T0201244

Earth's Climate Evolution

The Geological Time Scale for the Phanerozoic Aeon

Era	Period	Epoch	Base Age (Ma)
Cenozoic[a]	Quaternary	Holocene	0.0117
		Pleistocene	2.6
	Neogene	Pliocene	5.3
		Miocene	23
	Palaeogene	Oligocene	33.9
		Eocene	55.8
		Palaeocene	65.5
Mesozoic	Cretaceous	Upper	99.6
		Lower	145.5
	Jurassic	Upper	161.2
		Middle	175.6
		Lower	199.6
	Triassic	Upper	228.7
		Middle	245.9
		Lower	251
Palaeozoic	Permian	Upper	260.4
		Middle	270.6
		Lower	299
	Carboniferous	Upper	318.1
		Lower	359.2
	Devonian	Upper	385.3
		Middle	397.5
		Lower	416
	Silurian		443.7
	Ordovician		488.3
	Cambrian		542

[a]During the 19th century, geological time was divided into Primary, Secondary, Tertiary and Quaternary Eras. Mesozoic and Palaeozoic strata were regarded as belonging to the Secondary Era. The Tertiary was equivalent to the Cenozoic Era, but without the Quaternary. These older designations were done away with in the latter part of the 20th century, although 'Tertiary Era' is often misused for 'Cenozoic Era'. Of the older terms, 'Quaternary' has managed to hang on in the form of a geological period. From the International Stratigraphic Chart for 2010 (see http://www.stratigraphy.org/index.php/ics-chart-timescale, last accessed 29 January 2015).

Earth's Climate Evolution

COLIN P. SUMMERHAYES

Published in association with the Scott Polar Research Institute

Scott Polar Research Institute
University of Cambridge

WILEY Blackwell

Library of Congress Cataloging-in-Publication Data

Summerhayes, C. P.
    Earths evolving climate : a geological perspective / Colin Summerhayes.
        pages cm
    Includes bibliographical references and index.
    ISBN 978-1-118-89739-3 (cloth)
    1. Atmospheric carbon dioxide. 2. Climatic changes–Research. 3. Geological carbon sequestration. 4. Paleoclimatology.
5. Ice cores. I. Title.
    QC879.8.S86 2015
    551.609′01–dc23

                                                                                        2015007793

A catalogue record for this book is available from the British Library.

To my grandchildren, Reid, Torrin and Jove Cockrell and Zoe and Phoebe Summerhayes, in the hope that you can work towards freeing the future from the negative aspects of anthropogenic climate change.

Contents

Author Biography

Colin P. Summerhayes is an emeritus associate of the Scott Polar Research Institute of Cambridge University. He has carried out research on past climate change in both academia and industry: at Imperial College London; the University of Cape Town; the Woods Hole Oceanographic Institution; the United Kingdom's Institute of Oceanographic Sciences Deacon Laboratory; the United Kingdom's Southampton (now National) Oceanography Centre; the Exxon Production Research Company; and the BP Research Company. He has managed research programmes on climate change for the United Kingdom's Natural Environment Research Council, the Intergovernmental Oceanographic Commission of UNESCO and the Scientific Committee on Antarctic Research of the International Council for Science. He has co-edited several books relating to aspects of past or modern climate, including *North Atlantic Palaeoceanography* (1986), *Upwelling Systems: Evolution Since the Early Miocene* (1992), *Upwelling in the Oceans* (1995), *Oceanography: An Illustrated Guide* (1996), *Understanding the Oceans* (2001), *Oceans 2020: Science, Trends and the Challenge of Sustainability* (2002), *Antarctic Climate Change and the Environment* (2009) and *Understanding Earth's Polar Challenges: International Polar Year 2007–2008* (2011). Photo courtesy of the author; taken amid the snows of the Lofoten Islands, Norway, April 2009.

Foreword

Climate change is becoming increasingly obvious, through melting glaciers, extreme weather events and rising insurance premiums. Research on the topic is reported and reviewed more thoroughly than any other aspect of the world we live in, and yet we allow the principal cause, greenhouse gas emissions, to continue to rise.

In the last few years, many eminent climate scientists have shifted their focus from seeking new knowledge to reviewing what we already know of our warming world and what our followers will have to cope with in the future. All conclude with a call for action. What makes this book different is its multi-million-year perspective, looking at the climate of the past. Surprisingly, it turns out not only to be relevant for appreciating what we will be facing in coming decades and centuries, but also to add to the urgency of the need for action.

Colin has had a remarkable career, beginning in the 1960s as a scientist in the early days of the plate tectonics revolution, making discoveries in ocean circulation in the 1970s, and then in the 1980s moving into petroleum exploration to reconstruct geography and environments in the distant past to help find more oil. Since then, he has worked with UNESCO's Intergovernmental Oceanographic Commission and the World Meteorological Organization on the contribution of the oceans to modern climate change, going on to the Scientific Committee on Antarctic Research to oversee the development of Antarctic multidisciplinary studies of climate change and its effects on all time scales from the distant past to the future. His stories remind us that scientists are also human.

The real stimulus for this book came recently, through his realisation that many colleagues were still climate change 'sceptics', actively persuading the public that changes in climate in recent decades were either not significant or not related to greenhouse gas emissions, or both. First, he led a group within the Geological Society of London to develop a position paper for the Society on the issue. This paved the way for taking the case to the public through this book.

The story is a fascinating one, for a number of reasons. It reveals how much of our current understanding of Earth's climate history and the role of atmospheric CO_2 has been known for well over a century. In the 1830s, Charles Lyell, Father of Geology, described the great cooling of the last 50 million years, leading to the Ice Age of the last 2 million years. By 1896, Svente Arrhenius, at the behest of a geological colleague, had estimated the climatic consequences of increasing CO_2 levels in the atmosphere. Since then, this basic understanding has been improved upon and verified in remarkable detail through advances in imaging strata beneath the Earth's surface, and in determining environmental conditions (including temperature and atmospheric CO_2 levels) at various times and places in the past from ice and sediment cores going back tens of millions of years.

Colin also includes in his story the most recent scientific tool of all, numerical simulation of Earth's climate through computer modelling of various interactions involving atmosphere, water on land and in the oceans, snow and ice and the living world. These models are of course the only means we have for making projections of future climate, providing a rational basis for assessing possible consequences for both the physical and biological worlds. After 40 years of development, and astonishing advances in computer power, we now find broad agreement between model estimates of past climate and geological knowledge of the same periods, but also some mismatches, as well as significant differences between results from different modelling groups. As you'll see from Colin's overview of the field, the crucial issues are now in finding ways of increasing the robustness of the models for projecting regional consequences of climate change. However, critical issues, such as how fast these changes will be, have yet to be resolved, with ice loss and sea level rise a key concern.

Three aspects of the book are especially significant. The first is the extensive knowledge of the details of Earth's climate and its interaction with the ocean. These are not only captured from observations over the last 150 years and modelling in the last 30, but now include similar studies covering the period since the Last Glacial Maximum 20 000 years ago, and into the stable warm climate of the last 10 000 years, which led to the development of agriculture and our present society. We also get confirmation of the temperature–CO_2 link from much warmer times millions of years ago, reflecting Earth's future climate, to which all species (not just us) will have to adapt by 2100 if present emission rates continue. While our understanding of the Earth system is not complete, it is nevertheless huge, and fully justifies our confidence in acting on this new knowledge. The second aspect is the abundant evidence that Earth is now warming beyond the 'natural envelope' of the Ice Age glacial–interglacial climate cycles of the last 2.6 million years, a development that is becoming increasing significant for all life on Earth. The third is our growing appreciation that there is a lag between increasing greenhouse gas levels in the atmosphere and the response of warming of the atmosphere (more or less instantaneous), of the oceans (in decades to centuries) and of the ice sheets (decades to millennia). On the bright side, this gives us some time to act, but our geological knowledge shows us the ultimate consequences of not changing our present course. We might be able to cope with warmer temperatures in most places, but sea level rise of 10–20 metres in several hundred years will be more difficult. That prospect now seems inevitable, though we can still delay the worst if we reduce our emissions in coming decades. Earth has been there before, but change came slowly. Do we want to get there in a geological instant?

Beyond the message from climate science itself, Colin also provides intriguing glimpses of how scientists in the past were regarded by their contemporaries, and the context in which they worked. Some were very effective networkers long before the Internet age! I hope readers will also enjoy discovering from these pages how science makes progress, despite the human limitations to which we are all subject – occasionally pausing, but in the end always self-correcting.

P.J. Barrett
Fellow of the Royal Society of New Zealand
Holder of the NZ Antarctic Medal
Honorary Fellow, Geological Society of London
Emeritus Professor of Geology, Victoria University of Wellington
Wellington, New Zealand

Acknowledgements

Research for an all-embracing book like this is impossible without the help of many people. I thank Julian Dowdeswell for facilitating my research by making me an emeritus associate of the Scott Polar Research Institute.

The concept for the book emerged from the workings of the drafting group that produced the policy statement on climate change for the Geological Society of London (GSL) (www.geolsoc.org.uk/climaterecord). I am grateful to a former president of the GSL, Lynne Frostick, for appointing me in January 2010 to chair that group. My understanding of Earth's climate evolution was broadened through stimulating discussions with the members of the drafting group that produced that statement in November 2010 and its addendum in 2013, including Jane Francis, Alan Haywood, Joe Cann, Anthony Cohen, Rob Larter, Eric Wolff, John Lowe, Nick McCave, Paul Pearson and Paul Valdes, aided by Edmund Nickless, Nic Bilham and Sarah Day. The GSL's librarians, Michael McKimm, Wendy Cawthorne and Paul Johnson, helped me find a number of obscure publications.

Many individuals donated time and effort to helping me with advice or materials or discussions along the way. They included Ian Jamieson (Guildford), Vicki Hammond (Edinburgh), Peter Barrett (Wellington), Mike Sparrow, Eric Wolff, Marie Edmonds, Bryan Lovell, John Turner and Tom Bracegirdle (Cambridge), Peter Liss and Andy Watson (Norwich), Jane Francis (Leeds), Paul Mayewski (Maine), Martin Siegert and Cherry Lewis (Bristol), Valérie Masson-Delmotte (Gif-sur-Yvette), Phil Woodworth (Liverpool), Bob Berner (Yale), Terrence Gerlach (US Geological Survey), Peter Dexter (Melbourne), Chris Scotese (Arlington, Texas), Judy Parrish (Idaho), Cornelia Lüdecke (Munich), Jörn Thiede (Kiel), Heinz Wanner (Bern), David Bottjer (Los Angeles), Jim Kennett (Santa Barbara), Alan Lord (London), Iain Stewart (Plymouth), Mike Arthur (Penn State), Chris Rapley (UCL), Pieter Tans (NOAA), Malcolm Newell and Ralph Rayner (IMarEST, London), Keith Alverson (UNESCO, Paris), Ed Sarukhanian (WMO, Geneva), John Gould (Southampton), Eric Steig (Washington State), Bob Binschadler (NASA), Emily Shuckburgh (Oxford), Bryan Storey (Christchurch), John Lewis (NOAA), André Berger (Louvain) and David Archer (Chicago). I apologise if I have inadvertently left anyone out. Any errors in the text are my own. I have drawn extensively on findings published in the scientific literature, which is a vast storehouse of accumulated learning about all aspects of our climate history. Those writings, integrated together, tell a compelling story that I have tried to put into a form that anyone with a basic scientific education can understand. My gratitude goes to the thousands of scientists who have dedicated themselves to making the individual bricks in this impressive wall. I would also like to thank the many audiences in schools, universities, womens' groups, and mens' business groups, on whom I have tried out my ideas, and who have stimulated me with their probing questions. That list includes the many participants of Antarctic cruises on 'Le Boreal'.

Last but not least, I would like to thank my grandson, Reid Cockrell, for culling articles on global warming from my back issues of *Nature* and *New Scientist* and for indexing the book, and my long-suffering wife, Diana, for putting up with my mental absences on Planet Climate. Yes, that's where I was in those moments when my eyes were glazed and I didn't hear the question.

1

Introduction

In almost every churchyard, you'll find gravestones so old that their inscriptions have disappeared. Over the years, drop after drop of a mild acid has eaten away the stone from which many old gravestones were carved, obliterating the names of those long gone. We know this mild acid as rainwater, formed by the condensation of water vapour containing traces of atmospheric gases like carbon dioxide (CO_2) and sulphur dioxide (SO_2). It's the gases that make it acid. Rain eats rock by weathering.

Weathering is fundamental to climate change. Over time, it moves mountains. Freezing and thawing cracks new mountain rocks apart. Roots penetrate cracks as plants grow. Rainwater penetrates surfaces, dissolving as it goes. The CO_2 in the dissolved products of weathering eventually reaches the sea, where it forms food for plankton, and the seabed, in the remains of dead organisms. Once there, it goes on to form the limestones and hydrocarbons of the future; one day, volcanoes will spew that CO_2 back into the atmosphere and the cycle will begin all over again.

The carbon cycle includes the actions of land plants, which extract CO_2 from the air by photosynthesis. When plants die, they rot, returning their CO_2 to the air. Some are buried, preserving their carbon from that same fate, until heat from the Earth's interior turns them back into CO_2, which returns to the air. This natural cycle has been in balance for millions of years. We have disturbed it by burning fossil carbon in the form of coal, oil or gas.

This book is the story of climate change as revealed by the geological record of the past 450 million years (450 Ma). It is a story of curiosity about how the world works and of ingenuity in tackling the almost unimaginably large challenge of understanding climate change.

The task is complicated by the erratic nature of the geological record. Geology is like a book whose pages recount tales of the Earth's history. Each copy of this book has some pages missing. Fortunately, the American, African, Asian, Australasian and European editions all miss different pages. Combining them lets us assemble a good picture of how Earth's climate has changed through time. Year by year, the picture becomes clearer, as researchers develop new methods to probe its secrets.

As we explore the evolution of Earth's climate, we will follow the guidance of one of the giants of 18th-century science, Alexander von Humboldt, who wrote in 1788, '*The most important result of research is to recognize oneness in multiplicity, to grasp comprehensively all individual constituents, and to analyze critically the details without being overwhelmed by their massiveness*'[1]. All too often, those who seek to deny the reality of modern climate change ignore his integrative approach to understanding nature by focusing on just one or two aspects where the evidence seems, at the moment, to be less than compelling.

Can the history of Earth's climate tell us anything about how it might evolve if we go on emitting gigatonnes (Gt) of CO_2 and other greenhouse gases into the atmosphere? That is the key question behind the title to this book. I wrote it because I have spent most of my career working on past climate change, and it worries me that few of the results of the growing body of research on that topic reach the general public. Even many professional Earth scientists I meet, from both academia and industry, know little of what the most up-to-date Earth science studies tell us about climate change and global warming. For the most

Earth's Climate Evolution, First Edition. Colin P. Summerhayes.
© 2015 John Wiley & Sons, Ltd. Published 2015 by John Wiley & Sons, Ltd.

part, they have specialised in those aspects of the Earth sciences that were relevant to their careers. Unfortunately, their undoubted expertise in these topics does not prevent some of them from displaying their ignorance of developments in the study of past climate change by trotting out the brainless mantra, 'the climate is always changing'. Well, of course it is, but that ignores the all-important question: Why?

What we really need to know is in what ways the climate has been changing, at what rates, with what regional variability, and in response to what driving forces. With these facts, we can establish with reasonable certainty the natural variability of Earth's climate, and determine how it is most likely to evolve as we pump greenhouse gases into the atmosphere. This book attempts to address these issues in a way that should be readily understandable to anyone with a basic scientific education. It describes a voyage of discovery by scientists obsessed with exposing the deepest secrets of our changing climate through time. I hope that readers will find the tale as fascinating as I found the research that went into it.

The drive to understand climate change is an integral part of the basic human urge to understand our surroundings. As in all fields of science, the knowledge necessary to underpin that understanding accumulates gradually. At first we see dimly, but eventually the subject matter becomes clear. The process is a journey through time, in which each generation makes a contribution. Imagination and creativity play their parts. The road is punctuated by intellectual leaps. Exciting discoveries change its course from time to time. No one person could have discovered in his or her lifetime what we now know about the workings of the climate system. Thousands of scientists have added their pieces to the puzzle. Developing our present picture of how the climate system works has required contributions from an extraordinary range of different scientific disciplines, from astronomy to zoology. The breadth of topics that must be understood in order for us to have a complete picture has made the journey slow, and still makes full understanding of climate change and global warming difficult to grasp for those not committed to serious investigation of a very wide-ranging literature. The pace of advance is relentless, and for many it is difficult to keep up. And yet, as with most fields of scientific enquiry, there is still much to learn – mostly, these days, about progressively finer levels of detail. Uncertainties remain. We will never know everything. But we do know enough to make reasonably confident statements about what is happening now and what is likely to happen next. Looking

back at the progress that has been made is like watching a timelapse film of the opening of a flower. Knowledge of the climate system unfolds through time, until we find ourselves at the doorstep of the present day and looking at the future.

While the story of Earth's climate evolution has a great deal to teach us, it is largely ignored in the ongoing debate on global warming. The idea of examining the past in order to discover what the future may hold is not a new one. It was first articulated in 1795 by one of the 'fathers of geology', James Hutton. But it is not something the general public hears much about when it comes to understanding global warming. This book is a wake-up call, introducing the reader to what the geological record tells us.

Information about the climate of the past is referred to as 'palaeoclimate data' (American spelling drops the second 'a'). As it has mushroomed in recent years, it has come to claim more attention from Working Group I of the Intergovernmental Panel on Climate Change (IPCC). The Working Group comprises an international group of scientists, which surveys the published literature every 5 years or so to come up with a view on the current state of climate science. It has been reporting roughly every 5 years since its first report in 1990. Each of its past two reports, in 2007[2] and 2013[3], incorporated a chapter on palaeoclimate data. The Working Group's report is referred to as a 'consensus', meaning the broad agreement of the group of scientists who worked on it. Just one chapter in a 1000-page report does not constitute a major review of Earth's climate evolution: the subject deserves a book of its own, and there are several, as you will see from the Appendix to the present book.

The study of past climates used to be the exclusive province of geologists. They would interpret past climate from the character of rocks: coals represented humid climates; polished three-sided pebbles and cross-bedded red-stained sands represented deserts; grooved rocks indicated the passage of glaciers; corals indicated tropical conditions; and so on. Since the 1950s, we have come to rely as well on geochemists using oxygen isotopes and the ratios of elements such as magnesium to calcium (Mg/Ca) to tell us about past ocean temperatures. And in recent years we have come to realise that cores of ice contain detailed records of past climate change, as well as bubbles of fossil air; glaciologists have joined the ranks.

Climate modellers have also contributed. Since the 1950s, our ability to use computers has advanced apace. We now use them not only to process palaeoclimate data and find correlations, but also to run numerical models of past climate systems, testing the results against data from the rock record. Applying numerical models to past climates that were much colder or much warmer than today's has an additional benefit: it helps climate modellers to test the robustness of the models they use to analyse today's climate and to project change into the future. One of my reasons for writing this book is to underscore how research into past climates by both of these research streams, the practical and the theoretical, adds to our confidence in understanding the workings of Earth's climate system and in predicting its likely future.

My take on the evolution of Earth's climate is coloured by my experience. Early in my geological career, I applied knowledge of how oceans and atmospheres work to interpret the role of past climates in governing the distribution of the phosphatic sediments that form the basis for much of the fertiliser industry. That work broadened into a study of how climate affects runoff from large rivers like the Nile and the Amazon, as well as the accumulation of sediments on the world's continental shelves. Working for Exxon Production Research Company (EPRCo) in Houston, Texas, in the mid-1970s, I developed a model for how climate controlled the distribution of petroleum source beds: rocks rich in organic remains that, when cooked deep in the subsurface, yield oil or gas. Explorers tested my model's predictions by drilling. Later, with the BP Research Company (BPRCo), I studied how the changing positions of past continents, along with changing sea levels and climates, affected the distribution and character of sources and reservoirs of oil and gas, as the basis for developing predictions for explorers to test by drilling.

In the late 1980s to mid-1990s, as director of the UK's main deep-sea research centre, the Institute of Oceanographic Sciences Deacon Laboratory, I learned a great deal more from my physical, biological and chemical oceanographer colleagues about the ocean's role in climate change. I applied that knowledge to analysing the response of the upwelling currents off Namibia and Portugal to the glacial-to-interglacial climate changes of the last Ice Age.

In order to develop accurate forecasts of climate change, one has to have an observing system, much like that used for weather forecasting. In 1997, I joined UNESCO's Intergovernmental Oceanographic Commission (IOC) to direct a programme aimed at developing a Global Ocean Observing System (GOOS), which would provide the ocean component of a Global Climate Observing System (GCOS). The task further broadened my understanding of climate science. Then, from 2004 to 2010, I directed the Antarctic research activities of the International Council for Science (ICSU), while based at the Scott Polar Research Institute of the University of Cambridge. There I was awarded emeritus status, starting in 2010. These recent appointments exposed me to the thinking of the polar science community about the role of ice in the climate system. Few people can have been as fortunate as I in being exposed to the current state of knowledge about the operations of the climate system from the perspectives of the ocean, the atmosphere, the ice and the geological record.

Because of that diverse background, I was asked to advise the Geological Society of London on climate change. Many of the world's major scientific bodies, including the US National Academy of Sciences and the UK's Royal Society, have felt moved in recent years to publish statements on the science of global warming as part of their remit to inform the public and policy makers about advances in science. The Geological Society of London became interested in 2009 in developing such a statement, and its then president, Professor Lynne Frostick, invited me to chair the group that would draft it. Entitled 'Climate Change: Evidence from the Geological Record', the statement was published on the Society's Web page[4] and in its magazine, *Geoscientist*, towards the end of 2010[5]. I led basically the same team in writing an addendum to the statement in 2013, to show what advances had been made in the intervening 3 years and to provide a palaeoclimate-based statement that could be evaluated alongside the 5th Assessment Report of the IPCC's Working Group I, published in September 2013. We operated independently of the IPCC, and drew our own conclusions. The Society published the addendum in December 2013[4].

As the Society's statement was being developed, I realised that it did not allow the space to reveal either the human stories behind the long development of modern climate science or the full extent of advances emerging through palaeoclimatic research. So I resolved to write a book about climate change from the palaeoclimate perspective – the 'long' view – drawing on the Society's statement and summarising the history and knowledge of climate processes that took place over millions of years under conditions very different from today's. This book is the result. It shows how the climate record of past times is the key to understanding the natural variability of our climate, and explains why that knowledge is a necessary complement to what we learn from meteorologists and modern climatologists focusing on the instrumental records of the past 150 years.

The book focuses on the past 450 Ma or so of Earth's climate, starting with the period when land plants first emerged, because plants play an important role in tying up carbon on land. For most of the past 450 Ma, our planet has been a lot warmer than it is now. Our climate is usually of the greenhouse variety, with abundant CO_2 warming the part of the atmosphere in which we live: the troposphere. This long history of warmth is not widely recognised, because in the past 50 Ma Earth's atmosphere has lost much of its CO_2 and moved into an icehouse climate, characterised by cool conditions and polar ice. That cooling has intensified to the point where, over the past 2.6 Ma, Earth has developed large ice sheets in both polar regions. This period has earned a popular title: the Ice Age. We are living in a geologically brief warm interlude within that Ice Age. Before an ice sheet formed on Antarctica, 40–50 Ma ago, global temperatures were warmer by 4–6 °C than they are today. Where will our climate go next? Will we stay in the icehouse or move back into the greenhouse? The latest news from NASA's Jet Propulsion Laboratory, dated 12 May 2014, is that the West Antarctic Ice Sheet has begun an irreversible decline, making it likely that we are now moving away from the icehouse and towards the greenhouse[6].

Increasing scrutiny of the palaeoclimate record over the past few decades has helped us to explain why our present climate is the way it is. Most of the fluctuation from warm to cold climates through time takes place because of changes in the balance of Earth's interior processes. Changes over millions of years involve periods of excessive volcanic activity, associated with the break up and drift of continents, which fills the air with CO_2 and keeps the climate warm, and periods of continental collision, which build mountains and encourage the chemical weathering of exposed terrain, sucking CO_2 out of the atmosphere and keeping the climate cool. Continental drift moves continents through climatic zones, sometimes leaving them in the tropics, sometimes at the poles. It also changes the locations of the ocean currents that transport heat and salt around the globe. Individual volcanic eruptions large enough to eject dust into the stratosphere provide short-term change from time to time, while the equally erratic but more persistent volcanic activity of large igneous provinces, involving the eruption of millions of cubic metres of lava over a period of a million years or so, can change the climate for longer periods; at times, they may have done so enough to cause substantial biological extinctions.

External changes are important, too. The Sun is the climate's main source of energy. Orbital variations in the Earth's path around the Sun, combined with regular changes in the tilt of the Earth's axis, superimpose additional change on these millions-of-years-long changes, through cycles lasting 20 000 to 400 000 years (20–400 Ka). Variations in the Sun's output superimpose yet another series of changes, with variability at millennial, centennial and decadal scales. Examples include the 11-year sunspot cycle and its occasional failure. The best-known such failure is the Maunder Minimum between 1645 and 1715 AD, at the heart of the Little Ice Age. Large but rare meteorite impacts have had similar, albeit temporary, effects.

Internal oscillations within the ocean–atmosphere system, like El Niño events and the North Atlantic Oscillation, cause further changes at high frequencies but low amplitudes, and are usually regional in scope. Whatever the climate at any one time, it is modified by internal processes like those oscillations, and by the behaviour of the atmosphere in redistributing heat and moisture rapidly, by the ocean in redistributing heat and salt slowly and by the biota. An example of the latter is the 'biological pump', in which plankton take CO_2 out of surface water and transfer it to deep water and, eventually, to sediments, when they die. These processes can make attribution of climate change difficult, as can the smearing of the annual record in deep-water sediments by burrowing organisms.

In spite of the potential for considerable variation in our climate, close inspection shows that at any one time the climate is constrained within a well-defined natural envelope of variability. Excursions beyond that natural envelope demand specific explanation. As we shall see, one such excursion is the warming of our climate since late in the last century.

This book looks at these various processes and puts them into perspective in their proper historical context. Chapter 2 follows the evolution of thinking about climate change by natural scientists, philosophers and early geologists from the late 1700s on. It touches on the debates of the early 1800s on the virtues of gradual versus sudden change and highlights the growing realisation that the world cooled towards an Ice Age in geologically recent times. Chapter 3 takes us into the minds of 19th-century students of the Ice Age and examines the astonishing discovery that its climate cycles were probably controlled by metronomic variations in the behaviour of the Earth's orbit as it responded to the gravitational influences of the great gas planets, Venus and Jupiter.

The arrival of new technologies on the scene, often from different disciplines, changes the way in which science works; think of the effect of the telescope on Galileo's perception of astronomy. Geology is no exception. In Chapter 4, we explore the extraordinary mid-19th-century discovery of the absorptive properties of what we now know as the greenhouse gases, such as water vapour, carbon dioxide and methane, which changed the way we view past climates. At the end of that century, a Swedish chemist, Svante Arrhenius, made the first calculations of what emissions of CO_2 would do to the climate. Few people realise that he did so at the urging of a geological colleague, to try to see if variations in atmospheric CO_2 might explain the fluctuations in temperature of the Ice Age. An American geologist, Thomas Chamberlin, used Arrhenius's findings to construct an elegant hypothesis as to how CO_2 controlled climate, but it was soon forgotten for lack of data. Much of what he had to say on the subject has since been proved correct.

In Chapter 5, we examine the evolution of ideas in the early part of the 20th century about the way in which the continents move relative to one another through continental drift, which geophysicists discovered in the 1960s was driven by the process of plate tectonics. Once again, new technologies played a key role: in this case, the echo-sounder and the magnetometer. Knowing the past positions of the continents provides us with the maps of past geography – the palaeogeographic base maps – needed to determine the past locations of sedimentary deposits that are sensitive to climate, like coal swamps and salt pans. Along the way, we see how studies of past climates benefited from access to the accurate dating of rocks, minerals and fossils at the smallest possible intervals of time. Once again, a new technology was key: radiometric dating by the use of natural radioactivity.

Chapter 6 describes how the new science of palaeoclimatology developed, with Earth scientists plotting their indicators of past climates on maps, using yet another new technology – oxygen isotopes – to determine the temperature of past seawater. Geologists investigated the origins of sedimentary cycles, coming up with hypotheses explaining the evolution of climate from the Carboniferous glaciation roughly 300 Ma ago to the end of the Cretaceous at 65 Ma ago. Yet another new technology changed the picture again, this time in the shape of numerical models of the climate system, which capitalised on the rapid development of the computer. We see early attempts to use numerical models to find out why the Cretaceous Period was so

warm, and note that until the mid-1980s, the analysis of palaeoclimates virtually ignored CO_2.

Chapter 7 takes us into the Cenozoic Era, which includes what used to be known as the Tertiary, between 65 and 2.6 Ma ago, and the Quaternary, lasting from 2.6 Ma ago to the present. Here we follow the cooling of our climate from the warmth of the Cretaceous seas that flooded western Europe and central North America 60–100 Ma ago to the current Ice Age, which characterises the Pleistocene Period (2.6 Ma to 11.7 Ka ago) and the present Holocene Period (starting 11.7 Ka ago). We look at how climate changed, and at how our knowledge of climate change was dramatically expanded by drilling into the largely undisturbed sediments of the deep ocean floor. As we saw in Chapter 6, many of the theories explaining the changes in climate of the Cenozoic Era prior to the 1980s developed in the absence of substantial knowledge about the past composition of the air.

A clear understanding of the roles of greenhouse gases in the climate system demands an ability to measure those gases and examine their properties: capabilities that were limited until the mid 1950s, and which then took another 30 years to penetrate the world of geological thought. Chapter 8 explores the massive strides made over the past 50 years in enhancing that knowledge base and in formulating theories to explain how greenhouse gases behave within the air and ocean. Along with that understanding came the realisation that, in order to understand the climate problem, we must see our planet holistically – as a whole – and not in a reductionist way. Humboldt was right: everything is connected. One key consequence was the development of a new field of scientific endeavour, biogeochemistry, which has proved especially important for understanding how the carbon cycle works. Answering questions about the evolution of the climate system also came to involve a more international approach, in which national scientists increasingly worked with each other across borders on major scientific issues such as climate change that were not susceptible to resolution by individual investigators or even individual nations.

Chapter 9 reminds us of the amazing discovery that ice cores contain bubbles of fossil air holding pristine samples of CO_2 and other greenhouse gases. We also see how palaeoclimatologists eventually learned how to measure the amount of CO_2 in the atmosphere in the ages before the oldest ice cores (which span the past 800 Ka) using fossil leaves, tree rings, planktonic remains, soils, corals and cave deposits. These data are being used to check numerical models of past climates and to test the theory that the warm periods of the past occurred when CO_2 was most abundant.

Our planet's climate has experienced large cycles through time. Chapter 10 explores how these cycles relate to changes in plate tectonic processes, sea level, emissions of CO_2 and the weathering of emerging mountain chains as continents collided. It investigates the evidence for changes to our climate, and the creation of major biological extinctions, caused by occasional meteorite impacts and/or massive eruptions of plateau basalts.

In Chapter 11, we examine the evidence for how CO_2 and climate changed together through the Mesozoic and Cenozoic Eras, and explore two case histories. The first is from the Palaeocene–Eocene boundary 55 Ma ago, when a massive injection of carbon into the air caused dramatic warming, which at the same time made the seas more acid. It took the Earth 100 Ka to recover – now, there's a lesson from the past! The second is from the mid-Pliocene, about 3 Ma ago, when CO_2 levels rose to levels much like today's, but when temperatures were warmer and the sea level was higher: another lesson from the past. These periods are not precise analogues for today, because the world was configured slightly differently then. But they can teach us something about what is happening now and what might happen in the future.

Chapter 12 begins our exploration of the Ice Age of the past 2.6 Ma, noting how much of what we know comes from cores of sediment extracted with great difficulty from the ocean bed. It was a big surprise in 1976 when it emerged that marine sediment cores display signs of change in the Earth's orbit and the tilt of the Earth's axis through time. These cores also display unexpected millennial signals.

Our exploration of Ice Age climate continues in Chapter 13, where we examine the contribution made by ice cores collected in recent decades. We see what the records tell us from Greenland and from Antarctica, and explore the linkages between the poles. The latest research shows that during the warming from the Last Glacial Maximum, CO_2 in the Antarctic region rose synchronously with temperature, not after it, as had been thought. The chapter ends with a survey of plausible explanations for the fluctuations of the Ice Age, concluding that CO_2 played a crucial role in the changes from glacial to interglacial and back over the past 800 Ka.

In Chapter 14, we focus on the changes that took place over the past 11.7 Ka, forming the latest interglacial: the Holocene. Insolation – the amount of heat received due to the motions of the Earth's orbit and the tilt of the

Earth's axis – was greatest in the Northern Hemisphere at the beginning of the Holocene, but the great North American and Scandinavian ice sheets kept the Northern Hemisphere cool until they had completely melted by the middle Holocene. All that while, Northern Hemisphere insolation was in decline, moving Earth's climate towards a Neoglacial Period, the peak of which we reached in the Little Ice Age of the past few hundred years. CO_2 played no active part in this cooling.

Chapter 15 focuses on the end of the Holocene – the past 2000 years, up to the present – reviewing cyclical changes in solar output. It explores the development and extent of the Medieval Warm Period centred on 1100 AD and the subsequent Little Ice Age, and includes a review of the 'Hockey Stick' controversy. Multiple sources of palaeoclimatic data now make it abundantly clear that the years since 1970 were the warmest of the past 2000. Yet astronomical calculations show that despite variations in the sun's output, our climate should still be like that of the Little Ice Age. Only by adding our emissions of greenhouse gases like CO_2 to palaeoclimate models can we recreate the climate that we see today.

The concluding chapter, Chapter 16, provides an overview of Earth's climate evolution, concluding that, from the evidence of previous chapters, we should expect to see sea level rises of 6–9 m as temperatures rise 2–3 °C above the 'preindustrial' levels typical of the years before the Industrial Revolution. Those conditions were typical of recent interglacials, which were warmer than our own. We will not see such rises in sea level this century, because it takes a long time for the Earth system to arrive at an equilibrium, in which the ocean is heated as fully as it can be for a given level of atmospheric CO_2 and no more ice will melt.

As in any other field of science, the 200-year history of past climate studies has been punctuated with arguments and disagreements, but the influence of CO_2 on climate eventually emerged as highly significant. The exciting developments documented in this book revolutionised the way Earth science is done as much as did the discovery of plate tectonics. The demands of climate science now require sedimentologists and palaeontologists to become familiar with the host of related disciplines that deal with processes taking place on and above the Earth, and to take a holistic approach to interpreting their data. Due to the rapid evolution of these topics and techniques, including the use of computers to model palaeoclimate behaviour, much of what we now know is quite recent, and little publicised except in scientific journals.

In brief, the geological evidence now suggests that emitting further large amounts of CO_2 into the atmosphere over time will almost certainly push our climate from icehouse to greenhouse, something not experienced since the late Eocene about 40 Ma ago. We now have a strong enough base of geological evidence to agree that '*In the light of the evidence presented here it is reasonable to conclude that emitting further large amounts of CO_2 into the atmosphere over time is likely to be unwise, uncomfortable though that fact may be*'[4]. The evidence emerging from the past gives much the same answers about the nature of our future climate as those emerging from a different scientific community, the IPCC's Working Group I.

Hasn't all this been said before, in classical texts on palaeoclimatology? No, in the sense that my approach combines a depiction of the science with a study of its evolution and of the role of individuals and their imagination in reaching our current understanding of Earth's climate system. But there is growing appreciation that '*evidence from the Quaternary stratigraphic record provides key baseline data for predictions of future climate change*'[7].

Agreeing with Nate Silver[8], I argue that the way to test research findings like those laid out here is to see whether or not they make accurate predictions in the real world. Our ability to predict well is a measure of our scientific progress. If you start with an absolute belief that humans do not cause global warming then, following Bayes's Theorem, no amount of evidence will persuade you otherwise. But you have to recognise that what you hold is a belief, not scientific understanding.

One thing you will need to consider carefully is context. In this book you will see evidence that CO_2 does correlate with temperature. Correlation is not causation, but that is a trite observation that ignores context. When you know that CO_2 is a greenhouse gas that both absorbs and re-emits radiation, you should expect a correlation with temperature from that context. That's the prediction and it's easy to test. What then becomes interesting are the instances when the two do *not* correlate, for which we have to find alternative hypotheses. We have to think! Thus far, nobody has managed to explain what, if not our emissions of greenhouse gases and related feedbacks, has caused the global warming since 1970.

I will leave this introduction with two key questions for you to consider as you read on: **Can what we see of climate in the geological record tell us anything about what might happen if we go on emitting more and more carbon dioxide and other greenhouse gases into the atmosphere?** and **What are the chances that our increasing use of fossil fuels will drive Earth's climate out of the icehouse, where it has been stuck for several million years, and back into the greenhouse – the dominant climate mode for much of the past 450 million years?** We will revisit these questions at the end of the book.

References

1. Reported in Lettau, H. (1978) Explaining the world's driest climate. In: *Exploring the World's Driest Climate* (eds H. Lettau and K. Lettau). Center for Climatic Research, Institute for Environmental Studies, University of Wisconsin-Madison, Madison, WI.
2. Jansen, E., Overpeck, J., Briffa, K.R., Duplessy, J.-C., Joos, F., Masson-Delmotte, V., *et al.* (2007) Chapter 6. Palaeoclimate. In: *Climate Change 2007: The Physical Science Basis* (eds S. Solomon, D. Qin, M. Manning, Z. Chen, M. Marquis, K.B. Averyt, *et al.*). Cambridge University Press, Cambridge, 433–497.
3. Masson-Delmotte, V., Schulz, M., Abe-Ouchi, A., Beer, J., Ganopolski, A., Rouco, J.F.G., *et al.* (2013) Chapter 5. Information from paleoclimate archives. In: *Climate Change 2013: The Physical Science Basis* (eds T. Stocker, D. Qin and G.-K. Plattner). Cambridge University Press, Cambridge.
4. Climate change: evidence from the geological record. Available from: http://www.geolsoc.org.uk/climatechange (last accessed 29 January 2015).
5. Summerhayes. C.P., Cann, J., Cohen, A., Francis, J., Haywood, A., Larter, R., *et al.* (2010) Climate change: evidence from the geological record. *Geoscientist* **20** (11), 10 and 24–27.
6. West Antarctic Glacier Loss Appears Unstoppable. Available from: http://www.jpl.nasa.gov/news/news.php?release=2014-148 (last accessed 29 January 2015).
7. Walker, M. and Lowe, J. (2007) Quaternary science 2007: a 50-year retrospective. *Journal of the Geological Society, London* **164**, 1073–1092.
8. Silver, N. (2012) *The Signal and the Noise: The Art and Science of Prediction.* Penguin, London.

2

The Great Cooling

2.1 The Founding Fathers

Geologists have known for over 200 years that climate is
one of the main controls on the accumulations of minerals
and organic remains that end up as sedimentary rocks
and fossils. As early as 1686, Robert Hooke, a fellow
of London's Royal Society living in Freshwater on the
Isle of Wight, deduced from fossils discovered at Port-
land that the climate there had once been tropical[1]. His
perceptive observation remained unremarked upon until
the keeper of France's Royal Botanical Gardens – the
Jardin des Plantes – in Paris realised that differences in
climate might explain the differences between living and
fossil organisms found at the same place. This was the
naturalist Georges-Louis Leclerc, the Comte de Buffon
(1707–1788)[2], friend to Voltaire, and a member of both the
French Academy of Sciences and the literary Académie
Française.

Buffon planned to take his place in history with a
vast 50-volume encyclopaedia: the *Histoire Naturelle,
Générale et Particuliére*. The 36 volumes that he actually
produced were among the most widely read publications
of the time. His reconstruction of geological history
appeared in 1788 in *Époques de la Nature*, the supplement
to Volume 5. Buffon realised that each geographical region
had its own distinctive plants, animals and climate – a
basic principle of what we now call biogeography. Finding
in Siberia and Europe the fossil remains of animals that
now inhabit the tropics, he deduced that the climate there
must have been warmer in the past.

Buffon thought that the temperature of the air reflected
the temperature of the Earth, rather than the heat from

the Sun, and interpreted animal remains to show that the
Earth was cooling from its original molten state[2]. Sir
Humphry Davy, FRS, discoverer of sodium and potassium
and inventor of the coal miners' safety lamp, was another
who shared this popular notion, penning it in 1829 in his
Consolations in Travel, or the Last Days of a Philosopher,
shortly before he died. Measurements of the temperature
of the Earth and its atmosphere by the French scientist
Joseph Fourier had knocked this idea on its head in 1824,
however – an advance that Davy overlooked.

Buffon and other savants of the late 18th century considered that Earth's past history must be explained with reference to what is happening now[3]. Among them was James
Hutton (1726–1797) (Figure 2.1), a Scot who profoundly
influenced geological thought[4,5]. Born in Edinburgh, Hutton studied medicine and chemistry there and in Paris and
Leyden. Taking up farming, first in Norfolk, then on his
paternal acres in Berwickshire, he developed an interest in
geology, and exploited his chemical knowledge to become
partner in a profitable sal ammoniac business. By 1768,
he was established in Edinburgh, pursuing his geological
interests. In 1785 he published a *Theory of the Earth* in
the first volume of the Transactions of the Royal Society
of Edinburgh. Encouraged to seek observations to support
his theory, he found several telling examples, enabling him
in 1795 to expand his ideas into a two-volume book: *Theory of the Earth with Proofs and Illustrations*. His friend,
John Playfair, brought Hutton's ideas to a wider audience
in *Illustrations of the Huttonian View of the World*, published in 1802.

Hutton popularised the notion that '*the present is the
key to the past*'[6]. As he put it in his book, '*In examining

Earth's Climate Evolution, First Edition. Colin P. Summerhayes.
© 2015 John Wiley & Sons, Ltd. Published 2015 by John Wiley & Sons, Ltd.

Figure 2.1   *James Hutton.*

things present, we have data from which to reason with regard to what has been'[6]. In following that approach, Hutton echoed Isaac Newton's dictum that '*We are to admit no more causes of natural things than such as are both true and sufficient to explain their appearances*'[6]. But that's not all. Hutton went on to say, '*and, from what has actually been, we have data for concluding with regard to that which is to happen hereafter*'[6]. Here was an extraordinary notion – that examples from the past preserved in the geological record could provide examples of what might happen on Earth in future if the same guiding conditions were repeated.

Observing the processes at work on his farm and in the surroundings, Hutton saw that today's hills and mountains are far from being everlasting. They were sculpted by the slow forces of erosion. The eroded materials were transported by rivers and dumped in the ocean, where they accumulated to great thickness before being raised into mountains. The process then began again, in yet another great geological cycle. Hutton's idea that the ruins of earlier worlds lay beneath our feet was demonstrated by younger and undisturbed strata resting uncomfortably on older folded and eroded beds, notably at Siccar Point on Scotland's east coast. To some degree, Hutton's concept repeats ideas proposed originally in the notebooks of Leonardo Da Vinci in about 1500[7]. Perhaps this is an early example of the convergent evolution of ideas. The slowness of geological processes led Hutton to conclude that Earth's history was unimaginably long. Indeed, in dramatic contrast to biblical scholars, he found that Earth's history showed '*no vestige of a beginning, – no prospect of an end*'[6].

He was wrong in one respect: not all operations of nature are equable and steady. Earthquakes and volcanic eruptions are sudden, as are meteorite strikes of the kind that cratered the surface of the moon. Even so, he realised that earthquakes and volcanic eruptions, although discontinuous, are recurrent. Neither he nor many other savants of his time knew about the kind of catastrophic meteorite impact that we now believe led to the great extinction and loss of the dinosaurs 65 Ma ago.

When Buffon died, Georges Cuvier (1769–1832) (Figure 2.2) took his place as France's leading natural historian. Cuvier was a key figure in establishing the scientific fields of comparative anatomy and palaeontology. He was elected a member of the Academy of Sciences in 1795, professor of natural history at the College de France in 1799 and professor at the Jardin des Plantes in 1802. He also became a foreign member of the Royal Society of London in 1806 and was ennobled Baron Cuvier in 1819.

Cuvier used his knowledge of anatomy to identify fossil species and their likely interrelationships. While Buffon thought that Siberian fossils of woolly rhinoceros and elephant were the remains of animals still living, Cuvier showed that they were extinct, and identified the elephant remains as mammoths[3]. Both men knew that these animals were found frozen into the tundra '*with their skin, their fur and their flesh*'[8,9]. Unlike Hutton, Cuvier was keen on the moulding of geological history by catastrophic events.

Figure 2.2   *Georges Cuvier.*

He thus attributed this freezing to an environmental catastrophe: '*this event took place instantaneously, without any gradation ... [and] rendered their country glacial*'[3, 10]. Here we have the first inkling of the idea of the Ice Age.

Cuvier's senior colleague Jean-Baptiste de Monet, Chevalier de la Marck, commonly known as Lamarck (1744–1829), challenged his call for catastrophic change. Studying the sequence of fossil molluscs from the region around Paris, he concluded in 1802 that many of them belonged to genera that are now tropical and that they represented a slow change of climate with time[3].

At about the same time, in the late 1790s, William Smith (1769–1839), a land surveyor engaged in building the network of canals that now cross the English countryside, began using distinctive fossils to identify and map the occurrence of particular strata. This led him to publish in 1801 a 'prospectus' for the production of a geological map of England,[3] something he achieved in 1815. He had invented the science of 'stratigraphy' – the use of fossil remains to establish the succession of strata – which now underpins our appreciation of changes in climate through geological time.

French geoscientists were quick to seize upon this new approach to geohistory. In 1802, Alexandre Brongniart (1770–1847), the newly appointed young director of the porcelain factory at Sèvres, near Paris, visited England to find out more about the mass production of ceramics by the Wedgwood factory[3]. In London, he dined with fellows of the Royal Society, where is likely to have become aware of the novel ideas and unpublished maps of William Smith. Searching for new deposits of clay, and working with Cuvier to identify fossils, Brongniart began a systematic survey of the Parisian region. Much like Smith, Brongniart and Cuvier used fossils to determine the order of the layers of sedimentary rock of the Paris Basin and map the outcrops of the strata. They concluded that the area had been submerged at times by the sea and at times by freshwater – a first indication that environmental conditions could change with time in a relatively small area, and something that went beyond anything attempted by Smith in its high level of detail. In 1808, they delivered a preliminary report of their paper on the Paris Basin, with an accompanying draft geological map that was eventually published in 1811[3].

Brongniart and Cuvier were not the first to map the sedimentary divisions of the Paris Basin. The famous French chemist Antoine-Laurent Lavoisier (1743–1794), who had discovered oxygen and hydrogen, and was guillotined during the French Revolution, beat them to it.

Lavoisier's 1789 memoir on the topic[11] was brought to light by sedimentologist Albert Carozzi of the University of Illinois in 1965[12]. Lavoisier saw in the alternating deep- and shallow-water (littoral) deposits of the Paris Basin evidence for a succession of transgressions (floodings) and regressions (retreats) of the sea. His vision of how these packages of sediment were built up through time by the alternating rising and falling of sea level is like the modern understanding of the origin of sedimentary cycles. It involved '*a very slow oscillatory movement of the sea ... [each oscillation] requiring several hundred of thousand years for completion*'[11, 12]. Lavoisier's cross-sections of the Basin provide an outline for the correct classification of its Tertiary deposits. He was a man far ahead of his time.

These parallel French and English efforts were major developments in the evolution of palaeontology and geology. They provided an essential platform for the development of palaeoclimatic studies and influenced the thinking of those who followed.

Along with Cuvier, one of the most influential scientific men in Europe in the early 1800s was the German naturalist Baron Alexander von Humboldt (1769–1859) (Figure 2.3, Box 2.1)[13–18]. By 1797 Humboldt was planning an overseas expedition, learning to use a wide range of scientific and navigational instruments, and visiting experts in Vienna and Paris. While in Paris, Humboldt met the botanist Aimé Bonpland (1774–1854), who was to be his travelling companion. Humboldt focused on physical geography, geology, geomorphology and climatology, while Bonpland focused on flora and fauna. Visiting Madrid, they obtained royal assent to scientifically examine Spain's American territories as a contribution to understanding the physical make-up of the world. They sailed from La Coruña on 5 June 1799 and visited South and Central America, the West Indies and the United States, returning to France in August 1804. The major scientific outcomes were Humboldt's seminal *Essay on the Geography of Plants*, published in 1805, and his treatise on *Isotherms and the Distribution of Heat over the Earth's Surface*, published in German in 1816. Wider recognition followed publication of his more general works: *Views of Nature* in 1808 and the travelogue *Personal Narrative of a Journey to the Equinoctial Regions of the New Continent* in three volumes in 1814, 1819 and 1825. Further travels, to Russia and Siberia, in 1829 led to the publication of *Fragments of the Geology and Climatology of Asia*

Figure 2.3 *Alexander von Humboldt working on his botanical specimens.*

in 1831. These works laid the foundations for the study of physical geography, biogeography, meteorology and climatology.

Box 2.1 Baron Alexander von Humboldt.

Humboldt was born in Tegel, now the location of Berlin's major airport. At the age of 19, he developed a lifelong interest in botany, whcih led him to investigate the laws that govern not only the diversity of plant life, but also everything that impinged on the environment. Entering the University of Göttingen to study natural sciences in 1789, he travelled to Mainz to meet Georg Forster, the naturalist from Captain Cook's second voyage. Forster encouraged Humboldt to study the basalts of the Rhine, a topic that Humboldt wrote up in his first book, in 1790. Next year, the two men travelled to England together, visiting Sir Joseph Banks, who had been the naturalist on Cook's first expedition, and Captain William Bligh, who had been on Cook's third. These encounters gave Humboldt a desire to travel and study regions not yet explored scientifically and – like Forster – to combine science and travel writing. In England, he met the physicist Henry Cavendish, who introduced him to the work of Antoine Lavoisier. Humboldt's study of Lavoisier convinced him of the importance of measurement and experimentation and of the value of scientific cooperation and the exchange of ideas. Scientific networking is not new. In June 1791, Humboldt joined Freiburg's School of Mining, run by one of the great men of geological science, Abraham Gottlob Werner (1750–1817)[19]. Werner led the so-called Neptunists, who thought that all rocks were once precipitates in the ocean. Humboldt initially followed Werner on this, for example in his work on the Rhine basalts, but eventually joined the so-called Vulcanist or Plutonist school led by James Hutton, who showed that granites were created from molten rock. Despite his mining studies, Humboldt found time to continue research on plant life, winning the Saxon gold medal for his work. In 1792, aged 22, he joined the Prussian Mining Service, rising to become inspector of mines. During his early twenties, Humboldt dreamt of writing a *Physique du Monde*, a total description of the physics of the world. His dream would come to fruition in his five-volume work *Cosmos: A Sketch for a Physical Description of the Universe*, starting with a first volume in 1845. In recognition of his outstanding contributions, many geographical features are named after him, including the Humboldt Current off the coast of Peru, as well as numerous towns, forests, streets, parks, universities, colleges and schools, a lunar crater and several plants and animals. Humboldt was awarded the Copley Medal by London's Royal Society in 1852.

Humboldt's view of nature was holistic. He saw that its parts were intimately related and were only understandable with reference to the whole, with plants growing where they did in response to relationships between biology (plants, animals and soils), meteorology (temperature, winds, humidity and cloudiness), geography (altitude, latitude and distance from coast) and geology.

2.2   Charles Lyell, 'Father of Palaeoclimatology'

The ideas of Buffon, Hutton, Humboldt, Cuvier and Brongniart had a considerable influence on a young Scottish geologist, Charles Lyell (1797–1875) (Figure 2.4, Box 2.2). Lyell was famed for turning Hutton's big idea into a fundamental geological principle that has stood the test of time, albeit with certain modifications. He was destined to become the greatest geologist of his age[20, 21]. It seems oddly fitting that he was born in 1797, the year that Hutton died.

Figure 2.4   *Charles Lyell.*

Box 2.2   Charles Lyell[20, 21].

Born at Kinnordy, near Dundee, Scotland, Lyell was brought up at Bartley Lodge in England's New Forest. The son of a wealthy naturalist after whom the plant *Lyellia* was named, he was fascinated by natural history. Studying classics at Oxford between 1816 and 1819, he attended lectures in geology given by William Buckland. Deciding to become a lawyer, he entering Lincoln's Inn in London in 1820, but his interests drew him into the emerging science of geology. Lyell rose to fame with the publication of his *Principles of Geology* in three volumes between 1830 and 1833. This was the first comprehensive geological textbook. Its 12th edition was published in 1875, just after his death. His reputation was further enhanced by the publication in 1838 of a companion volume, *Elements of Geology*. Originally intended as a supplement to the *Principles*, this formed an independent practical guide to the new science of geology. Together, the two books put the study of geology on a firm footing. Lyell's influence was further assured with his naming of a number of geological periods: the Recent (now the Holocene), the Pleistocene, the Pliocene, the Miocene and the Eocene. From 1831 to 1833, Lyell was the first professor of geology at London's fledgling Kings College, but he later earned his living as a geological writer. His influence stretched far and wide, through publications, lectures and his association with the Geological Society of London. Having been elected a fellow of the Society in 1819, and after publishing his first paper there in 1823, he became one of its joint secretaries from 1823 to 1826, its foreign secretary from 1829 to 1835, a vice president for 20 sessions and its president in both 1835–37 and 1849–51. His talent was recognised early. He was elected a fellow of the Royal Society in 1826 and received its Copley Medal in 1858. The Geological Society awarded him its Wollaston Medal in 1866. Recognising his huge contribution to understanding of the Earth, he was knighted by Queen Victoria in 1848, at the age of 51, and made a Baronet in 1864. The year he died, the Geological Society inaugurated the prestigious Lyell Medal. Lyell has a crater on the Moon and a crater on Mars named after him, along with an Antarctic glacier and several mountains. He was buried in Westminster Abbey, an honour reserved for few scientists. His burial memorial reports that, '*For upwards of half a century he has exercised a most important influence on the progress of geological science, and for the last twenty-five years he has been the most prominent geologist in the world.*'

The key to Lyell's understanding of the Earth lies in the subtitle to his *Principles of Geology*, namely '*An attempt to explain the former changes of the Earth's surface by causes now in operation*', which demonstrates the influence of Hutton on his thinking. Lyell's conception that the same natural laws and processes that operate in the universe now have always operated, and that they apply everywhere, was later named 'uniformitarianism' by William Whewell (1794–1866). In effect, Lyell took Hutton's ideas and magnified them a hundredfold, showing how they applied to the many different aspects of geology, from fossil life to volcanoes. In doing so, he was labouring to overcome the catastrophist theories of scientists like Cuvier. Lyell believed that what appeared from the geological record to be the results of catastrophic events could instead have arisen through the slow and steady action of processes observable today. Like Hutton, he thought immense periods of time were required to wear down the land and deposit the sediments eventually represented by various uplifted strata. This would not endear him to strict interpreters of Genesis.

Lyell drew heavily on contemporary geological literature to produce the *Principles*[20–22]. Particularly influential was *Conchiologia fossile subapennina* (*The Fossil Seashells of the SubAppenines*), published in 1814 by the Italian geologist Giovanni Battista Brocchi (1772–1826), curator of the Museum of Natural History in Milan[3, 22]. Lyell was fascinated by Italy. Besides honeymooning there, he studied the geology with local guides, read the Italian literature and met local specialists. He may have read Brocchi's work in Italian, or else the English translation made in 1816 from a copy given to William Buckland during a visit to Milan in that year[23].

An expert on the fossil seashells of the Apennines, Brocchi used the change in the percentage of living forms in fossil assemblages with time as a means of dating relatively their encasing formations. Using this approach, he produced a definitive study of the historical geology of Italy, an advance comparable to that made by Smith in England and Brongniart and Cuvier in France. Comparing modern and ancient molluscs, he noticed that the recent species of older Tertiary strata now inhabit warmer climates, suggesting, much as Lamarck had seen in the Parisian region, that the world was cooling. Lyell took note both of the approach and its conclusion.

Young Lyell hoped to meet Cuvier during his first visit to the continent on a tour with his family in June 1818. Cuvier was away, however, so Lyell peeked into his office, looked at some of his fossil specimens and read his paper on the 'Geology of the Country around Paris'. He went on to climb the glaciers around Chamonix and the Grindelwald glacier in Switzerland, which gave him an inkling of the power of ice. This was the first of many visits to all parts of the United Kingdom, to much of Europe and to North America, which would make him the best-travelled of the geologists of his generation. Seeing the most rocks is one

route to becoming an excellent geologist, and Lyell saw plenty. Equally important is becoming fully submerged in the world of ideas about the subject, which Lyell managed by meeting and corresponding with all of the major geological figures of his time in Europe and North America.

Lyell eventually met Cuvier, along with Brongniart and Brongniart's former student Constant Prévost (1787–1856), when visiting Paris in 1823 to improve his French[20]. He was impressed to find that young Prévost, unlike Cuvier, thought that the changes in strata in the Paris Basin had come about gradually, not as the result of a series of catastrophic events. Others, like Karl von Hoff (1771–1837) in Germany, also concluded that, given enough time, ordinary agencies could effect major changes. Over the years, Lyell and Prévost worked closely together, recognising strong similarities between the Mesozoic strata of Normandy and of southern England.

Lyell became expert at identifying fossil molluscs. By 1828, following Brocchi, he had used the percentages of modern molluscs in each epoch, and the relations of strata to one another, to subdivide the Tertiary Period into several geological Epochs. This statistical approach was a novelty at the time. Perhaps Lyell was following Humboldt's dictum that all science should be based on numbers. The following year, he met Gérard Deshayes (1795–1875), a French palaeontologist with an even larger collection of fossil molluscs, who had arrived at similar views. He persuaded Deshayes to expand on his work and combine it with Lyell's own, publishing the results in the *Principles*, where he named the four periods of the Tertiary as the Eocene ('dawn of the recent', with 3.5% modern species), Miocene (with 17% modern species), Early Pliocene (with 35–50% modern species) and Late Pliocene (with 90–95% modern species).

Later, Lyell worked closely with the Danish palaeontologist Henrick Beck (1799–1863) to extract yet more information from fossil molluscs, finding that Europe's Eocene had a tropical climate, its Pliocene had a climate more like today's and the Miocene lay in between. In Chapter 10 of Volume 2 of the *Principles*, he established that there was '*a great body of evidence, derived from independent sources, that the general temperature has been cooling down during the epochs which immediately preceded our own.*' Later palaeobotanical work confirmed this. Large pointed leaves with many stomata and thin cuticles typical of warm humid climates characterised Europe's early Tertiary and the tropical rainforest flora of the Eocene London Clay[24].

Lyell was much influenced by Humboldt, whom he met in Paris in 1823 and again in Potsdam in 1850. He was particularly taken with Humboldt's holistic view of nature and his observations of the way in which the distribution of plants reflected both the geography and the climate. Lyell was also among the first to appreciate the geological significance of Humboldt's 'isothermal lines': lines of equal temperature that could be used to divide the world into climatic zones[14]. Observing that the positions and sizes of continents and the development of mountain ranges distorted those lines and climatic zones, he made a crucial intellectual leap: recognising that that many of Europe's older rocks had been deposited in much warmer climates than today's, he deduced that if the Earth's climate zones had not changed, then the land must have moved – the geography must have changed with time (Figure 2.5). Writing to Gideon Mantell (1790–1852) in February 1830, and swearing him to secrecy, he said, '*I*

will give you a receipt [i.e. recipe] for growing tree ferns at the pole, or if it suits me, pines at the equator; walruses under the line [the Equator], and crocodiles in the arctic circle'[25]. This exciting new idea profoundly changed the way people thought about the distant past.

Lyell acknowledged his debt to Humboldt in a letter to his geological friend George Poulett Scrope (1797–1876): '*Give Humboldt due credit for his beautiful essay on isothermal lines: the geological application of it is mine, and the coincidence of time 'twixt geographical and zoological changes is mine, right or wrong*'[25]. Would his theory hold the test of time? In the same letter, Lyell confessed '*That all my theory of temperature will hold, I am not so sanguine as to dream. It is new, bran new [at that time, the term 'bran new' was interchangeable with 'brand new']*'[25].

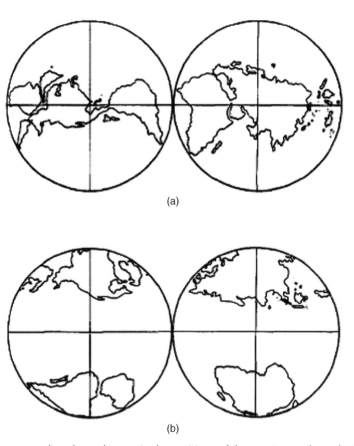

(a)

(b)

Figure 2.5 *Lyell's attempt to show how changes in the positions of the continents through time might contribute to extremes of (a) heat or (b) cold.*

I focus on Lyell because he was the first scientist to concentrate intently on the geological record of past climates, and it would be fair to call him the 'father of palaeoclimatology'. He devoted three chapters of Volume I of the *Principles* to showing how the climates of past times could be recognised from the types and distributions of sedimentary rocks and their enclosed fossils – especially the seashells he so enjoyed studying. Not only that, but he also incorporated seven chapters on 'aqueous causes', under which he listed rivers, torrents, springs, currents, tides and icebergs as agents of change in the inorganic world, all of which were likely to change as climate did. Lyell began his climate chapters by recapping the approach he took in a scientific paper in 1825–26, where he deduced the likely conditions of deposition of fossil freshwater limestones[26]. His chapters on climate rehearsed the standard arguments for climate change from fossil evidence. He agreed with Buffon, Cuvier and Brocchi that the fossil evidence showed that Europe's climate was much warmer in former times. Unlike Cuvier, he found no need for some catastrophe to explain the cooling, and, like Fourier, he thought Buffon wrong to suggest that this was due to the solid Earth having been hotter in former times.

Lyell's thinking on the geographical control of climate matured as he gathered more data – especially from his visits to North America in the 1840s. For example, in the 12th edition of his *Principles*, published in 1875, anticipating later notions of the break-up of formerly continuous continents, he observed that '*If we go back ... to the Eocene period ... we find such a mixture of forms now having their nearest living allies in the most distant parts of the globe, that we cannot doubt that the distribution of land and sea bore scarcely any resemblance to that now established*'[27]. Along the same lines, he noted that '*In the case of the great Ohio or Appalachian coal-field ... it seems clear that the uplands drained by one or more great rivers were chiefly to the eastward, or occupied a space now covered by the Atlantic Ocean*'[27]. Nothing he had discovered in 45 years of publishing the *Principles* detracted from his conclusion that '*Continents therefore, although permanent for whole geological epochs, shift their positions entirely in the course of ages*'[27]. He was well aware of how geography – manifested as the positions of continents, their coasts and their topography – modified climatic zones, observing that, '*on these geographical conditions the temperature of the atmosphere and of the ocean in any given region and at any given period must mainly depend*'[27].

Moving the continents around was dramatic stuff at the time, and, unfortunately for Lyell, he had no other information than his climate theory to back him up. It would be left to others to prove him correct as more data arrived. In due course, Lyell's uniformitarian assumption that the Earth's average temperature had remained more or less constant through time would be proved wrong, but many of his other assumptions about climate remain valid, including the notion that the continents had changed position through time.

Lyell broke with tradition in abandoning the theory in vogue in the early 19th century, and embraced by – among others – his old Oxford tutor, William Buckland (1784–1856) (Figure 2.6), that the erratic blocks of rock littering the British landscape were the relics of Noah's flood. Like Hutton, Lyell was keen to take the scripture out of geology and the geology out of scripture. Buckland's attempts to relate geology to scripture are hardly surprising, given that he was an Anglican clergyman. An influential man, he was twice elected president of the Geological Society, in 1824–26 and 1839–41. But he had an open mind, and eventually abandoned the 'deluge' hypothesis.

Just as Humboldt had influenced Lyell, so Lyell too influenced a younger man with a big future, Charles Darwin (1809–1882), who took Volume 1 of Lyell's *Principles* with him when he sailed in late 1831 on his scientific voyage around the world on HMS *Beagle*. Darwin became a fellow of the Geological Society of London

Figure 2.6   *William Buckland.*

in 1836, and, like Lyell, served the Society as an officer, being a member of council from 1837 to 1850, one of the two joint secretaries from 1838 to 1841 and a vice president from 1844 to 1845. Like Lyell, he was made a fellow of the Royal Society, was awarded the Geological Society's Wollaston Medal (1859), was awarded the Royal Society's Copley Medal (1864) and was buried in Westminster Abbey.

Lyell's notion of significant change arising from slow processes operating steadily over the eons of geological time provided Darwin with the long periods through which tiny natural variations, which we now understand as genetic mutations, could accumulate and give rise to different species through natural selection. The two men met in October 1836, shortly after Darwin's return, and became friends. Lyell nominated Darwin to the Council of the Geological Society, and later helped to ensure that on 1 July 1858 Darwin's paper on natural selection was read at the Linnaean Society in London, alongside that of Alfred Russell Wallace, who had reached the idea independently. Darwin was also influenced by Humboldt, whom he met in London in 1842. As Richard Holmes pointed out in 2008 in *The Age of Wonder*, '*Science is truly a relay race with*

each discovery handed on to the next generation … and the world of modern science begins to rush towards us'[28].

Lyell's *Principles* considerably influenced thought in Victorian times. As James Secord points out in his introduction to the Penguin Classic version in 1997, the *Principles* was '*a manifesto for fundamental change in the organisation of intellectual life*', capping the campaign by Lyell and others '*to sever all links of geology to a theology based in scripture*'[29]. After Lyell, geologists no longer accepted the biblical flood as having any worth in analysing Earth's history. For a modern view of Lyell's merits, we may turn to *Time's Arrow, Time's Cycle*, the 1987 work by palaeontologist Stephen Jay Gould, for whom Lyell '*doth bestride my world of work like a colossus*'[30].

Nevertheless, the *Principles* had its flaws. For example, the first volumes, published in 1830–32, did not mention the Ice Age. Like James Hutton, Lyell knew that mountain glaciers transport rock debris, which is dumped in piles called moraines where the glaciers melt, and that rivers sweep the glacial sand and mud away to the sea. But he also knew that polar mariners had seen drifting icebergs transporting large amounts of rock (Figure 2.7). This observation led him to speculate that melting icebergs would dump their loads on the seabed to '*offer perplexing*

(a)

Figure 2.7 *Ice transporting rocks at sea: (a) stranded iceberg carrying a load of rocks in the Fridtjof Channel, Antarctic Peninsula; (b) ice floe carrying a load of rocks in the Erebus and Terror Gulf, Antarctic Peninsula.*

(b)

Figure 2.7 *(continued)*

problems to future geologists[27]. In Volume 3 of the first edition of the *Principles,* he speculated that the huge erratic blocks of rock littering the landscape in the Alps and the Jura had been transported by floating ice, not by ice sheets sliding over land, as some Swiss geologists thought at the time. At that time, he did not connect the erratic blocks littering the Swiss landscape with those of the United Kingdom.

2.3 Agassiz Discovers the Ice Age

One great mind is never enough – it requires many for us to get to grips with how the world works. And so it was left to another geological genius, Jean Louis Rodolphe Agassiz (1807–1873) (Figure 2.8, Box 2.3), to point out how important ice may have been in the geological history of Europe and North America.

Like Lyell, Agassiz upset the comfortable world of established thought. In the summer of 1837, he turned the world of geological ideas upside down with a proposal made at the annual meeting of the Swiss Society of Natural Sciences, of which he was the new president: he thought

Figure 2.8 *Louis Agassiz drawing radiates, 1872.*

a vast ice sheet must have carried erratic blocks across Europe in a recent Ice Age. Living in Switzerland, he knew that far from the snouts of glaciers, the rock surfaces were

Box 2.3 Jean Louis Rodolphe Agassiz.

Agassiz was born in Switzerland in 1807. He studied medicine and natural sciences at Zurich, Heidelberg and Munich before moving to Paris, where he studied with Humboldt and Cuvier. In 1832, he was appointed professor of natural history at Neuchatel in the Swiss Jura. He published five volumes of research on fossil fish between 1833 and 1843. In 1846, he visited the United States and was invited to stay there, becoming head of the Lowell Scientific School of Harvard University in 1847. Harvard made him a professor of zoology and of geology, and he founded the Museum of Comparative Zoology there in 1859, serving as its director until he died. He was awarded the Geological Society of London's Wollaston Medal in 1836 for his work on fossil fish, and was elected one of the Society's foreign fellows in 1841. He thought highly of Charles Lyell, and named an ancient jawless fish after him: *Cephalaspis lyelli*, which lived in Scottish lakes in Old Red Standstone times in the Silurian and late Devonian Periods between 420 and 360 Ma ago. Like Lyell, Humboldt and Darwin, Agassiz became one of the world's best-known scientists of the 1800s. Mountains, glaciers and a Martian crater are named after him, as are several fish and beetles, a fly and a tortoise. A fossil glacial lake is also named after him, as is the Agassiz Glacier in the United States' Glacier National Park.

scratched and smoothed, and strewn with boulders and rubble like that still being carried and deposited by the ice. His new idea countered the suggestion of the first scientific explorer of the Alps, Horace Bénédicte de Saussure, in the late 1700s, that fast rushing streams deposited these boulders in catastrophic events. Hutton disagreed with Saussure, proposing in his 1795 *Theory of the Earth* that a former extension of Swiss glaciers accounted for the distribution of boulders of Mont Blanc granite, which de Saussure attributed to a deluge. Hutton wrote, '*There would then have been immense valleys of ice sliding down in all directions towards the lower country, and carrying large blocks of granite to great distance, where they would be variously deposited and many of them remain an object of admiration to after ages, conjecturing from whence or how they came*'[6].

Later Swiss observers agreed with Hutton. In 1818, Jean Pierre Perraudin, a guide and chamois hunter, interpreted the gouges in hard unweathered rock as indicating the former widespread extent of Alpine glaciers. His remarks came to the attention of Ignace Venetz, chief engineer for the Swiss Canton du Valais, who in 1821 deduced from the positions of old terminal moraines downslope from present glacier terminations that the climate had warmed and the glaciers shrunk. He presented this finding to the annual meeting of the Swiss Society of Natural Sciences in 1829, suggesting that glaciers had once extended over the Jura and into the European plain. Jean de Charpentier (1786–1855), director of mines of the Canton de Vaud, applied Venetz's observations more widely, proposing to the annual meeting of the same Society in 1834 that widespread erratic blocks and moraines had been deposited by ice – Swiss glaciers had formerly been much more extensive.

Agassiz, who attended Charpentier's lecture, had been one of his students. A trip into the field with Charpentier in 1836 to study the evidence for glacial transport convinced him. We'll let Elizabeth Agassiz tell us about her husband's ensuing lecture to the same Society in 1837: '*In this address he announced his conviction that a great ice-period, due to a temporary oscillation of the temperature of the globe, had covered the surface of the earth with a sheet of ice, extending at least from the north pole to Central Europe and Asia … "Siberian winter," he says, "established itself for a time over a world previously covered with a rich vegetation and peopled with large mammalia, similar to those now inhabiting the warm regions of India and Africa. Death enveloped all nature in a shroud, and the cold, having reached its highest degree, gave to this mass of ice, at the maximum of tension, the greatest possible hardness". In this novel presentation the distribution of erratic boulders, instead of being classed among local phenomena, was considered "as one of the accidents accompanying the vast change occasioned by the fall of the temperature of our globe before the commencement of our epoch" … This was, indeed, throwing the gauntlet down to the old expounders of erratic phenomena upon the principle of floods, freshets, and floating ice*'[31].

Much astonishment and not a little ridicule greeted Agassiz's proposal that an ice sheet like that now covering Greenland formerly covered much of northwest Europe as far south as the Mediterranean. The great German geologist Leopold von Buch (1774–1853) attended the meeting and could hardly conceal his indignation and contempt for this young upstart[31]. It was von Buch who had first identified the erratic blocks littering the north German plain as having come from Scandinavia, by some unknown means. Even Humboldt, who knew Agassiz from the time they had spent together in Paris, counselled his young friend to abandon '*these general considerations (a little icy besides) on the revolutions of the primitive world – considerations which, as you well know, convince only those who give them birth*'[31].

Some of the reluctance to accept Agassiz's idea stemmed from the fact that very few European scientists knew anything about the extent of ice sheets. The vast extent of the Antarctic ice sheet would not be fully appreciated until after the visits to the Ross Sea and Ross Ice Shelf with HMS *Erebus* and HMS *Terror* in 1841 and 1842 by James Clark Ross (1800–1862), whose book on his expedition was not published until 1847. The icy mass of East Antarctica had been only glimpsed before, by Von Bellingshausen in 1820 and by Dumont d'Urville and Charles Wilkes in 1840. It was not even known that a vast continuous ice sheet covered Greenland. Even so, Agassiz had leapt ahead of himself by claiming that ice extended as far as the Mediterranean, when glacial erratic blocks were actually confined to the Alps and northernmost Europe.

Undeterred, Agassiz wrote up his ideas for an English-speaking audience in a short paper published in 1838[32], in which he set out the evidence for glacial activity, and noted that grooved and polished rocks beneath Swiss glaciers are usually overlain by fine sand, followed by rounded pebbles and then by angular blocks – the opposite of the sequence expected from transport by currents. The fine sand came from the disintegration of rock fragments and most likely caused the polishing. He called for research to see whether this same relationship applied in the polar regions.

Agassiz pulled his theories together in his 1840 book *Etudes sur les Glaciers*[33], using the distribution of boulders that could only have been transported by ice and the gouges in the smoothed surfaces of rocks over which glaciers had passed, along with other features, to divine the former existence of ice sheets and glaciers where none now existed, and even to propose that ice sheets could move erratic blocks up hill, for example to the top of the Jura Mountains near his university, where the 3000 ton *Pierre-a-Bot* (toadstone) occurred. Charpentier published his own *Essay on the Glaciers and the Erratic Terrain of the Rhone Basin* in 1841. His Swiss-scale vista was eclipsed by Agassiz's Europe-wide vision, which became global after his arrival in North America in 1846,

where he discovered further evidence for former ice sheets. Agassiz started a major research programme on Alpine glaciers, spending a decade working at an alpine research station and climbing all over the Alps with fellow researchers and students, literally starting the study of glaciology. He published the results in 1847 in *Système Glaciare*.

Agassiz was keen to convert William Buckland, the leading proponent of the biblical deluge hypothesis for explaining erratic blocks. Buckland was intrigued enough by Agassiz's theory to visit Switzerland and see the evidence. Becoming convinced that Agassiz might be right, he invited Agassiz to visit Britain to see whether evidence of a past ice sheet could be found there too. Agassiz duly arrived on this mission in August 1840, visiting Scotland and lecturing on his new theory at the meeting of the British Association for the Advancement of Science in Glasgow. Lyell attended the meeting, but remained unconvinced.

Touring Scotland and other parts of Britain, Agassiz and Buckland found the evidence they were looking for: moraines, erratic blocks and polished and grooved rocks showed that great sheets of ice like that covering Greenland must formerly have covered the mountainous areas of Great Britain. Buckland even managed to convince Lyell that the piles of rocks near Lyell's Scottish home were moraines deposited at the edge of this former ice sheet[34]. In November and December 1840, Agassiz, Buckland and Lyell gave lectures at the meetings of the Geological Society of London on their discoveries of evidence for former British ice sheets[37]. The initial reaction was hostile[25]. Buckland concluded the 1840 meeting in high spirits by condemning to '*the pains of eternal itch without the privilege of scratching*'[36] anyone who challenged the evidence supporting the Ice Age theory. But although the papers were read at the Society's meetings, and précis were published by the Society's secretaries to convey the main points to the readers of the Society's Proceedings, the full papers were never published[34]. Part of the problem was that another Scottish lion of the British geological scene, Roderick Murchison (1792–1871), who had visited Scotland with Agassiz and Buckland, was unconvinced. Murchison had been president of the Geological Society in 1831–33, and was again in 1841–43. During his presidential address to the Society in the latter term, he chose to attack the Ice Age theory. He did not back away from this stance until 1862, when he finally recanted in an address to the Geological Society of London. Sending a copy of his 1862 paper to Agassiz, he wrote: '*I have the sincerest pleasure in avowing that I was wrong in opposing as I did your grand and original idea of my native mountains. Yes! I am now convinced that glaciers did descend from the mountains to the plains as they do now in Greenland*'[37]. The evidence had mounted.

In his 1840 paper[37], Agassiz explained that rivers draining the massive ice sheet that had brought the erratic boulders to the plains of northern Europe had also given rise to widespread outwash gravels, for which there was no other explanation. The existence of this ice sheet indicated that a period of intense cold – an Ice Age – had intervened between the warm conditions of the Tertiary period and those of today. Modern mountain glaciers were the remnants of that former ice sheet. Having found polished rocks, erratic blocks and outwash gravels across much of Scotland, Ireland and the north of England, along with rounded hillocks of ice-cut rock named '*roches moutonnés*', he deduced that an ice sheet had covered these areas too. The distribution of erratic blocks suggested that they had moved in all directions away from 'centres of dispersion', which would not be expected for deposition from floating ice. The main centres of dispersion in the British Isles were the mountains of Ben Nevis, the Grampians, Ayrshire, the English Lake District, Wales, Antrim, Wicklow and the west of Ireland. Floating ice from Scandinavia explained the origin of erratic blocks on the east coast of England.

2.4   Lyell Defends Icebergs

Lyell met Charpentier in 1832 in Switzerland, while on his honeymoon, and so was exposed to Charpentier's ideas[20]. He also met Agassiz several times during the 1830s, and they worked together on fossil fish for a while[20]. They met again when Agassiz visited Buckland in 1840, and together presented their papers on glaciers to Geological Society meetings late that year. But Lyell was a hard man to convince, and the extent to which he accepted the notion that sheets of ice had transported boulders was distinctly limited. Having seen moraines in the Alps, he was not going to deny the role of mountain glaciers in transporting erratic blocks. But – what happened beyond the mountains? Lyell thought icebergs had done the work.

Lyell used the word 'till', a Scottish farmers' term, to describe the widespread unstratified jumbled mass of erratic blocks, pebbles and clay covering parts of the British Isles, and which we now call 'boulder clay'. He lumped 'till' together with other deposits from the glacial era (like Agassiz's outwash fans of gravel) into what he called the 'glacial drift', a term chosen on the one hand to support his iceberg theory and on the other to replace the former term 'diluvium', which came from the biblically inspired 'flood' hypothesis formerly used to explain the distribution of this recent debris[20].

Agassiz was unable to explain in detail the origin of till, but assumed that it was derived from the ice sheet in some way, not least because the boulders in the till were typically striated and gouged like ice-transported rocks. He imagined that boulders now found as erratic blocks might have slid down the Alpine slopes and out across his proposed European ice sheet in some catastrophic fashion, to be left behind when the ice beneath them melted. Lyell, in contrast, offered a noncatastrophic 'steady-state' mechanism: the supply of rocks, pebbles and rock flour from floating icebergs. We now know more than both Lyell and Agassiz. The fine-grained clayey element of 'boulder clay' is rock 'flour' or powdered rock, derived from rock fragments interacting with each other and with the surrounding country rock as ice sheets move over the ground.

In his 1840 paper, Lyell suggested that '*the assumed glacial epoch*'[37] had arisen as Scottish glaciers first advanced to the sea, as they did in South Georgia, then remained stationary while the intervening hollows filled with snow and ice, on which boulders slid to their present positions – much as Agassiz was suggesting for the erratic blocks of western Europe. The ice then retreated, leaving moraines and debris behind. To explain the origin of boulder clay or till away from mountainous areas where glaciers could provide a means of transport, Lyell called, as he had in his *Principles*, on the transport of rocks and sediment by floating ice[37]. He rejected Agassiz's idea that some catastrophic event had caused boulders to fall off the Alps and slide out over the European ice sheet, because he believed that the Alps rose gradually, in consistency with his uniformitarian principles.

Although he was exposed back in 1832 to Charpentier's observation that Swiss glaciers moved boulders, Lyell was equally impressed during his visit to Sweden in 1834, where he saw along the coast granite boulders that appeared to have been carried by floating ice[10]. He was also impressed by accounts from mariners of boulders carried on icebergs. Not long after his excursion with Agassiz, he was told of similar observations made by Joseph Hooker on the James Clark Ross expedition to Antarctica in 1839–41. Darwin too had reported rocks being carried out to sea by icebergs broken off from glaciers in southern

Chile, as Lyell reported in his *Elements of Geology*. Lyell just missed observing this phenomenon for himself when he was crossing the Atlantic on his way to and from the Americas in the mid to late 1840s[38]. Writing to his sister Carry from the steamship *Britannia* in June 1846, he reported, '*We passed fifty icebergs or more in daylight ... One iceberg ... which came close to us when I was below, had a large rock twelve feet square on the top and as much gravel and dark sand on its side*'[25]. While this confirms my own observation that such occurrences are rare, it was enough to make Lyell stick to icebergs as accounting for the distribution of erratic blocks away from mountainous glaciated regions like Scotland and Scandinavia, no matter what Agassiz said.

Like Agassiz, Lyell also saw glacial erratics in North America. During his visit to the United States in 1853, his host James Hall took him to see trains of erratic boulders in the Berkshire Hills of western Massachusetts. Lyell's biographer explains[39], '*The boulders were distributed in long parallel rows, extending in nearly straight lines across ridges and valleys from their starting points on the Canaan Ridge. Their direction was nearly at right angles to the lines of the ridges and bore no relation to the direction of the streams and rivers. The boulders were rounded like the glacial boulders called in Switzerland roches moutonnés ... one of the larger boulders ... [near the meeting house in Richmond] was fifty-two feet long, forty feet wide, and, although partially buried, fifteen feet high*'[40]. *The boulders rested on a deposit resembling the European "northern drift". Where the underlying rock was exposed, its surface was polished, striated, and furrowed, with the furrows running in the same direction as the trains of boulders. Lyell thought that the trains of boulders must have been transported by floating ice at a time when the Berkshire hills stood at a much lower level, with only their highest ridges protruding above the sea. He thought their transport could not be explained by glaciation, because if glaciers had transported the boulders, the trains of boulders should have been distributed down the valleys instead of across them. In fact, the boulders had been transported by glaciers, but by continental glaciers rather than by mountain glaciers, the only ones with which Lyell was familiar.*' Nowadays, we would say 'transported by continental ice sheets' rather than by continental glaciers.

The contrast between Lyell and Agassiz was one of vision. Lyell stuck to what he knew to be true: glaciers occupied valleys and carried boulders down them, and, where they met the ocean, icebergs carrying boulders might break off and carry their burden of rocks out to sea.

Agassiz could envision a merging of mountain glaciers into great sheets of ice covering entire landscapes, ploughing across and shaping the land and dumping clay and boulders en route. 'God's Great Plough', he called the ice sheet.

Although in later years he would back away from Lyell's adherence to transport by icebergs, Charles Darwin initially followed Lyell's line closely in a paper on the glaciers of Caernarvonshire, in Wales[41]. Investigating the moraines near Lakes Ogwyn and Idwell in the Welsh mountains, he deduced that the glaciers from the valleys in which those lakes now sat had formerly united and plunged down the valley of Nant-Francon towards Bethesda, where they had dumped in the sea a whitish earth full of rounded and angular boulders that were deeply scored like the rocks over which a glacier had passed. Following Lyell's line, he assumed that the boulders had been dropped into this mud from floating icebergs, and that the land had since been uplifted. '*By this means*', he said, '*we may suppose that the great angular blocks of Welch [sic] rocks scattered over the central counties of England were transported*'[41]. He concluded '*that the whole of this part of England was, at the period of the floating ice, deeply submerged ... I do not doubt that at this same period the central parts of Scotland stood at least 1300 feet beneath the present level, and that its emergence has since been very slow. The mountains at this period must have formed islands, separated from each other by rivers of ice, and surrounded by the sea*'[41]. Lyell would have approved.

Like Lyell, Darwin accepted that there must also have been vast thicknesses of land ice locally, as a source for floating icebergs. His letter to W.H. Fitton[42] is a reminder that one may often not be able to 'see' what is under one's nose. On a field trip to Capel Curig in North Wales, he wrote: '*the valley about here, & the Inn, at which I am now writing, must once have been covered by at least 800 or 1000 ft in thickness of solid Ice! – Eleven years ago, I spent a whole day in the valley, where yesterday every thing but the Ice of the Glacier was palpably clear to me, and then I saw nothing but plain water, and bare Rock. These glaciers have been grand agencies*'[42]. But he then went on to extol the virtues of the power of drifting icebergs to distribute erratic blocks: '*I am the more pleased with what I have seen in N. Wales, as it convinces me that my views, on the distribution of the boulders on the S. American plains having been effected by floating Ice, are correct*'[42]. It would take a lot for Darwin to withdraw support from Lyell.

Lyell stuck to the iceberg theory more or less unchanged throughout his life. In the second edition of his *Elements of*

Geology, published in 1841, he admitted that small glaciers might once have existed in Scotland, but dismissed the theory that the widespread British deposits of 'glacial till', comprising mixed boulders and clay, had been deposited beneath an ice sheet, preferring still to think of them as deposited from floating icebergs. By the time he published *Antiquity of Man* in 1863, he had accepted the refrigeration of the climate in the post-Tertiary Pleistocene that Agassiz had postulated, and that this had led to large areas of Britain and northwestern Europe becoming covered by 'glacial drift'. Lyell's hypothesis that boulder clay was deposited from floating icebergs required that much of England north of a line joining the estuaries of the Thames in the east and the Severn in the west, as well as much of the northwest European plain, had been submerged. He explained away the grooves carved into exposed rocks on hillsides as having been made by stones embedded in the bottoms of icebergs, rather than – as Agassiz would have it – by stones embedded in a moving ice sheet.

Lyell accepted that the ice originated in glacial dispersion centres on highlands in Scandinavia, Scotland, Wales and the English Lake District. But he thought that those centres were limited in extent and discharged their ice into a surrounding ocean, rather than into a surrounding ice sheet like that of Greenland. Lyell also agreed with Agassiz's suggestion that within those distribution centres, glacial lakes dammed by ice were locally important, the beaches of different lake levels explaining the terraces or 'parallel roads' around Scotland's Glen Roy. His interpretation of the terraces around Glen Roy was not original: it had first been proposed by the Scottish geologist John MacCulloch in 1817, when he was president of the Geological Society of London.

Where Lyell and Agassiz differed profoundly was in explaining the origin of the glacial drift. In *Antiquity of Man*, Lyell expanded on his marine glacial theory, suggesting that during the Ice Age much of England and northwest Europe must have been submerged to depths of more than 600 feet, Scotland to depths of as much as 2000 feet and Wales to a depth of 1350 feet. By 1875, in the 12th edition of *Principles*, these figures had changed to 'perhaps' 500 feet in Scotland and 2000 feet in Wales. As is clear from that edition, much of his argument for submergence rested upon the occurrence of seashells at high altitudes among the boulder clay. For someone who denied any role for catastrophism in geology, Lyell was sailing close to the wind in invoking unexplained forces that could periodically lift the United Kingdom and Europe above the sea and then submerge them, during

the small amount of geological time represented by the Ice Age.

By 1848, Darwin began to realise that sticking to the Lyellian view required some contorted thinking[43]. Trying to answer a common criticism of the time – that floating ice could not carry erratic blocks from a lower to a higher level – he suggested that, with repeated subsidence of the land, floating ice could gradually deposit boulders at progressively higher levels. Special pleading, indeed! The subsidence would have to have been significant and more or less immediate, something for which there was no apparent mechanism, and to have been continually repeated. A certain Mr Nicol objected '*that when the parent rock was once submerged, no further supply of boulders could be derived from it*'[43]. Darwin confessed, '*this appears to me an objection of some force*'[43]. Well he might! Not to be deterred, he argued that the piling up of ice by storms along a shore would raise boulders above their original level. Rather a weak response, considering that erratic boulders of immense size occurred 900–1000 feet above the strata from which they had been carved. He would recant, as we see in Chapter 3.

Enter Archibald Geikie (1835–1924) (Figure 2.9, Box 2.4), a young Scottish geologist, who roundly criticised Lyell's iceberg transport theory early in the 1860s.

From his detailed examination of the Ice Age geology of Scotland, Geikie concluded that the land must have been shaped by the actions of a giant ice sheet, the remains of which mantled its surface as 'drift' deposits[44]. Geikie called for the iceberg theory to be abandoned forthwith. He said that he hoped he might have convinced Lyell of

Figure 2.9 *Archibald Geikie.*

the correctness of his conclusions. Lyell did read Geikie's book, writing to his wife in May 1863 that *'Geikie's book on the Glacial Period in Scotland is well done … '*[25]. Nevertheless, that same year – 1863 – Lyell published *Antiquity of Man*, with its illustrations showing Great Britain drowned beneath an iceberg-flooded sea! In the 12th edition of his *Principles*, Lyell continued with his ice-flooded sea, but conceded a little ground to Geikie, noting that in Scotland *'some examples of this … striation may have been due to the friction of icebergs on the bed of the sea during a period of submergence; others to a second advance of land glaciers over moraines of older date'*[27].

Lyell did reverse his conclusion about seaborne transport in one case. In the 12th edition of *Principles*, he reported that, on a visit to Switzerland in 1857, the local geologists had convinced him that an ice sheet had filled the Valley of Switzerland between the Alps and the Jura and transported down into it and up the other side the erratic blocks now found 50 miles away from the Alps, atop the Jura Mountains. Writing to his father-in-law, Leonard Horner, from Zurich in 1857, he said, *'If the hypothesis now adopted here to account for the drift and erratics of Switzerland, the Jura, and the Alps be not all a dream, we must apply the same to Scotland, or to the parts of it that I know best. All that I said in May 1841 on the old glaciers of Forfarshire … I must reaffirm'*[25]. In a letter to J.W. Dawson in February 1858, he went further, calling for glaciers (not icebergs) to transport erratics and drift on to the plains of the River Po in northern Italy[25].

By the 12th edition of the *Principles*, Lyell's conversion to the Ice Age cause was more or less complete. He recalled seeing that many of the rocky surfaces exposed in Switzerland were *'smoothed and polished, and scored with parallel furrows, or with lines and scratches produced by hard minerals … The discovery of such markings at heights far above the surface of the existing glaciers, and for miles beyond their present terminations,'* he said, *'affords geological evidence of the former extension of the ice beyond its present limits in Switzerland and other countries'*[27]. Although this meant that Agassiz had been right all along about the Swiss erratics, Lyell could not accept that Agassiz's theory could be extended beyond the Swiss region, except in mountainous places like Scotland (and presumably Scandinavia).

Next into the lists was yet another Scottish geologist, James Geikie (1839–1915) (Figure 2.10, Box 2.5), younger brother of the more famous Sir Archibald.

Following in his illustrious brother's footsteps, James amassed a vast storehouse of knowledge of the geology of the glacial and interglacial periods of the Ice Age from all over the world, publishing his tome *The Great Ice Age* in 1874[45]. The comprehensive 3rd edition, published in 1894, included a chapter on the glaciations of North America by the great American geologist T.C. Chamberlin. James's most telling fact came from Nansen's observation that ground moraines beneath the Greenland ice sheet were visible in arches and tunnels under the ice front, where one

Figure 2.10 *James Geikie.*

Box 2.5 James Geikie.

James Geikie was born in Edinburgh and educated at the university there. He served on the Geological Survey from 1862 to 1882, when he succeeded his brother Archibald as Murchison Professor of Geology and Mineralogy at Edinburgh University. He was elected a fellow of the Royal Society in 1875 and awarded the Geological Society of London's Murchison Medal in 1889. He published a standard textbook, *Outlines of Geology*, in 1886. An Alaskan glacier is named after him.

could see a bluish clay charged with blunted and scratched boulders. Geikie confirmed, '*Most ... icebergs [are] free from inclusions [meaning rocks, pebbles or clay] of any kind ... Nothing has been observed to lead us to believe that parallel striations and markings, like those produced by glaciers, are ever the result of iceberg action*'[45].

Having examined glaciers in the field, he went on to note that '*The finer material – the "flour of rocks" resulting from this action [glaciers with rock debris embedded in their bases and grinding away at the rocks beneath] ... renders the glacial rivers turbid and milky*'[45]. If there were no river to wash the 'flour' out, it would accumulate beneath the ice. A further argument against the iceberg hypothesis was that it should lead to deposits that were sorted, the coarser material being deposited first, the finer having settled last – not at all like the higgledy-piggledy nature of boulder clay.

A final telling point in his mind was that the topography itself was witness to the passage of an ice sheet, with humpbacked features like *roches moutonnées* (carved from underlying rock), drumlins (mounds made of squeezed up boulder clay), kames (collapsed piles of gravel), eskers (long, winding ridges of sand and gravel deposited by streams beneath the ice) and kettles (potholes eroded beneath the ice and now filled with water – witness the many glacial lakes of Canada and Finland). From the difference between river-cut and ice-carved topography, James Geikie deduced that the British ice sheet had covered the northern Pennine Hills – the backbone of England – but not the southern part.

On to the scene in 1875 strode yet another Scottish geologist, James Croll, at the time about to join the Geikies and Lyell as a fellow of the Royal Society. We'll see more of him in Chapter 3. Like the Geikies,

Croll could not find any evidence that icebergs produced striations when ploughing through the seabed, but he found ample evidence for ice sheets having done so when passing over land[46]. Reexamining the travellers' tales that Lyell had used to support his theory that icebergs transported stones, Croll found that such icebergs were rare. Besides, there were no reports of icebergs carrying clay. My own observations support his conclusion. Only one of the many icebergs I have seen around Greenland or in the Antarctic carried rocks, as did one ice floe among a myriad others that I sailed through in Erebus and Terror Gulf in the western Weddell Sea (Figure 2.7). Even so, Ice Age iceberg drift does explain the occurrence of blocks of granite and related continental rocks dredged from the Mid-Atlantic Ridge in the northern North Atlantic[47] and the African continental shelf south of Cape Town[48].

Croll agreed with the Geikies that the clay of boulder clay was eroded from the ground by the action of ice sheets on land and left behind as the ice melted. Where the boulder clay contained marine shells, for example in Caithness in Scotland, Croll showed that they were as fragmented and striated as the adjacent boulders. He inferred that the Scandinavian ice sheet had extended across the North Sea, and had ripped marine clays and shells from the bed of the sea en route, depositing them on Scotland.

Modern geologists confirm that the boulder clay of Europe's glacial 'drift' is the 'ground moraine' or 'bottom moraine' of the thick ice sheets that covered much of Europe and North America. It is quite different from 'end moraines': the piles of debris dumped at the terminations of glaciers or ice sheets. They also agree that ice sheets scouring the bed of the North Sea en route from Scandinavia, or scouring the Irish Sea en route to Wales, incorporated marine shells from the seabed and subsequently plastered them on Britain as part of the boulder clay when the ice melted. This obviates the need for Britain and northwest Europe to have oscillated up and down to accommodate Lyell's glacial theory. As in Switzerland, ice sheets can move both down and up. The gods, they say, have feet of clay – in Lyell's case, it was boulder clay.

Ironically, we are left with the word 'drift' as a descriptor of glacial deposits deposited by moving ice sheets, despite its original application to explain deposits from drifting icebergs. Lyell's influence lingers long! But he was not alone. Other influential figures, like Roderick Impey Murchison, one time president of the Geological Society, agreed with Lyell (for example, in Murchison's presidential addresses to the Society in 1842 and 1843), which helps to explain why we had to wait until the immaculate fieldwork and compilations of the brothers Geikie in the 1860s and 1870s to expose the fallacy of believing that the 'drift' of Europe and North America originated from icebergs.

Initially, neither Lyell nor Agassiz seem to have realised that Agassiz's theory, calling for the formation of a vast ice sheet on land, implied a large drop in sea level, something pointed out as early as 1842 by Charles MacLaren[38, 49]. Eventually, Lyell recognised the link between the formation of an ice sheet and a drop in sea level, and drew attention to it in the 10th and later editions of *Principles*.

Could parts of the British Isles have been submerged, by several hundred feet, as Lyell imagined to explain raised beaches and terraces and raised deposits of unbroken marine shells? Opposing Lyell, James Geikie[45] favoured the explanation provided by a Mr Jamieson[50] that the immense weight of the ice sheet had depressed the land surface, which then rebounded slowly as the ice melted. This process, called isostatic adjustment, which Lyell knew little about, would have raised coastal deposits and terraces. James seems to have forgotten that his own brother, Archibald, had already suggested more or less the same process as Jamieson in 1863 to explain the elevation of evidently marine deposits[44]. We now know that when ice sheets are abundant, sea level is low, and the land is depressed beneath the weight of ice. As the ice melts, relatively rapidly, the sea rises, quickly flooding the margins of the depressed landmass, which itself rises more slowly, allowing time for shell beds to form on the flooded margins. When all the ice sheets have melted, sea level stabilises but the submerged lands continue to rise, eventually raising the shell beds above sea level. This process might have helped to delude Lyell into thinking that the land had been submerged beneath an iceberg-flooded sea, rather than pushed down by an ice sheet. Scotland and Scandinavia are still slowly rising, as Lyell knew, having visited Sweden in 1834 and delivered the Bakerian Lecture on this topic to the Royal Society in the winter of that year[20]. In contrast, the south of England, which had bulged up south of the front of the ice sheet as the north of Britain sank, is now slowly sinking, while Scotland rises.

James Geikie's monumental tome shows that the great ice sheets of the glacial period extended over most of the north European plain, the southern boundary being a roughly east–west line starting in the west between the Severn and the Thames estuaries at around 51° 30′ N. In North America, the main Laurentide ice sheet centred over Hudson's Bay covered most of Canada and the northern United States east of the Rockies, extending down to 37° 35′ N, close to the junction of the Ohio and Mississippi Rivers in Illinois.

Continuing his study of what he called the 'glacial epoch', and what we know as the Ice Age, Lyell reported in his 1863 volume, *Antiquity of Man*, that Swiss geologists considered that there had been at least two phases of glacial action in the Alps, the first carrying erratics to the Jura Mountains before the glaciers retreated, the second filling Lake Geneva with ice but not reaching the Jura before retreating again. Lyell saw some parallels with the geology of the UK, associating the first retreat of the Alpine glaciers with the period of his supposed submergence of England and the deposition of boulder clay from icebergs. Being a canny Scot, he realised that such chronological comparisons were 'very conjectural'. He was confident, however, that when the ice advanced most, the sea would have been at its coldest, and that this would be reflected in the species of seashells – a prescient observation.

James Geikie's study of the Ice Age greatly extended that of Lyell, identifying up to six glacial periods, with intervening interglacials, since the end of Pliocene time[45]. Chamberlin had also identified several glacial periods in North America. Geikie included the Alps, where by the end of the 19th century geologists agreed that there had been at least four periods of advance of the ice – the Gunz, Mindel, Riss and Würm glaciations – separated by warm interglacial episodes, when the climate may have been slightly warmer than it is today[51]. The difficulty in establishing a chronology for the successive events of the Ice Age lay partly in the fact that each succeeding advance of the ice sheet tended to obliterate the evidence deposited by its predecessors – ice sheets were indeed, as Agassiz put it, 'God's Great Plough'.

The realisation that there were several pulses of ice advance during the Ice Age leads us on to consider what the drivers or such change might be, in Chapter 3.

References

1. Lamb, H.H. (1966) Britain's climate in the past. In: *The Changing Climate – Selected Papers*. Methuen, London, pp. 170–195.

2. Georges-Louis Leclerc, Comte de Buffon (1749–1788). *Histoire Naturelle, Générale et Particuliére*, in 36 volumes. See http://www.buffon.cnrs.fr/ice (last accessed 29 January 2015).

3. Rudwick, M.J.S. (2005) *Bursting the Limits of Time: The Reconstruction of Geohistory in the Age of Reason.* University of Chicago Press, Chicago and London.

4. Geikie, A. (1897) *Founders of Geology.* Macmillan, London.

5. Baxter, S. (2004) *Revolutions in the Earth: James Hutton and the True Age of the World.* Phoenix, Orion Books, London.

6. Hutton, J. (1795) *Theory of the Earth, with Proofs and Illustrations.* Edinburgh.

7. Capra, F. (2013) *Learning from Leonardo: Decoding the Notebooks of a Genius.* Berrett-Koehler, San Francisco, CA.

8. Cuvier, G. (1812) Discourse sur les révolutions de la surface du globe et sur les changements qu'elles ont produits dans la règne animal. In: *Recherches sur les Ossememnts Fossiles,* 4 vols, 1st edn. Deterville, Paris (new edn Christian Bourgeois, Paris, 1985, in French).

9. Cuvier, G. (1818) *Essay on the Theory of the Earth.* Kirk & Mercein, New York (5th edn T. Cadell, London, 1827).

10. Bard, E (2004) Greenhouse effect and ice ages: historical perspective, *Comptes Rendus Geoscience* **336**, 603–638.

11. Lavoisier, A. (1789) Observations generates sur les couches horizontales, qui ont ete deposees par la mer, et sur les consequences qu'on peut tirer de leurs dispositions, relativement a l'anciennete du globe terrestre. *Mémoires de l'Académie des Sciences* 351–371.

12. Carozzi, A.V. (1965) Lavoisier's fundamental contribution to stratigraphy. *The Ohio Journal of Science* **65** (2), 71–85.

13. McCrory, D. (2010) *Natures Interpreter – The Life and Times of Alexander von Humboldt.* Lutterworth, Cambridge.

14. von Humboldt, A. (1817) Von den isothermen linien und der verteilung der wärme auf dem erdkörper. *Mémoires de Physique et de Chimie de la Société d'Arceuil* **3**, 462–602; reprinted in von Humboldt, A. (1989) *Studienausgabe, Sieben Bände, hrsg. von Hanno Beck.* Wissenschaftliche Buchgesellschaft, Darmstadt, Band VI, 18–97; English translation (1820-21) On isothermal lines, and the distribution of heat over the globe. *Edinburgh Philosophical Journal* **III** (5), 2–20; **III** (6), 256–274; **IV** (7), 23–37; **IV** (8), 262–280; **V** (9), 28–38.

15. von Humboldt, A. (1805) *Essai sur la Géographie des Plantes.* Levrault Schoell et Cie, Strasbourg.

16. von Humboldt, A. (1849) *Aspects of Nature in Different Lands and Different Climates, with Scientific Elucidations* (trans. Sabine, M.). Lea and Blanchard, Philadelphia, PA (first published as *Ansichten der Natur* in 1808).

17. von Humboldt, A. (1907) *Personal Narrative of a Journey to the Equinoctial Regions of the New Continent during the Years 1799-1804,* 3 vols (trans. Ross, T.) George Bell & Sons, London.

18. von Humboldt, A. (1845–1861) *Cosmos: A Sketch of a Physical Description of the Universe,* 5 vols (trans. Otte, E.C.). Henry Bohn, London.

19. http://www.humboldt-foundation.de/web/kosmos-view-onto -germany-93-1.html (last accessed 29 January 2015).

20. Wilson, L.G. (1972) *Charles Lyell; the Years to 1841: The Revolution in Geology.* Yale University Press, New Haven, CT and London.

21. Bonney, T.G. (1895) *Charles Lyell and Modern Geology.* Macmillan, New York.

22. Rudwick, M.J.S. (1998) Lyell and the principles of geology. In: *Lyell: The Past is the Key to the Present* (eds D.J. Blundell and A.C. Scott). Geological Society Special Publication, London, pp. 3–15.

23. Vai, G.B. (2009) Light and shadow: the status of Italian Geology around 1807. In: *The Making of the Geological Society of London* (eds C.L.E. Lewis and S.J. Knell). Geological Society Special Publication, London, pp. 179–202.

24. Kräusel, R. (1961) Palaeobotanical evidence of climate. In: *Descriptive Palaeoclimatology* (ed. A.E.M. Nairn). Interscience Publishers, New York, pp. 227–254.

25. Lyell, K.M. (1881) *Life, Letters and Journals of Sir Charles Lyell, Bart,* vols 1 and 2. John Murray, London.

26. Lyell, C. (1826) On a recent formation of freshwater limestone in Forfarshire, and on some recent deposits of freshwater marl; with a comparison of recent with ancient freshwater formations; and an Appendix on the gyrogonite or seed-vessel of the chara. *Transactions of the Geological Society of London* **2**, 73–96.

27. Lyell, C. (1875) *Principles of Geology,* 12th edn. John Murray, London.

28. Holmes, R. (2008) *The Age of Wonder: The Romantic Generation and the Discovery of the Beauty and Terror of Science.* HarperPress, London.

29. Secord, J. (1997) Introduction. In: *Principles of Geology* (Lyell, C.). Penguin Classics, London.

30. Gould, S.J. (1987) *Time's Arrow, Time's Cycle: Myth and Metaphor in the Discovery of Geological Time.* Harvard University Press, Harvard, CT.

31. Agassiz, E.C. (1885) *Louis Agassiz, his Life and Correspondence* (ed. Agassiz, E.C.). Houghton Mifflin, Boston and New York.

32. Agassiz, L. (1838) On the erratic blocks of the Jura. *Edinburgh New Philosophical Journal* **24**, 176–179.

33. Agassiz, L. (1840) *Etudes sur les Glaciers.* Jent and Gassman, Neuchâtel. See also Agassiz, A. (1967) *Studies on Glaciers. Preceded by the Discourse of Neuchatel* (trans. and ed. A. Carozzi). Hafner, New York.

34. Davies, G.H. (2007) *Whatever Is Under the Earth.* Geological Society, London.

35. Agassiz, L. (1840) On the evidence of the former existence of glaciers in Scotland, Ireland and England. *Proceedings of the Geological Society of London* **III** (II, 72), 327–332. See also Buckland, pp. 332–337 and 345–348, and Lyell, pp. 337–345, in the same issue. Lyell applied to the Society on 5 May 1841 to have his complete paper withdrawn, and Buckland did the same in June 1842.

36. Woodward, H.B. (1908) *The History of the Geological Society of London*. Longmans, Green & Co., London.

37. McPhee, J. (1998) *Annals of the Former World*. Farrar, Straus, and Giroux, New York.

38. Dott, R.H. (1998) Charles Lyell's debt to North America: his lectures and travels from 1841 to 1853. In: *Lyell: The Past is the Key to the Present* (eds D.J. Blundell and A.C. Scott). Geological Society Special Publication, London, pp. 53–69.

39. Wilson, L.G. (1998) *Lyell in America – Transatlantic Geology 1841–1853*. Johns Hopkins University Press, Baltimore, MD.

40. Lyell, C. (1854) On certain trains of erratic blocks on the western borders of Massachusetts, United States. *Notices of the Proceedings of the Royal Institution of Great Britain* **2** (1854–1858), 86–97.

41. Darwin, C.R. (1842) Notes on the effects produced by the ancient glaciers of Caernarvonshire, and on the boulders transported by floating ice. *London, Edinburgh, Dublin Philosophical Magazine* **21**, 180–188.

42. Darwin, C.R. (1842) Letter to W.H. Fitton, 28 June 1842. Letter 632, Darwin Correspondence Project. Available from: http://www.darwinproject.ac.uk/letter/entry-632 (last accessed 29 January 2015).

43. Darwin, C.R. (1848) On the transport of erratic boulders from a lower to a higher level. *Quarterly Journal of the Geological Society of London* **4**, 315–323.

44. Geikie, A. (1863) On the phenomena of the glacial drift of Scotland. *Transactions of the Geological Society of Glasgow* **I** (II). Available from: http://digital.nls.uk/early-gaelic-book-collections/pageturner.cfm?id=77409534 (last accessed 29 January 2015).

45. Geikie, J. (1874) *The Great Ice Age, and its Relation to the Antiquity of Man*, 3rd edn. Stanford, London.

46. Croll, J. (1875) *Climate and Time in their Geological Relations: A Theory of Secular Changes of the Earth's Climate*. Appleton and Co, New York.

47. Huggett, Q.J. and Kidd, R.B. (1983/84) Identification of ice-rafted and other exotic material in deep-sea dredge hauls. *Geo-Marine Letters* **3**, 23–29.

48. Needham, H.D. (1962) Ice-rafted rocks from the Atlantic Ocean of the coast of the Cape of Good Hope. *Deep Sea Research* **9** (11–12), 475–486.

49. MacLaren, C. (1842) Review of the Glacial Theory of Prof. Agassiz. *American Journal of Science* **42**, 346–365.

50. Jamieson, J. (1882) *Geological Magazine*, p. 400. Cited in Geikie, J. (1894) *The Great Ice Age*. University of Chicago Press, Chicago, IL.

51. Penck, A. and Bruckner, E. (1901–09) *Die Alpen im Eiszeitalter*, 3 vols. Chr. Herm. Tauchnitz Verleger, Leipzig; summarised by Sollas, W.J. (1911) *Ancient Hunters*. Macmillan, London, esp. pp. 18–28.

3

Ice Age Cycles

3.1 The Astronomical Theory of Climate Change

Back in 1830, when Lyell was writing volume 1 of his *Principles*, he knew that glaciers had advanced and retreated. That led him to wonder if significant changes in our climate through time might be driven by some regular astronomical control on the amount of sunlight Earth received. To address this question, Lyell turned to his friend, the English astronomer John Herschel (1792–1871), another Fellow of the Royal Society, knighted in 1831. Herschel explained that the amount of sunlight falling anywhere on the Earth's surface varies with regular changes in the 'eccentricity' of the Earth's orbit around the Sun, in the 'precession of the equinoxes' and in the tilt (or 'obliquity') of the Earth's axis.

Starting with eccentricity, it seems self-evident that if Earth were the only planet orbiting the Sun, it would follow a circular orbit. But it is not and does not. Earth's orbit is influenced by the gravitational pull of its giant sister planets, which converts our orbit into an ellipse with the Sun off-centre, a state called 'eccentric'. The eccentricity of the Earth's orbit slowly changes from more or less circular, or 'centric' (with the Sun at the centre), to 'eccentric' and back following cycles of about 400 and 100 Ka. That affects climate, because when the orbit is at its most elliptical, the amount of radiation the Earth gets from the Sun at perihelion (the point on the orbit closest to the Sun) is 23% more than it gets at aphelion (the point on the orbit furthest from the Sun).

The way in which that radiation is distributed over the planet depends on the fact that the position of the Earth at any given point in time (e.g. the spring equinox (time of equal hours of day and night) in March) migrates slowly around the Earth's orbit in a cycle lasting about 21 Ka, known as the 'precession of the equinoxes' (for more details, see Appendix 1 of Reference 1). This migration has little effect on the climate when the Earth's orbit is more or less circular, but a large effect when the orbit is eccentric. Currently, the Earth is closest to the sun (at perihelion) in mid-winter, making Northern Hemisphere winters warm. In about 10.5 Ka, Earth will be furthest from the sun (at aphelion) in mid-winter, making Northern Hemisphere winters cool. The Earth's orbit is currently close to circular, so these differences have little effect at present. The effect will increase as the orbit becomes more eccentric with time. The effects of precession are greatest in the tropics.

Most people know that the Earth's spin axis is tilted at an angle of around 23° to the plane of the Earth's orbit, which accounts for the seasons. If the axis were upright, we would still have climatic zones – with more heat at the Equator and less at the poles – but no seasons. The tilt of the axis fluctuates from around 21.5° to around 24.5° and back on a cycle of 40 Ka. The higher the angle, the more the seasonal difference, with summers receiving more energy from the Sun, and winters less. The effects of the tilt cycle are greatest in temperate and polar latitudes, where snow accumulates when tilt is high and winters are cold.

Interaction between these three cycles controls the amount of incoming solar radiation – 'insolation' – received at any point on the Earth's surface through time. These cycles are so regular that astronomers can use them to calculate the amount of insolation anywhere on the Earth's surface over periods of millions of years.

Earth's Climate Evolution, First Edition. Colin P. Summerhayes.
© 2015 John Wiley & Sons, Ltd. Published 2015 by John Wiley & Sons, Ltd.

The Sun being the major driver of Earth's climate, these calculations provide geologists with a first-order means of estimating climate change.

Herschel presented his ideas on the astronomical theory of climate change to the Geological Society in December 1830[2]. He speculated that extreme variations in the eccentricity of the Earth's orbit might exaggerate the difference between summer and winter temperature, and that these might combine with the effect of precession of the equinoxes to produce '*periodical fluctuations in the quantity of solar heat received by the earth, every such fluctuation being of course accompanied with a corresponding alteration of climates; and therefore, if sufficiently extensive and continued, giving room for variation in the animal and vegetable productions of the same region at different and widely remote epochs.*' Herschel thought that the variation in the tilt of the Earth's axis was insufficient to affect the climate. He regarded '*the excentricity [sic] as the only element whose variation can possibly have any effect of the kind in view.*' He went on to say, '*by reason of the precession of the equinoxes combined with the motion of the apogee of the earth's orbit, the two hemispheres would alternately be placed in climates of a very opposite nature, the one approaching a perpetual spring, the other to extreme vicissitudes of a burning summer and a rigorous winter.*' Obscure prose indeed, but I'm sure you get the drift. Observing that the Earth's orbit was becoming more centric, he thought that this might indicate a cooling. We have to remember that these were early days and that Herschel was arguing without the benefit of detailed calculations of orbital changes, although they could have been determined from the work of the great French astronomer, Laplace.

Based on his discussions with Herschel, Lyell observed in volume 1 of *Principles* that it is '*of importance to the geologist to bear in mind that in consequence of the precession of the equinoxes, the two hemispheres receive alternately, each for a period of upwards of 10 000 years, a greater share of solar light and heat. This cause may sometimes tend to counterbalance inequalities resulting from other circumstances of a far more influential nature; but, on the other hand, it must sometimes tend to increase the extremes of deviation, which certain combinations of causes produce at distant epochs*'[2]. This dense prose hid the fact that Lyell suspected that these astronomical changes might affect climate enough to be detectable in the geological record, but had no idea if this were the case.

At the time, in 1830, Agassiz had not yet 'discovered' the Ice Age. The honour of being the first to conclude that astronomical forces controlled what happened during the Ice Age thus goes to French mathematician Joseph Alphonse Adhémar (1797–1862), who had the brilliant intuition that Agassiz's glaciations must be periodic and thus controlled by celestial mechanics, publishing his ideas in 1842[4]. He calculated that periods of cooling and warming would correspond to the 21 Ka precession cycle, and that they would alternate between the two hemispheres.

3.2 James Croll Develops the Theory

The first scientist to investigate the astronomical theory of climate change in detail was Scottish physicist-cum-geologist James Croll (1821–1890) (Figure 3.1, Box 3.1), the man who disagreed with Lyell's ideas about boulder clay.

Figure 3.1 *James Croll.*

Box 3.1 James Croll.

Croll's scientific career began late in life. Born on a farm near Wolfhill in Perthshire, Scotland in 1821, he was largely self-educated in mathematics and astronomy. Leaving school at age 13, he worked as a millwright, a carpenter, a tea merchant, the keeper of a temperance hotel, a life-insurance salesman and a writer for a temperance newspaper, before becoming caretaker of the museum at Anderson

College in Glasgow in 1859, at the age of 38. That gave him access to a fine scientific library, exposing him to the work of Herschel and Adhémar and enabling him to develop his own ideas about one of the 'hot' topics of the era: the origin of the Ice Age. His findings led to him being invited to work at the Geological Survey of Scotland, which he did from 1867 to 1881. For his Ice Age research, he was elected a Fellow of the Royal Society in 1876 and awarded an Honorary Degree by the University of St Andrews. He retired due to ill health in 1880 and died in 1890. The Quaternary Research Association now awards the James Croll Medal.

Thinking about the Ice Age, Croll felt sure that *'The recurrence of colder and warmer periods evidently points to some great, fixed, and continuously operating cosmic law'*[5]. Immersing himself in the studies of celestial mechanics by Frenchmen Pierre Simon de Laplace and Urbain Leverrier, discoverer of the planet Neptune, he worked out a more complex theory to explain the effect of seasonal contrasts of insolation, publishing his results in 1864[5]. His paper refined predictions of the timing of the glacial epochs of the Ice Age and showed how changes in the Earth's orbit provided a periodic extraterrestrial mechanism for initiating *multiple* glacial epochs. These would occur every 22 Ka, and when there was a glacial period in the north, there would be an interglacial in the south, as suggested by Adhémar. Unlike Adhémar, Croll thought that eccentricity was important, because the accumulation of snow would be encouraged by decreased sunlight and longer winters when the orbit was most elliptical and the Earth was furthest from the sun (at aphelion).

One of the most important and overlooked aspects of Croll's analysis of the astronomical theory of climate change was his realisation that the change in heat received by the Earth due to orbital changes was *not enough by itself* to cause glaciation. He concluded correctly that *'glacial cycles may not arise directly from cosmical causes, they may do so indirectly!'*[6] Feedbacks were needed. The effect of decreased insolation was amplified by increasing accumulation of snow and ice, which increased the reflectiveness of the Earth's surface – its albedo. This increased the development of mists and fogs, reflecting yet more solar energy.

Croll also considered how changes in insolation would affect winds and ocean currents[5]. Cooling of the poles in glacial periods would steepen the thermal gradient in the air between pole and Equator, making winds stronger. Glaciation in the Northern Hemisphere would weaken the Gulf Stream, which carried warm water from the Equator to the Arctic. It would also strengthen the Northeast Trade Winds, which would force the Equatorial Current south, limiting the supply of warm water to the Gulf Stream from the South Atlantic; *'The Gulf-stream would consequently be greatly diminished, if not altogether stopped'*[5]. If that were so, he speculated that the climate of northern Europe would resemble that of Greenland[7]. This is more or less what occurred during the Ice Age.

Lyell was intrigued, and began corresponding with Croll. This interaction led to Croll being appointed in 1867 to a clerical position as keeper of maps and correspondence in the Geological Survey of Scotland, where the director, Sir Archibald Geikie, encouraged his research. In 1875, Croll summarised his research findings in an influential book, *Climate and Time*[7] (Figure 3.2).

Given the importance he attached to the decrease of heat from the Gulf Stream as one of the positive feedbacks enhancing the glaciation of the north, Croll was keen to find out more about the nature of ocean circulation, then a topic of much speculation[8]. He closely monitored the work of physiologist William Carpenter (1813–1885), a scientist keen to test the notion of Edward Forbes (1815–1854) that there was no life in the deep ocean – it was 'azoic'. As registrar of the University of London and vice-president of the Royal Society, Carpenter used his influence to access a Royal Navy ship, HMS *Porcupine*, which, in 1870, dredged numerous creatures from the deep sea and so killed off Forbes's azoic theory.

Carpenter's work with the *Porcupine* showed most of the deep North Atlantic to be extremely cold. He thought this meant that a deep current originating in the Arctic carried cold water south into the interior of the Atlantic, and *'embarked on the development of his "magnificent generalization" that the cold temperatures were part of a large-scale general ocean circulation'*[8]. In his conceptual model, this deep current replaces the warmer surface water that flows from the Equator towards the poles. He saw *'this flow of water toward the equator and its eventual return towards the pole [as] just as much a physical necessity as that interchange of air which has so large a part in the production of winds'*[8]. Density was an important driver in Carpenter's model of the general circulation, with lighter, warmer water at the surface moving north connected to denser, colder water at depth moving south and eventually returning to the surface near the Equator. These were important insights into ocean circulation, which helps to

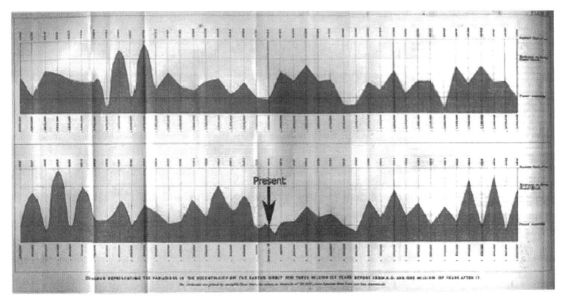

Figure 3.2 *Croll's orbital variations. Variations in the earth's orbit for 3 Ma before 1800 AD and 1 Ma thereafter. Each division on the horizontal axis represents 50,000 years. Croll thought that when the orbit was most eccentric (i.e. elliptical, with high values along the vertical axis) there would be longer winters allowing more snow to accumulate. In contrast, low values of eccentricity would equate with warm climates. The arrow indicating 'present' conditions denotes the position of 1800 AD, when eccentricity was close to its lowest value.*

regulate Earth's climate, although we now know that the return to surface takes place around Antarctica, not at the Equator.

Croll disagreed with Carpenter's model, setting out his own conception of how ocean currents contributed to glaciations[7, 8]. There just wasn't enough oceanographic information to enable them to resolve their differences, however[8]. The controversy did have one happy outcome, in providing a *raison d'être* for the world-encircling oceanographic expedition of HMS *Challenger* in 1872–76, which would create a much clearer picture of ocean circulation than was available to either man.

Although Croll focused his attention on precession and eccentricity, he suspected that changes in the tilt of the Earth's axis might also affect the climate, especially at the poles, where the longer summers at times of maximum tilt would melt more snow and ice than at other times. He surmised that particular combinations of eccentricity, precession and axial tilt would lead to periods of ice melt that would raise sea level, and that oscillations of sea level should be associated with changes from interglacial to glacial conditions – another prescient conclusion. Lacking calculations of the changes in tilt through time, Croll

could do little more than speculate about their effects. We now know that changes in axial tilt do have a strong effect in the polar regions. Croll was on the right track, and that got geologists thinking in the right direction. Noting that tilt was at a maximum 11.7 Ka ago, he speculated that this might have led to a rise in sea level that would explain the occurrence of raised beaches from about that period in Scotland and Scandinavia. This was a perceptive observation, although it ignored the effect of the isostatic upward adjustment of the land in response to the removal of the last ice sheet.

Croll was also one of the first to note the occurrence beneath the Scottish boulder clay of buried river channels, the depths of whose beds showed that they must have been cut when sea level was much lower. The cutting of deep channels implied that the sea level had dropped, thus steepening the gradient of the rivers' beds. While Croll thought the channels might have been cut during warm periods, it seems more likely that they were cut when ice sheets were extensive and sea level was low, both as the climate was cooling and as it was warming. Channel cutting would cease when the ice sheets advanced to the edge of the continental shelf.

In one particularly perceptive leap of the imagination, Croll deduced[7] that '*If the glacial epoch resulted from a high condition of eccentricity, we have not only a means of determining the positive date of that epoch, but we also have a means of determining geological time in absolute measure.*' This turned out to be the case, although not in precisely the way Croll imagined, as we shall see later. Following this leap to its logical conclusion, he calculated eccentricity not only for 3 Ma into the past, but also for 1 Ma into the future (Figure 3.2), making him the first to take a mathematical approach to estimating future climate change.

3.3   Lyell Responds

Having read Croll's 1864 paper[5] and corresponded with him, Lyell modified the last three editions of his *Principles of Geology* (numbers 10 in 1866, 11 in 1872 and 12 in 1875) to introduce a new chapter, 13, on 'Vicissitudes in Climate – How Far Influenced by Astronomical Changes'[9–11]. It explained how changes in the eccentricity of the Earth's orbit in combination with the precession of the equinoxes would cause alternations of climate with a period of around 21 Ka, repeated thousands of times throughout the geological past, which might explain '*some of the indications of widely different climates in former times*'[11]. Lyell attributed to Croll the observation that winters would be at their coldest when the Earth was at aphelion (farthest from the sun) and the orbit was at its most eccentric. The development of large amounts of ice at those times, Lyell went on to note, '*must have given rise at certain periods to some differences in the ocean's level*'[11]. This is a rather late acknowledgment of the notion that ice ages might be times of lowered sea level, which Charles Maclaren had proposed as early as 1842[12].

Despite the attractive features of the astronomical theory of climate change, Lyell thought that Croll had given insufficient credence to Lyell's own principle that '*abnormal geographical conditions*' – meaning the existence of land near or over the poles – were '*far the most influential in the production of great cold*'[11]. He reminded his readers that '*The simple fact that totally different climates exist now in the same hemisphere and under the same latitude would alone suffice to prove that their occurrence cannot be exclusively due to astronomical influence*'. For example, '*the climates of South Georgia and Tierra del Fuego are at present so different that the former might be supposed to belong to a glacial period, while the latter, by its flowers and humming-birds in the winter, and the genera of marine molluscs in the adjoining sea, might indicate to the traveller, as well as to some future geologist, such a temperature as has been spoken of as perpetual spring. This contrast is due to geographical causes.*' The confusion in Lyell's mind arose because the movements of continents, which could indeed cause changes in climate, took place on a time scale of millions of years, while the variations of the astronomical theory were measured in tens or hundreds of thousands of years.

Although Lyell considered '*the former changes of climate and the quantity of ice now stored up in polar latitudes to have been governed chiefly by geographical conditions*', he accepted that the combination of a large excess of polar land with maximum eccentricity of the Earth's orbit '*would produce an exaggeration of cold in both hemispheres*'[11]. To see when maximum eccentricity occurred, he had colleagues draw up a table showing the eccentricity of the Earth's orbit over the last 1 Ma. The table, on page 285 of the 12th edition of *Principles*, shows that major eccentricity occurred at intervals of about 100 Ka, more or less in agreement with modern calculations, with the greatest eccentricity within relatively recent times occurring some 200–210 Ka ago[11].

Changes in the tilt of the Earth's axis, Lyell agreed[11], might also have some effect on the climate, greater tilt causing colder winters at the poles. If that condition were combined with maximum eccentricity and abundant polar land, '*this would favour a glacial epoch*'[11].

Despite their differences, Croll and Lyell agreed in one key respect: as Croll noted, '*the geological agents are chiefly the ordinary climatic agents. Consequently, the main principles of geology must be the laws of the climatic agents, or some logical deductions from them. It therefore follows that, in order to [pursue] a purely scientific geology,* **the grand problem must be one of geological climate** [my emphasis]. *It is through geological climate that we can hope to arrive ultimately at principles which will afford a rational explanation of the multifarious facts which have been accumulating during the past century.*'[7] Where the two men differed was in emphasis, Lyell stressing the pre-eminence of geography, and Croll that the existence of warm interglacial periods goes '*to prove that the long epoch known as the Glacial was not one of continuous cold, but consisted of a succession of cold and warm periods [which] is utterly inexplicable on every issue of the cause of the glacial epoch which has hitherto been advanced*'[7].

Lyell was concerned enough about Croll's astronomical challenge to his geographical theory of climate change that he wrote to Herschel and the Astronomer Royal, Sir George Biddell Airey. This correspondence led him tentatively to accept Croll's theory as a minor cause of climate change[6]. He must have felt he had an edge over Croll in that Croll had concluded that, following his theory, ice ages should have recurred through time. It did not disturb Croll that no evidence for them had been found in the warmer Tertiary; after all, the inadequacy of the geological record could well explain their absence[7]. Lyell disagreed, considering that eccentricity alone could not be the cause of the post-Pliocene ice ages, because there was no evidence for ice ages in the Tertiary or the Cretaceous formations, or indeed back to the Carboniferous[11]. *'This absence of recurrent periods of cold is perfectly explicable'*, he went on, *'if I am right in concluding that they can only be brought about by an abnormal quantity of land in high latitudes'*[11].

While we might think this a stubborn adherence to what might be becoming an outdated idea, Lyell regarded the geographical principle of climate change as one of his major contributions to the science of geology, and it is hard to let go of your favourite ideas, especially when compelling evidence for the competing theory is weak or absent. Besides, we now know that they may both have been right: land in the polar regions does help to build up substantial accumulations of ice, in accordance with Lyell's view, and Earth's orbital changes do modify the climate, in accordance with Croll's, although not to the extent of forming glaciations in warm periods like the early Tertiary and Cretaceous, as Croll thought. That is where greenhouse gases come in, as we shall see later. Neither man considered them.

Croll's work influenced many eminent scientists, including Lyell, Darwin and James Geikie[6, 13]. Darwin was more forthcoming than Lyell in his praise for Croll's theory, *'in part because it provided a valuable mechanism for speciation'*[6]. He wrote to Croll on 24 November 1868 that *'I have never, I think, in my life, been so deeply interested by any geological discussion'*, agreeing with Croll that the advocates of the iceberg theory (such as Lyell) had formed *'too extravagant notions regarding the potency of floating ice as a striating agent'*[6] and that *'scored rocks throughout the more level parts of the United States result from true glacier action'*[14].

3.4   Croll Defends his Position

Croll realised that one barrier to getting his astronomical theory accepted was that scientists tend to find what they are looking for[7]. Before publication of his astronomical theory in 1864, there was no impetus to seek evidence for interglacial periods. That the evidence existed, he was sure, recalling reading some years previously a paper describing fossiliferous sediments found between deposits of glacial till, which contained *'rootlets and stems of trees, nuts, and other remains showing that it had evidently been an old interglacial land surface.'*[7] After 1864, evidence for interglacial deposits began to emerge[13], including evidence that warm interglacial conditions had extended as far north as 75° 32' N in the Arctic[7]. While Croll took these data on board[7], Lyell largely ignored in his last edition of 1875[11] the detailed findings on glacial and interglacial geology that James Geike began publishing in 1874[13]. By then, Lyell was at the end of his life. From 1875 on, Croll and Geikie carried the day, their major opponent having vanished from the scene.

Unfortunately for the astronomical theorists, while orbital variations could be calculated fairly accurately, 19th-century geologists could date rocks only crudely. It was impossible to precisely relate sedimentary sequences to astronomical variables, or to test Croll's notion that glaciations alternated between the hemispheres. All that Croll could do was use rates of erosion and deposition to suggest that the glacial epoch ended around 80 Ka ago[7]. In fact, it ended around 20 Ka ago, but this was a good best guess for those times.

James Geikie supported Croll's theory, with the caveat that *'it must be confessed that a complete solution of the [Pleistocene Ice Age] problem has not yet been found'*[13]. He realised that Lyell's requirement for land in the polar regions to explain glaciations did not address the real issue: the complex alternation of cold and warm epochs. It was unreasonable to suppose that land moved in and out of the polar regions sufficiently rapidly to explain the origin of glacial to interglacial cycles. Lyell had also called on increases in the elevation of land to explain the origin of cold periods. Geikie took issue with that too, considering it unlikely that highlands had popped up and down fast enough to account for the observed cycles[13].

Croll's work was widely discussed but generally disregarded, not least because of geologists' continuing inability to date variations at a fine enough scale. By the time Croll died, geologists realised that glacial conditions had persisted much later than his proposed peak glaciation

80 Ka ago, and many thought his theory must be wrong. James Geikie hoped that '*some modification of his views will eventually clear up the mystery. But for the present we must be content to work and wait*'[13]. Advances in the theory and in dating rocks were needed to revive his work.

3.5   Even More Ancient Ice Ages

Agassiz's discovery changed the way field geologists thought about the rocks they saw. One of the first to respond, in 1855, was Professor Andrew Constable Ramsay (1814–1891), another of those Scottish geological fellows of the illustrious Royal Society of London. He was president of the Geological Society of London, and in 1872 became director-general of the Geological Survey of Britain. He was awarded the Wollaston Medal of the Geological Society in 1871 and the Royal Medal of the Royal Society in 1880, and was knighted in 1881.

Ramsay had a special interest in the effects produced by ice, and applied it to interpreting the origin of the Permian breccias skirting the English and Welsh coal fields. The breccias comprise many large, polished and striated angular fragments of rock stuck in a '*marly paste*'[15]. He deduced '*that they are chiefly formed of the moraine matter of glaciers, drifted and scattered in the Permian sea by the agency of icebergs*'. The Permian strata overlie the older Carboniferous coal measures, in which the coal was assumed to be the remains of swamps and forests growing in '*a moist, equable, and temperate climate, possibly such as that of New Zealand*'[15]. Croll was delighted to see evidence for a major glaciation far earlier in time than the relatively recent Ice Age, because it supported his notion that regular changes in the Earth's orbit should have led to glaciations in the distant past[5] and meant that Earth's climate could not be explained simply in terms of a gradual cooling.

By the time he wrote his extensive work on climate and time[7], Croll knew that the great coal formations of the Carboniferous Period had been deposited in cycles, with thin coal beds made from the remains of forests formed on land alternating with thin beds of marine clay, suggesting a succession of geologically rapid changes from a terrestrial to a marine environment and back. Croll interpreted this succession as possibly representing repeated changes from warm interglacial periods, when the sea level was low and coal forests grew, to cold glacial periods, when the sea level was high and marine clays were deposited.

Although alternations between warm and cold conditions were consistent with his astronomical theory, he missed the point that during cold glacial conditions the sea level should have been low rather than high, with more water being trapped in ice at those times. We will explore the origin of cyclical deposits of the Carboniferous more in Chapter 6.

3.6   Not Everyone Agrees

By the end of the 19th century, a lot more was known about how the modern climate system worked than when Lyell first set down his *Principles*. Other geologists had begun to write treatises about the climates of the geological past. One was the Dutch palaeo-anthropologist and geologist Marie Eugène François Thomas Dubois (1858–1940) (Box 3.2), famed for discovering 'Java Man' (*Pithecanthropus erectus*, later named *Homo erectus*).

Box 3.2 Marie Eugène François Thomas Dubois.

Raised in Eijsden in Limburg, The Netherlands, Dubois was fascinated by natural history. He studied medicine at the University of Amsterdam, obtaining his degree in 1844. Specialising in comparative anatomy, he developed an interest in human evolution and the link between humans and apes. In 1887, he joined the Dutch army to get himself posted to the Dutch East Indies – now Indonesia – because he felt sure that the 'missing link' between apes and humans lay in the tropics. There, he searched caves on Sumatra and Java, finding 'Java Man' in 1891. He returned to Europe in 1895 and was awarded an honorary doctorate by the University of Amsterdam in 1897, becoming professor of geology there in 1899. He also served from 1897 to 1928 as keeper of palaeontology, geology and mineralogy at Teyler's Museum in Haarlem.

Aside from his fascination with the link between humans and apes, Dubois was intrigued by the climatic changes of the past, publishing an expanded essay on this topic in 1895[16]. His article brought to an English-speaking audience a wide range of references to the growing understanding of past climate change, especially in Germany. Notable among these was the work of Melchior Neumayer (1845–1890)[17], who had shown the biogeographical link between Brazil and Africa across the South Atlantic, and that of Swiss geologist and naturalist Oswald Heer (1809–1883)[18–20], whom Dubois considered to be the 'father of palaeoclimatology', noting that, '*To him we owe nearly all the information which we possess about ancient climates*'[16]. Heer was Professor of Botany at the University of Zurich. For his services to science, the Geological Society of London awarded him its Wollaston Medal in 1874.

Dubois drew attention to the growing evidence from fossil plants and animals that the climate of the Arctic had declined from almost tropical in the early Tertiary to the cold conditions of today. There was ample evidence from the early Tertiary and former times – especially the Jurassic and the Cretaceous – that, while they had been warmer than today, their heat had been distributed in climate zones between the Equator and the pole as they presently exist. No displacement of the poles seemed necessary to explain the distribution of these zones. That appeared to knock on the head, at least temporarily, Lyell's theory of continental displacement as a cause of climate change. For earlier times, such as the Permo-Carboniferous, where there was evidence for the action of ice far from the present poles, Dubois preferred to explain their occurrence as caused by increased precipitation rather than by excessive cold, citing as an example the occurrence of glaciers in temperate New Zealand today.

How, then, could the Cenozoic cooling be explained? Dubois believed that the answer must lie with the energy emitted by the Sun, which must have declined with time, but he also recognised that ocean currents play an important role in transporting heat away from the tropics. He thought that periodic changes in the Sun's output, like those causing the sunspot cycle, might explain the warm interglacial periods of the Ice Age. While the astronomical variations called upon by James Croll might explain some of that variation in Pleistocene times, they were inadequate to explain the general cooling of the Earth in Cenozoic times. '*It now seems to be completely proved*' said Dubois '*that no other source of heat than the sun can ever have exercised an appreciable influence either on the meteorological condition of the Earth, or on the climates. To any very considerable changes of the solar heat we must, therefore, look for the cause of the geological changes of climate*'[16].

Dubois can perhaps be taken as typical of the geologists of the time. He was unaware of the effects of greenhouse gases in modifying the Sun's heat and neglected the possibility that the continents might have changed their positions with relation to the poles and to each other. Breakthroughs in the sciences would be needed before the prevailing paradigm could change. We turn in the next chapter to the efforts of 19th-century scientists to explore the possibility that changes in the Earth's climate resulted from changes in the composition of the atmosphere.

References

1. Maslin, M., Seidov, D. and Lowe, J. (2001) Synthesis of the nature and causes of rapid climate transitions during the Quaternary, in *The Oceans and Rapid Climate Change: Past, Present and Future* (eds D. Seidov, B.J. Haupt and M. Maslin). American Geophysical Union, Wahington, DC.

2. Herschel, J. (1830) On the astronomical causes which may influence geological phenomena, read before the Geological Society, 15 December 1830. *Transactions of the Geological Society* **III**, 293, second series 244.

3. Lyell, C. (1830) *Principles of Geology*, vol. **1**. John Murray, London (followed by Vol. 2 in 1832 and Vol. 3 in 1833).

4. Adhémar, J. (1842) *Révolutions de la Mer, Déluges Périodique*. Carilian-Goeury et V. Dalmont, Paris (2nd edn: Lacroix-Comon, Paris, 1860). See review in Bard, E. (2004) Greenhouse effect and ice ages: historical perspective. *C.R. Geoscience* **336**, 603–638.

5. Croll, J. (1864) On the physical cause of the change of climate during geological epochs. *London, Edinburgh and Dublin Philosophical Magazine and Journal of Science* **XXVIII**, 4th series, August, 121–136.

6. Fleming, J.R. (2006) James Croll in context: the encounter between climate dynamics and geology in the second half of the nineteenth century. *History of Meteorology* **3**, 43–53.

7. Croll, J. (1875) *Climate and Time in their Geological Relations: a Theory of Secular Changes of the Earth's Climate*. Appleton and Co, New York.

8. Mills, E.L. (2009) *The Fluid Envelope of Our Planet – How the Study of Ocean Currents became a Science*. Toronto University Press, Toronto, ON.

9. Lyell, C. (1866) *Principles of Geology*, 10th edn. John Murray, London.

10. Lyell, C. (1872) *Principles of Geology*, 11th edn. John Murray, London.

11. Lyell, C. (1875) *Principles of Geology*, 12th edn. John Murray, London.

12. MacLaren, C. (1842) Review of the glacial theory of Prof. Agassiz. *American Journal of Science* **42**, 346–365.

13. Geikie, J. (1894) *The Great Ice Age, and its Relation to the Antiquity of Man*, 3rd edn. Stanford, London.

14. Fleming, J.R. (1998) Charles Lyell and climatic change: speculation and certainty. *Geological Society Special Publication* **143**, 161–169.

15. Ramsay, A.C. (1855) On the occurrence of angular, subangular, polished and striated fragments and boulders in the Permian breccia of Shropshire, Worcestershire, etc; and on the probable existence of glaciers and icebergs in the Permian Epoch. *Proceedings of the Geological Society of London* **11**, 185–205.

16. Dubois, E. (1895) *The Climates of the Geological Past and their Relation to the Evolution of the Sun*. Sonnenschein, London. This was a translation of an essay published in German in 1893 as *Die Klimate der geologischen Vergangenheit und ihre Beziehung zur Entwickelungsgeschichte der Sonne*, H.C.A. Thieme, Nijmegen, which in turn was an expansion of a paper initially published in Dutch in 1891 as De klimaten der voorwereld en de geschiedenis der zon, *Natuurkundig Tijdschrift voor Nederlandsch-Indie, Batavia* **51**, 37–92.

17. Neumayr, M. (1887) *Erdgeschichte*, vol. **II**, Leipzig; and his 1889 lecture: *Die klimatischen Verhaltnisse der Vorzeit*, Schriften des Vereins zur Verbreitung naturwiss. Kenntnisse. Wien. A translation of this lecture entitled 'The Climates of Past Ages' appeared in 1890 in *Nature* **42**, 148, 175.

18. Heer, O. (1864) *Die Urwelt der Schweiz [Lieferungen 7-11]*. Friedrich Schulthess, Zürich, pp. 289–496.

19. Heer, O. (1879) Die Urwelt der Schweiz. Zweite, umgearbeitete und vermehrte Auflage. Friedrich Schulthess, Zürich.

20. Heer, O. (1868–1882) *Flora Fossilis Arctica – Die fossile Flora der Polarländer*, Zweiter Band enthaltend: 1. *Fossile Flora der Bären-Insel*. 2. *Flora fossilis Alaskana*. 3. *Die Miocene Flora und Fauna Spitzbergens*. 4. *Contributions to the Fossil Flora of North-Greenland*, Mit 59 Tafeln, Winterthur. Verlag von Wurster & Company.

4

Trace Gases Warm the Planet

4.1 De Saussure's Hot Box

In the late 18th and early 19th centuries, physicists were puzzling over what it was that kept Earth's atmosphere warm. Among them was Horace Bénédicte de Saussure (1740–1799) (Box 4.1), a Genevan aristocrat, brilliant scientist and author of the widely known *Voyages dans les Alpes*[1].

Box 4.1 Horace Bénédicte de Saussure.

Born in Conches, near Geneva, Saussure was a candidate for a professorship of mathematics by the age of 20, and at the age of 22 obtained one in philosophy. Fascinated by the Alps, in 1787 he made the second ever ascent of Mt Blanc, and his scientific measurements there captured the popular imagination. Rector of Geneva University from 1774 to 1776, he was made a Fellow of the Royal Society of London and a foreign member of the Academies of Sciences of both France and Sweden. Well connected, he visited Buffon in Paris and discussed electricity with Benjamin Franklin in London. A genus of high alpine plant – the *Saussurea* – is named after him, as is the mineral Saussurite, and he appeared on the Swiss 20 Franc note (1979–1995). Saussure was the first to break the abhorrence of the English-speaking public for the dangers of the Alps and to foster the rise of mountaineering. He also popularised the use of the

term 'geology'[2]. As a geologically minded natural philosopher, he gave Hutton ideas about the former extent of glaciers. His *Voyages dans les Alpes* supplied Hutton with many of the illustrations of geological processes on which he based his *Theory of the Earth*.

Proximity to the Alps enabled Saussure to investigate the behaviour of the atmosphere. Carrying barometers and thermometers to the summits of Alpine peaks, he measured the temperature and relative humidity of the air and the strength of solar radiation at different heights. In 1767, to test the idea that the air is colder on peaks because the sunlight is weaker there, he invented the 'heliothermometer', a solar energy collector or solar oven comprising a black-lined, well-insulated box with a lid made up of three layers of glass with air between them (Figure 4.1). As the temperatures in his 'hot box' reached 110 °C regardless of altitude or outside air temperature, he concluded that sunlight was constant everywhere within a given locality, and that some other physical process must cause the air to cool upwards. Radiation was involved, but what kind, and from where?

4.2 William Herschel's Accidental Discovery

The explanation starts with visible light. When the Sun shines, tiny rainbows flit about the rooms in my house.

Earth's Climate Evolution, First Edition. Colin P. Summerhayes.
© 2015 John Wiley & Sons, Ltd. Published 2015 by John Wiley & Sons, Ltd.

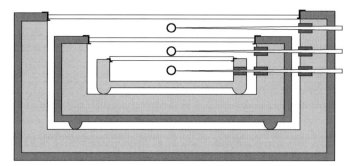

Figure 4.1   *Saussure's hot box: the heliothermometer, comprising several boxes encased one inside another, each of whose sides are glazed. Each case is isolated thermally from its neighbour by cork, and its bottom is painted black to minimise heat losses by reflection. Mercury thermometers, placed on the glass windows, make it possible to read the temperatures inside the various encased boxes.*

They come from crystals hanging in the windows. Physicists had long puzzled over the origin of the colours emerging from crystals and seen in rainbows, before, in 1665, young Isaac Newton used a prism to show that white light, or visible radiation, is made of a spectrum of colours mixed together. It was 135 years before another young genius extended Newton's work by discovering infrared radiation. This was the German-born musician and astronomer Frederick William Herschel (1738–1822) (Box 4.2), father of Sir John Herschel.

Box 4.2 Frederick William Herschel.

William Herschel was born in Hanover and moved to England aged 19, as a musician and composer. There, he turned his attention to astronomy, building his own telescopes and grinding his own lenses. He discovered the planet Uranus in 1781 and two of its major moons in 1787, as well as two of the moons of Saturn in 1789. He spent much of his life close to his observatory in Slough, in what was then Buckinghamshire, where I went to school. One of our school's houses was named after him. That same year, he was elected a fellow of the Royal Society and awarded their Copley Medal, and in 1782 he was appointed the 'King's Astronomer', not to be confused with the Astronomer Royal. He was knighted in 1816.

In 1800, the inquisitive Herschel thought it would be interesting to measure the temperatures of the different colours of the spectrum emerging from a prism placed in a beam of white light. Much to his amazement, he found the highest temperatures just beyond the visible red end of the spectrum. It was an accident: he had placed one thermometer just outside the visible spectrum emerging from his prism, to represent the ambient temperature of the environment. He realised that he had discovered an invisible form of light beyond the visible spectrum. Because it vibrates at a longer (i.e. lower) wavelength than does visible red light, we call it **infra**red light.

All warm objects emit heat as invisible infrared radiation. If you heat them enough, they give off visible radiation – we speak of things becoming red hot and, at extreme temperatures, white hot. You can use an infrared light detector to see people and objects at night, because they emit heat. Engineers use this principle to find heat leaks from buildings. You use infrared rays to operate your TV by remote control. Warmed by the visible and infrared radiation from the Sun, the surface of the Earth and the ocean emit invisible infrared radiation.

4.3   Discovering Carbon Dioxide

The man who introduced the term 'gas' to chemistry, Flemish chemist Jan Baptist van Helmont (1579–1644), was the first to combine carbon with air to produce a gas. Burning charcoal in a closed vessel, he found that the weight of the ash was less than the original weight of charcoal. Part of the charcoal was transmuted into an invisible substance that he named 'gas sylvestre'. He thought it was the same as the gas produced by fermenting fruit juice, which rendered the air of caves unbreathable.

One hundred years later, the Scottish chemist Joseph Black (1728–1799), a friend of James Hutton, discovered that treating limestone with acids or heat yielded a gas he called 'fixed air'. Fixed air was denser than air, did not burn and did not support animal life. When bubbled through limewater, it precipitated calcium carbonate. He proved Helmont right – the gas was produced by fermentation – and he showed that it was also respired by animals. Black also discovered latent heat. Asking the question, 'Why doesn't ice melt on a sunny day?', he found that when ice is heated, its temperature increases to the freezing point and then stays there until all the ice has melted. Similarly, if you boil water, its temperature stays the same until it has all evaporated. A specific quantity of heat is needed to make these changes. This 'lost' or 'hidden' heat he called 'latent heat'. It is fundamental to the science of thermodynamics[3].

4.4 Fourier, the 'Newton of Heat', Discovers the 'Greenhouse Effect'

Enter Jean Baptiste Joseph Fourier (1768–1830) (Figure 4.2), whom we met in Chapter 2 when he concluded that the Earth's internal heat did not affect the temperature of the atmosphere. He originated Fourier's Law regarding the rate of transfer of heat through a material, and the Fourier Transform, commonly used to transform a mathematical function of time into a new function, such as frequency. Fourier's contributions to the study of heat earned him fellowship in the Royal Society of London in 1823. The Université Joseph Fourier in Grenoble, in the Department of Isère – of which he was made prefect by Napoleon – is named after him.

Fourier made planetary temperature a proper object of study in physics, setting the stage for the development of that field over much of the 19th century[4]. Thinking about Saussure's 'hot box', he knew that glass was transparent to sunlight and opaque to infrared radiation, which might contribute to heating the oven. But knowing that convection redistributes heat, he reasoned that the glass lid of Saussure's 'hot box' would contribute to warming the oven by preventing the heat from blowing away[4]. Fourier pointed out that infrared radiation (which he called

Figure 4.2   *Joseph Baptiste Joseph Fourier.*

'nonluminous heat') is the only means by which a planet loses heat, and that while the air is largely transparent to sunlight, it is relatively opaque to infrared radiation from the Earth's surface, which keeps the atmosphere warmer than it would be if it were transparent to infrared. The same applies to the ocean.

In effect, Fourier, who regarded himself as 'the Newton of heat'[5], had discovered the 'greenhouse effect'. He was right, but he could not explain how the atmosphere absorbed infrared radiation to become warm. He did not coin the term 'greenhouse effect', although in some of his earlier work he did discuss the principles governing the temperature of a greenhouse. Greenhouses actually work both by preventing the escape of infrared radiation *and* by preventing heat from dissipating by convection. Using the term 'greenhouse effect' to describe what greenhouse gases do to the atmosphere is thus a misnomer, but one we are now stuck with.

Fourier had been studying heat for some 20 years before summarising his work in his 1822 magnum opus, *Theorie Analytique de Chaleur*[5]. His ideas on the atmosphere were presented to the Academie Royale des Sciences in 1824, published that same year in the *Annales de Chimie et de Physique* (as well as later, in 1827) and translated into English in the *American Journal of Science* in 1837[5]. The oft-quoted 1827 paper is reproduced in translation in *The Warming Papers*[4]. Science historian J.R. Fleming tells us that so few people seem to have read Fourier's various papers on the subject that '*We may ... safely say that ... [Fourier's contribution to the history of the greenhouse effect] ... is not well known [even] inside the atmospheric sciences*'[5].

4.5   Tyndall Shows How the 'Greenhouse Effect' Works

Fourier's 'nonluminous' or 'radiant' heat became the focus of attention of another scientific genius of the 19th century, John Tyndall (1822–1893) (Figure 4.3, Box 4.3), the man who discovered why the sky is blue.

Figure 4.3   *John Tyndall.*

Box 4.3 John Tyndall.

Born in County Carlow, Ireland, Tyndall moved to England to work as a land surveyor in 1844, then to Germany in 1848 to study mathematics at the University of Marburg, where he was taught by the great German chemist Robert Bunsen (1811–1899). Following the award of his PhD in 1850, he returned to London, where he gained the support of Michael Faraday (1791–1867). His work on magnetism led in 1853 to his being appointed professor of natural philosophy (physics) at the Royal Institution in Albemarle Street. As an experimental physicist, he became well known for his public lectures at the Institution, and for writing popular books on heat, sound and light. Among other things, he was fascinated by the effects of pressure on ice, which led him to study Alpine glaciers. Spending several summers in the Alps, he became an expert mountaineer, twice ascending the highest and second-highest peaks, Mont

Blanc and Monte Rosa, and becoming the first to ascend the Weisshorn, in 1861. Elected a fellow of the Royal Society in 1852, he was awarded its Royal Medal in 1853 for his work on magnetism, but declined the honour. He accepted the Royal Society's Rumford Medal in 1864 for his work on the absorption and radiation of heat by gases. Like Lyell, he has a Tasmanian mountain named after him. Tyndall met an untimely death when his wife accidentally administered him an overdose of the chloral hydrate he took to treat his insomnia.

In 1859, having become aware of the work of Saussure and Fourier, Tyndall began experimenting on the absorption and radiation of heat by gases. A meticulous experimenter, he used a spectrophotometer of his own design, whose tube could be filled with different mixtures of gases at variable pressures. The work required an almost superhuman effort, in which he spent many hours improving his apparatus. Following a brief synopsis of his results in his *The Glaciers of the Alps* in 1860[6], he presented a seminal paper on radiation and gases to the Royal Society in their Bakerian Lecture in February 1861[7]. In this lecture, he explained how he measured the relative powers of air, hydrogen, oxygen, nitrogen, water vapour, ozone, carbon dioxide (then called carbonic acid) and other compounds to absorb infrared radiation (radiant heat). His apparatus is also described in his *Heat and Mode of Motion*[8].

Tyndall was awestruck by his findings. Of the components in air, he found that water vapour absorbed the most radiant heat, and oxygen and nitrogen almost none, with ozone and carbon dioxide absorbing moderate amounts. Using a second apparatus, equipped with thermometers, he found that gases that absorb infrared radiation re-emit it as heat. Here was the proof of the greenhouse effect.

Using an improved apparatus, Tyndall confirmed his previous findings, reporting his results to the Royal Society in January 1862[9]. This time, he included methane (marsh gas), which he found to be around 4.5 times more absorbent than CO_2, and human breath. He continued to improve his apparatus (Figure 4.4), summarising his results in the Rede Lecture 'On Radiation' in May 1865[10], in which he confirmed that water vapour absorbs a great deal of infrared radiation. Recognising that the Sun's energy warmed the Earth's surface, causing it to emit heat, Tyndall deduced that even though water vapour makes up a mere 0.5% of the atmosphere, its enormous powers of

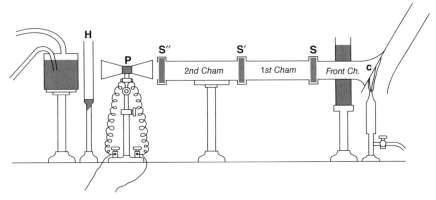

Figure 4.4 *Tyndall's revised apparatus, 1863–64. The new apparatus was a 49.4″ long tube consisting of three chambers. A heat source at one end played on a copper plate. Heat passed through the first chamber, full of dry air, then through a plate of rock salt into two more chambers filled with the gas or vapour to be measured, separated by a second plate of rock salt, then through a final plate of rock salt to a detector. The detector comprised a thermoelectric pile, in which the heating of different metal strips created a current that could be read by a galvanometer. Beyond the pile was a compensating cube to neutralise the radiation coming from the heat source, the cube being separated from the detector by an adjusting screen that helped to adjust for the compensation.*

absorption of radiant heat means that it has a huge influence on the temperature of the atmosphere, '*protecting its surface from the deadly chill which it would otherwise sustain … In consequence of this differential action upon solar and terrestrial heat, the mean temperature of our planet is higher than is due to its distance from the sun*'[10].

It was clear to Tyndall that water vapour absorbs much of the infrared radiation emitted by the Earth's solid or watery surface, re-emitting it in all directions to keep the atmosphere warm. '*Similar remarks*', he wrote, '*would apply to the carbonic acid [CO_2] diffused through the air*'[7,8]. Every variation of these minor constituents, he thought, must produce a change of climate, and could have produced '*all the mutations of climate which the researches of geologists reveal … they constitute true causes, the extent alone of the operation remaining doubtful*'[7,8].

As Tyndall explained, water vapour has an important effect on atmospheric temperatures. Its absence from the dry air of deserts makes them extremely cold at night. I experienced that cold at first hand on a camping trip to the Big Bend National Park in the Chihuahua Desert in west Texas at Thanksgiving in November 1979. It was so cold at night that my girlfriend and I had to don every article of clothing we had brought with us and squeeze into our double sleeping bag for warmth, and when we woke up the following morning the dishes were solidly frozen into the water in the washing-up bowl outside the tent.

Not everyone agreed with Tyndall about the overwhelming power of water vapour as a major forcing agent in warming the atmosphere. Beginning with Arrhenius (see Section 4.6), climate scientists realised that the abundance of water vapour in the atmosphere is limited by the hydrological cycle: if the air gets too humid, it rains. In effect, water vapour acts as a secondary agent in global warming, by amplifying the warming caused by increases in CO_2, which has no such limit. Increasing CO_2 warms the atmosphere, causing more water vapour to evaporate from oceans and lakes, so further increasing the temperature through positive feedback.

Tyndall showed that while carbon dioxide (CO_2) is 'in general' a weak absorber of radiant heat, it absorbs practically all of the radiant heat from a carbonic oxide flame, suggesting what we now know: that, because of its particular chemical structure, CO_2 preferentially absorbs heat in specific parts of the electromagnetic spectrum. He was unable to measure the true strength of the absorption of infrared radiation by the various gases, which turns out to be dependent on the exact wavelength of the radiation[4], as that example suggests.

We now know that the CO_2 absorption spectrum comprises a collection of very narrow peaks at set wavelengths, which broaden and coalesce with increasing pressure[4]. This helps, in part, to explain why Tyndall's results showed that when a gas like CO_2 is very dilute or at a low enough pressure, the absorption increases linearly with the concentration of the gas, but that when the concentration of gas is sufficiently high, it absorbs all or most of the radiant heat – a phenomenon known today as 'band saturation'. As a result, *'even today, a detailed calculation of the absorption and emission of IR [infrared radiation] by a column of atmosphere is not trivial; it can be done by computer models known as line-by-line codes, based on megabytes of detailed spectral information for the various greenhouse gases, but these calculations are too computationally expensive, that is to say slow, to be done in the full climate models that are used to predict things such as, say, the climate sensitivity or global warming forecasts'*[4].

Comparing the lack of absorption of infrared radiation by single-element gases like hydrogen (H_2), oxygen (O_2) and nitrogen (N_2) with its abundant absorption by compound gases like carbon dioxide (CO_2) and water (H_2O), Tyndall concluded that something about the chemical bonds in the compound gases encouraged the absorption of radiant heat. We now know that the structure of these compound gases encourages vibration, enabling them to capture and re-emit infrared radiation. Thanks to his work, we now use infrared radiation detectors and spectrometers to measure the abundance of CO_2 in the air and in the atmosphere of other planets. Hospitals still use Tyndall's system for measuring the carbon dioxide in human breath to monitor the health of patients under anaesthetics during surgical operations.

Although Tyndall gets the credit for having demonstrated the role of CO_2 in absorbing and re-emitting radiation, Edouard Bard reminded us that *'Jacques Joseph Ebelmen (1814–1852), professor at the "École des mines" in Paris and Director of the Royal Works of porcelains in Sèvres, was the first to suggest that past changes in the carbon cycle could have changed the atmospheric concentration of "carbonic acid" and, as a direct consequence, the climate of the Earth'*[11]. Ebelmen was an eminent French scientist in his day. His was one of the 72 names of famous scientists inscribed on the Eiffel Tower at the time of its construction, and the Ebelmen Award of the International Association of Geochemistry is named in his honour. Bob Berner and Kirk Maasch[12] tell us that that before his untimely death at the age of 37, Ebelmen had developed essentially all of the fundamental concepts not only of the geochemical carbon and sulphur cycles

and their effects on atmospheric CO_2 and O_2, but also on the whole process of chemical weathering; for example, with volcanic emissions supplying CO_2 to the atmosphere and the weathering of magnesium silicates removing it. He realised that a close balance was needed between CO_2 uptake by silicate weathering and CO_2 release by volcanism, to prevent a build up of its concentration in the atmosphere. He also recognised that changes in the rates of uptake and release could change CO_2 levels in the atmosphere over geological time. Ebelmen wrote in 1845 (translation by Berner and Maasch[12]) that *'many circumstances … tend to prove that in ancient geologic epochs the atmosphere was denser and richer in carbonic acid and perhaps oxygen, than at present. To a greater weight of the gaseous envelope should correspond a stronger condensation of solar heat and some atmospheric phenomena of a greater intensity'*[13]. Ebelmen's theoretical work was brought to the attention of the wider community by T.S. Hunt in 1880[14], but largely ignored[12]. Given the vibrancy of the intellectual exchange networks of the 19th century, even in the absence of the Internet, it seems surprising that apparently neither Lyell nor Tyndall knew anything of Ebelmen's work. But it seems quite likely that, by Tyndall's time, Ebelmen's ideas were in the scientific air.

Basically, then, the natural greenhouse effect works like this: some of the outgoing infrared radiation from the Earth's surface escapes directly to space, but much is absorbed by the atmosphere and re-emitted in all directions by trace gases in the form of complex molecules such as ozone, water vapour, CO_2 and methane (CH_4). The absorption of infrared energy makes these molecules vibrate and thus warm the air. This natural warming raises the average temperature of the atmosphere at the surface of the Earth by some $33\,°C$. Without it, we would be shivering at an unpleasant $-18\,°C$ average instead of a comfortable $+15\,°C$. As Bill Hay points out, if the oven in your kitchen has a light in it and a glass pane in its door, you will be familiar with the fact that glass allows visible radiation to pass through but blocks the heat radiating at longer (infrared) wavelengths[15]. Greenhouses and blankets work in much the same way as the glass, although both also work by preventing convection. Convection ensures that CO_2 and heat are widely mixed throughout the atmosphere.

Convection also moves water vapour around, but as Gilbert Plass pointed out[16], the distribution of water vapour is subject to the limits imposed by temperature. Very cold air contains no moisture, so water vapour tends to be concentrated within the troposphere, while CO_2 extends into the stratosphere; there is also little water vapour at the poles. Water rains out rapidly from the atmosphere, while CO_2 can continue to increase. Water vapour's main effect is to reinforce temperature change by positive feedback. CO_2 increases temperature, hence evaporation, hence water vapour, which further increases temperature, which warms surface waters so that they release more CO_2. CO_2, in contrast, tends to be present everywhere and at all altitudes. While, one-for-one, water is a less effective greenhouse gas than CO_2, with a global warming potential of 0.28 to CO_2's 1.0, the greater abundance of water vapour in the atmosphere means that its potency is higher (4.0 as against 1.42 for CO_2)[15]. In contrast, the main atmospheric gases – oxygen and nitrogen – have no effect on infrared radiation, as they lack the absorptive properties of the more complex trace gases. Under these circumstances, adding or subtracting CO_2 would be expected to raise or lower the temperature of the atmosphere, as we see next, not least by helping to increase the amount of water vapour present in it.

4.6 Arrhenius Calculates How CO_2 Affects Air Temperature

Someone had to take Tyndall's findings further, and calculate by how much CO_2 might change the temperature of the atmosphere. That person was Swedish chemist Svante August Arrhenius (1859–1927) (Figure 4.5, Box 4.4).

Stimulated by the work of an eminent geologist colleague, Arvid Gustaf Högbom, who had worked on the

Figure 4.5   *Svante August Arrhenius.*

Box 4.4 Svante August Arrhenius.

Arrhenius was born near Uppsala, Sweden. He studied at Uppsala University and at the Institute of Physics of the Swedish Academy of Sciences in Stockholm. In 1884, he published a dissertation on the chemical theory of electrolytes, and he worked on electrochemistry in several continental laboratories before returning to a lectureship in physics at the Stockholm Hogskola in 1891. There, he was promoted to professor and served as rector before becoming the first director of the Nobel Institute for Physical Chemistry. He was elected to the Swedish Academy in 1901, awarded the Nobel Prize for Chemistry in 1903 and elected a foreign member of the British Royal Society in 1911.

---

geochemistry of carbon, and being aware of Tyndall's observations on the radiative properties of gases, Arrhenius became intrigued by the question of what caused the onset of glacial and interglacial periods. In 1895, he presented to the Swedish Physical Society a paper published the following year suggesting that changes of the order of 40% in CO_2, a minor atmospheric constituent, might trigger feedback phenomena that could account for glacial advances and retreats[17]. '*A simple calculation,*' he said, '*shows that the temperature in the Arctic regions would rise about 8° to 9°C, if the carbonic acid [CO_2] increased to 2.5 to 3 times its present value. In order to get the temperature of the ice age between the 40th and 50th parallels, the carbonic acid in the air should sink to 0.65–0.55 of its present value (lowering of temperature 4°–5°C)*'[17].

To get that result, Arrhenius had to invent the field of climate modelling[11]. As explained by David Archer and Raymond Pierrehumbert[4], his energy-budget model took into account the radiative energy exchanges between the atmosphere and space, between the atmosphere and the ground and between the ground and space due to the transmission of infrared radiation through the atmosphere. It also took into account the increase in water vapour in the atmosphere as the atmosphere warms and the change in albedo (reflection of solar energy) as snow and ice melt with increased warming. As in modern climate models, his calculations assumed a four-dimensional space, with a grid based on average temperatures for every 10th degree of latitude, extending vertically through the atmosphere and repeated for every season. To keep the calculations manageable, he considered the entire atmosphere as a single layer whose radiative properties were characterised by a single vertically averaged temperature. To further simplify his calculations, he assumed that the transport of heat by winds and ocean currents remained unchanged with time, as did the extent of cloud cover, which allowed him to focus on the variation of the temperature with the transparency of the air. Even so, without the aid of a computer, this was a long and tedious process to say the least – but perhaps helped by those long Swedish nights and the absence of distractions like television.

Arrhenius did the best he could, given the limitations of the data at the time. But the infrared spectral data that he had at his disposal only went up to wavelengths of 13 microns (μ) – not far enough for him to pick up the major CO_2 absorption band near 15 μ, which is an important feature for global warming[12]. Moreover, his absorption data were largely at intervals of 0.5–1.0 μ, whereas individual absorption spectra for both water vapour and CO_2 are now known to occur in multiple spectral lines, each around 1/100 μ wide or less[18]. This was unknown to Arrhenius and his contemporaries, and would remain so for some decades. As mentioned earlier, Arrhenius knew of the limitations on water vapour caused by rainfall at 100% humidity. His model required that whatever the temperature, the relative humidity should stay at around 80%.

Arrhenius predicted that doubling the CO_2 content of the atmosphere from its present value, for which he used 300 ppm, would raise the temperature of the surface of the Earth by about 6 °C. Most modern estimates put the likely warming for a doubling of CO_2 at 2.5–4.0 °C. He got lucky, because his model contained two significant sources of error: first, in not having the full spectrum of absorption of CO_2, which biased the result on the high side; and second, in using just one layer for the entire atmosphere, which biased the result on the low side. These two errors cancelled one another[4]. As Archer and Pierrehumbert point out, '*the genius in the work of Arrhenius is that he turned Fourier's rather amorphous and unquantified notion of planetary temperature into exactly the correct conceptual framework, even going so far as to get the notion of water vapour feedback right. Most importantly he correctly identified the importance of satisfying the energy balance both at the top of the atmosphere and at the surface … [C]orrect spectroscopy was not brought together with a correct conceptual framework [like that of Arrhenius] in a multilevel model until*

the seminal work of Manabe in the early 1960s'[4]. The importance of Arrhenius's contribution is that '*While we can now compute the effects of CO_2 on climate at a level of detail and confidence that Arrhenius could hardly have dreamed of, we are basically doing the same energy book-keeping as Arrhenius … [but] in vastly elaborated detail with vastly better fundamental spectroscopic data*'[4].

Arrhenius predicted that, in order to get the temperature drop of 4–5 °C thought likely for the last glaciation, the CO_2 concentration of the atmosphere must have fallen to between 165 and 186 ppm. We now know it was about 180 ppm, so he was not far off. Nevertheless, he was unable to provide a mechanism for the changes in CO_2 with time between glacial and interglacial periods. His theory implied that the whole Earth should have undergone about the same variations of temperature at the same time, the atmosphere being well mixed, meaning that glacial periods should have been simultaneous in both hemispheres, contrary to Croll's notion that glacial and interglacial periods alternated between the hemispheres. Arrhenius disputed the notion that orbital variations drove variations in the climate as Croll required, but offered no alternative in its place. As we see later, the fluctuations of the Ice Age require a combination of the mechanisms of Croll, Arrhenius and Lyell[11].

Arrhenius was well aware that during Tertiary times, Arctic vegetation had been like that of the mid latitudes, implying that temperatures must have been 8–9 °C above what they are there today. This, he calculated, would have required an increase in CO_2 of 2.5–3.0 times above the current value of 300 ppm.

To underscore the relevance of his model to geology, Arrhenius reprinted in his paper a lengthy summary of the views of his geological colleague, Högbom. Högbom calculated that the burning of coal would supply about 1/1000th part of the carbonic acid (CO_2) in the atmosphere, '*completely compensating the quantity of carbonic acid that is consumed in the formation of limestone*'[17], and thus not building up in the atmosphere. At the time, Arrhenius went along with that, showing no concern about humanity's influence on the climate.

Just after the turn of the century, in 1907, the date of the preface to his 1908 book *Worlds in the Making*, Arrhenius looked into the question of how changes in CO_2 might affect our present climate. By then, he was referring to Tyndall's theory as the 'hot-house' (i.e. greenhouse) theory[19]. He first calculated by how much the temperature of the Earth's atmosphere would fall if deprived of its

CO_2, and what the added effect would be of depriving it of its water vapour, showing that '*comparatively unimportant variations in the composition of the air have a very great influence*'[19]. His new calculations showed that '*doubling of the percentage of carbon dioxide in the air [from its value then] would raise the temperature of the Earth's surface by 4°*'[19]. Recognising that we humans were now burning significant amounts of coal, and assuming that it was burned at the current rate, he suggested: '*the slight percentage of carbonic acid in the atmosphere may by the advances of industry be changed to a noticeable degree in the course of a few centuries*'[19]. But, being an astute fellow, he realised that this change in climate would be significantly more rapid if the rate of consumption of fossil fuels increased – as we now know it did.

Knowing that CO_2 was good for plants, he suggested that increasing the CO_2 content of the atmosphere by burning fossil fuels might be beneficial, making the Earth's climates warmer and more equable, stimulating plant growth, providing more food for a larger population and even preventing the recurrence of another glacial period. Arrhenius also knew that in the normal course of events, the content of CO_2 in the atmosphere did not continually increase as a result of volcanic activity, because it was taken out of the atmosphere by weathering of minerals and assimilation by plants. That is what we may call the 'slow cycle' of carbon on geological time scales, in cotrast to the burning of fossil fuel, which is happening much more rapidly in what we may call the 'fast cycle'.

4.7 Chamberlin's Theory of Gases and Ice Ages

It seems odd in retrospect that neither Lyell nor Croll, both publishing their key works in 1875, seem to have picked up on the significance of Tyndall's discoveries, reported in the 1860s and early 1870s, of the importance of water vapour and CO_2 as greenhouse gases capable of influencing the climate. Lyell only mentions CO_2, in the 12th edition of *Principles*, to dismiss the idea of some geologists that it was an abundance of CO_2 in the air that encouraged the flourishing of vegetation that led to the coal deposits of the Carboniferous. That he knew about Tyndall's discoveries is evident from a letter to Joseph Hooker dated October 1866, in which he said, '*I suspect that the vapour to which you allude, and on which Tyndall has written so much,*

may equalize the heat and cold caused by greater proximity to, and distance from, the sun'[20]. Fleming reminds us that in his 1866 edition of *Principles*, Lyell had taken some of Tyndall's discoveries into account in noting that plants could be saved from extinction during an Ice Age '*by the heat of the earth's surface … being prevented from radiating off freely into space by a blanket of aqueous vapour caused by the melting of snow and ice*'[21]. But, as Fleming points out, '*He was grasping at straws, aware of new theoretical problems, but taking from them only the aspects that reinforced his own preconceptions*'[21]. Perhaps this lack of connection between the frontiers of physics and geology is a sign of the increasing specialisation of the sciences at the time.

The geologist who did pick up Tyndall's findings and run with them, albeit indirectly, was Thomas Chrowder Chamberlin (1843–1928) (Figure 4.6, Box 4.5).

Rather than following Tyndall directly, Chamberlin was following Arrhenius, who had been stimulated by Tyndall's work. Like Lyell, Tyndall, Croll and Arrhenius, he too seems to have been unaware of the groundbreaking work of Ebelmen.

In 1899, having seen Arrhenius's seminal paper in 1896, and attracted by the idea that CO_2 might have helped to control the climates of former times, Chamberlin set about developing a working hypothesis to explain what its role might have been[22, 23]. He surmised that CO_2 in the air combines with water in rain to form a weak acid that decomposes the silicate minerals of shales,

Figure 4.6 *Thomas Chrowder Chamberlin.*

Box 4.5 Thomas Chrowder Chamberlin.

Chamberlin was born in Mattoon, Illinois, and educated at Beloit College, Wisconsin and the University of Michigan. On the faculty at Beloit, he mapped the glacial deposits of Wisconsin and devised the terminology used to describe the North American Pleistocene. He joined the Geological Survey of Wisconsin in 1873 and became its chief geologist in 1876. He was appointed head of the Glaciological Division of the US Geological Survey in 1881, president of the University of Wisconsin in 1887, professor of geology at the University of Chicago in 1892 and president of the Chicago Academy of Sciences in 1898. In 1893, he launched the *Journal of Geology*, and in 1904 a college textbook, *Geology*. He was awarded the Penrose Gold Medal of the Society of Economic Geologists in 1924 and the Penrose Medal of the Geological Society of America in 1927. He has craters named after him on the Moon and Mars.

sandstones and volcanic rocks, thus supplying mineral salts, including carbonate ions (CO_3^{2-}), to the ocean and leaving behind residual silicates and oxides[22, 23]. This 'weathering' sucks CO_2 out of the atmosphere slowly over long periods of time. When the dissolved carbonate ions arrive in the ocean, they are used by marine creatures to form shells of calcium and magnesium carbonate, which fall to the seabed to accumulate over time and be compressed into limestones and dolomites. CO_2 is also lost from the atmosphere through photosynthesis by plants, then trapped as carbonaceous compounds in the organic matter disseminated widely in small amounts in marine sediments and in the organic-rich deposits that ultimately form coal, bitumen, oil and natural gas. This loss cools the atmosphere. Compensating for this loss over the aeons of time is the supply of CO_2 to the atmosphere from volcanoes, and also from the decomposition of organic matter. Chamberlin concluded, '*the state of the atmosphere at any time is dependent upon the relative rates of loss and gain. [The] agencies of permanent supply [volcanoes and the decomposition of organic matter] and of permanent loss [weathering and the deposition of carbonates] ... are ... rather slow in action, and ... on the whole mutually compensatory ... but these relations are* believed to be subject to sufficient fluctuation to give a basis for pronounced climatic change'[22, 23]. His hypothesis reads remarkably like Ebelmen's in several respects, probably due to the convergent evolution of ideas, which can lead to independent generation of the same concept – as Darwin and Wallace found out in 1858.

Chamberlin was the first geologist to fully appreciate the role of the ocean in controlling the amount of CO_2 in the atmosphere. By his time, chemists knew that CO_2 dissolves in the ocean, and that the CO_2 dissolved in the ocean should be in equilibrium with that diffused in the air[22, 23]. Reducing the concentration or partial pressure of the CO_2 in the air would force some of the CO_2 in the ocean to diffuse into the air to restore the equilibrium. Chemists also knew that warm water contains less gas than cold water. So Chamberlin appreciated that a warming ocean would release more CO_2 to the air, while a cooling ocean would take more CO_2 from it. If CO_2 were lost from the air by weathering faster than it could be replaced by diffusion from the ocean, the atmosphere and the ocean would cool, and the cooler ocean would hold more CO_2, preventing its escape to the air. Bearing these various new facts in mind, Chamberlin was the first to suggest that '*the ocean, during a glacial episode instead of resupplying the atmosphere ... would withhold its carbon dioxide to a certain extent ... [and] when the temperature is rising after a glacial episode ... the ocean gives forth its carbon dioxide at an increased rate, and thereby assists in the amelioration of climate*'[22, 23].

In a nutshell, Chamberlin's theory of CO_2 and climate change runs like this: When land is raised and extended by mountain building, the rate of weathering and dissolution of silicate and carbonate rocks increases, consuming CO_2 from the atmosphere. The dissolved constituents are carried to the sea to form limestones, which keep CO_2 out of the atmosphere. The extension of the land reduces the area available for marine life (which Chamberlin assumes is more productive than terrestrial life), which in turn reduces the resupply of CO_2 to the atmosphere from decomposing marine organic remains. The lessened CO_2 concentration in the atmosphere cools the planet. The cooling ocean absorbs CO_2, further reducing its supply to the atmosphere. Over time, erosion reduces the land level, and weathering is reduced, permitting a rise in the CO_2 concentration in the atmosphere, supplied by the ongoing emission of volcanic gases. Warming ensues. The warming ocean floods the margins of the continents, further reducing weathering and enhancing CO_2 in the atmosphere. The warming evaporates water vapour from the ocean, warming the air yet further, which releases yet more CO_2 from the ocean. While increasingly abundant marine life takes CO_2 out of the atmosphere by photosynthesis, Chamberlin assumes that this is balanced by decomposition, which puts CO_2 back into the atmosphere. *'As a result … geological history has been accentuated by an alternation of climatic episodes embracing, on the one hand, periods of mild, equable, moist climate nearly uniform for the whole globe, and on the other, periods when there were extremes of aridity and precipitation, and of heat and cold, these last denoted … occasionally by glaciation'*[22, 23].

In many ways, Chamberlin's theory is much like that in vogue today. Applying it to the Ice Age of the past 2 Ma or so[24], he surmised that as CO_2 began to decrease in the atmosphere, starting the cooling process, ice sheets and snowfields expanded. These reflected more solar energy, accentuating cooling – the positive feedback that Croll had identified. Areas of frozen ground – permafrost – expanded, too. Combined, these effects limited the area exposed to chemical weathering, which in any case would have been reduced by the increasingly cold conditions, thus limiting the consumption of CO_2. Bearing in mind that the concentration of CO_2 in the atmosphere reflected a balance between supply and consumption, there would be a tendency for the continued supply of CO_2 from volcanic gases and other sources to restore the initial equilibrium, by building up the CO_2 concentration in the atmosphere and thus warming the ocean, which would then release more CO_2, causing further warming, which would evaporate water vapour, causing still further warming, leading eventually to the retreat of the ice sheets. Chemical weathering would then be renewed, starting the cycle over again. Chamberlin thought that these patterns would be influenced and modified, but not controlled, by variations in solar heating caused by orbital changes of the kind suggested by Croll.

In his later work, Chamberlin considered that *'a vital factor in initiating a glacial period is a reversal of the deep-sea circulation'*[5]. In interglacials like the present, he surmised, deep-ocean currents transfer heat to high latitudes, keeping the poles relatively warm. Reversal of this circulation cools the polar oceans, leading them to take up CO_2 at the expense of the atmosphere. Cooling puts less water vapour into the atmosphere, which accentuates the fall in temperature. Eventually, cold water occupies the ocean floor. In due course, this cold water reaches the surface in the tropics, warms, and starts to release more CO_2

than the polar waters are absorbing, so ending the glacial cycle. These ideas are rather like those of Carpenter and Croll, which we read about in Chapter 3, and which probably stimulated Chamberlin's thinking in this direction. The notion of the reversal of deep-ocean circulation is now integral to our understanding of the glacial–interglacial transition. What Chamberlin lacked was the trigger for it.

Chamberlin's work was eventually largely forgotten, but as Fleming tells us[5], he gave his contemporaries an almost modern understanding of multiple glaciations, of the role of the atmosphere as a geological agent and of the role of deep-ocean circulation in the ice age. His work *'was filled with fundamentally sound insights and represents a surprisingly modern voice from the past'*[5].

With time, Chamberlin regretted having been quite so willing to accept Arrhenius's results. He thought that the role of CO_2 in the atmosphere had been overemphasised, and that not enough attention had been given to the role of the ocean, which he considered his distinct contribution to the subject. We now know that Chamberlin's dilemma should not have been a question of choice between atmospheric CO_2 and deep-ocean circulation: both play key roles in the transitions from glacial to interglacial period and back. He simply lacked the data to support his CO_2 theory, a state of affairs that has long since been remedied.

Like all geologists in the 19th century, one of Chamberlin's greatest problems was the absence of any ready means of accurately dating the boundaries between glacial and interglacial deposits. Taking a conservative view, he estimated that the last glaciation might have ended some 20 Ka ago – *'the time since the last ice retired from the site of Niagara River'*[24] – and that interglacial intervals lasted 20–30 Ka[24]. Not a bad guess, as we shall see in later chapters

In Chamberlin's final paper on the causes of ice ages, he addressed the enigma of the glaciation in the Permo-Carboniferous period 300 Ma ago[25]. Evidence for the occurrence of extensive glaciation in India, Brazil, Australia and South Africa had grown throughout the last half of the 19th century. A Dr Blandford discovered a late-Carboniferous boulder bed of glacial origin at Talchir, India, in 1856. Tillites (essentially beds of boulder clay) of about the same age were identified in South Australia in 1859, in South Africa in 1870 and in Brazil in 1888[26]. Establishing the nature and cause of this earlier glaciation proved difficult in the face of inexact dates and the absence of information about contemporaneous conditions in other parts of the world. The distribution of the tillites was extraordinary; they were chiefly found within the tropics. Another confusing factor was the association of glacial remains with coal deposits, which seemed to imply tropical conditions and high concentrations of CO_2 in the atmosphere. Lyell had noted in the 12th edition of his *Principles* '*That the air was charged with an excess of carbonic acid in the Coal period has long been a favourite theory with many geologists, who have attributed partly to that cause an exuberant growth of plants*'[27]. He thought this inference '*most questionable*', because although carbonic acid was supplied to the atmosphere from volcanoes and other sources, there were ample causes in action to prevent it building up in the atmosphere, notably the burial of plant remains. Chamberlin was reduced to suggesting on flimsy grounds that the widespread deposition of carbonates and coals in the Carboniferous may have depleted the air of CO_2, so encouraging development of a glaciation. He was not far from the truth, as it happens. The various glacial deposits were judged to belong to the Gondwana Series – chiefly land and freshwater deposits whose distribution around the Indian Ocean implied unusual connections of some kind between the different continents.

Chamberlin was not alone in his speculations about past geography and climate. In the second half of the 19th century, geologists began to realise from the distributions of fossil plants and animals that there must have been close links between different continents. One of the first to point this out for the southern continents was Eduard Suess (1831–1914) (Figure 4.7, Box 4.6).

Noticing that fossils of the extinct order of seed ferns known as *Glossopteris* were common to southern Africa, India and Australia, Suess postulated in Volume II of his book *The Face of the Earth*[28] that they had formerly been connected in a supercontinent that he named 'Gondwanaland' – 'Land of the Gonds' – after an ancient people of India. Suess thought that the present separation of the fragments of Gondwanaland was due to flooding of its low parts by the ocean, not to the fragments moving apart from one another. He was the first to recognise that an ocean, which he named Tethys, must have connected the Atlantic and Pacific Oceans more or less along the line of the Mediterranean Sea and Persian Gulf in the distant geological past, a significant geological development that must have affected the regional climate. Suess also believed that the rises and falls of the sea (transgressions and regressions) must be connected from continent to continent and should be correlatable across the Earth. We will revisit that important concept in later chapters.

Figure 4.7   *Eduard Suess in 1869.*

Box 4.6 Eduard Suess.

Born in London to a German merchant family, Eduard Suess moved to Vienna as a teenager and spent much of his career as professor of geology at the university there. In his magnum opus, *Das Antlitz der Erde* (*The Face of the Earth*), Suess pointed out that organic life was limited to a narrow zone at the surface of the Earth's crust or lithosphere, which he named the 'biosphere'. Highly respected, and considered one of the fathers of the science of ecology, he was elected a member of the Royal Swedish Academy of Sciences in 1895, awarded the Copley Medal by the British Royal Society in 1903 and the Wollaston Medal of the Geological Society of London in 1896 and had craters named after him on the Moon and Mars.

When setting down his ideas between 1897 and 1899, Chamberlin might not have known about Suess's concept of the unity of the southern continents in Gondwanaland, because the three volumes of Suess's book were not translated into French (*La Face de la Terre*, in four volumes) until between 1897 and 1918, or into English (in five volumes) until between 1904 and 1924.

The idea that the continents had been connected by former 'land bridges' now sunk into the ocean became popular among palaeontologists seeking to explain the distribution of similar species on widely separated land masses. We may laugh at it now, with the benefit of hindsight, but in those days little was known about the composition of the deep ocean floor, so it did not seem unreasonable to assume that former land bridges simply sank to form ocean basins. So influential was Suess that, as an undergraduate in the early 1960s, I still had to write essays on the possible existence of 'land bridges'. But I was never taught about Chamberlin or CO_2.

The tree fern *Glossopteris* (or 'tongue fern' in Greek), which would help in coming years to establish the reality of a southern supercontinent, was named by botanist Adolphe Brongniart (1801–1876), the son of Alexandre Brongniart, whom we met as a friend of Lyell's in Chapter 2. Adolphe's first memoir on fossil plants appeared in 1822, and his *Histoire des Végétaux Fossiles* was published between 1828 and 1837, earning him the title 'father of palaeobotany'[29]. The boundary between the Gondwanan flora, in which *Glossopteris* plays such an important role – especially in the Permian – and that of the northern continents has been described as '*the most profound floristic boundary in the history of land vegetation*'[30]. Captain Scott of Antarctic fame would unwittingly identify Antarctica as a former part of Gondwana by collecting a rock sample containing fossil *Glossopteris* leaves from Mt Buckley on the Beardmore Glacier on his tragic return from the South Pole in February 1912. Scott's leafy rock sample was analysed in 1914 by the eminent palaeobotanist Albert Charles Seward (1863–1941), who later became president of the Geological Society of London (1922–24) and won the Society's Wollaston Medal in 1930[31]. Suess would have been pleased to see the results, although not with the human cost.

As following chapters show, we would have to wait until the late 1950s and beyond before science advanced far enough to test the ideas put forward by 19th-century pioneers like Tyndall, Arrhenius and Chamberlin concerning CO_2 in the atmosphere and its role in the climate system. Atmospheric physicists and chemists would have to understand much more about how the greenhouse effect worked, which would require the development of more powerful analytical techniques, along with advanced computers for the processing of data. And palaeoclimatologists would have to follow Chamberlin and learn much more about atmospheric and ocean chemistry and circulation, as well as the role of the greenhouse gases, in order to

appreciate how the Earth's climate system worked in the past. As the migration of continents is fundamental to understanding the change in their climates through time, it is to continental drift that we turn next.

References

1. De Saussure, H.B. (1779, 1786, 1796) *Voyages Dans les Alpes, Précédés d'un Essai sur l'Histoire Naturelle des Environs de Genève*, **3** vols. Fauche Barde, Manget, et Fauche-Borel, Neuchatel.

2. Geikie, A. (1897) *Founders of Geology*. Macmillan, London.

3. Uglow, J. (2002) *The Lunar Men*. Faber & Faber, London.

4. Archer, D. and Pierrehumbert, R. (2011) *The Warming Papers*. Wiley-Blackwell, Chichester.

5. Fleming, J.R. (1998) *Historical Perspectives on Climate Change*. Oxford University Press, Oxford.

6. Tyndall, J. (1860) *The Glaciers of the Alps*. Longmans, Green, London.

7. Tyndall, J. (1861) On the absorption and radiation of heat by gases and vapours, and on the physical connexion of radiation, absorption and conduction. *London, Edinburgh and Dublin Philosophical Magazine and Journal of Science*. Reproduced in Tyndall, J. (1868) *Contributions to Molecular Physics in the Domain of Radiant Heat*. D. Appleton, New York; also reproduced with comments in Archer, D. and Pierrehumbert, R., (2011) *The Warming Papers*. Wiley-Blackwell, Chichester, pp. 21–44.

8. Tyndall, J. (1863) *Heat and Mode of Motion*, 1st edn, D. Appleton, New York. (7th edn Longmans, Green, London, 1887).

9. Tyndall, J. (1868) Further researches on the absorption and radiation of heat by gaseous material. In: *Contributions to Molecular Physics in the Domain of Radiant Heat* (ed. J. Tyndall). D. Appleton, New York.

10. Tyndall, J. (1868) *On Radiation: The 'Rede' Lecture, Delivered in the Senate-House before the University of Cambridge on Tuesday, May 16, 1865*. D. Appleton, New York. Also included in Tyndall, J. (1871) *Fragments of Science*. Longmans, Green, London.

11. Bard, E. (2004) Greenhouse effect and ice ages: historical perspective. *Comptes Rendus Geoscience* **336**, 603–638.

12. Berner, R.A. and Maasch, K.A. (1996) Chemical weathering and controls on atmospheric O_2 and CO_2: fundamental principles were enunciated by J.J. Ebelmen in 1845. *Geochimica et Cosmochimica Acta* **60** (9), 1633–1637.

13. Ebelmen, J.J. (1845) Sur les produits de la décomposition des espèces minérals de la famille des silicates. *Annales des Mines* **7**, 3–66.

14. Hunt, T.S. (1880) The chemical and geological relations of the atmosphere. *American Journal of Science* **19**, 349–363.

15. Hay, W.W. (2013) *Experimenting on a Small Planet: A Scholarly Entertainment*. Springer, New York.

16. Plass, G.N. (1961) The influence of infrared absorptive molecules on the climate. *New York Academy of Science* **95**, 61–71.

17. Arrhenius, S. (1896) On the influence of carbonic acid in the air upon the temperature of the ground. *London, Edinburgh, and Dublin Philosophical Magazine and Journal of Science*, Series 5, **41** (251), 39.

18. Pierrehumbert, R.T. (2010) *Principles of Planetary Climate*. Cambridge University Press, Cambridge, p. 221.

19. Arrhenius, G. (1908) *Worlds in the Making: The Evolution of the Universe*. Harper, New York.

20. Lyell, K.M. (1881) *Life, Letters and Journals of Sir Charles Lyell, Bart*, vols 1 and 2. John Murray, London.

21. Fleming, J.R. (1998) *Charles Lyell and Climatic Change: Speculation and Certainty*. Geological Society Special Publication, London.

22. Chamberlin, T.C. (1897) A group of hypotheses bearing on climate changes. *Journal of Geology* **5**, 653–683.

23. Chamberlin, T.C. (1899) An attempt to frame a working hypothesis of cause of glacial periods on an atmospheric basis. *Journal of Geology* **7**, 545–584.

24. Chamberlin, T.C. (1899) An attempt to frame a working hypothesis of cause of glacial periods on an atmospheric basis. Part II, special application of the hypothesis to the known glacial periods. *Journal of Geology* **7**, 667–685.

25. Chamberlin, T.C. (1899) An attempt to frame a working hypothesis of cause of glacial periods on an atmospheric basis. Part III, localisation of glaciation. *Journal of Geology* **7**, 751–787.

26. Holmes, A. (1965) *Principles of Physical Geology*. Nelson, London.

27. Lyell, C. (1875) *Principles of Geology*, 12th edn. John Murray, London.

28. Suess, E. (1888) *Das Antlitz der Erde*, vol. II. F. Tempsky, Vienna.

29. Seward, A.C. (1892) *Fossil Plants as Tests of Climate; Being the Sedgwick Prize Essay for the Year 1892*. C.J. Clay & Sons, London.

30. Schopf, J.M. (1968) Distribution of the Glossopteris-Gangamopteris flora. *Geological Society of America Annual Meeting*, Mexico City, p. 267.

31. Seward, A.C. (1914) *Antarctic Fossil Plants*. British Museum (Natural History), London.

5

Moving Continents and Dating Rocks

5.1 The Continents Drift

Climate is intimately linked to geography, because the amount of energy received at any point on the Earth's surface varies with latitude. As information from the rock record concerning the climates of the past grew, geologists realised that there were large variations in climate through the aeons of geological time in any given location. One favoured explanation was that the continents were fixed, and displacement of the Earth's pole and axis caused climate shifts. An alternative was that the pole and axis were fixed, and the Earth's thin, hard crust slid over a fluid subcrust in relation to them. The result was the same: displacement of the poles in relation to the continents, or 'polar wandering'. The problem with applying the polar wandering concept was that evidence for glaciation in the Carboniferous and Permian around 300 Ma ago was spread across the widely separate southern continents and India. Where might the pole have been then?

The matter was complicated by Eduard Suess's notion that the Earth was shrinking, making the continental crust collapse in places to form the ocean basins[1]. He thought the world's mountain belts arose where the edges of floating pieces of crust crept up over sunken pieces. Suess's notion of continental creep was taken up and modified by the American geologist Frank Bursley Taylor (1860–1938), who lectured on it to a Geological Society of America meeting in Baltimore, Maryland in December 1908, and published his ideas in 1910[2]. Taylor saw the Tertiary mountain ranges of Europe and Asia as '*as the product of southward creep of the entire crustal sheet of Eurasia*'[2], which extended into Alaska. To account

for the southward bend of the mountain ranges of South East Asia around the northern edge of India, Suess and Taylor required Africa and India to remain still while southward-moving Eurasia moulded itself around them to form the Alps, the Himalayas and the South East Asian ranges. Taylor saw North America as creeping toward the west or southwest and away from the fault-bounded margins of Greenland.

Combining these notions with the observation that the American Appalachian Mountains looked as if they had once been continuous with the Caledonide Mountains running from Norway to Ireland, Taylor wrote that '*it seems probable that a considerable part of the present [North Atlantic] oceanic interval is due to Tertiary and perhaps to older crustal movements which divided the original chain near Greenland and carried the parts away on divergent lines – to the northeast and the southwest*'[2]. Noting the remarkable size and shape of the Mid-Atlantic Ridge and its position midway between the Americas and Africa, he suggested, '*The ridge is a submerged mountain range of a different type and origin from any other on the earth … It is apparently a sort of horst ridge – a residual ridge along a line of parting or rifting – the earth-crust having moved away from it on both sides … It is probably much nearer the truth to suppose that the mid-Atlantic ridge has remained unmoved, while the two continents on opposite sides of it have crept away in nearly parallel and opposite directions*'[2]. Taylor thought Africa moved east away from the Mid-Atlantic Ridge before the Mesozoic Era, while South America moved away to the west in the Tertiary. Elsewhere in the Southern Hemisphere, Antarctica appeared to have held fast, while Australia

Earth's Climate Evolution, First Edition. Colin P. Summerhayes.
© 2015 John Wiley & Sons, Ltd. Published 2015 by John Wiley & Sons, Ltd.

moved north. Observing that '*both polar areas were areas of crustal dispersion or spreading, and the continental sheets, excepting Africa, all crept toward the equatorial zone*', Taylor declined '*to attempt any discussion of the ultimate cause of the Tertiary mountain making*'[2].

Here, then, we have the suggestion that the continents may move apart, forming mountain ranges at their leading edges, where they meet other continental fragments (as between Europe and Africa, or Asia and India) or are thrust over the ocean basins (as on the western coast of North America), and leaving some new kind of seabed behind them in newly opened ocean basins: a prescient observation, although it caused barely a ripple at the time. Taylor's concept implied that the residual ridge down the middle of the Atlantic was a remnant of continental crust.

The suggestion that the Americas had moved away from Africa and Europe was not new. Noticing the nice 'fit' between west Africa and eastern South America, several eminent thinkers and scientists – starting with Abraham Ortelius in 1596 – suggested that these continents had been torn apart. The French geographer Antonio Snider-Pellegrini (1802–1885) argued in 1859 that this separation was sudden and had something to do with Noah's flood[3]. He also imagined that Australia might formerly have been connected to Kenya, before moving away to form the Indian Ocean. Geological evidence played no part in his speculations.

Examining the fit between South America and Africa, a 32-year-old German meteorologist, Alfred Wegener (1880–1930) (Figure 5.1, Box 5.1), was led in 1912 to propose his own controversial notion of 'continental drift'.

Wegener's musings were triggered by an atlas[4]. In her 1960 biography of Alfred, his wife Else quoted from a letter he wrote in January 1911: '*My roommate … got Andree's Handatlas (Scobel 1910) as a Christmas gift. For hours we admired the magnificent maps. A thought occurred to me: … Does not the east coast of South America fit the west coast of Africa as though they had been contiguous in the past? Even better is the fit seen on the bathymetric map of the Atlantic when comparing … the break-off into the deep sea. I must follow this up*'[5]. So we can date the spark of Wegener's revolutionary idea to Christmas 1910[4], just as Frank Taylor's idea was being published in the United States, in yet another convergence of ideas.

Wegener scoured the geological literature for evidence to support his idea. He linked continents together by matching the shapes of opposing continental shelves, by matching geological structures and fossil groups across

Figure 5.1   *Alfred Wegener.*

continental boundaries and by bringing together continental fragments with similar palaeoclimatic information, such as corals, coals, desert dunes and glacial deposits.

Box 5.1 Alfred Wegener.

Born in Berlin, Wegener studied natural sciences at the university there, receiving a doctorate in astronomy in 1904. He then pursued a career in meteorology, focusing on the upper atmosphere. He pioneered the use of balloons to study currents of air. He and his brother, Kurt, broke the world endurance record for hot air balloons, staying aloft for 52 hours in 1906. He was the meteorologist on the 1906 Danish expedition to Greenland, before taking up a post at the University of Marburg, where he lectured on meteorology, astronomy and navigation for explorers. In 1911, his collected meteorology lectures were published as *The Thermodynamics of the Atmosphere*. Wegener also worked at the German Naval Observatory and the University of Hamburg. He was the first to explain the formation of raindrops and of the haloes around the Sun, formed by ice crystals in Arctic air. He was back in Greenland in 1912, becoming the first explorer to over-winter on the ice cap. In due course, he became a world expert on polar meteorology and glaciology, and in 1924 was appointed professor of meteorology and geophysics at the

University of Graz in Austria. He died on the ice cap in Greenland in November 1930. The Alfred Wegener Institute for Polar and Marine Research, in Bremerhaven, Germany, is named after him, as are craters on the Moon and Mars. The Institute awards a Wegener Medal, and the European Union of Geosciences awards an Alfred Wegener Medal.

This formed the basis for his claim that about 300 Ma ago the continents had formed a single mass, which he called the 'Urkontinent', and that this had subsequently split, with the individual fragments moving gradually to their present positions. His theory explained the odd distribution of Cretaceous corals and Carboniferous coals. And by bringing the Southern Hemisphere continents together, he could show that the Carboniferous–Permian glaciation 300 Ma ago was centred about a South Pole located near Durban (Figure 5.2).

Wegener startled his audience with this sweeping proposal at a meeting of the Geological Association in Frankfurt am Main on 6 January 1912, and repeated it 4 days later at a meeting of the Society for the Advancement of Natural Science in Marberg. He titled his Frankfurt talk 'The Geophysical Basis of the Evolution of the Large-Scale Features of the Earth's Crust (Continents and Oceans)', and it was published the same year, in German, as 'The Origins of Continents'[6].

Like others, Wegener saw that the Earth's topography has a bimodal distribution, with the continents standing consistently higher than the ocean floors. He realised from newly acquired gravity data that the ocean floors differed geologically from the continents, in being made of dense rocks like basalt, full of heavy iron and magnesium silicates, on which less dense continental rocks like granite, full of relatively light quartz and feldspar, would float – rather like icebergs in an ocean – following the principle of isostasy, or hydrostatic equilibrium. The dense oceanic rocks he called 'sima', using Suess's term for the dense global subcrust of basaltic rocks rich in silica and magnesium; the light ones he called 'sial', modified from Suess's term 'sal' for the less dense granitic rocks of the continents rich in silica and aluminium. As an example of isostasy, he cited the northern polar regions, where the land was now rising, owing to the removal of the weight of the great ice sheets. Clearly, given enough time, the Earth's solid crust could flow.

Wegener thought hard about what might drive continental drift. In 1912, he wrote that the Mid-Atlantic Ridge represented a zone in which the floor of the Atlantic, as

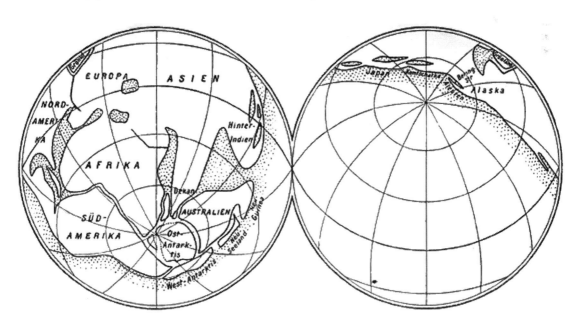

Figure 5.2 *Wegener's continental palaeoreconstruction for the Carboniferous.*

it keeps widening, is continuously tearing open and making space for fresh, relatively fluid and hot basaltic material (sima) rising from depth; a prescient observation, but one that he did not follow up. Later, influenced by his Arctic experience, he dropped this early rendition of 'seafloor spreading', replacing it with the idea that the lighter rocks of the continents ploughed through the denser rocks of the ocean basins like icebergs through the ocean, under the influence of 'accidental currents in the globe' – later called 'convection currents'.

Wegener's aim of writing something more substantial to explain his ideas was delayed by his participation in an expedition to Greenland in 1912. There, he made the longest traverse of the ice cap ever undertaken, a 750 mile crossing at the widest point, and was the first scientist to winter over on the ice cap. A further delay came with the advent of the First World War, in which he served as an army officer. Convalescing from a wound, he expanded on his theory to write one of the most influential and controversial books in the history of science: *The Origins of Continents and Oceans*, published in German in 1915, and revised in 1920, 1922 and 1929[7]. Due to the war, the book was not appreciated widely beyond Germany until the third German edition (1922) was translated into English in 1924[8].

In his 1920 text, Wegener provided three reconstructions of past continental positions: one for the Carboniferous (300 Ma ago), with the South Pole estimated to lie just off the coast of Durban in South Africa; one for the Eocene (50 Ma ago); and one for the Ice Age (Quaternary). By 1924, he had expanded his suite of maps to nine geological periods, an example of which is given in Figure 5.3[9]. They showed the assembly of continental fragments in the Palaeozoic and their subsequent disintegration and dispersion in Mesozoic and Cenozoic times.

Wegener's 'Urkontinent' included Suess's 'Gondwanaland' (the southern continents plus India) at its southern end, with its separate continental pieces migrated back to their 'correct' initial positions instead of being connected by sunken land bridges. He was still referring to 'Gondwanaland' in his 1920 book, but by 1922 he had started to use the term 'Pangaea' (meaning 'all the Earth' in Greek) for his 'Urkontinent' and had dropped 'Gondwanaland' entirely[8].

Wegener gave us a first crude way of looking at Earth's palaeogeography: the locations of the continental fragments on which climate-sensitive sediments were deposited through time. It was a boon to palaeoclimatologists seeking to interpret past climate change from

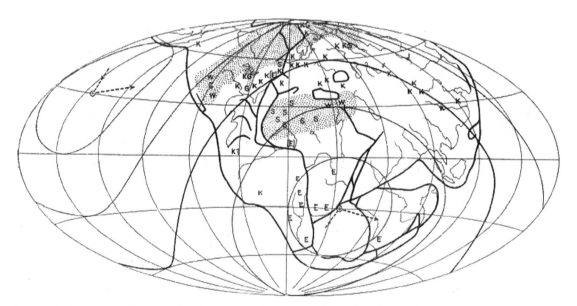

Figure 5.3   *Köppen and Wegener's palaeogeographic map for the Carboniferous, showing regions of ice, marshes and deserts. E = signs of ice; K = coal; S = salt; G = gypsum; W = desert sands; dotted area = dry regions.*

the sedimentary record. Wegener himself used the climate record of the rocks to estimate probable past polar positions and improve the 'polar wandering curve': the line joining the successive estimated positions of past poles from the Carboniferous to today[10]. Unfortunately, he lacked the magnetic data that would help later investigators improve on his reconstructions of past continental positions. Even so, his maps were not vastly different from modern ones. One major difference is that while he showed India in its more or less correct southern location in the Carboniferous between Africa, Antarctica and Australia, he created an imaginary land bridge between India and Asia, across what we now know was the ocean that Suess named 'Tethys'.

Wegener can claim several 'firsts': first use of the expression 'continental drift' to describe the movement of the continents; first reconstruction of continental positions based on the edges of the continental shelf, rather than the coastline; first amassing of large volumes of geological and palaeontological evidence to support his reconstructions; and first statement of the proposal that his supercontinent broke apart due to rifting, like that forming today's East African Rift Valley. By considering the changes in flora and fauna with time on the different continents, he was even able to suggest when they broke apart: Australia and Africa in the Jurassic; South America and Africa in the Lower to Middle Cretaceous; and India and Madagascar at the end of the Cretaceous[8,9]. There was no longer any need for Suess's land bridges.

Geologists are a conservative bunch, and many could not accept the radical idea that the continents might formerly have been joined together, not least because its advocate was a meteorologist[10]. Everyone knew that land bridges explained how similar plants and animals came to be found on the different continents. And anyway, Wegener could not explain precisely what moved the continents. '*Utter, damned rot!*' said the president of the American Philosophical Society[10]. But not everyone was so dismissive. Some recalled that even Humboldt had noted in his five-volume *Cosmos* that there were striking similarities between the rocks on either side of the South Atlantic. But the lack of a plausible mechanism for continental drift would consign Wegener's theory to the dustbin in the minds of his many critics. He would eventually be proved right and they wrong, but not until long after he was dead.

Was criticism of Wegener's background justified? In fact, he was far from being 'just a meteorologist'. He had

trained in the natural sciences, including geology, held a PhD in astronomy, practised as a meteorologist and had considerable experience of glaciology from expeditions to the Greenland ice cap, all of which led him to adopt a highly interdisciplinary approach to scientific questions that was unusual at the time. His observations of the behaviour of ice floes and icebergs around Greenland probably helped him come to his drift theory.

A major problem in getting his theory accepted lay in the absence of a plausible mechanism for the movement of the continents; geologists just did not know about the processes taking place deep beneath the Earth's crust. As he put it, '*Relative continental movements have been determined by purely empirical means … [but] the Newton of continental drift has not yet appeared [to explain how they took place]*'[4]. He was adamant that '*Continental drift, divergence and convergence, earthquakes, volcanism, transgressions and polar wander are interconnected in a grand causative scheme*'[4]. How right he was.

A plausible scheme by which to explain the drifting of the continents came in 1928 from an English geologist, Arthur Holmes (1890–1965) (Figure 5.4, Box 5.2).

Holmes suggested that the continents were split apart by rifting induced by convection currents within the warm, viscous rocks of the Earth's mantle, which stretches 35–3000 km deep beneath the lithosphere, Earth's rigid outer crust. As the convection currents transported continental fragments away from one another, new deep-ocean floor formed between them through the crystallisation of basalt magma welling up from below, beneath the mid-ocean ridge[11]. Elaborating on this idea in the first edition of his *Principles of Physical Geology*,

Figure 5.4 *Arthur Holmes.*

Box 5.2 Arthur Holmes.

Arthur Holmes, from Gateshead, Northumberland, had a significant influence on the development of geology. A precocious pioneer in geochronology, he carried out the first uranium-lead dating while still an undergraduate at the Royal School of Mines in London and published *The Age of the Earth* in 1913, when just 23 years old. He taught geology at Durham University before becoming head of the geology department at the University of Edinburgh. He was widely known for his 1944 textbook, *Principles of Physical Geology*, in whose title we see an attempt to emulate Lyell. I learned a lot from the massive 1288-page second edition of this book (1965). Holmes was much honoured by his peers. He won the Geological Society's Murchison Medal in 1940 and its Wollaston Medal in 1956. He became a fellow of the Royal Society in 1942 and won its Copley Medal, as well as the Geological Society of America's Penrose Medal. He has a crater on Mars named after him. The European Geosciences Union awards the Holmes Medal.

Box 5.3 Alexander Du Toit.

Du Toit was born in Cape Town and educated there and at the Royal Technical College in Glasgow. Joining the Geological Commission of the Cape of Good Hope led him to map much of South Africa and, in 1926, to write a book on its geology. In 1923, the Carnegie Institute of Washington gave him a grant to study the geology of Argentina, Paraguay and Brazil, to see how it matched that of the west coast of Africa, as a test of continental drift. He ended his career as a consulting geologist for the mining house De Beers, retiring in 1941. The Geological Society of London awarded him its Murchison Medal in 1933, and he was elected a fellow of the Royal Society in 1943. Like Wegener, he has a crater on Mars named after him.

drifting under the influence of convection currents in the Earth's subcrust.

5.2 The Seafloor Spreads

In the absence of geophysical evidence for the movement of the continents or the opening of the ocean basins, geologists organised themselves into different camps. The 'drifters' followed Wegener and Du Toit; the 'fixists' thought the ocean basins permanent; the 'contracters' thought the Earth was shrinking, with mountains forced up as the space for rocks diminished; and the 'expanders' thought the continents were being forced apart as the Earth expanded. This diversity of opinion provided me with ample meat for arm-waving undergraduate essays in the early 1960s.

Growing evidence for movement of the continents through time came from the striking discovery by Achille Delesse in 1849 that magnetic minerals like magnetite became aligned parallel to the Earth's magnetic field in solidifying volcanic lavas. But due to the lack of a suitably sensitive magnetometer, it was 100 years before the study of the magnetic field trapped in ancient rocks could begin in earnest. The experimental physicist Patrick Maynard Stuart Blackett (1897–1974), of Imperial College London, invented the appropriate device. Blackett won the Nobel Prize in Physics in 1948 for his work on cosmic rays, became president of the Royal Society in 1965 and was

Holmes explained that '*the basaltic layer becomes a kind of endless travelling belt on the top of which a continent can be carried along*'[11]. Where the fronts of advancing continents met old ocean floor, it sank beneath them into the Earth's interior, along ocean deeps like the Peru–Chile Trench. Advancing continental fronts were folded into mountains.

Another early supporter of continental drift was Alexander Du Toit (1878–1949) (Box 5.3), who presented detailed geological evidence for it in 1937 in his influential book *Our Wandering Continents*, which he dedicated to Wegener.

Pointing out that Wegener's '*illuminating hypothesis … can be tested on the basis of prediction*', Du Toit went on to say that he had made predictions on the basis of the theory and '*verified [them] by field work*'[12]. His detailed studies led him to propose that 'Pangaea' was an amalgamation of the southern continents, grouped together as 'Gondwanaland', and of the northern continents, grouped together as 'Laurasia', separated at times by Suess's Tethys Ocean, which disappeared as the southern continents collided with the northern ones in Cenozoic times (during the past 50 Ma). He favoured Holmes's concept of continents made a life peer, Baron Blackett of Chelsea, in 1969. He also has a crater on the Moon named after him.

Blackett's magnetometer triggered many sophisticated palaeomagnetic studies in the late 1950s, by the British scientists Stanley Keith Runcorn (1922–1995)[13] and Edward (Ted) Irving (1927–2014)[14], among others. Like Blackett, both Runcorn and Irving received several honours for their contributions, including being made fellows of the Royal Society. Their studies pointed to what appeared to be successive changes in the geographic position of the magnetic North and South Poles through time. In due course, they reasoned that it was not the magnetic poles that had moved – at least, not much – but the continents on which the sampled rocks were found. A sequence of measurements of magnetic polar position from rocks of different ages on each continent could define the 'polar wandering curve' for that continent. In effect, it showed the palaeo**latitude** of the measured rocks in relation to the magnetic pole through time. From the difference between the positions of the polar wandering paths in rocks of the same age on different continents, the **relative** palaeo**longitude** of those rocks could be obtained. The differences in the polar wandering paths of Europe and North America disappeared when the Atlantic was closed, proving that the continents had drifted apart. At last, scientists could determine with some confidence where the continents had been in relation to one another through time. Blackett was not to be left out, publishing his own comparisons of palaeolatitudes with the evidence of ancient climates from geological data[15]. Thus, by the end of the 1950s, palaeomagnetic data had vindicated Wegener: the continents had moved, although how they had done so was still unknown.

This revolution in the understanding of the way the Earth works picked up pace with a landmark paper widely circulated in 1960 by Harry Hammond Hess (1906–1969) (Figure 5.5, Box 5.4), professor of geology at Princeton, which was eventually published in 1962 as 'History of Ocean Basins'[16].

Hess's knowledge of the shape of the ocean basins came from his vast collection of echo-soundings. Before the Second World War, few research vessels had an echo-sounder, and none had continuous depth recorders printing out the depth as a line on a rolling strip of paper. Widespread use of echo-sounders and line printers by US Navy vessels during the war and by research ships after it ended provided abundant continuous depth records, or bathymetric profiles, enabling the shape of the ocean floor to be mapped in much more detail than could be

Figure 5.5   *Harry Hammond Hess.*

Box 5.4 Harry Hammond Hess.

Born in New York City, Hess taught for a year at Rutgers University (1932–33) and spent a year at the Geophysical Laboratory in Washington, DC, before joining Princeton in 1934, where he remained, becoming chairman of the geology department from 1950 to 1966. As a naval officer in the Second World War, Hess, who ended up as a rear admiral in the US Naval Reserve, was made captain of the USS *Cape Johnson*, an attack transport ship equipped with a relatively new tool in oceanography and navigation: the echo-sounder. Hess used its soundings to map the deep-ocean floor in more detail than had been achieved before, which deepened his understanding of sea-floor processes and led him to articulate novel ideas about the origin of the ocean basins. He was honoured with the establishment of the Harry H. Hess medal by the American Geophysical Union.

achieved by lead-line soundings. In 1957, Bruce Heezen (1924–1977), a graduate student under Maurice Ewing (1906–1974) at the Lamont Geological Observatory of Columbia University, working with the geologist and cartographer Marie Tharp (1920–2006), used these new data to make the first physiographic map of the North Atlantic. This map confirmed that there was a deep V-shaped rift running down the axis of the Mid-Atlantic Ridge. By 1977, Heezen and Tharp had expanded their map to cover all the ocean basins. Heezen was honoured by having a US Navy survey ship named after him – the USNS *Bruce C Heezen* – in 1999.

Maurice Ewing, meanwhile, had begun using seismic techniques to build up a picture of the Earth's crust beneath the ocean basins, showing it to be just a few kilometres thick. He had also dredged the Mid-Atlantic Ridge, in 1947, finding it to be made of basalt and other igneous rocks[17]. For his contributions to marine science, Ewing was awarded fellowship of the US National Academy of Science, the US National Medal of Science in 1973, the Vetlesen Prize in 1960, the Wollaston Medal of the Geological Society of London in 1969 and fellowship in the Royal Society in 1972.

Mapping the focal points of earthquakes showed the deep ones to be concentrated in the deep trenches around the Pacific margin. Heezen showed that the shallow ones tended to be concentrated along the axes of the mid-ocean ridges. Sir Edward Bullard (1907–1980) of Cambridge University, a fellow of the Royal Society and the Geological Society of London's Wollaston Medallist for 1967, found that heat flow was also high along those axes.

Thinking about the significance of this growing body of information, and building on Holmes's ideas, Hess surmised that molten rock oozes up from the Earth's interior beneath the world-encircling belt of mid-ocean ridges, creating new seafloor that spreads away on either side from the active zone at the ridge crest and eventually cools and sinks to form the deep-ocean basins. Recalling the fate of Wegener's ideas about continental drift, Hess was cautious, describing his as 'an essay in geopoetry'[16]. In a paper in the journal *Nature* in 1961, Robert (Bob) Sinclair Dietz (1914–1995), a scientist with the US Coast and Geodetic Survey who had read the widely circulated draft of Hess's paper, named Hess's process the 'seafloor spreading hypothesis'[18]. It provided the missing mechanism for Wegener's continental drift, and followed the seafloor spreading model that Wegener had originally proposed. This theory, which still needed testing, was the trigger for the astonishing development of the theory of plate tectonics in the late 1960s. Dietz was honoured for his achievements in marine science with the Penrose Medal of the Geological Society of America in 1988.

The first acid test of Hess's geopoetry came from the world of magnetics. Back in 1906, Bernhard Brunhes (1867–1910) had discovered that some rocks are magnetised in opposition to the Earth's present magnetic field: the compass needle points south when it should be pointing north, or vice versa. By the early 1960s, we knew that Earth's magnetic field had reversed polarity at least

170 times in the past 80 Ma, probably due to changes in the molten core. The distinctive pattern of these reversals back through time provided a magnetic polarity time scale that could be used to date rocks in the absence of other means.

In 1962, surveying the Carlsberg Ridge, part of the mid-ocean ridge in the Indian Ocean, from HMS *Owen*, graduate student Fred Vine (1939–) and his supervisor Drummond Matthews (1931–1997), both from Cambridge University, found the ridge's crest to be paralleled on either side by puzzling stripes of different magnetic character. Making a profound intellectual leap, they surmised in a paper in *Nature* in 1963 that, if Hess and Dietz were right about seafloor spreading, younger rocks along the mid-ocean ridge crest should have today's magnetic polarity, with the magnetic and geographic north poles being in the same direction, while older rocks further from the ridge crest should have the opposite magnetic polarity, having formed at the ridge crest when the magnetic field was reversed[19]. Given the multiple changes in magnetic polarity with time, the sideways-moving seafloor should show a pattern of ridge-parallel magnetic stripes increasing in age away from the ridge, with the pattern on one side of the ridge being an exact replica of the pattern on the opposite side. By matching the magnetic patterns on the seafloor to the magnetic polarity time scale, one could in theory date the seafloor far from the ridge crest. Vine and Matthews were right, and for that insight both were made fellows of the Royal Society. In due course, the Geological Society rewarded Matthews with the Bigsby Medal, in 1975, and the Wollaston Medal, in 1989, and awarded Vine the Bigsby Medal, in 1971, and the Prestwich Medal, in 2007.

Many scientific discoveries are 'ideas waiting to happen', and so it proved with the Vine and Matthews hypothesis. Their idea had been hit upon independently by Canadian geophysicist Lawrence Whittaker Morley (1920–2013), director of the geophysics branch of the Canadian Geological Survey in Ottawa (1950–69). He too had submitted a paper to *Nature* in 1963, proposing much the same theory. As reported in Morley's obituary in the *Canadian Globe and Mail*, a reviewer gave his opinion that the paper might be interesting for '*talk at cocktail parties, but it is not the sort of thing that ought to be published under serious scientific aegis*'[20]. Another geologist later described Morley's paper as '*probably the most significant paper in the Earth sciences ever to be denied publication*'[20]. *Nature* published the Vine and Matthews paper a few months after Morley's submission.

But Morley was not forgotten: eventually, the seafloor spreading hypothesis, confirming the reality of continental drift, became known as the Vine–Matthews–Morley hypothesis. Morley was duly honoured with the Royal Canadian Geographical Society's Gold Medal in 1995 and was made an officer of the Order of Canada in 1999.

By 1965, Fred Vine, working with Canadian geologist John Tuzo Wilson (1908–1993) (Figure 5.6, Box 5.5) from the University of Toronto, showed that these patterns were widespread in the ocean basins and could be predicted to occur where they were not yet found. Proof of the correctness of their theory came from dating of rocks drilled by the Deep-Sea Drilling Project's *Glomar Challenger* from differently magnetised zones on either side of the mid-ocean ridge, starting in 1968 (more about that in Chapter 6). The model had a further advantage: if the ages of the magnetic stripes were known then the timing of the separation of the continents and the opening of the ocean basins could be determined. This would be a boon to palaeoclimatologists.

Capitalising on the new information about the age of the seafloor, Sir Edward Bullard worked with colleagues at Cambridge, including young Alan Gilbert Smith (1937–), to show in 1965 that, by applying the principles of seafloor spreading, they could fit together the edges of the margins of the continents – the continental shelves – to close the Atlantic tightly, demonstrating the validity of Wegener's theory.

By late 1967 to early 1968, geophysicists like Dan McKenzie (1942–) at Cambridge and Jason Morgan (1935–) at Princeton had realised that the Earth's crust consists of a set of thin but rigid giant plates apparently being moved about by underlying convection in the hot

Figure 5.6 *John Tuzo Wilson.*

mantle. These 'tectonic plates' are thinnest beneath the oceans and thickest beneath the continents. Some, like the Pacific Plate, consist entirely of oceanic crust, with a thin surface veneer of relatively young sediments. Others, like the North American Plate, comprise a mixture of continental parts (North America) and oceanic parts (the North Atlantic west of the Mid-Atlantic Ridge). Where the edges of plates moving in opposite directions meet, one sinks beneath the other in a process known as 'subduction', which is accompanied by deep earthquakes. Great ocean trenches like those around the edges of the Pacific form as the downgoing or subducting plate buckles. The solid, subducting plate melts deep in the Earth's hot and semi-molten mantle beneath the thin and rigid crust, and the freshly melted material rises as magma, which spews out as lava through volcanoes in island arcs, such as the Aleutians, next to deep-ocean trenches. Much of the Pacific Ocean comprises an oceanic plate that is being subducted beneath the surrounding continents, with the resulting ring of volcanoes around its margin forming the so-called 'Pacific ring of fire'.

Sometimes, whole mountain ranges rise up over a subducting oceanic plate, as in the Andes, where the Pacific and related plates dive under South America. Where continents meet, the overriding one is thrust up as mountain ranges, like the Alps, between the European and African plates, or the Himalayas, between the Indian and Asian plates. Within the ocean basins, as McKenzie predicted and John Sclater confirmed, heat flow declines as the seabed moves away from the narrow zone of crustal formation at the ridge crest. As a result, the ocean crust cools, becomes denser and sinks, which explains why the deepest parts of the ocean basins contain the oldest (hence, coldest) ocean crust. Convection of some kind is assumed to 'close the circle' between magma rising beneath ridges, plates moving sideways and ocean crust descending into the Earth's mantle. It is accepted that there had to be some kind of connection between the rising material and the sinking material. As you might imagine, mapping convection currents is impractical without the ability to drill into the hot and ductile asthenosphere – the upper part of the Earth's mantle – beneath the rigid crust. What we do know of what goes on down there is somewhat fuzzy, and based on crude geophysical imaging of the subcrust, earthquake focal depths, the occasional rocks that rise from extreme depths – like Kimberlite pipes containing diamonds – and subtle differences in the geochemical makeup of volcanic rocks in island arcs, on volcanic islands and at mid-ocean ridges. We need not concern ourselves here with that level of detail, because we are interested in what happens on the surface under the control of climate.

Plate tectonics revolutionised the Earth sciences by explaining how and why everything was connected: the locations of the continents, mountain ranges, ocean trenches, water depths, deep and shallow earthquakes, regions of high heat flow, magnetic patterns and volcanoes. Plate tectonics also allowed geoscientists to reconstruct past geographies, one of the key controls of climate, and enabled geophysicists to use the dated magnetic stripes on the seafloor, along with the past positions of the poles as determined by palaeomagnetic analysis of continental rocks, to reconstruct the motions of the plates – and hence the continents – through time with a fair amount of precision. Among other things, it became possible to calculate how rapidly the continents were moving apart: the North Atlantic is opening at a rate of around 2.5 cm/year – the rate a fingernail grows.

Plate tectonics provided us not just with past geographies, but also with mechanisms for supplying CO_2 through volcanic activity and for removing CO_2 by weathering in newly uplifted mountains.

We use magnetic lineations on the seabed and palaeomagnetic data from the land to reconstruct continental positions since the breakup of Pangaea began 180 Ma ago. Making reconstructions prior to that is more difficult, because there is no older seafloor anywhere in the ocean basins. To position the continents in pre-Jurassic time, we use palaeomagnetic data. Given the 'palaeogeography' of past continental fragments, the original geographic position of climatically sensitive sediments on those fragments

can now be established with a high degree of certainty, confirming Lyell's notion that climatic zones much like those of today have prevailed through time, although modified by geographic changes such as the development of seaways and mountain belts, as he suspected. Plate tectonic theory fully vindicated Wegener, whose ideas have now been brought in from the cold, as beautifully illustrated by Robin Muir Wood[21].

By 1972, Runcorn and Donald Harvey Tarling (1935–) could report at a NATO Advanced Study Institute at the University of Newcastle-upon-Tyne that '*These developments have … essentially ended the long debate about whether or not the classical lines of geological evidence, palaeoclimatic, palaeontological distributions, global tectonic patterns and lithological relationships, support or refute drift*'[22]. Kenneth Creer, professor of magnetism at the same university from 1966 to 1972, used palaeomagnetic data to compile a suite of maps of past continental positions for the main geological periods: Cambrian, Ordovician, Silurian to early Devonian, Devonian, lower to middle Carboniferous, upper Carboniferous to Permian, Triassic, Jurassic and early Cretaceous[23]. Others demonstrated how plate tectonics explained the opening of the Atlantic and Indian Oceans[24–26].

These were exciting times for the Earth sciences – and for me, as, shortly after leaving Oxford University at the end of 1964 to join the New Zealand Oceanographic Institute as a marine geologist, I began using the seafloor spreading concept to analyse the origins of seafloor ridges, trenches and plateaux around New Zealand. Operating from the New Zealand Navy's HMNZS *Endeavour* in the Southern Ocean south of New Zealand, I surveyed the submerged continental crust of the Campbell Plateau, the eastern margin of which had broken off from Antarctica's Marie Byrd Land and the Ross Sea area some 80 Ma ago. My research on the adjacent Macquarie Ridge, running south from New Zealand's Fjordland, showed in 1967 that it was an island arc with an associated trench[27]. Just before that, in 1965, Tuzo Wilson proposed that plate tectonic theory required major faults crossing the mid-ocean ridges to be interpreted as a new class of fault – the 'transform fault' – and that transform faults also connected the oppositely directed ends of island arcs. Putting two and two together, I deduced that the famous Alpine Fault cutting across New Zealand had to be a transform fault connecting the west-facing Macquarie–Fjordland island arc in the south to the east-facing Tonga–Kermadec island arc in the north[27]. Wilson was pleased to see this result when he visited our research institute over the lollipop factory on Thorndon Quay in Wellington in 1967. My New Zealand colleagues were not all as kind to me as Wilson; many of the older kiwi geologists resented a young upstart from the United Kingdom revising the history of their great Alpine Fault. In those days, their notions of geology stopped at the coast.

I was also lucky enough to be assigned as the New Zealand marine geology representative to the Scripps Institution of Oceanography's Nova Expedition to investigate the submarine geology of the Southwestern Pacific in 1967, aboard the research vessel *Argo*. The expedition was under the leadership of Henry William (Bill) Menard (1920–1986) (Figure 5.7), author of one of my bibles of the time, the 1964 text *Marine Geology of the Pacific*[28]. Menard later became the 10th director of the US Geological Survey (1978–81), a member of the US National Academy of Sciences and a winner of the Geological Society of America's Penrose Medal.

Menard was a great teacher, and I had some fascinating company to bounce ideas off, including students who would later become leaders in their own right in the fast-evolving field of plate tectonics. Sailing from Noumea in New Caledonia to Auckland, New Zealand, via Brisbane in Australia, we surveyed several of the features expected from the seafloor spreading hypothesis: first a spreading seafloor – the Tasman Sea; next the submerged thinned continental fragments pushed away from Australia by the opening of that sea – the Lord Howe Rise and Norfolk Ridge; and then a back-arc basin – the South Fiji Basin; this last lay behind an active island arc and trench – the Tonga-Kermadec Ridge and Kermadec Trench[29]. I learned a great deal.

Figure 5.7 *Henry William Menard.*

Later, in October 1990, I was able to study a mid-ocean ridge at first hand during a research cruise to the Reykjanes Ridge, from which Iceland emerges in the North Atlantic[30]. Lindsay Parson of the United Kingdom's Institute of Oceanographic Sciences Deacon Laboratory, of which I was then director, led the cruise, and I shared a cabin with Sir Anthony Laughton, FRS, my predecessor and an expert on mid-ocean ridges, who had been awarded the Geological Society of London's Murchison Medal in 1989. We sailed from Bergen to New York aboard the MV *Maurice Ewing*, of the Lamont-Doherty Earth Observatory, on a storm-tossed sea. The occasional equinoctial gale had our captain steaming for gentler seas from time to time, which rather interrupted our seabed studies. Losing scarce research time to storms is all part of the trials and tribulations of open ocean research.

My involvement in the early days of plate tectonics would stand me in good stead as the topic developed in future years to underpin the palaeogeographic and palaeoclimatic studies in which I became increasingly involved.

5.3 The Dating Game

Palaeoclimatologists also needed to know precisely when climatic changes occurred. For that, they needed a means of dating rocks much more precisely than by using fossils. The breakthrough in rock dating came about through the discovery of radioactivity in 1896 by Antoine Henri Berquerel (1852–1908). Ernest Rutherford (1871–1937), a New Zealander working at McGill University in Montreal, proved that radioactivity involved an alchemist's dream – the transmutation of one chemical element into another – and that it did so at a set rate, the 'half-life', which was unique to each element. The half-life is the time by which half of the initial element, say uranium, has decayed into its 'daughter' product, which in the case of uranium is the element lead (Pb). Rutherford, who won the Nobel Prize in Chemistry in 1908 for his work on radioactive decay, realised that the decay of uranium into lead could be used to date rock samples. Bertram Boltwood (1870–1927) applied this understanding to date some rocks in practice in 1907. His samples appeared to range in age from 92 to 570 Ma old, but improvements to his technique changed their age range to 250–1300 Ma. In 1911, 21-year-old Arthur Holmes further developed the uranium–lead method, dating a rock from Ceylon to 1.6 billion years of age[31]. Because each radioactive element has its own half-life, geologists can use them as clocks to date past events.

The following year, Frederick Soddy (1877–1956) found that elements could occur in more than one form, with identical properties except for their atomic weight. He coined the word '*i*sotope' (meaning 'one place') for these identical entities of different mass. We now know that chemical elements contain nuclei made up of positively charged protons and particles of the same size but no charge: neutrons. Individual chemical elements are characterised by their number of protons. They usually contain the same number of neutrons. Isotopes occur where there are more or fewer neutrons than protons. Within any one element – say, potassium – there may be both stable and radioactive isotopes. For example, potassium has three naturally occurring isotopes, two of which are stable (^{39}K and ^{41}K) and one of which (^{40}K) is not. The superscript numbers refer to the numbers of protons and neutrons in the nucleus. Radioactive ^{40}K, with a half-life of 1.3 billion years, decays into the stable argon isotope ^{40}Ar. Similarly, the radioactive rubidium isotope ^{87}Rb, with a half-life of 49 billion years, decays into the stable strontium isotope ^{87}Sr. When I was a student at Oxford in 1963–64, I used a mass spectrometer to measure rubidium and strontium isotopes in order to date the igneous intrusion of Garabal Hill on the edge of Loch Fyne in the Scottish Highlands. It was 392 Ma old.

Radiometric dating, as it is called, has pinned down the finer details of the geological time scale, commonly by dating volcanic lavas and ash embedded within fossiliferous sediments. The minerals in igneous rocks like lavas and granites contain the dates of their formation, unlike the minerals in sedimentary rocks, which contain the dates of formation of the rocks from which they were weathered.

We have now entered the world of 'geochemistry'. Chemists had been examining geological materials as chemical objects in their own right for perhaps 200 years or more before it was realised that the tools and principles of chemistry could be used to understand and explain geological processes and the origins of the Earth's atmosphere, ocean and crust. This new way of doing things spawned a new subfield of geology. Stimulated by the development of new analytical techniques in the Second World War and the arrival on the scene of new analytical instruments capable of analysing smaller and smaller amounts of material with greater and greater accuracy, this new geological subdivision grew rapidly, attracting its own journal, *Geochimica et Cosmochimica Acta*, by 1950, and its own society, the Geochemical Society, by 1955. The

geochemist's bible, *Geochemistry*, emerged in 1954 from the pen of one of the founders of this new subject, Victor Moritz Goldschmidt (1888–1947), regarded by some as the 'father of modern geochemistry'. My interest in this new field led to my earning a PhD in it at Imperial College London in 1970. As we follow the development of the science of past climate change, we will see an increasingly prominent role played by geochemists.

By 1949, Willard Libby (1908–1980) of the University of Chicago had discovered that a naturally occurring radioactive isotope of carbon could be used to date organic remains up to about 60 Ka old. The abundant naturally occurring stable isotopes of carbon are ^{12}C, with six protons and six neutrons, and ^{13}C, with six protons and seven neutrons. Radioactive ^{14}C (or carbon-14), discovered in 1940, has eight neutrons, forms naturally in the atmosphere and decays with a half-life of 5730 years to the nitrogen isotope ^{14}N. Both fossil plants and calcareous skeletons made from calcium carbonate contain carbon and so can be dated by radiocarbon dating, provided they are young enough. Libby won the Nobel Prize in Chemistry in 1960 for his contributions. Among his colleagues was a young Austrian chemist, Hans Seuss (1909–1993), a grandson of the famous Eduard Suess. In 1950, Hans emigrated to the United States to work at the University of Chicago. Libby showed him how to use ^{14}C to date Earth materials[32]. Suess soon established a radiocarbon laboratory at the nearby US Geological Survey office, publishing his first dates, on fossil wood, in 1954.

New techniques and refinements to old ones continue to provide measurements of smaller and smaller intervals of time. Just to cite one example, a variant of the K-Ar dating method was developed in the 1970s to ascertain the ratio between two argon isotopes, ^{40}Ar and ^{39}Ar, in volcanic rocks. It allowed ages as young as 2 Ka to be determined.

With these various astonishing developments, geologists could accurately date the ages of magnetic reversals and fossil zones, enabling palaeoclimatologists to pin down with increasing confidence precisely when climate changes had occurred. Dating also helped to refine plate tectonic reconstructions of past continental positions, another aid in palaeoclimatology.

5.4   Base Maps for Palaeoclimatology

By the early 1970s, geologists were producing suites of continental reconstruction maps. Among them was a much-admired set compiled in 1973 by Alan Smith, Jim

Briden and G.E. Drewry, from Cambridge[33]. Smith's maps led to his being awarded the Geological Society's Bigsby Medal in 1981 and the Lyell Medal in 2008.

Reconstructing past continental positions is of critical interest not only to palaeoclimatologists, but also to petroleum companies wanting to reduce the risk involved in the increasingly expensive drilling required to find commercial quantities of oil and gas. Plotting information about past geography and environments on maps of past continental positions from different geological periods provides exploration geologists with an accurate regional perspective with which to determine oil and gas prospectivity ahead of the drill in frontier areas. It tells the explorers where to expect the organic-rich sediments that will eventually form petroleum source rocks, the porous sandy deposits that can form reservoirs for oil and gas and the clay minerals that can clog those pores. Source rocks, reservoir sands and pore-clogging clays result in many different ways from the climate of their depositional environment.

Recognising the potential value of palaeogeographic mapping, in 1981 the Exploration Division of BPRCo at Sunbury-on-Thames, under the leadership of Peter Llewellyn, set up extramural research awards with Smith's group in the Earth sciences department at Cambridge, to supply palaeoreconstruction maps, and with Professor Brian Funnell's group in the school of environmental sciences at the University of East Anglia (UEA), to provide information on past mountainous areas, shorelines, depositional environments and climate-sensitive sediment types, which they superimposed on Smith's maps[34] to provide BP's explorers with a technical edge in their worldwide search for oil and gas. Richard Tyson would do much of the legwork under Funnell's supervision at UEA.

BP's Global Palaeoreconstruction Group, located in BP Research's Palaeontology Branch (later renamed Stratigraphy Branch), managed the project. I led that group from September 1982 until 1985, following Ian Hoskin, the leader from 1981–82, and followed by David Smith, who led the project to its completion in 1987. We planned to produce maps for each of the 50 geological stages of the past 245 Ma. In some cases, there were not enough data to plot a single stage, so it was combined with another; we thus created 31 maps through time, each representing a period averaging about 8 Ma long. In 1994, Alan Smith, David Smith and Brian Funnell published the reconstructed continental positions and shorelines as *Atlas of Mesozoic and Cenozoic Coastlines*[35] (Figure 5.8). The oldest reconstruction was of the early

Triassic (Scythian stage) at 245 Ma. Tyson and Funnell published the palaeogeographic environmental data in a number of research papers[36, 37]. This was '*the first known attempt to draw a continuous series of global maps with such a fine resolution*'[35].

A major difference between Wegener's and later reconstructions is that later groups knew that Pangaea took the shape of the letter C, with its east side open to the Tethys Ocean, separating eastern Laurasia in the north from eastern Gondwanaland in the south (Figure 5.8). Wegener also thought that Greenland had remained connected to Europe until the Quaternary, but its separation actually began in the Eocene.

The main uncertainties in today's reconstruction maps are the positions of continental fragments that were compressed into mountain ranges at times between those for which each map was made[35]. This applies not only to pieces involved in building the Alps and the Himalayas, but also to Pacific continental margins, which accumulated small continental fragments that had migrated considerable distances along major transcurrent faults. Palaeomagnetic data will continue to help in determining the original positions and pathways of those fragments, as well as with unravelling the geology of complex areas like South East Asia. Fortunately, these uncertainties involve quite small areas and do not affect the main elements of the reconstructions. These days, new reconstruction maps refine rather than greatly change the reconstructions made in the early 1970s.

To check on the results of this in-house project, I got BPRCo to join a multicompany consortium funding the work of the Palaeogeographic Atlas Project, led by Alfred M. Ziegler (1938–) at the University of Chicago, which began in 1975[38]. Ziegler's proposal for this project was written '*in the expectation of stimulating the interest and support of the petroleum industry … The time has come to apply the findings of plate tectonics to the field of paleogeography*'[38]. Work began in 1975 with the help of seed money from the Shell Development Company. The US National Science Foundation provided co-funding from 1977[39], and the project continued with funding from some seven additional companies into the 2000s, with the latest publication listed on the Internet as 2007[40].

I was pleased to be working with Ziegler, having been a fellow student of his in the department of geology and mineralogy at Oxford in 1963–64. A palaeontologist, Ziegler was working closely with geologist Chris Scotese (1953–), Chicago's equivalent to Alan Smith at Cambridge. As part of his PhD (1976–83), Scotese

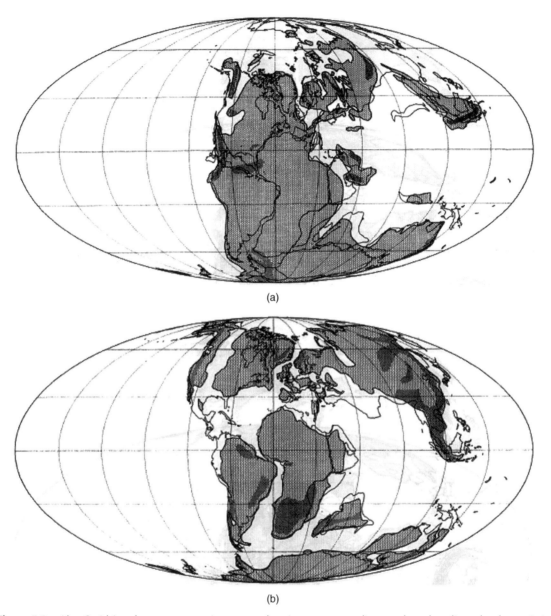

(a)

(b)

Figure 5.8 *Alan Smith's palaeoreconstruction maps, showing current outlines, palaeoshorelines, land area (stippled) and highland (shaded). (a) Early Triassic, 245 Ma ago. (b) Mid Cretaceous (Albian), 105 Ma ago.*

produced seven palaeoreconstruction maps for the Palaeo-zoic (late Cambrian (Franconian), middle Ordovician (Llandeilo-Caradoc), middle Silurian (Wenlock); middle Devonian (Emsian); lower Carboniferous (Visean); upper Carboniferous (Westphalian); late Permian (Kazanian))[41] and seven for the Mesozoic and Cenozoic (earliest Triassic (Induan), early Jurassic (Pliensbachian), latest Jurassic (Volgian), middle Cretaceous (Cenomanian), latest Creta-ceous (Maestrichtian), middle Eocene (Lutetian), middle Miocene (Vindobonian))[42], showing shallow seas, high-lands and lowlands. Ziegler's team would plot on to refined versions of those maps[43, 44] palaeogeographic and palaeoclimatic data derived from the literature. Ultimately, they hoped to produce one map for every 16 Ma during the Mesozoic and Cenozoic[39].

The differences between the reconstructions of Smith from Cambridge and Scotese from Chicago were small. The main one lay in time interval: Cambridge produced maps at 8 Ma intervals for Mesozoic and Cenozoic times from the Triassic to the present, but no maps for the Palaeozoic (pre-Triassic), while Chicago produced maps at 16 Ma intervals for Palaeozoic, Mesozoic and Cenozoic times. To ensure a close match between the approaches of the Cambridge and Chicago schools, BPRCo invited Scotese to spend a sabbatical at Sunbury-on-Thames in 1985, working with me on applications of palaeore-constructions – more on that in Chapter 6. At the time of writing (summer 2014), Alan Smith was still active at Cambridge and Chris Scotese was supplying palae-oreconstruction maps through his PALEOMAP Project (www.scotese.com).

5.5   The Evolution of the Modern World

Modern geological and geophysical methods confirm that during the late Palaeozoic the various scattered fragments of continental crust were in the process of being swept together to form a single supercontinent: 'Pangaea'. By early Carboniferous times, around 360 Ma ago, the Paleozoic ocean between Gondwanaland in the south and the fragments that would become Europe and North America in the north had shrunk to a narrow seaway. A similarly narrow seaway separated the North Ameri-can and European block from the Asian block. By late Carboniferous–early Permian times, around 300 Ma ago, these seaways had closed, making the supercontinent

a reality. Closure took place in the central upright part of the C-shaped supercontinent, which still enclosed a broad Tethyan Ocean between its two eastern-facing arms (Figure 5.8a).

This configuration would not last. In the middle Jurassic (200–180 Ma ago), Pangaea began to break apart. Linear domes grew over rising magma. Huge cracks developed along the crests of the domes, then widened to form gigan-tic linear rift valleys like that in East Africa today. The core of these rift valleys, or 'grabens' (after the German for trench), was a down-dropped block of the Earth's crust bordered on either side by steep faults commonly associated with volcanoes, like Mount Kilimanjaro in East Africa. Lakes like East Africa's Lake Nyasa (now Lake Malawi) grew in the down-dropped blocks. Rifting was accompanied by the intrusion of thin vertical dykes of basalt. Jurassic basaltic dykes and sills – horizontal sheets tens to hundreds of metres thick and extending over thousands of square kilometres – are widespread as precursors of eventual continental separation, for example in Antarctica, South Africa and Tasmania.

In most places, rifting proceeded to the point where the continent split and basalt magma welled up from below to form new ocean floor between the now separate continen-tal parts. This second stage is just beginning in today's Red Sea, where the continental rocks on either side of the origi-nal rift valley have thinned and sunk, letting in the sea, and new basaltic ocean floor is just beginning to emerge along the axis. The Gulf of California is an example of the third stage: the continental sides of the rift have moved apart, the deep basin between them is floored by new fresh basaltic deep-ocean floor and seafloor spreading has begun. If the East African rift were to progress over a few million years to stage three, we would see Tanzania, Kenya and Soma-lia moving east to form a large new island in the Indian Ocean. Where rifting aborted before reaching stage two, we are left with failed rifts like the one beneath the North Sea, holding the North Sea's oil and gas fields, and the Rhine Graben, through which the Rhine flows on its way to the sea.

Pangaea's first major break, at about 160 Ma, split the South America/Africa block from the North America block, opening a seaway comprising the central North Atlantic and a proto Gulf of Mexico and Caribbean region. That break connected the Tethys Ocean in the east with the Pacific Ocean in the west. To the north lay Laurasia, comprising North America, Europe and Asia. To the south lay Gondwana, containing the rest of today's continental bits and pieces.

Laurasia began to break up in the late Jurassic, with rift basins forming between Europe and Greenland and between Greenland and North America. The aborted North Sea rift formed at that time. Figure 5.8b provides a representative view. Starting in the late Cretaceous, around 90 Ma ago, the North Atlantic Ocean began to open like a zipper, from south to north. The process was extremely slow, with new ocean floor not forming between Britain and Greenland until some 60 Ma ago, in Palaeocene times[45]. Widespread volcanic activity, including outpourings of plateau basalts and the injection of swarms of basaltic dykes, accompanied the opening of this northern section between 60 and 50 Ma ago in Britain and Greenland. This activity formed the columnar basaltic wonders of the Giant's Causeway in Northern Ireland and of Fingal's Cave on Scotland's Isle of Staffa. Following a visit in 1829, the sound of the waves in the cave stimulated Felix Mendelssohn to produce Hebrides Overture Opus 26, known as the Fingal's Cave Overture.

Extensive volcanic activity continues in Iceland, over a plume of molten material from deep within the Earth's mantle, which started life with the opening of the Norwegian-Greenland Sea 60 Ma ago. The slowly opening zipper finally created a deep-water connection between the North Atlantic and the Arctic through Fram Strait between Greenland and Svalbard only 12 Ma ago, in Miocene times.

Meanwhile, in the south, Gondwana began slowly disintegrating. About 180 Ma ago, it split into two. The Africa and South America block slid sedately north, while the other half (Madagascar, India, Antarctica, Australia and New Guinea) stood still. This latter block started moving south about 170 Ma ago, then began to break apart in the lower Cretaceous, between 140 and 130 Ma ago (Figure 5.8b). At that time, Antarctica and Australia continued moving south away from Madagascar and India. At about the same time, South America began to move west away from Africa. Rifting between Antarctica and Australia began around 110 Ma ago, in the mid Cretaceous, as Antarctic moved south and Australia stayed still. About 85 Ma ago, New Zealand split away from Antarctica and India split away from Madagascar. India then moved rapidly north to collide with Tibet around 50 Ma ago, creating the Himalayas. Australia started to move slowly north at about the same time, and the Australia–New Guinea block collided with the Indonesian island arc around 30 Ma ago. Antarctica has been where it is now for about 90 Ma.

The northward movement of Africa, Arabia, India and Australia to collide with the southern margin of Laurasia swallowed up the Tethys Ocean between Gondwana and Laurasia. Remnants of Tethys exist today as the Mediterranean, Black and Caspian Seas. With the opening of the North Atlantic, the Tethys did for a while extend west from the Indian Ocean through the Mediterranean to the Pacific (Figure 5.8b), but that connection eventually closed at both ends, first with the collision of Africa and Iberia beginning in the late Cretaceous and second with the rising of the central American isthmus to close the central American seaway and create the Caribbean Sea in the late Pliocene, 2.8–2.5 Ma ago. These changes radically changed ocean circulation and hence climate. Today's Indian Ocean contains nothing of Tethys – it formed as a new ocean basin in the wake of the northward movement of India.

The creation of 2000 km wide and 2 km high ridges of hot rock in the middle of the ocean basins inevitably displaced seawater on to the edges of the adjacent continental margins. Sea levels rose steadily from about 200 Ma ago to a peak around 90 Ma ago as seafloor spreading expanded and caused ridges to grow, then declined as seafloor spreading declined and caused ridges to cool and subside. Investigating the ways in which different strata lapped on to continental margins, several of my colleagues at EPRCo in Houston, Texas, led by Peter Vail (1930–), defined a curve of eustatic sea level through time (Figure 5.9)[46]. As a colleague of Vail's, I was much influenced by his thinking when I worked for EPRCo between 1976 and 1982. Another close colleague of mine, Bilal Haq (1942–), later worked with Vail to refine his efforts, concluding that Cretaceous sea level stood 100–250 m above today's[47]. Their results, derived from seismic data, paralleled those derived by Walt Pitman (1931–) of Lamont from rates of seafloor spreading and resultant changes in mid-ocean ridge volume[48]. Similar results were obtained by Tony Hallam (1933–) of Birmingham University, a winner of the Geological Society's Lyell Medal in 1990[49] (Figure 5.9). Later, Ken Miller and colleagues from Rutgers University derived a new global sea level curve based on a combination of oxygen isotope data, stratigraphic data and the subsidence history of the New Jersey continental margin, concluding that late Cretaceous sea levels were only 20–25 m above today's[50].

Dietmar Müller of Sydney University and colleagues used a comprehensive analysis of the volumes of

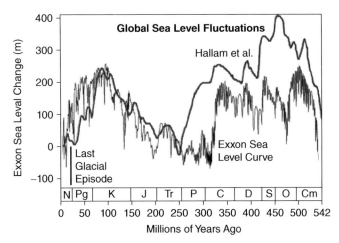

Figure 5.9 *Sea level curves through time. From Vail (Exxon) and Hallam (University of Birmingham), prepared by Robert A. Rohde and made available through Global Warming Art at http://commons .wikimedia.org/wiki/File:Phanerozoic_Sea_Level.png (last accessed 29 January 2015).*

mid-ocean ridges through time, along with a reanalysis of Miller's data from the New Jersey margin, to conclude that Mesozoic sea levels lay midway between the high estimates of Haq and the low estimates of Miller[51]. These high sea levels help to explain the widespread distribution of Cretaceous chalk over western Europe, for example (Figure 5.8b). The various studies, including geochemical analyses by Sean McCauley and Don DePaulo of the Center for Isotope Geochemistry of the Lawrence Berkeley National Laboratory[52], suggest that the rate of production of mid-ocean ridges declined by about 50% over the past 100 Ma.

Studies of seamounts in the southwestern Pacific produced further evidence for an uplift of the ocean floor of the kind that would have made sea level rise during the Cretaceous. From his wartime naval echo-soundings, Harry Hess discovered submerged flat-topped seamounts all over the Pacific Ocean, which he named 'guyots' after the 19th-century Swiss-American geographer Arnold Henry Guyot (1807–1884)[53]. Guyot, a friend of Louis Agassiz, emigrated to America at Agassiz's urging in 1848, ending up as professor of physical geography and geology at Princeton; hence the connection to Hess, who was a professor there. Hess concluded that the guyots' flat tops were eroded at sea level before sinking to their present depths of 1–2 km. His conclusion echoed that of Charles Darwin, who had deduced that atolls sat atop sunken volcanoes and had reasoned from the preponderance of

atolls in the region that a vast area of the southwestern Pacific must have subsided[54]. Mapping the distribution of atolls and guyots, Bill Menard realised that they occupied an elongate region with its northwest–southeast trending axis about halfway between New Zealand and Hawaii. As the amounts of subsidence indicated by the guyots increased towards the axis, Menard deduced that the rise must have arisen as a broad upward bulge in the Earth's mantle – a kind of super-swell – which he named the Darwin Rise[28]. While it does not appear to have been a mid-ocean ridge associated with seafloor spreading, it does seem to have been a centre for mid-plate or intra-plate volcanism between 100 and 60 Ma ago. Atolls and guyots are abundant there, so this must have been a major source for some of the CO_2 that filled the Cretaceous atmosphere. More recently, the chains of island and guyots that define the Darwin Rise have come to be seen as the result of the Pacific plate moving northwest over multiple fixed plumes or 'plumelets' of vertically rising magma[55, 56], much in the same way that the Hawaiian island chain originated.

In 2002 and 2008, David Rowley of the University of Chicago calculated that the average rate of ridge production evident from the visible distribution of ocean crust had been more or less constant at around 3.4 km^2/year[57, 58]. That led him to question Müller's hypothesis that rates of ridge production, and hence the sizes of ridges, were greater in the Cretaceous than since[51]. Rowley was not arguing that there was no possibility that rates of spreading were higher in the past, but was criticising what he regarded as unjustifiable assumptions back beyond the period for which we had accurate seafloor data. He made no attempt to evaluate the evidence for rates of seafloor spreading derived from the flooding of continental margins (à la Vail or Haq or Pitman or Hallam[46–49]), nor evidence derived from the geochemical record (à la Miller or McCauley or DePaulo[50,52]). Why does this matter? Because rates of seafloor spreading affect rates of production of CO_2, hence climate. In geology, we have to take all factors into consideration. Going solely with Rowley is to ignore multiple alternative strands of evidence; more on that later.

With these various advances, and despite the fact that we still do not know precisely how plate tectonics works[59], we have a sound palaeogeographic framework for the study of climate change. In Chapter 6, we explore what the distribution of climate-sensitive strata through time on the different continents has to tell us about the changes in Earth's climate with time.

References

1. Suess, E. (1885–1908) *Das Antlitz der Erde*, vols 1–3. F. Tempsky, Vienna. Translated by H.B.C. Sollas as Suess, E. (1904–1924) *The Face of the Earth*. Clarendon Press, Oxford.
2. Taylor, F.B. (1910) Bearing of the Tertiary mountain belt in the origin of the Earth's plan. *Bulletin of the Geological Society of America* 21, 179–227.
3. Snider-Pellegrini, A. (1859) *La Creation et ses Mystères Dévoiles*. Librarie A. Frank, Paris. (See Figures 9 (before) and 10 (after) between his pp. 314 and 315.)
4. Jacoby, W. (2012) Alfred Wegener – 100 years of mobilism. *Geoscientist* 22 (9), 12–17.
5. Wegener, E. (1960) *Alfred Wegener: Tagebücher, Briefe, Erinnerungen*. F.A. Brockhaus, Wiesbaden.
6. Wegener, A. (1912) Die Enstehung der Kontinente, *Petermanns Geographische Mitteilungen* 58 (1), 185–195, 253–256, 305–309; and (somewhat abbreviated) in *Geologische Rundschau* 3 (4), 276–292.
7. Wegener, A. (1920) *Die Entstehung der Kontinente und Ozeane*, 2nd edn. Vieweg und Sohn, Braunschweig.
8. Wegener, A. (1924) *The Origins of Continents and Oceans* (trans. of the 3rd (1922) edn). Methuen, London.
9. Köppen, W. and Wegener, A. (1924) *Die Klimate der Geologischen Vorzeit*. Borntraeger, Berlin. Note that one sees this commonly referred to as published in English as *Climates of the Geological Past* by Van Nostrad, London, 1963, which is an erroneous reference to a book with a similar title: Schwarzbach, M. (1963) *Climates of the Past: An Introduction to Paleoclimatology*. Van Nostrad, London, translated from the German 2nd edition by R. Muir. Köppen and Wegener's *The Climate of the Geological Past* will be reissued in German and English in 2015.
10. Demhardt, I.J. (2006) Alfred Wegener's hypothesis on continental drift and its discussion in Petermanns Geographische Mitteilungen. *Polarforschung* 75 (1), 29–35.
11. Holmes, A. (1928) Radioactivity and Earth movements. *Transactions of the Geological Society of Glasgow* 18, 559–606.
12. Du Toit, A.L. (1937) *Our Wandering Continents: An Hypothesis of Continental Drifting*. Oliver & Boyd, London.
13. Runcorn, S. K. (1959) Rock magnetism. *Science* 129, 1002–1011.
14. Irving, E. (1959) Paleomagnetic pole positions. *Royal Astronomical Society Geophysical Journal* 2, 51–77.
15. Blackett, P.M.S. (1961) Comparison of ancient climates with the ancient latitudes deduced from rock magnetic measurements. *Proceedings of the Royal Society A* 263 (1312), 1–30.
16. Hess, H.H. (1962) History of ocean basins. In: *Petrologic Studies: A Volume to Honor A.F. Buddington* (eds A.E.J. Engel, H.L. James and B.F. Leonard). Geological Society of America, New York, pp. 599-620.
17. Ewing, M. (1948) Exploring the mid Atlantic ridge. *National Geographic* XCIV (3), 275–294.
18. Dietz, R.S. (1961) Continent and ocean basin evolution by spreading of the sea floor. *Nature* 190, 854–857.
19. Vine, F.J. and Matthews, D.H. (1963) Magnetic anomalies over oceanic ridges. *Nature* 199, 947–949.
20. Csillag, R. (2013) Pioneering geophysicist Lawrence Morley broke new ground. *The Globe and Mail*, Tuesday, 14 May. Available from: http://www.theglobeandmail.com/news/national/pioneering-geophysicist-lawrence-morley-broke-new-ground/article11925771/ (last accessed 29 January 2015).
21. Wood, R.M. (1985) *The Dark Side of the Earth*. George Allan and Unwin, London.
22. Tarling, D.H. and Runcorn, S.K. (eds) (1973) *Implications of Continental Drift to the Earth Sciences*, Vol. 1. Academic Press, London (see 'Introduction', pp. ix–x).
23. Creer, K.M. (1973) A discussion of the arrangement of palaeomagnetic poles on the map of Pangaea for epochs in the Phanerozoic. In: *Implications of Continental Drift to the Earth Sciences*, Vol. 1 (eds D.H. Tarling and S.K. Runcorn). Academic Press, London, pp. 47–76.
24. Heirtzler, J.R. (1973) The evolution of the North Atlantic Ocean. In: *Implications of Continental Drift to the Earth Sciences*, Vol. 1 (eds D.H. Tarling and S.K. Runcorn). Academic Press, London, pp. 191–196.

25. Francheteau, J. (1973) Plate tectonic model of the opening of the Atlantic Ocean south of the Azores. In: *Implications of Continental Drift to the Earth Sciences*, Vol. 1 (eds D.H. Tarling and S.K. Runcorn). Academic Press, London, pp. 197–202.

26. Laughton, A.S., Sclater, J.G. and McKenzie, D.P. (1973) The structure and evolution of the Indian Ocean. In: *Implications of Continental Drift to the Earth Sciences*, Vol. 1 (eds D.H. Tarling and S.K. Runcorn). Academic Press, London, pp. 203–212.

27. Summerhayes, C.P. (1967) New Zealand region volcanism and structure. *Nature* **215**, 610–611.

28. Menard, H.W. (1964) *Marine Geology of the Pacific*. McGraw-Hill, New York.

29. Menard, H.W. (1969) *Anatomy of an Expedition*. McGraw Hill, New York.

30. Parson, L.M., Murton, B.J., Searle, R.C., *et al.* (1993) En echelon volcanic ridges at the Reykjanes Ridge: a life cycle of volcanism and tectonics. *Earth and Planetary Science Letters* **117**, 73–87.

31. Holmes, A. (1911) The association of lead with uranium in rock-minerals and its application to the measurement of geological time. *Proceedings of the Royal Society A* **85**, 248–256. See also the biography of Holmes by Lewis, C. (2002) *The Dating Game: One Man's Search for the Age of the Earth*. Cambridge University Press, Cambridge.

32. Waencke, H. and Arnold, J.R. (2005) Hans E. Suess, 1909–1993, a biographical memoir. Biographical Memoirs, Vol. 87. National Academy of Sciences. Available from: http://www.nasonline.org/publications/biographical-memoirs/memoir-pdfs/suess-hans.pdf (last accessed 29 January 2015).

33. Smith, A.G., Briden, J.C. and Drewry, G.E. (1973) Phanerozoic world maps. In: *Organisms and Continents Through Time* (ed. N.F. Hughes). Special Papers on Palaeontology, No. 12. Palaeontological Association, London, pp. 1–42.

34. Smith, A.G., Hurley, A.M. and Briden, J.C. (1981) *Phanerozoic Palaeocontinental World Maps*. Cambridge University Press, Cambridge.

35. Smith, D.G., Smith, A.G. and Funnell, B.M. (1994) *Atlas of Mesozoic and Cenozoic Coastlines*. Cambridge University Press, Cambridge.

36. Tyson, R.V. and Funnell, B.M. (1987) European Cretaceous shorelines, stage by stage. *Palaeogeography, Palaeoclimatology, Palaeoecology* **59**, 69–91.

37. Funnell, B.M. (1990) Global and European shorelines, stage by stage. In: *Cretaceous Resources, Events and Rhythms* (eds R.N. Ginsburg and B. Beaudoin). Kluwer, Netherlands, pp. 221–235.

38. Ziegler, A.M. (1975) A proposal to produce an atlas of paleogeographic maps. Unpublished ms, Department of Geophysical Sciences, University of Chicago.

39. Ziegler, A.M. (1982) The University of Chicago Paleogeographic Atlas Project: Background – Current Status – Future Plans. Unpublished ms, Department of Geophysical Sciences, University of Chicago.

40. The Paleogeographic Atlas Project in 2002. Available from: http://www.geo.arizona.edu/~rees/PGAP2000-1.html (last accessed 29 January 2015).

41. Scotese, C.R., Bambach, R.K., Barton, C., *et al.* (1979) Paleozoic base maps. *Journal of Geology* **87**, 217–277.

42. Ziegler, A.M., Scotese, C.R. and Barrett, S.F. (1983) Mesozoic and Cenozoic paleogeographic maps. In: *Tidal Friction and the Earth's Rotation II* (eds P. Brosche and J. Sundermann). Springer-Verlag, Berlin, pp. 240–252.

43. Scotese, C.R., Gahagan, L.M. and Larson, R.L. (1988) Plate tectonic reconstructions of the Cretaceous and Cenozoic ocean basins. *Tectonophysics* **155**, 27–48.

44. Scotese, C.R., Boucot, A.J. and McKerrow, W.S. (1999) Gondwanan palaeogeography and palaeoclimatology. *Journal of African Earth Sciences* **28** (1), 99–114.

45. Bott, M.H.P. (1973) The evolution of the Atlantic North of the Faeroe Islands. In: *Implications of Continental Drift to the Earth Sciences*, Vol. 1 (eds D.H. Tarling and S.K. Runcorn), Academic Press, London, pp. 175–189.

46. Vail, P.R. and Mitchum, R.M. (1977) Seismic stratigraphy and global changes of sea level. *American Association of Petroleum Geologists Memoir* **26**, 49–212.

47. Haq, B.U., Hardenbol, J. and Vail, P.R. (1988) Mesozoic and Cenozoic chronostratigraphy and cycles of sea-level change. *SEPM Special Publication* **42**, 71–108.

48. Pitman, W.C. (1978) Relationship between eustasy and stratigraphic sequences of passive margins. *Geological Society of America Bulletin* **89**, 1389–1403.

49. Hallam, A. (1984) Pre-Quaternary sea level changes. *Annual Review of Earth and Planetary Sciences* **12**, 205–243.

50. Miller, K.G., Kominz, J.V., Browning, J.V., *et al.* (2005) The Phanerozoic record of global sea-level change. *Science* **312**, 1293–1298.

51. Müller, R.D., Sdrolias, M., Gaina, C., *et al.* (2008) Long-term sea-level fluctuations driven by ocean basin dynamics. *Science* **319**, 1357–1362.

52. McCauley, S.E. and DePaolo, D.J. (1997) The marine $^{87}Sr/^{86}Sr$ and $\delta^{18}O$ records, Himalayan alkalinity fluxes, and Cenozoic climate models. In: *Tectonic Uplift and Climate Change* (ed. W.F. Ruddiman). Plenum Press, New York, pp. 427–467.

53. Hess, H.H. (1946) Drowned ancient islands of the Pacific basin. *American Journal of Science* **244**, 772–791.

54. Darwin, C. (1896) *The Structure and Distribution of Coral Reefs*, 3rd edn. Smith and Elder, London.

55. Janney, P.E. and Castillo, P.R. (1999) Isotope geochemistry of the Darwin Rise seamounts and mantle dynamics beneath the south central Pacific. *Journal of Geophysical Research* **104** (B5), 10 571–10 590.

56. Foulger, G.R. (2010) *Plates vs Plumes: A Geological Controversy*. John Wiley & Sons, Chichester.
57. Rowley, D.B. (2002) Rate of plate creation and destruction: 180 Ma to present. *Bulletin of the Geological Society of America* **114** (8), 927–933.
58. Rowley, D.B. (2008) Extrapolating oceanic age distributions: lessons from the Pacific Region. *Journal of Geology* **116**, 587–598.
59. Ananthaswamy, A. (2012) Earthly powers. *New Scientist* **2878**, 38–41.

6

Mapping Past Climates

6.1 Climate Indicators

Because many climatically sensitive deposits have well defined latitudinal ranges, and climate zones are reasonably well defined, we can make crude forecasts about what kinds of sediments we might expect in different geographical locations[1–4]. Coral reefs tend to be concentrated between the 30th parallels. Thick accumulations of sediments rich in the calcium carbonate remains of marine organisms tend to accumulate in warm seas – as did the chalk of the White Cliffs of Dover – although muds from tropical rivers may locally mask this tendency[5]. Glacial deposits or glacial striae (scratches made on rocks by glaciers carrying boulders) tend to occur at high latitudes and/or at high elevations, along with boulder clays or tillites, moraines and glacial landforms like drumlins, kames and eskers[6]. Salt (halite) and gypsum formed by evaporation normally occur in mid- to low-latitude arid areas with other 'evaporite' deposits[7]. They may be associated with dunes and other indications of deserts, including 'dreikanters': pyramid-shaped pebbles facetted by the wind[8]. Sand grains blow up the long fore-slope and avalanche down the steep lee-slope of dunes, making them advance downwind and providing internal structure known as dune bedding or 'cross bedding' at an angle of ~30° to the desert surface. We use dune bedding to estimate past wind directions[9]. Iron oxide is common in hot deserts, making them reddish; their pebbles show a haematite (iron oxide) glaze, or 'desert varnish'.

Of course, not all deserts are hot, and not all dunes form in deserts. As Alan Eben Mackenzie Nairn (1927–2007)

reminded us in his book *Descriptive Palaeoclimatology* in 1961, '*the association of one or more criteria, such as evaporite deposits representing hot dessicating conditions with the dune bedded sandstone, may remove the ambiguity*'[10]. A Scottish palaeomagnetist and stratigrapher, Nairn was a fellow of the University of Durham and King's College, Newcastle-upon-Tyne. In 1965, he was a co-founder of the journal *Palaeogeography, Palaeoclimatology and Palaeoecology*. By 1991, his book was being described as '*the first modern book on palaeoclimatology*'[11]. As a sign of the times, Nairn confessed he was not fond of the astronomical theory of climate change, writing: '*the effect of the varying distance between the earth and the sun from perihelion to aphelion, the basis of Croll's theory of the origin of ice ages, is not now thought to be a significant factor in climate*'[10]. By 1976, he would be shown to be completely wrong, as we shall see in Chapter 12.

Nairn's book contains articles on ancient deserts, evaporites, red beds, cold climates and fossils as climatic indicators. Arid conditions can be recognised from mud cracks. Peats tend to form in mid- to high-latitude bogs. Organic-rich deposits also form by the accumulation of terrestrial plant remains in humid tropical settings. Both may give rise to coals. Marine sediments rich in the organic remains of plankton may occur along coasts where winds run parallel to the shore, especially at mid latitudes off desert coasts, often in association with phosphate-rich phosphorite rock. Laterites are red soils rich in iron and aluminium that form in hot, wet areas. Extreme tropical weathering converts them to bauxite, a mixture of iron and

Earth's Climate Evolution, First Edition. Colin P. Summerhayes.
© 2015 John Wiley & Sons, Ltd. Published 2015 by John Wiley & Sons, Ltd.

aluminium oxides and hydroxides. In arid and semi-arid regions, soils may acquire a hard crust rich in calcium, forming caliche, also known as hardpan or calcrete.

Types of animals and plants also provide climate signals; think of crocodiles and penguins, for instance, or fir trees and banana plants. Even such lowly creatures as the marine plankton may signal oceanic climates, some species preferring warm and others cold seas. Siliceous oozes made of the remains of radiolaria may be common beneath the tropical ocean, while those made of the remains of diatoms may be common beneath high-latitude seas[1–3]. Clay minerals give away their climate zones: kaolinite comes from tropical weathering and chlorite from mechanical weathering in polar regions. Illite tends to predominate in between[1–3]. Seasonal variation is signalled by the growth rings in trees and corals, and by 'varves' (alternating light and dark layers, representing the change from summer to winter deposition) in lakes and closed marine basins. We can also use a variety of chemical indicators to simulate past temperatures. The range of indicators has grown in recent years[12–14]; they help to establish the likely palaeolatitude of the environment of deposition at the time a rock was formed. Ultimately, the truest analysis of past climates comes from combining many different lines of evidence[4, 10].

6.2 Palaeoclimatologists Get to Work

Wegener's concept of drifting continents provided a testable means of predicting where past climate zones were. Although he didn't put the locations of climatic indicators on to past continental positions in his 1920 maps[15], he did use them to support his theory and to indicate where lines of latitude probably lay. Most of the coal deposits of the Carboniferous lay along what he thought was the palaeo-equator, extending from modern Texas through Germany to China (Figure 5.3). Salt and gypsum deposits typical of arid climates in low to mid latitudes lay just north and south of this equatorial zone. He thought that the *Glossopteris* ferns from the Southern Hemisphere were deposited in subpolar peat bogs, because they surrounded a region carrying evidence for glaciation (Figure 6.1). Based on his biogeographic analysis and the distribution of samples indicating glaciation, he concluded that the South Pole lay near Durban, South Africa. He

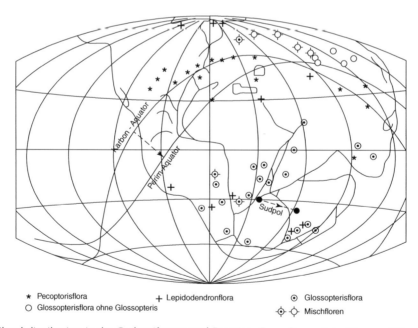

| ★ Pecoptorisflora | + Lepidodendronflora | ⊙ Glossopterisflora |
| ○ Glossopterisflora ohne Glossopteris | | ⊙ ⟡ Mischfloren |

Figure 6.1 *Floral distribution in the Carboniferous and Permian. From Figure 8 in Köppen, W. and Wegener, A. (1924)* Die Klimate der geologischen Vorzeit. *Borntraeger, Berlin, pp. 1–255.*

Figure 6.2   *Wladimir Köppen in 1875.*

plotted some of these biogeographical features on his first palaeoclimate map in 1922[16].

Wegener expanded his understanding of past climates by collaborating with Wladimir Köppen (1846–1940) (Figure 6.2), a Russian climatologist of German extraction who was one of the founders of modern meteorology and climatology. In 1875, Köppen was appointed to the newly formed Deutsche Seewarte (the German Marine Observatory) in Hamburg. Much as Matthew Fountaine Maury (1806–1873) had begun doing in the United States in the 1850s and 1860s, Köppen began using ships' reports to map the winds over the ocean, contributing to the Seewarte's sailing handbooks for the Atlantic, Pacific and Indian Oceans[17]. Continuing to develop his ideas on climatology, in 1884 he published the first comprehensive map of global climate zones. This formed the basis for the Köppen climate classification system, which first appeared in 1901[18], was later expanded[19] and is still in use today[20].

Köppen saw that the combination of dryness and temperature driven by the distribution of radiation and precipitation creates largely latitudinal climatic zones characterised by their vegetation. His system divides the land surface up on the basis of annual and monthly temperatures and precipitation and the seasonality of precipitation, in such a way as to coincide as much as possible with the world's patterns of vegetation and soils (Figure 6.3). It comprises six major climate types, designated by capital letters (Box 6.1).

Wegener's association with Köppen was close. He married Köppen's daughter Else in 1913. Köppen liked his son-in-law's theory enough to publish a paper in support of it in 1921[21].

A correction seems in order here: it has been suggested that the polar wandering positions that Wegener gave to Köppen for his 1921 paper were based on palaeomagnetic data[22], but the use of palaeomagnetic data to establish past polar positions was not developed until the 1950s.

In 1924, the two men published *Die Klimate der geologischen Vorzeit* (*The Climate of the Geological Past*)[23]. As I write, the book is being translated into English, with publication expected in mid 2015 (see Appendix). Its central feature was an innovative suite of crude palaeoclimatic maps for selected time periods between the Devonian and the Pleistocene (e.g. Figure 5.3). These were the first comprehensive global palaeoclimatic maps. They featured Wegener's continental reconstructions, the distributions of climate-sensitive indicators and selected geographic features: the positions of the North and South Poles, the Equator and the 30° and 60° lines of latitude. Other maps showed the flora of the Carboniferous and Permian (Figure 6.1), the flooded areas of the continents in the Jurassic and the corals of the Cretaceous.

The two compared their data (Figure 5.3) with Köppen's conceptual model of the climate system (Figure 6.3). The salt and gypsum deposits occurred where such evaporites are found today, in the arid belts north and south of the Equator. Cretaceous corals occurred in the equatorial zone between the 30th parallels, more or less like today. Glacial indications occurred around the poles. Coals formed under temperate humid conditions, as well as in the humid tropics. These findings vindicated Lyell's suggestion that a shifting of the continents through time might explain the global distribution of fossils and the location of past climate-sensitive deposits. The genius of Wegener was to leap beyond Lyell in determining where and when the continents had moved.

By 1937, the South African geologist Alexander Du Toit[24] was advising his readers to consult Köppen and Wegener's maps, while providing additional supporting data of his own. He was repaying the compliment: they had used his published comparisons of the geology of Africa and South America to support their continental

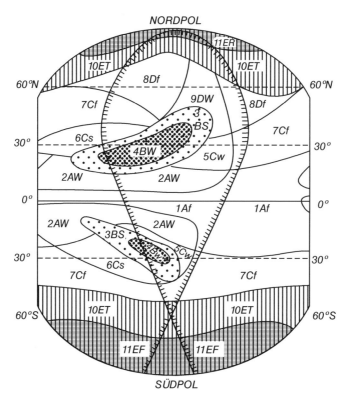

Figure 6.3   *Köppen's climate classification system. The heavy serrated line denotes a large and a small southern continent. See Box 6.1 for explanation of symbols. f = constantly humid; s = dry summers; w = dry winters.*

Box 6.1 Koppen's classification system.

A. Moist tropical climates with high temperature and rainfall; average temperature of the coldest months > 18 °C.
B. Dry climates with little rain and a large daily temperature range; this category is divided into S = semi-arid or steppe and W = arid or desert.
C. Humid mid-latitude climates with warm, dry summers and cool, wet winters.
D. Continental climates in the interiors of large land masses, with low overall precipitation and a wide range of seasonal temperature; snow and forest with warmest month > 10 °C and coldest month < −3 °C.
E. Cold climates, where permanent ice and tundra are present and temperatures are below freezing for most of the year; warmest month < 10 °C.
F. Polar with warmest month < 0 °C; T = tundra.

These types are divided into subgroups designated by lower-case letters. For example: Af = tropical rainforest; Aw = savanna; Bs = grassland; Cf = deciduous forest; Dfc = boreal forest (taiga). An additional localised climate type is H = cold Alpine climate, which is important in mid latitudes for water storage (snow in winter) and release (spring thaw).

reconstructions. In turn, Du Toit influenced the England-born, New Zealand-trained geologist Lester King (1907–1989), who became professor of geology at the University of Natal in Durban in 1935. King deduced that the Gondwanaland glaciation passed from west to east through time, starting with early Carboniferous deposits in western Argentina, moving through upper Carboniferous deposits in South Africa and early Permian deposits in India and finishing with mid-Permian tillites in Australia[25], presumably reflecting migration of Gond-wanaland across the pole. Within the glacial deposits, King found evidence for multiple advances and retreats, like those of the Quaternary Ice Age. Consistent with the notion that Gondwanaland travelled '*through a succession of climatic girdles*', King found that the main phase of coal formation in Gondwanaland ranged from late Car-boniferous in Brazil, through early Permian in Africa and India to late Permian in eastern Australia[25].

These were the authors I was exposed to when, as an undergraduate student in geology at University College London (UCL) in 1960–63, I learned about palaeoclimates from our head of department, Professor Sydney Holling-worth, winner of the Geological Society's Murchison Medal in 1959. Hollingworth's presidential address to the Geological Society of London in 1961 dwelt upon 'The Climate Factor in the Geological Record'. Under his tutelage, and with urging from the sedimentology lecturer Alec Smith and geology lecturer Eric Robinson, I became fascinated by the prospect of divining past climates from the geological record. Following in the footsteps of Lyell, Wegener, Köppen, Du Toit and King, we students learned how climate-sensitive sediments and fossils occurred in distinct climatic zones. What we needed to know was the palaeogeography: where had the continental fragments on which those sediments were deposited been located through time? Thanks to the pioneers of continental drift, we had some idea, but much of what they had to say was dismissed by the geological community. Ahead of the Vine and Matthews era, we were reduced to writing arm-waving essays like 'Continental Drift – Pros and Cons'.

6.3 Palaeomagneticians Enter the Field

Forty years after the publication of Köppen and Wegener's book, another seminal palaeoclimatic publication appeared, stimulated by the tremendous advances in palaeomagnetic studies of continental rocks that had

been made in the late 1950s and the very early 1960s. Edited by A.E.M. Nairn, it contained the proceedings of a NATO-funded conference at the University of Newcastle upon Tyne in January 1963, which brought together an eclectic mix of palaeomagnetists, palaeontologists and palaeoclimatologists[26]. The NATO meeting was in many respects a follow-up to Nairn's 1961 book on *Descriptive Palaeoclimatology*.

It is worth bearing in mind that a conference held in January 1963 would predominantly review research results from earlier times – mostly no later than the middle of the preceding year, 1962 – so it is not surprising that only 1 of the 54 papers at that meeting, by Australian geologist Rhodes Fairbridge (1914–2006), referred to Hess's 1962 'geopoetry' paper on seafloor spreading. Indeed, the conference preceded by 9 months the proof of the seafloor spreading concept by Vine and Mathews. What a difference those 9 months would make! Nairn's 1964 volume[26] contained almost no reconstructions of past continental positions. Its successor – the report of the 1972 NATO Advanced Study Institute at Newcastle University, published almost a decade later, in 1973 – contained several.

The paper from the 1963 NATO conference that is most remembered in palaeoclimate circles is the classic by Jim Briden and Ed Irving[27], which posed the question: '*with reference to palaeoclimatology, has the balance of rainfall and temperature and their gradients been the same in the past as they are today?*' In other words, did Lyell's uni-formitarian views hold water when the details of past cli-mates were examined? This question could be examined by recognising '*some feature, which may be called a palaeo-climatic indicator, and which may reasonably be assumed to indicate the occurrence of a particular climatic condi-tion, say heavy rainfall or low temperature, at the time it was formed … [and by] the use of some model of past cli-matic zonation of the Earth, so that the indicator can be placed in its correct palaeoclimatic zone*'[27]. This is what Köppen and Wegener had done, but as Briden and Irving pointed out, there was a drawback to using modern ana-logues to determine past climates. For example, while the spread of modern corals was limited by the 18 °C isotherm, past corals may not have had the same limit. Equally, the position of the 18 °C isotherm may have varied through time with respect to the Equator. '*Palaeomagnetism*', they affirmed, '*affords the means of estimating numerically the palaeolatitude spectra of palaeoclimatic indicators in a manner which is not subject to these fluctuations in the climatic model, being based on an entirely different type of observation and analysis*[27].

Their palaeomagnetic data enabled Briden and Irving to plot accurately for the first time the palaeolatitudes for the main geological periods of the past 540 Ma on North America, Eurasia and Australia (see Frontispiece), superimposing on each map (more or less as had Köppen and Wegener) the distribution of selected palaeoclimatic indicators: red beds, desert sandstones, evaporates, glacial beds and coal. Unlike Köppen and Wegener, they did not reconstruct past continental positions, nor did they show any data from Africa or South America.

Their other novel contribution was to determine the past latitudinal distributions for their various palaeoclimatic indicators. Most carbonates clustered between the 40th parallels, with the bulk between the 30th parallels, as they do today. Most fossil coral reefs also occurred between the 30th parallels, like modern reefs. Red beds indicative of arid environments occupied similar latitudes. Dune-bedded sandstones occur today between latitudes 18° and 40°; in the past they occurred between 20° N and 30° S, while in the Permian they occurred within 10° of the Equator – *'much lower latitudes than is common at present'*[27]. Most fossil evaporites (primarily salt and gypsum) also occurred within 30° of the palaeo-equator, whereas modern terrestrial evaporites show maxima at 25° S and 40° N; the discrepancy may be explained by some of the fossil evaporites being marine rather than terrestrial. Fossil coals were bimodal. Most occurred in tropical and temperate humid zones; very few occurred at palaeolatitudes between 15° and 30°, indicating the presence of an arid zone there. Figure 6.4 is a recent update of the palaeolatitudinal zonation of climate-sensitive deposits.

Briden and Irving found that, while ancient marine carbonates and coral reefs displayed strikingly similar distributions to their modern counterparts, this was less true of the indicators of arid conditions, which were concentrated much closer to the Equator than at present, especially in Palaeozoic times, suggesting some disturbance to latitudinal climate zoning by the distribution of land. There was also a rather abrupt change from low-latitude coals in the Carboniferous to temperate-latitude coals in later times. Briden and Irving speculated that these patterns came about because of the creation of a large land area (Wegener's Pangaea) in low latitudes, encouraging the initial development of equatorial coals and later development of dry conditions all across its interior. In due course, the Geological Society of London rewarded the pair for their efforts, with the Murchison Medal for Briden in 1984 and the Wollaston Medal for Irving in 2005. Irving was also elected a fellow of the Royal Societies of London and Canada and a member of the US National Academy of Sciences, and was honoured with several other medals.

Another attendee at the meeting, Walter Bucher of Columbia University, was not impressed. *'The main conflict'*, he pointed out, *'arises from the implicit assumption that the width of the latitudinal climatic belts has always been essentially the same as at present. Yet, during the Cenozoic it has certainly undergone drastic changes in width ... Within the first third of that time, warm temperate floras grew on Ellesmere Island, Greenland, Iceland and Spitsbergen ... and conditions were favourable for limestone deposition [there] ... The change to present conditions started slowly, speeded up during the Middle Miocene, and led at an accelerating rate to the glacial conditions in the shadow of which we still live ... Why should the present width of climatic zones and the conditions it implies for temperature, wind direction and rainfall be applied to the fossil record?'*[28]

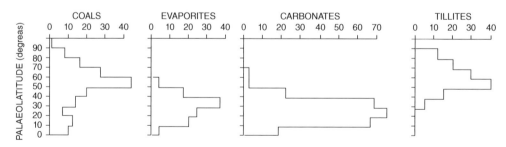

Figure 6.4 *Palaeolatitudinal zonation of climate-sensitive deposits. Frequency in number of deposits against palaeolatitude.*

Reinforcing Bucher's message, another palaeontologist, Erling Dorf of Princeton University, showed that the Arctic and temperate forests of the Cenozoic migrated from around 50–65° N in the Eocene to 35–45° N in the late Pliocene. He reminded his audience that tropical floras characterised the Eocene London clay, affirming that '*the present epoch in geological history is rather abnormal in many ways, but especially in its climatic characteristics*', because it is just an interglacial within the recent Ice Age[29]. Yet another participant, Curt Teichert of the US Geological Survey, observed in that same report that '*relationships between climate and coral-reef growth are not very straightforward, because coral evolution seems to have been influenced more by intrinsic biologic factors than by climate*'[29].

The Bucher–Dorf–Teichert message from the palaeontologists was that one could not apply rigidly the Huttonian–Lyellian dictum that the present is the key to the past. Nevertheless, there had to be something to it, or the likes of Wegener, Köppen, Du Toit, Briden and Irving would not have been able to confirm that most climatic indicators were broadly where one would expect to find them if past climate zones resembled present ones. Briden and Irving did recognise exceptions: notably, the abundant development of tropical coals in the Carboniferous, but not later, and the widespread development of arid deposits in the interior of Pangaea. The solution to the riddle would not come until palaeoclimate data were plotted on accurate reconstructions of continental positions at rather fine geological intervals, and until accurate means were found to determine past temperatures.

6.4 Oxygen Isotopes to the Rescue

That last requirement was in the process of being met. The man who won a Nobel Prize in 1934 for discovering deuterium, American chemist Harold Urey (1893–1981), discovered at the University of Chicago in the late 1940s that the isotopes of oxygen measured in seashells are related to the temperature of the seawater in which they grow[30–32]. It works like this: While oxygen carries eight protons in its nucleus, the number of neutrons varies between eight and ten, thus giving rise to stable isotopes known as oxygen-16 (or ^{16}O), with eight protons and eight neutrons, and oxygen-18 (or ^{18}O), with eight protons and ten neutrons. As the ocean warms, water molecules carrying the light isotope, ^{16}O, evaporate preferentially,

enriching warm surface waters in ^{18}O. Planktonic organisms such as foraminifera, growing in the water, use that oxygen to construct their skeletons, which thus reflect the isotopic composition, and hence the temperature, of the water. Urey published his oxygen isotopic temperature scale in 1951, opening a magnificent new vista for studies of the changes in climate with time. Of this discovery, it has been said, '*The measurement of the paleotemperatures of the ancient oceans stands as one of the great developments of the earth sciences; a truly remarkable scientific and intellectual achievement*'[33]. Urey was showered with honours during his career, among them election to the Royal Society of London in 1947 and the US National Medal of Science in 1964. He also has a lunar crater and an asteroid named after him.

In practice, the widespread use of stable isotopes in palaeoclimate studies awaited the development of the isotope ratio mass spectrometer to provide the necessary accuracy and precision, something that was achieved around 1950[34, 35]. In due course, the relation between oxygen isotopes and temperature turned out to be not as simple as was first supposed, because ice volume also affects this ratio, although only at times when there were large volumes of ice on Earth – as we see in later chapters.

Analyses of oxygen isotopes in fossil shells, together with studies of climate-sensitive fossil plants and animals, confirmed Lyell's observation that global temperatures fluctuated through time. They were relatively warm between 540 and 340 Ma ago, cold during the Permo-Carboniferous glaciation between 340 and 260 Ma ago, warm again between 260 and 40 Ma ago and cold from 40 Ma ago to the present. To some extent, these patterns reflected the influence on climate of the changing positions of the continents. But other factors also affected temperature, including the concentrations of greenhouse gases in the air, as we see later.

Heinz Lowenstam (1912–1993) (Box 6.2), who had been part of Urey's Chicago University group but had moved to Caltech, presented oxygen isotope data from the Permian and the Cretaceous to the 1963 NATO conference in Newcastle[36].

Box 6.2 Heinz Lowenstam.

Lowenstam was born in Germany. He started out studying palaeontology at the Universities of Frankfurt and Munich, but unfortunately fell foul in 1936 of a new Nazi law prohibiting the awarding of doctorates to Jews, which was passed the week before his PhD thesis defence. He and his wife Ilse emigrated to the United States in 1937. There, his prior work was accepted by the University of Chicago, which gave him a doctorate in 1939.

Within the Cretaceous, Lowenstam found that while temperatures were similar to those found today in the tropics, they declined less rapidly towards the North Pole. While the 18 °C isotherm was shifted northward from about 32 to 60° N in the Santonian (86–84 Ma ago), it progressed slightly southward in the Albian (112–100 Ma ago), Cenomanian (100–94 Ma ago) and Maastrichtian (70–65 Ma ago). Estimated average temperatures for polar waters of 10 °C for the Cenomanian, 15 °C for the Albian and 16–17 °C for the Santonian are in sharp contrast to those of today – around 0 °C, which *'points towards a considerably more uniform temperature distribution of the oceanic surface waters during the Cretaceous periods as compared with today'* [36]. Lowenstam estimated Cretaceous deep-water temperatures to also be around 10 °C in the Cenomanian, 15 °C in the Albian and 16–17 °C in the Santonian, implying that the Cretaceous oceans were considerably more uniform than they are today, where bottom waters average between 1.5 and 4.0 °C. These various findings underscore the limits on the application of modern climate zones to ancient environments.

Isotopic evidence for both cool and warm temperatures in the Permian of Australia, and for significant temperature variations between the different ages of the Cretaceous within specific areas like Europe, led Lowenstam to stress *'that palaeobiogeographic studies must be limited to short time-stratigraphic intervals to serve as a meaningful palaeoclimatological tool'* [36].

Another member of Urey's team was the Italian geologist Cesare Emiliani (1922–1995), who we shall come across again in later chapters. In 1961, Emiliani analysed the ratio of ^{16}O to ^{18}O in the benthic (bottom-dwelling) foraminifera collected from cores of deep-sea sediment from the early Cenozoic, and found that the bottom waters in which those creatures grew were significantly warmer than they are today – further proof that climate and ocean circulation had changed profoundly [37].

6.5 Cycles and Astronomy

As we saw in Chapter 3, James Croll thought that periodic changes in the Earth's orbit might have caused not only the fluctuations of the Ice Age, but also cycles earlier in Earth's history. Following up on Croll's ideas, in 1895, the prominent American geologist Grove Karl Gilbert (1843–1918) (Box 6.3) thought that variations in the Earth's orbital behaviour might also explain oscillations in the carbonate content of Cretaceous marls in Colorado. Later, Wilmot Hyde (Bill) Bradley (1899–1979) (Box 6.3) of the US Geological Survey suggested in 1929 that cycles in the oil shales of the Eocene Green River Formation in Wyoming might also have been caused by variations in orbital precession [38].

Box 6.3 Grove Karl Gilbert and Wilmot Hyde Bradley.

G.K. Gilbert's talents were widely recognised. He is the only geologist to be elected twice as president of the Geological Society of America (in 1892 and 1909). Craters have been named after him on the Moon and Mars, and he was awarded the Geological Society of London's Wollaston Medal in 1900 and the Charles P. Daly Medal of the American Geographical Society in 1910.

Bill Bradley was chief geologist of the US Geological Survey from 1944 to 1959, president of the Geological Society of America in 1965 and winner of the Society's Penrose Medal in 1972.

These pioneering approaches seem to have been largely ignored or forgotten when an explanation was sought in the first half of the 20th century for the so-called 'cyclothems' of coal-rich Carboniferous strata like those of the British coal measures. Cyclotherms are repeated sedimentary cycles several metres thick, comprising coal formed in a swamp forest, then shallow marine shales, lagoonal deposits and deltaic sands, capped with mudstone and clay containing rootlets from the next coal seam. How might they have originated? The authors of several of my undergraduate textbooks, written in the late 1950s, invoked unexplained tectonic processes to alternately lift and lower the land, enabling the sea to flood the coastal plain and then retreat, so giving rise to these cycles [39, 40]. Having the land surface raise and lower hundreds of times might seem realistic in a tectonically active setting, but not on the stable continental margins where most Carboniferous cyclothems were found. Nevertheless, these geologists were in illustrious company, since – as we saw in Chapter 2 – Lyell too had called upon large and unexplained changes in land level to enable his icebergs to drop glacial erratics on English highlands, and even Darwin had followed him down that same illusive path.

In due course, sedimentologists realised that these cyclothems were deposited on flat plains in slowly subsiding basins that eventually accumulated thick piles of sediment[41]. Wet and swampy environments on the plains encouraged the accumulation of organic matter away from the oxidising conditions that would otherwise have encouraged decomposition of organic matter. Across these plains, rivers and their associated deltas migrated and lakes formed from time to time, leading to a geological record of alternating coal seams and mudstones[42]. The sea invaded the subsiding basins at times, leading to the deposition of marine clays. What we see is thus the end result of an interplay between the tectonic processes of Earth movement causing basins to subside, not necessarily uniformly; sedimentary processes causing the lateral migration of river channels and deltas to shut off coal formation temporarily at one site and move it laterally to another; and eustatic changes in sea level, reflecting changes in the volume of water in the oceans, caused either by glacial–interglacial fluctuations in some distant polar region or by alternate warming (i.e. expanding) and cooling (i.e. shrinking) of the ocean's mass[43]. Croll knew the basins were subsiding, but he did not cater for the effects of river systems swinging back and forth across the flat and swampy plains through time. His insight that some of the cycles between coal and mud were due to elevations or depressions in sea level caused by glacial–interglacial changes driven by variations in the Earth's orbit was well ahead of its time, even though he got the association the wrong way around (see Chapter 3). By 1977, the idea that cyclic deposition of sedimentary sequences at all scales was probably controlled by eustatic rather than tectonic changes in sea level was being widely promoted by EPRCo researchers, led by Pete Vail[44]. Nowadays, it is accepted that cyclothems are millennial-scale sedimentary cycles controlled by the rhythms of Earth's orbit and their effects on climate and sea level[45] (although local tectonics may influence the pattern).

Support for this leap in the imagination required a significant advance on the work of James Croll, which we read about in Chapter 3. Ludwig Pilgrim, a German mathematician whose efforts have long been overlooked, kicked off the necessary work in 1904[46]. He calculated in minute detail the changes through time expected in the eccentricity of the Earth's orbit, the precession of the equinoxes and the tilt of the Earth's axis, and linked them to the probable chronology of the ice ages.

Next on the scene was the man who would 'solve' the mystery of the ice ages, Serbian engineer Milutin Milankovitch (1879–1958) (Figure 6.5, Box 6.4).

Figure 6.5　*Milutin Milankovitch.*

Box 6.4 Milutin Milankovitch.

Milankovitch was born into an affluent family that owned extensive farms and vineyards in Serbia. Being more inclined towards science and engineering than to managing the family estates, he went off to attend the University of Vienna, where he earned a PhD in engineering in 1904. After some years as a civil engineer, building bridges and dams in Vienna, he returned to his native land to take up a post at the University of Belgrade, where he lectured on mechanics, theoretical physics and astronomy. But, like all young men, he needed a challenge – a way to make his mark on the world. Starting in 1911, he chose climate, deciding to develop a mathematical theory that would enable him to determine the temperature of the Earth at different times and places, as well as the temperatures of the other planets in the solar system. An ambitious goal! Milankovitch was a reserve army officer, and when war broke out in 1914 he was interned for a while. At the urging of a Hungarian university professor who was familiar with his work, he was eventually paroled and allowed to work in Budapest, where he could access the library of the Hungarian Academy of Sciences. He spent the war years refining his theory for predicting the world's climates through time and describing the climates of Mars and Venus, publishing his work in 1920 as *Mathematical Theory of Heat Phenomena Produced by Solar Radiation*[47]. In 1941, Milankovitch synthesised all of his results into a magnum opus known as *The Canon*[48-50]. Published first in German, it was translated into Serbian in 1977, then into English in 1969, and again in 1998[50]. Aleksander Grubic of Belgrade University published the key elements of Milankovitch's theory, from a study of the 1998 version of *The Canon*, in 2006[51]. For his applications of celestial mechanics to climatology, Milankovitch is often regarded as the founder of cosmic climatology. His efforts were rewarded in the naming of craters on the Moon and Mars and in the establishment of the Milutin Milankovitch Medal, for climatological investigations, by the European Geophysical Union, in 1993, among other accolades.

Milankovitch realised that Croll lacked the detail needed to solve the problem and that Pilgrim lacked the understanding of the operation of the climate system. He was happy enough with Pilgrim's work, however, to use the German's figures to make his own calculations. Before the First World War, he published several papers documenting the emerging results of his theory, which he refined during the war (see Box 6.4). His theory showed how astronomical changes altering the amount of solar radiation could account for the glaciations of the Ice Age, as we shall see later.

Wladimir Köppen was struck by the similarity between Milankovitch's curves and the sequence of glaciations established for Europe by the geographers Albrecht Penck (1858–1945), from Germany, and Eduard Bruckner (1862–1927), from Austria[52], which seemed to confirm Milankovitch's theory. He was so impressed by Milankovitch's conclusions that he invited him to contribute to *Climates of the Geological Past*, the book that

he and Wegener were then writing[23]. Milankovitch was much influenced by Köppen, who told him that it was long periods of low summer temperature that produced glaciation[48-51]; that contradicted Croll, who thought that long winters were the key factor.

Milankovitch recognised that '*Köppen, with his ingenious insight, was the first to discover the connection between the secular march of insolation explored mathematically and the proved historical climates of the Earth*'[48-51]. At Köppen's urging, he produced for *Climates of the Geological Past* a set of graphs showing the variation in summer radiation with time at middle latitudes between 55 and 65° N over the past 600 Ka[23] (Figure 6.6). These showed four cold periods, which Köppen recognised as the four glacial periods of the Penck–Bruckner scheme (Günz, Mindel, Riss and Würm), identified from studies of gravels in river terraces north of the Alps[52]. Milankovitch's graph was a great leap forward. It provided a time calibration for glacial events and explained their occurrence[51]. At last, we had an Ice Age calendar with which to date glacial epochs. Milankovitch's contribution to the Köppen and Wegener book drew his work to the attention of a wide audience.

Climatologists now appreciate that long, cool summers could be critical for the inception of glaciation[10]. They might also help to explain cyclic sedimentation in periods when there were no major ice sheets, by affecting the thermal expansion of seawater, and hence eustatic change in sea level.

Others refined Milankovitch's theory, such as the Belgian climatologist André Berger (1942–), starting in the mid 1970s, and the French astronomer Jacques Laskar (1955–). In 2004, Laskar presented a new solution for the astronomical calculation of Earth's insolation due to changing orbital properties over the past 250 Ma. He and his team expect it to be useful for calibrating palaeoclimate data back to 50 or even 65 Ma ago[53]. Their solutions were improved in 2011[54].

Fascinating though this topic is, it will not detain us further until we get to the Ice Age in later chapters. This is partly because less of the geological record is preserved in older strata, and age control declines in older strata, and partly because I want to focus attention on the geological record closest to the period in which we live now. For more information about the role of Milankovitch cycles in driving climate change, a good source is the chapter on 'Orbital Forcing' in Andrew Miall's book *The Geology of Stratigraphic Sequences*[55].

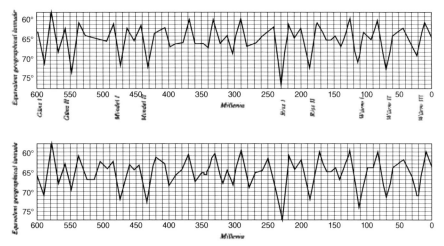

Figure 6.6   *Variation in insolation at 65° N. Amplitudes of the secular variations of the summer radiation at the northern latitude of 65°. Top, the old graph; bottom, the new graph. The vertical axes represent the equivalent geographical latitude for the variation in radiation at 65°N prior to 1800 AD over the past 600,000 years. Where the radiation curve moved down (i.e. towards latitude 70°N), conditions were colder than normal, while when the curve moved up (i.e. towards latitude 60°N), conditions were warmer than normal. The glacial advances recorded in the field are labelled from left to right as Gauss I and II, Mindel I and II, Riss I and II, and Wurm I, II and III.*

6.6   Pangaean Palaeoclimates (Carboniferous, Permian, Triassic)

In April 1972, experts at a NATO Advanced Study Institute at the University of Newcastle upon Tyne reviewed evidence for the relationship between continental positions and past climates[56]. Among them was Pamela Lamplugh Robinson (1919–1994) (Box 6.5) of the zoology department of UCL.

Box 6.5   Pamela Lamplugh Robinson.

Robinson was a vertebrate palaeontologist and an expert in the fossil vertebrates of Gondwanaland[57]. She had a somewhat unusual career. Her university studies as a pre-med student at the University of Hamburg in 1938 were interrupted by the threat of war and she returned to England, where she worked in a munitions factory until 1945. She finally registered as a geology undergraduate at University College London in 1947. Graduating in 1951, she moved to the zoology department to study a giant Triassic lizard for her PhD. She finished her career

in the same department, as reader in palaeozoology in 1982. She was well known for her benchmark review 'The Indian Gondwana Formations', published in the *First Symposium on Gondwana Stratigraphy* (1967). She was Alexander Agassiz Visiting Professor at Harvard University in autumn 1972, and in 1973 was awarded the Wollaston Fund of the Geological Society of London for her work in India. Her biography describes her as '*an excellent, if demanding, teacher, with an immense breadth and depth of knowledge of biology and geology. She could be patient, helpful, charming, and thoroughly entertaining, but also intimidating, imperious, and quite terrifying*'[57]. I can vouch for the accuracy of that description, having been taught by Pamela during my undergraduate days in the Geology Department at UCL. Pamela smoked, and one of her colleagues, Tom Barnard, the professor of micropalaeontology, hated smoking. Alan Lord, a former UCL colleague, told me, '*They would stand in the lab until she finished her cigarette. Tom would then invite her to his office, whereupon she would light a new cigarette just to annoy him.*' She was quite a character.

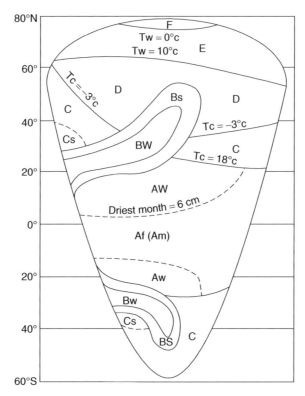

Figure 6.7   *Robinson's conceptual climate model. Distribution of climatic regions on a hypothetical continent of low and uniform relief, after Köppen (compare Figure 6.3).*

Robinson used Köppen's conceptual model of climate zones (Figure 6.3) to demonstrate the likely distribution of climatic regions on a hypothetical continent of low and uniform relief (Figure 6.7). She then used an idealised diagram of world wind and pressure systems to show the likely distribution of annual precipitation on such a continent, which could represent modern Africa or ancient Pangaea[58]. She realised that both the climate zones and the precipitation zones integrated the operations of the atmosphere and ocean and were essentially controlled by latitude. This meant that shifting the hypothetical continent north or south would change the location of the climate and precipitation zones on its surface: they would stay fixed while it moved. Her conceptual model ignored the effects of topography and of ocean currents like the Gulf Stream. Nevertheless, her approach helped to demonstrate how meridional movement of a continent to north or south would lead to changes in

the sequences of climate-sensitive sediments at any one location along that journey. Uplift to form mountains would complicate the picture by inviting rainfall on the windward side and aridity in the rain shadow on the leeward side.

Robinson's climate zone model (Figure 6.7) reminds us that, while aridity is common at around latitudes 20–30° on western coasts (e.g. the Sahara in the Northern Hemisphere), its latitudinal position rises poleward as one progresses inland eastward to around 45° (e.g. Mongolia in the Northern Hemisphere); thus, one can have the same kind of aridity under two quite different temperature regimes. These patterns explain what led Köppen to stipple certain areas to denote aridity on the maps that he had produced with Wegener 50 years earlier. Later, we'll examine the validity of the Robinson–Köppen assumption about the location of arid zones.

Robinson applied her conceptual climate-modelling approach to a suite of continental reconstruction maps rather like those of Alan Smith (Figure 5.8a)[58]. On each, she plotted the likely positions of high-pressure maxima, winds and the Inter-Tropical Convergence Zone (ITCZ) for the northern summer (July) and winter (January) seasons, for the late Triassic (235–200 Ma ago) and late Permian (260–250 Ma ago). Applying first principles, she deduced which regions were likely to have been dry year round, which had sharply seasonal (monsoonal) rainfall and which were likely to be humid at high latitude. To test her predictions, she compared them with the distribution of climate-sensitive sedimentary rocks.

Starting with the Triassic, she suggested that, during the northern summer (July), the warming of the land-mass would have led to a major centre of low pressure developing over northeastern Pangaea, which would have deflected the ITCZ northward over the coast of eastern Laurasia. As the ITCZ is the boundary between the north-east Trades and the southeast Trades, this displacement would have sucked in wet air from the south over the Tethyan Ocean, causing summer monsoonal rains to fall over the northern coasts of Tethys, much as happens in southern Asia today. In the Southern Hemisphere, the winter cooling of southern Pangaea (Gondwanaland) would have formed a high-pressure maximum there, creating a dry winter season. The winds blowing from that centre across land towards western Laurasia (North America) would have led to dry summers in the latter region. In January, these conditions would have reversed, with the ITCZ being pushed far to the south over eastern Gondwanaland, bringing monsoonal rains to the southern margins of Tethys, in what is now Arabia and northern India. Robinson thought that smaller high-pressure cells would have developed over both poles. Today's polar high-pressure cells are surrounded by low-pressure zones, which, if they occurred in the same way in the past, would have brought seasonal rains to places like Alaska and Japan in the northern summer and to coastal Australia, Antarctica and southern South America in the southern summer.

Robinson considered that conditions would have been slightly different in Permian times, because – compared with the younger Triassic period – the Equator lay some 10° further north, the North Pole lay 10° north of the coast of Laurasia and the South Pole still lay in Antarctica and close to Africa. This meant that there would have been less divergence between the northern and southern

extremes of the ITCZ. The arid conditions of the interior would have shifted north, covering most of North America and Greenland; monsoon rains would still have characterised the northern and southern coasts of Tethys; and the humid temperate conditions at the southern end of Gondwanaland would have extended further into the continent. At that time, she thought, the more central position of the Equator within Tethys would have encouraged development of a warm ocean current flowing east along the coasts of India and Australia at the northern margin of Gondwanaland, increasing the chances of heavy rains along those coasts.

Did her model work? She found evaporites where her model predicted that climates were dry year round and coals where the climate was humid, so *'On the whole agreement between the model and the pattern of distribution of the four types of "climate-sensitive" rocks is a good one'*, although she accepted that *'there are some anomalies'*[58]. Why were there no equatorial coals in the Permian and Triassic like there were in the Carboniferous? Robinson reminded us that, in the Carboniferous, the northern and southern components of Pangaea were still in the process of coming together, and so were separated by an equatorial seaway, on either side of which monsoonal conditions would have provided the rainfall necessary to sustain extensive coal swamps. That Carboniferous seaway, which Köppen and Wegener had not included in their own maps (Figure 5.3), had disappeared by Permian and Triassic times (Figure 5.8a), and the monsoon rains could not penetrate far enough into the arid hinterland to support the vegetation necessary to form equatorial coal deposits where Carboniferous ones had formed along the palaeo-equator.

The 1972 conference clarified other aspects of the climatic history of Palaeozoic times. For instance, much of the discussion about past climate change prior to the conference was rather confused because many geologists thought that coal must have formed in a tropical climate. Coal deposits first became widespread during the late Carboniferous. They contain beautifully preserved structures of a wide variety of terrestrial plants that once formed parts of a swamp community. The lack of herbaceous plants and the abundance of tree ferns or lianas with giant leaf fans, along with the remains of trees with smooth cortex and little bark, show that they formed in rainforests, which may have been tropical or subtropical[59]. At the 1972 conference, the palaeobotanist Bill Chaloner (1928–) from Royal Holloway College, near London, who was to be elected a fellow of the Royal Society in 1979 and was awarded the Geological Society's Lyell Medal in 1994, showed that trees that grew at temperate latitudes differed considerably from those in tropical locations. Temperate tree carried rings representing seasonal change, while tropical trees did not[60]. Most of the Carboniferous and Permian coals of North America and Europe, near Köppen and Wegener's palaeo-equator, lacked tree rings; they were tropical. Those of the same age from southern Gondwanaland, including Antarctica, which formed near Köppen and Wegener's South Pole, carried tree rings; they were temperate[60]. Problem solved. Tying coals lacking trees with rings to the palaeo-equator removed the necessity for Carboniferous coals to signify global warmth. Coals could just as well have formed in cool, humid environments, which would be signalled by trees with rings. We no longer had to think of the Carboniferous as a period that was especially warm *globally*. Zonal conditions ruled, much as Lyell suspected.

Like Humboldt and Köppen, Chaloner saw that '*climate is the overriding influence controlling the distribution of plant communities*'[61]. Hence, palaeoclimatic information could be extracted from fossil plant remains by observing what climate zones their nearest living relatives inhabited, the shape of their leaves or the character of their wood – especially the presence or absence of rings, representing seasonality[61]. Seeing that Ian Woodward, then at Cambridge University, had discovered in 1987 that the frequency of stomata (pores) on leaves was proportional to the abundance of CO_2 in the atmosphere[62], Chaloner noted in 1990 that this '*offers promise for direct palaeobotanical evidence for past changes in the level of this climatically significant atmospheric constituent*'[61], something we follow up on later.

Robinson concurred with Wegener and Du Toit that, during the Carboniferous and Permian, today's southern continents were clustered over a South Pole in South Africa, where we find the extensive Dwyka Tillite. Glacial conditions covered South Africa, Antarctica, India and much of southern Australia and South America. Lyell would probably have been pleased to see the association of cooling with high-latitude land (see Figure 2.5).

We now know more about the Permo-Carboniferous glacial period. Emerging evidence suggests that global ice volume reached a peak at the Carboniferous–Permian boundary, causing a significant global fall of sea level at about 300 Ma ago. In due course, that was followed by a rise in sea level, manifest as a global transgression during the following Sakmarian stage (295–290 Ma ago),

signifying the beginning of the major deglaciation of Gondwana[63].

Gondwana continued to warm through the Permian and into the Triassic, as the supercontinent moved north away from the South Pole. During this time, while Pangaea's maritime margins were humid, much of its immense interior was arid and desert-like; imagine a gigantic version of modern Australia. Where evaporation exceeded precipitation, vast deposits of salt accumulated in the Permian of western Europe. By late Permian (Zechstein) times (270–250 Ma ago), salts were being deposited in a basin extending from west central Poland to northeastern England and from Denmark to southern Germany[64]. Deposition began in the early Permian, around 280 Ma ago, and extended up into Triassic time, diminishing towards its end at 200 Ma ago. Laminations within the deposits suggest climate cycles of more or less aridity. The salt may have been deposited during particularly arid times, rather than continuously. Most past evaporites were deposited in warm, arid regions between 45 °N and 40 °S, with a peak in the desert regions centred on about latitude 32°[64].

Knowing how the continents were distributed through time was a boon to palaeontologists, who could now begin to understand, rather than just guess, why the fossils of animals and plants were distributed in the way that they were across today's continents[65]. It was simple: the break-up of Pangaea disrupted former land links. Knowing the timing of the different breaks, they could understand the divergence of fossil lineages from one another on today's different continents.

As we saw in Chapter 5, one of those palaeontologists, Fred Ziegler of the University of Chicago, realised that it would benefit the wider community to construct an accurate series of palaeogeographic maps to show how fossil plants and animals and climate-sensitive deposits had been distributed through time, which led to the inception of the Paleogeographic Atlas Project. In 1979, Ziegler and his colleagues publish a suite of seven continental reconstruction maps for the Palaeozoic, on to which were plotted the locations of climate-sensitive sediments[66]. They found that '*The distribution of climatically sensitive sediments shown on our reconstructions for the Paleozoic is in good agreement with expectations based on the model of the Earth's present atmospheric and oceanic circulation patterns*'[66]. They went on to explain: '*We do not mean to imply that climate has been constant through time. The proportion of land, and its latitudinal array, must have been very important in controlling world temperature and*

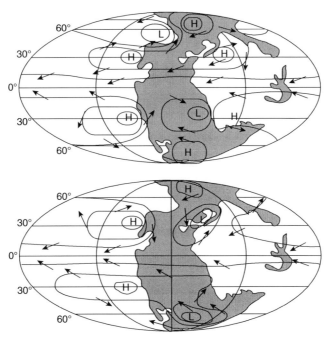

Figure 6.8 *Past distribution of atmospheric pressure for the earliest Triassic, with northern winter above and northern summer below. Heavy solid lines are isobars; arrows represent wind directions. H = high-pressure centre; L = low-pressure centre.*

precipitation. The heat derived from solar radiation is absorbed and redistributed in the oceans, and by contrast, lost over land areas during the nights and the winters. From this, one would expect that the world climate of periods like the Recent, the Permo-Carboniferous and the late PreCambrian, with much land in high latitudes, would be generally cool and this is confirmed by glaciations of these times. At the other extreme were times like the early Paleozoic and the late Mesozoic with large expanses of shelf seas associated with relatively low latitude continents. The occurrence during such times of carbonates in higher latitudes than present may be evidence of more uniform temperature conditions[66]. Lyell would have been pleased to see the emphasis on latitude as a controlling factor in climate.

One of the co-authors of the 1979 paper on 'Paleozoic Paleogeography' was Ziegler's former PhD supervisor from Oxford, W.S. (Stuart) McKerrow (1922–2004), a palaeo-ecologist and the Geological Society of London's Lyell Medallist for 1981. McKerrow went on, with Chris Scotese, to write about palaeogeography and palaeoclimatology, notably in Africa, making ample use of the usual climatic indicators[67]. Among their cold-climate indicators were glendonites: carbonate pseudomorphs of ikaite, a calcium carbonate hexahydrate ($CaCO_3.6H_2O$) that forms in organic-rich marine or brackish sediments at near-freezing temperatures and which decomposes when the temperature rises above 5 °C.

In 1977–78, the Paleogeographic Atlas Project was expanded to apply the likely circulation patterns of the atmosphere and ocean (à la Robinson) to the continental reconstruction maps. Judith (Judy) Totman Parrish was invited to supervise this part of the programme[68]. In due course, Parrish would rise through the ranks to become president of the Geological Society of America in 2008–09. She built upon and expanded Robinson's conceptual approach to palaeoclimatology in a set of landmark papers published in 1982/83. Basically, she superimposed conceptual distributions of likely past air pressure on continental reconstruction maps for different time slices (Figure 6.8) and used the pressure maps to determine likely palaeo-wind directions and areas of high or low rainfall, to compare with palaeoclimate data.

Where the winds blow parallel to the coast, the surface waters move offshore. Nutrient-rich subsurface waters well up to replace them, stimulating high productivity. Under appropriate conditions, this can lead to the deposition of the organic-rich rocks that are the source rocks for oil, and of rocks rich in phosphorus: phosphorites, which can be mined for fertiliser. Parrish produced a set of papers predicting where upwelling might have occurred in the Palaeozoic[70, 71] and in the Mesozoic and Cenozoic[69], and where rainfall might have been high, perhaps leading to coal deposition, in Mesozoic and Cenozoic times[72]. As we shall see later, her predictions correlated well with palaeoclimatic data from the field.

Compared with what Robinson had to offer, Parrish benefitted by having improved reconstruction maps, more time slices and more data on the distribution of climate-sensitive deposits. She found that the present-day rainy zones around 55° N and 55° S and at the Equator also existed in the past: '*from this, it can be concluded that atmospheric circulation has not been radically different from its present configuration, despite some apparently great differences in some climatic parameters such as the equator-to-pole temperature gradient*'[72]. She agreed with Robinson that the Triassic world was generally dry, with seasonal rainfall on eastern coasts. Sea level was low at the time.

Applying Parrish's conceptual palaeoclimatic model helped to refine the Chicago group's analysis of climate change in the Carboniferous[73]. Her maps showed that the collision between Gondwana and Laurasia changed the climate from mainly zonal to mainly monsoonal, causing increasing asymmetry of climate patterns from east to west. That dried out the equatorial region, leading to the demise of formerly flourishing coal swamps along an equatorial seaway, as mentioned earlier, and increased seasonality. Formation of a single large land mass dried the interior and deflected to both north and south the former through-flowing warm equatorial currents, which then carried heat to high latitudes along the east coasts of Pangaea. The mountain belt created along the suture line between Gondwana and Laurasia would have interacted with the monsoonal circulation much as the Tibetan Plateau does today. As noted by Briden and Irving[27], coals forming after the suture were abundant at high latitudes, not at the Equator.

Neither Robinson nor Parrish had much to say about the likely range of temperature from the coast to the interior of Pangaea. Numerical modelling experiments suggest that mean monthly summer temperatures there exceeded 35 °C:

more than 6 °C above today's maximum temperatures for the interior[11]. Indeed, daytime highs may have approached 45–50 °C. Numerical climate models agree that most of the interior would have been dry, especially between 40° N and 40° S[11].

Certain caveats must be applied in palaeoclimate studies. Bruce Sellwood and Brian Price remind us that, despite enthusiasm for the use of sedimentary facies (or types) as indicators of past climate, the data have to be interpreted with care[74]. The most climatically informative sediments are tills, laterites, evaporates and aeolianites (e.g. dunes). Other criteria provide supplementary evidence of climate. But, because sedimentary rocks are subject to post-depositional change (diagenesis), they seldom faithfully record subtle climate signals. They are imperfect receivers of the climate signal, although certain settings (e.g. deserts and ice caps) preserve such signals better than others[66].

Among the best preservers of climate signals are fossil plants. As Ziegler and his team pointed out, they occupy realms with pronounced climate signals, are sedentary (no seasonal migrations), are not subject to diagenetic alteration (unlike isotopes), represent ground truth (unlike model outputs) and are abundantly preserved in many places[75]. The team assigned the fossil vegetation of Eurasian floras from the Triassic and Jurassic periods to one or other of 10 biological zones (biomes). Most plants fell into the dry subtropical, warm temperate and cool temperate biomes. There was a general absence of tropical rainforests. Tropical coal swamps disappeared in the early Triassic, except locally in the Asian monsoon region, and the equatorial belt became arid in the Triassic. Coal swamps emerged at mid to high latitudes during the late mid Triassic. Warm temperate floras reached above 70° N in the Triassic and up to 70° N in the Jurassic, and there was no hint of the cold temperate (Arctic or glacial) climates of today. Triassic warmth contrasted with the glacial conditions of the southern continents in the Carboniferous and Permian, reflecting the drift of Gondwana north away from the South Pole.

During the Triassic, generally arid conditions prevailed over North America and Europe within 5–50° north of the Equator. They were interrupted in the late Triassic Carnian period (228–216 Ma ago) by a warm, wet monsoonal phase[76]. Substantial changes occurred within the marine invertebrate fauna at the end of the early Carnian, and there was a major change in the terrestrial biota at the end of the Carnian. Michael Sims of Trinity College, Dublin and Alastair Ruffell of UCL interpreted these developments to suggest that the final coalescence of Gondwana and Laurasia to form Pangaea was followed in the mid Carnian by rifting preceding the break-up of the supercontinent. This rifting would have been associated with volcanism and the emission of CO_2, which might have led to sufficient warming to have caused the development of the monsoonal conditions[75], along the lines suggested by the Australian geologist John Veevers[77].

Among those reviewing the relationship between continental positions and past climates was Lawrence A. Frakes (1930–) (Box 6.6), who produced a series of papers on Palaeozoic glaciation in Gondwanaland, starting in 1969[78].

As Frakes pointed out in 1981[79], the idea that variations in the age and distribution of late Palaeozoic glacial deposits on Gondwanaland resulted from the drift of the supercontinent over the pole was first elaborated in 1937 by Du Toit[24], then in 1961 by Lester King[25], and again in 1970 by Crowell and Frakes[80]. Palaeomagnetic studies had established by 1981 that South America and South Africa were the first parts of Gondwanaland to cross the pole, and that Australia was the last[79]. It was not entirely obvious to Frakes why glaciation ceased by the early late Permian, as Gondwanaland remained at fairly high latitudes then, as did its southernmost fragments during the continental break-up that followed in the early Mesozoic. One possibility was that more of Gondwanaland now lay at or closer to latitude 65° S, where conditions were warm enough to melt ice and prevent its further accumulation. Global warming of unspecified cause – and an associated decrease in albedo – might account for these changes, along with a decrease in the precipitation required to build an ice sheet[79]. A decrease in the requisite precipitation might have resulted from the gradual shift of the continents or from shifts in the locations of warm ocean currents. We have to remember that, at the time, CO_2 had only just been discovered in ice cores, and nothing much was known about its past distribution.

Box 6.6 Lawrence A. Frakes.

Larry Frakes was born in the United States and started his career with John Crowell at the University of California, Los Angeles (1964–71), studying late Palaeozoic glaciations on Gondwana fragments. Later, at Florida State University, he worked with Elizabeth Kemp on global reconstructions of Eocene–Oligocene palaeotemperatures, making an early contribution to climate modelling,

and publishing key findings in the journal *Nature* in 1972. Working with Jane Francis and Neville Alley at Adelaide University (1987–99), where he was appointed the Foundation Douglas Mawson Professor of Geology and Geophysics (1985), his research overturned the concept of a uniformly warm Cretaceous through discoveries of evidence for glacial activity. He and Jane Francis found evidence for glaciation in most periods of the Phanerozoic (e.g. through the occurrence of dropstones and related criteria), culminating in a paper in *Nature* in 1988. Frakes was awarded the Antarctic Service Medal by the US National Science Foundation and has a mountain named after him in Marie Byrd Land, Antarctica.

6.7   Post-Break-Up Palaeoclimates (Jurassic, Cretaceous)

With seafloor spreading taking place in all of the new seaways, as well as in the pre-existing Pacific Ocean, the rate of production of new ocean crust increased significantly, forming several new mid-ocean ridges during late Jurassic and Cretaceous times[81, 82]. These massive new upstanding ridges displaced ocean water, thereby raising sea level (Figure 5.9) and drowning low-standing parts of the former fragments of Pangaea, creating warm, shallow seas in North America and Europe (Figure 5.8b)[83]. Following Peter Vail's lead (Figure 5.9), Parrish thought that Cretaceous sea levels stood on average 170 m higher than today[12]. Sea level began to fall from these high levels when the Izanagi Plate, with its associated mid-ocean ridge, in the northeast Pacific was subducted beneath East Asia around 60 Ma ago[81]. The new seaways changed the pattern of ocean currents, introducing a new element into the story of climate change. For example, the creation of a north–south passage by the opening of the Atlantic Ocean increased the opportunity for oceanic transport of heat from the tropics to the poles, making polar glaciations less likely.

The climate of the Jurassic has been described as 'equable', in the sense of warm but with low variability. Warm it was, compared with today, but there were strong seasonal contrasts in continental interiors, where, during the early Jurassic (195 Ma ago), the annual range of temperature was up to 40 °C in Eurasia, at about 60° N, and more than 45 °C in Gondwana, at about 60° S[11]. Hardly 'equable'!

Tony Hallam is a fount of knowledge about Jurassic climate[84, 85]. He tells us that there were no significant polar ice caps then, but that dropstones indicate the presence of seasonal ice in the mid Jurassic of Siberia, where winter temperatures probably hovered close to 0 °C. Most of Africa, Madagascar, India, South America, North America south of the Canadian border, western Europe and western Asia would have been dry. Monsoons would have made the margins of these Pangean fragments seasonally wet. Year-round humidity characterised high latitudes, South East Asia and southernmost South America. Coral reefs were confined to a tropical belt mostly between the 30th parallels.

Parrish's palaeoclimate models and Ziegler's data showed that monsoonal circulation with extended wet and dry periods allowed evaporites to form seasonally in equatorial regions in the Mesozoic[72], as Briden and Irving also found[27]. North Africa and northern South America became wetter with time as the North Atlantic opened.

In a seminal study of the climate of the past 540 Ma – the Phanerozoic Aeon – Larry Frakes and his colleagues from the University of Adelaide in South Australia, Jane Francis and Joseph Sytkus, reported in 1992 that sea surface temperatures in the low latitudes of the mid to late Jurassic were 26–28 °C, while bottom waters were about 17 °C[86]. Their oxygen isotope data showed that water temperatures cooled towards the lower Cretaceous. The Frakes team provided evidence for transport by ice at high latitudes in the late Jurassic and early Cretaceous; chiefly the occurrence of boulders and dropstones of exotic rock types embedded in fine-grained mudstones, harking back to Lyell's ice-rafting (see Section 2.2 and Figure 2.7). But, in the absence of glacial deposits such as tillites, it seemed likely that the northern polar environment was periglacial, with seasonal winter ice forming on rivers and shorelines and incorporating exotic materials from the banks and bases of rivers and from cliffs. Seaward transport of floating ice explained the occurrence of dropstones offshore[86]. Lyell's theory that cool conditions would result from the polar locations of continents did not work all the time. Something operated against it.

While evidence for episodes of cooler conditions in the early Cretaceous has been proposed, a recent detailed study of palaeotemperatures from the lower Cretaceous (Berriasian–Barremian, between 145 and 125 Ma ago) showed that sea surface temperatures were much warmer than today, averaging 26 °C at 53° S and 32 °C

at 15–20° N[87]. It seems that the climate was warm and stable, with a weaker meridional temperature gradient (0.2–0.3 °C per degree of latitude) than we have today (0.5 °C per degree). The temperatures appear to be no different from those of the late Cretaceous Cenomanian and Turonian periods (100–88 Ma ago). If there were cool or cold periods in the early Cretaceous, as was formerly supposed, they may have been just short cold snaps or seasonal extremes[87]. One such early Cretaceous cold snap gave rise to glacial tillites in the Flinders Range of Australia[88].

Parrish's conceptual model implied that Cretaceous opening of the entire Atlantic (Figures 5.8b and 6.9) brought rainfall to the formerly dry east coasts of both North and South America, while the interiors of Asia and Africa remained dry. Much the same applied in the late Cretaceous (Maastrichtian, 70–65 Ma ago), but the widening North Atlantic would have encouraged the development of westerlies, bringing rain to western Europe. An equatorial Tethyan current separated the northern and southern fragments of Pangaea (Figures 5.8b and 6.9), and there was probably a proto Gulf Stream in the North Atlantic. When the break-up was well underway, sea level was at a maximum (Figure 5.9), with flooded continental margins (Figure 5.8b and 6.9)[72].

Parrish interpreted her climate model to suggest where winds blew parallel to the coast, generating upwelling currents (e.g. in the mid Cretaceous) (Figure 6.9)[69]. These are locations where one might expect organic-rich rocks to form.

Based on the studies of phosphorite that I carried out off the coast of northwest Africa for my PhD (1967–70) and off southwest Africa while at the University of Cape Town (1970–72), I independently developed a similar approach to Parrish's for predicting the likely occurrence of organic-rich rocks. Whereas she focused on the winds, I focused on what was happening within the body of the ocean – especially on the depletion of oxygen in the oxygen minimum zone. This zone arises because the sinking and decomposing remains of dead plankton consume oxygen at intermediate depths at rates faster than it can be replenished by mixing from well-oxygenated surface waters and bottom waters, making the ocean into an oxygen sandwich. Where oxygen depletion is extreme, conditions may become anoxic (zero oxygen), and sediments reaching the seabed there may be well preserved, leading to organic enrichment, especially on continental margins – as off Peru, California and Namibia today. I applied this model extensively in my research

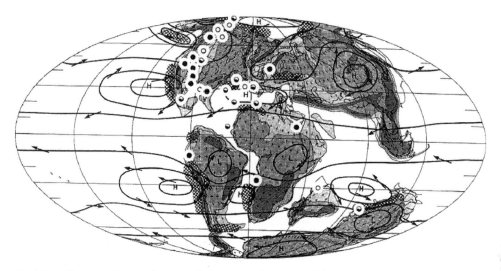

Figure 6.9   *Upwelling predictions from qualitative circulation models: Cenomanian (99.6–93.6 Ma ago). The map shows highland in dark shading, lowland in medium shading and flooded continental edges in light shading. Lines represent isobars, with H = high pressure and L = low pressure. Upwelling indicated by cross-hatching along continental margins. Dots = locations of samples of organic rich rocks.*

for EPRCo in Houston (1976–82), with the object of predicting where explorers might find oil-rich source rocks in ancient basins. Parrish and I presented papers on our complementary approaches at a NATO meeting organised by Jörn Thiede and Erwin Suess in September 1981 in Villamoura, Portugal, on the topic of 'Coastal Upwelling – Its Sediment Record'[71,89].

When Chris Scotese from Ziegler's group joined me for a sabbatical at BPRCo in the mid 1980s, we devised a method for quantifying the Robinson–Parrish approach to climate modelling. The end result was a set of palaeogeographic maps, complete with isobars, that I used to show where upwelling currents may have formed organic-rich deposits in past times[90]. While our results did not differ much from Parrish's, we felt that quantifying the principles made them more credible.

Knowing that plants are strongly related to climate[75], Bob Spicer of Oxford University and colleagues used Cretaceous fossil plant remains to show that cool temperate rain forests in polar coastal areas were conifer-dominated and deciduous[91]. At high latitudes and in continental interiors, winter temperatures likely fell below freezing, but some plants retained leaves year round, with reduced leaf size and thick cuticles. At mid latitudes, conifers, ferns and cycads dominated open-canopy woodlands and forests, giving way in the late Cretaceous to broadleaved

angiosperms, including shrubs and small trees. Forests were patchy at low latitudes.

Working with Parrish, Spicer suggested that late Cretaceous–early Cenozoic floras from high palaeolatitudes (75–85° N) experienced a similar light regime to that at present. Their plant data suggested that mean annual air temperatures at sea level there were 10 °C in the Cenomanian (100–94 Ma), rising to 13 °C in the Coniacian (88–86 Ma) and dropping to 5 °C in the Maastrichtian (71–65 Ma), then rising again to 6–7 °C in the Palaeocene (65–55 Ma)[92]. They thought that polar winter temperatures were freezing in the Maastrichtian and that '*Permanent ice was likely above 1700 m at 75° N in the Cenomanian, and above 1000 m at 85° N in the Maastrichtian*'[92]

Jane Francis (1956–) (Figure 6.10, Box 6.7), an eminent palaeobotanist, specialises in using the fossil plants of the polar regions as indicators of past climates.

Francis likes to point out 'the Antarctic paradox', which is that '*despite the continent being the most inhospitable … on Earth with its freezing climate and a 4-km thick ice [sheet], some of the most common fossils preserved in its rock record are those of ancient plants. These fossils testify to a different world of warm and*

Figure 6.10   Jane Francis.

Box 6.7 Jane Francis.

Jane Francis was a palaeobotanist with the British Antarctic Survey from 1984 to 1986, before spending 5 years as a post-doctoral researcher with Larry Frakes at the University of Adelaide in Australia. Returning to the United Kingdom in 1991, she joined Leeds University, where she rose to become professor of palaeoclimatology in the school of earth and environment, and director of the Centre for Polar Science in 2004. For her polar research, she was awarded the US Antarctic Service Medal, the US Navy Antarctic Medal and, in 2002, the UK's Polar Medal – only the fourth woman to receive that honour. She received the President's Award of the international Paleontological Society in the 1980s, became president of the UK's Palaeontological Association for 2010–12, was appointed to head the British Antarctic Survey in October 2013 and was awarded the Coke Medal of the Geological Society of London in 2014. Francis leads the United Kingdom's involvement in ANDRILL, the international Antarctica drilling programme.

---

ice-free climates, where dense vegetation was able to survive very close to the poles. The fossil plants are an

important source of information about terrestrial climates in high latitudes, the regions on Earth most sensitive to climate change'[93]. Francis and her team found that fossil plants from the mid Cretaceous were abundant on Alexander Island on the west side of the Antarctic Peninsula, at around 70° S. Conifers, tree ferns and ginkgos were abundant there, with shrubs, mosses and liverworts in the rich undergrowth (Figure 6.11). Ginkgo trees were common in the distant past but are rare today. You may have come across one: the Maidenhair tree, *Ginkgo biloba*, used in traditional Asian medicine. Evidence from the plants and their associated soils showed that the climate was warm and humid; probably dry in summer and wet in winter. This was around the time when Antarctica reached the South Pole, 90 Ma ago. Dinosaurs roamed the woods. See if you can find the one in Figure 6.11.

Younger Cretaceous strata are preserved on the opposite, eastern, side of the Antarctic Peninsula, on James Ross Island and Seymour Island, at about 64° S. Flowering plants (*angiosperms*) were abundant. Their modern equivalents live in warm temperate or subtropical conditions, including wet tropical mountain rainforests and cool temperate rainforests. Analysing the shapes of the margins of leaves, which are related to temperature, told Francis and her team that mean annual temperatures in this part of Antarctica, then some 2000 km from the pole, averaged around 17–19 °C. Winter temperatures must have been above freezing, and rainfall ranged from around 600 to 2400 mm/year, with peaks in the growing season.

Most of the perceptions of climate change that we have examined so far come from sediments and fossils found on land. But ocean sediments also have something to tell us. Here we benefit from the application of oil company technology to the solution of fundamental science questions. Many of the advances in our understanding of the evolution of our climate following the break-up of Pangaea come from the use of a floating drill rig to sample the sediments deep beneath the 72% of the planet's surface covered by the ocean. Drill cores obtained through the Deep Sea Drilling Project (DSDP) and its successors (Figure 6.12, Box 6.8) extend as far back as the early Jurassic in a few places: the age of the oldest known deep marine sediments formed since the break-up of Pangaea. We have many more Cretaceous deep-ocean drill cores, and yet more from the Cenozoic, as we'll see in Chapter 7. The marine microfossils from these cores tell us a great deal about past climates[94].

Figure 6.11   Cretaceous vegetation on Alexander Island, Antarctica 100 Ma ago.

Integrating data from marine and terrestrial sources, Larry Frakes and his team tell us that '*the period from the mid-Cretaceous (mid-Albian) to the mid-early Eocene … (105–55 Ma) was one of the warmest times in the late Phanerozoic*'[86]. The average global temperature then was probably at least 6 °C higher than today, the poles were likely free of permanent ice and there was no evidence for seasonal ice-rafting. As the high-latitude oceans were warm, the Equator–pole temperature gradient was low, resulting in relatively weak atmospheric and oceanic circulation. '*Temperate climates extended right up to the poles during the Cretaceous and early Tertiary, allowing the growth of forest vegetation at high latitudes. The plants were able to tolerate the rather extreme light regime that they would have experienced*'[86].

This analysis seems to neglect Parrish and Spicer's conclusion that the high latitudes of the Maastrichtian (70–65 Ma ago) were rather cold[92]. More recent data from deep-sea sites confirm that bottom waters were about 12 °C during the mid Cretaceous, reached 20 °C during

the latest Cenomanian (100–94 Ma ago) and Turonian (94–88 Ma ago), and cooled to 9 °C by the Maastrichtian at the end of the Cretaceous[97]. Kenneth MacLeod of the University of Missouri and colleagues confirmed in 2013 that Turonian seas were particularly warm, with surface water temperatures of 30–35 °C and bottom temperatures of 18–25 °C[98].

Fossil leaves from Alaska during the Albian (112–100 Ma ago) and Cenomanian (100–94 Ma ago) suggest temperatures of around 10 °C, warming in the Coniacian (89–86 Ma ago) to about 13 °C then cooling during the Campanian–Maastrichtian (84–65 Ma ago) to around 2–8 °C. Winter temperatures may have declined below freezing there. Under these conditions, dinosaurs thrived in Arctic deltas among mild to cold temperate forests of deciduous conifers and broad-leaved trees. At the other end of the world, there were rainforests on the Antarctic Peninsula and in Tierra del Fuego. The climate there was like that of New Zealand and Tasmania today[93]. The continental interiors, like central Asia, remained very dry[86].

Figure 6.12   *Deep-ocean drilling vessel* DV JOIDES Resolution *(1989–).*

Box 6.8 Deep-ocean drilling.

The Deep Sea Drilling Project (DSDP) on the 120 m-long drilling vessel *Glomar Challenger* was run by the US National Science Foundation (NSF) and collected samples from the summer of 1968 through 1972. It was followed from 1975 to 1985 by the International Phase of Ocean Drilling (IPOD) on the same ship, funded by the NSF, Germany, France, the United Kingdom, Japan and the Soviet Union. A new phase, the Ocean Drilling Program (ODP), began with the advent of a larger drill ship, the DV *JOIDES Resolution*, and ran from 1985 to 2003, when it was replaced by the Integrated Ocean Drilling Program (IODP). From 2012, this became the International Ocean Discovery Program (also IODP), which involves the United States, 17 European countries, India, China and South Korea. It employs two ships: the *JOIDES Resolution* and Japan's *Chikyu*, which is equipped with a 'riser' – a device for preventing blowouts – thus enabling drilling into deep sediment sections on continental margins where natural gas may be a potential hazard. As Bill Hay tells it[95], the development of the DSDP benefitted from the efforts of Cesare Emiliani in obtaining long cores with which to study the history of the ocean. In 1963, he submitted his LOCO (long cores) proposal to the NSF, and the drill ship *Submarex* duly collected some test drill cores of late Tertiary and Quaternary age from the Nicaragua Rise late that year. In 1964, the major US oceanographic institutions Lamont, Woods Hole and Scripps (of which more in Chapter 7) formed the Joint Oceanographic Institutions for Deep Earth Sampling (JOIDES) and used the drill ship *Caldrill* for a drilling campaign on the Blake Plateau off Florida in 1965. The successes of *Submarex* and *Caldrill* led JOIDES to propose to the NSF in 1966 that there be an 18-month programme of ocean drilling – the DSDP – and a uniquely outfitted research drill ship, the *Glomar Challenger*, was commissioned. Complete with a dynamic positioning system, the *Glomar Challenger* was named after the first major oceanographic survey ship, HMS *Challenger*. In recent years, a comparable programme to the DSDP has developed for the continents: the International Continental Scientific Drilling Programme (ICDP)[96].

Myriads of tiny planktonic plants, the *coccolithophoridae*, flourished in the warm shallow seas that flooded western Europe during the Cretaceous period between 145 and 65 Ma ago. The remains of their calcium carbonate skeletons sank to the shallow seabed to form white ooze, now consolidated and uplifted as the chalk of the White Cliffs of Dover and the French coast. When you rub a piece of chalk between your fingers, the dust that comes off is made of the miniscule platelets, or coccoliths, that covered these tiny creatures. Strange to think that while watching my science teacher scribble on the blackboard, I was seeing fossil coccoliths scrawled on slate. And I doubt my mother ever knew that she was dusting her face with fossils when powdering her nose.

As with the Jurassic, in the past geologists thought the Cretaceous had a warm, equable climate with a lack of seasonal extremes. Nowadays, Hallam tells us that the mid-latitude Cretaceous climate was probably seasonal and the concept of an equable climate belongs on the scrap heap[84]. Besides that, Frakes and his team found that while the mid to late Cretaceous was generally warm, abundant evidence of cyclic sedimentation showed that the warmth was interrupted by cool periods lasting from a few thousand to 2 million years, which might be related to variations in the Earth's orbit of the kind identified by Croll and Milankovitch[86]. Mid Cretaceous sea surface temperatures in Israel, then located at about 10° N, were between 29 and 31 °C, but may have dropped to 21 °C in the late Campanian (84–71 Ma ago). Equatorial bottom waters were around 10 °C cooler than surface waters, but in restricted basins like the South Atlantic they were as warm as 22 °C[86]. The high-latitude ocean was cooler, with Antarctic shelf waters ranging from 9 to 16 °C[99].

The tropical ocean was significantly warmer during parts of the Cretaceous than it is today, notably during the Turonian (94–89 Ma ago), when sea surface temperatures in the equatorial Atlantic reached 33–42 °C[96]. But, as Frakes and his team noted, conditions were not permanently warm. There were periodic coolings of 1–3 °C[100] and the fluctuations in sea level identified from seismic records, for example by Pete Vail and colleagues from EPRCo[101, 102], strongly suggest the fluctuating presence of at least small ice caps in the polar regions during Jurassic, Cretaceous and early Eocene times, despite their warm greenhouse climates[103]. In later chapters, we will address the question of what caused the cooling of the late Cretaceous: changes in oceanic heat transport or declining concentrations of atmospheric CO_2?

More evidence has emerged from a study of the remains of the dinoflagellate cyst *Impletosphaeridium clavus*, peaks in whose abundance occur in the muds of the Maastrichtian (latest Cretaceous) and Danian (earliest Palaeocene) of Seymour Island, Antarctica[104]. Such peaks suggest the existence of blooms of this species, which typically occur at high latitudes in association with the melting of winter sea ice. If winter sea ice was forming along the eastern side of the Antarctic Peninsula at 65° S 70–60 Ma ago then conditions may well have been suitable to allow land ice to form on the highlands of the continental interior as well. And if such ice was present, and periodically melted under the influence of orbital variations in insolation, then we have a mechanism for the periodic changing of sea level.

Planet Earth was quite different in warm Cretaceous times from how it is now. It was a world largely without polar ice. The difference in temperature between pole and Equator was about 20 °C, while today it is about 33 °C. The weaker Equator–pole thermal gradient would have weakened westerly winds like the jet stream. Polar climates would have become much more seasonal than they now are. As Bill Hay points out, '*If there were no perennial ice in the Polar Regions, the temperatures there could alternate between cold in winter and warm in summer, and that means that the polar atmospheric pressure systems would change between summer and winter*'[95]. Hay supposed that these changes meant that the Hadley cell that governs the positions of the westerlies could have expanded poleward and that the westerly and easterly winds at high latitudes would have become seasonal and disorganised. This would have had a knock-on effect on the circulation of the surface ocean, which is driven by the winds. While the Trade Winds and east–west flowing ocean currents beneath them would still have existed in the tropics, Hay suspected that at higher latitudes there would have been '*a chaotic pattern of giant eddies generated by storms*'[95]. Without the steady westerly winds of the mid latitudes, the vertical structure of the ocean that we are familiar with would have broken down: '*no great surface gyres, no subtropical and polar frontal systems, no clear separation between surface and deep waters, no "Great Conveyor"*'[95]. Hay went on to suppose that '*upwelling would have depended on the development of cyclonic eddies, which pump water upward*'[95]. That situation would have persisted until the cooling towards the modern ocean that took place at the end of the Eocene, of which more in Chapter 7. Hay's vision begs the question: By how much might the jet streams have moved position?

Crowley and North used numerical models to suggest that such displacements were probably minor[11].

Surprising though it may seem, Hay's apocalyptic vision of ocean and climate change did not occur to Robinson, or Parrish or Scotese and me when we constructed our maps of past climate, since we basically used annual pressure models derived from the modern climate to derive our palaeoclimatic maps. There was a danger in that approach, because the Cretaceous world was so different from today's. As ever, concepts evolve with time. Proof of the pudding would lie not in mental concepts, which might be based on unsound premises, but in the development of numerical models of the atmosphere and ocean based on sound physical principles.

6.8 Numerical Models Make their Appearance

By the very early 1980s, climatologists were using a brand new tool – the numerical general circulation model (GCM) – to simulate the behaviour of the present climate system. Such models can also be used to explore the relationships between climate and geology in the past[105]. Some global warming contrarians like to portray these numerical models as computer games. Games are designed to allow you to pit your wits against a series of known obstacles in order to win. GCMs are different. Climate scientists use them to find out how the climate system works and to discover what hidden properties of the system emerge when they are run for long periods, such as whether the climate tips from one stable state to another as warming continues.

I became familiar with the operation of numerical models when I joined the United Kingdom's Institute of Oceanographic Sciences Deacon Laboratory as its director in 1988 and found myself responsible, among other things, for oversight of the Southern Ocean modelling project FRAM (the Fine Resolution Antarctic Model). This was the first high-resolution, ocean-scale model capable of simulating typical oceanic eddies of no more than about 100 km across. There was no way at the time that we could gain a comprehensive understanding of how the Southern Ocean worked from the mere 100 years' worth of scattered ocean data points we possessed. In that remote region, they were far too sparsely distributed in time and space. But, given those data points and certain other starting conditions, FRAM could apply natural laws, such as the first law of thermodynamics and Newton's three laws of

motion, at closely spaced points on a 27 km grid, and at several levels down through the ocean, to show precisely how the Southern Ocean worked at all levels through time. It was as if the static school atlas of ocean currents had suddenly come alive. We could see in real time the sinuous motions of currents and the spinning of eddies[106]. Comparing the output to sea surface temperatures as seen from an ocean-observing satellite showed that the model results were very close to the real world. FRAM really did show how the Southern Ocean worked[107]. It was a breakthrough.

Such models, of the ocean, the atmosphere or both combined, provide us with a unique and verifiable means of connecting widely scattered data points and of understanding why the data are distributed the way they are. More than that, they tell us where to go to test ideas about how the ocean circulates or how the climate system works. They are vital aids. For instance, trying to sample every square metre of the ocean so as to understand its circulation is simply impossible. It can only be done from expensive research ships or through a massive and costly collection of autonomous floats and data buoys, and we have to remember that the ocean covers 72% of the surface of the planet! Satellites alone will not do the trick, because they cannot see below the ocean's surface.

Michael Crucifix of the Institut d'Astronomie et de Géophysique G. Lemaître, at the Université Catholique de Louvain, tells us, '*The aim of climate modeling is to understand past changes in climate that are currently unexplained and to be able to predict successfully the future evolution of climate*'[108]. It is a myth that the system is too chaotic for us to do that. As John Barrow explains· '*The standard folklore about chaotic systems is that they are unpredictable ... [but in fact] classical ... chaotic systems are not in any sense intrinsically random or unpredictable ... An important feature of chaotic systems is that, although they become unpredictable when you try to determine the future from a particular uncertain starting value, there may be a particular stable statistical spread of outcomes after a long time, regardless of how you started out. The most important thing to appreciate about these stable statistical distributions of events is that they often have very stable and predictable average behaviours*'[109]. For an example, look at Boyle's law, **PV/T = a constant**, where P is pressure, V volume and T temperature. These are the average properties of a confined gas, comprising a number of molecules whose interactions are unpredictable. '*The lesson of this simple example is that chaotic* *systems can have stable, predictable, long-term, average behaviours'* [109].

Crucifix explained how we use this understanding in modelling climate: *'The ocean-atmosphere-cryosphere-biosphere system is a complex system in the sense that it is made of different components that may interact with each other on a very wide range of time-scales … These interactions are generally nonlinear, that is the response is not proportionate to the amplitude of the excitation. A physical system with at least three components interacting nonlinearly with each other may be chaotic … In other words, its evolution cannot be predicted accurately beyond a certain time horizon because any error on the initial conditions grows exponentially with time. The atmosphere is chaotic. This is the reason we cannot forecast weather much beyond about 6 days. Yet, we can predict global warming. Indeed, conservation of energy, heat, and momentum makes it possible to predict the general evolution of a chaotic system in statistical terms. This statistical description of weather is nothing but the definition of climate'* [108].

For many geologists, this is a new world, brought to us courtesy of the massive increase in computing power since early 1980s. As John Barrow reminds us *'The advent of small, inexpensive, powerful computers with good interactive graphics has enabled large, complex, and disordered situations to be studied observationally – by looking at a computer monitor. Experimental mathematics is a new tool. A computer can be programmed to simulate the evolution of complicated systems, and their long-term behaviour observed, studied, modified and replayed. By these means, the study of chaos and complexity has become a multidisciplinary subculture within science. The study of the traditional, exactly soluble problems of science has been augmented by a growing appreciation of the vast complexity expected in situations where many competing influences are at work'* [109]; for example, in the climate system. **Mathematics is essential to understanding the complexities of the climate system.**

Mathematical models of the climate system are not reality. Nor are they perfect. But they are useful. Uncertainties arise for several reasons. First, they encompass the interaction between components of very different time scales, ranging from cloud formation and precipitation, on the scale of hours, to long-lived ice sheets. Second, in order to be addressed efficiently, the operations of the different elements of the climate system must be simplified. Third, the horizontal and vertical spacing of the points on the global grid, dictated by the capacity of

the computer, restricts the resolution of the outputs. Early GCMs were also limited because computer power was too small to simulate the circulation of both the ocean and the atmosphere.

Given these limitations, model outputs should be seen as aids to understanding how past climate systems worked [105]. They are tools grounded in the application of fundamental laws. We can verify their outputs by comparing them to climatic indicators from the sedimentary record. Where there are discrepancies between record and output, the challenge is to work out which is wrong. For example, a disagreement between model output and oxygen isotope data may come about because the isotope data come from species that live deep in the water column, while the model represents sea surface temperatures.

GCMs have an advantage over the conceptual models of Robinson and Parrish in that numerical modellers can tweak the controlling parameters of their model to see what effect each has. From that, they can evaluate the sensitivity of the climate to different controls, providing insights into the operation of the climate system through time. Without coupled ocean–atmosphere GCMs, it would be impossible for geologists to relate knowledge about meteorological and oceanographic processes to geological information about past climates. In in the remainder of this chapter, we will ignore the effects on climate of CO_2, which we will explore in detail later.

Apparently, the first to use GCMs in analysing Cretaceous climate was Eric James Barron (1951–) (Box 6.9).

As is obvious from the titles of his papers in 1981 ('Ice-Free Cretaceous Climate? Results from Model Simulations' [110]) and 1983 ('A Warm Equable Cretaceous: The Nature of the Problem' [111]), Barron was fascinated with the question of why the Cretaceous was warm and ice-free. Could it have to do with the distribution of land masses, with the opening of oceanic gateways (allowing warm water to penetrate poleward) or with the growth of mountain chains as continents continued to split apart? In 1984, together with Warren Washington, Barron used an atmospheric GCM, combined with a simple energy-balance model, to suggest that Cretaceous geography alone could have warmed the global surface temperature by $4.8\,°C$ [112]. The actual temperatures were warmer, so geography was not the sole forcing factor. The results might have been slightly different had he coupled his atmospheric model to an ocean model, but that required more computing power than was available at the time.

Box 6.9 Eric James Barron.

Getting his PhD in Oceanography from the University of Miami in 1980, Barron went to sea in July and August that same year with palaeoclimatology expert William (Bill) Hay (1934–), then of the Joint Oceanographic Institutions Office and later of the University of Colorado. They were participants in *Glomar Challenger*'s shipboard scientific party on leg 75 of the DSDP, which looked at the accumulation of organic matter in the Cretaceous and Cenozoic sediments of the Angola Basin and the Walvis Ridge in the southeast Atlantic. That same year, Barron joined the US National Center for Atmospheric Research (NCAR), in Boulder, CO, where he was able to combine his palaeogeographic and palaeoclimatic interests and test his ideas on Cretaceous climate with GCMs. In due course, he would become NCAR's director (2008–09), the president of Florida State University (2009–14), and finally the president of Penn State University. As a young scientist, Barron benefitted from working closely with two scientific luminaries. One was Bill Hay, whose many scientific contributions were recognised by the award of the Twenhofel Medal of the Society of Sedimentary Geology in 2006. The other was leading-edge climate modeller Steve Schneider (1945–2010), who was then the deputy-director of NCAR. Barron co-wrote with Judy Parrish a useful primer on 'Paleoclimates and Economic Geology'[3], introducing geologists to the potential applications of GCMs to palaeoclimate studies.

Investigating the effect of the opening mid Cretaceous seaways, in 1986 Barron used a slightly more advanced GCM to show that a well-developed zonal tropical rain belt, located 10° south of the Equator in January and 20° north of it in July, developed when the ITCZ was located over the zonal Tethyan Ocean. '*Clearly*', he said, '*the zonal subtropical ocean has influenced the general circulation pattern through the importance of latent heating in driving the circulation*'[105]. The creation of seaways, like the zonal equatorial ocean evident in Figure 6.9, provided a source of rain, which then fell on Pangaea's formerly dry interiors. His model showed that there would have been a mid-latitude rain belt at around 40° north and south of the Equator, and that it would have extended to 50° in both hemispheres in their respective summers. Palaeoclimatic evidence in the form of widespread laterites, bauxite and kaolin-rich clays confirms an increase in precipitation in mid latitudes at the time, as do the widespread coals of that age in North America[105].

Evaluating the effects of topography on atmospheric circulation and precipitation patterns, Barron found that '*The greater the starting temperature and relative humidity, the more topography will influence temperature and evaporation-precipitation patterns … If the topographic expression is high enough, evaporites can be formed in the lee of a mountain range at <u>any</u> latitude. During warmer climatic periods, the role of topography may be accentuated. The greatest potential for extensive evaporites is in the lee of a mountain range in the tropics where temperatures and relative humidities are high*'[105].

Like Parrish and myself, Barron tried to see if his numerical models could predict the likely location of upwelling conditions, and hence of organic rich rocks. Indeed, they could. Even so, Barron accepted that his GCM had several limitations for that purpose, including crude coastlines, a lack of seasons, poor representation of ocean circulation and a lack of bathymetry. Of course, the same limitations applied to the outputs of the Parrish model[69] and the Scotese and Summerhayes model[90].

Barron was also keen to use a GCM to test Pamela Robinson's notion that Köppen's arid zones would increase in latitude from west to east (Figure 6.7), just as happens today in Asia (Figure 6.3)[113]. Carrying out a number of experiments with Bill Hay, he found that mountainous areas and high plateaus distort zonal climate boundaries. In none of their experiments did zonal climate boundaries slope steeply poleward from east to west, '*suggesting that the "standard climatic pattern" attributed to Köppen may be an artefact of the topography and disposition of contemporary continents, and more related to the Tibetan-Himalayan and North American uplifts and to the configuration of South America than to the general atmospheric circulation*'[113]. Without the Tibetan Plateau, '*Climates may have been more zonal in the past than they are today*'[113].

This tends to confirm what is known from the distribution of palaeoclimatic indicators (Figures 5.3 and 6.1)[27]. What of Bill Hay's suggestion, then, that the Hadley

Cell could have expanded its range northward when the Equator–pole thermal gradient was much lower in the warm climates of the mid Cretaceous[95]? To test that idea, H. Hasegawa of the University of Tokyo and colleagues mapped the distribution of palaeo-desert deposits and palaeo-wind directions. Where air descending from high altitude in the northern limb of the Hadley Cell reaches the surface, it diverges to the north, forming the mid-latitude westerlies, and to the south, forming the easterly Trade Winds. Hasegawa's maps showed that the zone of divergence shifted poleward during the early and late Cretaceous, suggesting poleward expansion of the Hadley Cell[114]. However, it shifted equatorward during the hot mid-Cretaceous 'super greenhouse' period (of which more later), suggesting shrinkage of the Hadley Cell at that time, against Hay's expectation.

Fred Ziegler also realised the importance of including such factors as topography in a GCM. He presented a paper on this topic with co-author John Kutzbach at a Royal Society discussion meeting in London in 1993[115]. Kutzbach hails from the Center for Climatic Research at the Gaylord Nelson Institute for Environmental Studies in Madison, WI. His contributions were recognised by election to the US National Academy of Sciences in 2006 and by the award of the American Geophysical Union's Revelle Medal in 2006 and of the European Geophysical Society's Milankovitch Medal in 2001.

Using a GCM to model the climate of the late Permian, Ziegler and Kutzbach found that adding mountains, plateaus, inland seas and large lakes to palaeogeographic base maps produced outcomes different from model simulations lacking such features. Mountains and plateaus became focal points for enhanced precipitation, intensifying monsoonal circulation. Inland seas and lakes damped the seasonal range of mid-continental temperature. Agreement between Kutzbach's model results and Ziegler's palaeoclimate data was much better for the model with lakes and inland seas than for the one without. Failure to include lakes and inland seas may explain the extreme seasonality of continental interiors suggested by models lacking those constraints.

Anticipating that topography was likely to have modified climate in the past, much as it does today, a team from Chevron experimented with different topographies in their palaeoclimate model runs in 1992[116]. Their most convincing results emerged from the model containing mountain ranges with variable heights up to 3 km. Inputting more

simplified (lower) topography produced more simplified global circulation patterns, much as Ziegler and Kutzbach found. Evidently, palaeotopography provides an important boundary condition in applying numerical models.

Nevertheless, uncertainties in modelling remain; not least, for example, in establishing past palaeogeography and palaeotopography accurately. Equally, global circulation models suffer from having a high spatial scale (e.g. a 300 km grid), which makes them incapable of resolving regional climatic patterns accurately. Although GCMs have evolved to the point of being able to consider feedback from both land ice and vegetation, making them more like Earth system models and allowing more realistic coupling between the physical climate system and the biosphere, assumptions based on present-day vegetation are not likely to have been applicable to times prior to the evolution of angiosperms (130 Ma age) and the expansion of grasses (34 Ma ago)[29].

This finding underscores Barron's observation that we should see models as 'thought experiments'. As the palaeoclimate modeller Paul Valdes of the University of Bristol points out, even the most comprehensive numerical model will not include every detail of the ocean and atmosphere that might be important for the climate, especially at regional and local scales[117]. Then, too, the models may not include all likely forcing factors (such as atmospheric composition) appropriately. Finally, because the geological record itself is more gap than record, it may be difficult to find the field data required to test (or, in model speak, 'verify') the results supplied.

The shortcomings of palaeoclimate modelling have lessened with time, as the modellers' skills have improved. The topic was reviewed in 2011 by a committee of the US National Research Council, led by Isabel Montañez of the University of California at Davis and Richard Norris of Scripps Institution of Oceanography[103]. Remarking on the abundant evidence for anomalous polar warmth during past greenhouse periods (e.g. the middle Cretaceous to Eocene and the Pliocene), they pointed out that '*To date, climate models have not been able to simulate this warmth without invoking greenhouse gas concentrations that are notably higher than proxy estimates … [prompting] modeling efforts to explain high latitude warmth through vegetation … clouds … intensified heat transport by the ocean … and increased tropical cyclone activity … The ability to successfully model a reduced latitudinal temperature gradient state, including anomalous polar warmth, presents a first-order check on the efficacy of climate models as the basis for predicting future greenhouse conditions*[103]. Mismatches between model outputs, modern observations and palaeoclimate records may suggest important deficiencies in scientific knowledge of climate and the construction of climate models[103]. Past analyses of warm worlds might help to resolve these disparities.

The Montañez-Norris committee was concerned about tropical climate stability and the answer to the question, were the tropics warmer in the past than now, or do they have some natural 'thermostat' that keeps them to a limit of about 30–32 °C, like the tropical Pacific temperatures of today[103]? They found that tropical ocean temperatures during past greenhouse times were much warmer than modern tropical maxima – possibly as high as 42 °C – and that the temperatures of tropical continental interiors were anomalously high (30–34 °C). A tropical thermostat seemed unlikely under the circumstances. Clearly, we need to know more about the tropical climates of these past warm worlds in order not only to better understand Earth's climate system, but also to know what to expect of a warmer world in the future.

Climate models have advanced a good deal since Barron's day, as has the power of computers. Nowadays, climate models incorporate a very wide range of parameters that might have some effect on the climate system, including soil dynamics, vegetation dynamics, ocean biogeochemistry, ice sheet dynamics and river hydrology[108]. Although the coarse scale of global grid points in GCMs (300 km) prevents them from resolving regional details, the outputs of a global model can be used to drive regional dynamical models that resolve much smaller length scales (30–50 km), enabling details of the climates of specific regions to be obtained.

To assist in understanding past climate change, and to test the ability of climate models to match climate data, various climate models have been applied to past time slices for which palaeoclimate data were both reliable and abundant. The international modelling community has pooled its efforts through the Paleoclimate Modeling Intercomparison Project (PMIP), which simulated the climates of the mid Holocene (6 Ka ago) and the Last Glacial Maximum (21 Ka ago) with different atmospheric GCMs and organised systematic comparisons between model outputs and palaeoclimate data[118]. PMIP showed *'that climate models correctly reproduce a number of observed features of the mid-Holocene climate'*[108].

There is now an abundant literature on the newly emergent discipline of climate modelling and its application to past climates, which concerns us most here. In effect, it forms a new subfield of science: theoretical palaeoclimatology. Thomas Crowley and Gerald North, then resident at Texas A&M University, explained in 1991 that, *'For the first time, physicists and atmosphere and ocean scientists are applying quantitative climate models to interpret many fascinating observations uncovered by geologists. In some cases we have a good understanding of the observed changes. In other cases there is a considerable gap between models and data'*[11]. It is the task of the palaeoclimatologist to resolve such discrepancies. *'Responsible use of climate models can help in the formulation of theories of climate change and in some cases such models can lead the geologist to collect and analyze data in new ways'*[11]. Far from being just 'computer games', palaeoclimate models provide valuable physical insights into how past climates may have operated.

Despite that, palaeoclimate models do have one interesting problem, as Paul Valdes reminds us: *'If anything, the models are underestimating change, compared with the geological record. According to evidence from the past, the Earth's climate is sensitive to small changes, whereas the climate models seem to require a much bigger disturbance to produce abrupt change'*[119]. This means that *'Simulations of the coming century with the current generation of complex models may be giving us a false sense of security'*[119]. Caveat emptor.

6.9   From Wegener to Barron

Why did it take so long to reach agreement on how the climate had changed with time? Geologists were not prepared to agree that Köppen and Wegener might have been right until the palaeomagnetists entered the arena in the late 1950s and early 1960s, with geophysicists like Briden and Irving confirming that the climate-sensitive deposits of the past did indeed fall into climate zones much like those of today. The advent of the plate tectonics and seafloor spreading paradigm in the late 1960s radically changed peoples' view of what was possible – and, indeed, probable. That helps to explain why it took almost 50 years until the landmark NATO meeting in 1972, when Robinson broke new ground by using conceptual palaeoclimate models to show how continental displacement affected climate-sensitive deposits on land, basically confirming the validity of the Köppen and Wegener model and applying it to 'modern' plate tectonic reconstructions for the Permian and Triassic. Her work implied that in order for geologists to understand past climates, they first had to learn and apply the principles of meteorology. As we shall see later, they also had to learn and apply the principles of oceanography.

New techniques, like oxygen isotope palaeothermometry, added another new dimension to palaeoclimate studies from the 1950s on. Our understanding of orbital controls on climate grew with the work of Milankovitch. We began amassing palaeoclimatic data from land, with Ziegler's Paleogeographic Atlas Project, from the mid 1970s on, and from the ocean, with the advent of the Deep Sea Drilling Project and its successors, in 1968. By the mid 1980s, 60 years after Köppen and Wegener, the geography of past continental fragments and the original geographic position of climatically sensitive sediments could be established with a fair degree of certainty. The advent of numerical modelling of the climate system introduced a thoroughly modern understanding of how the climate system works, further improving our appreciation of how past climates evolved. Nevertheless, there is more to learn about how those deposits formed, as we shall see.

Much of the work reviewed in this chapter dates from before the mid 1980s, when there was little or no discussion in the geological community of the possible role of CO_2 as a modifier of Earth's climate through time. Influenced by Lyell's thinking, most geologists simply attributed the glaciation of the Permo-Carboniferous and the warmth and high sea levels of the mid Cretaceous to the changing positions of the continents. More land over the pole led to cooling; more land over the Equator led to warming. To many of them, the results of increasingly detailed palaeoclimatic studies summarised in this chapter confirmed Lyell's notion that climatic zones much like those of today must have prevailed through time. Much effort was devoted to considering how those climatic zones might have been modified by geographic changes like the development of seaways and mountain belts, with their attendant effects on winds and ocean currents. Little if any thought was given to the possible role of the changing composition of the atmosphere. The question of why the Cretaceous was so warm remained unanswered.

Having examined the evidence for climate change from the Carboniferous to the Cretaceous, it is now time to turn our attention to the Cenozoic: the past 65 Ma. During much of this time, the Earth cooled towards the Ice Age of the Pleistocene. In Chapter 7, we review that change.

References

1. Lisitzin, A.P. (1972) *Sedimentation in the World Ocean* Society for Sedimentary Geology, Tulsa, OK.
2. Kennett. J.P. (1982) *Marine Geology*. Prentice-Hall, Upper Saddle River, NJ.
3. Parrish, J.T. and Barron, E.J. (1986) *Paleoclimates and Economic Geology*. Society for Sedimentary Geology, Tulsa, OK.
4. Nairn, A.E.M. (1961) The scope of palaeoclimatology. In: *Descriptive Palaeoclimatology* (ed A.E.M. Nairn). Interscience Publishers, New York, pp. 1–7.
5. Fairbridge, R.W. (1964) The importance of limestone and its Ca/Mg content to palaeoclimatology. In: *Problems in Palaeoclimatology* (ed. A.E.M. Nairn). Proceedings of the NATO Palaeoclimates Conference, University of Newcastle-upon-Tyne, January 1963. Interscience Publishers, London, pp. 431–477.
6. Flint, R.F. (1961) Geological evidence of cold climate. In: *Descriptive Palaeoclimatology* (ed. A.E.M. Nairn). Interscience Publishers, New York, 140–155.
7. Green, R. (1961) Palaeoclimatic significance of evaporites. In: *Descriptive Palaeoclimatology* (ed. A.E.M. Nairn). Interscience Publishers, New York, pp. 61–88.
8. Opdyke, N.D. (1961) The palaeoclimatological significance of desert sandstone. In: *Descriptive Palaeoclimatology* (ed. A.E.M. Nairn). Interscience Publishers, New York, pp. 45–60.
9. Runcorn, S.K. (1964) Paleowind directions and palaeomagnetic latitudes. In: *Problems in Palaeoclimatology* (ed. A.E.M. Nairn). Proceedings of the NATO Palaeoclimates Conference, University of Newcastle-upon-Tyne, January 1963. Interscience Publishers, London, pp. 409–419.
10. Nairn, A.E.M. (1961) *Descriptive Palaeoclimatology*. Interscience Publishers, London, New York.
11. Crowley, T.J. and North, G.R. (1991) *Paleoclimatology*. Oxford Monographs on Geology and Geophysics 18. Oxford University Press, Oxford.
12. Parrish, J.T. (1998) *Interpreting Pre-Quaternary Climate from the Geologic Record*. Columbia University Press, New York.
13. Vaughan, A.P.M. (2007) Climate and geology – a Phanerozoic perspective. In: *Deep-Time Perspectives on Climate Change: Marrying the Signal from Computer Models and*

Biological Proxies (eds M. Williams, A.M. Haywood, F.J. Gregory and D.N. Schmidt). Micropalaeontological Society Special Publication, The Geological Society, London, pp. 5–59.

14. Markwick, P.J. (2007) The palaeogeographic and palaeoclimatic significance of climate proxies for data-model comparisons. In: *Deep-Time Perspectives on Climate Change: Marrying the Signal from Computer Models and Biological Proxies* (eds M. Williams, A.M. Haywood, F.J. Gregory and D.N. Schmidt). Micropalaeontological Society Special Publication, The Geological Society, London, pp. 251–312.

15. Wegener, A. (1920) *Die Entstehung der Kontinente und Ozeane*, 2nd edn. Die Wissenschaft, Band 66. Vieweg und Sohn, Braunschweig.

16. Wegener, A. (1924) *The Origin of Continents and Oceans*. Methuen, London.

17. Lewis, J.M. (1996) Winds over the world sea: Maury and Köppen. *Bulletin of the American Meteorological Society* **77** (5), 935–952.

18. Greene, M.T. (2008) Köppen, Wladimir Peter. In: *Complete Dictionary of Scientific Biography*. Available from: http://www.encyclopedia.com/topic/Wladimir_Peter _Koppen.aspx (last accessed 29 January 2015).

19. Köppen, W.P. (1923) *Klimate der Erde*. Walter de Gruyter, Berlin and Leipzig; revised 2nd edn (1931) *Grundriss der Klimakunde*. Walter de Gruyter, Berlin and Leipzig.

20. Peel, M.C., Finlayson, B.L. and McMahon, T.A. (2007) Updated world map of the Koppen-Geiger climate classification. *Hydrological Earth Systems Science* **11**, 1633–1644 (with free download of the current map of climate zones).

21. Köppen, W. (1921) [Polar wandering, movement of the continents and climate history]. *Dr A. Petermanns Mitteilungen aus Justus Perthes' Geographischer Anstalt* **67** (1–8), 57–63.

22. Demhardt, I.J. (2006) Alfred Wegener's hypothesis on continental drift and its discussion in Petermanns Geographische Mitteilungen. *Polarforschung* **75** (1), 29–35.

23. Köppen, W. and Wegener, A. (1924) *Die Klimate der geologischen Vorzeit*. Borntraeger, Berlin. Note that one sees this commonly referred to as published in English as *Climates of the Geological Past* by Van Nostrad, London, 1963, which is an erroneous reference to a book with a similar title: Schwarzbach, M. (1963) *Climates of the Past: An Introduction to Paleoclimatology*. Van Nostrad, London, translated from the German 2nd edition by R. Muir. Köppen and Wegener's *The Climate of the Geological Past* will be reissued in German and English in 2015.

24. Du Toit, A.L. (1937) *Our Wandering Continents: An Hypothesis of Continental Drifting*. Oliver and Boyd, London.

25. King, L.C. (1961) The Palaeoclimatology of Gondwanaland during the Palaeozoic and Mesozoic Eras. In: *Descriptive Palaeoclimatology* (ed. A.E.M. Nairn). Interscience Publishers, New York, pp. 307–331.

26. Nairn, A.E.M. (1964) *Problems in Palaeoclimatology*. Proceedings of the NATO Palaeoclimates Conference, University of Newcastle upon Tyne, January 1963. Interscience Publishers, London.

27. Briden, J.C. and Irving, E. (1964) Palaeolatitude spectra of sedimentary palaeoclimatic indicators. In: *Problems in Palaeoclimatology* (ed. A.E.M. Nairn). Proceedings of the NATO Palaeoclimates Conference, University of Newcastle upon Tyne, January 1963. Interscience Publishers, London, pp. 199–224.

28. Bucher, W.H. (1964) The third confrontation. In: Problems in Palaeoclimatology (ed. A.E.M. Nairn). Proceedings of the NATO Palaeoclimates Conference, University of Newcastle upon Tyne, January 1963. Interscience Publishers, London, pp. 3–9.

29. Dorf, E. (1964) The use of fossil plants in palaeoclimatic intepretation. In: *Problems in Palaeoclimatology* (ed. A.E.M. Nairn). Proceedings of the NATO Palaeoclimates Conference, University of Newcastle upon Tyne, January 1963. Interscience Publishers, London, pp. 13–31.

30. Urey, H.C. (1947) The thermodynamic properties of isotopic substances. *Journal of the Chemical Society* **61**, 562–581.

31. Urey H.C. (1948) Oxygen isotopes in nature and in the laboratory. *Science* **108**, 489–496.

32. Urey, H.C., Lowenstam, H.A., Epstein, S. and McKinney, C.R. (1951) Measurement of paleotemperatures and temperatures of the Upper Cretaceous of England, Denmark, and the southeastern United States. *Bulletin of the Geological Society of America* **62** (4), 399–416.

33. Arnold, J.R., Bigeleisen, J. and Hutchison, C.A. (1995) *Harold Clayton Urey 1893–1981, Biographical Memoir*. National Academies Press, Washington, DC, pp. 363–411.

34. Nier, A.O. (1947) A mass spectrometer for isotope and gas analysis. *Review of Scientific Instruments* **19** (6), 398–411.

35. McKinney, C.R., McCrea J.M., Epstein, S., Allen, H.A. and Urey, H.C. (1950) Improvements in mass spectrometers for the measurement of small differences in isotope abundance ratios. *Review of Scientific Instruments* **21** (8), 724–230.

36. Lowenstam, H.A. (1964) Palaeotemperatures of the Permian and Cretaceous Periods. In: *Problems in Palaeoclimatology* (ed. A.E.M. Nairn). Proceedings of the NATO Palaeoclimates Conference, University of Newcastle upon Tyne, January 1963. Interscience Publishers, London, pp. 227–248.

37. Emiliani, C. (1961) The temperature decrease of surface sea-water in high latitudes and of abyssal-hadal water in

open oceanic basins during the past 75 million years. *Deep Sea Research* **8**, 144–147.

38. Bradley, W.H. (1929) The varves and climate of the Green River Epoch. *US Geological Survey Professional Paper* **158**, 87–110.

39. Wells, A.K. and Kirkaldy, J.F. (1959) *Outline of Historical Geology*. Thomas Murby, London.

40. Krumbein, W.C. and Sloss, L.S. (1958) *Stratigraphy and Sedimentation*. W.H. Freeman, San Francisco, CA.

41. Klein, G. de V. and Willard, D. (1989) Origin of the Pennsylvanian coal-bearing cyclothems of North America. *Geology* **17** (2), 152–155.

42. Leeder, M.R. and Strudwick, A.E. (1987) Delta-marine interactions: a discussion of sedimentary models for Yoredal-type cyclicity in the Dinantian of Northern England. In: *European Dinantian Environments* (eds J. Miller, A.E. Adams and V.P. Wright). John Wiley & Sons, Chichester, pp. 115–130.

43. Wanless, H.R. and Shepard, F.P. (1936) Sea level and climatic changes related to late Paleozoic cycles. *Bulletin of the Geological Society of America* **47**, 1177–1206.

44. Vail, P.R. and Mitchum, R.M. (1977) Seismic stratigraphy and global changes of sea level. *American Association of Petroleum Geologists Memoir* **26**, 49–212.

45. Tucker, M.E., Gallagher, J. and Leng, M.J. (2009) Are beds in shelf carbonates millennial-scale cycles? An example from the mid-Carboniferous of Northern England. *Sedimentary Geology* **214**, 19–34.

46. Pilgrim, L. (1904) *Versuch einer rechnerischen behandlung des eiszeitenproblems*. Jahreshefte fur Vaterlandische Naturkunde in Wurtemberg **60**.

47. Milankovitch, M. (1920) *Theorie Mathématique des Phénoménes Thermiques Produits par la Radiation Solaire*. Paris, Gauthier-Villars.

48. Milankovitch, M. (1941) *Kanon der erdbestrahlung und seine anwendung auf das eiszeitenproblem*. Special Publications 132, Section of Mathematics and Natural Sciences 33. Königliche Serbische Akademie, Belgrade.

49. Milankovitch, M. (1969) *Canon of Insolation and the Ice Age Problem*. Israel Program for Scientific Translations, Jerusalem.

50. Milankovitch, M. (1998) *Canon of Insolation and the Ice-Age Problem*. Zavod za udzbenike i nastavna sredstva, Belgrade.

51. Grubic, A. (2006) The astronomic theory of climatic changes of Milutin Milankovich. *Episodes* **29** (3), 197–203.

52. Penck, A. and Bruckner, E. (1909–11) *Die Alpen im Eiszeitalter*, Vols 1–3. Tauchnitz, Leipzig.

53. Laskar, J., Robutel, P., Joutel, F., Gastineau, M., Correia, A.C.M. and Levrard, B. (2004) A long-term numerical solution for the insolation quantities of the Earth. *Astronomy and Astrophysics* **428**, 261–285.

54. Laskar, J., Fienga, A., Gastineau, M. and Manche, H. (2010) A new orbital solution for the long-term motion of the Earth. *Astronomy and Astrophysics* **532**, A 89.

55. Miall, A.D. (2010) *The Geology of Stratigraphic Sequences*, 2nd edn. Springer, New York.

56. Tarling, D.H. and Runcorn, S.K. (1973) *Implications of Continental Drift to the Earth Sciences*, Vol. 1. Academic Press, London.

57. Milner, A.C. (2004) Robinson, Pamela Lamplugh (1919–1994). *Oxford Dictionary of National Biography*. Oxford University Press, Oxford.

58. Robinson, P.L. (1973) Palaeoclimatology and continental drift. In: *Implications of Continental Drift to the Earth Sciences*, Vol. 1 (eds D.H. Tarling, and S.K. Runcorn). Academic Press, London, pp. 451–476.

59. Kräusel, R. (1961) Palaeobotanical evidence of climate. In: *Descriptive Palaeoclimatology* (ed. A.E.M. Nairn). Interscience Publishers, New York, pp. 227–254.

60. Chaloner, W.G. and Creber, G.T. (1973) Growth rings in fossil woods as evidence of past climates. In: *Implications of Continental Drift to the Earth Sciences*, Vol. 1 (eds D.H. Tarling and S.K. Runcorn). Academic Press, London, pp. 425–437.

61. Chaloner, W.G. and Creber, G.T. (1990) Do fossil plants give a climate signal? *Journal of the Geological Society of London* **147**, 343–350.

62. Woodward, F.I. (1987) Stomatal numbers are sensitive to increases in CO_2 from pre-industrial levels. *Nature* **327**, 617–618.

63. Koch, J.T. and Frank, T.D. (2011) The Pennsylvanian-Permian transition in the low-latitude carbonate record and the onset of major Gondwanan glaciation. *Palaeogeography, Palaeoclimatology, Palaeoecology* **308**, 362–372.

64. Green, R. (1961) Palaeoclimatic significance of evaporates. In: *Descriptive Palaeoclimatology* (ed. A.E.M. Nairn). Interscience Publishers, New York, pp. 61–88.

65. Cox, C.B. (1974) Vertebrate palaeodistributional patterns and continental drift. *Journal of Biogeography* **1**, 75–94.

66. Ziegler, A.M., Scotese, C.R., McKerrow, W.S., Johnson, M.E. and Bambach, R.K. (1979) Paleozoic paleogeography. *Annual Review of Earth and Planetary Science* **7**, 473–503.

67. Scotese, C.R., Boucot, A.J. and McKerrow, W.S. (1999) Gondwanan palaeogeogaphy and palaeoclimatology. *Journal of African Earth Sciences* **28** (1), 99–114.

68. Ziegler, A.M. (1982) *The University of Chicago Paleogeographic Atlas Project: Background – Current Status – Future Plans*. Unpublished ms, Department of Geophysical Sciences, University of Chicago.

69. Parrish, J.T. and Curtis, R.L. (1982) Atmospheric circulation, upwelling, and organic-rich rocks in the Mesozoic and Cenozoic. *Palaeogeography, Palaeoclimatology, Palaeoecology* **40**, 31–66.

70. Parrish, J.T. (1982) Upwelling and petroleum source beds, with reference to the Palaeozoic. *American Association of Petroleum Geologists Bulletin* **66**, 750–754.

71. Parrish, J.T., Ziegler, A.M. and Humphreville, R.G. (1983) Upwelling in the Paleozoic Era. In: *Coastal Upwelling – Its Sediment Record, Part B, Sedimentary Records of Ancient Coastal Upwelling* (eds J. Thiede and E. Suess). NATO Conference Series IV, Marine Sciences. Plenum, New York and London, pp. 553–578.

72. Parrish, J.T., Ziegler, A.M. and Scotese, C.R. (1982) Rainfall patterns and the distribution of coals and evaporates in the Mesozoic and Cenozoic. *Palaeogeography, Palaeoclimatology, Palaeoecology* **40**, 67–101.

73. Rowley, D.B., Raymond, A.R., Parrish, J.T., Lottes, A.L., Scotese, C.R. and Ziegler, A.M. (1985) Carboniferous paleogeographic, phytogeographic and paleoclimatic reconstructions. *International Journal of Coal Geology* **5**, 7–42.

74. Sellwood, B.W. and Price, G.D. (1994) Sedimentary facies as indicators of Mesozoic palaeoclimate. In: *Palaeoclimates and their Modelling – with Special Reference to the Mesozoic Era* (eds J.R.L. Allen, B.J. Hoskins, B.W. Sellwood, R.A. Spicer and P.J. Valdes). Chapman & Hall, London, pp. 17–25 (originally published as *Philosophical Transactions of the Royal Society, Series B* **342**, 203–343 (1993); based on a discussion meeting held on 24–25 February 1993).

75. Ziegler, A.M., Parrish, J.M., Jiping, Y., Gyllenhaal, E.D., Rowley, D.B., Parrish, J.T., *et al.* (1994) Early Mesozoic phytogeography and climate. In: *Palaeoclimates and their Modelling – with Special Reference to the Mesozoic Era* (eds J.R.L. Allen, B.J. Hoskins, B.W. Sellwood, R.A. Spicer and P.J. Valdes). Chapman & Hall, London, pp. 89–97 (originally published as *Philosophical Transactions of the Royal Society, Series B* **342**, 203–343 (1993); based on a discussion meeting held on 24–25 February 1993).

76. Sims, M.J. and Ruffell, A.H. (1990) Climatic and biotic change in the late Triassic. *Journal of the Geological Society of London* **147**, 321–327.

77. Veevers, J.J. (1989) Middle/Late Triassic (230 ± 5 Ma) singularity in the stratigraphic and magmatic history of the Pangaean heat anomaly. *Geology* **17**, 784–787.

78. Frakes, L.A. and Crowell, J.C. (1969) Late Paleozoic glaciation, I. South America. *Bulletin of the Geological Society of America* **80**, 1007–1042.

79. Frakes, L.A. (1981) Late Paleozoic paleoclimatology. In: *Paleoreconstruction of the Continents* (eds M.W. McElhinny and D.A. Valencio). Geodynamics Series 2. American Geophysical Union, Washington, DC, pp. 39–44.

80. Crowell, J.C. and Frakes, L.A. (1970) Phanerozoic glaciation and the causes of ice ages. *American Journal of Science* **2688**, 193–224.

81. Müller, R.D., Sdrolias, M., Gaina, C., Steinberger, B. and Heine, C. (2008) Long-term sea-level fluctuations driven by ocean basin dynamics. *Science* **319**, 1357–1362.

82. Pitman, W. (1978) Relationship between eustacy and stratigraphic sequences of passive margins. *Bulletin of the Geological Society of America* **89**, 1389–1403.

83. Smith, D.G., Smith, A.G. and Funnell, B.M. (1994) *Atlas of Mesozoic and Cenozoic Coastlines*. Cambridge University Press, Cambridge.

84. Hallam, A. (1985) A review of Mesozoic climates. *Journal of the Geological Society of London* **142**, 433–445.

85. Hallam, A. (1994) Jurassic climates as inferred from the sedimentary and fossil record. In: *Palaeoclimates and their Modelling – with Special Reference to the Mesozoic Era* (eds J.R.L. Allen, B.J. Hoskins, B.W. Sellwood, R.A. Spicer and P.J. Valdes). Chapman & Hall, London, pp. 79–88 (originally published as *Philosophical Transactions of the Royal Society, Series B* **342**, 203–343 (1993); based on a discussion meeting held on 24–25 February 1993).

86. Frakes, L.A., Francis, J.E. and Syktus, J.I. (1992) *Climate Modes of the Phanerozoic – the History of the Earth's Climate over the Past 600 Million Years*. Cambridge University Press, Cambridge.

87. Littler, K., Robinson, S.A., Brown, P.R., Nederbragt, A.J. and Pancost, R.D. (2011) High sea surface temperatures during the early Cretaceous Epoch. *Nature Geoscience* **4**, 169–172.

88. Alley, N.F. and Frakes, L.A. (2003) First known Cretaceous glaciation: Livingston tillite member of the Cadna-Owie Formation, South Australia. *Australian Journal of Earth Science* **50**, 139–144.

89. Summerhayes, C.P. (1983) Sedimentation of organic matter in upwelling regimes. In: *Coastal Upwelling – Its Sediment Record, Part B, Sedimentary Records of Ancient Coastal Upwelling* (eds J. Thiede and E. Suess). NATO Conference Series IV, Marine Sciences. Plenum, New York and London, pp. 29–72.

90. Scotese, C.R. and Summerhayes, C.P. (1986) Computer model of paleoclimate predicts coastal upwelling in the Mesozoic and Cenozoic. *Geobyte* **1** (3), 28–44, 94.

91. Spicer, R.A., Rees, P.M. and Chapman, J.L. (1994) Cretaceous phytogeography and climate signals. In: *Palaeoclimates and their Modelling – with Special Reference to the Mesozoic Era* (eds J.R.L. Allen, B.J. Hoskins, B.W. Sellwood, R.A. Spicer and P.J. Valdes). Chapman & Hall, London, pp. 69–78 (originally published as *Philosophical Transactions of the Royal Society, Series B* **342**, 203–343 (1993); based on a discussion meeting held on 24–25 February 1993).

92. Spicer, R.A. and Parrish, J.T. (1990) Late Cretaceous-early Tertiary palaeoclimates of northern high latitudes: a quantitative view. *Journal of the Geological Society of London* **147**, 329–341.

93. Francis, J.E., Ashworth, A., Cantrill, D.J., Crame, J.A., Howe, J., Stephens, R., *et al.* (2008) 100 million years of Antarctic climate evolution: evidence from fossil plants. In: *Antarctica: A Keystone in a Changing World* (eds A.K. Cooper, P.J. Barrett, H. Stagg, B. Storey, E. Stump and W. Wise). National Academies Press, Washington, DC, pp. 19–27.

94. Kennett, J.P. (1982) *Marine Geology*. Prentice-Hall, Englewood Cliffs, NJ.

95. Hay, W.W. (2013) *Experimenting on a Small Planet: A Scholarly Entertainment*. Springer, New York.

96. Cronin, T.M. (2010) *Paleoclimates: Understanding Climate Change Past and Present*. Columbia University Press, Cambridge.

97. Huber, B.T., Norris, R.D. and MacLeod, K.G. (2002) Deep-sea paleotemperature record of extreme warmth during the Cretaceous. *Geology* **30** (2), 123–126.

98. MacLeod, K.G., Huber, B.T., Berrocoso, A.J. and Wendler, I. (2013) A stable and hot Turonian without glacial $\delta^{18}O$ excursions is indicated by exquisitely preserved Tanzanian foraminifera. *Geology* **41** (10), 1083–1086.

99. Cramer, B.S., Miller, K.G., Barrett, P.J. and Wright, J.D. (2011) Late Cretaceous-Neogene trends in deep ocean temperature and continental ice volume: reconciling records of benthic foraminiferal geochemistry ($\delta^{18}O$ and Mg/Ca) with sea level history. *Journal of Geophysical Research* **116**, C12023.

100. Forster, A., Schouten, S., Baas, M. and Damste, J.S.S. (2007) Mid-Cretaceous (Albian–Santonian) sea surface temperature record of the tropical Atlantic Ocean. *Geology* **35**, 919–922.

101. Vail, P.R., Mitchum, R.W. and Thompson, S. (1977) Seismic stratigraphy and global changes of sea level 4. global cycles of relative changes of sea level. *American Association of Petroleum Geologists Memoir* **26**, 82–97.

102. Haq, B.U., Hardenbol, J. and Vail, P.R. (1987) Chronology of fluctuating sea levels since the Triassic. *Science* **235**, 1156–1167.

103. Montañez, I.P., Norris, R.D., Algeo, T., Chandler, M.A., Johnson, K.R., Kennedy, M.J., *et al.* (2011) *Understanding Earth's Deep Past: Lessons for Our Climate Future*. National Academy Press, Washington, DC.

104. Bowman, V.C., Francis, J.E. and Riding, J.B. (2013) Late Cretaceous winter sea ice in Antarctica? *Geology* **41**, 1227–1230.

105. Barron, E.J. (1986) Mathematical climate models: insights into the relationship between climate and economic sedimentary deposits. In: *Paleoclimates and Economic Geology* (eds J.T. Parrish and E.J. Barron). Lecture Notes for Short Course 18. Society for Sedimentary Geology, Tulsa, OK, pp. 31–83.

106. Webb, D.J., Killworth, P.D., Coward, A.C. and Thompson, S.R. (1991) *The FRAM Atlas of the Southern Ocean*. Natural Environment Research Council, Swindon.

107. Sparrow, M.D., Heywood, K.J. and Brown, J. (1996) Current structure of the South Indian Ocean. *Journal of Geophysical Research: Oceans* **101** (C3), 6377–6391.

108. Crucifix, M. (2009) Modeling the climate of the Holocene. In: *Natural Climate Variability and Global Warming: A Holocene Perspective* (eds R.W. Battarbee and H.A. Binney). Wiley-Blackwell, Chichester and Oxford, pp. 98–122.

109. Barrow, J.D. (2010) Simple really: from simplicity to complexity – and back again. In: *Seeing Further: The Story of Science and the Royal Society* (ed. B. Bryson). Harper Press, London, pp. 360–383.

110. Barron, E.J., Thompson, S.L. and Schneider, S.H. (1981) Ice-free Cretaceous climate? Results from model simulations. *Science* **212**, 501–508.

111. Barron, E.J. (1983) A warm, equable Cretaceous: the nature of the problem. *Earth Science Reviews* **19**, 305–338.

112. Barron, E.J. and Washington, W.M. (1984) The role of geographic variables in explaining paleoclimates: results from Cretaceous climate model sensitivity studies. *Journal of Geophysical Research* **89**, 1267–1279.

113. Hay, W.W., Barron, E.J. and Washington S.L. (1990) Results of global atmospheric circulation experiments on an Earth with a meridional pole-to-pole continent. *Journal of the Geological Society of London* **147**, 385–392.

114. Hasegawa, H., Tada, R., Jiang, X., Suganuma, Y., Imsamut, S., Charusiri, P., *et al.* (2012) Drastic shrinking of the Hadley circulation during the mid-Cretaceous supergreenhouse. *Climate of the Past* **8**, 1323–1337.

115. Kutzbach, J.E. and Ziegler, A.M. (1994) Simulation of late Permian climate and biomes with an atmosphere-ocean model: comparisons with observations. In: *Palaeoclimates and their Modelling – with Special Reference to the Mesozoic Era* (eds J.R.L. Allen, B.J. Hoskins, B.W. Sellwood, R.A. Spicer and P.J. Valdes). Chapman & Hall, London, pp. 119–132 (originally published as *Philosophical Transactions of the Royal Society, Series B* **342**, 203–343 (1993); based on a discussion meeting held on 24–25 February 1993).

116. Moore, G.T., Sloan, L.C., Hayashida, D.N. and Umrigar, N.P. (1992) Paleoclimate of the Kimmeridgian/Tithonian (Late Jurassic) world: II. Sensitivity tests comparing three different paleotopographic settings. *Palaeogeography, Palaeoclimatology, Palaeoecology* **95**, 229–252.

117. Valdes, P.J. (1994) Atmospheric general circulation models of the Jurassic. In: *Palaeoclimates and their*

Modelling – with Special Reference to the Mesozoic Era (eds J.R.L. Allen, B.J. Hoskins, B.W. Sellwood, R.A. Spicer and P.J. Valdes). Chapman & Hall, London, pp. 109–118 (originally published as *Philosophical Transactions of the Royal Society, Series B* **342**, 203–343 (1993); based on a discussion meeting held on 24–25 February 1993).

118. Joussaume, S. and Taylor, K.E. (2007) Status of the Paleoclimate Modeling Intercomparison Project (PMIP). Available from: http://pmip.lsce.ipsl.fr/ (last accessed 29 January 2015).

119. Valdes, P. (2011) Built for stability. *Nature Geoscience* **4**, 414–416.

7

Into the Icehouse

7.1 Climate Clues from the Deep Ocean

Much of what we now know about climate change through the Cenozoic comes from studying the climate record in deep-ocean sediments. That's not entirely surprising, because the oceans are an important driver of climate change. They cover 72% of the Earth's surface and store huge amounts of solar energy, with the top 3.5 m containing as much heat as the entire atmosphere. They move this heat around the planet in giant ocean currents that girdle the Earth, transporting significant amounts of heat from the Equator to the poles. Europe is much warmer than Labrador, despite both lying at the same latitude, because of the heat transported north by the Gulf Stream. In effect, the ocean is the flywheel of the climate system, storing solar energy and moving it around much more slowly than would be the case if the Earth's surface were all land. Deep-ocean sediments carry the record of that oceanic behaviour, and hence of climate change.

We call students of the ocean 'oceanographers'. The science of oceanography came of age with the publication in 1942 of a magnificent reference work, *The Oceans*, by scientists at Scripps Institution of Oceanography: Harald Ulrik Sverdrup (1888–1957), an import from Norway, and Martin Johnson and Richard Fleming. While *The Oceans* helped to explain how the oceans worked, it also showed that we still didn't know much more about the distribution of sediments on the deep-ocean floor than had the scientists on the round-the-world expedition of HMS *Challenger* (1872–76). Much the same applied at the end of 1964, when I began my career as a marine geologist with the New Zealand Oceanographic Institute. That lack

of knowledge largely reflected the costs and difficulties of sampling the ocean floor: on average, it is about 4 km deep, and all that water gets in the way. Any breakthrough in our understanding of the role of the ocean in past climate change called for new technologies with which to sample the deep-sea floor.

Before the Second World War, weighted tubes known as gravity corers were dropped from ships to retrieve cores of deep-sea sediment of up to about 3 m long. Inspecting cores collected from the South Atlantic by the German *Meteor* Expedition of 1925–27, Wolfgang Schott (1905–1989) identified three distinct layers characterised by different species of planktonic foraminifera[1]: the upper layer contained abundant *Globorotalia menardii*, a species typical of warm surface waters; the second layer contained cold-water types; and the deepest layer contained *menardii* again. Schott deduced that the middle layer represented the last glaciation, when the Atlantic was cooler, and that the other two layers represented warm interglacials, the last one being the Holocene. Changes in planktonic foraminifera could serve as a sort of thermometer for past times.

A simple technological breakthrough 20 years later enabled us to expand on Schott's findings, when the Swedish geologist Börje Kullenberg (1906–1991) had the bright idea of adding an internal piston to avoid the compression of sedimentary layers under the weight of gravity corers. This allowed much longer cores to be collected without the layers being compressed. The piston corer became the tool of choice for studying past climates in ocean sediments. First deployed in Sweden's Gullmar Fjord in 1945, it obtained an undisturbed core of

---

Earth's Climate Evolution, First Edition. Colin P. Summerhayes.
© 2015 John Wiley & Sons, Ltd. Published 2015 by John Wiley & Sons, Ltd.

sediment 20.3 m long. Following successful trials in the Mediterranean, the device was used aboard the *Albatross* on the round-the-world Swedish Deep Sea Expedition of 1947–48, which obtained 57 cores, totalling 500 m in length.

Widespread use of the Kullenberg piston corer began to expose the secrets of climate change hidden in deep-ocean sediments. Many cores were collected by the worldwide research expeditions of the ships of three great American oceanographic institutions. The Scripps Institution of Oceanography, based in the village of La Jolla near San Diego, California, began life as the Marine Biological Association of San Diego in 1903, before becoming Scripps in 1925. The Woods Hole Oceanographic Institution (WHOI), on Cape Cod, Massachusetts, where I worked in the 1970s, formed in 1930. The Lamont Geological Observatory of New York's Columbia University, known as Lamont, was established in 1949 and renamed the Lamont–Doherty Earth Observatory in 1993. The comparable major research institutions of the United Kingdom, Germany and France also began to collect piston cores across the world's oceans. The National Institute of Oceanography at Wormley, in Surrey, was created in 1948, morphing with time into the Institute of Oceanographic Sciences Deacon Laboratory, before moving to Southampton in 1995 to become the heart of the Southampton (now National) Oceanography Centre. In France, the Institut français de recherche pour l'exploitation de la mer (IFREMER) was formed in 1984 and the Institut polaire français Paul-Émile Victor (IPEV) in 1992. In Germany, the Alfred Wegener Institute for Marine and Polar Research (AWI) was formed in 1980.

Piston corers themselves evolved. In the early 1970s, when I was an Assistant Scientist at WHOI, my colleague Charlie Hollister (1936–1999) developed the giant piston corer. This monster could easily collect large-diameter cores up to around 20 m long, some 8 m longer than the average piston corer, providing much larger samples for analysis[2]. My last marine research expedition, between the Azores and Marseilles in July 1995, was aboard the brand new French research vessel *Marion DuFresne*. We used the latest French giant piston corer to collect cores as long as 39 m off the coast of Portugal. Before we boarded the ship in the Azores, its coring crew had collected cores of up to 52 m long on the Bermuda Rise. The mind boggles!

Opening up the next dimension, rather like a bird watcher graduating from binoculars to a telescope, called for deep-ocean drilling, profiled in Chapter 6. While piston corers can be launched from medium-sized to large research ships, deep-sea drilling requires a specialised vessel, of which there is usually only one available to the global research community at any one time. The international Deep Sea Drilling Project (DSDP) and its successors have supplied such drilling platforms since late 1968, enabling us to penetrate the older geological record that is buried too deep to be reached by piston corers. Even there, piston coring has enhanced recovery, through operation of the hydraulic piston corer, developed in 1979; this allows the recovery of undisturbed cores of 10–100 m long in the soft uppermost sediments of the seafloor, which are poorly sampled by the rotary drilling process.

Piston coring and deep-sea drilling underpin the new science of palaeoceanography, which has unlocked many of the secret files of past climate change. I learned all about piston coring early on in my career, when I was lucky enough to join the Scripps research vessel *Argo* in the southwest Pacific in 1967, as mentioned in Chapter 5[3]. We collected a piston core every day. Lessons learned on that cruise would come in handy when I began collecting my own piston cores for my PhD at Imperial College London, while on expeditions down the Moroccan coast. The *Argo* cruise also exposed me to some of the hazards of ocean research, when the cylindrical 1 ton weight that drove the piston corer into the sediment broke free of its lashings on the after deck in a big storm and began to roll from one side of the ship to the other, smashing everything in its path. Stopping it was rather like trying to lasso a raging bull, and Bill Menard pulled a tendon trying to leap over the rolling monster when it unexpectedly headed his way. We had to go to Brisbane for repairs (as an unexpected bonus, that visit introduced me to Australian aboriginal art). Later, when I worked for EPRCo in Houston, I was much involved with the DSDP, analysing drill cores for their concentration of organic matter and serving as an advisor on its Organic Geochemistry Panel.

7.2   Palaeoceanography

Palaeoceanography is the study of how the ocean behaved in the past in response to changes in the positions of the continents, the patterns of the winds, the presence or absence of polar ice and changes in solar radiation determined by variations in the Earth's orbit[4]. It can tell us about the distribution of the water masses, the average positions of major currents and oceanic fronts, the regional patterns of sea surface temperatures, changes in global ice volume and extent, changes in the vertical thermal structure of the ocean, changes in bottom currents and bottom water masses, changes in the positions and intensities of the Trade Winds and past changes in the biogeography (distribution) of different ocean-dwelling organisms. Changes with time reflect changes in controlling processes, enabling us to decipher the history of ocean and atmospheric circulation and ice and to understand its effects on global climate.

The earliest use of the word 'palaeoceanography' that I can find was by William (Bill) R. Riedel (1927–), an Australian specialist in studies of radiolaria (a form of siliceous plankton), who joined Scripps in 1951. Riedel used it in a paper submitted in August 1960 and published in 1963[5]. The New Zealander Jim Kennett (1940–) (Figure 7.1, Box 7.1) used the term in his 1971 paper on the Cenozoic palaeoglacial history of Antarctica, referring to 'paleo-oceanography'[6], and again in the report of DSDP's Leg 29 (exploring the climate history of the Southern Ocean south of New Zealand), which he co-led with Bob Houtz of Lamont in March and April 1973[7]. 'Palaeo-oceanography' was soon to lose the hyphen. Kennett was one of the initial developers of this new field, in the early 1970s, although the title of 'father of palaeoceanography' is generally bestowed on the Italian isotope geochemist Cesare Emiliani for his pioneering work in the 1950s.

Like all new research fields, this one led to the formation of a new scientific journal: *Paleoceanography* (note the American spelling). Jim Kennett helped establish the new journal in 1986. It also led to a new scientific conference

Figure 7.1   *Jim Kennett.*

Box 7.1 Jim Kennett.

Kennett was one of the primary developers of palaeoceanography. As a youngster, he spent much of his spare time studying geology at the National Museum and the New Zealand Geological Survey. He was particularly taken with the usefulness of foraminifera as recorders of ocean history, and completed a PhD in Wellington in 1965 on Neogene foraminifera across the New Zealand region. The two of us shared an office at the New Zealand Oceanographic Institute for a while, and I was much influenced by his growing awareness of the as yet unnamed field of past ocean studies. With the advent of the DSDP, many of those involved in interpreting the deep-sea sediment record began to see obvious connections between the oceans, the ice, the atmosphere and the biosphere. Interpreting these connections required a broader, more integrative Earth systems approach to geology than was taught in most geology schools. Kennett was among the first to take such an approach, which permeates his textbook *Marine Geology*, published in 1982[4]. This was the first comprehensive textbook on ocean history and processes following the plate tectonics revolution, and it remains an authoritative source book. I was thrilled to see it appear, and I envied the new generation of marine geologists who would kick off their studies and careers with such a comprehensive text to guide them. In 1987, Kennett was appointed director of the Marine Science Institute of the University of California at Santa Barbara. There he used deep-ocean drilling to provide a 160 000 year high-resolution record of climate change in the Santa Barbara Basin, showing that it was synchronous with that in Greenland ice cores. He is well known as an imaginative thinker, having proposed that Quaternary climate shifts (deglaciations) may have been accentuated by the destabilisation of seafloor methane hydrates and that the abrupt cooling at the onset of the cold Younger Dryas episode 13 000 years ago was caused by a meteorite impact. He was awarded the Shephard Medal of the Society of Economic Palaeontologists and Mineralogists in 2002.

series: I attended the first International Conference on Paleoceanography in Zurich in 1983, and it continues today.

Among the pioneers of palaeoceanography was the Swedish scientist Gustaf Arrhenius (1922–), grandson of Svante Arrhenius of CO_2 fame. Arrhenius is a member of the Royal Swedish Academy of Sciences and the 1998 recipient of its Hans Pettersson Gold Medal. In 1952, after sailing on the *Albatross* cruise and completing his PhD at the University of Stockholm, he joined Scripps. Among his *Albatross* cores were several from the eastern equatorial Pacific, in which he found that productivity, measured by the accumulation of planktonic skeletal remains on the deep-sea floor, was high during what appeared to be the glacial stages of the Ice Age. This increase in productivity, he proposed, came about when the increased temperature gradient between the Equator and the poles during glacial periods strengthened the Trade Winds, thus enhancing the upwelling currents that bring nutrients to the surface at the Equator[8, 9]. This was a fundamental advance in our understanding of the workings of the atmosphere–ocean system in an Ice Age world.

Another pioneer was Cesare Emiliani (1922–1995) (Figure 7.2, Box 7.2), whom we met in Chapter 6, working with Harold Urey at the University of Chicago.

Emiliani realised early on, from studies of deep-sea benthic foraminifera, that the bottom waters of the Miocene and Oligocene were warmer than those of today, although

Figure 7.2 *Cesar Emiliani. Taken during the early 1950s, when he was doing his pioneering research at the University of Chicago.*

Box 7.2 Cesare Emiliani.

Emiliani was born in Bologna, Italy, and gained his PhD in geology there in 1945, specialising in micropalaeontology, before moving to Chicago and gaining a second PhD in 1950, specialising in isotopic palaeoclimatology. He worked in Urey's geochemistry lab until he moved in 1957 to the Institute of Marine Sciences at the University of Miami, which gave him ready access to the ships he needed in order to sample deep-sea sediments for their foraminifera. In 1967, he became director of the Institute's Geology and Geophysics Division. In due course, the Institute became the Rosenstiel School of Marine and Atmospheric Science. Emiliani was honoured by having the coccolithophorid genus *Emiliana* named after him, and was awarded the Swedish Vega Medal and the Alexander Agassiz Medal of the US National Academy of Sciences.

they fell from 10 °C in the mid Oligocene to 2 °C in the late Pliocene in the eastern equatorial Pacific, a sign of cooling driven by the sinking of cold water in high latitudes[10, 11]. In contrast, his studies of pelagic foraminifera showed that ocean surface temperatures in the equatorial Atlantic were rather uniform through Oligocene and Miocene times[11]. Later research confirmed this picture. In addition, he made substantial contributions to the study of the record of the Ice Age in deep-ocean sediments, as we'll see later.

Emiliani was also one of the first to apply oxygen isotopes to the study of the temperatures of past surface and bottom waters. By 1965, the British physicist Nicholas J. Shackleton (1937–2006) (Figure 7.3, Box 7.3), from the Godwin Laboratory of the University of Cambridge, had radically improved the technology of oxygen isotopic analysis. Realising that substantial advances could be made if many more samples were available, he rectified inadequacies in the mass spectrometric instruments used to analyse the proportions of oxygen isotopes, paving the way for multiple oxygen isotopic analyses on very small samples[12]. Shackleton became a major force in palaeoceanography over the next 40 years.

I first met Shackleton at WHOI when I was working there in the early 1970s. We discussed the results emerging from the DSDP on many occasions. As the DSDP drew to a close, we co-organised a Geological

Figure 7.3 *Nicholas J. Shackleton.*

Box 7.3 Nicholas J. Shackleton.

Nick Shackleton was the leading British oxygen isotope analyst and one of the stars of the new and growing field of palaeoceanography. A distant relative of polar explorer Sir Ernest Shackleton, in due course he became famous in his own right. He was also owner of the largest collection of antique clarinets in the world and himself a talented clarinettist. His renown in scientific circles came from his innovative and pioneering research. As his Cambridge colleagues Nick McCave and Harry Elderfield said, '*his lifetime achievements define the emergence of our understanding of the operation of Earth's natural climate system*'[13]. Among many honours for his services to science, the Geological Society of London awarded him its Lyell Medal in 1987 and its prestigious Wollaston Medal in 1996. He was also made a fellow of the Royal Society in 1985, received its Royal Medal in 2003 and was knighted in 1998 – the ultimate British accolade.

Society of London conference on the Palaeoceanography of the North Atlantic to review the advances that had come from deep-sea drilling. The meeting, in November 1984, showed how climate had influenced ocean circulation to create the Atlantic sedimentary record over the past 180 Ma[14]. It helped to ensure the success of the bid to replace the DSDP with the next phase of ocean drilling.

While I was working for BPRCo, in the mid 1980s, I persuaded Shackleton to come and talk to us about isotope stratigraphy. At the time, I was responsible for the micropalaeontologists who provided BP explorers with the ages of the fossils in drill cuttings and cores and explained what they told us about the depositional environments and climates in which they had lived, in order to help the explorers evaluate the petroleum potential of their sedimentary basins. I thought we could use Nick's isotopic stratigraphy techniques to tell us yet more about past environments. It was amusing to see him turning heads in the senior dining room at BP's research centre. The senior managers in their pinstriped suits were not used to seeing hippie-looking men with shoulder-length hair, open-necked shirts and sandals in their carpeted and waitress-serviced inner sanctum. Only once did I see Nick wearing a tie: at a conference at the Royal Society in London. The horizontal black and white stripes spaced irregularly down this neckpiece turned out to be the magnetic polarity time scale used to interpret the ages of the magnetic stripes on the deep-sea floor on either side of mid-ocean ridges. Typical!

In 1988, I was invited to organise the Geological Society's 3rd Lyell Meeting, on Palaeoclimates, which took place in February 1989[15]. The papers presented there examined in depth the several different ways we then had of looking at past climates: the biostratigraphic approach, using the fossils of marine plankton and land plants; the geochemical approach, using oxygen isotopes and alkenones (more on that later); high-resolution studies, based on the astronomical approach; and a new kid on the block, numerical modelling (reviewed at the end of Chapter 6). The conference concluded that deep-ocean drilling into ancient marine sediments had an increasingly important role to play in showing us how climate had changed with time and in providing the data needed to test global circulation models. Papers from the meeting were published in Volume 147 of the *Journal of the Geological Society* in 1990.

Looking for a chemical method independent of oxygen isotopes by which to estimate the temperatures of past surface waters, a Bristol University group led by the chemist Geoff Eglinton (1927–) discovered in 1985 that the relative proportions of organic chemical compounds known as alkenones, derived from marine algae, correlated with the temperature of ocean surface waters[16]. This was one of the most important discoveries in organic geochemistry in the last 50 years. For his work in applying the tools of organic chemistry to the understanding of geological problems, Eglinton was made a fellow of the Royal Society. He received the Society's Royal Medal in 1997 and the Geological Society's Wollaston Medal in 2004.

Alkenones are lipids. They are long-chained organic compounds belonging to a class known as ketones and come from cell membranes in a few species of haptophyte algae, such as the widespread marine coccolithophorid *Emiliana huxleyi*[17]. In effect, they are fossil organic molecules. They are distinguished from one another by the number of carbon atoms in their chain (C_{37}, C_{38}, C_{39}), their degree of unsaturation (the number of double bonds they contain) and the structure of the group at the end of their chain (methyl or ethyl). The compounds of interest for palaeothermometry are the C_{37} methyl ketones containing two or three double bonds[18]. The temperature of the surface waters in which they grew is determined from the $U^{k'}_{37}$ index, measured as the ratio of $C_{37.2}/(C_{37.2} + C_{37.3})$, where $C_{37.2}$ represents the amount of the diunsaturated ketone and $C_{37.3}$ the amount of the triunsaturated form[19]. The beauty of this technique is that it provides results even where the skeletons of the organisms have been dissolved.

Another new tool in palaeothermometry capitalised on the discovery that the ratio of magnesium to calcium (Mg/Ca) in the calcium carbonate skeletons of planktonic foraminifera was directly related to temperature[20]. The first to observe a tantalising relationship between the Mg content of biogenic carbonates and their growth temperature was the American geochemist Frank Wigglesworth Clarke (1847–1931), after whom the Geochemical Society's F.W. Clarke Award is named[21]. This relationship only came to fruition 70 years later, in the mid 1990s, through the application of powerful modern analytical techniques like the electron microprobe, which allowed the trace-element concentration of a single foraminiferal chamber to be measured to high accuracy[22, 23]. Progress has been rapid since then[21]. One of the scientists who developed Mg/Ca palaeothermometry was my former fellow student and bridge-playing opponent at Imperial College, Henry (Harry) Elderfield (1943–)[20]. For his many outstanding geochemical contributions, Elderfield was made a fellow of the Royal Society in 2001, was awarded the Geological Society's Prestwich Medal in 1993 and its Lyell Medal in 2003 and was given the Urey Medal of the European Association of Geochemistry in 2007.

Using Mg/Ca ratios to detect past temperature change is especially helpful in the tropics, where glacial–interglacial changes are small (<5 °C) and thus more difficult to detect by other means. The Mg/Ca method works well there because its relationship to temperature is exponential, steepening towards higher temperatures. It indicates tropical cooling of around 3 °C in glacial times, as do data from alkenones[21]. The Mg/Ca ratio changes only slightly at temperatures typical of glacial–interglacial bottom waters, so it is not so useful at demonstrating change there.

Elderfield and his students also applied the Mg/Ca technique to pre-Quaternary marine sediments. In 2000, they used Mg/Ca ratios in benthic foraminifera to produce a deep-sea temperature record for the past 50 Ma[24]. This closely replicated the $\partial^{18}O$ curve, defined a cooling of about 12 °C and showed that when ice first formed on Antarctica 34 Ma ago, the event was not accompanied by an abrupt decrease in deep-sea temperature. One of Elderfield's students, Aradhina Tripati, used this technique to investigate changes in tropical sea surface temperatures in the late Palaeocene through middle Eocene (59–40 Ma ago) from an Ocean Drilling Program site on Allison Guyot in the western central Pacific[25]. They observed long-term warming into the early Eocene – reaching a peak between 51 and 48 Ma ago, with values slightly warmer than those in the present tropical Pacific – followed by cooling of 4 °C over the next 10 Ma. The Mg/Ca temperatures were warmer than those calculated from oxygen isotopes, which had suggested a cool tropics for that period. Tripati concluded that '*absolute SST [sea surface temperature] values inferred from the $\partial^{18}O$ of mixed-layer planktonic foraminifera in deep-sea sediments are biased towards colder values due to (1) secondary calcification, and (2) assumptions about surface water $\partial^{18}O$*'[25]. They went on to estimate that there had been '*reduced equator-to-pole thermal gradients during the early Paleogene, with minimum gradients coincident with peak tropical warmth during the early Eocene. Gradients increase during the middle Eocene, with tropical SST cooling by 3–4 °C and high-latitude SST cooling by several degrees. This finding is consistent with a weaker ocean circulation and heat transport during the early Paleogene, relative to modern*'[25].

As we will see later in this chapter, additional evidence confirms that the deep ocean cooled gradually from the mid Eocene onwards, while ice growth occurred rapidly at certain periods, starting in the earliest Oligocene[26].

The apparent lack of warming of the Eocene tropics derived from oxygen isotope data led to what Eric Barron called the 'cool tropics paradox'[27]. Puzzling over this paradox, scientists eventually came to realise that recrystallisation on the deep-sea floor distorted the original isotopic signal from tropical surface waters, biasing the record towards cold temperatures[28, 29]. Science continually advances, and as Tripati and colleagues found, the latest techniques exposed the problem and made the paradox disappear.

In 1986, a new organic molecular proxy for sea surface temperature was discovered by researchers on the Dutch island of Texel, in the Netherlands. They named it TEX_{86}, shorthand for the Tetraether index based on the 86-tetraether-carbon chain in lipids derived from the cell membranes of the marine picoplankton known as the Archaea (picoplankton are just $0.2–2.0\,\mu$ in diameter and Archaea are single-celled microorganisms lacking a nucleus; they are similar in size to bacteria). The membrane lipids are glycerol dialkyl glycerol tetraethers, known by the shorthand GDGT. TEX_{86} data confirmed that Eocene tropical temperatures were 5–10 °C warmer than previous reconstructions based on $\partial^{18}O$ had shown[28–30]. This tells us that in the warmer worlds of the past, both the tropics and the poles were warm, and thus the Equator–pole temperature gradient was less than in past cool periods. Under such conditions, wind strengths would be weaker than today, weakening oceanic circulation patterns.

By the early 2000s, we thus had four independent geochemical tools with which to map past climate and ocean temperature in deep-sea sediment cores: oxygen isotope ratios ($\partial^{18}O$), which also provided a means of documenting changes in ice volume; alkenones; Mg/Ca ratios; and TEX_{86}. These were supplemented by calculations of sea surface temperature based on the assemblages of planktonic microfossils, a technique developed by John Imbrie, which we explore in Chapter 12. As time went by, we were also able to use carbon isotope ratios ($\partial^{13}C$) and the ratios of cadmium to calcium (Cd/Ca) as tracers by which to document the changing characteristics of deep waters with time (developed by Ed Boyle at MIT[31]). Robert Anderson of Lamont showed how an isotope of thorium (^{230}Th) could be used to measure the flux of sediment particles to the seabed as a guide to past ocean productivity[32]. The striking advances represented by new technologies enabled us to investigate climate variability in marine sediment cores from the Cenozoic in unprecedented detail.

When new techniques are added to the toolkit, they have to be tested in a wide range of environments before they will be widely accepted. TEX_{86} is no exception. Concerned that proxy-based estimates of sea surface temperature for the southwest Pacific in the Eocene were too high, a team led by Christopher Hollis of the department of palaeontology of GNS Science in Lower Hutt, New Zealand recently tested the TEX_{86}, $\partial^{18}O$ and Mg/Ca proxies for sea surface temperature against one another and against numerical model results[33]. They were worried that proxies based on lipid organic material (e.g. the GDGTs underpinning the TEX_{86} palaeothermometer) might respond differently to inorganic palaeothermometers such as $\partial^{18}O$ and Mg/Ca, based on chemical analyses of fossil shells. They found that the high temperatures estimated for surface and bottom waters could be reconciled with numerical model outputs if the proxies were biased towards summer seasonal maxima. Temperatures may also have been influenced by the warm East Australian Current. Evidently, multiple approaches to evaluating palaeotemperatures are advisable.

Application of these different tools demonstrates that the cycles of orbital insolation occur throughout the Cenozoic. Indeed, they form the 'heartbeat' of Oligocene time[34]. It appears that particular combinations of insolation minima in both eccentricity and obliquity led to sustained cooling accompanied by the build-up of ice at the Oligocene–Miocene boundary (23 Ma ago) and may have been responsible for the occurrence of other such 'thresholds' in climate change through time[34]. More on that later.

7.3   The World's Freezer

It was not only the remote floor of the oceans that we needed to study to find out more about the history of our climate. We also needed much more information from another remote spot: the coldest place on Earth, Antarctica. The world's freezer, Antarctica sucks in heat from the tropics and pours out cold. The world's biggest ice cube, at close to 30 million km^3, it holds 70% of the world's freshwater in frozen form. The latest satellite data, from 2010, show that up on the 3000 m-high Polar Plateau, temperatures can be as low as −93.2 °C (data from National Snow and Ice Data Center, reported at an American Geophysical Union meeting, December 2013).

Rather like the deep ocean, Antarctica has only been open to detailed and systematic scientific investigation since the Second World War. This expansion in scientific knowledge began with the International Geophysical Year (IGY) of 1957–58, when a dozen nations put research stations on the continent. Many of them are still in operation, and they have since been joined by the research stations of several other nations. Much of the research carried out in the Antarctic today focuses on climate change.

Like the deep ocean, Antarctica is hard to get to and a dangerous place to be. In the ocean, you can fall off a ship and drown. In Antarctica, you can fall into a crevasse and die. In both places, you are subject to fierce storms. Living conditions for researchers are basic, both on the ice and at sea. And the speed of research on the ground around the pole is about the same as it is in the deep ocean: caterpillar-tracked trucks and one-man skidoos move at about the speed you can pedal a bicycle, just like research ships. Antarctica is probably a worse environment, in that for 6 months of the year it is dark, which is a real limitation for climate researchers.

To see how Antarctica fits into the climate story, we are going to have to become better acquainted with it (Figure 7.4). It helps to think of the continent as two distinct pieces. The largest, East Antarctica, lies mostly east of the Greenwich Meridian and south of Africa and Australia. It forms 75% of the continent and is rimmed by mountains. Buried beneath its ice are the Gamburtsev Mountains, with peaks up to 3500 m above sea level. More than 98% of East Antarctica, and all of the interior, is covered with the largest mass of ice on the planet: up to 4776 m thick, averaging around 1800 m. If this 24 million km^3 of ice melted, it would raise sea level 52 m. The weight of ice has depressed the land beneath by about 600 m. The base of the ice is mostly above sea level, but locally it can be as much as 2 km below sea level.

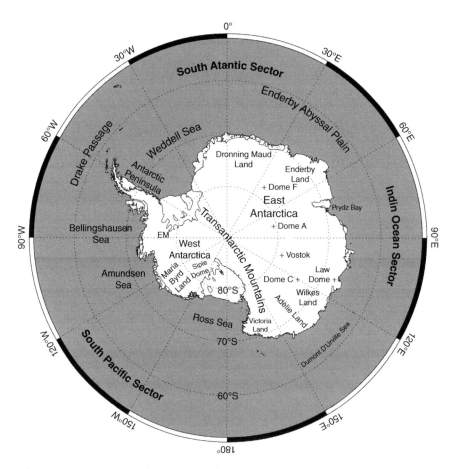

Figure 7.4 *Antarctica. Showing main features and locations of main ice core sites.*

The smaller piece, West Antarctica, lies south of the Pacific Ocean and South America. Although it forms 25% of the continent, it has only 10% of the ice: around 3 million km^3, equivalent to 5 m of sea level rise. West Antarctica has a mean elevation of just 850 m. It is really an archipelago of large islands. Most of its ice sheet is grounded below sea level, by over 2 km in places, making it inherently vulnerable to melting. The main peak, in the Ellsworth Mountains, rises to 4892 m.

The top of the East Antarctic Ice Sheet is the windswept Polar Plateau, the almost featureless white plain that Scott, Amundsen and Shackleton crossed on their way towards the South Pole early in the 20th century. They slogged it on foot and by sled and ski. Nowadays you can fly there in 3 hours from McMurdo Station, the US base on Ross Island in the Ross Sea, on a four-engined, ski-equipped Lockheed C-130 Hercules. How times change!

The Polar Plateau rises slightly towards a central crest with three broad peaks: Dome Argus, or Dome A, at 4093 m; Dome Fuji, or Dome F, at 3810 m; and Dome Charlie, or Dome C, at 3233 m. The Chinese recently built a base at Dome A, and the French and the Italians have a base at Dome C. The Americans built their Amundsen-Scott base at the South Pole itself, which lies on one edge of the plateau, at 2835 m. It is not the geographic centre of the continent, which lies at the Pole of Inaccessibility at 85° 50′ S, 65° 47′ E. The Soviet Union built its Mirny base there, but the major Soviet (now Russian) base was Vostok, now famous for supplying a 400 Ka record of climate change and for the lake beneath the ice.

The height of the continent keeps it cool, because cooling increases with height by roughly 6.5 °C per 1000 m in humid conditions, and 10 °C per 1000 m if it's dry. East Antarctica is 3000 m high on average and has an average annual temperature of −50 °C. It is so cold that the air is extremely dry. Precipitation is low enough for it to be classified as a desert. With its mean elevation of just 850 m, West Antarctica is warmer and wetter, with an average temperature of only −20 °C, and snowfall averaging 50–100 cm per year, mostly on the coast.

The East Antarctic Ice Sheet is bounded on the west by the Transantarctic Mountains, which cross the continent from the Ross Sea to the Weddell Sea. It lies atop an ancient piece of continental crust that migrated to its present position during the break-up of Gondwanaland, as we saw in Chapter 5. Marked by the formation of new ocean floor, the break-up started with the separation from Africa in the early Cretaceous around 140 Ma ago,

continued with that from India around 125 Ma ago and ended with that from Australia around 40 Ma ago and that from South America about 20 Ma ago. The leading edge of Antarctica reached the South Pole about 100 Ma ago. Since then, the continent has moved slowly back and forth across the pole, which has been within about 2000 km of its present position for the last 90 Ma[35].

East Antarctica's ancient crustal core was bordered on the side facing the Pacific Ocean by small continental fragments merged into one of the volcanic 'island arcs' that form when ocean crust is 'subducted' beneath the advancing edge of a moving continental fragment. This island arc was a southward continuation of the Andes of South America. It continues into Antarctica through the Antarctic Peninsula. The peninsula separated from Patagonia about 23 Ma ago, forming the Drake Passage, which is named after its discoverer, the 16th-century English freebooter Sir Francis Drake, who found it when blown off course in 1578, while trying to reach the west coast of North America and steal gold from Spanish galleons out of Lima.

The islands beneath West Antarctica are separated from the Transantarctic Mountains of East Antarctica by a 1000 m-deep oceanic trough, the opposite ends of which are the Weddell Sea and the Ross Sea. This fault-bounded depression is a back-arc basin. The island arc running from the Peninsula through West Antarctica originally continued into New Zealand, which migrated away around 80 Ma ago.

Examining the geological record for evidence of climate change in and around Antarctica is extremely difficult and costly, not least because only 46 000 km^2 of rock, or about 0.33% of the continent, is exposed to the geologist's hammer or drill. Captain Scott's team collected numerous geological samples from Antarctica, including coal from the cliffs of the Beardmore Glacier. As we saw in Chapter 4, among the plant remains from the Beardmore Glacier was the first recorded occurrence from the continent of the fossil Gondwanan seed fern *Glossopteris*. It was the recovery of geological specimens and scientific data that ultimately distinguished Scott's expedition from Amundsen's.

During and after the IGY of 1957–58, palaeontologists scoured the few peaks protruding through the ice for fossils that could tell us about the icy continent's past climate. All that these rocks provided were a few snapshots of climate change through time. Reviewing the status of Antarctic glacial geology in 1964, the geologist R.L. Nichols was moved to state, '*this writer believes that as yet there is no good evidence for Tertiary Antarctic glaciation*'[36] – it was all thought to be Quaternary, as in the Northern Hemisphere. Alas for him, in the same published collection of papers, Cam Craddock (1930–2006), then at the University of Minnesota, published radiometric potassium–argon (K-Ar) dates showing that a glaciated surface overlain by basalt in the Jones Mountains of West Antarctica was older than 10 Ma[37]. Not long after that, small basaltic cones overlain and underlain by evidence of glaciation yielded ages ranging between 2.8 and 3.6 Ma in Taylor Valley, west of McMurdo Sound. Evidently, Antarctic ice had a significant pre-Quaternary history[38, 39].

Through his studies of the foraminifera of the New Zealand region, Kennett recognised a significant cooling in the late Miocene that corresponded to a sea level fall. Knowing about Craddock's 1964 paper[37], he surmised that the cooling resulted from expansion of the Antarctic ice sheet[40]. Kennett broadened his studies when he emigrated to the United States in 1966 to work with Stan Margolis at Florida State University on piston cores from the Southern Ocean. By 1970, their work had pushed back the inception of Antarctic ice-rafting to the early and middle Eocene[41]. The following year, their core studies led them to conclude that the Southern Ocean had been cool through much of the Cenozoic, with periodic major cooling during the early Eocene, late middle Eocene and Oligocene, causing '*considerable ice-rafting of continental sediments to present-day Subantarctic regions*'[6]. They argued that there had been a warming and a reduction of ice-rafted sands in the lower and middle Miocene, followed by a cooling trend in the late Miocene[6]. The discovery that the Southern Ocean had cooled over the past 40 Ma tied in nicely with Emiliani's finding that deep waters had cooled at low latitudes over the same period. A connection via the deep-ocean circulation was implied.

Back in the late 1960s–early 1970s, Kennett was intrigued by the revolution in plate tectonics that showed how New Zealand had migrated away from Antarctica. He teamed up with the palaeomagnetist Norman Watkins from the University of Rhode Island to examine the spreading of New Zealand and Australia away from Antarctica. Comparing Kennett's palaeontological data with Watkins' palaeomagnetic data, they realised in 1971 that the Antarctic Circumpolar Current only developed after Australia separated from Antarctica sometime in the early Cenozoic[42].

The problem shared by geologists keen to understand the climatic history of Antarctica was that the ice was in the way. The solution was obvious: they had to drill through

it into the rocks beneath. That would be very expensive, as well as logistically challenging. Then came the Eureka moment: a clever way around the problem would be to sample the sediments that glaciers and ice streams had dumped into the ocean, and which had been spread out over the adjacent ocean floor by currents. Sampling the ocean sediments close to the continent would have the added benefit of revealing microfossils that could tell us about the behaviour of the Southern Ocean through time, another key part of the climate story. Sampling could be carried out either by drilling through the ice shelves adjacent to the land or by deep-ocean drilling further offshore. Land drilling in ice-free areas was also possible, for example in the McMurdo Dry Valleys discovered by Captain Scott in 1903.

7.4   The Drill Bit Turns

Drilling through strata on land in Antarctica began in 1971, in the international Dry Valley Drilling Project (DVDP)[43], and was extended seawards by drilling through the 'fast ice' close to shore. Drilling further offshore, from the DSDP's *Glomar Challenger*, began shortly afterwards, on DSDP Leg 28, between December 1972 and February 1973[44]. Although the onshore and fast ice drilling took place in parallel with the DSDP drilling deep offshore, for convenience in telling the story I'll first review the results of the drilling on land and from the fast ice.

The ice-free McMurdo Dry Valleys were explored and mapped in 1957–58 by Victoria University of Wellington students Barrie McKelvey and Peter Webb, as part of New Zealand's programme for the IGY. Initiated by the US Office of Polar Programs, the DVDP project was led by Lyle McGinnis, a geophysicist at the University of Northern Illinois, and valley mapper Peter Webb, by then chairman of geology at Northern Illinois. The project was eventually sponsored and staffed by New Zealand, the United States and Japan, continuing until 1976[45]. In November 1973, the drill was tested on the volcanic rocks at McMurdo Station on Ross Island, reaching a depth of over 300 m. It was then taken to the Antarctic mainland, near the mouth of Taylor Dry Valley, where it cored through 300 m of glacial sediment, recovering material as old as 6 Ma. By now, Peter Barrett (1940–) (Figure 7.5, Box 7.4), a young palaeoclimatologist from Victoria University in Wellington, was involved in the project.

Figure 7.5 *Peter Barrett.*

Having been to sea on DSDP Leg 28 (more on that later), Barrett and Webb were convinced that in order to reach further back in time than the 6 Ma sampled by the DVDP, they would have to use the sea ice as a drilling platform. Drilling from the fast ice (sea ice frozen to the land, as opposed to drifting pack ice) would enable them to penetrate hundreds of metres of sediment beneath the seafloor off the coast. This would be a better option than waiting several years until the DSDP returned to the Ross Sea. Selecting a site 12 km seaward of the mainland, they set up their drill rig on 2 m of sea ice over 120 m of water in November 1975. The 50 m of core recovered just scratched the surface, but it proved the concept and led Barrett, Webb and colleagues to create further multinational drilling projects in the area, including the McMurdo Sediment and Tectonic Studies (MSSTS) programme in 1976, the Cenozoic Investigation in the Western Ross Sea (CIROS) programme in 1984–86 and the Cape Roberts Project in 1997–99. The latter provided a continuous core, with a remarkable 98% recovery, through 1600 m of strata beneath the sea ice, and a record of coastal tundra and oscillating warm ice sheets from 33 to 17 Ma ago. Unfortunately, the oldest glacial strata lay more or less directly atop 330 Ma-old Beacon Sandstone, and strata had been eroded off the top, so another drilling project was needed to fill in both the older and younger ice and climate stories[46, 47].

By now, the scientific value of Antarctic offshore drilling had been established. Following the creation of a new international drilling consortium (ANDRILL),

Box 7.4 Peter Barrett.

Barrett's interest in geology came from exploring and mapping caves in his high school years in the Waitomo district of New Zealand. He went to Ohio State University in the early 1960s to help map the strata of the Transantarctic Mountains, and discovered the first four-legged vertebrate dinosaur fossil in Antarctica, a 200 Ma-old amphibian. That helped to confirm that Antarctica had once been part of Gondwana. After completing a PhD on the geological history of the central Transantarctic Mountains, he joined Victoria University of Wellington. There he helped to establish the stratigraphy of Antarctica's Gondwana sequence, comprising Devonian quartz sandstone, Carboniferous–early Permian glacial deposits and Permian–Triassic coal measures. In the early 1970s, Barrett became much involved with drilling, both through the ice adjacent to the Antarctic coast and through the DSDP, leading several projects that revealed much about Antarctica's climate history. Through the Scientific Committee on Antarctic Research (SCAR), he promoted studies of the history of the Antarctic ice sheet, and he was a leader of SCAR's major research programme on Antarctic Climate Evolution from 2004 to 2010. SCAR has provided the forum for planning international drilling activities in the Ross Sea region since the early 1980s, with Barrett providing leadership and advice. Professor of geology at Victoria University and founding director of the Antarctic Research Centre there (1972–2007), Barrett was instrumental in establishing the NZ Climate Change Research Institute at Victoria University in 2008. He was awarded fellowship in the Royal Society of New Zealand in 1993, the Marsden Medal by the NZ Association of Scientists in 2004, the first SCAR President's Medal in 2006, the NZ Antarctic Medal in 2010 and honorary fellowship of the Geological Society of London in 2011.

Tim Naish of Victoria University of Wellington and Ross Powell of the University of Northern Illinois led the drilling of the first ANDRILL hole, AND-1B, in 2006–07. Penetrating the 85 m-thick McMurdo Ice Shelf and the underlying 870 m of ocean, it cored 1284 m into the sediments beneath, getting the geologists back to 13 Ma ago[48]. The top 80 m was made up of cycles, in which sediments with strongly glacial and weakly glacial character matched the eight major climate cycles of the last 800 Ka. Further back in time, only a few cycles were preserved, most having been eroded. But those few provided useful snapshots of Antarctic climate, with weaker glacial–interglacial cycles at intervals of 40 Ka. There was also one big surprise: between 2 and 5 Ma ago, in the Pliocene, warm conditions melted the Ross Ice Shelf, replacing it with open water full of planktonic siliceous diatoms[49]. Might this tell us something about our modern global warming? We shall see as we go on.

Later, Dave Harwood of the University of Nebraska and Fabio Florindo of Rome's Institute of Geophysics drilled a second hole, AND-2B, from the sea ice seaward of the dry valleys, reaching a depth of 1139 m below the seafloor and finding evidence for a warm, pulsating East Antarctic Ice Sheet 15–20 Ma ago[50].

Four decades of research in mapping and dating the landscape of the McMurdo Dry Valleys by George Denton of the University of Maine, David Sugden of the University of Edinburgh and their students shows that the region was cold and dry for the last 14 Ma, refrigerated by the huge East Antarctic Ice Sheet[46, 47]. The last vestiges of shrubby vegetation were found there recently by Adam Lewis, a student at Boston University working with Dave Marchant, a student of Denton and Sugden in the 1990s. The remains occurred in thin, scattered moraines at 1700 m above sea level, dated from volcanic ash layers at 14.0 and 13.6 Ma ago, respectively. They marked the transition from warm to cold glaciers. The evidence for a cold stable ice sheet on East Antarctica co-existing with the warm unstable ice sheet on West Antarctica might give us clues about what could happen as global warming continues.

At least drilling from the 'fast ice' and from the ice shelf eliminated any possibility of the drill rig being damaged by icebergs. Drilling offshore might be a gamble, in contrast, given that Antarctica's continually moving ice streams, glaciers and ice shelves hive off icebergs, which move west in the easterly coastal current. Would the DSDP's *Glomar Challenger* be able to operate at high latitudes, where the weather was severe and icebergs were common? The acid test took place between December 1972 and February 1973 on DSDP Leg 28[44], led by Dennis Hayes of Lamont and Larry Frakes (whom we met in Chapter 6), then at Florida State University and soon to move to Monash University in Australia. Also aboard were Barrett and Webb.

Coring from south of Fremantle to the Antarctic margin, they established that Australia had been moving away from Antarctica for the last 55 Ma. Coring the Ross continental shelf up to 78° S, they found evidence of grounded ice on the continent as far back as 25 Ma ago. Recovering some 1405 m of core from various sites, they deduced from the distribution of ice-rafted fragments of rock and the fossils of microorganisms that major continental glaciation had begun at least by the late Oligocene, some 25 Ma ago. The climate began deteriorating then, and cool waters pushed north until the end of the Miocene, around 5 Ma ago, when there was a rapid build-up and subsequent retreat of the ice sheet.

Between March and April 1973, *Glomar Challenger* was back in Antarctic waters on DSDP Leg 29, led by Jim Kennett, then at the Graduate School of Oceanography in Rhode Island, and Bob Houtz of Lamont[7]. Earlier, Kennett had participated as a micropalaeontologist on DSDP Leg 21 to the southwest Pacific (November 1971–January 1972), where he and his colleagues had found a vast regional unconformity that had developed in deep water during the Oligocene[51]. They deduced that this represented massive erosion caused by the development of fast-moving bottom waters in the Ross Sea area in response to extensive glaciation. They also thought that the movement apart of Australia and Antarctica had allowed the inception of the Antarctic Circumpolar Current in the late Oligocene, and that there must be a connection of some kind between that current and the deep waters that caused the erosion[51].

The Leg 29 team showed that the Antarctic Circumpolar Current must have developed in the middle to late Oligocene, close to 30 Ma ago[52]. Although the slow separation of Australia from Antarctica began much earlier, around 80–90 Ma ago, the Tasman Rise, a ridge extending south from Tasmania, blocked development of a full ocean connection between the Indian Ocean and Pacific Ocean south of Australia until 30 Ma ago. The team confirmed that the onset of substantial glaciation near the Eocene–Oligocene boundary dramatically increased the volume of ocean-bottom water, causing massive worldwide erosion in the deep sea. '*The separation of Australia from Antarctica*', they considered, '*led to a fundamental change in the world's oceanic circulation and its climate that marks the onset of the modern climate regime*'[52]. Subsequent deep-ocean drilling around Antarctica refined that picture.

Appreciating the value of the growing field of oxygen isotope stratigraphy, Kennett persuaded Shackleton to analyse the oxygen isotopes in benthic calcareous micro-fossils from Leg 29. In 1975, they concluded that sea surface temperatures on the Campbell Plateau south of New Zealand had declined from 19 °C in the early Eocene to 11 °C by the late Eocene and 7 °C by the Oligocene, with a dramatic drop of 1–2 °C at the Eocene–Oligocene boundary, some 34 Ma ago[53]. In deeper water nearby, on the Macquarie Ridge, early Oligocene bottom-water temperatures were as low as those of the present day, implying that the mean annual temperature in high southern latitudes must have been near freezing by the beginning of the Oligocene, and that deep-ocean bottom waters were being created then around Antarctica. From that time, glaciers descended to sea level and sea ice was abundant. The isotopic evidence suggested that a small ice sheet formed during the Oligocene and a large one in the middle Miocene, at around 14 Ma ago[53].

Kennett integrated micropalaeontological and oxygen isotope results from deep-sea drilling around the continent on Legs 28, 29, 35 and 36 to infer in 1977 that the Antarctic ice sheet first formed in the earliest Oligocene[54]. His summary of Antarctic climate history, in which continental shifts changed ocean currents, which changed climate and even affected the evolutionary pathways of life in the ocean, was highly influential. Cores from the Ocean Drilling Program in the 1980s from the Prydz Bay Shelf and the Kerguelen Plateau confirmed the timing of the first continental ice sheet at 34 Ma ago[46]. A key contributor was Jim Zachos of the University of California, Santa Cruz[55].

Using oxygen isotopes from surface-dwelling and benthic foraminifera in deep-sea drill samples from the Southern Ocean (Figure 7.6), Zachos reported in 1994 that sea surface temperatures in high southern latitudes rose from around 10–12 °C in the late Palaeocene to as high as 15 °C in the early Eocene, declined back to 10–12 °C in the early middle Eocene, then fell to 6 °C in the late Eocene and to 4 °C in the early Oligocene[56]. Deep-water temperatures followed sea surface temperatures but were around 2 °C cooler. Tropical and subtropical sea surface temperatures averaged around 24–25 °C in the early Palaeogene, much like today, while deep-water temperatures then were around 3 °C warmer there than around Antarctica.

Fossil plants confirmed the picture emerging from drilling and from oxygen isotopes. Jane Francis found botanical evidence from Seymour Island showing that after the peak warmth of the mid to late Cretaceous, Antarctica

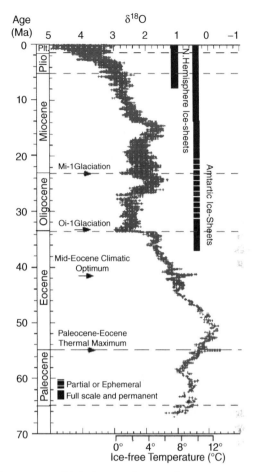

Figure 7.6 *Global compilation of oxygen isotope records for the Cenozoic. Solid bars span intervals of ice-sheet activity in the Antarctic and Northern Hemisphere.*

began to cool[57,58]. Warmth returned during the Palaeocene and early Eocene, between 65 and 50 Ma ago, when there were moist, cool, temperate rainforests like those of southern Chile today, before cooling began again in the mid to late Eocene. According to Francis, 'Richly fossiliferous Eocene sediments have yielded the only fossils of land mammals in the whole Antarctic continent … along with fossil wood, fossil leaves, a rare flower, plus … giant penguins'[58]. Analyses of leaf margin shapes suggested that between the late Palaeocene and the middle Eocene, average Antarctic temperatures fell from around 13 °C to 11 °C, ranging from around 1–2 °C in winter to 24–25 °C in summer. Independent confirmation came from oxygen isotopes in fossil Antarctic molluscs, which suggested that annual average temperatures fell to around 10 °C by the late Eocene. Cooling in the late Eocene was confirmed by the growing abundance of relatives of today's monkey puzzle conifer trees, along with widespread growth of the southern beech, *Nothofagus*, which became progressively more dominant as floral diversity declined in the middle Eocene, around 45 Ma ago. Declining diversity is another sign of cooling.

Confirmation of Antarctica's early Eocene warmth came in 2012 from a core drilled off the Wilkes Land coast at a palaeolatitude of 70 °S, where scientists from Integrated Ocean Drilling Program (IODP) expedition 318 analysed fossil pollen and spores and used organic geochemical proxies to show that the coastal lowlands of the time supported a highly diverse, near-tropical forest, including palms[59]. Winters were warmer than 10 °C and essentially frost-free despite polar darkness.

The pattern of cooling suggests that ice may have been present on the continent, at least during winter, by the middle Eocene, 47–44 Ma ago, although not as a major ice sheet like that of today[57]. Ice-rafted debris suggests the presence of valley glaciers during the Eocene, and '*by the end of the Eocene it is possible that an ice sheet extended over much of the peninsula, although average Seymour Island temperatures did not reach below zero*'[58].

After 3 decades of surveying the Antarctic continental shelf, John B. Anderson of Rice University developed the SHADRIL project to take a series of short (~100 m) cores from gently dipping marine strata of Eocene–Oligocene age beneath the continental shelf on the east side of the Antarctic Peninsula. He and his colleagues found that the onset of full glacial conditions from a beginning near the Eocene–Oligocene boundary was a long, slow process that took several million years to go from the development of mountain glaciers to a full-bodied ice sheet[60]. This slow development, he pointed out, would tend to favour the Kennett hypothesis for the gradual development of ice on the continent. Was he right? We will see further on.

The first big Antarctic ice sheets were similar in extent to modern Antarctic ice sheets, but warmer, thinner and more highly dynamic[61]. Independent confirmation of the changing thermal conditions on the continent came from the ages of glacial tillites on King George Island, west of the Peninsula, dated at 45–41 Ma ago. It seems likely that middle Eocene ice caps would then have been present on the continent above elevations of 1000 m, with perhaps a few valley glaciers reaching sea level.[57, 62]

In 1999, Peter Barrett summarised the results of earlier work, noting that '*The earliest evidence of glaciers forming on the Antarctic continent since the Gondwana glaciation of Carboniferous-early Permian times comes as sand grains in fine-grained uppermost lower Eocene and younger deep-sea sediments from the South Pacific, with isolated sand grains interpreted to record ice-rafting events centred on 51, 48 and 41 Ma. Cores from Maud Rise and Kerguelen Plateau have similar sand grains at around 15 Ma … Lonestones in cores of deep-water late Eocene to early Oligocene mudstones in the CIROS-1 drill hole indicate that glaciers (but not necessarily ice sheets) were discharging ice at sea level through those times … The eustatic sea-level curve … indicates that over this period there were several short periods of substantial sea level change (nominally ~100 m, but more likely a few tens, at 58, 55, 49, 42 and 39 Ma), suggesting the growth and collapse of ice sheets thousands of km across … [although] isotopic data provide no support for such variations in ice volume beyond around 34 Ma*'[63].

The earliest Oligocene, when the first major ice sheet formed, was marked by an ice-rafting event on the Kerguelen Plateau and a change in the mineralogy of clays from smectite to illite and chlorite on the Kerguelen Plateau and Maud Rise, representing a shift from chemical to physical weathering of the Antarctic continent as a consequence of the progressive cooling on land[63]. In Prydz Bay, on the Antarctic coast south of India, '*grounded ice first deposited debris beyond the coast from earliest Oligocene (or possibly late Eocene) times … The ice sheet margin extended to the edge of the continental shelf, implying that the ice sheet of that time was of similar size and extent to the present ice sheet in its expanded form*'[63]. Things were different in the Ross Sea, where '*close to the Transantarctic Mountains glaciers were calving at sea level in Eocene times*'[63]. Drilling in 1999 found evidence for glaciers grounding at sea level in the Ross Sea area from 33 Ma ago onwards. Ice cover increased substantially around 15 Ma ago, since when the East Antarctic Ice Sheet has been a more or less permanent feature, with temperatures close to present values.

Despite Antarctica's icing up, it took a while for terrestrial plant life to be wiped out. As Jane Francis says, '*Even though many warmth-loving plant taxa disappeared during the mid-Eocene, floras dominated by Nothofagus [the southern beech] remained for many millions of years*'[57, 58]. One of the most remarkable Antarctic floras, comprising twigs of *Nothofagus* along with moss cushions, liverworts and the fruits, stems and seeds of several other plants, is preserved near the head of the Beardmore Glacier at 85°S in the Sirius Group of sediments at Oliver Bluffs in the Transantarctic Mountains. This suggests a tundra environment, with a mean annual temperature of around −12°C, summers up to +5°C and winters below freezing. This 'interglacial' environment existed at a time when glaciers temporarily retreated from the bluffs, allowing dwarf shrubs to colonise a tundra surface only about 500 km from the pole. These deposits are regarded as frozen remnants from the time of the warmer ice sheet that pulsated on the continent between 34 and 14 Ma ago. The plants and insect remains from Oliver Bluffs are like the 14 Ma-old remnants discovered by Adam Lewis in the McMurdo Dry Valleys[64].

One of the fascinating problems in reconstructing Cenozoic climate was that the global oxygen isotope data sets and data on sea level suggested that there must have been more ice around than was predicted by general circulation models (GCMs) for the Eocene–Oligocene boundary 34 Ma ago. Could that ice have been located in the Northern Hemisphere? The then current model took it for granted that the area of West Antarctica was relatively small at the time. Was that correct? Doug Wilson and Bruce Luyendyk of the Marine Science Unit of the University of California at Santa Barbara thought not. From the mass of Oligocene sediment found in nearby deep-sea drilling cores, they realised that a significant highland must have existed in the vicinity of West Antarctica. This led them to question the geophysical models for rifting and subsidence of the Ross Sea and uplift of the Transantarctic Mountains, which led in turn to the realisation that, 34 Ma ago, West Antarctica was part of a mountain belt, of which the Transantarctic Mountains are a remnant. There was thus a much larger land area than had been supposed available for the growth of ice on a proto-West Antarctica 34 Ma ago. Having a larger land area available provided the basis for a larger ice sheet, which matched the requirements based on the sea level and isotope data[65]. A Northern Hemisphere ice sheet was not necessary after all.

Does the evidence accumulated since the mid 1970s bear out Kennett's theory that the first ice sheets formed about 34 Ma ago in response to the opening of the Drake Passage and Tasman Rise and the formation of an Antarctic Circumpolar Current that isolated Antarctica thermally by preventing warm currents from impinging on the continent? This was still the prevailing view in the early 1980s[4, 66]. While John Anderson's data from the Peninsula region suggested that ice had built up gradually during the

mid Cenozoic[60], Jim Zachos's isotope data (Figure 7.6) showed that there must have been a major cooling event[55], likely associated with a sudden increase in ice volume at the Eocene–Oligocene boundary. Ken Miller of Rutgers University and his colleagues documented a global sea level fall of some 55 m at that boundary, consistent with the idea of a rapid growth of ice[67]. Even so, before we can answer the question with confidence, we must examine the data for global cooling.

7.5   Global Cooling

So much for Antarctica, which contains the record of Earth's most extreme cooling during the Cenozoic Era. What about the rest of the world? In the 19th century, geologists like Charles Lyell, Eugène Dubois and T.C. Chamberlin were well aware that the world had cooled towards the Ice Age throughout the Cenozoic, just as we saw happening at the regional scale in and around Antarctica. Close inspection shows that the picture was not quite that simple. The Palaeocene, which intervened between the Cretaceous and the Eocene from 65 to 55 Ma ago, had been about as warm as the mid to late Cretaceous, but the Eocene was warmer.

Back in April 1972, at the NATO Palaeoclimate meeting in Newcastle upon Tyne, the warming through the Palaeocene towards the Eocene intrigued Larry Frakes, then at Florida State University, and his colleague Elizabeth Kemp. They were keen to explain '*the poleward expansion of a zone of warm and humid climate during the early part of the Tertiary*' and '*the gradual decrease in temperature which followed*'[68]. This warming, recognised in the 19th century by the palaeobotanist Oswald Heer, as we saw in Chapter 3, was represented by the Eocene temperate forests at the high latitudes of Greenland, Spitzbergen, the high Canadian Arctic and Alaska. The occurrence of such fossil floras, at 81°N in Grinnell Land on Canada's Ellesmere Island, for example, '*points to the absence of a polar ice cap in the Early Tertiary*'[68], according to Frakes and Kemp, echoing for the Arctic what Margolis and Kennet had discovered in the Southern Ocean in 1971[6]. Jan Zalasiewicz and Mark Williams of the University of Leicester later reminded us that the ancient forest buried on Ellesmere Island was discovered during the First International Polar Year of 1882–83 by '*the handsomely whiskered First Lieutenant Adolphus Washington Greely (1844–1935) of the United States Army*'[69]. Greely's experiences were horrific. Stuck in the ice, his party survived by eating their boots and their dead companions. A relief expedition 2 years later found just six men left alive, including Greely; a reminder that in the early days, new geological information came at a sometimes terrible price.

The Norwegians followed the Americans, led by Otto Sverdrup (1854–1930), a pal of fellow Arctic explorer Fridtjof Nansen (1861–1930). Sverdrup had commanded Nansen's *Fram* on its celebrated drift across the Arctic Ocean in 1893–96. In 1898, while trying to circumnavigate Greenland in the *Fram*, he was caught by the ice and had to winter over on Ellesmere Island, something he did again three times between 1899 and 1902. The expedition's geologist, Per Schei, sampled the Eocene forest. His specimens, including the 60 m-tall redwood *Metasequoia*, showed the forest to have been lush and the climate humid. Summer temperatures reached 15 °C, falling to 0 °C in the winter, no more than 700 km from the pole[69].

Plant fossils confirmed that warm, humid conditions typical of the tropics were widespread at around 60 °N in the Eocene, and there was growing evidence that such conditions also prevailed in the Southern Hemisphere, including in the Antarctic Peninsula[68]. Eocene laterites indicating tropical warmth and high rainfall extended from 55 °N to 45 °S, while today they are confined to Köppen's 'type A' climate zones (wet and tropical) between the 30th parallels. By the mid Eocene, coals were forming both in equatorial regions and at high latitudes, making Eocene coals more widespread than Cretaceous ones[62]. Evaporites formed at mid latitudes. Rainfall continued to be high along the high-latitude western margins of the Pacific[62].

In contrast, plant fossils indicated widespread cooling in the Oligocene, especially in North America between 34 and 31 Ma ago, although there was disagreement about how sharp the decline was globally[68]. Laterites indicating warm, humid conditions were less geographically extensive in the Oligocene. Sparse oxygen isotope data confirmed this picture, with shallow water invertebrate shells from Australia and New Zealand showing a drop from 17–22 °C in the Eocene to 12 °C in the Oligocene.

Frakes and Kemp wondered wehther continental displacements might have something to do with these patterns of warming and cooling. To answer that question, they reconstructed continental positions for the Eocene and Oligocene[68]. This was most likely done in 1971 for the April 1972 conference, so predated the definitive suite of reconstruction maps by Smith, Briden and Drewry that

became available in 1973[70]. As there was no significant latitudinal difference of the Arctic between the two time periods, changing proximity to the pole did not account for the cooling.

By the mid 1980s, we knew much more about the distribution of plants during the Cenozoic than did Heer in the late 1800s or Frakes and Kemp in 1972. Jack Albert Wolfe (1936–2005) of the US Geological Survey office in Denver, Colorado argued that if the boundary between tropical and paratropical rain forests bore the same relationship to temperature in the past as it does today in East Asia, where the boundary today lies at 50 °N, then that boundary must have lain at 70 °N in the early Eocene. Broad-leaved evergreens would then have extended north of that to the pole[71]. As more data arrived, it seemed that climate zones could shift through time; they were not as static as might be supposed from the earlier work of Briden and Irving, which was based on a limited suite of samples[72].

Wolfe pointed out that the great poleward expansion of broad-leaved evergreen vegetation in the warm Palaeocene and Eocene must have produced a different carbon cycle from that operating today. Yet another version of the carbon cycle would have operated during the Oligocene and Miocene, when broad-leaved deciduous forests covered large parts of the Northern Hemisphere at middle latitudes, evergreen forests were more restricted than in earlier times and there were no extensive grasslands or deserts. The grasslands and deserts that we see today are late developments in the Cenozoic, something we return to in Chapter 11.

Large leaves with continuous (i.e. smooth) margins are typical of tropical plants, while small leaves with serrated edges indicate cooler climes. Based on those modern relationships, one can infer past temperatures from leaf shape. Accepting that assumption, Wolfe plotted the likely zonal distribution of vegetation on the reconstruction maps of Alan Smith for the late Palaeocene–early Eocene (55–50 Ma ago), for warm intervals of the middle Eocene, for cool intervals of the middle to late Eocene (50–46 and 41–37 Ma ago) and for the early Miocene (22–18 Ma ago). The maps showed the gradual origin and spread of vegetation types of low biomass, including savanna (incorporating woodland and scrub), steppe, taiga, tundra and true desert, all of which seemed to be Miocene or younger[71]. Tundra vegetation consists of dwarf shrubs, sedges and grasses, mosses and lichens, locally with scattered and stunted trees. Taiga (coniferous forest) covers much of Canada, Russia and Siberia south of the tundra. These low-biomass types of vegetation originated and spread in response to prolonged cooling and the steepening of the Equator–pole thermal gradient. The steeper thermal gradient intensified the subtropical high-pressure systems, and the associated summer drought along the western sides of the continents caused deserts to develop there. Savanna spread at low latitudes, where temperature increased and precipitation decreased. As a result, the mass of carbon in land plants decreased during the Cenozoic[73]. Later, we explore the relation of these developments to atmospheric CO_2.

Marine plankton also changed in response to the Cenozoic cooling[62]. The four plankton provinces characterising the Eocene ocean expanded to six by the Oligocene, as two cold polar regions became established, signalling the development of a more differentiated latitudinal temperature gradient. Globigerinid planktonic foraminifera became restricted to mid- and low-latitude sites, and nannofossils moved equatorward.

To see if the Eocene to Oligocene cooling might be related to the separation of the continents, Frakes and Kemp pondered the implications of the five main changes that took place in continental positions during that time. First, India continued moving north, shrinking the Tethyan Ocean between India and Tibet and expanding the Indian Ocean in its wake. Warming began to decline from a peak in the early Eocene 50 Ma ago, when the advancing edge of the Indian fragment of Gondwana, much of it now subducted beneath Tibet, collided with the southern edge of Laurasia, much of it now crumpled up to form the Himalayas. That collision closed the east–west Tethyan Ocean connecting the Atlantic and Pacific through the tropics. Removal of the Tethyan seaway between Laurasia and Gondwana dried the climate along the Mediterranean–Caspian corridor and its surrounding lands. Second, the Australia–New Guinea–New Zealand complex separated from Antarctica and the Drake Passage opened between South America and Antarctica, allowing the creation of the Antarctic Circumpolar Current between 50 and 60° S. As Judy Parrish noted[74], the new wind and ocean current regime brought moisture to west Australia. Equatorial easterlies made east Africa wet, but there would not yet have been a substantial monsoonal regime over Asia, because that would require massive (later) uplift to form the Tibetan Plateau. Third, the northward movement of Australia–New Guinea narrowed the space for the Pacific Equatorial Current between Australia and South East Asia, diminishing the supply of warm water to the widening Indian Ocean. Fourth, continued westward migration of the Americas shrank the Pacific. Fifth, a slight widening of the narrow gap between Greenland and Scandinavia offered the prospect of more oceanic exchange between the Arctic and Atlantic. These opening or closing 'oceanic gateways' ought to have influenced the climate.

Knowing that today's easterly Trade Winds blow warm water west to accumulate in the Pacific Warm Pool just north and east of New Guinea, Frakes and Kemp suggested that during the Eocene, when the Pacific Ocean was wide, the warm pool accumulated more heat than during the Oligocene, when the ocean was narrower[68]. The north- and south-moving branches of the Eocene equatorial current would have moved that warm water towards high latitudes along the coasts of Asia and Australasia, thus warming the polar regions more than would have been possible during the Oligocene. The Earth would also have been slightly warmer than it is now, because there was no substantial polar ice to reflect the Sun's radiation.

Analysing the rather limited and scattered oxygen isotope data from the North Pacific, Australia and New Zealand, the pair went on to suggest that in the Oligocene, '*the extensive cooling indicated by the oxygen isotope curves could have led to the development of the first extensive glaciers on Antarctica*'[68], where evidence for glaciation was beginning to appear. '*We favour a model which allows limited or regional ice formation in the Eocene, construction of the bulk of the Antarctic ice sheet in the Oligocene, and culmination of this process in the Late Miocene by the development of extensive ice shelves … Ice formed during the Oligocene probably exceeded the amount built up during the Eocene because the principal source, the new seaway south of Australia … was much closer to the interior of the old polar continent*'[68].

This begs the question, what caused the extensive cooling from the Eocene into the Oligocene? Was it, as Frakes and Kemp suggested[68], the narrowing of the Pacific Ocean, which weakened the supply of warm water to the polar regions? Their analysis ignored two potentially important controls on global heat exchange. One was a change in the global content of CO_2 in the atmosphere, which we examine in Chapter 9. The other was the potential of the newly formed Antarctic Circumpolar Current to isolate Antarctica thermally from the rest of the world, as Kennett suggested. We'll look at the implications of that next.

By 1975, as more and more cores arrived from the DSDP, which was then only 7 years old, the analysts of oxygen isotopes began to derive a comprehensive picture of global climate change for the past 70 Ma. One of the leading analysts was Sam Savin, from Case Western Reserve University in Cleveland, Ohio[75]. A striking feature of his new $\partial^{18}O$ record was its fall from the end of the Palaeocene to the present in a series of major steps indicative of cooling. These steps were far more apparent in the polar ocean samples analysed by Shackleton and Kennett[53] than in the tropical ocean samples analysed by Savin and his team[75]. One of these cooling steps occurred at the boundary between the Eocene and the Oligocene. As we saw earlier, in the mid 1970s when these data started to become available, that cooling was thought to represent the change in ocean circulation caused by the separation of Australia and South America from Antarctica. This opening allowed the formation of an Antarctic Circumpolar Current and associated strong westerly winds that isolated Antarctica from the warmer climes to the north, encouraging the formation of a major ice sheet, which in turn led to the formation of cold, dense bottom water capable of changing the global circulation of the world ocean[53].

Refinements to the curve of changing oxygen isotopes continued with time. One was made in 1987 by Ken Miller (1956–) from Lamont, whom I remember as the fast-talking, cigar-smoking PhD student I played poker with when I was on the staff at WHOI in the 1970s. Miller and his team compiled oxygen isotope data from planktonic foraminifera and benthic foraminifera collected by deep-ocean drilling to suggest that ice-free conditions characterised the Palaeocene and Eocene, 65–34 Ma ago, and that continental ice sheets characterised the Oligocene and later world from 34 Ma ago to the present[76]. Evidently, the positive isotopic shifts seen in earliest Oligocene and middle Miocene sediments of the Southern Ocean by Shackleton and Kennett in 1975[53] were indeed global.

One significant refinement to the interpretation of the curve of changing $\partial^{18}O$ with time came from Shackleton's realisation that only about half of the oxygen isotope signal from the base of the Oligocene onwards represented the cooling of the ocean, while the rest represented the trapping of water on land as ice[77, 78]. We now know that the $\partial^{18}O$ trend in benthic foraminifera is a proxy for two properties of the climate: first, deep-ocean temperatures, which represent the temperatures of surface waters at the high latitudes where those waters formed; and second, the seawater $\partial^{18}O$ value, which reflects how much water is tied up in ice on land[26].

This picture was further refined by Jim Zachos and colleagues, who confirmed in 2001 that temperatures had risen from warm conditions in the late Cretaceous

towards a climatic optimum in the early Eocene, between 54 and 50 Ma ago, when deep-ocean temperatures reached 10–12 °C, as we saw in Figure 7.6[79]. Temperatures then fell to around 5–6 °C at the Eocene–Oligocene boundary 34 Ma ago, when a major ice sheet first formed on Antarctica.

Measurements of oxygen isotope data from benthic foraminifera of late middle Eocene to early Oligocene age (38–28 Ma ago) by Miriam Katz of Rensselaer Polytechnic Institute and her team also showed the jump at the Eocene–Oligocene boundary, representing the cooling that accompanied formation of the Antarctic ice sheet (as in Figure 7.6)[80]. Katz's carbon isotopic data showed that a large $\partial^{13}C$ offset developed between mid-depth waters (about 600 m deep) and waters deeper than 1000 m in the western North Atlantic in the early Oligocene, indicating the development of a low-$\partial^{13}C$ zone at intermediate water depths at around 31–30 Ma ago[80]. Low $\partial^{13}C$ means abundant ^{12}C organic matter, which is typically abundant due to decomposition in the oxygen minimum zone. In the modern ocean, this is about the depth of Antarctic Intermediate Water (AAIW), which sinks at the polar front. Katz deduced that '*ventilation by AAIW leads to a relatively low O_2 and low $\partial^{13}C$ layer ($\sim$700–1000 m) ... At the same time, the ocean's coldest waters became restricted to south of the ACC [Antarctic Circumpolar Current], probably forming a bottom-ocean layer, as in the modern ocean*'[80].

Katz deduced that the modern four-layer ocean structure of surface, intermediate, deep and bottom waters probably developed in the Oligocene as a consequence of the development of the Antarctic Circumpolar Current and its associated frontal structures. In effect, '*intermediate water circulation today is a consequence of the ACC, which blocks warm surface waters entrained in subtropical gyres from reaching Antarctica; this thermally isolates the continent and the surrounding ocean, allowing large-scale ice sheets to persist.*[80]' As an aside, that helps explain what keeps Antarctica cold today. The Arctic lacks this thermal isolation, receiving warm subsurface water from the North Atlantic via the Norwegian-Greenland Sea.

The Antarctic Circumpolar Current began to develop in the middle Eocene, Katz thought, with shallow through-flow via the Drake Passage. It strengthened and deepened with the opening of the Tasman Gateway between Australia and Antarctica in the late Eocene–early Oligocene and the deepening of the Drake Passage in the Oligocene. While it was thought that the Antarctic Circumpolar Current did not develop until the late Oligocene, even a shallow Antarctic Circumpolar Current would have thermally isolated Antarctica from the rest of the ocean from the late Eocene onwards[80]. Several lines of evidence suggest the progressive development of a deep Antarctic Circumpolar Current from the middle Eocene through to the late Oligocene.

Katz's team's data from the Atlantic continental slope, and their comparison of one ocean basin with another, showed that the thermal structure and circulation of the ocean were affected at an early stage of the opening of the Drake Passage. Their results suggested that '*changes in tectonic gateways affected the middle Eocene to early Miocene ocean circulation, which in turn affected global climate*'[80].

How do Katz's conclusions square with other interpretations? Some other data do not entirely support development of the Antarctic Circumpolar Current as the sole cause of the build-up of ice on the continent[81]. By 2009, several pieces of evidence would change the picture[58]. It now appears that narrow gaps opened between Antarctica and Tasmania and between Antarctica and South America in the mid to late Eocene, before the first ice sheets developed at the Eocene–Oligocene boundary. However, the dinoflagellate phytoplankton population around the Antarctic coast, especially in the narrow gaps between Tasmania and Antarctica and between South America and Antarctica, proved to be highly endemic (i.e. local), precluding the operation of a major circumpolar current connecting these environments in the late Eocene. Numerical modelling of the behaviour of the Southern Ocean suggests instead that during the Eocene it comprised two major clockwise gyres: one on the South Pacific side and one on the South Atlantic and Indian Ocean sides of the continent, with no major connection between them. In addition, Peter Barker and Ellen Thomas found that the inception of siliceous biogenic sedimentation near 34 Ma ago did not necessarily indicate deep-water connections and circum-Antarctic through-flow[81].

Clearly, the geographical isolation of Antarctica by an eventual Antarctic Circumpolar Current driven by powerful westerly winds would, as Kennett first suggested, have contributed to Antarctica's thermal isolation. Nevertheless, as we shall see in Chapter 9, the icing up of Antarctica may also have been driven, at least in part, by cooling forced by the gradual decline in the CO_2 content of the air.

Zachos's $\partial^{18}O$ data showed that short warm periods occurred in the late Oligocene and middle Miocene (Figure 7.6), before another isotopic shift heralded the formation of a major ice sheet on East Antarctica in the mid Miocene (15–14 Ma ago)[79]. In the mid 1980s, this shift was thought to have initiated the modern ocean–atmosphere system, with its cold polar regions, strong Equator–pole temperature gradient and well-developed thermocline (warm waters sitting over cold)[4, 82]. A corollary of this concept was that the southern Subtropical Convergence separating subtropical waters from the sub-Antarctic waters of the Antarctic Circumpolar Current developed then, further isolating Antarctica from tropical warmth[62]. The subsequent work of Katz's team suggests that while those changes may have been accentuated in the mid Miocene, they had been initiated in the Oligocene[80].

High-latitude bottom waters were as cool as about 5 °C in the mid to late Miocene but had dropped to 2.5 °C by the Pleistocene, reflecting the further development of the world's cold deep water. The circulation of North Atlantic Deep Water gradually strengthened through that period, intensifying at around 2.75 Ma ago[62]. We will explore the details of Arctic change in the next section.

By the mid Miocene, the rising of the Tibetan Plateau, caused by the collision of India with Asia, would have strengthened monsoon circulation[74]. Sea level was dropping, and east Africa had become increasingly arid.

There was a further cooling in the latest Miocene, 6–5 Ma ago. Coinciding with the effects of the ongoing collision between Africa (Morocco) and Europe (Spain), which was gradually closing the equatorial Tethyan Ocean at the Strait of Gibraltar, that cooling created sufficient land ice to lower sea level enough to cut the connection between the Atlantic and the Mediterranean. The Mediterranean Sea almost completely dried out, leading to the deposition on its bed of thick accumulations of salt in a number of discrete basins[83, 84]. It is unclear whether the floor of the Mediterranean comprised one large brackish lake or a series of smaller ones, in which salts might precipitate in relatively deep water. In the eastern Mediterranean, the River Nile subsequently dumped vast masses of sediment on top of these Miocene salt deposits to form a giant submarine fan, which mantles the continental slope seawards of the Nile Delta. Its weight has squeezed the salt into vertical finger-like intrusions up to a kilometre or so across, like those beneath the continental slope of the Gulf of Mexico[85]. As Harry Elderfield and I found during a WHOI research cruise to the eastern Mediterranean in 1975, the salt is rising through the thick pile of sediments dumped on the continental slope by the outflow of the Nile[86].

Besides these major changes, the oxygen isotope record (Figure 7.6) reveals much low-level, short-term variability, most of it reflecting changes in the amount of insolation received from the Sun due to regular changes in the tilt of the Earth's axis[79]. These changes have a period of around 40 Ka and persisted through much of the Neogene and Oligocene.

Can we use the new techniques of palaeoceanography to separate the influence of temperature and ice volume in the Cenozoic? The answer is likely to be 'yes'. For example, comparing Mg/Ca-based temperatures with measured $\partial^{18}O$ across the Eocene–Oligocene boundary confirmed that the $\partial^{18}O$ shifts were dominated by changes in global ice volume[21]. In 2011, a team led by Benjamin Cramer of Rutgers University used Mg/Ca ratios and $\partial^{18}O$ records from benthic foraminifera, along with a record of sea level change as a proxy for ice volume, to evaluate trends in ice volume and deep-ocean temperature for the past 108 million years (Figure 7.7)[26]. They showed that ice volume and temperature did not change proportionately through time. Cenozoic deep-ocean cooling

occurred in two prolonged intervals: a gradual decline in temperature from about 14 °C in the early Eocene to about 6 °C in the late Eocene and a gradual decline from about 6 °C in the late Miocene to 0 °C in the Pliocene. In contrast, the Cenozoic increase in ice volume (and decrease in sea level) took place in three steps: one at the Eocene–Oligocene boundary, one in the middle Miocene and one in the Plio-Pleistocene. According to Cramer, '*These differences are consistent with climate models that imply that temperatures ... should change only gradually on timescales >2 [Ma], but growth of continental ice sheets may be rapid in response to climate thresholds due to feedbacks that are not yet fully understood*'[26].

Can numerical models help to explain the cooling? In 2000, a team led by Karen Bice of WHOI used atmosphere and ocean GCMs with fixed atmospheric CO_2 to see whether changes in palaeogeography could affect heat transport and so explain the high-latitude and deep-water warmth of the early Cenozoic and its subsequent decline[87]. Using an atmospheric GCM connected to a slab ocean, they found little difference between different time slices

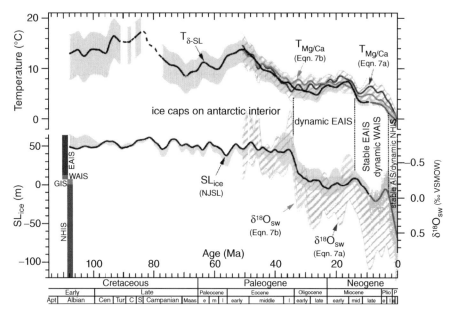

Figure 7.7 *Variations in bottom-water temperature and sea level for the past 108 Ma. Data smoothed to show only variations on >5 Ma timescales. Vertical bars indicate the approximate cumulative range of sea level variations due to ice sheets. EAIS = East Antarctic Ice Sheet, WAIS = West Antarctic Ice Sheet, GIS = Greenland Ice Sheet, NHIS = Northern Hemisphere ice sheets, NJSL = New Jersey sea level. Shading indicates confidence levels; cross-hatching indicates uncertainties in either temperature (top) or seawater $\partial^{18}O$ (bottom) estimates. The latter do not necessarily reflect realistic uncertainty in ice-volume calculations.*

(early Eocene, middle Eocene, early Oligocene, early Miocene and middle Miocene), suggesting that variations in geography were not enough by themselves to explain the global Cenozoic cooling trend. In contrast, a full three-dimensional ocean model forced by the output of the atmospheric GCM produced substantial differences in the magnitude of heat transport from the Equator to the poles between the different time periods.

Given the tropical conditions typical of the Eocene in the Arctic, Bice's model outputs are quite instructive. At that time, there was much higher poleward heat transport in the Southern Hemisphere than in the Northern. This is because warm equatorial water could flow south to Antarctica in the absence of an Antarctic Circumpolar Current, the Drake Passage being closed. In contrast, heat transport north of the Equator was dominated by a strong east–west current, taking warm equatorial waters from the Pacific through the narrowing Tethyan Ocean between India and Asia and across the Atlantic into the Pacific via the Central American Seaway. Links to the Arctic were weak, due to the slow opening of the northernmost Atlantic. So what made Ellesmere Island more or less tropical in the Eocene? The answer seems to be the atmosphere. Sea levels were high, warm conditions caused increased evaporation and the atmosphere transported significant latent heat poleward in humid air, to be released in the Arctic. All that would be required to achieve this would be higher than modern equatorial temperatures.

Over time, the Equator–pole heat transport in the Southern Hemisphere reduced with the gradual opening of the Southern Ocean's gateways, as the distance widened between Antarctica and both South America and Australia. In the Northern Hemisphere, the gradual closing of the Tethyan gateway by the collision of India and Asia, along with the closing off of the Mediterranean and the narrowing of the central American and Indonesian–Pacific gateways, blocked east–west equatorial transport and diverted warm currents towards the Arctic. Over time, warm, salty water reaching the Norwegian-Greenland Sea eventually cooled and became dense enough to sink and move south at depth as North Atlantic Deep Water, strengthening the Meridional Overturning Circulation between 55 and 14 Ma ago, as the northern North Atlantic basins opened. Northern Hemisphere overturning and heat transport increased further as the central American seaway closed and the Indonesian–Pacific seaway further narrowed towards modern times. Clearly, geography plays a key role in controlling heat transport by the ocean[87].

But as we shall see later, changes in CO_2 may be equally important, if not more so.

7.6   Arctic Glaciation

In contrast to the major development of ice on Antarctica starting 34 Ma ago, it was long thought that the ice sheets of the Northern Hemisphere did not develop until the Pleistocene, beginning around 2.6 Ma ago. But by the early 1990s, reports of so-called 'drop-stones' in sediments from Svalbard suggested that there might have been some seasonal winter ice there in Eocene times, while ice-rafted debris was recovered in sediments of late Miocene age (5 Ma ago) from the central Arctic Ocean[62]. Alaskan tillites suggested intermittent glaciation in Alaska's Wrangel Mountains over the past 10 Ma[62]. Pack ice may have been present intermittently in the Arctic basin since 13–10 Ma ago, and there were signs of ice-rafting in the Russian sector of the Pacific from the late Oligocene onwards[62]. This and related evidence convinced Ana Christina Ravelo of the University of California, Santa Cruz and her colleagues in 2007 that *'small ice sheets were present in the Miocene and earliest Pliocene, the marked increase in global ice volume, or the onset of significant Northern Hemisphere glaciation, occurred gradually from 3.5–2.5 Ma'*[88]. By 2.6 Ma ago, vast regions of the Northern Hemisphere were in the grip of huge ice sheets that covered large areas of the Eurasian Arctic, northeast Asia and Alaska, and, by 2.5 Ma ago, North America (more on that in Chapter 13)[61].

Coming at the question of Arctic glaciation from a different perspective, Aradhina Tripati and her colleagues thought that the changes in the deep-sea oxygen isotope ($\partial^{18}O$) record during the Cenozoic were larger than could be explained by the formation of ice on land just in Antarctica, and concluded, *'there was ice stored on Antarctica and in the Northern Hemisphere at 44.5 Ma, 42 Ma, 38 Ma, and after 34 Ma'*[89]. If true, perhaps this meant that the transition from greenhouse to icehouse was not as sudden as had been initially suggested, although the transition from mountain ice caps to ice sheets on Antarctica may well have taken place over a short period 34 Ma ago. Was Tripati right? No. As we saw earlier, by 2009 Wilson and Luyendyk had deduced that the missing ice had not been in the Northern Hemisphere, as Tripati thought, but on West Antarctica[65].

The only way to test ideas about the timing of Arctic glaciation was to drill close to the North Pole. This was done in 2004, on the Arctic Ocean Coring Expedition (ACEX), Leg 302 of the IODP, under the leadership of Ted Moore[90]. The expedition drilled close to the North Pole on the Lomonosov Ridge, using an ice-strengthened drill ship, the *Vidar Viking*, protected by icebreakers: the *Sovetskyi Soyuz* from Russia and the *Oden* from Sweden.

The results were stunning, as Jan Backman and Kathryn Moran explain[91]. During the Eocene, the Arctic was an isolated deep-water basin containing brackish water with a salinity of 21–25‰. From time to time, its surface became fresh enough to support blooms of the freshwater fern *Azolla*, which formed vast organic mats. The main *Azolla* event lasted for about 1 Ma between the late early Eocene and the early middle Eocene. Isolation from the adjacent Pacific and Atlantic Oceans ended about 45 Ma ago, but there was no substantial deep-water connection until the Miocene (see later). Starting about 46 Ma ago, conditions cooled to the point where sea ice formed a seasonal cover, transporting ice-rafted debris, mainly from the Eurasian margin. The rate of supply of ice-rafted sediment increased with time into the Miocene, sea ice becoming permanent about 13 Ma ago, shortly after the mid Miocene warm climatic optimum 15 Ma ago. At that time, Arctic Intermediate Water began to form on the Eurasian shelf, presumably in response to increased density caused by the excretion of salt into subsurface waters as sea ice formed at the surface.

The full-scale connection of the Arctic Ocean to the North Atlantic was delayed because, although seafloor spreading began in the Norwegian-Greenland Sea in the Eocene around 54 Ma ago, the Fram Strait between Svalbard and Greenland did not open fully until sometime between the early Miocene (25–20 Ma ago) and the late Miocene (9.8 Ma ago)[92]. The ACEX results show that the ventilation of the Arctic by deep Atlantic waters began 17.5 Ma ago, at about the same time as Arctic outflow to the Atlantic began[91].

In 2010, Jörn Thiede and colleagues brought together all of the evidence on Arctic ice-rafting to paint a picture of the history of Northern Hemisphere glaciation for the Cenozoic[93]. Thiede is the former director of Germany's Alfred Wegener Institute, and currently holds positions with the Academy of Sciences and Literature of Mainz, in Germany, and the St Petersburg State University, in Russia. For his contributions to the study of climate change, and especially Arctic climate change, he has received several awards, among them the Murchison Medal of the Geological Society of London, presented in 1994.

Thiede's analysis shows that we cannot say anything about the sea ice cover of the Arctic Ocean for the late Eocene to the early Miocene (44.4–18.2 Ma ago) or for much of the Miocene (11.6–9.4 Ma ago) because sediment is missing in the cores from the Lomonosov Ridge. But it is plain that there was abundant sea ice in the Arctic even in the late Eocene and that it was a continuous feature of the Arctic winter from possibly 47.5 Ma ago onwards[93].

Thiede's group also examined the evidence from deep-ocean drilling for ice-rafting around Greenland, concluding that there must have been an ice sheet there much earlier than was formerly supposed. Ice-rafting in this area was most likely from icebergs rather than, as in the Arctic Ocean, from sea ice. The data '*suggest the more or less continuous existence of the Greenland ice sheet since 18 Ma, maybe much longer*'[93]. That is consistent with evidence cited by Backman and Moran for some glaciation on Greenland as early as 38 Ma ago[91].

What might explain the development of conditions cold enough for sea ice to form in the Arctic Ocean 47 Ma ago? Thiede and his team '*speculate that this may be related to a reorganization of Siberian drainage patterns as a result of the tectonic changes subsequent to the collision of the plate carrying the Indian subcontinent northwards, once it collided with the southern margins of the Eurasian plate*'[93]. That collision began about 50 Ma ago and may have caused Siberia to tilt down to the north, away from Tibet, increasing the runoff of freshwater from central Asia to the Arctic, making conditions more suitable for the formation of sea ice there[93].

These new data paint a different picture from the old one. We now have evidence for ice at both poles dating back into the Eocene. Ice was evidently beginning to be present in the Arctic around 46 Ma ago, and, as Barrett told us, there is evidence for ice-rafting events in the South Pacific at 51, 48 and 41 Ma ago[63]. Once you have ice, the Earth's albedo increases, thus providing a positive feedback to whatever caused the late Eocene cooling[91].

What sort of temperatures were we dealing with? The global isotopic and related data sets and their translation into global temperature records (Figures 7.6, and 7.7)[79] are matched by palaeoclimate data from the Arctic. A multinational team led by Gifford Miller found in 2010 that late Cretaceous Arctic temperatures averaged 15 °C, and that temperatures there were still at about 10 °C in the Eocene 50 Ma ago, when *Azolla* dominated the vegetation of the relatively fresh Arctic surface waters[94]. Temperatures rose by 3 °C in the later Eocene, when forests of *Metasequoia* dominated an Arctic landscape of organic-rich floodplains and wetlands. Despite the cooling and floral turnover at the Eocene–Oligocene transition 34 Ma ago, warm conditions persisted into the early Miocene 23–16 Ma ago, when the central Canadian islands were covered in mixed conifer-hardwood forests like those of southern maritime Canada and New England today. *Metasequoia* was still present, although less abundant than in the Eocene. This was the period of winter sea ice, according to the ACEX teams[91].

Late Miocene–Pliocene riverside forests of pine, birch and spruce populated the Canadian Arctic islands after a pronounced mid Miocene cooling[94]. Miller's palaeo-climate data from Ellesmere Island suggest mean annual temperatures there were 14 °C warmer than today as recently as the mid Pliocene warm period. Given that the Arctic Ocean was rather fresh in the Eocene, we cannot call on poleward heat transport by the ocean to keep conditions warm at that time. Miller and his team pointed out that *'Taken very broadly, the Arctic changes parallel global changes during the Cenozoic, except that the changes in the Arctic were larger than those globally averaged … In general, global and Arctic temperature trends parallel changing atmospheric CO_2 concentrations, which is the likely cause for most of the temperature changes'*[94]. We will look at the role of CO_2 in later chapters.

Once again, we see the slow accumulation of fragments of knowledge filling out the spaces in the jigsaw puzzle. Some pieces don't fit and have to be discarded; others fit well. The Cenozoic palaeoclimate data confirm that, as in the Cretaceous (Chapter 6), despite the fact that there was land in the polar regions, the world did not become cool as a consequence of Antarctica's arrival at the South Pole 90 Ma ago, contrary to what would be implied by Lyell's theory. It took 45 Ma before a major ice sheet developed on Antarctica. The data also confirm that, while there is considerable evidence for the persistence of climate zones more or less unchanged through time, they expanded during periods of extreme warmth such as the mid to late Cretaceous and the Eocene, and began to contract with the cooling that began in the late Eocene. The warm world of the Cretaceous and early Cenozoic gave way after 50 Ma ago to a cool mid to late Cenozoic world, in which sea ice formed at both poles and large ice sheets then formed apparently first on Antarctica and then in the Arctic, although the jury is still out on the dating of the first Greenland glaciation. Much the same level of global cooling affected the late Cenozoic as had occurred in the Carboniferous and early Permian.

The cooling from the middle Eocene (50 Ma) onwards was a puzzle. It was reasonable to explain the small changes from warm to cold that were superimposed on the broad envelope of Cenozoic cooling as being the product of Milankovitch's orbital variations on time scales of 10–100 Ka, but these did not explain the underlying cooling trend. Could the Sun be at fault? There was no evidence to suggest that the trend was the product of variations in solar output. If we only had the Sun's output to consider then – as Jim Hansen and Makiko Sato pointed out[95] – the planet should have warmed, because solar luminosity increases as the Sun continues to burn hydro-gen to form helium by nuclear fusion, and so the Sun is slowly getting brighter. Solar physics models suggest that during this time, the Sun should have brightened by about 0.4%, equivalent to a forcing of about 1 W/m². We should have seen modest warming throughout the Cenozoic Era.

Most geologists prior to the mid 1980s looked for physical explanations for the cooling, like changes in continental shape and position. Remember that Frakes and Kemp thought the world cooled because the Pacific shrank, diminishing the supply of heat to the polar regions. Shack-leton and Kennett thought that the separation of Australia and Antarctica created the Antarctic Circumpolar Current, causing ice to form on Antarctica and cold deep water to cool the ocean.

What was really going on? Why were there warm forests at high latitudes in the early Cenozoic? Why did sea ice begin forming at both poles in the late Eocene? Could we blame changes in ocean circulation? Or did CO_2 play an unexpected role? How had atmospheric chemistry changed? Was that what geologists had missed from their thinking? As we shall see in later chapters, in order to fully understand climate change, geologists would have to break away from their conservative bastion and let in the light being shed by discoveries in other disciplines, especially oceanography, meteorology, glaciology and atmospheric and ocean chemistry. Above all, they would have to consider the behaviour of the carbon cycle. Hav-ing established how palaeoclimatology developed as a discipline, we now turn to the discoveries being made outside geology about the nature and influence on climate of the greenhouse gases.

References

1. Schott, W. (1935) Wissensch. *Ergebn. Deutschen Atlantis-chen Expedition Vermess. Forschungsschiff 'Meteor' 1925–1927* **3** (3), 43–134.

2. Hollister, C.D., Silva, A.J. and Driscoll, A. (1973) A giant piston corer. *Ocean Engineering* **2**, 159–168.

3. Menard, H.W. (1967) *Anatomy of an Expedition*. McGraw Gill, New York.

4. Kennett, J.P. (1982) *Marine Geology*. Prentice-Hall, New York.

5. Riedel, W.R. (1963) The present record: paleontology of pelagic sediments. In: *The Sea: Ideas and Observations on Progress in the Study of the Seas, Vol. 3, The Earth Beneath the Sea – History* (ed M.N. Hill). John Wiley & Sons, Chicherster, pp. 866–887.

6. Margolis, S. and Kennett, J.P. (1971) Cenozoic paleoglacial history of Antarctica recorded in subantarctic deep-sea cores. *American Journal of Science* **271**, pp. 1–36.

7. Kennett, J.P., Houtz, R.E., Andrews, P.B., Edwards, A.R., Gostin, V.A., Hajós, M., *et al.* (1975) *Initial Reports of the Deep Sea Drilling Project 29*. US Government Printing Office, Washington, DC.

8. Arrhenius, G. (1952) Sediment cores from the East Pacific: properties of the sediment. *Reports of the Swedish Deep-Sea Expedition 1947–48* **1**, 1–288.

9. Arrhenius, G. (1966) Sedimentary record of long-period phenomena. In: *Advances in Earth Science* (ed. P.M. Hurley). Contributions to the International Conference on the Earth Sciences, MIT, 1964. MIT Press, Camridge, MA, pp. 155–174.

10. Emiliani, C. (1954) Temperatures of Pacific bottom waters and polar superficial waters during the Tertiary. *Science* **119**, 853–855.

11. Emiliani, C. (1956) Oligocene and Miocene temperature of the equatorial and subtropical Atlantic Ocean. *Journal of Geology* **64**, 281–288.

12. Shackleton, N.J. (1965) The high-precision isotopic analysis of oxygen and carbon in carbon dioxide. *Journal of Scientific Instruments* **42**, 689–692.

13. McCave, I.N. and Elderfield, H. (2011) Sir Nicholas John Shackleton. 23 June 1937–24 January 2006. Biographical Memoirs of Fellows of the Royal Society. Available from: http://rsbm.royalsocietypublishing.org/content/57/435 (last accessed 29 January 2015).

14. Summerhayes, C.P. and Shackleton, N.J. (eds.) (1986) *North Atlantic Palaeoceanography*. Geological Society and Blackwell, London.

15. Summerhayes, C.P. (1990) Palaeoclimates, *Journal of the Geological Society of London* **147**, 315–320.

16. Brassell, S.C., Eglinton, G., Marlowe, I.T., Pflaumann, U. and Sarnthein, M. (1986) Molecular stratigraphy: a new tool for climatic assessment. *Nature* **320**, 129–133.

17. Lawrence, K.T., Herbert, T.D., Dekens, P.S. and Ravelo, A.C. (2007) The application of the alkenone organic proxy to the study of Plio-Pleistocene climate. In: *Deep-Time Perspectives on Climate Change: Marrying the Signal from Computer Models and Biological Proxies* (eds M. Williams, A.M. Haywood, F.J. Gregory and D.N. Schmidt). The Geological Society, London, pp. 539–562.

18. Herbert, T.D. (2003) Alkenone paleotemperature determinations. In: *The Oceans and Marine Geochemistry, Vol. 6, Treatise on Geochemistry* (eds H.D. Holland and K.K. Turekian). Elsevier-Pergamon, Oxford, pp. 391–432.

19. Prahl, F.G. and Wakeham, S.G. (1987) Calibration of unsaturation patterns in long-chain ketone compositions for paleotemperature assessment. *Nature* **330**, 367–369.

20. Elderfield, H. and Ganssen, G. (2000) past temperature and ∂^{18}o of surface ocean waters inferred from foraminiferal Mg/Ca ratios. *Nature* **405**, 442–445.

21. Lea, D.W. (2003) Elemental and isotopic proxies of past ocean temperatures In: *The Oceans and Marine Geochemistry, Vol. 6, Treatise on Geochemistry* (eds H.D. Holland and K.K. Turekian). Elsevier-Pergamon, Oxford, pp. 365–390.

22. Nürnberg, D., Bijma, J. and Hemleben, C. (1996) Assessing the reliability of magnesium in foraminiferal calcite as a proxy for water mass temperatures. *Geochimica et Cosmochimica Acta* **60** (5), 803–814.

23. Rosenthal, Y., Boyle, E.A. and Slowey, N. (1997) Temperature control on the incorporation of Mg, Sr, Fe, and Cd into benthic foraminiferal shells from the Little Bahama Bank: prospects for thermocline paleoceanography. *Geochimica et Cosmochimica Acta* **61** (17), 3622–3643.

24. Lear, C.H., Elderfield, H. and Wilson, P.A. (2000) Cenozoic deep-sea temperatures and global ice volumes from Mg/Ca in benthic foraminiferal calcite. *Science* **287**, 269–272.

25. Tripati, A.K., Delaney, M.L., Zachos, J.C., Anderson, L.D., Kelly, D.C. and Elderfield, H. (2003) Tropical sea-surface temperature reconstruction of the early Paleogene using Mg/Ca ratios of planktonic foraminifera, *Paleoceanography* **18** (4), 1101.

26. Cramer, B.S., Miller, K.G., Barrett, P.J. and Wright, J.D. (2011) Late Cretaceous–Neogene trends in deep ocean temperature and continental ice volume: reconciling records of benthic foraminiferal geochemistry (∂^{18}O and Mg/Ca) with sea level history. *Journal of Geophysical Research* **116**, C12023.

27. Barron, E.J. (1987) Eocene equator-to-pole surface ocean temperatures: a significant climate problem? *Paleoceanography* **2**, 729–739.

28. Pearson, P.N., Ditchfield, P.W., Singano, J., Harcourt-Brown, K.G., Nicholas, C.J., Olsson, R.K., *et al.* (2001) Warm tropical sea surface temperatures in the late Cretaceous and Eocene epochs. *Nature* **413**, 481–487.

29. Pearson, P.N., van Dongen, B.E., Nicholas, C.J., Pancost, R.D., Schouten, S., Singano, J.M. and Wade, B.S. (2007) Stable warm tropical climate through the Eocene epoch. *Geology* **35**, 211–214.

30. Montañez, I.P., Norris, R.D., Algeo, T., Chandler, M.A., Johnson, K.R., Kennedy, M.J., *et al.* (2011) *Understanding Earth's Deep Past: Lessons for Our Climate Future*. National Academies Press, Washington, DC.

31. Boyle, E.A. and Keigwin, L.D. (1982) Deep circulation of the North Atlantic over the last 200,000 years: geochemical evidence. *Science* **218** (4574), 784–787.

32. Anderson, R.F. (2003) Chemical tracers of particle transport. In: *The Oceans and Marine Geochemistry, Vol. 6, Treatise on Geochemistry* (eds H.D. Holland and K.K. Turekian). Elsevier-Pergamon, Oxford, pp. 247–291.

33. Hollis, C.J., Taylor, K.W.R., Handley, L., Pancost, R.D., Huber, M., Creech, J.B., *et al.* (2012) Early Paleocene temperature history of southwest Pacific Ocean: reconciling proxies and models. *Earth and Planetary Science Letters* **349–350**, 53–66.

34. Cronin, T.M. (2010) *Paleoclimates: Understanding Climate Change Past and Present*. Columbia University Press, New York.

35. Di Venere, V.J., Kent, D.V. and Dalziel, I.W.D. (1994) Mid-Cretaceous paleomagnetic results from Marie Byrd Land, West Antarctica: a test of post-100 ma relative motion between East and West Antarctica. *Journal of Geophysical Research* **99** (B8), 15 115–15 139.

36. Nichols, R.L. (1964) Present status of Antarctic glacial geology. In: *Antarctic Geology* (ed. R.J. Adie). North-Holland, Amsterdam, pp. 123–137.

37. Craddock, C., Bastien, T.W. and Rutford, R.H. (1964) Geology of the Jones Mountains area. In: *Antarctic Geology* (ed. R.J. Adie). North-Holland, Amsterdam, pp. 171–187.

38. Armstrong, R.L., Hamilton, W. and Denton, G.H. (1968) Glaciation in Taylor Valley, Antarctica, older than 2.7 million years. *Science* **159**, 187–89.

39. Denton, G.H., Armstrong, R.L. and Stuiver, M. (1970) Late Cenozoic glaciation in Antarctica the record in the McMurdo Sound region. *Antarctic Journal of the United States* **5**, 15–21.

40. Kennett, J.P. (1967) Recognition and correlation of the Kapitean Stage (upper Miocene, New Zealand). *New Zealand Journal of Geology and Geophysics* **10**, 1051–1063.

41. Kennett, J.P. and Margolis S.V. (1970) Antarctic glaciation during the Tertiary recorded in sub-Antarctic deep-sea cores. *Science* **170**, 1085–1087.

42. Watkins, N.D. and Kennett, J.P. (1971) Antarctic bottom water: major change in velocity during the late Cenozoic between Australia and Antarctica. *Science* **173**, 813–818.

43. http://arf.fsu.edu/publications/documents/DVDP_01.pdf (last accessed 29 January 2015).

44. Hayes, D.E., Frakes, L.A., Barrett, P.J., Burns, D.A., Chen, P.-H., Ford, A.B., *et al.* (1973) *Initial Reports of the Deep Sea Drilling Project, Vol. 28, Freemantle, Australia to Christchurch, New Zealand, December 1972–February 1973*. US Government Print Office, Washington, DC.

45. McGinnis, L.D. (1981) *Dry Valley Drilling Project*. Antarctic Research Series 81. American Geophysical Union, Washington, DC.

46. Barrett, P.J. (2007) Cenozoic climate and sea level history from glacimarine strata off the Victoria Land Coast, Cape Roberts Project, Antarctica. In: *Glacial Processes and Products* (eds M. J. Hambrey, P. Christoffersen, N.F. Glasser and B. Hubbart). International Association Of Sedimentologists, Gent, pp. 259–287.

47. Barrett, P.J. (2009) A history of Antarctic Cenozoic glaciation – view from the margin. In: *Antarctic Climate Evolution* (eds F. Florindo and M. Siegert). Developments in Earth and Environmental Sciences 8. Elsevier, Amsterdam, pp. 33–83.

48. Naish T.R., Powell, R.D., Barrett, P.J., Levy, R.H., Henrys, S., Wilson, G.S., *et al.* (2008) Late Cenozoic climate history of the Ross Embayment from the AND-1B drill hole: culmination of three decades of Antarctic margin drilling. In: *Antarctica: A Keystone in a Changing World* (eds A.K. Cooper, P.J. Barrett, H. Stagg, B. Storey, E. Stump and W. Wise). Proceedings of the 10th Inernational Symposium on Antarctic Earth Science, Santa Barbara, California, 26 August–1 September 2007, pp. 71–82.

49. Naish, T.R., Powell, R., Levy, R., Wilson, G., Scherer, R., Talarico, F., *et al.* (2009) Obliquity-paced Pliocene West Antarctic Ice Sheet oscillations. *Nature* **458**, 322–329.

50. Harwood, D., Florindo, F., Talarico, F. and Levy, R.H. (2008–09) Studies from the ANDRILL Southern McMurdo Sound Project, Antarctica – initial science report on AND-2A: *Terra Antartica* **15** (1).

51. Kennett, J.P., Burns, R.E., Andrews, J.E., Churkin, M., Davies, T.A., Dumitrica, P., *et al.* (1972) Australian–Antarctic continental drift, palaeocirculation changes and Oligocene deep-sea erosion. *Nature* **239** (91), 51–55.

52. Kennett, J.P., Houtz, R.E., Andrews, P.B., Edwards, A.R., Gostin, V.A., Hajós, M, *et al.* (1975) *Cenozoic paleoceanography in the Southwest Pacific Ocean, Antarctic glaciation, and the development of the Circum-Antarctic Current*. Initial Reports of the DSDP 29, US Government Print Office, Washington, DC, pp. 1155–1169.

53. Shackleton, N.J. and Kennett, J.P. (1975) *Paleotemperature history of the Cenozoic and the initiation of Antarctic glaciation: oxygen and carbon isotope analyses in DSDP sites 277, 279, and 280*. Initial Reports of the DSDP 29, US Government Print Office, Washington, DC, pp. 743–755.

54. Kennett, J.P. (1977) Cenozoic evolution of Antarctic glaciation. *Journal of Geophysical Research* **82**, 3843–3860.

55. Zachos, J.C., Breza, J. and Wise, S.W. (1992) Early Oligocene ice sheet expansion on Antarctica, sedimentological and isotopic evidence from the Kerguelen Plateau. *Geology* **20**, 569–573.

56. Zachos, J.C., Stott, L.D. and Lohmann, K.C. (1994) Evolution of early Cenozoic marine temperatures. *Paleoceanography* **9** (2), 353–387.

57. Francis, J.E., Ashworth, A., Cantrill, D.J., Crame, J.A., Howe, J., Stephens, R., *et al.* (2008) 100 million years of Antarctic climate evolution: evidence from fossil plants. In: *Antarctica: A Keystone in a Changing World* (eds A.K. Cooper, P.J. Barrett, H. Stagg, B. Storey, E. Stump and W. Wise). Proceedings of the 10th Inernational Symposium on Antarctic Earth Science, Santa Barbara, California, 26 August–1 September 2007, pp. 19–27.

58. Francis, J.E., Marenssi, S., Levy, R., Hambrey, M., Thorn, V.C., Mohr, B., *et al.* (2009) From Greenhouse to Icehouse – The Eocene/Oligocene in Antarctica. In: *Antarctic Climate Evolution* (eds F. Florindo and M. Siegert). Developments in Earth and Environmental Sciences 8. Elsevier, Amsterdam, pp. 309–368.

59. Pross, J., Contreras, L., Bijl, P.K., Greenwood, D.R., Bohaty, S.M., Schouten, S., *et al.* (2012) Persistent near-tropical warmth on the Antarctic continent during the early Eocene epoch. *Nature* **488**, 73–77.

60. Quoted in Schultz, C. (2013) Tectonic, climate and cryospheric evolution of the Antarctic Peninsula. *EOS* **94** (23), 210. (A review of Anderson, J.B., and Wellner, J.S. (2011) *Tectonic, climate and cryospheric evolution of the Antarctic Peninsula.* AGU Special Publication 63).

61. Bertler, N.A.N. and Barrett, P.J. (2010) Vanishing polar ice sheets. In: *Changing Climates, Earth Systems and Society, International Year of Planet Earth* (ed J. Dodson). Springer, Berlin, pp. 49–83.

62. Frakes, L.A., Francis, J.E. and Syktus, J.I. (1992) *Climate Modes of the Phanerozoic – the History of the Earth's Climate Over the Past 600 Million Years.* Cambridge University Press, Cambridge.

63. Barrett, P. (1999) Antarctic climate history over the last 100 million years. *Proceedings of the Workshop on Geological Records of Global and Planetary Changes. Terra Antarctica Reports* **3**, 53–72.

64. Barrett, P.J. (2013) Resolving views on Antarctic Neogene glacial history – the Sirius debate. *Transactions of the Royal Society of Edinburgh* **104** (1), 31–53.

65. Wilson, D.S. and Luyendyk, B.P. (2009) West Antarctic paleotopography estimated at the Eocene-Oligocene climate transition. *Geophysical Research Letters* **36**, L16302.

66. Kennett, J.P. (1980) Paleoceanographic and biogeographic evolution of the Southern Ocean During the Cenozoic, and Cenozoic microfossil datums. *Palaeogeography, Palaeoclimatology, Palaeoecology* **31**, 123–152.

67. Miller, K.G., Browning, J.V., Aubry, M.-P., Wade, B.S., Katz, M.E., Kulpecz, A.A. and Wright, J.D. (2008) Eocene-Oligocene global climate and sea-level changes: St. Stephens Quarry, Alabama. *Bulletin of the Geological Society of America* **120** (1–2), 34–53.

68. Frakes, L.A. and Kemp, M.E. (1973) Palaeogene continental position and evolution of climate. In: *Implications of Continental Drift to the Earth Sciences*, Vol. 1 (eds D.H. Tarling and S.K. Runcorn). Academic Press, London, pp. 539–558.

69. Zalasiewicz, J. and Williams, M. (2012) *The Goldilocks Planet – the Four Billion Year Story of Earth's Climate.* Oxford University Press, Oxford.

70. Smith, A.G., Briden, J.C. and Drewry, G.E. (1973) Phanerozoic World Maps. In: *Organisms and Continents through Time* (ed. N.F. Hughes). Special Paper in Palaeontology 12. Palaeontological Association of London, pp. 1–43.

71. Wolfe, J.A. (1985) Distribution of major vegetational types during the Tertiary. In: *The Carbon Cycle and Atmospheric CO_2: Natural Variations Archean to Present* (eds E.T. Sundquist and W.S. Broecker). Geophysics Monographs 32. American Geophysical Union, Washington, DC, pp. 357–375.

72. Briden, J.C. and Irving, E. (1964) Palaeolatitude spectra of sedimentary palaeoclimatic indicators. In: *Problems in Palaeoclimatology* (ed. A.E.M. Nairn). Proceedings of the NATO Palaeoclimates Conference, University of Newcastle upon Tyne, January 1963. Interscience, London, pp. 199–224.

73. Olsen, J.S. (1985) Cenozoic fluctuations in biotic parts of the global carbon cycle. In: *The Carbon Cycle and Atmospheric CO_2: Natural Variations Archean to Present* (eds E.T. Sundquist and W.S. Broecker). Geophysics Monograms 32. American Geophysical Union, Washington, DC, pp. 377–396.

74. Parrish, J.T., Ziegler, A.M. and Scotese, C.R. (1982) Rainfall patterns and the distribution of coals and evaporites in the Mesozoic and Cenozoic. *Palaeogeography, Palaeoclimatology, Palaeoecology* **40**, 67–101.

75. Savin, S.M., Douglas, R.G. and Stehli, F.G. (1975) Tertiary marine paleoemperatures. *Bulletin of the Geological Society of America* **86**, 1499–1510.

76. Miller, K.G., Fairbanks, R.G. and Mountain, G.S. (1987) Tertiary oxygen isotope synthesis, sea level history, and continental margin erosion. *Paleoceanography* **2**, 1–19.

77. Shackleton, N.J. and Opdyke, N.D. (1973) Oxygen isotope and palaeomagnetic stratigraphy of equatorial Pacific core V28-238: oxygen isotope temperatures and ice volumes on a 10^5 year and a 10^6 year scale. *Quaternary Research* **3**, 39–55.

78. Shackleton, N.J. (1987) Oxygen isotopes, ice volume and sea level. *Quaternary Science Reviews* **6**, 183–190.

79. Zachos, J., Pagani, M., Sloan, L., Thomas, E. and Billups, K. (2001) Trends, rhythms and aberrations in global climate 65 Ma to present. *Science* **292**, 686–693.

80. Katz, M.E., Cramer, B.S., Toggweiler, J.R., Esmay, G., Liu, C., Miller, K.G., *et al.* (2011) Impact of Antarctic Circumpolar Current development on late Paleogene ocean structure. *Science* **332**, 1076–1078.

81. Barker, P.F. and Thomas, E. (2004) Origin, signature and paleoclimatic influence of the Antarctic Circumpolar Current. *Earth Science Reviews* **66**, 143–166.

82. Vincent, E. and Berger, W.H. (1985) Carbon dioxide and polar cooling in the Miocene: the Monterey hypothesis. In: *The Carbon Cycle and Atmospheric CO_2: Natural Variations Archean to Present* (eds E.T. Sundquist and W.S. Broecker). Geophysics Monographs 32. American Geophysical Union, Washington, DC, pp. 455–468.

83. Hodell, D.A., Elmstrom, K.M. and Kennett, J.P. (1986) Late Miocene benthic $\delta^{18}O$ changes, global ice volume, sea level and the 'Messinian Salinity Crisis'. *Nature* **320**, 411–414.

84. Cita, M.B. and McKenzie, J.A. (1986) The terminal Miocene event. In: *Mesozoic and Cenozoic Oceans* (ed. K.J. Hsu). Geodynamics Series 15. American Geophysical Union, Washington, DC, pp. 123–140.

85. Ross, D.A., Uchupi, E., Summerhayes, C.P., et al. (1978) Sedimentation and structure of the Nile Cone and Levant Platform area. In: *Sedimentation in Submarine Canyons, Fans, and Trenches* (eds D.J. Stanley and G. Kelling). Dowden, Hutchinson and Ross, Stroudsburg, PA, pp. 261–275.

86. Elderfield, H. and Summerhayes, C.P. (1978) Salt diffusion in eastern Mediterranean Sea sediments. *Deep Sea Research* **25**, 837–841.

87. Bice, K.L., Scotese, C.R., Seidov, D. and Barron, E.J. (2000) Quantifying the role of geographic change in Cenozoic ocean heat transport using uncoupled atmosphere and ocean models. *Palaeogeography, Palaeoclimatology Palaeoecology* **161**, 295–310.

88. Ravelo, A.C., Billups, K., Dekens, P.S., Herbert, T.D. and Lawrence, K.T. (2007) Onto the ice ages: proxy evidence for the onset of Northern Hemisphere glaciation. In: *Deep-Time Perspectives on Climate Change: Marrying the Signal from Computer Models and Biological Proxies* (eds M. Williams, A.M. Haywood, F.J. Gregory and D.N. Schmidt, D. N.). Micropaleological Society Special Publication. The Geological Society, London, pp. 563–573.

89. Tripati, A.K., Dawber, C.F., Ferretti, P., Backman, J., Elderfield, H. and Macintyre, H. (2007) Evidence for synchronous glaciation of Antarctica and the Northern Hemisphere during the Eocene and Oligocene: insights from Pacific records of the oxygen isotopic composition of seawater. 10th International Symposium on Earth Science. Extended Abstract 186. US Geological Survey and The National Academies, USGS OF-2007-1047.

90. Moore, T.C. and the Expedition 302 Scientists (2006) Sedimentation and subsidence history of the Lomonosov Ridge. In: *Proceedings of the Integrated Ocean Drilling Program, Vol. 302: Edinburgh* (eds J. Backman , K. Moran, D.B. McInroy, L.A. Mayer and the Expedition 302 Scientists). Available from: http://publications.iodp.org/proceedings/302/105/105 _.htm (last accessed 29 January 2015).

91. Backman, J. and Moran, K. (2008) Introduction to special section on Cenozoic paleoceanography of the Central Arctic Ocean. *Paleoceanography* **23**, PA1SO1.

92. Engen, O., Faleide, J.I. and Dyreng, T.K. (2008) Opening of the Fram Strait gateway: a review of plate tectonic constraints. *Tectonophysics* **450**, 51–69.

93. Thiede, J., Jessen, C., Knutz, P., Kuijpers, A., Mikkelsen, N., Nørgaard-Pedersen and Spielhagen, R.F. (2010) Millions of years of Greenland ice sheet history recorded in ocean sediments. *Polarforschung* **80** (3), 141–159.

94. Miller, G.H., Brigham-Grette, J., Anderson, L., Bauch, H., Douglas, M.A., Edwards, M.E., *et al.* (2010) Temperature and precipitation history of the Arctic. *Quaternary Science Review* **29**, 1679–1715.

95. Hansen, J.E. and Sato, M. (2012) Paleoclimate implications for human-made climate change. In: *Climate Change: Inferences from Paleoclimate and Regional Aspects* (eds A. Berger, F. Mesinger and D. Šijački). Springer, Berlin, pp. 21–48.

8

The Greenhouse Gas Theory Matures

Before palaeoclimatologists could understand the role of CO_2 in the climates of the past, they first needed to know what its role was in the modern climate system. And until the 1950s, even meteorologists did not know that. Progress had to wait until our ability to measure CO_2 in the ocean and atmosphere evolved. We will review these developments here, as they set the scene for later chapters.

8.1 CO_2 in the Atmosphere and Ocean (1930–1955)

In the late 1930s, the steam engineer, inventor and amateur meteorologist Guy Callendar (1898–1964) started thinking about how CO_2 might have affected climate[1,2]. His pronouncements fell on deaf ears: he was about as welcome to meteorologists as Wegener was to geologists. Besides, his calculations contained some errors. Like Arrhenius, he assumed wrongly that the burning of fossil fuels and emission of CO_2 would increase linearly, while in fact they increased exponentially. And he failed to take into account the effect on the climate of feedback from water vapour[3].

What held CO_2 science back? First, spectroscopists couldn't measure variation across the infrared radiation spectrum with the necessary refinement. They didn't know where in the spectrum CO_2 had the most effect. Second, they lacked computers with which to calculate the flux of infrared radiation at different wavelengths. Three communities pushed for improvement. Astronomers wanted better measures of the absorption and emission spectra of gases in the Earth's atmosphere, to correct for their effect on the infrared spectra of light from other planets. Meteorologists wanted to know how the absorption and emission of infrared radiation affected the vertical temperature structure of the atmosphere, to improve weather forecasting. But the biggest driver for change after the Second World War was the military, which wanted to detect heat sources like fighter jet engines and air-to-air and ground-to-air missiles. It was important to be able to determine where in the spectrum infrared radiation was blocked by absorption by gases like CO_2 and where the spectrum was transparent, letting radiation pass.

By the early 1950s, spectroscopists knew that the spectrum of infrared light emitted by Earth's Sun-warmed surface ranged from wavelengths of 5 to 100 μ, with about 50% peaking between 8.5 and 28 μ, and that CO_2 preferentially absorbed infrared radiation at wavelengths of 13–18 μ, right in the middle of that peak. Hence, absorption of infrared light by CO_2 must significantly impede the transmission of infrared radiation through the atmosphere, causing it to warm. If you want to see that absorption, just point an infrared spectrometer at the sky and squirt a jet of CO_2 from a standard CO_2 fire extinguisher across the light path.

Military funding was needed to build spectrometers of very high resolution, in order to precisely establish the spectrum of CO_2 as the basis for accurate calculations of its absorptive effects. In 1950, the US Office of Naval Research (ONR) funded the design of an ultra-high-resolution spectrometer at the Laboratory of Astrophysics and Physical Meteorology of Johns Hopkins University in Baltimore, Maryland, where scientists like Robert P. Madden did the research[4]. By then, the

Earth's Climate Evolution, First Edition. Colin P. Summerhayes.
© 2015 John Wiley & Sons, Ltd. Published 2015 by John Wiley & Sons, Ltd.

absorption spectrum of CO_2 between 13 and 18 μ was known to contain clusters of hundreds and perhaps thousands of individual absorption lines, between which the spectrum was transparent, allowing light from a source to pass without being absorbed. An ultra-high-resolution instrument was needed to separate the individual lines so that the amount of light absorbed by each could be calculated accurately[5]. High resolution demanded a long light path: 14 m in the case of Madden's instrument. Later, ONR funded the development of a more advanced spectrometer, with a light path of 182 m[6]. Data began to emerge as early as 1952[7], although Madden's report was not declassified and published until 1961[8]. Aside from ascertaining the precise spectrum of CO_2, the military needed to calculate absorption by CO_2 at different wavelengths along different paths through the atmosphere, so as to improve their detection of heat from the engines of enemy aircraft. ONR funded research on that too, in 1957[9]. Here, then, we have a practical application of greenhouse gas theory.

The spectroscopic data from these various military-funded experiments were collected by the US Air Force at their Cambridge (Massachusetts) Research Laboratories in the late 1960s, and are now available through the **hi**gh-resolution **tran**smission molecular absorption (HITRAN) database, which was further developed at the Atomic and Molecular Physics Division of the Harvard-Smithsonian Center for Astrophysics[10]. NASA uses HITRAN's data to study the changing properties of the atmosphere, while numerical modellers use the data to simulate and predict the effects of CO_2 and other greenhouse gases in the atmosphere. This is another example of a spinoff from military research adding value in the civilian science sphere.

Calculation was exceptionally tedious before computers arrived on the scene. The marriage of high-resolution spectroscopy and the computer constituted a major technical breakthrough, rather like Galileo using his new-fangled telescope to discover the four moons of Jupiter, which ultimately led him to prove Copernicus right, to the dismay of the conservative church.

Among those working on how to measure infrared radiation in relation to the temperature of the atmosphere in the 1950s and 1960s was John (later Sir John) Houghton (1931–), then a research scientist at Oxford, later to become director of the United Kingdom's Meteorological Office and leader of the scientific panel (Working Group 1) of the Intergovernmental Panel on Climate Change (IPCC). A much respected scientist and fellow of the United Kingdom's Royal Society, Houghton was the

recipient in 2006 of the Japan Prize, the equivalent in the environmental sciences of the Nobel Prize. In the mid 1950s, he designed the first airborne infrared spectrometers, later refining them so that they could be deployed on NASA's Nimbus 4 satellite in 1970. This enabled us to measure Earth's infrared radiation from space for the first time, a breakthrough reliant on the development of three new technologies: spectrometers, to make the measurements; computers, to process the data; and satellites, to carry the payload. The tale is told in Houghton's 2013 autobiography, *In the Eye of the Storm*[11]. His work on infrared radiation is ably summarised in his 1966 book with Des Smith, *Infra-red Physics*[12], a must-read for anyone interested in CO_2 and infrared radiation. Summarising decades of experimental work, they remind us that '*a large proportion of [the infra-red radiation emitted by the Earth's surface] … [is] absorbed and in turn re-emitted by different layers of the atmosphere, in amount dependent on the concentration of the absorber, the temperature, and the absorption coefficient, which itself depends on pressure and temperature*'[12]. This is not theory, but measurement.

8.2 CO_2 in the Atmosphere and Ocean (1955–1979)

One of the key people working on the absorption and emission of infrared radiation with the new data emerging from Johns Hopkins was the Canadian physicist Gilbert Norman Plass (1920–2004) (Figure 8.1). In 1956, Plass capitalised on improvements in spectroscopy and computing to recalculate the radiation flux due to CO_2 in the atmosphere[13–17]. He knew that each of the many thousands of closely spaced absorption lines in the CO_2 spectrum broadened as the concentration of CO_2 increased, eventually eliminating the transparent gaps through which infrared energy could pass between them. Furthermore, when the CO_2 concentration was sufficiently high, even its weaker absorption bands became effective, enabling a greater amount of radiation to be absorbed.

Plass calculated that doubling the amount of CO_2 in the atmosphere would cause a rise in the average temperature of the atmosphere at the Earth's surface to 3.6 °C, to restore equilibrium in the balance between solar radiation received at the top of the atmosphere and that emitted back into space[13]. He qualified this estimate in 1961 by pointing out that '*The actual temperature changes for the earth may even be somewhat larger than these values because of the action of water vapour in reinforcing temperature*

Figure 8.1 *Gilbert Norman Plass.*

changes caused by other factors'[17]. Like Callendar and Arrhenius, he made some erroneous assumptions[3], but his findings woke people to the potential effects of an exponential increase of CO_2 in the atmosphere.

To get the calculations right, we needed to map the spectroscopy of water vapour, represent the effects of convection on the structure of the atmosphere and take account of the energy balance at the top of the atmosphere[3]. These goals were reached by 1967, allowing Syukuro Manabe and Richard Wetherald of the Geophysical Fluid Dynamics Laboratory in Washington, DC to provide the first fully sound estimate of the warming that would arise from a doubling of CO_2[3, 18]. They were well aware of the effects of water vapour, calculating that doubling CO_2 would increase global temperatures by 1.3 °C if the water vapour content of the atmosphere remained constant, but by 2.4 °C if water vapour increased to maintain the same relative humidity. Their calculations would be further refined with time[19].

Looking to the future, Plass calculated that, *'If no other factors change, man's activities are increasing the average temperature by 1.1 °C per century'*[13]. Considering the effect of the absorption of CO_2 by the ocean, and taking account of its slow circulation, Plass calculated that *'it would probably take at least 10,000 years for the atmosphere-ocean system to come to equilibrium after a change in the atmospheric CO_2 amount'*[13]. He warned, *'the influence of the extra CO_2 on the climate will become increasingly important in the near future as continuously greater amounts of CO_2 are released into the atmosphere by man's activities'*[13]. Remember, this was in 1956: it's not something new.

The oceanographers of the mid 1950s were just as interested as Plass in how CO_2 was distributed. They knew that about 90 billion tonnes (1 billion tonnes = 1 gigatonne (Gt)) of carbon were routinely exchanged per year as part of the chemical equilibrium between the ocean (with a carbon reservoir of 38 000 Gt) and the atmosphere (with a carbon reservoir of 780 Gt). One, Roger Revelle (1909–1991) (Figure 8.2), the director of Scripps, thought that studying the radioactive isotope of carbon, ^{14}C, in the atmosphere and ocean might help him to understand the behaviour of the marine part of the global carbon cycle. He needed to hire an expert on [14]C. The key person at the leading edge of applying ^{14}C to the study of natural materials was Hans Suess (Figure 8.3), whom we met in Chapter 5. Suess had already deduced that ^{14}C in the atmosphere was being diluted by the ^{12}C derived from the burning of fossil fuels[20]. In 1955, Revelle hired Suess to work with him.

The two men knew that fossil fuel combustion produced CO_2 at a rate two orders of magnitude greater than its usual rate of production from volcanoes. Echoing Plass, they concluded in a landmark paper in 1957 that *'human beings are now carrying out a large scale geophysical experiment of a kind that could not have happened in the past nor be reproduced in the future. Within a few centuries we are returning to the atmosphere and oceans the concentrated organic carbon stored in sedimentary rocks over hundreds of millions of years. This experiment, if adequately documented, may yield a far-reaching insight into the processes determining weather and climate. It therefore becomes of prime importance to attempt to*

Figure 8.2 *Roger Revelle in August 1952.*

Figure 8.3   *Hans Suess in 1972.*

determine the way in which carbon dioxide is partitioned between the atmosphere, the oceans, the biosphere and the lithosphere[21].

Revelle hinted at the need for funds: '*An opportunity exists during the International Geophysical Year [the IGY of 1957–58] to obtain much of the necessary information*'[21]. His wish would be granted. During the IGY, he got funds to employ Charles David Keeling (1928–2005) to set up a CO_2 measuring station 3000 m up on the Mauna Loa volcano in Hawaii. Keeling began measuring CO_2 in well-mixed oceanic air in 1958[22], using a system he developed while a post-doctoral fellow in geochemistry at the California Institute of Technology. His measurements, which continue at Mauna Loa today, established the steady rise in CO_2 with time, as well as its seasonal ups and downs. They are also now made at numerous other sites, including one at the South Pole. Why Mauna Loa? Keeling saw that the content of CO_2 in the atmosphere near ground level varies widely from place to place, depending on proximity to major human sources like cities and factories. He could only obtain the well-mixed 'background' concentration in remote areas. The coasts of open ocean areas with onshore breezes were ideal. Wouldn't the Mauna Loa volcano emit CO_2? Yes, but the recording station was upwind of any volcanic emissions. Measurements at multiple sites confirm that the Mauna Loa record is not corrupted.

After the development of mass spectrometric analytical methods in the 1950s, geochemists began measuring the distribution of isotopes of various kinds in rocks, sediments, biological materials, the oceans and the atmosphere[23]. Keeling measured the carbon isotopes in atmospheric CO_2 as a way of characterising this molecule. Carbon has two stable isotopes: 98.9% is ^{12}C (with six protons and six neutrons) and 1.1% is ^{13}C (with one extra neutron). During photosynthesis, the enzyme *Rubisco* discriminates against the heavier of the two isotopes, giving land plants a higher proportion of ^{12}C and hence a more negative $^{13}C/^{12}C$ (or $\partial^{13}C$) ratio than the air. Back in 1960, the air from Mauna Loa and the South Pole contained $\partial^{13}C$ ratios typical of land plants[22]. The ratio decreased in the summer (i.e. values were more negative, with less ^{13}C), along with CO_2, when photosynthetic activity was greatest, and increased in the winter (i.e. had more ^{13}C), along with CO_2, when plants were no longer active. The ratio is opposite in the Northern Hemisphere compared with the Southern because of the opposition of the seasons, but the limited area of land plant production in the south means that the northern seasonal signal dominates the global signal.

Keeling also found that atmospheric CO_2 was more abundant in the Northern Hemisphere, where most industrial sources were located, than in the Southern[24]. The amount of CO_2 released to the atmosphere annually from the burning of fossil fuels can be calculated from records of fuel production compiled internationally, for example by the Oak Ridge National Laboratory in Tennessee[25]. The estimated addition closely matches the observed amount. The $\partial^{13}C$ ratio decreases (becomes more negative, with less ^{13}C) in the same direction, north, confirming the source for this pattern as the burning of fossil fuel, which is depleted in ^{13}C because it originates from dead ^{12}C-rich plant carbon. Thus, while short-term variations in $\partial^{13}C$ reflect annual changes in the biosphere, the long-term underlying trend of a decline in atmospheric $\partial^{13}C$ of around 1.5‰ over the past 100 years represents the contribution from burning fossil fuels[23, 26]. From the carbon isotopic composition of the burned fuels, one can estimate their likely effect on the $\partial^{13}C$ ratio of the CO_2 in the atmosphere. It closely matches the observed values[27]. The ^{14}C content of atmospheric CO_2 shows a similar effect, decreasing at the same time as the content of CO_2 increases, which is to be expected if the ^{14}C is being diluted by CO_2 from fossil fuels in which all the original ^{14}C has decayed away. Finally, the combustion of fuel removes oxygen from the atmosphere, as is shown by the inverse relation between growing CO_2 and declining O_2[28].

Keeling's son, Ralph (1959–), the present director of the Scripps CO_2 Program, continues his father's work. CO_2 science has become a family business!

Callendar was still going strong in 1957. For comparison with the measurements that would be made during the IGY, he reviewed the published analyses of CO_2 in the air from around the world[29]. Like Keeling, he noticed that high values were reported near towns. His equivalent of Keeling's 'background air' was what he called 'free air'. To obtain free-air CO_2, he devised a system for rejecting samples collected in or near towns and within buildings. He also concluded that measurements made before 1870 were unreliable. The resulting data showed that free air in the northeast Atlantic region contained 290 ppm CO_2 in 1900, which rose to 326 ppm over Scandinavia in the mid 1950s. Those figures compare favourably with the global average of 315 ppm reported by Keeling in 1958. The rising CO_2 closely followed the rise in the consumption of fossil fuels[29].

Was that rise real? As we shall see, it matches well with other sources of data. But in 2007, it was questioned by Ernst-Georg Beck (1948–2010), a biology teacher at the Merian technical grammar school in Freiburg, and co-founder of the European Institute for Climate and Energy (EIKE). Beck compiled the results of some 90 000 CO_2 analyses made since 1812 and used them in an attempt to demonstrate that the CO_2 content of air had fluctuated considerably since then[30]. Unfortunately, he ignored the simple and straightforward Keeling/Callendar sampling protocol, and the reasons for it. To cite just one example, he accepted for his global compilation of atmospheric data the CO_2 measurements made at Giessen, in Germany, where CO_2 values ranged from lows of around 300 to highs of up to 550 ppm. There is no doubt that the Giessen measurements were made well, but that's not the point. Giessen lies about 140 km southeast of the industrial centres of the Ruhr Valley, in an area of prevailing westerly winds, and just 50 km north of another major industrial centre: Frankfurt. It is not an ideal site at which to determine the atmospheric background level of CO_2. Values recorded there most likely represent contamination by downwind transport from industrial sources and population centres. Beck's misleading analysis was roundly criticised[31,32], but it still has adherents on the Internet.

Although Revelle's and Suess's landmark 1957 paper[21] is much quoted, it contains two statements in apparent contradiction. Their calculations showed that the average lifetime of a CO_2 molecule in the atmosphere before it is dissolved into the sea is about 10 years, from which one might conclude that any new emission of CO_2 would be quickly taken up by the ocean rather than staying in the atmosphere. But, as Spencer Weart points out, the paper also contains a much overlooked implication: '*While it was true that most of the CO_2 molecules added to the atmosphere would wind up in the oceans within a few years, most of these molecules (or others already in the oceans) would promptly be evaporated out*' by the normal processes of ocean–atmosphere exchange[33]. Bearing that ongoing exchange in mind, Revelle and Suess stated that '*In contemplating the probably large increase in CO_2 production by fossil fuel combustion in coming decades we conclude that a total increase of 20 to 40% in atmospheric CO_2 can be anticipated*'[21]. In other words, much of the extra CO_2 pumped into the atmosphere by our emissions would stay there, with obvious implications for Earth's climate[3].

The first part of that paragraph reminds me that there is a popular misconception that CO_2 does not last long in the atmosphere. This is based on evidence from the behaviour of CO_2 containing the radioactive isotope ^{14}C, which was created by the nuclear bomb tests of the 1950s and 1960s, and was found to have a relatively short residence time of 5–10 years in the air. While this confirms that any individual molecule of CO_2 is likely to have a short residence time in the air, it has no bearing whatsoever on what the bulk of the CO_2 is doing over time. This is because, as implied in the second part of the preceding paragraph, CO_2 is constantly being exchanged between the tiny reservoir of carbon in the atmosphere (720 Gt) and the vast reservoir of carbon in the ocean (38 400 Gt), both of which are dominated by the stable isotopes of carbon: mainly ^{12}C, with a little ^{13}C. Every year, some 90 Gt of carbon dominated by the ^{12}C isotope goes from the atmosphere into the ocean, and more or less the same amount comes out in the other direction. The tiny fraction of CO_2 containing ^{14}C is lost in the noise of this massive exchange. Some of it is mixed down into the ocean interior above the thermocline and some disappears through the sinking of dense cold surface waters down into the ocean depths, where the ^{14}C decays. In effect, it is massively diluted out of the exchange equation by the vast amounts of ^{12}C and its little sister, ^{13}C, and so its abundance and brief residence time have no bearing on the overall and very long residence time of total CO_2 in the atmosphere. Exchange is the key factor.

The Swedish meteorologists Bert Bolin (1925–2007) and Erik Eriksson of the International Meteorological Institute in Stockholm would spell these points out brilliantly in 1958, in yet another landmark paper[34]. Their work took into account the slow mixing time of the ocean, the impact of the ocean's carbonate buffer chemistry and the accelerating emission of CO_2 with time[3, 34]. In summary: we put excess CO_2 into the atmosphere; it constantly exchanges with the CO_2 in the ocean; some stays in the ocean because physical and biological processes prevent it getting back into the atmosphere; some is trapped in land plants and soils; but the constant exchange keeps the atmospheric concentration high; and much of the excess CO_2 causes the atmospheric concentration to rise. This was widely accepted by knowledgeable scientists at the time[35].

Thus, we have known a great deal about the likely effects of rising CO_2 on our climate since scientists started taking a close look at the matter in the mid to late 1950s, some 30 years before the founding of the IPCC.

Independent of these studies of CO_2, meteorologists were compiling meteorological data in order to see whether the world was really warming. Among them was Hubert Horace Lamb (1913–1997) (Figure 8.4, Box 8.1), an English climatologist.

One of Lamb's tasks at the Met Office was to answer the question, is the climate changing? Temperatures had risen from 1900 to 1940, then declined into the 1960s. Considering, in 1963, the greater frequency of severe winter weather in the United Kingdom in recent decades, Lamb asked whether this were '*(i) Only a temporary lapse from the warm climate attained in the early 20th century, possibly to be followed by a renewed trend to still greater warmth? (ii) A return to conditions normal in the past century or two? Or (iii) the beginning of a climatic decline to still harsher conditions? [i.e. towards the next*

Figure 8.4   *Hubert Horace Lamb.*

Box 8.1 Hubert Horace Lamb.

Lamb started his working life as a meteorologist for the United Kingdom's Met Office, where he was responsible for long-range weather forecasting and studies of climate change[36]. In 1972, he founded the Climate Research Unit (CRU) of the University of East Anglia's school of environmental sciences, based in Norwich. In August 2006, the CRU building was renamed the Hubert Lamb Building in his honour. That same year, in a report listing the '*top 100 world-changing discoveries, innovations and research projects to come out of the UK universities in the last 50 years*', Lamb was hailed as '*instrumental in establishing the study of climate change as a serious research subject*'[36]. For his contributions to climate science, Lamb won several awards, among them the Symons Gold Medal of the Royal Meteorological Society and the Murchison Medal of the Royal Geographical Society.

Ice Age]'[37]. At the time, he thought (i) unlikely and (iii) more likely, concluding, '*it seems prudent to assume that the longer-term temperature trend is at present on balance downward and likely to remain so*'[37].

Although some other scientists also agreed that the world might be cooling towards the next Ice Age, it is a myth that there was some kind of scientific consensus on the subject[38]. In fact, only around 10% of scientific papers on climate change published between 1965 and 1979 predicted global cooling; the others all predicted warming[38]. The media, looking as usual for scary things to lead with, backed the wrong horse and banged the cooling drum.

Lamb eventually accepted that CO_2 played a key role in governing climate change[39]. Recognising that the orbital geometry calculations of André Berger suggested that over the long term we were heading towards a colder climate, he concluded that our future climate would reflect the balance between orbital effects and human emissions, modulated by shorter-term phenomena such as El Niño events and volcanic eruptions like that of Mt Pinatubo: '*Man may be obliged in the future either to seek to avert, or slow down, the onset of a new ice age by deliberately increasing the CO_2 in the atmosphere*'[39], he wrote; an intriguing geoengineering notion. He did not think that urban heat islands significantly distorted global temperature records, because that would not explain warming in remote locations, nor the melting of glaciers.

Another prominent climate scientist to enter the ring in the late 1960s was the Soviet meteorologist Mikhail Ivanovich Budyko (1920–2001) (Figure 8.5, Box 8.2).

Much like Lamb, Budyko found that the annual temperature of the Northern Hemisphere had risen by 0.6 °C from 1900 to 1940, then fallen by 0.2 °C by 1960[40]. Puzzling over what caused the decline, he found that it correlated with a decrease in direct solar radiation measured under a cloudless sky. Thinking that this was due to rising industrial output, he wrote, '*the decrease in radiation after 1940 could … depend on the increase in dust in the atmosphere due to man's activity*'[40]. By 1977, he calculated that direct solar radiation had decreased by 6% between the late 1950s and the late 1960s, due to our growing emission of fine particles (aerosols), which could have reduced the global temperature by around 0.5 °C (Figure 8.6)[41].

Being aware that CO_2 was a greenhouse gas, in 1977 Budyko agreed that adding CO_2 emissions to the atmosphere would be expected to warm the climate[42]. He calculated that a continued rise of CO_2 would cause temperatures to rise by 0.6–0.7 °C by the end of the century[41]; a good guess. By 1979, he concluded that '*man's influence on the atmosphere has a considerable significance … As a result of the combustion of steadily increasing amounts of coal, oil, and other forms of fossil fuel as well as the reduction in the amount of carbon in living organisms (from the felling of forests) and a reduction in the amount of carbon in soil humus, the CO_2 level in the atmosphere has recently*

Box 8.2   Mikhail Ivanovich Budyko.

A brilliant and original thinker, Mikhail Budyko was one of the founders of modern climatology, well known for his calculations of the planetary 'albedo': the amount of solar radiation reflected back into space by light-coloured surfaces such as ice, snow and deserts. He recognised that more ice and snow would reflect more energy, cooling the Earth, which would lead to more snow and ice and so on: a positive feedback. Conversely, warming would lead to less snow and ice and greater exposure of the darker ocean, meaning less reflected energy and more energy absorbed by the ocean: another positive feedback. Budyko was head of the Division for Climate Change Research of the State Hydrological Institute in Leningrad (now St Petersburg), a member of the Russian Academy of Sciences and a 1994 recipient of the Robert E. Horton Medal of the American Geophysical Union.

been rising. It is assumed that this mass might increase by a factor of 6–8 within 100–200 years. The important point is that in such a case the CO_2 level would attain the values typical of the average for the Phanerozoic. Such a change in the atmospheric composition would rule out any further cooling of the climate and will probably result in the return to warmer climatic conditions typical of the Pre-Quaternary period'[43]. In bringing in the geological dimension, he was influenced by his co-author, geologist Alexander Borisovitch Ronov (1913–1996) of the V.I. Vernadsky Institute of the USSR Academy of Sciences.

Budyko realised that warming due to increased CO_2, plus the increase in CO_2 itself, might provide more favourable conditions for the plant growth needed to feed the growing human population[44]. Besides, warming would expand the area of northern Russia and Siberia capable of growing wheat. It would also reduce Arctic sea ice, which could be a boon to shipping[41].

Budyko's observations make it clear that air temperature is modified by two anthropogenic effects: emissions of CO_2 and emissions of aerosols. His Soviet data help to explaining the hiatus in global warming in the 1950s and 1960s: the effects of aerosols then declined as nations introduced legislation to limit air

Figure 8.5   *Mikhail Ivanovich Budyko.*

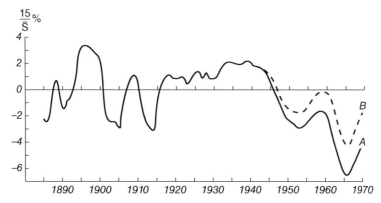

Figure 8.6 *Budyko's measurements show the effects of growing aerosol concentration. A = actual measurement; B = adjustment for cloudless sky. 5 year running means of the anomalies of direct radiation expressed in per cent of its normal amount, calculated from the data observed at a number of stations. Attenuation is equal to about 6%.*

pollution from particulates, spurred by such events as the smogs in London in 1952 and in New York in 1953[a].

Oceanographers like Revelle and climatologists like Budyko were not alone in agreeing with Plass. By 1972, the United Kingdom's Met Office had become convinced by Keeling's data of the dangers inherent in continued emissions of CO_2[45]. Others waited until numerical models could incorporate the annual hydrological cycle, rather than just the assumption of fixed relative humidity that Manabe and Wetherald had used[18]. Developing a general circulation model (GCM) for '*solving the three-dimensional fluid dynamical and thermodynamical equations governing transfer of heat, moisture and momentum around the world*'[3] was a very big step. It would tax the powers of the biggest computers of the time. In 1975, Manabe and Wetherald proved equal to the task[46], confirming among other things that high latitudes (especially the Arctic) warm more than low latitudes; this is the 'polar amplification' predicted by Arrhenius. Their model confirmed that land warms faster than ocean, and that global precipitation increases as the world warms

because warming evaporates water vapour from the ocean[3].

Even so, in 1976 the meteorological community was not in complete agreement that CO_2 was contributing or might in future contribute to climate change. At that time, the climatologist Steve Schneider, referring to the 0.5 °C rise in global temperature over the previous 75 years, wrote, '*climatic theory is still too primitive to prove with much certainty whether the relatively small increases in CO_2 and aerosols up to 1975 were responsible for this climate change*'[47, 48]. So we should not be surprised that most geologists of the 1970s and before did not consider the CO_2 theory of climate change worth a second thought.

Before the CO_2 model could be accepted, it was necessary to quantify the errors and uncertainties built into the radiative–convective models that had been used to calculate climate change. Tommy Augustsson, of the NASA Langley Research Centre of George Washington University, and Veerabhadran (Ram) Ramanathan, of the National Center for Atmospheric Research (NCAR) at Boulder, Colorado, carried out that task in 1977[49]. They calculated that while most of the absorption of infrared radiation by CO_2 occurred in the main spectral bands attributed to CO_2 vibrations centred on about 15 μ, about 30% occurred in the so-called 'weak' spectral bands attributed to CO_2 vibrations between 12 and 18, 9 and 10 and 7 and 8 μ. They found that while the increase in temperature tailed off logarithmically as CO_2 increased in the 15 μ band, it was almost linear with increasing CO_2 in the weak bands. Hence, '*the warming effect of CO_2 on*

---

[a] New evidence shows a rise in the emission of sulfur aerosols after the Second World War. It paralleled the rise in CO_2 and is likely to have prevented the warming signal due to CO_2 from becoming manifest between 1945 and 1970, much as implied by Budyko's data. The decline in these aerosols after 1970 allowed the warming signal due to CO_2 to emerge (Smith, S.K., Van Aardenne, J., Klimont, Z., et al. (2011) Anthropogenic sulfur dioxide emissions:1850–2005. Atmos. Chem. Phys., 11, 1101–1116).

the global surface temperature may never saturate out even for large increases in CO_2 concentrations'[49]. They estimated that doubling CO_2 raised temperature between 1.98 and 3.20 °C.

It was also important to evaluate the latitudinal and seasonal changes that would be likely upon doubling of the CO_2 content of the atmosphere, which Ramanathan did in 1979[50]. He showed that there were two reasons why the Antarctic should be colder than the Arctic: first, Antarctica is land surrounded by ocean, while the Arctic is ocean surrounded by land; and second, because the atmosphere in the Southern Hemisphere is thinner than that in the Northern, the surface pressure is 20% lower at 80° S than at 80° N, and as the opacity of the atmosphere to CO_2 at infrared wavelengths is proportional to atmospheric pressure, the atmosphere is less opaque in the south at comparable latitudes, and so warms less. Differences between the tropics and the poles arise because of the distribution of water vapour. Because the CO_2 absorption bands at 12–18 μ overlap H_2O absorption bands, where both occur together in the tropics enhanced emission by CO_2 is partly absorbed by H_2O, enhancing tropospheric heating. At high latitudes, where there is little H_2O, CO_2 emissions warm the surface. Polar warming is also amplified by decreases in the albedo resulting from melting sea ice.

Ramanathan found that '*At 75° N the CO_2-induced enhancement in surface temperature is roughly 2 times as great in summer as compared with winter, while at 85° N this factor is greater than 3*'[50]. He noted that '*This effect is not, however, due to seasonal variability of the CO_2 radiative heating, but is solely the result of ice albedo feedback*'[50]. This led him to observe that '*The high-latitude summer enhancement in surface temperature could be of considerable importance with regard to the stability of arctic sea ice*'[50]. He calculated that, with $1.33 \times CO_2$ (which was then at 372 ppm), the mean increase in surface temperature for the Northern Hemisphere would be 1.45 °C, whereas for June at 85° N it would be 6.5 °C[50]. This polar-amplification effect led him to conclude that a significant influence of CO_2 on the climate would first appear during Arctic summers, not Antarctic summers, because heating rates would be 50% lower in the latter. Looking back, we can see that his CO_2 signal did indeed appear, in the very year he published his paper, in the shape of the beginning of a remarkable decline in summer Arctic sea ice. Clever fellow! From the geological perspective, his observations about the

differences between the two poles will be important to our evaluation of changes in ice cover with time.

The growing understanding of the role of CO_2 in the air and ocean outlined in this section began to trigger public warnings. As early as 1965, one of President Lyndon B. Johnson's Science Advisory Committees warned that burning fossil fuels would lead to a rise in CO_2 of 25% by the year 2000, which might be expected '*to produce measurable and perhaps marked changes in climate ... [that could be] deleterious from the point of view of human beings*'[51]. Revelle wrote that section, aided by Wally Broecker of Lamont and Charles Keeling, among others. The actual rise was close to 40%. This opinion was reiterated in the Club of Rome's report, *Limits to Growth*, in 1972, which forecast a rise to 380 ppm by the year 2000[52].

By 1975, the marine geochemist Wallace Smith Broecker (1931–) (Figure 8.7, Box 8.3) was ready to speak out, publishing a paper with the subtitle 'Are We on the Brink of a Pronounced Global Warming?'[53]. Sometimes referred to as 'the dean of climate scientists', Broecker is credited with popularising the phrase 'global warming', a term he first used in his 1975 article, which predicted that the recent cooling trend would give way to a rise in temperature as CO_2 continued to rise. In a memorable phrase from the 1990s, Broecker described the world's climate as an 'angry beast' that we're continuing to poke through growing emissions of CO_2, without being sure how it might respond[54]. In 1978, a similar warning came from John Mercer of Ohio University's Institute of Polar Studies, expressing the concern that polar

Figure 8.7 *Wallace Smith Broecker.*

warming would potentially destabilise the West Antarctic Ice Sheet and lead to a global rise in sea level[55].

Broecker was keen to see if the dissolution of deep-ocean carbonate sediments might ultimately neutralise the CO_2 generated by burning fossil fuels, a project he worked on with Taro Takahashi[56]. They calculated that the process would take 1500 years or longer. Others agreed. James Walker of the University of Michigan and James Kasting of Penn State University concluded that '*dissolution of pelagic carbonates is not likely to influence the levels of atmospheric carbon dioxide before several hundred years have elapsed*'[57].

We still needed to know more about the ocean's role in the CO_2 story, and Roger Revelle led the charge. In 1960, on the occasion of the founding of UNESCO's Intergovernmental Oceanographic Commission (IOC), based in Paris, Revelle highlighted '*the present attempt to determine the total carbon dioxide content in the atmosphere and the change in this content with time as a result of the input from fossil fuel combustion and the loss to the ocean and biosphere. One of the questions we are asking is: Where is the carbon dioxide absorbed by the ocean? Does it remain in the surface layers or does it extend throughout the ocean volume?*'[58]. Part of the answer would come from improved measurements of the global distribution of carbon in the ocean made by Broecker and others in the Geochemical Ocean Sections Study (GEOSECS), launched by the United States' National Science Foundation as part of its International Decade of Ocean Exploration (IODE) in 1970. GEOSECS ran from 1971–78, making some 6000 measurements of dissolved organic carbon and total ocean alkalinity. However, still more accurate measurements would be required before

the anthropogenic contribution to oceanic CO_2 could be established with confidence[59].

To progress understanding of CO_2 in the ocean, the IOC and the Scientific Committee on Oceanic Research (SCOR) of the International Council for Science (ICSU), appointed Revelle to chair their new Committee on Climate Change and the Ocean. By 1984, this committee had established a CO_2 Advisory Panel under Revelle's chairmanship. His panel recommended an observation programme and a sampling strategy to determine the global oceanic inventory of CO_2 to an accuracy of 10–20 petagrams of carbon (PgC)[59] (1 Pg is equivalent to 1 Gt). Later, we'll see where that recommendation led.

8.3 CO_2 in the Atmosphere and Ocean (1979–1983)

One of the first public statements on CO_2 and climate from the heart of the meteorological community was a 1977 report to the National Research Council (NRC), a branch of the US National Academy of Sciences, by Robert M. White (1923–), who headed the newly formed National Oceanic and Atmospheric Administration (NOAA) from 1970 to 1977[60]. White was a member of the US Academy of Sciences, and would become president of the National Academy of Engineering (1983–95) and, in 1980, president of the American Meteorological Society. His report concluded that increasing CO_2 would warm the planet.

A team of experts led by meteorologist Jule Gregory Charney (1917–1981) of the Massachusetts Institute of Technology (MIT) made a second report along the same lines for the NRC 2 years later[61]. Charney, considered the father of modern dynamical meteorology, was awarded the Carl-Gustaf Rossby Research Medal of the American Meteorological Society for his outstanding contributions to atmospheric science. His team agreed that doubling atmospheric CO_2 would increase global temperature by about 3 °C.

By 1979, interest in the CO_2 question was beginning to move out of national laboratories and into the international arena. In February 1979, the first World Climate Conference took place in Geneva, Switzerland, organised by the UN's World Meteorological Organization (WMO), which is headquartered there[62]. Other UN agencies provided assistance, among them UNESCO, the UN Environment Programme (UNEP), the UN's Food and Agricultural Organization (FAO) and the UN's World Health Organization (WHO).

CO_2 was not a primary focus for the meeting. The preceding decade had seen one of the strongest El Niño events of the 20th century (in 1972–73), a 5-year Sahelian drought, failure of the Indian monsoon in 1974, damage to the Brazilian coffee crop from cold waves in 1975, drought in Europe and cold waves in the United States. Recognising the importance of variations in climate, the WMO had decided to initiate a World Climate Programme (WCP), which would include a World Climate Research Programme. The conference was designed to provide a solid platform for the launch of these two initiatives, by assessing the state of our knowledge of our climate, considering the effects of climate variability and change on human society and addressing the issue of possible human influences on climate.

Robert White kicked off the meeting, reminding his audience, '*If natural disasters had not been enough to motivate governments and the scientific community to action, the ominous possibilities for man-induced climate changes would have triggered our presence here … In recent years, we have come to appreciate that the activities of humanity can and do affect climate*'[63]. As to what could be done about that, he went on, '*It is essential that we join together to consider what we can do collectively and individually about climatic issues in the interests of all … The possibility that actions by individual nations may influence the climates of others may demand new types of international action … We thus see emerging a need for some mechanism to develop global environmental impact assessments that will be accepted by all nations … Let us hope that this conference marks the commencement of a new level of collaboration for the protection and productive use of climatic resources*'[63]. Note the call for a 'mechanism', and use of the word 'assessments' – we shall explore their implications later.

Among the experts making presentations was the German climatologist Hermann Flohn (1912–1997) (Box 8.4), of the University of Bonn, who had discussed this topic at a Dahlem Workshop in Berlin in 1976[64].

Flohn thought that the probability of artificially induced future global warming was high and would increase with time, such that '*soon after the turn of the century, a level may possibly be reached that exceeds all warm periods of the last 1000–1200 years … This risk must be avoided even at very high costs*'[65]. He reminded the conference that global changes are larger by a factor of about three in the polar regions, hence, '*The most fascinating, and also the most controversial problem of the future evolution of our climate is the possibility of a complete disappearance*

Box 8.4 Hermann Flohn.

Hermann Flohn is well known for having produced the first north–south cross-section through the atmosphere, with widespread implications for our understanding of atmospheric processes at the global scale. His analysis of the atmospheric circulation of the Southern Hemisphere contributed to his becoming one of the world's most famous climatologists after the Second World War. He won the WMO International Meteorological Association Prize in 1986 and was a member of the Leopoldina (the German Academy of Sciences) and the Royal Belgium Academy.

---

of the drifting ice of the Arctic Ocean … an increase of the atmospheric CO_2 content might lead rather rapidly to an ice-free Arctic Ocean'[65]. Budyko had reached that same conclusion in 1977[41]. The disappearance of ice, Flohn thought, would lead to '*an increase of cold season snowfall along the northern coasts of the continents and arctic islands*', along with '*reduction and northward displacement of the winter-rain belts in the Mediterranean, Near East, and southwestern North America, together with frequent summer droughts in the belt 45°–50° N and an extension of the subtropical dry areas towards north*'[65]. These are things we have now begun to see with increasing frequency. The news was not all bad: '*The possibility of a significant melting of the polar ice caps is … small*', he reminded his audience[65].

Not everyone agreed with Flohn's diagnosis. One of the experts, F.K. Hare of the University of Toronto, thought that the climate system was highly variable and that we did not yet know enough about it to be able to predict with confidence whether temperatures would rise or fall[66]. R.E. Munn of the University of Toronto and L. Machta of NOAA agreed, noting that as well as predictions of warming, '*There are predictions for naturally occurring cooling trends during the next few decades*'[67]. If the climate cooled, they thought, '*the CO_2 induced global warming would then be welcome*'[67]. They recognised that '*few predictions call for significant climatic effects before 2000 AD*', and, further, that '*Few, if any, scientists believe the CO_2 problem in itself justifies a curb, today, in the use of fossil fuels or deforestation*'[67]. Nevertheless, given the exponential rise in the consumption of fossil fuels and emissions of CO_2, they concluded that

'*There is therefore a sense of urgency in determining whether there is likely to be any real environmental or socio-economic threat from growing atmospheric CO_2*'[67]. First, we needed more information: '*Studies of the climate impacts of an increase in the concentrations of greenhouse gases, and of the resulting impacts on society, should be pursued internationally with great vigour*'[67].

Summing up the results of this groundbreaking meeting, the secretary-general of the WMO, David Arthur Davies (1913–1990) of the United Kingdom, said in the foreword to the report, '*This publication may safely be considered as the most profound and comprehensive review of climate and of climate in relation to mankind yet published*'[68]; in effect, it could be seen as the first climate-change textbook. On the basis of their deliberations, the experts adopted 'The Declaration of the World Climate Conference'[62], which agreed, '*it appears plausible that an increased amount of carbon dioxide in the atmosphere can contribute to a gradual warming of the lower atmosphere, especially at high latitudes ... [which] may be detectable before the end of this century and become significant before the middle of the next century*'[62]. As a result, the declaration went on more controversially, in order to avoid serious environmental problems it might become necessary to '*redirect ... many aspects of the world economy*'[62]. We might have to *do* something about this warming.

Enter the US National Aeronautics and Space Administration (NASA). NASA's atmospheric physicists, located at its Institute for Space Studies at the Goddard Space Flight Centre in New York, had been studying the atmospheres of other planets in the solar system. Knowing that the greenhouse effect explained the atmospheric temperatures of Mars (low H_2O and low CO_2 = cold) and Venus (high H_2O and high CO_2 = hot), they realised that it ought also to be able to explain the atmosphere of the home planet, Earth. Stimulated by Charney's 1979 paper for the NRC, and led by Jim Hansen, they published in 1981 a compelling analysis of the effects of increasing CO_2 on Earth's climate[69]. Their model included feedbacks from clouds and from changes in the surface albedo caused by snow and ice, and suggested a rise of 2.8 °C for a doubling of atmospheric CO_2. That rise is known as the 'climate sensitivity', defined as the equilibrium change in global mean surface temperature for a doubling of atmospheric CO_2 above preindustrial values.

Hansen's team showed that exchange of CO_2 with the ocean would slow the full impact of warming and

that warming could be masked for short periods (1–2 years) by the ejection of reflective materials into the stratosphere by large volcanoes. Human emissions of fine particles – aerosols – could also have an effect, although, while some reflected radiation and so cooled the air, others absorbed radiation and so warmed it. Other trace gases, such as ozone (O_3), could also warm the planet, and were increasing. Warming was broadly consistent with the human output of CO_2, especially in the Southern Hemisphere[69].

Other factors affected the warming pattern in the Northern Hemisphere, where temperatures fell between 1940 and 1970. Taking into account the rise in CO_2 plus volcanic activity plus solar variability, Hansen was able to recreate the global temperature curve from 1880 to 1980, giving some confidence to their methods. Projecting the amount of likely energy use to 2100, he suggested that the signal of CO_2 warming should be detectable (reaching values more than 1 standard deviation from the mean trend) above the noise level of natural climate variability by the mid to late 1980s, and significantly detectable (reaching values more than 2 standard deviations from the mean trend) by the mid to late 1990s. The prediction was accurate, so it is wrong to say, as doubters do, that climate models have not been tested. By now, the predictions of these various early modelling efforts have been tested several times, and are still holding well.

Much like Arrhenius, Hansen's team calculated that the climate would warm further and faster at the poles than elsewhere. With a 2 °C rise in global temperature, Antarctic temperatures would rise by up to 5 °C, which could perhaps trigger the decay of the West Antarctic Ice Sheet and cause a rise in sea level of 5 m, possibly over a century or less[69], much as Mercer suggested in 1978. The 5 to 10 °C Arctic warming expected for doubled CO_2 in the 21st century should eventually melt all the sea ice in summer, opening the Northwest and Northeast Passages to navigation. Predicted temperatures would likely exceed those of the last interglacial.

Echoing Plass and Revelle, Hansen concluded that '*The climate change induced by anthropogenic release of CO_2 is likely to be the most fascinating global geophysical experiment that man will ever conduct*'[69], requiring challenging efforts in global observation and climate analysis. And the implications were becoming more stark. Given past evidence that it takes several decades to complete a major change in fuel use, Hansen thought that substantial climate change was almost inevitable, depending on the rate of growth of energy use and the mix of fuels. He thought that full exploitation of coal resources might become undesirable and it would be wise to conserve energy and develop alternative energy sources. These messages have rumbled on with little change since then.

In October 1982, with Taro Takahashi, Hansen convened the 4th Maurice Ewing Symposium, at Lamont, to discuss 'Climate Processes and Climate Sensitivity'[70]. The meeting was supported by a grant from the Exxon Research and Engineering Company, whose president, E.E. David, provided the first paper in the published volume of the proceedings[71]. '*Few people doubt*', he wrote, '*that the world has entered an energy transition away from dependence upon fossil fuels and toward some mix of renewable resources that will not pose problems of CO$_2$ accumulation. The question is how do we get from here to there while preserving the health of our political, economic and environmental support systems*'[71]. Mr David was '*upbeat about the chances of coming through this most adventurous of human experiments with the ecosystem*'[71.] Good for Exxon!

Much as in his 1981 paper, Hansen's contribution to the Ewing Symposium explored the feedbacks within the system that conspired to boost temperature increases beyond those attributable to CO$_2$ alone, such as that from water vapour[72]. But following in the footsteps of Arrhenius and Plass, he also investigated the likely change in climate due to the lesser amounts of CO$_2$ in the atmosphere of the last glacial maximum, 20 Ka ago. We shall explore his findings on that topic in Chapter 13.

Wally Broecker contributed two papers exploring possible causes for the rise in atmospheric CO$_2$ at the end of the last glaciation – and especially the ocean's role – concluding that CO$_2$ acts as a feedback amplifier[73, 74]. The ocean's precise role in controlling atmospheric CO$_2$ remained somewhat obscure at the time. That same year, Broecker pointed out in his classic *Tracers in the Sea*, co-written with Tsung-Hung Peng, that our current emissions '*will alter the ocean's chemistry even on geologic time scales*'[75]. How can that be? Consider this: The surface mixed layer equilibrates with the atmosphere in about a year. The waters of the main ocean thermocline equilibrate with the air on time scales of several tens of years. The waters of the deep sea equilibrate with the air on time scales of many hundreds of years. The calcite in marine sediments will equilibrate with the air on time scales of several thousands of years. Finally, the excess calcium dissolved in the sea as the result of the calcite–CO$_2$ reaction will be removed on the time scale of many tens of thousands of years[75].

8.4 Biogeochemistry: The Merging of Physics and Biology

By 1983, it was widely recognised that Earth's living and inanimate systems are inextricably intertwined, much as Humboldt told us in the early 1800s. To address these interactions in a more effective way, a whole new area of science sprang into being: biogeochemistry, the study of the chemical, physical, geological and biological processes and reactions that govern the composition of the natural environment. It focuses on chemical cycles that are driven by or have an impact on biological activity, with a particular emphasis on the cycles of carbon, nitrogen, sulphur and phosphorus, which play key roles in biological systems and their interactions with the Earth[76].

While some of the roots of biogeochemistry can be traced back to Humboldt, others lead to Eduard Suess, who invented the term 'biosphere', and James Hutton, who maintained that geological and biological systems were linked. In his 1926 book *The Biosphere*, the Soviet mineralogist and geochemist Vladimir Ivanovich Vernadsky (1863–1945), founder of the Ukrainian Academy of Sciences, now the National Academy of Sciences of the Ukraine, popularised the use of Suess's term by arguing that life is the geological force that shapes the Earth. He is often credited with the first use of the term 'biogeochemistry', but its use grew only gradually. Some of the impetus behind the development of this 'new' field came from surprising discoveries in the oceans; for example, that chemotrophic bacteria lived off hydrogen sulphide emitted from hot vents on mid-ocean ridges, and that microbes were altering sediment composition down to depths of 800–1000 m beneath the deep-sea floor, as found in cores collected by the Deep Sea Drilling Project (DSDP)[77].

How should the community proceed? One could not apply biogeochemical principles by keeping the physical and biological sciences separate, as had formerly been the trend; the IGY of 1957–58, for example, had been just that: geophysics with no biology. At the urging of the Swedish meteorologist Bert Bolin and the Dutch chemist Paul Crutzen, the International Council of Scientific Unions (ICSU), whose name (but not its acronym) changed in 1998 to the International Council for Science, decided that something must be done. A proposal was tabled for the development of an International Geosphere-Biosphere Programme (IGBP) linking geophysics, chemistry, biology and geology in a more 'holistic' framework than before, to address such topics as acid rain, ozone depletion, CO_2 and methane build-up and biogeochemical cycles. ICSU agreed to review this proposal during its 20th session, in Ottawa, Canada in September 1984. Meanwhile, the US NRC had considered the proposal at a meeting in 1983[78]. The topic was reviewed within the ICSU meeting, at a 1-day symposium on global change that would be informed by the US NRC report and by a collection of papers prepared for the meeting and eventually published in 1985[79]. The symposium recommended that ICSU develop an interdisciplinary programme to study interrelations between the geosphere and the biosphere, so as to understand global change over long time scales and the influence of humans on the environment. ICSU launched the IGBP in 1987.

This development put efforts to understand the functioning of the Earth as an integrated system into effect on a grand scale for the first time[80]. An important underlying context was the increasing scale and significance of humanity's role as an agent of global change and the need to use scientific knowledge to underpin management of our global life-support system, to enhance biological productivity and to respond to the needs of a growing population[81]. The creation of the IGBP recognised the urgency of improving '*understanding of the pathways and rates of exchange for the primary constituents of living organisms (carbon, nitrogen, phosphorus, sulfur, hydrogen, and oxygen) and their relation to the other great domains of planet Earth*'[82]. As Bill Fyfe of the University of Western Ontario put it in his concluding remarks to the Ottawa symposium, '*The Earth is changing and man is contributing to the change at an accelerating rate. We must acquire the fundamental data on geosphere interactions necessary to understand the impact of change on the biosphere*'[81]. At the urging of Thompson Webb, John Kutzbach and Alayne Street-Perrott[83], one of the core programmes of the IGBP would be PAGES, the Past Global Changes programme, dealing with past climate change.

Biogeochemistry lay at the heart of the IGBP. Recognising the birth of this new field of science, Springer launched the journal *Biogeochemistry* in 1984/85 and the American Geophysical Union launched *Global Biogeochemical Cycles* in 1987. By the mid 1990s, biogeochemistry was considered to be one of the core elements of the field of geochemistry, featuring in 2005 as Volume 8 of the nine-volume *Treatise on Geochemistry*[84]. W.H. Schlesinger, the editor of Volume 8, explained that we rely on biogeochemists '*to understand fully the full biogeochemical cycles of water and the various chemical elements and the human impacts on each of them ... Biogeochemistry must emerge as the critical scientific discipline that informs planetary stewardship through the rest of this century*'[84]. Discoveries in biogeochemistry have advanced our understanding of Earth's climate evolution.

8.5   The Carbon Cycle

What we are concerned with mostly in this book is the carbon cycle, which lies at the heart of biogeochemistry. There are in fact two interconnected carbon cycles, one fast and one slow. Many texts describe these, but here I draw on descriptions from the *Treatise on Geochemistry*[85, 86]. In the fast cycle, photosynthesis chemically reduces inorganic carbon (CO_2) to form organic matter plus oxygen:

$$CO_2 + H_2O \leftrightarrow CH_2O + O_2$$

When organic matter decomposes, the equation is reversed, and oxygen is used to produce CO_2 and water. This cycle operates on time scales of days to millennia.

Carbon as CO_2 is exchanged between a number of 'reservoirs' of different gigatonne sizes, including the atmosphere (780 Gt), vegetation (550 Gt), litter (300 Gt), soil (1200 Gt), the surface ocean (dissolved inorganic carbon, 700 Gt; dissolved organic carbon, 25 Gt), surface biota (3 Gt) and the deep ocean (dissolved inorganic carbon, 36 300 Gt; dissolved organic carbon, 975 Gt). As mentioned earlier, about 90 Gt is exchanged annually between the ocean and the air. Very little organic matter reaches the seafloor, where it may escape decay and become incorporated into sediment and thus eventually into the lithosphere, from whence it is transferred to the slow carbon cycle. Sediment accumulation of both organic and inorganic carbon in the sea is around 0.2 Gt/yr, a tiny amount given the vast size of the oceanic reservoir[86].

The slow carbon cycle (Figure 8.8) does not involve biology. It determines the abundance of CO_2 in the Earth's air and oceans on time scales of tens to hundreds of millions of years and is controlled by tectonics, volcanism and weathering. CO_2 is released from the mantle to the air and oceans via volcanic activity and seafloor spreading, and is removed from the air mainly by reaction with silicate minerals through weathering in mountains, forming

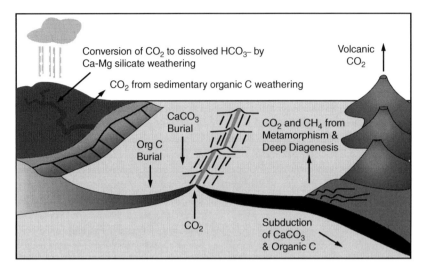

Figure 8.8 *The elements of the long-term carbon cycle. Readers interested in the equations for the various chemical reactions involved should consult Bob Berner's 1999 paper in GSA Today[87]. In effect, his work expands on that of Ebelmen in France in the 1840s[88].*

carbonate minerals that accumulate in deep sea-sediments and become part of the lithosphere. Most of the carbonates are eventually subducted into the Earth's mantle, where they are heated, releasing their carbon as CO_2 to the air and ocean, beginning the cycle again. The chemistry of the slow carbon cycle would operate whether or not there was life on the planet.

Volcanoes are an important natural source of CO_2. Terry Gerlach of the US Geological Survey showed in 2011 that present-day volcanoes emit rather modest amounts of CO_2 globally: about as much annually as all human activities in Florida[89]. Their CO_2 comes from the degassing of magma. In the absence of humans, volcanoes provide the main means of restoring CO_2 lost from the atmosphere and ocean by silicate weathering and the deposition and burial of sedimentary carbonates and organic carbon. Gerlach estimated that subaerial and submarine volcanoes emit between 0.18 and 0.44 Gt CO_2/year, the average being 0.26 Gt CO_2/year.

The human-induced emission of 35 Gt CO_2/year estimated for 2010 is 135 times greater than the 0.26 Gt CO_2/year estimated from volcanoes of all sorts, a ratio that Gerlach called the anthropogenic CO_2 multiplier (ACM): an index of the dominance of anthropogenic over volcanic CO_2 emissions. The ACM rose gradually from around 18 in 1900 to 38 by 1950, then rapidly to its 2010

level of 135, due the exponential rise in documented CO_2 emissions from the burning of fossil fuels.

Volcanoes do not erupt steadily, releasing a continuous stream of gas. They erupt periodically and unpredictably. Large, infrequent eruptions cause significant divergence from the global estimate of 0.26 Gt CO_2/year. But, as Gerlach noted, modern volcanic paroxysms are unlikely to have breached the upper limit of 0.44 Gt CO_2/year. When Mt St Helens blew up in May 1980, it contributed a mere 0.01 Gt CO_2. Even Mt Pinatubo, erupting in June 1991, only released around 0.05 Gt CO_2. Although the actual rate of CO_2 emission per hour during a big eruption may be about the same as what we humans emit over the same period, eruptions don't last long, so big eruptions have small effects averaged over the long term. As Gerlach noted, '*humanity's ceaseless emissions release an amount of CO_2 comparable to the 0.01 gigaton of the 1980 Mount St Helens paroxysm every 2.5 hours and the 0.05 gigaton of the 1991 Mount Pinatubo paroxysm every 12.5 hours*'[89].

Exploring further, Gerlach calculated that if the global volcanic output equalled or exceeded 35 Gt CO_2/year, the annual mass of volcanic CO_2 emissions would be more than three times the known annual mass of erupted magma. Such a supply would suggest that CO_2 makes up more than 30 wt % of global magma supply, which, if true, would make all volcanoes explosive. In reality, CO_2 concentrations are more like 1.5 wt % in magmas from ocean ridges, plumes and subduction zones. It is geologically implausible that magma production of the amount required to supply CO_2 to the atmosphere at present rates (more than 40 times the rate of current mid-ocean-ridge magma supply) comes from hitherto undiscovered magma sources on the seabed. Indeed, Gerlach pointed out, the release of such vast amounts of volcanic CO_2 into the ocean would turn it acid fast[89]. Production of the volumes of CO_2 required would call for magma output rates higher than known outputs of past massive continental plateau basalt eruptions, over much shorter times than they lasted.

Independent analyses by other volcanological experts confirmed Gerlach's findings and were taken into account in reaching his conclusions. For example, Marty and Tolstikhin calculated that the flux of CO_2 into the atmosphere and ocean from submarine volcanic sources (mid-ocean ridges, island arcs and plumes) is about 0.264 Gt/yr[90], while the flux from subaerial volcanoes is about 0.065 ± 0.046 Gt/yr, based on data from Williams and colleagues[91]. Some CO_2 is emitted from hydrothermal vents on mid-ocean ridges. As Chris German pointed out, while the first hydrothermal vent sites discovered on the deep-sea floor contained less than twice the CO_2 present in seawater, we now know that few vent sites have CO_2 levels less than or equal to those in seawater, while many sites have an order of magnitude more CO_2 than seawater, and a few have two orders of magnitude more[92]. However, these contributions are rapidly diluted away from immediate vent sources. Independent confirmation of the weakness of the submarine volcanic signal comes from the limited abundance in the ocean of the ^{3}He isotope, which is an expression of submarine volcanic activity. Helium-3 is modestly abundant at oceanic depths between about 1500 and 3000 m, decreasing rapidly away from mid-ocean ridge crests[93]. John Lupton of NOAA's Pacific Marine Environmental Laboratory (PMEL) in Washington state has found CO_2-rich fluid locally venting on to the seabed on the flanks of a submarine volcano in an island arc setting[94], but this is thought to be a highly localised phenomenon. It has also been found in a sedimentary setting in the Okinawa Trough.

An independent study by Giuseppe Etiope and Nils-Axel Mörner agreed that natural emissions of CO_2 lie far below those from human activities, but pointed out that geological seepage of light hydrocarbons from deep within the Earth, including methane, ethane and propane, could be a neglected source of greenhouse gases, which might prove to be more important than volcanic emissions of CO_2. Even

so, the amounts emitted are trivial compared with emissions of greenhouse gases from human activities[95].

Aside from these calculations of the amounts of CO_2 emitted, we can use the carbon isotopic composition of CO_2 to indicate its likely source. The CO_2 emitted from magmatic sources and uncontaminated by air usually has a carbon isotopic signature of around $-2\,\partial^{13}C$[96], which is quite different from that of air or petroleum. But the volume emitted is too small to have an effect on the composition of the air on human time scales, unlike the change caused by the burning of fossil fuels over the last century.

In summary, the data show that global emissions of CO_2 from within the crust by volcanic and related processes on land and beneath the ocean amount to less than 1% of the annual anthropogenic emissions of CO_2. There is no justification in the rock record for the unbelievably high volumes of magma production or unbelievably high concentrations of magmatic CO_2 required to support the notion that volcanoes have supplied the increases in atmospheric CO_2 since the start of the Industrial Revolution. Volcanic CO_2 is dwarfed by anthropogenic outputs.

8.6   Oceanic Carbon

On the millennial time scale, it is the vast amount of dissolved inorganic carbon in the ocean that determines the equilibrium amount of CO_2 in the atmosphere. CO_2 dissolves in the ocean to form carbonic acid (H_2CO_3), which dissociates to form bicarbonate ions (HCO_3^-) and carbonate ions (CO_3^{2-}), which are by far the main forms of dissolved inorganic carbon in seawater. Bicarbonate ions in turn dissociate to form hydrogen ions (H^+) and carbonate ions (CO_3^{2-}). The concentration of hydrogen ions dictates the ocean's acidity, or pH. As CO_2 dissolves in the ocean, it initially alters its carbonate chemistry, causing the ocean to become more acid and reducing its pH. This process cannot continue indefinitely, because the carbonate system in the ocean is buffered to the extent that the more acid it becomes, the less CO_2 it is able to absorb from the atmosphere. The buffering effect keeps the ocean's pH to a narrow range between pH 7.5 and 8.5[97]. While that may appear to be rather small, the pH scale is logarithmic, and the pH decrease of 0.1 unit between 1751 and 1994 is equivalent to an increase in acidity of 30%[86]. Because the buffer factor increases as dissolved CO_2 increases, when we put more CO_2 into the air, the ocean becomes increasingly resistant to taking it up, and more of it stays in the air.

This long-term equilibrium is modified in the short term by three key factors. One is ocean mixing, which moves CO_2 absorbed from the air by the ocean deeper in the water column with time, enabling more CO_2 to be adsorbed at the surface. An example is the sinking of relatively dense sub-Antarctic surface water from the surface of the Southern Ocean to form Antarctic Intermediate Water at an average depth of around 1000 m in the South Atlantic. The other two factors are 'pumps'. The 'solubility pump' operates because CO_2 is twice as soluble in cold water as it is in warm water. Warm the water and CO_2 comes out; cool it and CO_2 goes in. Thanks to this pump, the sinking of cold Arctic and Antarctic water takes CO_2 from the air down into the deep ocean, making the atmospheric concentration of CO_2 lower than it would otherwise be[86]. The 'biological pump' transfers CO_2 from surface waters to the deep ocean through the decomposition at depth of sinking dead organic matter that grew by using CO_2 in photosynthesis in the sunlit zone at the ocean's surface[98, 99]. This process also enriches deep waters in dissolved carbon, thus lowering the CO_2 content of surface waters and making less CO_2 available to the air. If these pumps were switched off, the preindustrial concentration of CO_2 in the atmosphere would likely have been 720 ppm rather than its measured 280 ppm[86].

Most primary production by ocean plankton is recycled as part of the fast carbon cycle, with the release of dissolved inorganic carbon to the upper ocean within the sunlit top 100 m of the water column[98]. This recycling process keeps most carbon out of the biological pump. Rates of primary production (in $gC/m^2/yr$) are highest (about 420) in coastal regions where upwelling currents bring nutrients to the surface, somewhat less (about 250) in other coastal areas and only about 130 in the open ocean. But because 90% of the ocean is open, 80% of carbon fixation occurs there, rather than in the more productive continental margins[98]. About 75% of production in upwelling and coastal regions is siliceous diatoms. These make up only 25% of open-ocean productivity, in which calcareous phytoplankton (coccolithophores) predominate away from the poles, where siliceous diatoms hold sway.

Organic matter commonly sinks at rates of around 100 m/day in the form of dense aggregates or clumps known as 'marine snow', some of which comprises the faecal pellets of zooplankton. The rates of decomposition en route are so high that the rate of accumulation of organic carbon in marine sediments is only 0.5% of the rate of carbon fixation at the ocean surface. Mapping organic matter on the seabed shows that around 94%

of the sedimentary organic carbon preserved there is buried on continental shelves and slopes, with only 6% accumulating in the open ocean. As the open ocean plays host to 80% of primary production, the accumulation of only 6% of organic carbon there suggests that this area has an overall preservation efficiency of 0.02%. On the continental shelves and slopes, the preservation efficiency is more like 1.4%[98]. To put it another way, only 10% of primary productivity escapes from the production zone in the upper 100 m; only 10% of that reaches the seabed at 4000 m; 90% of that is degraded at the seabed; leaving only 0.1% of the original production locked in sediments under the open ocean[100]. As implied by the efficiency numbers, significantly more sinking organic matter is trapped on continental margins. Some of the organic material deposited on the continental shelf and upper continental slope is resuspended by bottom currents and drifts downslope in turbid bottom waters. Some moves to the deep ocean much faster through turbidity currents. These lateral processes may mix organic matter from land with that from marine production in surface waters, on the deep-sea floor.

Because of the difficulty of measuring either changes in the ocean's inventory of carbon or the exchange of carbon across the air–sea interface, the uptake of anthropogenic carbon by the oceans was primarily calculated with models that simulate the chemistry of carbon in seawater, the air–sea exchanges of CO_2 and oceanic circulation[86]. However, with more than 4 million observations of the partial pressure of CO_2 in the oceans made in recent years, we can now calculate the uptake of CO_2 more directly[101]. Models and data are in good agreement.

8.7   Measuring CO_2 in the Oceans

By the mid 1980s, oceanographers recognised that we still knew far too little about how the ocean worked, not least because it was grossly under-sampled. Yet the ocean plays a key role in the climate system, by storing masses of heat and moving it slowly around the globe. As much heat is stored in the top 3.5 m of the ocean as in the entire atmosphere. To reduce uncertainties about global ocean dynamics and the rates of ventilation and mixing in the deep ocean, and to improve understanding of the role of the ocean in the global climate system, the World Climate Research Programme launched a major international global experiment – the World Ocean Circulation Experiment (WOCE) – conceived in the mid 1980s, with a field phase between 1990 and 1998 and a modelling phase extending to 2002[102].

Today, the oceans contain about 50 times as much CO_2 as the atmosphere, so small changes in the ocean carbon cycle can have large atmospheric consequences. Realising that it was essential to increase focus on the role of the ocean in the CO_2 story, the US scientific community began developing a strategy for large-scale, coordinated studies of oceanic biogeochemical processes. By 1987, several countries were following that lead, and the SCOR was persuaded to combine these efforts in the international Joint Global Ocean Flux Study (JGOFS)[59], which aimed to understand how physical, chemical and biological processes influence the exchange of CO_2 between the ocean and the air, and to elucidate the role of biological feedbacks in amplifying the chemical and physical effects. Its planners followed the recommendations of Revelle's CO_2 Advisory Panel in designing an observation programme to determine the global oceanic inventory of CO_2. In 1989, JGOFS would become a core project of the fledgling IGBP. Together, WOCE and JGOFS would improve our understanding of the workings of the ocean and its role in the carbon cycle. As director of the UK's major deep-sea oceanographic laboratory at Wormley between 1988 and 1995, I helped to oversee the development of the UK's substantial contribution to WOCE and my institute's contribution to JGOFS.

Reviewing Revelle's legacy, Chris Sabine from NOAA, Hugh Ducklow from WHOI and Maria Hood from the IOC had this to say (with my apologies for the acronym soup!): '*The first global survey of ocean CO_2 was carried out under the joint sponsorship of IOC and the Scientific Committee on Oceanic Research (SCOR) in the Joint Global Ocean Flux Study (JGOFS) and the World Ocean Circulation Experiment (WOCE) in the 1990s. With these programs and underway pCO$_2$ measuring systems on research vessels and ships of opportunity, ocean carbon data grew exponentially, reaching about a million total measurements by 2002 when Taro Takahashi (Lamont-Doherty Earth Observatory) and others provided the first robust mapping of surface ocean CO_2. Using a new approach developed by Nicolas Gruber (ETH Zurich) and colleagues with JGOFS-WOCE and other synthesized data sets, one of this article's authors (Sabine) with a host of coauthors estimated*[103] *that the total accumulation of anthropogenic CO_2 between 1800 and 1994 was $118 \pm 19 \, Pg \, C$, just within the uncertainty goals set by JGOFS and IOC prior to the global survey. Today, ocean carbon activities are coordinated through the International Ocean Carbon Coordination Project (IOCCP). Ocean carbon measurements now accumulate at a rate of over a million measurements per year – matching the total number achieved over the first three decades of ocean carbon studies. IOCCP is actively working to combine these data into uniform data sets that the community can use to better understand ocean carbon uptake and storage*'[59]. Through the IOCCP, which became a standing project of SCOR and the IOC in 2005, the '*IOC is achieving the international cooperation in ocean carbon observations that Roger Revelle was promoting back in 1960*'[59].

Observations confirm that the ocean is accumulating anthropogenic CO_2 in much the same way that Keeling showed was happening in the air. As of 1980, the oceans had absorbed around 40% of the emissions made to that time[86]. On a time scale of several thousand years, Dave Archer and colleagues calculated that around 90% of anthropogenic CO_2 emissions will end up in the oceans[104].

Sabine's study showed that our (anthropogenic) emission of CO_2 is not distributed uniformly through the oceans[103]. Most is found in the North Atlantic, where cold CO_2-rich surface water sinks into the depths to form North Atlantic Deep Water. There are substantial, although lesser, amounts in a belt along the north side of the Antarctic Circumpolar Current, where cold CO_2-rich surface waters sink to form Antarctic Intermediate Water and Sub-Antarctic Mode Water. These various sources make the Atlantic Ocean much richer in anthropogenic CO_2 than the Pacific Ocean. The surface waters least rich in anthropogenic CO_2 are the upwelling waters of the Southern Ocean south of the polar front, which bring old, naturally CO_2-rich water to the surface and supply it to the air. Thus, the Southern Ocean is a source for old CO_2 south of the front, and a sink for anthropogenic CO_2 north of it.

8.8   A Growing International Emphasis

For the purposes of my tale, the next big development after ICSU's 1984 Ottawa meeting was a conference convened by the UNEP, the WMO and ICSU, which took place in the autumn of 1985 in Villach, Austria, on the 'Assessment of the Role of Carbon Dioxide and of Other Greenhouse Gases in Climate Variations and Associated Impacts'[105]. The meeting participants concurred that '*we must place the CO_2 question in the context of a global picture which involves understanding the linkages of carbon* *to the biogeochemical cycles of nutrients, and treat these cycles in the context of climate feedbacks rather than as isolated, independent systems*'[106]. Much remained to be done, including clarifying the preindustrial concentration of CO_2, quantifying the oceanic sink for CO_2, studying the impact of El Niño events on CO_2 and studying the natural variability of CO_2 between glacial and interglacial cycles (which we examine in Chapters 12 and 13) and during climatic events like the Little Ice Age and Medieval Optimum (which we examine in Chapters 14 and 15). Evidently, geologists too would have to contribute to the growing understanding of the behaviour and effects of CO_2 within the climate system.

In tune with conclusions reviewed in previous sections, the Villach meeting concluded that '*As a result of the increasing concentrations of greenhouse gases, it is now believed that in the first half of the [21st century] a rise of global mean temperature could occur which is greater than any in man's history*'[105]. They considered that the effect of doubling CO_2 on temperature lay in the range 1.5–4.5 °C, and on sea level in the range 0.2–1.4 m. To address this problem, they believed, '*governments and funding agencies should increase research support and focus effort on crucial unsolved problems related to greenhouse gases and climate change*'[105]. Participants also suggested that governments should support periodic assessments of the state of scientific understanding and its practical implications. This recommendation would lead in 1988 to the formation of the IPCC.

By 1987, the 10th Congress of the WMO saw the need for an objective, balanced and internationally coordinated scientific assessment of the effects of increasing concentrations of greenhouse gases on the Earth's climate and of ways in which these changes might impact socioeconomic patterns. This mechanism took shape as the IPCC, a group of experts established in 1988 and first reporting in 1990, repeating its assessments at intervals of about 5 years. The compositions of the groups changed somewhat as time went by. The IPCC operates through three main working groups: Working Group I, which reports on the state of the climate system – the science of climate change; Working Group II, which reports on impacts, adaptation and vulnerability; and Working Group III, which reports on mitigation. In this book, where appropriate, we will review a few of the findings of Working Group I. Working Group I does not do original research. It reviews the pertinent published literature and produces reports of about 1000 pages long that in effect form textbooks on the state of climate science. These can be downloaded free from the IPCC Web site (www.ipcc.ch). The latest – the 5th Assessment Report – was made public in September 2013; it concluded, '*It is extremely likely that human influence has been the dominant cause of the observed warming since the mid-20th century*'[107]. In other words, the forecasts of Plass (1956), Revelle (1965, for Johnson's Science Advisory Committee), Manabe (1967), Broecker (1975), Charney (1979), Flohn (1979), Ramanathan (1979) and Hansen (1981 and 1982) had come to pass.

Do the IPCC's Working Group I reports constitute a scientific consensus 'across the board'? No. They represent the broad agreement of the experts who authored those reports. Parts of the reports may be inaccurate. For instance, in my Antarctic science community, we were unhappy with the climate modelling for Antarctica in the 2007 report, because the models used for the region did not properly represent the Antarctic climate system[108]. Similarly, glaciologists took exception to the sea level projections in the 2007 report because they did not include results from dynamic models for the mechanical decay of ice sheets (largely because those data were not yet in the public domain). Even so, at the time these reports are published, they are comprehensive reviews of the state of the art of climate science, and useful guides to what research needs to be done next. They do not, however, focus on palaeoclimates, which is why I have taken the trouble to write this book.

8.9   Reflection on Developments

Reflecting on developments since Plass's pronouncements in the late 1950s, a good beginning has been made towards understanding the operation of the planetary climate system and of the role of CO_2 within it. These advances owe a great deal to an extraordinary expansion in the numbers and kinds of satellites, and a comparable advance in the number of research ships, floats, buoys and other means of observing the ocean. For example, at the time of writing there were some 3500 *Argo* floats measuring the temperature and salinity of the upper ocean, and some 3000 satellites operating in space, many observing Earth's environment.

Following the launch of the first satellite, Sputnik, on 4 October 1957, the use of satellites to look at the Earth has mushroomed. TIROS-I was the first Earth-observing satellite, launched in 1960 to monitor the weather; now there are dozens doing the same. SEASAT was the first ocean-observing satellite, launched in 1978; there are now many more. Ships, planes and now people can know where they are through the use of the Global Positioning System (GPS), which has been operational since 1978 and generally available since 1994. The GPS is a boon to oceanographers trying to unravel the secrets of the motion of the oceans. And these days we all talk to each other and send messages via communication satellites, the first of which, Telstar, was launched in 1962. Instruments like *Argo* floats use communications satellites to send their data back to base and know where they are through the use of GPS.

Just as Earth-observing operations, both *in situ* and remotely from satellites, have grown dramatically over the years, so too has the research into what the operational data mean, and into the processes acting within the climate system. Research continues through ongoing activities of the World Climate Research Programme, the IGBP and groups like the IOCCP, mentioned earlier. One might justifiably say that the 1st World Climate Conference in 1979 had momentous outcomes. Climate science, like medical science, has become high-tech, and most of the developments in both fields have taken place since the 1950s.

So, by the late 1970s, some 20 years after publication of his key papers in 1956, one of Plass's key wishes had come true: CO_2 was being measured in the atmosphere and ocean. We now have abundant and ongoing measurements of CO_2 and other radiatively active gases in the air, and a comprehensive theoretical understanding of how the climate system works and of the roles of CO_2 and water vapour within it. Not everything is known with absolute certainty, however. Ramanathan was quick to point out back in 1988 that we did not know enough about the role of clouds, the tops of which could reflect radiation, and the moisture within which could trap it, and this is still an area of some difficulty[86]. But even before the IPCC began its deliberations in 1988, a great deal was known about CO_2 and climate and what the future might hold. And at about that time, development had begun on a global observing system capable of making an annual health check on the climate of the planet, and JGOFS had set out to make the quantum leap in measurements of CO_2 in the ocean. Momentous advances were made as we moved towards the end of the 20th century and into the 21st. We still do not know everything we need to. Some uncertainties remain, but they are steadily getting fewer and smaller.

One of the key elements I have not yet addressed, but which has a bearing on our understanding of how CO_2 may have behaved in the climate systems of the past, is the question of duration: How long will CO_2, once emitted, stay in the atmosphere? In 2007, a team led by Alvaro Montenegro of the University of British Columbia assessed the climate response on millennial time scales and calculated '*that 75% of the anthropogenic CO_2 has an average perturbation lifetime of ~1800 years with the remaining 25% having average lifetime much longer than 5000 years*'[109]. They showed that if all of the available fossil fuels (then estimated as ~5000 Gt C) were burned over, say, the next 200 years, global temperatures would rise by 6–8 °C and remain at least 5 °C higher than preindustrial levels for more than 5000 years. Later, in 2009, a subset of those same researchers refined their calculations to show that '*For emissions up to about 1000 [Gt], 50% of the CO_2 anomaly is taken up within 100 yr and another 30% is absorbed within 1000 yr, which is similar to IPCC estimates … Above 1000 [Gt], the time to absorb 50% of the emissions increases dramatically, and more than 2000 yr are needed to absorb half of a 5000-[Gt] perturbation*'[110,111]. In addition, due to the logarithmic relationship between CO_2 and its radiative forcing, '*Temperature anomalies may last much longer*'[110,111]: 10 000 years or more. These studies agree with those of Susan Solomon of the US NOAA and colleagues, who demonstrated that '*the climate change that takes place due to increases in carbon dioxide concentration is largely irreversible for 1000 years after emissions stop … Following cessation of emissions, removal of atmospheric carbon dioxide decreases radiative forcing, but is largely compensated by slower loss of heat to the ocean, so that atmospheric temperatures do not drop significantly for at least 1000 years*'[112].

In addition, one of the key things we have learned, which will have a bearing on interpretations of past climate change, is that water vapour is the dominant contributor to warming (~50% of the effect), followed by clouds (~25%) and CO_2 (~20%). According to Gavin Schmidt of NASA's Goddard Institute for Space Studies, these proportions remain about the same even with a doubling of CO_2 in the atmosphere[113]. This is likely to be typical of past climates, too.

The palaeoclimate data that we evaluate in following chapters will lead us to a view on future change. That view will not be based solely on the outputs of the IPCC, useful though they may be. It will reflect all that we have learned in these pages, starting with the work of Ebelmen in 1845 and Tyndall in 1859, about CO_2 in the atmosphere and its role in the climate system.

How would geologists apply this new knowledge to their interpretation of past climate change? And would they be able to grant Plass's second wish: '*to be able to determine atmospheric CO_2 concentration for past geological periods*'[13]? We shall explore these questions in Chapter 9.

References

1. Callendar, G.R. (1938) The artificial production of carbon dioxide and its influence on temperature. *Quarterly Journal of the Royal Meteorological Society* **64**, 223–237.
2. Callendar, G.R. (1949) Can carbon dioxide influence climate? *Weather* **4**, 310–314.
3. Archer, D. and Pierrehumbert, R.T. (2011) *The Warming Papers – The Scientific Foundation for the Climate Change Forecast*. Wiley-Blackwell, Chichester and Oxford.
4. Madden, R.P. (1957) *Study of CO_2 Absorption Spectra between 15 and 18 Microns*. Progress Report to Office of Naval Research, Contract 248(01), declassified. Available from Armed Forces Technical Information Agency, Arlington, Virginia.
5. See Czerny–Turner Monochromator diagram at http://en .wikipedia.org/wiki/Monochromator (last accessed 29 January 2015).
6. Strong, J. (1962) *Infrared Spectroscopy and Infrared Properties of Atmospheric Gases*. Fort Belvoir Defense Technical Information Center, Fort Belvoir, VA.
7. Cloud, W.H. (1952) *The 15 Micron Band of CO_2 Broadened by Nitrogen and Helium*. ONR Progress Report, Johns Hopkins University, Baltimore, MD.
8. Madden. R.P. (1961) A high-resolution study of CO_2 absorption spectra between 15 and 18 microns. *Journal of Chemical Physics* **35**, 2083.
9. Anding, D. (1967) *Band-Model Methods For Computing Atmospheric Slant-Path Molecular Absorption*. IRIA Technical Report. Available from NTIS (National Technical Information Service), Department of Commerce, Washington DC.
10. http://www.cfa.harvard.edu/hitran/ (last accessed 29 January 2015).
11. Houghton, J. (2013) *In the Eye of the Storm: The Autobiography of Sir John Houghton*. Lion Books, Oxford.
12. Houghton, J. and Smith, D. (1966) *Infra-red Physics*. Oxford University Press, Oxford.
13. Plass, G. N. (1956) The carbon dioxide theory of climate change. *Tellus* **VIII**, 140–154.
14. Plass, G.N. (1956) The influence of the 15 micron carbon dioxide band on the atmospheric infra-red cooling rate. *Quarterly Journal of the Royal Meteorological Society* **82**, 310–329.
15. Plass, G.N. (1956) Effect of carbon dioxide variations on climate. *American Journal of Physics* **24**(5), 376–387.
16. Plass, G.N. (1959) Carbon dioxide and climate. *Scientific American* **201**(1), 41–47.
17. Plass, G.N. (1961) The influence of infrared absorptive molecules on the climate. *New York Academy of Science* **95**, 61–71.
18. Manabe, S. and Wetherald, R.T. (1967) Thermal equilibrium of the atmosphere with a given distribution of relative humidity. *Journal of Atmospheric Sciences* **24** (3), 241–259.
19. Pierrehumbert, R.T. (2010) *Principles of Planetary Climate*. Cambridge University Press, Cambridge.
20. Keeling, C.D. (1980) The Suess effect: [13]carbon–[14]carbon interrelations. *Environment International* **2** (4–6), 229–300.
21. Revelle, R. and Suess, H.E. (1957) Carbon dioxide exchange between atmosphere and ocean and the question of an increase of atmospheric co_2 during the past decades. *Tellus* **IX** (1), 18–27.
22. Keeling, C.D. (1960) The concentration and isotopic abundance of carbon dioxide in the atmosphere. *Tellus* **XII** (2), 200–203.
23. Yakir, D. (2006) The stable isotopic composition of atmospheric CO_2. In: *The Atmosphere, Vol. 4, Treatise on Geochemistry* (eds H.D. Holland and K.K. Turekian). Elsevier-Pergamon, Oxford, pp. 175–213.
24. Keeling, C.D., Piper, S.C., Bacastow, R.B., Wahlen, M., Whorf, T.P., Heimann, M., *et al.* (2001) *Exchanges of Atmospheric CO_2 and $^{13}CO_2$ with the Terrestrial Biosphere and Oceans from 1978 to 2000. I. Global Aspects*. SIO Reference Series, No. 01–06 (revised from SIO Reference Series, No. 00–21). Scripps Institution of Oceanography, San Diego, CA.
25. Marland, G., Andres, R.J., Boden, T.A. and Johnston, C. (1998) Global, regional and national CO_2 emissions estimates from fossil fuel burning, cement production and gas flaring: 1751–1995 (revised 1998). Report ORNL/CDIAC NDP-030/Rb, available from: http://cdiac.esd.ornl.gov/ndps/ndp030.html (last accessed 29 January 2015).
26. Francey, R.J., Allison, C.E., Etheridge D.M., Trudinger, C.M., Enting, I.G., Leuenberger, M., *et al.* (1999) A 1000-year high precision record of $\partial^{13}C$ in atmospheric CO_2. *Tellus Series B: Chemical and Physical Meteorology* **51** (2), 170–193.
27. Andres, R.J., Marland, G., Boden, T. and Bischoff, S. (2000) Carbon dioxide emissions from fossil fuel consumption, 1751–1991, and an estimate of their isotopic composition and latitudinal distribution. In: *The Carbon Cycle* (eds T.M.L. Wigley and D. Schimel). Cambridge University Press, Cambridge, pp. 53–62.
28. Manning, A.C. and Keeling, R.F. (2006) Global oceanic and land biotic carbon sinks from the Scripps atmospheric oxygen flask sampling network. *Tellus* **58B**, 95–116.
29. Callendar, G.S. (1958) On the amount of carbon dioxide in the atmosphere. *Tellus* **10**, 243–248.

30. Beck, E.-G. (2007) 180 years of atmospheric CO_2 gas analysis by chemical methods. *Energy and Environment* **17** (2), 259–282.

31. Meijer, H.A.J. (2007) Comment on '180 years of atmospheric CO_2 gas analysis by chemical methods' by Ernst-Georg Beck. *Energy and Environment* **18** (5), 635–636.

32. Keeling, R.F. (2007) Comment on '180 years of atmospheric CO_2 gas analysis by chemical methods' by Ernst-Georg Beck. *Energy and Environment* **18** (5), 637–639.

33. Weart, S. (2008) *The Discovery of Global Warming*. Harvard University Press, Harvard, CT. 2011 expanded and updated edition available online courtesy of the American Institute of Physics at http://www.aip.org/history/climate/index.htm (last accessed 29 January 2015).

34. Bolin, B. and Eriksson, E. (1958) Changes in the carbon dioxide content of the atmosphere and sea due to fossil fuel consumption. In: *The Atmosphere and the Sea in Motion: Scientific Contributions to the Rossby Memorial Volume* (ed B. Bolin). Rockefeller Institute Press, New York, pp. 130–142.

35. Anon. (1957) Polar ice caps may melt with industrialisation. *Otago Daily Times, January* **23**, p. 5.

36. Elliot, L. and Gristock, J.M. (2006) *Eureka UK: 100 Discoveries and Developments in UK Universities that have Changed the World*. Universities UK, London.

37. Lamb, H.H. (1963) What can we find out about the trend of our climate? *Weather* **18**, 194–216; reprinted inLamb, H.H. (1966) *The Changing Climate*. Methuen, London, pp. 196-214.

38. Peterson, T., Connolley, W. and Fleck, J. (2008) The myth of the 1970s global cooling scientific consensus. *Bulletin of the American Meteorological Society* **89** (9), 1325–1337.

39. Lamb, H.H. (1995) *Climate History and the Modern World*, 2nd edn. Routledge, London.

40. Budyko, M.I. (1969) The effect of solar radiation variations on the climate of the Earth. *Tellus* **21**, 611–619.

41. Budyko, M.I. (1977) On present-day climatic changes. *Tellus* **29**, 193–204.

42. Budyko, M.I. (1971) *Klimat I Zhizn*. Gidrometeoizdat, Leningrad; translated asBudyko, M.I. (1974) *Climate and Life* (ed. D.H. Miller). Academic Press, New York and London.

43. Budyko, M.I. and Ronov, A.B. (1979) Chemical evolution of the atmosphere in the Phanerozoic. *Geochemistry International* **16** (3), 1–9; translated from *Geokhimiya* **5**, 643–653 (1979).

44. Budyko, M.I. (1996) Past changes in climate and societal adaptations. In: *Adapting to Climate Change* (eds J.B. Smith, N. Bhatti, G.V. Menzhuin, R. Benioff, M. Campos, B. Jallow, *et al.*). Springer, New York, pp. 16–26.

45. Sawyer, J.S. (1972) Man-made carbon dioxide and the 'greenhouse' effect. *Nature* **239**, 23–26.

46. Manabe, S. and Wetherald, R.T. (1975) The effects of doubling CO_2 concentration on the climate of a general circulation model. *Journal of Atmospheric Sciences* **32** (1), 3–15.

47. Schneider, S.H. (1996) *Laboratory Earth, the Planetary Gamble We Cannot Afford to Lose*. Weidenfeld and Nicolson, London.

48. Schneider, S.H. and Mesirow, L. (1976) *The Genesis Strategy: Climate and Global Survival*. Plenum, New York.

49. Augustsson, T. and Ramanathan, V. (1977) A radiative-convective model study of the CO2-climate problem. *Journal of Atmospheric Science* **34**, 448–451.

50. Ramanathan, V., Lian, M.S. and Cess, R.D. (1979) Increased atmospheric CO_2: zonal and seasonal estimates of the effect on the radiation energy balance and surface temperature. *Journal of Geophysical Research* **84** (C8), 4949–4958.

51. Revelle, R., Broecker, W., Keeling, C.D., Craig, H. and Smagorinsky, J. (1965) Atmospheric carbon dioxide. In: *Restoring the Quality of Our Environment: Report of the Environment Pollution Panel President's Science Advisory Committee*. The White House, Washington, DC, pp. 126–127.

52. Meadows, D.H., Meadows, D.L., Randers, J. and Behrens, W.W. (1972) *The Limits to Growth: A Report for the Club of Rome's Project on the Predicament of Mankind*. Universe Books, New York.

53. Broecker, W.S. (1975) Climatic change: are we on the brink of a pronounced global warming? *Science* **188**, 460–463.

54. Cross, T.P. (2012) Wallace Broecker '53 battles the angry climate beast. *Columbia College Today*, Summer 2012; available from http://www.college.columbia.edu/cct/summer12/features4 (last accessed 29 January 2015).

55. Mercer, J.H. (1978) West Antarctic ice sheet and CO_2 greenhouse effect: a threat of disaster. *Nature* **271**, 321–325.

56. Broecker, W.S. and Takahashi, T. (1978) Neutralization of fossil fuel CO_2 by marine calcium carbonate. In: *The Fate of Fossil Fuel CO_2 in the Oceans* (eds N.R. Andersen and A. Malahoff). Plenum Press, New York, pp. 213–248.

57. Walker, J.C.G. and Kasting, J.F. (1992) Effects of fuel and forest conservation on future levels of atmospheric carbon dioxide. *Global and Planetary Change* **5** (3), 151–189.

58. Revelle, R. (1960) Summary of statement on international cooperation in oceanography in Scripps Institution of Oceanography Archives. *Proceedings of Preparatory Meeting of the Intergovernmental Conference on Oceanographic Research, Paris, March 21–29, 1960, NS/2503/620*. UNESCO, Paris.

59. Sabine, C.L., Ducklow, H. and Hood, M. (2010) International carbon coordination: Roger Revelle's legacy in the Intergovernmental Oceanographic Commission. *Oceanography* **23** (3), 48–61.

60. White, R.M. (1978) Oceans and climate – introduction. *Oceanus* **21**, 2–3.

61. Charney, J.G., Arakawa, A., Baker, J.D., Bolin, B., Dickinson, R.E., Goody, R.M., *et al.* (1979) *Carbon Dioxide and*

Climate: A Scientific Assessment. National Academy of Sciences, Washington, DC.

62. WMO (1979) *First World Climate Conference, February 1979*. WMO, Geneva.

63. White, R.M. (1979) Climate at the millennium. *Proceedings Report of First World Climate Conference, February 1979*. WMO, Geneva.

64. Flohn, H. (1977) Man-induced changes in heat budget and possible effects on climate. In: *Global Chemical Cycles and their Alterations by Man*. Dahlem Workshop, Berlin, November 1976, pp. 207–224.

65. Flohn, H. (1979) A scenario of possible future climates – natural and man-made. *Proceedings Report of First World Climate Conference, February 1979*. WMO, Geneva.

66. Hare, F.K. (1979) Climatic variation and variability. *Proceedings Report of First World Climate Conference, February 1979*. WMO, Geneva.

67. Munn, R.E. and Machta, L. (1979) Human activities that affect climate. *Proceedings Report of First World Climate Conference, February 1979*. WMO, Geneva.

68. Davies, D.A. (1979) Foreword. *Proceedings Report of First World Climate Conference, February 1979*. WMO, Geneva.

69. Hansen, J., Johnson, D., Lacis, A., Lebedeff, S., Lee, P., Rind, D. and Russell, G. (1981) Climate impact of increasing atmospheric carbon dioxide. *Science* **213**, 957–966.

70. Hansen, J.E. and Takahashi, T. (1984) *Climate Processes and Climate Sensitivity*. Geophysics Monographs 29. American Geophysical Union, Washington, DC.

71. Davies, E.E. (1984) Inventing the future: energy and the CO_2 'greenhouse' effect. In: *Climate Processes and Climate Sensitivity* (eds J.E. Hansen and T. Takahashi). Geophysics Monographs 29. American Geophysical Union, Washington, DC, pp. 1–5.

72. Hansen, J.E., Lacis, A., Rind, D., Russell, G., Stone, P., Fung, I., *et al.* (1984) Climate sensitivity: analysis of feedback mechanisms. In: *Climate Processes and Climate Sensitivity* (eds J.E. Hansen and T. Takahashi). Geophysics Monographs 29. American Geophysical Union, Washington, DC, pp. 130–163.

73. Broecker, W.S. and Takahashi, T. (1984) Is there a tie between atmospheric CO_2 content and ocean circulation? In: *Climate Processes and Climate Sensitivity* (eds J.E. Hansen and T. Takahashi). Geophysics Monographs 29. American Geophysical Union, Washington, DC, pp. 314–326.

74. Broecker, W.S. and Peng, T.H. (1984) The climate-chemistry connection? In: *Climate Processes and Climate Sensitivity* (eds J.E. Hansen and T. Takahashi). Geophysics Monographs 29. American Geophysical Union, Washington, DC, pp. 327–336.

75. Broecker, W.S. and Peng, T.H. (1982) *Tracers in the Sea*. Lamont-Doherty Geological Observatory, New York.

76. Schlesinger, W.H. (1997) *Biogeochemistry: An Analysis of Global Change*, 2nd edn. Academic Press, San Diego, CA.

77. Parkes, R.J., Cragg, B.A. and Wellsbury. P. (2000) Recent studies on bacterial populations and processes in subseafloor sediments: a review. *Hydrogeology Journal* **8**, 11–28.

78. National Research Council (1983) *Toward an International Geosphere-Biosphere Program – A Study of Global Change*. National Academies Press, Washington, DC.

79. Malone, T.C. and Roederer, J.G. (1985) *Global Change*. Cambridge University Press, Cambridge.

80. Mooney, H.A., Duraiappah, A. and Larigauderie, A. (2013) Evolution of natural and social science interactions in global change research programs. *Proceedings of the National Academy of Sciences of the United States of America*. Available from http://www.pnas.org/cgi/doi/10.1073/pnas.1107484110 (last accessed 29 January 2015).

81. Fyfe, W.S. (1985) The international geosphere-biosphere program: global change. In: *Global Change* (eds T.S. Malone and J.G. Roederer). Cambridge University Press, Cambridge, pp. 499–508.

82. Malone, T.F. (1985) Preface. In: *Global Change* (eds T.S. Malone and J.G. Roederer). Cambridge University Press, Cambridge, pp. xi–xxi.

83. Webb, T., Kutzbach, J. and Street-Perrott (1985) 20 000 years of global climatic change: paleoclimate research plan. In: *Global Change* (eds T.S. Malone and J.G. Roederer). Cambridge University Press, Cambridge, pp. 182–218.

84. Schlesinger, W.H. (2005) Volume editor's introduction. In: *Biogeochemistry, Vol. 8, Treatise on Geochemistry* (eds H.D. Holland and K.K. Turekian). Elsevier-Pergamon, Oxford, pp. xv–xvii.

85. Falkowski, P.G. (2005) Biogeochemistry of primary production in the sea. In: In: *Biogeochemistry, Vol. 8, Treatise on Geochemistry* (eds H.D. Holland and K.K. Turekian). Elsevier-Pergamon, Oxford, pp. 185–213.

86. Houghton, R.A. (2005) The contemporary carbon cycle. In: *Biogeochemistry, Vol. 8, Treatise on Geochemistry* (eds H.D. Holland and K.K. Turekian). Elsevier-Pergamon, Oxford, pp. 473–513.

87. Berner, R.A. (1999) A new look at the long-term carbon cycle. *GSA Today* **11** (9), 1–6.

88. Ebelmen, J.J. (1845) Sur les produits de la décomposition des espèces minerales de la famille des silicates. *Annales des Mines* **7**, 3–66.

89. Gerlach, T. (2011) Volcanic versus anthropogenic carbon dioxide. *EOS* **92** (24), 201–202.

90. Marty, B. and Tolstikhin, I.N. (1998) CO_2 fluxes from mid-ocean ridges, arcs and plumes. *Chemical Geology* **145**, 233–248.

91. Williams, S.N., Schaeffer, S.J., Calvache, M.L. and Lopez, D. (1992) Global carbon dioxide emissions to the atmosphere by volcanoes. *Geochimica et Cosmochimica Acta* **56**, 1765–1770.

92. German, C.R. and Von Damm, K.L. (2003) Hydrothermal processes. In: *The Oceans and Marine Geochemistry,*

Vol. 6, Treatise on Geochemistry (eds H.D. Holland and K.K. Turekian). Elsevier-Pergamon, Oxford, pp. 181–222.

93. http://www.whoi.edu/sbl/liteSite.do?litesiteid=28952&articleId=47707 (last accessed 29 January 2015).

94. Lupton, J.E., Butterfield, D.A., Lilley, M., Evans, L., Nakamura, K., Chadwick, W. Jr,, *et al.* (2006) Submarine venting of liquid carbon dioxide on a Mariana Arc volcano. *Geochemistry, Geophysics, and Geosystems* **7**, Q08007.

95. Mörner, N.-A. and Etiope, G. (2002) Carbon degassing from the lithosphere. *Global and Planetary Change* **33** (1–2), 185–203.

96. Wardell, L.J., Kyle, P.R. and Campbell, A.R. (2003) Carbon dioxide emissions from fumarolic ice towers, Mount Erebus Volcano, Antarctica. *Geological Society of London Special Publication* **213**, 231–246.

97. Varney, M. (1996) The marine carbonate system. In: *Oceanography, an Illustrated Guide* (eds C.P. Summerhayes and S.A. Thorpe). Manson, London, pp. 182–194.

98. De La Rocha, C.L. (2003) The biological pump. In: *The Oceans and Marine Geochemistry, Vol. 6, Treatise on Geochemistry* (eds H.D. Holland and K.K. Turekian). Elsevier-Pergamon, Oxford, pp. 83–111.

99. Sigman, D.M. and Haug, G.H. (2003) Biological pump in the past. In: *The Oceans and Marine Geochemistry, Vol. 6, Treatise on Geochemistry* (eds H.D. Holland and K.K. Turekian). Elsevier-Pergamon, Oxford, pp. 491–528.

100. Eglinton, T.I. and Repeta, D.J. (2003) Organic matter in the contemporary ocean. In: *The Oceans and Marine Geochemistry, Vol. 6, Treatise on Geochemistry* (eds H.D. Holland and K.K. Turekian). Elsevier-Pergamon, Oxford, pp. 145–180.

101. Takahashi, T., Sutherland, S.C. and Kozyr. A. (2011) *Global Ocean Surface Water Partial Pressure of CO_2 Database: Measurements Performed During 1957–2010 (Version 2010)*. Report ORNL/CDIAC-159, NDP-088(V2010), Carbon Dioxide Information Analysis Center, Oak Ridge National Laboratory, US Department of Energy, Oak Ridge, TN. Available from 10.3334/CDIAC/otg.ndp088(V2010) (last accessed 29 January 2015).

102. Morel, P. (1989) The global climate system: current knowledge and uncertainties. In: *Carbon Dioxide and Other Greenhouse Gases: Climatic and Associated Impacts* (eds R. Fantechi and A. Ghazi). Proceedings of the European Commission Symposium, Brussels, 3–5 November 1986. Kluwer Academic Press, Dordrecht, pp. 5–10.

103. Sabine, C.L., Feely, R.A., Gruber, N., Key, R.M., Lee, K., Bullister, J.L., *et al.* (2004) The oceanic sink for anthropogenic CO_2. *Science* **305** (5682), 367–371.

104. Archer, D.E., Kheshgi, H. and Maier-Reimer, E. (1998) Dynamics of fossil fuel CO_2 neutralization by marine $CaCO_3$. *Global Biogeochemical Cycles* **12**, 259–276.

105. WMO (1986) Report of the International Conference on the assessment of the role of carbon dioxide and of other greenhouse gases in climate variations and associated impacts, Villach, Austria, 9–15 October 1985. WMO No. 661.

106. Oeschger, H., *et al.* (1986) Cycling of carbon and other radiatively active constituents. *Proceedings of Villach meeting, October 1985*. WMO Report 661, pp. 48–55.

107. IPCC (2013) *Summary for Policy Makers*. Working Group I Contribution to the IPCC Fifth Assessment Report Climate Change 2013. The Physical Science Basis, IPCC.

108. Turner, J., Bindschadler, R.A., Convey, P., Di Prisco, G., Fahrbach, E., Gutt, J., *et al.* (2009) *Antarctic Climate Change and the Environment*. SCAR, Cambridge.

109. Montenegro, A., Brovkin, V., Eby, M., Archer, D. and Weaver, A.J. (2007) Long term fate of anthropogenic carbon. *Geophysics Research Letters* **34**, L19707.

110. Eby, M., Zickfield, K., Montenegro, A., Archer, D., Meissner, K.J. and Weaver, A.J. (2009) Lifetime of anthropogenic climate change: millennial time scales of potential CO_2 and surface temperature perturbations. *Journal of Climate* **22**, 2501–2511.

111. Archer, D. and Brovkin, V. (2008) The millennial atmospheric lifetime of anthropogenic CO_2. *Climate Change* **90**, 283–297.

112. Solomon, S., Plattner, G.-K., Knutti, R. and Friedlingstein, P. (2009) Irreversible climate change due to carbon dioxide emissions. *Proceedings of the National Academy of Science* **106** (6), 1704–1709.

113. Schmidt, G.A., Ruedy, R.A., Miller, R.L. and Lacis, A.A. (2010) Attribution of the present-day total greenhouse effect. *Journal of Geophysical Research* **115**, D20106.

9

Measuring and Modelling CO₂ Back through Time

9.1 CO₂: The Palaeoclimate Perspective

By the late 1970s, then, the atmospheric and ocean sciences had provided a comprehensive theoretical understanding of the role of the greenhouse gases in the climate system. Subsequent work would strengthen it. How, and when, would geologists begin to apply this new knowledge to their interpretations of change in past climates? A first step, identified by Gilbert Plass in the 1950s, was '*to be able to determine atmospheric CO₂ concentration for past geological periods*'[1,2].

In the absence of those determinations, we had calculations to go by. Plass calculated that if weathering halved the CO₂ content of the atmosphere, it would lower average global temperature by 3.8 °C and could precipitate an ice age[3]. That convinced him (in agreement with Tyndall and Arrhenius) that CO₂ played a role in past climate change. And he knew that at least one geologist, Thomas Chamberlin, agreed that more CO₂ led to warm periods and less led to cool ones, thus helping to explain the fluctuations between glacial and interglacial periods of the recent Ice Age[1,2].

Looking back further in time, Plass knew that Chamberlin had proposed that the weathering of igneous rocks in mountain belts removed CO₂ from the atmosphere, which was added back by volcanic emissions[3] – an idea that Nobel chemist Harold Urey had recently explained in his 1952 book, *The Planets*[4]. Urey's chemical equations showed how the abundance of CO₂ in the atmosphere was controlled by the transformation of silicate rocks to

carbonate rocks by weathering and sedimentation, and their retransformation back into silicate rocks by metamorphism (the recrystallisation of sedimentary rocks into fine-grained schists, coarser grained gneisses or marble, under the high temperatures and pressures prevailing deep in the Earth's crust) and magmatism (the melting of silicate rocks deep in the Earth's crust to produce igneous rocks like granite).

From the fossil evidence for warmer times during much of the geological record, Plass deduced that the air then must have contained much more CO₂ than it does now[5], a prediction that subsequent data and models have show to be the case, as we shall see. He accepted Urey's argument that losses of CO₂ from the air by weathering would eventually be balanced by the addition of CO₂ emitted from volcanoes, which turns out to be reasonable on long time scales. And he surmised that the burial of masses of organic matter to form coal in the early Carboniferous reduced the level of CO₂ in the air sufficiently to generate the late Carboniferous–early Permian Ice Age[3], an idea that is now widely accepted. In contrast, the notion that changes in the supply of CO₂ by volcanoes and its extraction by weathering might explain the warming and cooling of the recent Ice Age has been played down. Chamberlin was closer to the mechanism, invoking the ocean's role in expelling CO₂ when warm and retaining it when cool, as we saw in Chapter 4.

Plass was far ahead of his time, and it would be years before the geological community caught up with him. CO₂, the greenhouse effect and the work of Tyndall, Arrhenius,

Earth's Climate Evolution, First Edition. Colin P. Summerhayes.
© 2015 John Wiley & Sons, Ltd. Published 2015 by John Wiley & Sons, Ltd.

Chamberlin and Plass rated not a single mention in the textbooks that I was recommended to read as a first-year geology undergraduate at University College London in 1960[6-9]. In his 1961 presidential address to the Geological Society of London on 'The Climate Factor in the Geological Record', my then head of department, Prof. Sydney Hollingworth of University College London, whom we met in Chapter 6, mentioned CO_2 just twice, and not as a major control on past climate[10]. That can be taken as typical of the geological thinking of the time.

Not everyone was quite so oblivious. From his 1965 landmark textbook *Principles of Physical Geology*, it is clear that Arthur Holmes had read Plass's 1959 paper in *Scientific American*[5]. He reminded his readers about the greenhouse effect and the implications of Tyndall's work of a century before, while saying little about the variation of CO_2 in past atmospheres[11]. But he went on to note that although sea ice had been melting and glaciers retreating in the early part of the 20th century, *'During the last decade a few glaciers have ceased to melt away and have begun to advance again'*[11]. He interpreted that correctly to mean that *'other factors besides the CO_2 content of the atmosphere are concerned with climatic changes, but what their relative effects may be still remains problematical'*[11]. Holmes was clearly aware of the theory, but was equally aware that, in the 1960s, it wasn't working quite the way it was supposed to. CO_2 was increasing, but the world seemed to be cooling, as we saw in Chapter 8. Even so, Holmes was concerned about what burning more fossil fuels would do to the climate, pointing out that *'The CO_2 effect has been mentioned here only to illustrate the remarkable consequences of burning fuel hundreds of times as fast as it took to accumulate'*[11,a].

Holmes's neglect of the possible role of CO_2 in affecting past climates was not surprising, because at that time there was no direct evidence for atmospheric CO_2 having been abundant in the past. Despite the widely known contributions of Tyndall, Arrhenius and Chamberlin, neither Köppen nor Wegener nor Du Toit seems to have considered the possible role of CO_2 in the climate story. Most

geologists were in the dark as far as greenhouse gases were concerned.

By the early 1970s, things had begun to change. In 1974, Mikhael Budyko, whom we met in Chapter 8, realised that the warmth of the Mesozoic and early Tertiary compared with later periods may have been caused by a decrease in *'the atmospheric transparency for long-wave radiation'*, caused by *'a high carbon dioxide content in the atmosphere'*[13]. He and his geochemical colleague Alexander Ronov tested this idea by using the known volumes of volcanic rock, carbonate sediment and the organic carbon content of sedimentary rocks to calculate the likely amounts of O_2 and CO_2 in the atmosphere over the past 550 Ma. Although their approach was crude, the results (published first in 1979[14] and then in a book in 1987[15]) confirmed that it was highly likely that changing atmospheric CO_2 had affected past climates.

Budyko and Ronov estimated that volcanism and CO_2 were abundant in the Devonian (400 Ma ago) and the Permian (275 Ma ago) and low during the Permo-Carboniferous glaciation (300 Ma ago) and at the Permo-Triassic boundary (250 Ma ago). They rose to intermediate levels from the mid Triassic, through the Jurassic and the Cretaceous, before declining to the low values of the Pliocene. Budyko and Ronov assumed – as Urey had suggested – that volcanism had a primary role in supplying CO_2 to the air, and that the fall in volcanism and CO_2 during the Cenozoic cooled the Earth and produced the Quaternary glaciations. They were puzzled by the fact that the amount of CO_2 that they had calculated for the Carboniferous atmosphere had not fallen as low as during the recent Ice Age. We now know that they overestimated the abundance of CO_2 for the Carboniferous. Despite their attempt, we still lacked measurements of past CO_2.

9.2   Fossil CO_2

A substantial advance in geological thinking about the climatic effects of changes in atmospheric CO_2 had to wait for a dramatic breakthrough: the discovery of fossil atmospheric CO_2 in 1978, by a pioneering Swiss team led by Hans Oeschger (1927–1998) (Figure 9.1), Head of the division of climate and environmental physics at the University of Bern[16]. Analysing fossil air from bubbles in the Camp Century ice core from Greenland, Oeschger's team found low values for atmospheric CO_2, consistent with Callendar's estimated late-19th-century value of 290 ppm. The idea of being able to analyse fossil air grabbed the

---

[a] Note added in press: In 1922, the geologist A.S. Woodward noted in the Foreword to *Man as a Geological Agent* by R.L. Sherlock *'the question whether man, by his prodigious combustion of coal and other carbonaceous substances, is producing more carbonic acid than can be eliminated by ordinary processes'*, which may eventually lead to *'an unwelcome change in his atmospheric surroundings'*[12]. He was commenting on Sherlock's support for the ideas of Arrhenius and Chamberlin, reviewed here in Chapter 4, which implied a linear increase in coal burning, rather than the exponential increase that we have experienced.

Figure 9.1   *Hans Oeschger.*

imagination of climate scientists and geologists. We could now test Plass's ideas with data.

Oeschger's team published more results in January 1984 at a Chapman Conference on 'Natural Variations in Carbon Dioxide and the Carbon Cycle', convened by Eric Sundquist and Wally Broecker in Tarpon Springs, Florida[17]. Oeschger showed that for the last 1000 years, atmospheric CO_2 had hovered around 260–270 ppm, with little fluctuation, and that by the late 19th century it had reached around 290 ppm, in agreement with Callendar's data[18]. It looked as if an association between CO_2 and temperature extended back through time, although the origin of the CO_2 signal was unclear and the details of the CO_2–temperature relationship had yet to be worked out.

Corroboration of Callendar and Oeschger's independent findings came from yet another source: tree rings, whose chemical composition is related to the air in which the tree lived. The relationship between the carbon isotopic composition of the atmosphere and its CO_2 content led to efforts to use the $\partial^{13}C$ signal in annual tree rings to determine past levels of atmospheric CO_2. The first to establish that link, in 1978, was the geochemist Minze Stuiver, from the University of Washington[19]. Building on Stuiver's $\partial^{13}C$ results, Tsung-Hung Peng of the Oak Ridge National Laboratory in Tennessee concluded in 1985 that the pre-1850 atmosphere contained about 266 ppm CO_2, consistent with the data from ice cores[20]. Plass would have been pleased to see these data emerge from the dim and distant past.

It was time to revisit Callendar's results. The arrival on the scene of the exciting new measurements of fossil air from ice cores triggered a meeting of experts under the aegis of the World Climate Research Programme. The results of the meeting in Boulder, Colorado in June 1983 were published as an obscure WMO Technical Report[21] and summarised in the scientific newsletter *EOS* by W. Elliott[22]. The experts reviewed a wider range of techniques than were available to Callendar for measuring the content of CO_2 in modern and fossil air. Much as Callendar had done (see Chapter 8), they rejected many of the pre-existing measurements of CO_2 as unrepresentative of background air. In addition, they analysed solar spectral data collected from 1902 to 1956 by the Smithsonian Solar Constant Program and CO_2 values derived by analysing the carbon isotopic composition of tree rings and the carbon chemistry of old ocean water. They concluded from this multidimensional research that the air in the mid 19th century most probably contained between 260 and 280 ppm CO_2, around the same as the levels found in ice cores.

Oeschger's team continued refining their measurements, and in November 1986 they showed that the CO_2 content of bubbles in ice from the Siple Dome core in the Ross Sea region increased exponentially from about 280 ppm in 1760 towards the values of the present day[23]. The measurements from the youngest ice mapped on to the measurements of CO_2 in air made since 1957 by Keeling and his team[23]. The ice record and the air record formed an uninterrupted time series of CO_2 increase that has since been verified by several groups of scientists in ice cores from both Antarctica and Greenland (Figure 9.2)[24]. Not only that, but the growth of CO_2 in the air matched both the growth in consumption of fossil fuels and the change in the $\partial^{13}C$ composition of the CO_2 in the air bubbles in the Siple Dome core and in the air from Mauna Loa[23]. As more data from ice cores became available, and as analytical techniques improved, the analytically determined preindustrial level of CO_2 converged on 280 ppm.

The rise in CO_2 since 1760 is not surprising. By then, steam engines were being used increasingly to power the Industrial Revolution. James Watt developed his new version of the steam engine between 1773 and 1775, patenting his design in 1769 and going into full production in 1776 – a momentous year in more ways than one. His new engine set off a demand for fossil fuel that continued to rise; for example, from 5.1 Gt/year in 2000 to 8.4 Gt/year in 2011[29].

Oeschger's team went one step further. They devised a numerical model of the carbon cycle to replicate their observations of atmospheric CO_2 and its carbon isotopic composition[23]. As to what controlled the natural

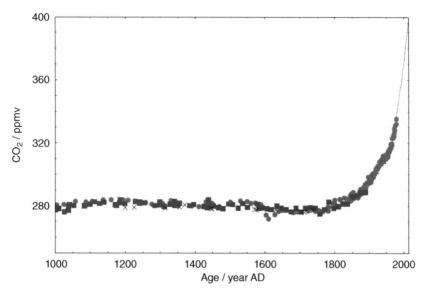

Figure 9.2   *Concentrations of CO$_2$ in the atmosphere from bubbles in ice cores over the past 1000 years (symbols) and from atmospheric measurements at Mauna Loa between 1957 and 2013 (solid line). Ice core data sourced from Law Dome[25,26], Siple Dome[27], EPICA Dronning Maud Land[28] and the South Pole[28].*

atmospheric concentration of CO$_2$ through time, the answer seemed to lie in the ocean, where – because the ocean contains about 60 times the amount of carbon in the atmosphere – small changes in productivity or circulation cause large changes in atmospheric CO$_2$.

CO$_2$ was not the only greenhouse gas showing signs of change with time. By 1986, it was clear from studies of fossil air in ice cores that the methane (CH$_4$) content was also increasing, partly due to an increase in rice growing and cattle production. CH$_4$ in ice core air showed little change for 1000 years before an exponential rise began in the late 1700s, doubling its abundance in the atmosphere[30]. The rising values for the past 1000 years mapped on to the air measurements of methane made since 1978[31], much as is shown for CO$_2$ in Figure 9.2.

As time went by, an important caveat emerged on the usefulness of CO$_2$ extracted from ice cores: high CO$_2$ concentrations may result from chemical reactions between rock dust rich in carbonate and acidic materials derived from volcanic eruptions. Oxidation of organic matter trapped in ice may also produce CO$_2$. This caveat applies mainly to Greenland, where the dust of glacial intervals is most abundant and most rich in carbonate. It does not

apply to Antarctica, where there is little dust and virtually none of it is carbonate or organic.

As Sundquist points out, ice core records are most compelling where they agree despite different locations, accumulation conditions and analytical procedures[24]. We now rely on Antarctic ice for changes in CO$_2$ through Ice Age times, but on both Greenland and Antarctica for changes in CH$_4$ over the same period, because methane is not subject to distortions arising from the chemistry of dust.

9.3   Measuring CO$_2$ Back through Time

So far, so good, and as we'll see in detail in Chapter 13, we can now measure CO$_2$ in bubbles of fossil air trapped in Antarctic ice cores back to 800 Ka ago. The key new question for geochemists was: How do we obtain hard data on the likely abundance of CO$_2$ in the atmosphere through time in the eras *before* fossil air became trapped in ice cores? As Crowley and North wrote in 1999, '*Documenting the magnitude of past CO$_2$ fluctuations is an important problem in paleoclimatology*'[32]. Geochemists have now cracked that nut.

Nowadays, we use several techniques to estimate the abundance of CO_2 in past atmospheres. Let's start with fossil leaves. Examining a leaf under a microscope, you will find its surface dotted with tiny pores, or stomata. These let in the CO_2 that plants use to build their bodies, and prevent loss of water. When CO_2 increases in the air, the number of pores per unit area of leaf (the stomatal density) decreases, as does the percentage of leaf cells that are pores (the stomatal index)[33]. The stomatal index provides a reliable yardstick for estimating CO_2 abundance back to the Ordovician (c. 450 Ma ago), when land plants evolved[34].

Dana Royer is expert at determining past levels of atmospheric CO_2 from fossil leaves, having published several papers on this topic in 2001 and 2002, which stemmed from his PhD studies at Yale University under the supervision of geochemist Bob Berner and palaeobotanist Leo Hickey[35–38]. Royer moved to Wesleyan University in 2005, where he used pore measurements from fossil leaves and $\partial^{13}C$ ratios from fossil soils to estimate how much CO_2 there was in the air at million-year intervals back to 450 Ma ago (Figures 9.3 and 9.4)[39]. He concluded, '*For periods with sufficient CO_2 coverage, all cool events are associated with CO_2 levels below 1000 ppm. A CO_2 threshold of below c. 500 ppm is suggested for the initiation of widespread, continental glaciations, although this threshold was likely higher during the Paleozoic due to a lower solar luminosity at that time … A pervasive, tight correlation between CO_2 and temperature is found both at coarse [10 million-year time scales] and fine resolutions up to the temporal limits of the data set (million-year timescales), indicating that CO_2, operating in combination with many other factors such as solar luminosity and paleogeography, has imparted strong control over global temperatures for much of the Phanerozoic*'[39]. Royer summarised his 2006 results as follows: '*the overarching pattern is of high CO_2 (4000+ ppm) during the early Paleozoic, a decline to present-day levels by the Pennsylvanian [upper Carboniferous, 318–299 Ma ago], a rise to high values (1000–3000 ppm) during the Mesozoic, then a decline to the present-day*'[39]. Although CO_2 was higher during the early Palaeozoic prior to 420 Ma ago than during the Jurassic and Cretaceous, it affected temperature less because the Sun's output was 6% less during the former period.

We can also estimate changes in CO_2 through time from changes in the carbon isotopic composition of carbonate sediments. Because organic matter selectively absorbs ^{12}C, the atmosphere and the ocean become relatively

enriched in ^{13}C when that organic matter is trapped in marine sediments or coals. By examining the $\partial^{13}C$ signal in carbonate sediments, we know when more organic matter was trapped, thereby reducing the abundance of CO_2 in the air. We can do this thanks to Jan Veizer of Ottawa-Carlson University in Canada, winner of the 1995 Logan Medal of the Geological Association of Canada, who compiled with his colleagues a global record of changes in $\partial^{13}C$ ratio in carbonate sediments[40]. There were substantial rises in $\partial^{13}C$, suggesting a lowering of CO_2 in the air, at the Silurian–Devonian boundary and in the Carboniferous and Permian: times of significant glaciation. Subsequent $\partial^{13}C$ levels were modest, such as those in the Devonian, although there were small peaks in the late Jurassic, mid Cretaceous (associated with the formation of organic-rich black shales) and mid Cenozoic.

The fall of atmospheric CO_2 from high levels in the early Palaeozoic to low values in the Carboniferous paralleled the rise of land plants, which increased extraction of CO_2 from the air by photosynthesis[41]. Mosses and liverworts emerged in the Ordovician some 470 Ma ago and accelerated the chemical weathering of rock and soil, which drew CO_2 out of the air[42]. As CO_2 fell, a brief ice age followed near the Ordovician–Silurian boundary about 440 Ma ago. The vascular land plants dominating today's land vegetation evolved in the Devonian, about 400 Ma ago. They absorbed CO_2 through their leaves, and their roots altered the chemistry of rocks and soils 10 times more effectively than did mosses and liverworts, making the substrate more susceptible to chemical weathering capable of extracting CO_2 from the air.

In the marine realm, we can estimate atmospheric CO_2 from $\partial^{13}C$ ratios in complex sedimentary organic molecules known as alkenones. As we saw in Chapter 7, these are long-chained organic compounds produced by a few haptophyte algae, such as *Emiliana huxleyi*[43]. Their carbon isotopic character can be used to estimate atmospheric CO_2 back to the Cretaceous[44]. Using this technique, Mark Pagani of Yale University, together with Jim Zachos of the University of Santa Cruz, whom we met in Chapter 7, and their colleagues, found in 2005 that the mid to late Eocene air contained 1000–1500 ppm CO_2, which declined to modern levels through the Oligocene[44]. Pagani and Zachos used oxygen isotopes to document the cooling of the planet after the middle Eocene (Figure 7.6)[45].

One of the problems with Pagani's alkenone-based reconstruction of past CO_2 levels was that it combined records from different localities to produce a single

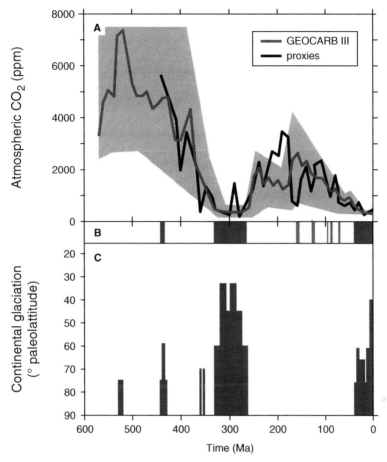

Figure 9.3 *Relation of cold and cool intervals to CO₂ through time. (a) Comparison of model predictions from GEOCARB III with proxy reconstructions of CO₂, at 10 Ma time-steps. (b) Intervals of glacial (dark) and cool (light) climates. (c) Latitudinal distribution of direct glacial evidence (tillites, striated bedrock, etc.) throughout the Phanerozoic.*

'stacked' record. Recognising that the process of stacking might have introduced regional biases in CO₂ trends and magnitudes, Pagani worked with Yi Ge Zhang of Yale and colleagues to derive a continuous alkenone-based CO₂ record at one deep-sea drilling site in the western equatorial Pacific. Their new CO₂ record spans the past 40 Ma[46]. While their results are broadly similar to other alkenone-based CO₂ reconstructions, they found CO₂ reaching 400–500 ppm in the middle Miocene. These values are higher than previous records from alkenones, but close to those estimated from boron isotope data and stomatal indices from leaves. They suggested that CO₂

levels were high during the middle Miocene climatic optimum 17–14 Ma ago, a period of global warmth following which there was a decline in both CO₂ and temperature.

The carbon isotopes in foraminifera from marine sediments can also tell us about past levels of CO₂ in the air. Nick Shackleton, whom we met in Chapter 7, pioneered their use in palaeoclimate studies of marine sediments. He knew that during photosynthesis, the phytoplankton, known as the 'grass of the sea', preferentially extract the smaller of the two carbon isotopes, ¹²C, from seawater, leaving surface waters relatively enriched in the heavier ¹³C. Dead phytoplankton sink and decompose, releasing

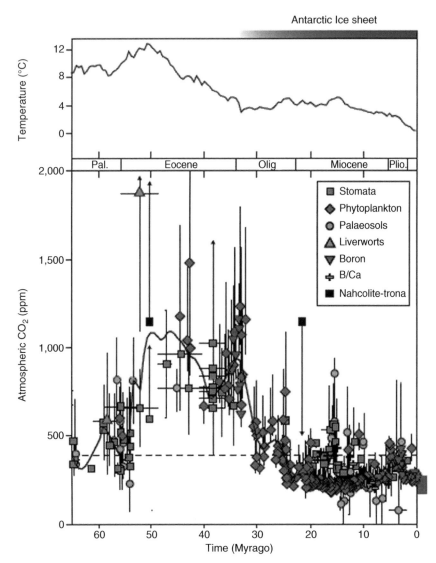

Figure 9.4 *Variation in atmospheric CO₂ through the past 65 Ma. Deep-sea temperatures (upper panel) generally track the estimates of atmospheric CO₂ (lower panel), reconstructed from terrestrial and marine proxies and showing error bars. Symbols with arrows indicate either upper or lower limits. Vertical grey bar at right indicates glacial–interglacial CO₂ range from ice cores. The top dark bar represents development of Antarctic ice sheet. Horizontal dashed line indicates present-day atmospheric CO₂ concentration in 2011 (390 ppm).*

their ^{12}C-rich carbon load to deep waters, which feed the organisms living on the deep-sea floor. As a result, the carbonate skeletons of the planktonic foraminifera that live near the ocean surface and eat phytoplankton are slightly enriched in ^{13}C, while the benthic foraminifera living on the deep-sea floor are slightly enriched in ^{12}C. In 1983, Shackleton used the gradient in ∂^{13}C between planktonic and benthic species of foraminifera from ancient sediments to infer changes in the CO_2 content of the air for the past 100 Ka. His predictions were close to the CO_2 values observed in the Vostok ice core from Antarctica[47], which we examine in Chapter 13.

Then there's soil. CO_2 enters soils from the air and through the production of CO_2 as organic matter decomposes, leading to the precipitation of soil carbonates. Like leaves and marine alkenones, these minerals contain the stable carbon isotopes ^{13}C and ^{12}C. The ratio between the two in soils depends on temperature and factors like depth, porosity, diffusion, rate of respiration and so on. Where these factors are well known and the processes are well understood, the ∂^{13}C ratio should reflect large-scale changes in atmospheric CO_2[48, 49]. CO_2 values from analyses of soil carbonates are referred to as 'pedogenic'. The data tend to show a fairly wide spread, but they may help where other data are unavailable.

Royer's CO_2 estimates confirm that the Carboniferous–Permian glaciation 326–267 Ma ago coincided with low levels of atmospheric CO_2 (Figure 9.3)[39]. The event occurred in two phases, with no evidence for ice from 310 to 300 Ma ago, when there was a pulse of high CO_2. As we know from Chapter 6, the glaciation occurred at a time when Antarctica and Australia drifted across the South Pole[50].

Royer realised in 2004[51] that his data conflicted with the interpretation of past oxygen isotope data by Jan Veizer, who suggested in the year 2000 that global temperature and CO_2 were decoupled through time[52]. Why the discrepancy? Royer concluded that Veizer had not corrected his temperature proxy – the ∂^{18}O ratio of ancient shallow-water carbonates – for the changing acidity of seawater – the pH – through time. Once that correction was applied, the temperature and CO_2 data agreed[39, 51]. This disposed of the idea that the cosmic ray flux, rather than the flux of CO_2 through time, controlled global temperature[53].

Various lines of geological and palaeontological evidence confirm, as we saw in Chapter 6, that the Earth was much warmer than today during most of Jurassic and Cretaceous time, when CO_2 levels were high (Figure 9.3). Nevertheless, there is increasing evidence at those times for cool pulses 1–3 Ma long. Several of them correlate with periods of low CO_2[39]. In some cases, increased burial of CO_2 in organic-rich deposits caused atmospheric CO_2 to decline, for example at times when ocean-bottom waters became oxygen-free, or anoxic, so preserving substantial accumulations of organic matter on the deep-sea floor. These unusual circumstances are designated 'oceanic anoxic events', and we will examine them more closely in Chapters 10 and 11. Accumulation of organic matter at those times temporarily reduced the concentration of CO_2 in the air.

Independent confirmation of Royer's results comes from many sources. For example, a team led by Gregory Retallack from the University of Oregon used stomatal data from fossil leaves and ∂^{13}C analyses to show that there were significant fluctuations in the levels of atmospheric CO_2 in the late Permian and early Triassic of Australia[54]. They found that '*Successive atmospheric CO₂ greenhouse crises coincided with unusually warm and wet paleoclimates for a paleolatitude of 61° S ... [which] punctuated long-term, cool, dry, and low-CO₂ conditions, and may account for the persistence of low diversity and small size in Early Triassic plants and animals*'[54]. The short-lived pulses of high CO_2 might have been caused by the intrusion of coals by flood basalts in Asia, they thought. Analyses of stomatal frequencies in the leaves of different plant groups from locations in Greenland and Ireland across the Triassic–Jurassic boundary suggested a rise of CO_2 from around 1000 to 2000–2500 ppm, agreeing with published ∂^{13}C records[55]. CO_2 values between 853 and 1033 ppm were calculated from stomatal data from leaves from the Messel Formation in Germany, which was deposited during the warm Eocene period[56].

Confirmation that CO_2 was abundant in the Eocene atmosphere comes from mineralogy. The sodium bicarbonate mineral **nahcolite** ($NaHCO_3$), which is abundant in the fossil muds of the 'Piceance Creek Member' of the Eocene 'Green River Formation' of the western United States, precipitates only under elevated levels of CO_2. Tim Lowenstein and Robert Demicco remind us that its co-precipitation there with common salt in the form of halite (NaCl) anchors minimum levels of CO_2 at >1125 ppm in the early Eocene atmosphere between 51 and 49 Ma ago[57]. They consider that the nahcolite and halite in the Piceance Creek Member probably precipitated at temperatures of between 25 and 35 °C, as happens today in the Dead Sea in Jordan. This suggests that atmospheric CO_2 ranged from 1125 to 2985 ppm, averaging about 2100 ppm. In similar but younger deposits, the main mineral is **trona** ($NaHCO_3 \cdot Na_2CO_3 \cdot 2H_2O$), which crystallises at temperatures above 25 °C in modern alkaline saline lakes. Trona precipitates at lower levels of CO_2 than nahcolite and is common at today's levels.

Finally, we can also use boron isotopes to estimate past oceanic acidity and, from that, atmospheric CO_2. As Ravizza and Zachos explain[58], boron has two stable isotopes, ^{10}B and ^{11}B, of which ^{10}B is the most abundant (80%). Variations in the two are calculated as the ratio $\partial^{11}B$, which varies with the pH of seawater. Boron is incorporated into calcite, so we can estimate pH from the $\partial^{11}B$ of marine carbonates – something first attempted in 1993 and more common today[59, 60]. Between around 60 and 20 Ma ago, oceanic pH increased from 7.4 to modern values (near 8.2) as atmospheric CO_2 fell from about 1000 to 200–400 ppm. This is similar to the levels calculated independently from alkenone data. These are early days in the use of $\partial^{11}B$; further work will refine its usefulness.

Proxy measurements do have their downsides. Alkenone and leaf stomatal proxies for past levels of atmospheric CO_2 have two key deficiencies[61]. First, the sensitivity of both declines above 1000 ppm, due to CO_2 saturation. Second, they use calibrations based on living taxa or their nearest living relatives, which may not have precisely the same relationship to their environment as their fossil equivalents did. Independent mineral-based methods like the $\partial^{11}B$ technique may be more reliable. Methods based on soil carbonate are less reliable, because of their high sensitivity to soil CO_2, which varies with soil moisture and productivity[61]. The nahcolite–trona transition is found in only two parts of the world to date: the southwestern United States and Henan Province in China. Still, it provides a well established, independent boundary for palaeo-CO_2 levels for the Eocene.

Equally, we need to be alert to the difficulties of determining palaeotemperatures accurately, especially in older geological intervals (Jurassic and earlier), where the carbonates used for $\partial^{18}O$ measurements may have been more subjected to diagenesis: post-depositional precipitation of carbonate, and thus overprinting of the original seawater signal. These difficulties have led to the development of independent measurements of palaeotemperature, such as the use of alkenones and the Mg/Ca ratio, as explained in Chapter 7; these have been particularly useful in analysing Pleistocene palaeotemperatures.

To conclude this section, we examine one more recorder of atmospheric CO_2 abundance: the carbonate compensation depth (CCD). This is the boundary at the deep-sea floor between sediment rich in the calcium carbonate skeletons of foraminifera and coccolithophorids and deeper sediment made of nonfossiliferous red clay. It is the point at which the rate of dissolution of the mineral calcite exceeds the rate of supply. The red clay is always present as a tiny fraction of deep-sea carbonate oozes, but below the CCD it is all that is left after the carbonate has gone. Major factors controlling the position of the CCD are the effects of increasing hydrostatic pressure with depth on the solubility of carbonate minerals (calcite and aragonite) and the increasing concentration of CO_2 in old bottom waters due to the decomposition of sinking organic matter[62, 63].

It was not until the Deep Sea Drilling Project (DSDP) began collecting core samples on transects across the CCD that geologists realised that the depth of this boundary changed with time. Because it was initially thought that the CCD was a pressure-solution horizon (reflecting ocean depth), it was believed that fluctuations in the CCD reflected alternating up or down movements of the seabed. The first person to realise that this was not the case was the University of Miami's Bill Hay, on Leg 4 of the DSDP (in the western Atlantic and Caribbean in early 1969). At the end of the Leg 4 report[64], Hay noted that there were in effect two CCDs, one being the lower limit of occurrence of pelagic foraminifera and the other the slightly deeper lower limit of the occurrence of the remains of calcareous nannofossils (coccolithophorids). Their vertical changes through time, which were of the order of 1 km, he said, '*are more easily explained by assuming that the only vertical motion of points on the ocean floor has been downward as they moved away from the crest of the Mid-Atlantic Ridge, but that the zone of calcium carbonate compensation has fluctuated through considerable distances in the water*'[64]. As he later said about this perceptive conclusion, tongue in cheek, '*Nobody read it*'[63].

Hay was right about the CCD, which meant that ocean chemists were wrong to assume that the chemistry of the ocean had remained more or less constant for millions of years. In his words, '*We were becoming aware of just how much we did not know*'[63]. What we are looking at is the effect of changing atmospheric CO_2 on ocean chemistry and on the stability of the main calcium carbonate minerals, calcite and aragonite. The solubility of these minerals depends not only on depth – representing pressure – but also on the acidity of subsurface water. As mentioned in Chapter 8, when CO_2 dissolves in the ocean, it reacts with water to produce either inorganic bicarbonate ions (HCO_3^-) or carbonate ions (CO_3^{2-}). In the process, hydrogen ions (H^+) are also produced; the greater the solution of CO_2, the lower the concentration of carbonate ions and the greater the concentration of hydrogen ions, and hence the lower the pH. Rising acidity dissolves $CaCO_3$, causing the CCD to shallow, decreasing the area covered by carbonate sediment and increasing the area covered by red clay. This 'ocean acidification' is 'the other CO_2 problem' (Box 9.1).

Box 9.1 Ocean acidification: the other CO₂ problem.

At times when the atmosphere contained more CO_2, the dissolution of CO_2 in the ocean produced more H^+ ions and fewer CO_3^{2-} ions. As a result, the ocean became slightly less alkaline. This process, which is going on today, is popularly described as 'ocean acidification'. We measure acidity by the pH scale, a measure of the free hydrogen (H^+) ions in the system. The scale runs from 1 to 14. Solutions are described as acid when their pH is less than 7. Applying these measures, we find that today's oceans are slightly alkaline, with a pH of around 8.2. As we put more CO_2 into the ocean, it moves towards the acid end of the pH scale. Between 1751 and 1994, the pH of the surface ocean is estimated to have changed from 8.25 to 8.14, a decrease of approximately 0.1 pH units. Because the pH scale is logarithmic (for readers with a technical bent, pH is the logarithm of the reciprocal of the hydrogen ion activity in a solution), this apparently tiny change is equivalent to a 30% increase in hydrogen ion concentration. Fortunately, as we saw in Chapter 8, seawater is chemically buffered against moving into the acid half of the scale (pH <7). The 'antacid' is the $CaCO_3$ sediment of the ocean floor. Unfortunately, the present rate of CO_2 rise is so rapid that it is faster than the normal process of neutralisation (buffering) by $CaCO_3$[65]. Acidification is beginning to erode the shells of living creatures that make their skeletons from $CaCO_3$, such as corals and planktonic organisms at the base of the food chain[66, 67].

Today, the concentrations of dissolved CO_2 and carbonate ions (CO_3^{2-}) are about the same in the surface waters of both the Atlantic and Pacific, but there is more dissolved CO_2 and less CO_3^{2-} in the deep Pacific than in the deep Atlantic. Hence, calcite particles sinking in the Atlantic dissolve below 5000 m, while in the Pacific they dissolve below 4200–4500 m. Aragonite dissolves at shallower depths than calcite, so it is calcite dissolution that defines the CCD. The process of dissolution is limited by the fact that when $CaCO_3$ dissolves, it uses up H+ ions, thus buffering the ocean against further acidification. Evidently, we can use changes in the CCD to tell us about past ocean chemistry and past atmospheric CO_2.

These various studies show that although, as Eric Sundqvist said in 2005, we lack '*a paleo-CO₂ "gold standard" analogous to the ice-cores of the Late Quaternary period*'[24], we do have considerable evidence, from a number of methods, for global trends in the abundance of CO_2 in the air through Phanerozoic time. Supporting evidence for the cycle of organic carbon production and consumption through time comes from the nature of the rock record; for example, from periods when carbonate sediments, coals or organically enriched black shales accumulated.

9.4 Modelling CO₂ and Climate

Enter the geochemical modellers. Along with the application of more powerful geochemical tools, the 1960s saw a significant increase in the application of chemical models through which to understand geological problems. Among the practitioners in this new field was the American geochemist Robert Minard Garrels (1916–1988) (Box 9.2).

Box 9.2 Bob Garrels.

Garrels revolutionised the field of aqueous geochemistry with the publication of his 1965 book *Solutions, Minerals and Equilibria*, based on applying experimental physical chemistry data to geological and geochemical problems. He was fascinated by the cycles of carbon and other elements, and developed several numerical models by which to explain in quantitative terms how these cycles worked to produce what we see in sedimentary rocks. During his career, he worked at various times for the US Geological Survey,

Northwestern University, Harvard, Scripps, the University of Hawaii and the University of South Florida. Garrels' many efforts were rewarded with election to the US National Academy of Sciences in 1961, presidency of the Geochemical Society in 1962 and several medals, among them the Penrose Medal of the Geological Society of America in 1978 and the Wollaston Medal of the Geological Society of London in 1981.

Building upon the work of Harold Urey[4], Garrels documented equations explaining how CO_2 in rainwater created the acidity needed to dissolve the silicate rocks of mountain belts, pointing out that the CO_2 extracted from the air in this way would eventually be restored, for instance by reactions in the oceans and sediments, not to mention via eruption from volcanoes[68]. There follows a typical equation for the effect of atmospheric CO_2 on silicate rocks in the presence of water. The magnesium silicate olivine in the presence of CO_2 and water dissolves to give magnesium and bicarbonate ions plus silicic acid in solution[69]:

$$Mg_2SiO_4 + 4CO_2 + 4H_2O \leftrightarrow 2\,Mg^{2+} + 4HCO_3^{-} + H_4SiO_4$$

One can replace the magnesium silicate with a calcium silicate, $CaSiO_3$, to produce calcium carbonate and silica, $CaCO_3 + SiO_2$[69]. These two equations are commonly called the 'Urey reactions' after their description by Harold Urey, as discussed earlier[4]. When the bicarbonate ions reach the ocean, they combine with calcium ions to produce calcium carbonate, water and carbon dioxide, thus:

$$4HCO_3^{-} + 2Ca^{2+} \rightarrow 2CaCO_3 + 2H_2O + 2CO_2$$

Marine organisms use the $CaCO_3$ to build skeletons made of calcite or aragonite, which ends up as carbonate ooze before hardening to limestone. Half of the original $4CO_2$ is thus prevented from returning rapidly to the atmosphere.

The rates of these reactions depend on temperature, doubling with a $10\,°C$ rise[63]. When CO_2 increases in the air, so does the temperature, increasing the rates of reaction so that more CO_2 is consumed by weathering. This keeps the CO_2 from expanding out of hand. Here we have Earth's natural thermostat. CO_2 is also removed from the air by photosynthesis, as we saw in Chapter 8. A proportion of the organic matter created in this way is trapped in sediments,

providing yet another path for the removal of CO_2 from the atmosphere.

CO_2 is returned to the atmosphere when organic matter decomposes or is burned. It is also returned when limestone ($CaCO_3$) is dissolved by rainwater, but with no net addition of CO_2 to the atmosphere:

$$CO_2 + H_2O \rightarrow H_2CO_3$$
$$H_2CO_3 + CaCO_3 \rightarrow Ca^{2+} + 2HCO_3^{-}$$

Then, in the ocean:

$$Ca^{2+} + 2HCO_3^{-} \rightarrow CaCO_3 + H_2O + CO_2$$

Note that there is no net change in the amount of CO_2 in the atmosphere: one molecule is consumed in these reactions, and one is returned. But the processes take a long time. There is a lag between the use and return of CO_2 molecules.

These equations conceal the voyage of a typical molecule of CO_2 from its emission in volcanic gas through its solution in a bead of rainwater, its impact on an exposed mountainside, its transfer through rivers to the sea, its absorption by plankton to form organic matter that is eaten by zooplankton and its incorporation into their $CaCO_3$ skeletons to its ending up as calcareous ooze on the ocean bed, awaiting transformation to limestone and the repeat of the cycle.

Some of Garrels' initial work on the carbon cycle was published in 1974, when he showed that the geochemical cycles of carbon and sulphur help to control the amount of oxygen in the atmosphere, the oxidation state of minerals in rocks and life processes in general[70]. A decade later, he showed that the cycles of carbon and sulphur were closely aligned: '*Reduction of carbonate carbon to organic carbon is ... mirrored in a roughly comparable oxidation of reduced sulfur to oxidized sulfur over time periods of tens of millions of years ... This relationship is a consequence of a remarkable coupling among the sedimentary reservoirs in a world that cannot afford wild fluctuations in the oxygen and carbon dioxide contents of the atmosphere without eliminating life*'[71]. There was an important implication: that the world of the past 550 Ma, the Phanerozoic, was not markedly different from the world of today in terms of rates of deposition and erosion. Here, then, we have a restraint on the bounds of physical conditions on the planet in former times. These processes act to keep the climate within a natural envelope.

While Garrels' work did not contribute directly to our understanding of the role of CO_2 in the air or its changes with time, things changed with the arrival of

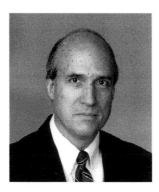

Figure 9.5 *Robert Arbuckle Berner.*

James Callan Gray Walker, from Yale, who was to win the F.W. Clarke Award of Geochemical Society for his contributions to geochemistry. In 1981, Walker and colleagues published a landmark paper suggesting that a negative feedback involving surface temperature, CO$_2$ and chemical weathering controlled the evolution of the Earth's climate through time by modulating the strength of the greenhouse effect[72]. That same year, another Yale geochemist, Robert Arbuckle (Bob) Berner (1935–2015) (Figure 9.5, Box 9.3), was trying to contribute to our understanding of the overall geochemical cycles of carbon and sulphur by determining their rates of removal from the modern oceans to form organic matter and pyrite in sediments[73]. One of his conclusions was that the concentration of organic carbon in modern bottom sediments was so low that this could not be where the bulk of carbon from human emissions was accumulating in the ocean. Most must be in dissolved form.

Box 9.3 Robert Arbuckle Berner.

Bob Berner received his PhD from Harvard in 1962. He was elected to the National Academy of Sciences and received numerous medals, among them the V.M. Goldschmitt Medal of the Geochemical Society in 1995, the Murchison Medal of the Geological Society of London in 1996 and the Vernadsky Medal of the International Association of Geochemisttry in 2012. Berner and Garrels knew one another reasonably well, Berner having taken courses from Garrels while studying for his PhD.

Walker and Berner's papers got Garrels thinking, and while he was at the University of South Florida in 1983, he got together with Berner, along with Antonio Lasaga from Penn State, to produce a classic paper on 'The Carbonate-Silicate Geochemical Cycle and its Effect on Atmospheric Carbon Dioxide over the Past 100 Million Years'[75]. Their quantitative model became known as the BLAG model, after the initials of the authors. Berner and his team built on Walker's notion by proposing that climate evolution was controlled by a negative-feedback loop balancing the supply of CO$_2$ to the air from the Earth's interior, via volcanoes, with the consumption of CO$_2$ from the air by chemical weathering of silicate rocks[75]. Supply and consumption must broadly match over long periods of time, or we would see a build-up of CO$_2$ leading to a runaway greenhouse effect like that on Venus, or a depletion to near zero and an icehouse effect like that on Mars.

In the BLAG model, high rates of seafloor spreading and volcanic activity in the Cretaceous – estimated from sea level curves that were thought to reflect rates of seafloor spreading, and hence volcanic activity – supplied CO$_2$ to strengthen the greenhouse effect and increase temperatures. In the inelegant jargon of the field, the supply of CO$_2$ from the Earth's subcrust (the mantle) to the air, via volcanic activity associated with seafloor spreading, was referred to as 'mantle degassing'. Warmth encouraged evaporation, increased rainfall and so increased chemical weathering, which acted as a negative feedback, preventing further build-up of CO$_2$ in the atmosphere. Rates of supply of CO$_2$ slowed when seafloor spreading slowed in the Cenozoic. The resulting cooling also slowed chemical weathering. The model showed Earth's thermostat at work. It made a valuable contribution, not only in making geologists think about the carbon cycle and the greenhouse effect for the first time, but also in providing testable predictions. Like all models of reality, it was bound to be deficient in some respects, but criticism – of the kind we explore further on – was welcome and spurred the team to introduce improvements. The model drew on the understanding of the slow carbon cycle portrayed in Figure 8.8, representing the geochemical theory proposed by Urey[4], Garrels[71], Walker[72], Berner[76] and Ebelmen[77].

By the time of the Tarpon Springs Chapman Conference in January 1984, the authors had refined the BLAG model to take into account the natural reservoirs of carbon, pyrite (iron sulphide), gypsum (calcium sulphate) and atmospheric oxygen, as well as factors such as the formation and oxidation of pyrite, the reduction of sulphate at mid-ocean ridges and the weathering and burial of organic carbon[78]. While this modified the original results, removing some of the model's problems, it did not change the prediction that the Cretaceous atmosphere 100 Ma ago was much richer than today in CO_2. It did, however, reduce the predicted amount of CO_2 in that atmosphere by a factor of 2, to 13 times as much as today. This inevitably reduced the predicted Cretaceous global temperature at that time from 25 to 23 °C, more in keeping with the evidence from oxygen isotopes. As in the 1983 version of the model, CO_2 and temperature changed together, the temperature falling to a low point (around 15 °C) by 55 Ma ago, rising to a peak (18.5 °C) around 43 Ma ago, falling to a new low (14.5 °C) around 18 Ma ago, rising to a new peak (15.5 °C) 5–3 Ma ago, before finally falling to Holocene preindustrial levels (on average about 14 °C).

One of the implications of these various studies is that land plants affect the levels of atmospheric CO_2 through a stabilising feedback loop[79]. The evolution of plants with strong roots enhanced the chemical weathering of rocks, thus helping to drive down the concentration of CO_2 in the air. The sequestering of CO_2 as skeletal remains in limestones, or as organic matter in coals, added to this effect.

9.5   The Critics Gather

All scientists are trained to be sceptical, and the BLAG model was not exempt from critiques. Some of them came to the fore in 1984 at the Tarpon Springs Chapman Conference in Florida, papers from which were published as Monograph 32 of the American Geophysical Union in 1985.

Based on studies of the $\partial^{13}C$ ratios in deep-ocean sediments, Nick Shackleton suggested that much more organic matter was trapped in Cenozoic sediments prior to the middle Miocene (around 14 Ma ago) than since[80]. He attributed the reduction in trapping of organic matter to the increasing oxygen content of bottom waters that came about as polar waters cooled from 10 to 2 °C. Polar cooling created more sea ice, which excreted salt and made surface waters dense enough to sink, taking oxygen with them. Shackleton concluded that the amount of dissolved CO_2 in the Palaeocene ocean was about the same as it is today, placing a constraint on the predictions of Berner's model[75, 78]. Clever though he was, he was wrong in this case. Royer's wide range of proxy data show that Berner was right: CO_2 was abundant in the Palaeocene air.

Somewhat more controversial was Shackleton's claim that CO_2 was not more than twice its present level in the Cretaceous – nothing like the factor of 13 claimed by the modellers. Behind his calculations lay the assumption that the carbon input to the ocean had been constant through time. If it had been, then his criticism would be justified. As we see further on, it was not. Besides this, more recently it was found that using $\partial^{13}C$ ratios from marine sediments as a proxy for atmospheric CO_2 could be suspect due to isotopic fractionation by phytoplankton in the so-called 'vital effects', such as changing growth rate[81].

Berner's assumptions about the Cretaceous were questioned by Michael Allen Arthur (Box 9.4) of the University of Rhode Island, along with Walter Dean from the US Geological Survey and Sy Schlanger from Northwestern University[82].

Box 9.4 Michael Allen Arthur.

Mike Arthur, professor of geosciences at Pennsylvania State University since 1991, was a student of Al Fischer (see Chapter 10) at Princeton, where he earned his PhD in 1979. Like his mentor, he is a very broad geological thinker. He was one of the first geoscientists to apply stable isotopic techniques to palaeoceanographic and palaeoclimatic problems. As a rigorous practitioner of the multiproxy approach to palaeoenvironmental analysis, he has made significant contributions to our understanding of the controls on organic carbon burial and the global carbon cycle, the expression of orbital forcing of climate in sedimentary systems, the biogeochemical cycles of sulphur, iron, nitrogen and phosphorus, and the relationship of these elements to the redox state of the oceans and atmosphere[83]. He was honoured with the F.P. Shepard Medal of the Society for Sedimentary Geology in 1996 and the Laurence L. Sloss Award of the Geological Society of America in 2007.

The BLAG model assumed that over the past 100 Ma the carbon cycle had been in balance and that the rate of burial of organic carbon had been small compared with rates of fixing of carbon by weathering and of supply of carbon by out-gassing from the Earth as a byproduct of metamorphism, volcanism and geothermal activity. Arthur and his team presented '*evidence that during the middle to late*

Cretaceous (110 to 70 m.y. ago), high rates of production and burial of Corg [= organic carbon], and intense and widespread volcanism … may have had profound effects on the global CO$_2$ cycle that were not accounted for by the [BLAG] model'[82]. Although Arthur's team knew that Berner and his team had made improvements to the BLAG model while they prepared their own presentation for the conference[78], they felt that even the revised BLAG model under-represented reality.

Both Arthur and Berner accepted the observation of Harvard's Heinrich Holland (1927–2012) that, prior to the Industrial Revolution, out-gassing by volcanoes was the major source for the replacement of atmospheric CO$_2$ drawn down by weathering[84]. As we saw in Chapter 5, the creation of numerous hot, young mid-ocean ridges starting 180 Ma ago with the break-up of Pangaea[85] undoubtedly pumped large quantities of CO$_2$ into the atmosphere during Jurassic and Cretaceous times. To understand how this might have affected the production of CO$_2$, it was important to know how much the rate of ridge production had changed through time. Initially, it was thought that it had changed a lot, with periods of rapid seafloor spreading and mid-ocean-ridge production causing flooding of the continents in the Jurassic and Cretaceous (Figures 5.8b and 6.9)[86, 87]. As we saw in Chapter 5, David Rowley called that long-held notion into question[88, 89], but Dietmar Müller and his team disagreed. Reconstituting the pattern of Mesozoic mid-ocean ridges from the magnetic anomaly patterns in the Pacific, they concluded that the highest rates of production of new crust occurred 140 Ma ago and declined towards the end of the Cretaceous[85], in keeping with what was understood from patterns of the change in sea level with time (Figure 5.9) as documented by Pete Vail[86], Tony Hallam[87] and Bilal Haq[90]. I take Vail and Haq's sea level curves as independent evidence for the likely correctness of Müller's reconstructions of past ridge volumes[85], and hence for likely changes in the rates of production of CO$_2$ by mantle degassing through time. Besides, as we shall see, Müller's data agree with changes in ocean chemistry.

Arthur and his team argued not only that there was more mid-ocean-ridge activity supplying CO$_2$ to the atmosphere during the Cretaceous, but also that there was widespread mid-plate volcanic activity right across the Pacific basin, contributing to what Bill Menard called the Darwin Rise (see Chapter 5)[91]. Arthur's team went further[82], suggesting that volcanic activity in the middle of ocean crustal areas away from mid-ocean ridge crests might also have been equally abundant on the Kula,

Farallon and Phoenix Plates, which formerly occupied much of the eastern Pacific seafloor but had since been subducted beneath the surrounding continents. We call this 'mid-plate volcanism'. In his 1985 paper, Arthur was anticipating the conclusions of Müller and his team in 2008[85]. If Arthur's team was right, then the supply of CO$_2$ to the atmosphere by volcanic activity in the Cretaceous must have been much greater than is supposed in Berner's models, which had not taken into account the Darwin Rise or contributions from long-subducted mid-ocean ridge crests.

Large amounts of the organic matter supplied to the ocean basins during Cretaceous times were buried during what became called oceanic anoxic events (more on those in Chapter 10). While the increased atmospheric CO$_2$ of the time may have been balanced to some extent by the increased rates of burial of organic carbon, it was feasible that some CO$_2$ had accumulated in the air as suggested by Berner's models. Reviewing evidence for causes of the constraint in the rise in atmospheric CO$_2$ in the Cretaceous, Arthur concluded that the CO$_2$ concentration in the air then was less than eight times present values[82]. Rates of burial of organic carbon were also unusually high during Carboniferous and Permian times 300–250 Ma ago, reflecting the rise in vascular land plants and the abundance of flooded and swampy lowlands[92].

By 1990, Berner had simplified his approach to modelling the carbon cycle[69]. His revised model's results showed that natural long-term processes involving carbon stored in rocks, brought about changes in atmospheric CO$_2$ with time that were gradual but large enough to have caused the climate to change through the operation of CO$_2$ as a greenhouse gas. These changes involved the slow carbon cycle, in which the long lag between the chemical weathering of silicate minerals in the Urey reactions (see earlier) and the thermal decarbonation of the carbonate products of that weathering over time by metamorphism could (and did) lead occasionally to a large imbalance in CO$_2$ fluxes that affected the Earth's climate, as could be seen in the palaeoclimate record. Hence, the Urey reactions must control the role of CO$_2$ in the climate system on long geological time scales. Like the BLAG model, the new one reproduced the low CO$_2$ concentrations of the Carboniferous and the late Cenozoic, leading Berner to conclude that climates should have been warm in the early Palaeozoic and Mesozoic and cool in the late Palaeozoic and late Cenozoic, in agreement with palaeoclimatic data. He was well aware that the model was no better than the data and assumptions that went into it, all of which needed refinement. The semi-quantitative trends emerging from the model were not hard facts. They showed how one could '*use the geochemical carbon cycle as a means of exploring the various factors affecting atmospheric CO₂ over long geological times*'[69].

By 1992, Berner's model was under fire for not paying enough attention to mountain building (orogenesis). The criticism came from Maureen Raymo (Box 9.5) and her former professor Bill Ruddiman (1943–) (Figure 9.6, Box 9.6), who thought that the uplift and chemical weathering of plateaus and mountains changed the global climate[93, 94].

Figure 9.6   *Bill Ruddiman.*

Box 9.5 Maureen Raymo.

Maureen Raymo received her Bachelors, Masters and PhD degrees at New York's Columbia University and has been a research professor at Lamont Doherty Earth Observatory since 2011, following periods at the University of California, Berkeley, MIT, Woods Hole Oceanographic Institution and Boston University. She has also been a visiting investigator with the National Institute of Oceanography in Goa, India and at the University of Melbourne, Australia. She has been much involved with deep-ocean drilling over the years, and chaired the Science Advisory Structure Executive Committee for the Integrated Ocean Drilling Program (2010–11). She has received several awards for her work, including the Cody Award in Ocean Sciences in 2002 and the Wollaston Medal of the Geological Society of London in 2014.

Box 9.6 Bill Ruddiman.

Trained as a marine geologist, much of Ruddiman's career has been spent examining one or other aspect of past climate change, a topic on which he has written several papers, edited various volumes and published three books. Ruddiman gained a PhD from Columbia University. After working for the US Naval Oceanographic Office, he returned to New York to join what was then Lamont-Doherty Observatory, where he spent most of his time until moving to the University of Virginia in 1991 as professor in the department of environmental sciences, of which he was chairman from 1993 to 1996. Having retired in 2001, he holds emeritus status with that university and continues publishing. During his career, he was director of the CLIMAP programme (see Chapter 12 for details) in 1982–83 and was on the executive committee of the COHMAP project (see chapter 14 for details) in 1982–90. In 2010, he won the Geological Society of London's Lyell Medal. Among other accolades, he received a Distinguished Career Award from the American Quaternary Association in 2012.

Back in 1988, Raymo and Ruddiman, together with Flip Froelich, thought that '*if the concept of a global temperature-weathering feedback was correct, then global chemical weathering rates should have dropped through the Cenozoic in concert with falling mantle degassing rates and surface temperature*'[93]. To determine these weathering rates, they suggested using the ocean-wide ratio of strontium isotopes $^{87}Sr/^{86}Sr$ in carbonate sediments. This ratio is high in continental granites and low in oceanic basalts. They inferred that the increase in this ratio in carbonate sediments over the past 40 Ma came from increased chemical weathering of continental rocks with high ratio values, driven by increased chemical weathering in the rapidly uplifting Himalaya-Tibet region, which caused atmospheric CO₂ values to fall following the collision of India with Tibet[94, 95].

By 1997, Ruddiman had put together a book on *Tectonic Uplift and Climate Change*[96]. He noted that tectonic uplift may control climate in one of two ways[97]. The **direct effect** works as follows: Through the planetary moist lapse rate of around 6.5 °C per 1000 m, uplift cools surfaces at high altitudes. The accumulation of snow and ice there further cools these areas by reflecting solar energy. The growth of frost and ice accentuates mechanical erosion. Mountains also encourage rainfall on their windward sides, leading to rain shadows on their lee sides, and they can create and intensify the seasonal monsoonal circulation, as happens over Tibet, which has a mean elevation of 5 km over an area of more than 4 million km^2. Heating of the Tibetan Plateau in summer leads to hot rising air, which is compensated for by the inflow of moist air from the ocean at sea level. The ensuing heavy rains enhance erosion on windward mountain slopes. In winter, cold dry air sinks over the plateau as it cools and becomes snow-covered, causing a reversal of the monsoon winds over the surrounding lowlands and ocean. This has been going on for the past 40 Ma or so, since the Tibetan Plateau began to rise with the collision between India and Asia. Large mountain and plateau areas can also cause and amplify meanders in the jet stream, affecting local climate. By changing the circulation of the winds, mountains can also change ocean circulation, as shown today by the action of monsoonal winds in the Indian Ocean. The **indirect effect** of tectonic uplift on climate can be seen in the action of rain on windward slopes, which chemically weathers silicate rocks, drawing down atmospheric CO_2 and lowering temperature.

Ruddiman suggested that the Himalayas and the Tibetan Plateau were uplifted gradually from about 40 Ma ago, and that there must have been significant relief and erosion 20 Ma ago to account for the massive accumulation of sediment in the Bay of Bengal[96]. The evidence suggested that '*the Tibetan Plateau reached a threshold size sufficient to induce the Asian monsoon about 8 million years ago*', he wrote[96]. Uplift led to the drying of the Asian interior through a rain-shadow effect, and the counter-clockwise circulation around the low-pressure cell over Tibet in summer brought hot dry air from central Asia south over Arabia, the Middle East and northeast Africa, causing a shift towards dry-adapted vegetation types.

Other uplifts had more local effects. African uplift, giving rise to a relatively high plateau in the east and south over what seemed to be mantle hot spots, began in the Miocene, probably about 8 Ma ago. Both that rise and the accelerated uplift of the Himalayas and Tibetan Plateau at that time may have been caused by global reorganisation of the world's tectonic plates. Uplift of the Andes is also relatively recent, beginning in the Eocene and accelerating with plate reorganisation in the Oligocene around

26 Ma ago. And both the Alps and the North American interior plateaus are Cenozoic features. All of these uplifts changed regional climate. But only the Tibetan uplift seems to have been extensive enough to have a global effect.

Ruddiman, Raymo and colleagues thought that the pervasive cooling of the late Cenozoic might have been enhanced by its own feedbacks. Decreasing CO_2 caused cooling, leading to formation of ice sheets and mountain glaciers, which accentuated mechanical erosion, exposing yet more silicate minerals to chemical weathering and thus further lowering CO_2[98]. They found no need to call upon changes in heat transport towards the Arctic by the ocean to explain the high-latitude late Cenozoic changes in vegetation that we explored in Chapter 7. The tectonic opening or closing of oceanic gateways was discontinuous through that period, with each having discrete beginnings and/or ends, and all occurring in scattered locations. Rather than being an ongoing source of global climate change, like the Himalayan-Tibetan uplift, Ruddiman's team thought that these gateways were more likely to have had important effects on the formation of deep and bottom water, and on the creation of regional perturbations to longer-term trends.

Raymo and Ruddiman's uplift-weathering hypothesis elevated the importance of chemical weathering to a major factor in reducing CO_2 levels and thus leading to global cooling. It assumed that increased chemical weathering in mountains must be a key factor controlling CO_2 and climate, and so challenged Berner's BLAG thermostat model of the climate system, in which mountains played little part. One of the implications of their hypothesis was that even steady-state plate motions could lead to non-steady-state effects on climate through variations in continental relief with time, since continental collisions resulting in the uplift of plateaus the size of Tibet were rare and episodic: '*Despite the continuous presence throughout geological history of high mountain terrain along the convergent margins of the world it may be the rarer occurrence of plateaux that can drive climate away from steady state and decouple rates of horizontal and vertical tectonic movement*'[99].

Geophysicists Peter Molnar and Phil England of MIT reversed this hypothesis, proposing that weathering drove uplift by reducing mountain mass, so allowing the Himalayas to rise through isostatic adjustment[100, 101]. It wasn't the uplift that drove the climate change, but the climate change that drove the uplift! Bearing that proposal in mind, Cronin concluded in 2010 that '*The jury is still* out on the hypothesis that Neogene climatic cooling was forced by Himalayan uplift, changes in the rate or intensity (or both) of continental erosion, and CO_2 drawdown, or whether climate changes drove plateau erosion'[102]. Even so, elsewhere evidence was accumulating to tie uplift (in Africa and the Andes) to Neogene changes in climate. What we need in order to resolve this issue are improvements in proxies for past elevation.

Could strontium isotopes be used in the way Raymo supposed? Joel Blum, of Dartmouth College, New Hampshire, thought not, noting that '*the Sr isotopic composition released by weathering of a single rock type changes dramatically with the age of the soil. Thus, the marine $^{87}Sr/^{86}Sr$ ratio cannot be considered a direct proxy for silicate weathering rates*'[103]. Others disagreed. John Edmond (1943–2001), a hard-drinking, red-bearded, Scottish geochemist and a fellow of the Royal Society from MIT, who was one of my seagoing expedition companions at Christmas 1973 en route from Dakar to Cape Town, sided with Raymo. He argued that '*tectonically active mountain belts are the loci for accelerated drawdown of atmospheric CO_2; hence their initiation and evolution have a direct influence on global climate*'[104]. Edmond concluded that given the lack of any increase in the rates of seafloor spreading and the volcanic activity associated with mid-ocean ridges over the past 50 Ma, the intensive mountain building of that period was sufficient not only to have prevented any runaway greenhouse effect, but also to have accounted for the observed climatic deterioration. '*The resulting CO_2 drawdown*', he said, '*is now close to complete, such that the Earth has entered into what will likely be a prolonged glacial epoch with the atmospheric CO_2 held at a kinetic minimum by the resulting exposure of aluminosilicate rock*'[104]. In other words, we are in for a rerun of the lengthy Carboniferous–Permian glaciation.

Berner fought back. In his own paper in Ruddiman's 1997 book, he concluded that Edmond's argument about strontium isotope ratios was flawed in its reliance on analyses of river chemistry that paid insufficient attention to the presence of carbonate minerals, which would distort the strontium isotope data[105]. He accepted that mountain uplift is an important control on rock weathering and atmospheric CO_2, agreeing that '*to have appreciable silicate weathering it is necessary to have sufficient relief so as to enable erosive removal of any protective clay overburden*'[105]. But, '*given sufficient relief*' (his italics)[105], he insisted that climate also plays an important role in weathering through its control on temperature, rainfall and vegetation. Agreeing that topography is important,

he incorporated a special term for topographic relief into the successor to the BLAG model. The outputs from his new GEOCARB model of the variation of CO_2 through time matched independent estimates of past levels of atmospheric CO_2 obtained from fossil soils (palaeosols), for example in showing that a major fall in CO_2 preceded the Carboniferous–Permian glaciation and was sustained throughout that glaciation. '*Focussing only on mountain uplift,*' he concluded, '*with neglect of ... other factors, is unnecessarily narrow. Earth system science requires that all factors be evaluated and first-order attempts at quantification at least be attempted*'[105]. Humboldt and Lyell would have agreed.

Together with Lee Kump of Penn State, Mike Arthur examined the assumptions about rates of seafloor spreading and CO_2 emission for the Cenozoic. They knew that atmospheric CO_2 declined from about 20 Ma ago, consistent with the uplift history of the Himalayas. They assumed that from the end of the Cretaceous, '*the rate of volcanism (and release of CO_2) has in general decreased, rather than increased, throughout the Cenozoic*'[106]. That assumption led them to '*reject the hypothesis of increased global weathering rates as the cause of Cenozoic cooling and of the Sr isotope trend toward more radiogenic values*'[106]. Instead they thought that while rates of chemical weathering and erosion probably increased as the Himalayas rose, mass balance considerations meant that chemical erosion rates must have decreased elsewhere, or else atmospheric CO_2 would have shown a steep decline as the mountains went up. They explained the increase with time in the marine $^{87}Sr/^{86}Sr$ ratio as reflecting the changing chemistry of the rocks exposed with time as the Himalayas were eroded and deeper, more crystalline basement rocks with higher $^{87}Sr/^{86}Sr$ ratios were gradually exposed. The ratio was not an index of the amount or rate of chemical weathering. Rather than having us consider rates of weathering, they thought we should be thinking about 'weatherability'.

For example, in 2000, they pointed out that as well as there being an increase in the rate of chemical weathering in the region of the Himalayas and the Tibetan Plateau, there would have been an increase in chemical erosion in areas of continental glaciation, especially Antarctica, where mechanical erosion exposed fresh rocks to chemical attack: the rocks became more weatherable[107]. They also highlighted the key role played by plants, whose root systems penetrate the soil, allowing acid-carrying fluids and CO_2 to permeate it. Plants also bind fine particles to the soil, thus increasing the surface area exposed to soil pore waters. Tree roots and associated fungi accelerate weathering by acidifying the weathering environment, hence affecting the concentration of CO_2 in the atmosphere, and thus global temperature[108]. Decomposition of soil organic matter also plays a role in controlling atmospheric CO_2 levels. While warming tends to increase the rate of decomposition, cooler temperatures will slow it. The rise of land plants 450 Ma ago increased the weatherability of silicates by about sevenfold, and, as a result, atmospheric CO_2 dropped from about twelve times preindustrial levels to about five times[108]. At the same time, the increasing burial of organic remains helped to draw down formerly high levels of CO_2. In developing the GEOCARB model, Berner considered both weatherability and the sequestration of organic matter in sediments to have been important.

Kump and Arthur concluded that a rise in CO_2 brought the Permian glaciation to an end. The widespread aridity of Permian Pangaea decreased the rate of chemical weathering and reduced the supply of nutrients to the ocean, hence reducing ocean productivity. Less weathering and less productivity meant less extraction of CO_2 from the air, allowing a long-term build-up of volcanic CO_2, which eventually led to global warming and a more active hydrological cycle, higher rates of weathering, more nutrient run-off and increased productivity[107].

In 2003, Ravizza and Zachos pointed out that if the curve of $^{87}Sr/^{86}Sr$ reflected the uplift and erosion of the Himalayas, one would expect changes in its slope to be related to events in Himalayan uplift, but they are not[109]. It may be no coincidence that the rapid rise in $^{87}Sr/^{86}Sr$ ratios in the Cenozoic began around 35 Ma ago, close to the initiation of Antarctic glaciation. Glaciation exposes large areas of fresh rock to weathering, which raises the $^{87}Sr/^{86}Sr$ ratio[109], much as Kump and Arthur suggested[107]. In summary, the $^{87}Sr/^{86}Sr$ curve doesn't tell us about the drawdown of CO_2.

Does mountain building lead to cooling? Consideration of the likely balance between rates of CO_2 drawdown by chemical weathering throughout the Cenozoic and the rates of supply of CO_2 to the atmosphere from volcanic and metamorphic sources tended, for a while, to suggest not[109–111]. Then again, cooling and ice growth during the Neogene are not associated with systematic decreases in atmospheric CO_2[44, 59, 60]. Ravizza and Zachos suggested that steady-state levels of CO_2 could be reduced without changes in global weathering flux by changes in 'weatherability'[109], as shown by Kump and Arthur[106, 107]. Tectonic and climatic factors could act to allow weathering

fluxes to remain high enough to balance CO_2 input in spite of overall cooling. This concept provides a defence for the causative link between Himalayan uplift and global cooling, since physical weathering can be argued to increase 'weatherability'. It provides the basis for arguing that Cenozoic cooling might have been enhanced by a limited decoupling of average global temperatures and weathering rates. Glaciation itself may have contributed by greatly increasing the area of fresh mineral surfaces available to chemical weathering. As we shall see in a moment, the 'orogenesis leads to cooling' theory is not dead yet, although it does not provide a single, unified explanation for Cenozoic climatic variation.

Bill Hay added another astute observation to the mix[63]. The rising of the Himalayas did not simply expose old rocks to chemical weathering that would tend to lower CO_2, but also led to massive deposition of eroded sediments in the enormous submarine fans of the Indus in the Arabian Sea and the Ganges in the Bay of Bengal. Even though individual layers of sediment in the fan contain rather small amounts of organic matter, the fans are so large that the vast quantity of matter locked up in them must have removed significant amounts of CO_2 from the system.

Royer's data from leaf stomata[39], Pagani's from marine alkenones[44] and Ekart's from soils[48] support each other in providing a convincing view of the variation of atmospheric CO_2 through time that is consistent with separately obtained geological data, palaeontological data and the backward predictions (or hindcasts, as we might call them) of Berner's GEOCARB model (Figure 9.7). The different data sets continue to be refined, the latest compilation of estimates of CO_2 levels in the Cenozoic being published in 2011 by David Beerling and Dana Royer[112].

A similar compilation of results from carbon isotopic analyses of fossil plant remains (liverworts from five continents) and from models was presented in 2007 for the Mesozoic and early Cenozoic by a team from the University of Sheffield led by Benjamin Fletcher and including David Beerling and Bob Berner[113]. Their results showed that atmospheric CO_2 rose from ~420 ppm in the Triassic (200 Ma ago) to a peak of ~1130 ppm in the mid Cretaceous (100 Ma ago), then declined to ~680 ppm by 60 Ma ago, coincident with Mesozoic climate change as determined from oxygen isotopes from marine carbonate fossils. Evidently, climate and CO_2 were not decoupled during this interval, contrary to Veizer's suggestion[52]. The team found that '*These reconstructed atmospheric CO₂ concentrations drop below the simulated threshold*

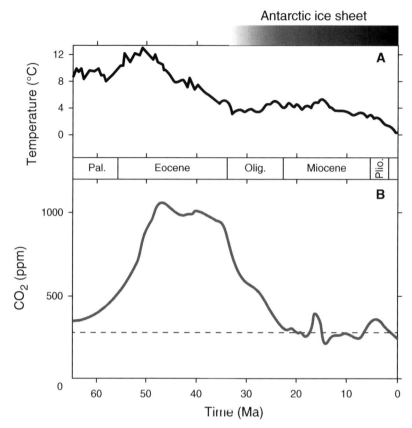

Figure 9.7   *Cenozoic CO₂ and temperature records, as of 2012. The temperature record is from Hansen, based on the benthic ∂¹⁸O record of Zachos, and is a rough minimum estimate of global surface temperature change relative to preindustrial conditions. The CO₂ record is updated from Beerling and Royer; the dashed line is the preindustrial value (280 ppm).*

for the initiation of glaciations on several occasions and therefore help explain the occurrence of cold intervals in a "greenhouse world"[113]. In view of their close correlation to the ∂¹⁸O record, fossil bryophytes (nonvascular plants, like mosses) provided a more accurate representation of past levels of CO_2 than did leaf stomatal indices or palaeosols, the latter being the least accurate. Given this association, the team said, we have a plausible *'explanation for Jurassic and Cretaceous climates without the need to invoke the influence of cosmic rays on cloud cover and planetary albedo'*[113].

The latest twist in the tale comes from a comprehensive re-evaluation of leaf–gas exchange, carbon isotopic composition and the stomatal characteristics of fossil leaves by a team including Dana Royer and David Beerling[114]. Their analysis shows that atmospheric CO_2 was constrained within an envelope ranging from 200 to 1000 ppm, and that this condition arose after the development of forests in the Devonian 390 Ma ago. Until that time, atmospheric CO_2 exceeded 1000 ppm. Forests then captured and sequestered vast amounts of CO_2 and, through their interactions with rocks, enhanced chemical weathering. Nevertheless, there are likely to have been short periods when CO_2 excursions overcame the climate system's ability to balance them with weathering, for example at the Triassic–Jurassic boundary, where CO_2 temporarily exceeded 1000 ppm.

These new data agree with progressive updates of Berner's GEOCARB model. By 1999, Berner was able to provide an elegant explanation of the workings of the slow and fast carbon cycles, pointing out how positive or negative feedbacks could arise[76]. He continued to revise the GEOCARB model, progressing from GEOCARB II to GEOCARB III, which emphasised the uptake of CO_2 by continental weathering[115]. '*The new results*', he said in 2001, '*exhibit considerably higher CO₂ values during the Mesozoic*'[115]. He continued to improve GEOCARB, which evolved into the GEOCARBSULF model, showing the variation in oxygen and CO_2 for the Phanerozoic[116].

Nevertheless, some concern was expressed that the chemical weathering associated with the acceleration of mountain building over the past 15 Ma should have lowered atmospheric CO_2 by more than the amount observed. Li and Elderfield[117] and Torres, West and Li[118] concluded recently that the loss of CO_2 by interaction with silicates would have been balanced to some extent by the weathering of iron sulphide (pyrite). Oxidation of sulphides produces sulphuric acid, which then attacks carbonate minerals ($CaCO_3$) to release CO_2. The balance between these two forms of chemical weathering explains better the observed patterns of atmospheric CO_2 over the past 15 Ma[117, 118].

With both the model and the data base having been refined, things now hang together well. Convergence between the very different approaches provided by geochemical modelling on the one hand and analyses of proxies for CO_2 on the other gives us some confidence that the science is on the right track and that CO_2 has played a significant role in controlling past climate. But CO_2 will not have acted alone. More water evaporates from the ocean in a warming world, and H_2O itself is a powerful greenhouse gas. There may also have been fluctuations in methane, another powerful greenhouse gas that is released from wetlands, which were probably more abundant in a warmer, wetter world. Unfortunately, we lack a way to determine past levels of CH_4 back beyond the range of ice cores, and it tends to be rapidly oxidised to CO_2 in the atmosphere.

Back in 1999, a review of the state of the art of palaeoclimatology concluded that, while it was reasonable to assume that CO_2 provided an important control on climate, '*A major problem with the CO₂ scenario is that insufficient proxy evidence has been gathered to rigorously support the CO₂ model*'[32]. In the 15 years since then, much of the missing proxy evidence has been gathered, thanks to the efforts of Dana Royer, David Beerling, Mark Pagani and Paul Pearson.

Further evidence is emerging to support the geochemical theory. In 2008, Richard Zeebe of the University of Hawaii and Ken Caldeira of the Carnegie Institution in Stanford California set out to test the notion that '*atmospheric CO₂ concentrations over millions of years are controlled by a CO₂-driven weathering feedback that maintains a mass balance between the CO₂ input to the atmosphere from volcanism, metamorphism and net organic matter oxidation, and its removal by silicate rock weathering and subsequent carbonate mineral burial*'[119]. Using the CO_2 concentrations from ice cores, they found no more than a 1–2% imbalance between the supply and uptake of CO_2 during the past 610 Ka, providing '*support for a weathering feedback driven by atmospheric CO₂ concentration that maintains the observed fine mass balance*'[119]. If this process did not take place, the release of CO_2 from volcanism and metamorphism would double the amount of CO_2 in the atmosphere within less than 600 Ka. These emissions are balanced by weathering acting as a stabilising feedback controlled by the concentration of atmospheric CO_2, much as Chamberlin concluded in the 1890s.

Bill Hay recently brought up an additional point, not yet comprehensively examined, that has a bearing on the long-term cycling of CO_2[63]. Coccolithophores and planktonic foraminifera that build their skeletons out of calcium carbonate evolved in *shallow* seas during the Jurassic and spread throughout the oceans during the Cretaceous. Whereas carbonate sediments were previously primarily deposited in shallow seas, they are now deposited right across the *deep ocean*, as carbonate oozes. Eventually, these deep-ocean carbonates will be subducted at ocean trenches, and their trapped CO_2 will return to the atmosphere via volcanic eruptions that tap the sedimentary loads of downgoing subducted slabs of ocean crust. But that is tens to hundreds of millions of years hence; for instance, there is virtually no subduction on the margins of the Atlantic, where most deep-sea carbonates are currently being deposited. The processes of deposition and destruction have not yet reached equilibrium. That may partly help to explain the slow decline in atmospheric CO_2 from the Cretaceous into the Cenozoic.

New data and ideas are constantly arriving, the latest to cross my desk being the discovery by Flip Froehlich and Sambuddha Misra that the isotope of lithium, ∂^7Li, in deep-ocean sediments can be used as an indicator of the weathering of continental rocks[120]. Froehlich and Misra found that the ∂^7Li chemistry of the Palaeocene–Eocene ocean '*indicates that continental relief during this period of the Early Cenozoic was one of peneplained (flat) conti-nents characterized by high chemical weathering intensity and slow physical and chemical weathering rates, yielding low river fluxes of suspended solids, dissolved cations, and clays delivered to the sea*[120]. Only when mountain building was reinitiated in the Oligocene–Miocene, fol-lowing the collision of India with Asia, *'did continental weathering take on modern characteristics of rivers with high suspended loads … with much of the cations released during weathering being sequestered into secondary clay minerals'*[120]. Hence, *'The early Cenozoic climatic opti-mum was a result of increased supply of carbon dioxide to the ocean-atmosphere system as well as diminished removal of carbon dioxide from the atmosphere through the weathering of silicate rocks'*[120]. What we are seeing is the absence of the negative-feedback mechanism pro-vided by weathering, rather than some steady increase in the supply of CO_2. The absence of a CO_2 sink explains how the hot Eocene climate persisted for millions of years[120]. This hypothesis also *'provides another piece of circumstantial evidence in support of the Late Cenozoic Uplift-Weathering Hypothesis'*[120].

I end this chapter with some quotes from a review led by Isabel Montañez and Richard Norris, whom we met in Chapter 6. They gave a useful summary of the net result of all these efforts over the years in a 2011 report to the US National Academy of Sciences, which concluded: *'The carbon fluxes in and out of the surface and sedimentary reservoirs over geological timescales are finely balanced, providing a planetary thermostat that regulates Earth's surface temperature. Initially, newly released CO_2 (e.g., from the combustion of hydrocarbons [or from volcanoes]) interacts and equilibrates with Earth's surface reservoirs of carbon on human timescales (decades to centuries). However, natural "sinks" for anthropogenic CO_2 [such as weathering] exist only on much longer timescales, and it is therefore possible to perturb climate for tens to hundreds of thousands of years … Transient (annual to century-scale) uptake by the terrestrial biosphere (includ-ing soils) is easily saturated within decades of the CO_2 increase, and therefore this component can switch from a sink to a source of atmospheric CO_2 … Most (60 to 80 percent) CO_2 is ultimately absorbed by the surface ocean, because of its efficiency as a sweeper of atmospheric CO_2, and is neutralized by reactions with calcium carbonate in the deep sea at timescales of oceanic mixing (1000 to 1500 years). The ocean's ability to sequester CO_2 decreases as it is acidified and the oceanic carbon buffer is depleted. The remaining CO_2 in the atmosphere is* sufficient to impact climate for thousands of years longer while awaiting sweeping by the "ultimate" CO_2 sink of the rock weathering cycle at timescales of tens to hundreds of thousands of years … Lessons from past hyperthermals [more on those in Chapter 11] suggest that the removal of greenhouse gases by weathering may be intensified in a warmer world but will still take more than 100 000 years to return to background values for an event the size of the Paleocene-Eocene Thermal Maximum'[61]. We'll explore the Paleocene–Eocene Thermal Maximum (PETM) in detail in Chapter 11.

I have spent some time on the seemingly convoluted development of thinking about the levels of CO_2 in past atmospheres, not least because it displays how science works. Science is not an authoritarian business. Propo-sitions are continually being tested. Arguments rage. Analytical methods improve. Data bases grow. Models improve. Knowledge and understanding evolve. The frontier is always being pushed forward. Old ideas are discarded as new ones take their place. There is always more research to do! But eventually, we converge on robust answers that stand the test of time. There should be no doubt now about the positive relationship between CO_2 and climate over long periods of geological time. We can think of the change in CO_2 with time as a forcing factor in the climate system. Over the course of the Cenozoic, atmospheric CO_2 fell from around 1000 ppm to as low as 170 ppm, equivalent to a climate forcing of -10 W/m^{2}[102]. No wonder the climate cooled! There was no comparable change in any other known forcing factor.

What emerges from these various studies of CO_2 is that throughout the 550 Ma of the Phanerozoic, warm periods with abundant atmospheric CO_2 alternated with cool peri-ods with much less CO_2. We now turn to look in more detail at the history of those fluctuations – the 'pulse of the Earth' – in Chapter 10.

References

1. Plass, G.N. (1956) The carbon dioxide theory of climate change. *Tellus* **VIII**, 140–154.
2. Plass, G.N. (1956) The influence of the 15 micron carbon dioxide band on the atmospheric infra-red cooling rate. *Quarterly Journal of the Royal Meteorological Society* **82**, 310–329.
3. Plass, G.N. (1961) The influence of infrared absorptive molecules on the climate. *New York Academy of Science* **95**, 61–71.
4. Urey, H.C. (1952) *The Planets: Their Origin and Develop-ment*. Yale University Press, New Haven, CT.

5. Plass, G.N. (1959) Carbon dioxide and climate. *Scientific American* **201** (1), 41–47.

6. Wells, A.K. (1959) *Outline of Historical Geology*, 4th edn. Thomas Murby and Co., London.

7. Zeuner, F.E. (1959) *The Pleistocene Period*. Hutchinson, London.

8. Dunbar, C.O. and Rogers, J. (1957) *Principles of Stratigraphy*. John Wiley and Sons, Chichester.

9. Krumbein, W.C. and Sloss, L.L. (1958) *Stratigraphy and Sedimentation*. W.H. Freeman and Co., San Francisco, CA.

10. Hollingworth, S.E. (1962) The climate factor in the geological record. *Quarterly Journal of the Geological Society of London* **118**, 1–21.

11. Holmes, A. (1965) *Principles of Physical Geology*. Nelson and Sons, London.

12. R.L. Sherlock (1922) *Man as a Geological Agent*. H.F. & G. Witherby, London.

13. Budyko, M.I. (1974) *Climate and Life*. Academic Press, New York and London; translation of a book published in Russian in 1971.

14. Budyko, M.I. and Ronov, A.B. (1979) Chemical evolution of the atmosphere in the Phanerozoic. *Geochemistry International* **16** (3) 1–9; translated from *Geokhimiya* **5**, 643–653, 1979.

15. Budyko, M.I., Ronov, A.B. and Yanshin, A.L. (1987) *History of the Earth's Atmosphere*. Springer-Verlag, New York.

16. Berner, W., Stauffer, B. and Oeschger, H. (1978) Past atmospheric composition and climate, gas parameters measured in ice cores. *Nature* **276**, 53–55.

17. Sundquist, E.T. and Broecker, W.S. (1985) *The Carbon Cycle and Atmospheric CO₂: Natural Variations Archean to Present*. Geophysics Monographs 32. American Geophysics Union, Washington, DC.

18. Oeschger, H., Stauffer, B., Finkel, R. and Langway, C.C. (1985) Variations of the CO₂ concentration of occluded air and of anions and dust in polar ice cores. In: *The Carbon Cycle and Atmospheric CO₂: Natural Variations Archean to Present* (eds E.T. Sundquist and W.S. Broecker). Geophysics Monographs 32. American Geophysics Union, Washington, DC, pp. 132–142.

19. Stuiver, M. (1978) Atmospheric CO₂ increases related to carbon reservoir changes. *Science* **199**, 253–258.

20. Peng, T.-H. 1985, Atmospheric CO₂ variations based on the tree-ring ¹³C record. In: *The Carbon Cycle and Atmospheric CO₂: Natural Variations Archean to Present* (eds E.T. Sundquist and W.S. Broecker). Geophysics Monographs 32. American Geophysics Union, Washington, DC, pp. 123–131.

21. World Climate Research Program (1983) *Report of the WMO (CAS) Meeting of Experts on the CO2 Concentrations from Pre-Industrial Times to IGY*. WCP-53, Report 10. WMO, Geneva.

22. Elliott, W. (1984) The pre-1958 atmospheric concentration of carbon dioxide. *EOS* **65** (26), 416–417.

23. Oeschger, H. (1989) Information on the history of atmospheric CO₂ and the carbon cycle from ice cores. In: *Carbon Dioxide and Other Greenhouse Gases: Climatic and Associated Impacts* (eds R. Fantechi and A. Ghazi). Proceedings of the European Commission Symposium, Brussels, 3–5 November 1986. Kluwer Academic Press, Dordrecht, pp. 40–54.

24. Sundquist, E.T. and Visser, K. (2005) The geologic history of the carbon cycle. In: *Biogeochemistry, Vol. 8, Treatise on Geochemistry* (eds H.D. Holland and K.K. Turekian). Elsevier-Pergamon, Oxford, pp. 425–513.

25. Etheridge, D.M., Steele, L.P., Langenfelds, R.L., Francey, R.J., Barnola, J.-M. and Morgan, V.I. (1996) Natural and anthropogenic changes in atmospheric CO₂ over the last 1000 years from air in Antarctic ice and firn. *Journal of Geophysical Research* **101**, 4115–4118.

26. MacFarling Meure, C., Etheridge, D., Trudinger, C., Steele, P., Langenfelds, R., van Ommen, T., *et al.* (2006) Law Dome CO₂, CH₄ and N₂O ice core records extended to 2,000 years BP. *Geophysical Research Letters* **33** (14), L14810.

27. Friedli, H., Lötscher, H., Oeschger, H., Siegenthaler, U. and Stauffer, B. (1986) Ice core record of the ¹³C/¹²C ratio of atmospheric CO₂ in the past two centuries. *Nature* **324**, 237–238.

28. Siegenthaler, U., Monnin, E., Kawamura, K., Spahni, R., Schwander, J., Stauffer, B., *et al.* (2005) Supporting evidence from the EPICA Dronning Maud Land ice core for atmospheric CO₂ changes during the past millennium. *Tellus B Chemical and Physical Meteorology* **57**, 51–57.

29. EIA (2012) *International Energy Statistics*. US Energy Information Administration. Available from: http://www.eia.gov/cfapps/ipdbproject/iedindex3.cfm?tid=1&pid=7&aid=1&cid=regions&syid=2000&eyid=2012&unit=TST (last accessed 29 January 2015).

30. Khalil, M.A.K. and Rasmussen, R.A. (1989) Trends of atmospheric methane: past, present and future. In: *Carbon Dioxide and Other Greenhouse Gases: Climatic and Associated Impacts* (eds R. Fantechi and A. Ghazi). Proceedings of the European Commission Symposium, Brussels, 3–5 November 1986. Kluwer Academic Press, Dordrecht, pp. 91–100.

31. Etheridge, D.M., Steele, L.P., Francey, R.J. and Langenfelds, R.L. (1998) Atmospheric methane between 1000 AD and present: evidence of anthropogenic emissions and climatic variability. *Journal of Geophysical Research Atmospheres* **103**, 15 979–15 993.

32. Crowley, T.J. and North, G.R. (2011) *Paleoclimatology*. Oxford Monographs on Geology and Geophysics 18. Oxford University Press, Oxford.

33. Woodward, F.I. (1987) Stomatal numbers are sensitive to increases in CO₂ from pre-industrial levels. *Nature* **327**, 617–618.

34. Beerling, D.J. (1999) Stomatal density and index: theory and application. In: *Fossil Plants and Spores: Modern*

Techniques (eds T.P. Jones and N.P. Rowe). Geological Society of London, London, pp. 251–256.

35. Royer D.L. (2001) Stomatal density and stomatal index as indicators of atmospheric CO_2 concentration. *Review of Palaeobotany and Palynology* **114**, 1–28.

36. Royer D.L., Berner R.A. and Beerling D.J. (2001) Phanerozoic atmospheric CO_2 change: evaluating geochemical and paleobiological approaches. *Earth-Science Reviews* **54**, 349–392.

37. Royer D.L., Wing S.C., Beerling D.J., Jolley, D.W., Koch, P.L., Hickey, L.J. and Berner, R.A. (2001) Palaeobotanical evidence for near present-day levels of atmospheric CO_2 during part of the Tertiary. *Science* **292**, 2310–2313.

38. Beerling, D.J. and Royer, D.L. (2002) Fossil plants as indicators of the Phanerozoic global carbon cycle. *Annual Reviews of Earth and Planetary Sciences* **30**, 527–556.

39. Royer, D.L. (2006) CO_2 – forced climate thresholds during the Phanerozoic. *Geochimica et Cosmochimica Acta* **70**, 5665–5675.

40. Veizer, J., Alab, D., Azmy, K., Bruckschen, P., Buhl, D., Bruhn, F., *et al.* (1999) $^{87}Sr/^{86}Sr$, ^{13}C and ^{18}O evolution of Phanerozoic seawater. *Chemical Geology* **161**, 59–88.

41. Lenton, T.M. and Watson, A. (2011) *Revolutions that Made the Earth*. Oxford University Press, Oxford.

42. Lenton, T.M., Crouch, M., Johnson, M., Pires, N. and Dolan, L. (2012) First plants cooled the Ordovician. *Nature Geoscience* **5**, 86–89.

43. Lawrence, K.T., Herbert, T.D., Dekens, P.S. and Ravelo, A.C. (2007) The application of the alkenone organic proxy to the study of Plio-Pleistocene climate. In: *Deep-Time Perspectives on Climate Change: Marrying the Signal from Computer Models and Biological Proxies* (eds M. Williams, A.M. Haywood, F.J. Gregory and D.N. Schmidt). Micropalaeontological Society Special Publication. The Geological Society, London, pp. 539–562.

44. Pagani, M., Zachos, J.C., Freeman, K.H., Tipple, B. and Bohaty, S. (2005) Marked decline in atmospheric carbon dioxide concentrations during the Paleogene. *Science* **309**, 600–603.

45. Zachos, J., Pagani, M., Sloan, L., Thomas, E. and Billups, K. (2001) Trends, rhythms and aberrations in global climate 65 Ma to present. *Science* **292**, 686–693.

46. Zhang, Y.G., Pagani, M., Liu, A., Bohaty, S.M. and Deconto, R. (2013) A 40-million-year history of atmospheric CO_2. In: *Warm Climates of the Past – A Lesson for the Future* (eds D.J. Lunt, H. Elderfield, R. Pancost and A. Ridgwell). *Philosophical Transactions of the Royal Society A* **371** (2001): 20130096.

47. Shackleton, N.J., Hall, M.A., Line, J. and Cang, S. (1983) Carbon isotope data in core V19-30 confirm reduced carbon dioxide concentration in the ice age atmosphere. *Nature* **306**, 319–322.

48. Ekart D.D., Cerling T.E., Montañez I.P. and Tabor N.J. (1999) A 400 million year carbon isotope record of pedogenic carbonate: implications for paleoatmospheric carbon dioxide. *American Journal of Science* **299**, 805–827.

49. Cerling, T.E. (1991) Carbon dioxide in the atmosphere: evidence from Cenozoic and Mesozoic paleosols. *American Journal of Science* **291**, 377–400.

50. Frakes, L.A., Francis, J.E. and Syktus, J.I. (1992) *Climate Modes of the Phanerozoic – The History of the Earth's Climate Over the Past 600 Million Years*. Cambridge University Press, Cambridge.

51. Royer, D.L., Berner, R.A., Montañez, I.P., Tabor, N.J. and Beerling, D.J. (2004) CO_2 as a primary driver of Phanerozoic climate. *GSA Today* **14** (3), 4–10.

52. Veizer, J., Godderis, Y. and François, L.M. (2000) Evidence for decoupling of atmospheric CO_2 and global climate during the Phanerozoic Eon. *Nature* **408**, 698–701.

53. Shaviv, N.J. and Veizer, J. (2003) Celestial driver of Phanerozoic climate? *GSA Today* **13** (7), 4–10.

54. Retallack, G.R., Sheldon, N.D., Carr, P.F., Fanning, M., Thompson, C.A., Williams, M.L., et al. (2011) Multiple early Triassic greenhouse crises impeded recovery from late Permian mass extinction. *Palaeogeography Palaeoclimatology Palaeoecology* **308**, 233–251.

55. Steinthorsdottir M., Jeram A.J. and McElwain, J.C. (2011) Extremely elevated CO_2 concentrations at the Triassic/Jurassic boundary. *Palaeogeography Palaeoclimatology Palaeoecology* **308**, 418–432.

56. Grein M., Konrad W., Wilde V., Utescher, T. and Roth-Nebelsick, A. (2011) Reconstruction of atmospheric CO_2 during the early middle Eocene by application of a gas exchange model to fossil plants from the Messel Formation, Germany. *Palaeogeography Palaeoclimatology Palaeoecology* **309**, 383–391.

57. Lowenstein, T.K. and Demicco, R.V. (2006) Elevated Eocene atmospheric CO_2 and its subsequent decline. *Science* **313** (5795), 1928.

58. Ravizza, G.E. and Zachos, J.C. (2003) Records of Cenozoic ocean chemistry. In: *The Oceans and Marine Geochemistry, Vol. 6, Treatise on Geochemistry* (eds H.D. Holland and K.K. Turekian). Elsevier-Pergamon, Oxford, pp. 551–581.

59. Pearson, P.N. and Palmer, M.R. (1999) Middle Eocene seawater pH and atmospheric carbon dioxide concentrations. *Science* **284** (5421), 1824–1826.

60. Pearson, P.N. and Palmer, M.R. (2000) Atmospheric carbon dioxide concentrations over the past 60 million years. *Nature* **406**, 695–699.

61. Montañez, I.P., Norris, R.D., Algeo, T., Chandler, M.A., Johnson, K.R., *et al.* (2011) *Understanding Earth's Deep Past: Lessons For Our Climate Future*. National Academy Press, Washington, DC.

62. Sverdrup, H.U., Johnson, M.W. and Fleming, R.H. (1942) *The Oceans, Their Physics, Chemistry and General Biology*. Prentice Hall, New York.

63. Hay, W.W. (2013) *Experimenting on a Small Planet: A Scholarly Entertainment*. Springer, New York.

64. Benson, W.E., Gerard, R.D. and Hay, W.W. (1970) Summary and Conclusions. Initial Reports of the Deep-Sea Drilling Project 4. US Government Print Office, Washington, DC, pp. 659–673.

65. Caldeira, K. and Wickett, M.E. (2003) Anthropogenic carbon and ocean pH. *Nature* **425**, 394–395.

66. Riebesell, U., Zondervan, I., Rost, B., Tortell, P.D., Zeebe, R.E. and Morel, F.M. (2000) Reduced calcification of marine plankton in response to increased atmospheric CO$_2$. *Nature* **407**, 364–367.

67. Doney, S.C., Ruckelshaus, M., Duffy, J.E., Barry, J.P., Chan, F., English, C.A., *et al.* (2012) Climate change impacts on marine ecosystems. *Annual Reviews in Marine Science* **4**, 11–37.

68. Garrels, R.M. and Mackenzie, F.T. (1971) *Evolution of Sedimentary Rocks.* Norton, New York.

69. Berner, R.A. (1990) Atmospheric carbon dioxide levels over Phanerozoic time. *Science* **249**, 1382–1386.

70. Garrels, R.M. and Perry, E.A. (1974) Cycling of carbon, sulfur and oxygen through geologic time. In: *The Sea* (ed. E. Goldberg). John Wiley and Sons, Chichester, pp. 303–336.

71. Garrels, R.M. and Lerman, A. (1984) Coupling of the sedimentary sulfur and carbon cycles – an improved model. *American Journal of Science* **284**, 989–1007.

72. Walker, J.C.G., Hays, P.B. and Kasting, J.F. (1981) A negative feedback mechanism for the long term stabilization of earth's surface temperature. *Journal of Geophysical Research* **86**, 9976–9982.

73. Berner, R.A. (1982) Burial of organic carbon and pyrite sulfur in the modern ocean: its geochemical and environmental significance. *American Journal of Science* **282**, 451–473.

74. Berner, R.A. (2013) From black mud to earth system science: a scientific autobiography. *American Journal of Science* **313**, 1–60.

75. Berner, R.A., Lasaga, A.C. and Garrels, R.M. (1983) The carbonate-silicate geochemical cycle and its effect on atmospheric carbon dioxide over the past 100 million years. *American Journal of Science* **283**, 641–683.

76. Berner, R.A. (1999) A new look at the long-term carbon cycle. *GSA Today* **11** (9), 1–6.

77. Ebelmen, J.J. (1845) Sur les produits de la décomposition des espèces minerales de la famille des silicates. *Annales des Mines* **7**, 3–66.

78. Lasaga, A.C., Berner, R.A. and Garrels, R.M. (1985) An improved geochemical model of atmospheric CO$_2$ fluctuations over the past 100 million years. In: *The Carbon Cycle and Atmospheric CO$_2$: Natural Variations Archean to Present* (eds E.T. Sundquist and W.S Broecker). Geophysics Monographs 32. American Geophysical Union, Washington, DC, pp. 397–411.

79. Cowie, J. (2013) *Climate Change: Biological and Human Aspects.* Cambridge University Press, Cambridge.

80. Shackleton, N.J. (1985) Oceanic carbon isotope constraints on oxygen and carbon dioxide in the Cenozoic atmosphere. In: *The Carbon Cycle and Atmospheric CO$_2$: Natural Variations Archean to Present* (eds E.T. Sundquist and W.S Broecker). Geophysics Monographs 32. American Geophysical Union, Washington, DC, pp. 412–417

81. Laws, E.A., Popp, B.N., Bidigare, R.B., Kennicutt, M.C. and Macko, S.A. (1995) Dependence of phytoplankton carbon isotopic composition on growth rate and $[CO_2]_{aq}$: theoretical considerations and experimental results. *Geochimica et Cosmochimica Acta* **59** (6), 1131–1138.

82. Arthur, M.A., Dean, W.E. and Schlanger, S.O. (1985) Variations in the global carbon cycle during the Cretaceous related to climate, volcanism, and changes in atmospheric CO$_2$. In: *The Carbon Cycle and Atmospheric CO$_2$: Natural Variations Archean to Present* (eds E.T. Sundquist and W.S Broecker). Geophysics Monographs 32. American Geophysical Union, Washington, DC, pp. 504–529.

83. http://www.geosociety.org/awards/07speeches/sloss.htm (last accessed 29 January 2015).

84. Holland, H.D. (1978) *The Chemistry of the Atmosphere.* John Wiley and Sons, New York.

85. Müller, R.D., Sdrolias, M., Gaina, C., Steinberger, B. and Heine, C. (2008) Long-term sea-level fluctuations driven by ocean basin dynamics. *Science* **319**, 1357–1362.

86. Vail, P.R., Mitchum, R.M. Jr., Todd, R.G., et al. (1977) Seismic stratigraphy and global changes of sea level. In: *Seismic Stratigraphy – Applications to Hydrocarbon Exploration* (ed. C.E. Payton). *American Association of Petroleum Geologists Memoir* **26**. 49–212.

87. Hallam, A. (1984) Pre-Quaternary sea level changes. *Annual Reviews of Earth and Planetary Sciences* **12**, 205–243.

88. Rowley, D.B. (2002) Rate of plate creation and destruction: 180 Ma to present. *Bulletin of the Geological Society of America* **114** (8), 927–933.

89. Rowley, D.B. (2008) Extrapolating oceanic age distributions: lessons from the Pacific region. *Journal of Geology* **116**, 587–598.

90. Haq, B.U., Hardenbol, J. and Vail, P.R. (1988) Mesozoic and Cenozoic chronostratigraphy and cycles of sea-level change. *Society of Economic Paleontologists and Mineralogists* **42**, 71–108.

91. Menard, H.W. (1964) *Marine Geology of the Pacific.* McGraw-Hill, New York.

92. Berner, R.A. (1989) Biogeochemical cycles of carbon and sulfur and their effect on atmospheric oxygen over Phanerozoic time. *Palaeogeography, Palaeoclimatology, Palaeoecology* **75**, 97–122.

93. Raymo, M.E., Ruddiman, W.F. and Froelich, P.N. (1988) Influence of late Cenozoic mountain building on ocean geochemical cycles. *Geology* **16**, 649–653.

94. Raymo, M.E. and Ruddiman, W.F. (1992) Tectonic forcing of late Cenozoic climate. *Nature* **359**, 117–122.

95. Raymo, M.E. (1997) Carbon cycle models: how strong are the constraints? In: *Tectonic Uplift and Climate Change* (ed. W.F. Ruddiman). Plenum, New York and London, pp. 367–381.

96. Ruddiman, W.F. (1997) *Tectonic Uplift and Climate Change*. Plenum, New York and London.

97. Ruddiman, W.F. and Prell, W.L. (1997) Introduction to the uplift-climate connection. In: *Tectonic Uplift and Climate Change* (ed. W.F. Ruddiman). Plenum, New York and London, pp. 3–15.

98. Ruddiman, W.F., Raymo, M.E., Prell, W.E. and Kutzbach, J.E. (1997) The uplift-climate connection: a synthesis. In: *Tectonic Uplift and Climate Change* (ed. W.F. Ruddiman). Plenum, New York and London, pp. 471–515.

99. Raymo, M.E. and Ruddiman, W.F. (1992) Tectonic forcing of late Cenozoic climate. *Nature* **359**, 117–122.

100. Molnar, P. and England, P. (1990) Late Cenozoic uplift of mountain ranges and global climate change: chicken or egg? *Nature* **346**, 29–34.

101. England, P. and Molnar, P. (1990) Surface uplift, uplift of rocks, and exhumation of rocks. *Geology* **18**, 1173–1177.

102. Cronin, T.M. (2010) *Paleoclimates: Understanding Climate Change Past and Present*. Columbia University Press, New York.

103. Blum, J.D. (1997) The effect of late Cenozoic glaciation and tectonic uplift on silicate weathering rates and the marine $^{87}Sr/^{86}Sr$ record. In: *Tectonic Uplift and Climate Change* (ed. W.F. Ruddiman). Plenum, New York and London, pp. 259–288.

104. Edmond, J.M. and Huh, Y. (1997) Chemical weathering yields from basement and orogenic terrains in hot and cold climates. In: *Tectonic Uplift and Climate Change* (ed. W.F. Ruddiman). Plenum, New York and London, pp. 329–351.

105. Berner, R.A. and Berner, E.K. (1997) Silicate weathering and climate. In: *Tectonic Uplift and Climate Change* (ed. W.F. Ruddiman). Plenum, New York and London, pp. 353–365.

106. Kump, L.E. and Arthur, M.A. (1997) Global chemical erosion during the Cenozoic: weatherability balances the budgets. In: *Tectonic Uplift and Climate Change* (ed. W.F. Ruddiman). Plenum, New York and London, pp. 399–426.

107. Kump, L.R., Brantley, S.L. and Arthur, M.A. (2000) Chemical weathering, atmospheric CO_2, and climate. *Annual Reviews of Earth and Planetary Sciences* **28**, 611–667.

108. Doughty, C.E., Taylor, L.L., Girardin, C.A.J., Malhi, Y. and Beerling, D.J. (2014) Montane forest root growth and soil organic layer depth as potential factors stabilizing Cenozoic global change. *Geophysical Research Letters* **41**, 983–990.

109. Ravizza, G.E. and Zachos, J.C. (2003) Records of Cenozoic ocean chemistry. In: *The Oceans and Marine Geochemistry, Vol. 6, Treatise on Geochemistry* (eds H.D. Holland and K.K. Turekian). Elsevier-Pergamon, Oxford, pp. 551–581.

110. Berner, R.A. and Caldeira, K. (1997) The need for mass balance and feedback in the geochemical carbon cycle. *Geology* **25** (10), 955–956.

111. Broecker, W.S. and Sanyal, A. (1998) Does atmospheric CO_2 police the rate of chemical weathering? *Global Biogeochemical Cycles* **12** (3), 403–408.

112. Beerling, D.J. and Royer, D.L. (2011) Convergent Cenozoic CO_2 history. *Nature Geoscience* **4**, 418–420.

113. Fletcher, B.J., Brentnall, S.J., Anderson, C.W., Berner, R.A. and Beerling, D.J. (2007) Atmospheric carbon dioxide linked with Mesozoic and early Cenozoic climate change. *Nature Geoscience* **1**, 43–48.

114. Franks, P.J., Royer, D.L., Beerling, D.J., Van de Water, P.K., Cantrill, D.J., Barbour, M.M. and Berry, J.A. (2014) New constraints on atmospheric CO_2 concentration for the Phanerozoic. *Geophysical Research Letters* **41** (13), 4685–4694.

115. Berner, R.A. (2006) GEOCARBSULF: a combined model for Phanerozoic atmospheric O_2 and CO_2. *Geochimica et Cosmochimica Acta* **70**, 5653–5664.

116. Berner, R.A. and Kothavala, Z. (2001) GEOCARB III: a revised model of atmospheric CO_2 over Phanerozoic time. *American Journal of Science* **301**, 182–204.

117. Li, G. and Elderfield, H. (2013) Evolution of carbon cycle over the past 100 million years. *Geochimica et Cosmochimica Acta* **103**, 11–25.

118. Torres, M.A., West, J.A. and Li, G. (2014) Sulphide oxidation and carbonate dissolution as a source of CO2 over geological timescales. *Nature* **507**, 346–349.

119. Zeebe, R.E. and Calderia, K. (2008) Close mass balance of long-term carbon fluxes from ice-core CO_2 and ocean chemistry records. *Nature Geoscience* **1**, 312–315.

120. Froehlich, P. and Misra, S. (2014) Was the late Paleocene-early Eocene hot because Earth was flat? *Oceanography* **27** (1), 36–49.

10

The Pulse of the Earth

10.1 Climate Cycles and Tectonic Forces

Rather like human history, Earth's history did not run smoothly. Patches where nothing much seemed to be happening apart from business as usual were punctuated by episodes of relatively rapid change. The Dutch geologist Johannes (Jan) Herman Frederick Umbgrove (1899–1954) (Box 10.1) dubbed this periodicity *The Pulse of the Earth*: the title for his classic textbook, first published in 1942[1], which I was still poring over as an undergraduate in 1960. Like Alexander Humboldt and Eduard Suess, Umbgrove thought of the Earth as a single integrated dynamic system.

Umbgrove ascribed the alternating periods of Earth history to '*deep-seated forces [that are] the paramount source of all subcrustal energy, which manifests itself with a [~250 Ma] periodicity observed in a whole series of phenomena in the earth's crust and on its surface … [including] the magmatic cycles and rhythmic cadence of world-wide transgressions and regressions [of the sea], the pulsation of climate, and – lastly – the pulse of life*'[1]. Writing before plate tectonic theory, and while Wegener's continental drift concept was still discredited, he could not ascribe a mechanism to his earthly pulsations. He favoured Suess's land bridges rather than Wegener's moving continents, and discounted the notions of Arrhenius and Chamberlin, concluding that things other than CO_2 primarily controlled Earth's climate. With the benefit of hindsight, it's easy to be critical, but like others of his generation he lacked the data to say anything much different. Even so, his concept of the Earth's pulse had certain attractions and lingered on.

Box 10.1 Johannes Herman Frederick Umbgrove.

A product of Leiden University, Umbgrove spent 3 years as a palaeontologist in what is now Indonesia with the Geological Survey of the Dutch East Indies, before returning to Leiden. In 1930, he became professor of stratigraphy and palaeontology at the University of Delft. He was an 'all-rounder', taking an interest in many different branches of geology and publishing on the palaeogeography of the Dutch East Indies, the palaeontology of corals and coral reefs, volcanology and the geology of the Netherlands. Among other honours, he was made an honorary fellow of the Royal Society of Edinburgh and of the New York Academy of Sciences, as well as being a fellow of the Royal Dutch Academy of Sciences.

The next key figure to pick up the baton in our story, in the late 1970s, was Princeton stratigrapher Alfred (Al) George Fischer (1921–) (Figure 10.1, Box 10.2).

In 1977, Fischer and his PhD student, Mike Arthur, whom we met in Chapter 9, drew wide-ranging conclusions from their study of the carbonate-rich, deep-marine, open-ocean (or pelagic) Cretaceous sediments from Gubbio in Italy's Apennine Mountains[2]. Their conclusions were bolstered by analyses of similar pelagic sections

Earth's Climate Evolution, First Edition. Colin P. Summerhayes.
© 2015 John Wiley & Sons, Ltd. Published 2015 by John Wiley & Sons, Ltd.

Figure 10.1   *Alfred George Fischer.*

Box 10.2 Alfred George Fischer.

Born in Germany, Al Fischer emigrated to the United States, where he obtained a PhD from Princeton in 1950. After sitting on the staff at Princeton from 1956 to 1984, he moved to the University of Southern California, where at the time of writing he is professor emeritus, living in Santa Barbara. As a 'biogeohistorical visionary', he would become one of the world's best-known stratigraphers. He became an expert on cyclic sedimentation in Mesozoic and Cenozoic sequences, and thus one of the pioneers of 'cyclostratigraphy'. His work in this field culminated in 2004 in the publication by the Society for Sedimentary Geology of *Cyclostratigraphy: Approaches and Case Histories*, containing several papers that he wrote or co-authored. His lasting contribution lay in linking variations in biodiversity, through changes in climate and the chemistry of oceans and atmospheres, to variations in rates of seafloor spreading, changes in sea level and fluctuations in continental igneous activity. Among his accolades, Fischer was awarded the Penrose Medal of the Geological Society of America, the Mary Clark Thompson Medal of the National Academy of Sciences in 2009 and the Geological Society of London's Lyell Medal in 1992. He was elected to the US National Academy of Sciences in 1994.

---

found on land or in deep-ocean drill cores, some collected by Fischer as stratigrapher and sedimentologist on DSDP Leg 1[3]. Stimulated by the new concept of plate tectonics, they reasoned that *'the episodic development of great glacial ages suggests that patterns of energy distribution and perhaps the energy budget as a whole have undergone fluctuations … One might expect, then, that variations in sedimentation through time reflect not only locally generated changes but also carry an overprint produced by shifts in the state of the oceans, of the biosphere, and perhaps of the earth as a whole – changes of a sort not considered by Hutton, Lyell, and other classical uniformitarianists'*[2]. Given that this 'overprint' was subtle, it had to be sought in sediments deposited in open-ocean pelagic realms far from the influence of land and tectonic 'noise'[2].

Their examination of pelagic sections convinced them that these sediments carried the signal of a cycle with a period of about 32 Ma[2]. This represented an oscillation between two oceanic states that they termed 'oligotaxic' and 'polytaxic'. Like all scientists, geologists are not immune to the curse of jargon. Oligotaxic times were periods of low global biodiversity, reduced complexity of biological communities and widespread extinction of free-swimming organisms, including plankton. They were also typically cool, like the Pleistocene and present seas, and associated with lowered sea level, marked by regression of the shoreline. In contrast, polytaxic times were periods of high rates of speciation and high biodiversity. They featured warm seas and a high sea level, marked by transgressions of the shoreline. They were also associated with weaker latitudinal temperature gradients, less vigorous ocean circulation, an intensified and expanded oxygen minimum zone, widespread deposition of organic-rich sediment, a net loss of CO_2 from the air with time and a rise in the carbonate compensation depth (CCD).

Cool (oligotaxic) episodes were centred on the Permo-Triassic boundary (250 Ma ago), the Triassic–Jurassic boundary (200 Ma ago), the Bathonian–Callovian boundary (165 Ma ago), the early Neocomian (~140 Ma ago), the Cenomanian (95 Ma ago), the early Paleocene (62 Ma ago), the mid Oligocene (30 Ma ago) and the Pleistocene–Holocene (2 Ma ago). Warm (polytaxic) episodes favourable to the accumulation of petroleum source beds were centred on the early Jurassic (190 Ma ago), the late Jurassic (155 Ma ago), the mid Cretaceous (110 Ma ago), the late Cretaceous (85 Ma ago), the Eocene (50 Ma ago) and the Miocene (15 Ma ago). These cycles lay within a broader climatic cycle of 200–300 Ma duration that tended to emphasise one state over the other, with, for example, pronounced warm (polytaxic) episodes during Jurassic–Cretaceous time and cool (oligotaxic) periods during the Cenozoic[2].

The cycles seemed to be unrelated to plate tectonics or magnetic reversals, but were related to cycles in terrestrial biodiversity, in sea level and in the $\partial^{13}C$ ratio – probably due to periodic burial of isotopically light (^{12}C-rich) organic matter, rather than to variation in productivity. Fischer and Arthur were unsure what made the polytaxic climates warm, suggesting that '*processes within the earth's interior influence sea levels by changing the earth's surface configuration, and may simultaneously affect the atmosphere and therefore climates through vulcanism*'[2]. Arguing from first principles, they thought that the warm phases likely had more atmospheric CO_2 (for which they had no direct evidence), which declined towards the cool phases[2].

By 1981, thinking deeply about these cycles, and considering the many advances being made by deep-ocean drilling, Fischer realised that new developments were dragging historical geology away from the uniformitarian view of Hutton and Lyell, in which the present state and functioning of the Earth were taken as the 'norm'[4,5]. The palaeontological record, for instance, he observed, is neither uniform nor gradual, but rather a record of sharp discontinuities. The discovery of ancient glaciations implied intervals of major climatic deterioration. Former warm periods might be explained by an increase in the abundance of greenhouses gases (like CO_2) in the atmosphere, as suggested by Arrhenius and Chamberlin. Orogenic events seemed to be periodic, as Umbgrove had suggested, as did changes in sea level. There were regular changes in the thermal structure and behaviour of the ocean. Clearly, the Earth had not persisted in an invariant state of which the present was representative, as Hutton and Lyell had implied.

Fischer stated that '*While most of this change has come about slowly – at what might be thought of as a uniformitarian pace – the role of catastrophe cannot be dismissed as it was by Lyell*'[5]. Unlike Lyell, he knew (as we see in more detail later) that an asteroid had hit the Earth at the end of the Cretaceous, killing off the dinosaurs, and that some regions had experienced long periods of massive volcanic eruptions, creating floods of basalt covering vast areas and no doubt filling the air with noxious fumes (such as the Deccan Traps of India and the flood basalts of Siberia and the Columbia River in the northwest United States). Fischer thought that '*We may now view earth history as a matter of evolution in which some changes*

are unidirectional (at least, in net effect), others are oscillatory or cyclic, and still others are random fluctuations, while the whole is punctuated by smaller or greater catastrophes. The prime tasks of modern historical geology are to separate the local signals from the global ones, to plot the relationships of global patterns both to time and to each other, and to search for the forces that drive these varied processes'[5]. There was a role here for a blend of Lyellian uniformitarianism and Cuverian catastrophism. Fischer's conclusion was echoed in 1993 by Derek V. Ager, one-time professor of geology at Imperial College London, former head of the department of geology and oceanography of the University College of Swansea and a former president of the United Kingdom's Geological Association[6]. Adapting Napoleon's aphorism, Ager concluded that '*the history of any one part of the earth, like the life of a soldier, consists of long periods of boredom and short periods of terror*'[6]. And so it is that today we find geological thought combining the uniformitarian and the catastrophic approaches. The ongoing mundaneness of the day-to-day can be interrupted by the special event. The two camps have merged.

Fischer identified two great tectonic–climatic cycles, with a periodicity of around 300 Ma, which he proposed were driven by cycles in mantle convection leading to cyclic changes in the abundance of CO_2 in the air[5]. These cycles began with the Ice Age of the late Proterozoic (around 650 Ma ago), continued with the inferred high CO_2 greenhouse conditions of the early-middle Paleozoic, the low CO_2 conditions of the Ice Age of the late Paleozoic (late Carboniferous to early Permian), and the high CO_2 of the Mesozoic greenhouse state, and ended in the low-CO_2 icehouse state of the late Cenozoic to the present. Volcanism was abundant and sea level was high in what Fischer called the 'greenhouse states', while both were low in what he called the 'icehouse states'. This was probably because the continents were dispersed and mid-ocean ridges were abundant and rapidly spreading during the greenhouse states, while continents tended to be aggregating and mid-ocean ridges to be less active in icehouse states. He considered that the continents tended to be thicker and to have less freeboard when aggregated in icehouse states, and that the post-Eocene drop in sea level might mark the start of the next phase of continental aggregation, which began with the collisions of India and Tibet, Australia and South East Asia and Africa and Europe[4,5].

Fischer thought that the primary source for the increase in atmospheric CO_2 in the greenhouse states was abundant volcanism when mid-ocean ridges were spreading fastest and granites were being emplaced beneath volcanoes at the leading edges of the moving continents. Large mid-ocean ridges made sea level high, flooding the continents and reducing the area susceptible to the weathering that would draw down the CO_2 content of the air. When volcanism declined and sea level was low, weathering would catch up, reduce CO_2 and lower the temperature, eventually leading to conditions suitable for glaciation in the icehouse state. Fischer credited Budyko and Ronov with drawing attention to the association between volcanism, weathering, sedimentation and CO_2. Citing Heinrich Holland's work on the carbon cycle[7], of which more later, he argued that '*carbon dioxide in the oceanic/atmospheric reservoir has a residence time of about 500 000 years [which is] not long in terms of the timescales here considered: an imbalance in input versus loss that continued over tens of millions of years could effect considerable changes*'[5]. That much was also evident to Chamberlin at the end of the 19th century, as we saw in Chapter 4.

Fischer was a great integrator. What we see in his work is the seepage into mainstream geology of ideas from the worlds of geochemistry (Walker and Holland), atmospheric chemistry and physics (Keeling and Plass) and ocean chemistry (Revelle, Suess and Broecker). And as we saw in Chapter 9, his conclusions are borne out by the later data of Royer and the models of Garrels and Berner.

Thomas Worsley of Ohio University published a refinement of Fischer's ideas in 1985[8]. Seeing that the flow of heat through the ocean crust from the Earth's interior is about six times what it is through the continents, Worsley concluded that the build-up of heat beneath supercontinents like Pangaea might account for their uplift and eventual rupture. Eventually, the passive margins of the newly expanding oceans would become sufficiently old and dense to spontaneously self-subduct, a process that seemed to him likely to occur within 200 Ma of initial rifting. In due course, Worsley argued, the Atlantic and the Indian Oceans should close to form a new supercontinent. Mad idea? Not when you realise that that the Atlantic Ocean has opened and closed before, more or less along the same lines. The Iapetus Ocean separated Europe from North America 600 to 400 Ma ago. Its closure thrust up the Caledonian mountain chain that runs from Norway through Scotland and Ireland and continues south in the Appalachians of North America. The present Atlantic is a new break. In Worsley's conceptual model, sea level was low on supercontinents and high during their break-up,

matching Vail's curves of sea level through time (Figure 5.9)[9]. The distributions of stable isotopes of carbon, sulphur and strontium through time matched the model, supporting the idea that episodic plate-tectonic processes drove biogeochemical cycles, including the slow carbon cycle and, by inference, CO_2.

Fischer's and Worsley's cycles followed the cycle of ocean basin evolution proposed in 1966 by Tuzo Wilson, whom we met in Chapter 5[10, 11].

Despite the realisation emerging in the geological community of the early 1980s that fluctuations in CO_2 had most likely played an important role in determining the changes in Earth's climate with time, many of the papers published on past global change in that era did so without mentioning CO_2 as a driver – such as a 1984 paper by Haq entitled 'Paleoceanography: A Synoptic Overview of 200 Million Years of Earth History'[12], to cite just one example. We should not be surprised. New concepts take time to work their way through the system. The BLAG model was first published in 1983; Fischer's key paper on the topic did not emerge until 1984. As we saw in Chapter 8, even within the climate science community, it was only in the very late 1970s and early 1980s that convincing studies of the role of CO_2 in the climate system began to emerge in papers, by the likes of Charney, Ramanathan, Hansen and Broecker.

Following Fischer's lead, the link between CO_2 and climate was now 'in the geological air'. Writing in 1986 and 1987, Bob Sheridan of the University of Delaware used the links between tectonics, CO_2 and climate to propose a theory of 'pulsation tectonics'[13, 14]. This involved plumes of hot magma periodically erupting from the boundary between the Earth's core and mantle, speeding seafloor spreading and causing mid-ocean ridges to grow, which increased sea level and displaced seawater on to the continents. The increased volcanic activity associated with seafloor spreading added CO_2 to the atmosphere, which warmed it and, in concert with the expanded ocean area, increased evaporation. Adding water vapour to the atmosphere further increased warming, leading to a warm, wet climate. Lessening of plume activity, on the other hand, lessened all the other factors, including the output of CO_2 and evaporation of seawater, leading to a cooler climate as weathering extracted CO_2 from the atmosphere.

In 1992, Fischer's cycles concept was adapted by Larry Frakes, whom we met in Chapter 6. With his colleagues Jane Francis and Josef Syktus, Frakes '*divided climate history into Warm Modes and Cool Modes, in a way not unlike Fischer's ... "Greenhouse" and "Icehouse" states, but our Modes are of shorter duration*'[15]. Frakes and his team '*questioned the theory that the Mesozoic climates were [uniformly] warm and ice free and instead propose[d] a Cool Mode in the Middle Mesozoic*'[15], listing the resulting modes as follows:

- **Cool Mode 5: early Eocene to present, 55–0 Ma ago**
- *Warm Mode 4: early Cretaceous to early Eocene, 105–55 Ma ago*
- **Cool Mode 4: late Jurassic to early Cretaceous, 167–105 Ma ago**
- *Warm Mode 3: latest Permian to middle Jurassic, 253–167 Ma ago*
- **Cool Mode 3: early Carboniferous to late Permian, 333–253 Ma ago**
- *Warm Mode 2: early Silurian to early Carboniferous, 436–333 Ma ago*
- **Cool Mode 2: late Ordovician to early Silurian, 445–436 Ma ago**
- *Warm Mode 1: earliest Cambrian to late Ordovician, 540–445 Ma ago*
- **Cool Mode 1: latest Precambrian to earliest Cambrian, 615–540 Ma ago**

The evidence for these modes comprised fossil animals and plants, along with sedimentary rock types and characteristics (tillites, carbonates, evaporates, aeolian sandstones, calcrete, kaolinite, coal and so on), as well as oxygen isotopes and the relative heights of sea level. Warm Mode 2 included a brief glaciation recognised in South America, and there was evidence for some cool periods within Warm Mode 4 and some warm periods during Cool Mode 4. This latter was identified as a cool mode from growing evidence of ice-rafted debris of that age at high latitudes. Like Fischer, the Frakes team accepted that CO_2 had some part to play as a driving force in changing climate, with more CO_2 being supplied when seafloor spreading was rapid than when it was slowed and mountains were built, although they did not go out of their way to provide details.

The cool modes were associated with low sea level and seemed to require a considerable extent of land at high latitudes, with long intervals of cooling leading to glaciation. They tended to be associated with a marked increase in $\partial^{13}C$, rising to a peak during extreme cooling, and with a marked decrease in the abundance of evaporites. While land at high latitudes seemed to be a necessary condition for the development of ice sheets, it was not sufficient. Global cooling, and especially the development of cool

summers to prevent snowmelt, was required too. This independently imposed cooling implied a decrease in atmospheric CO_2, possibly reflecting in turn the sequestration of large amounts of ^{12}C-rich organic carbon as coal on land, or as organic-rich black shales in the ocean, which would have increased $\partial^{13}C$ as cooling progressed.

The warm modes tended to appear quite suddenly but to end gradually. Their beginnings were usually preceded by an abrupt decrease in $\partial^{13}C$ (hence, an increase in CO_2), and their ends saw a gradual increase in $\partial^{13}C$ (hence, a decrease in CO_2). They were characterised by high sea level, a rise in volcanic activity and an increase in the accumulation of organic carbon with time, suggesting tectonic control in the form of an upsurge of seafloor spreading and volcanic emissions of CO_2 associated with sea level rise and warming, followed by the pulling down of CO_2 by the accumulation of organic carbon in sediments, leading eventually to cooling.

Clearly, the Frakes team was convinced of the strong link between plate tectonic activity and climate that had been alluded to in 1979 by Budyko and Ronov[16]. They noted that some 70% of above-average volcanism in tectonic regimes occurred in warm modes, '*consistent with the hypothesis that global climates are influenced, and perhaps forced, by volcanic outgassing*'[15]. This activity was usually most enhanced towards the middle of a warm mode. While orbital changes were obviously important controls on climate, especially within ice ages, the Frakes team considered that they did not contribute to the warm and cool modes, and that they did not initiate the late Cenozoic glaciation.

Looking towards the future, Frakes' team observed, '*The great variability of Phanerozoic climates has not seen a clear trend towards overall warming or cooling in the last 570 m.y., but rather can be characterized as alternating cool and warm intervals of long period*'[15]. Most ancient climates were relatively warm. With the increase in geological information, '*it has come to be recognized that climates have varied more often than previously accepted ... Further work may also reveal greater variability on both short and long wavelengths*'[15]. We will see plenty of evidence of that in later chapters.

As in all the sciences, nothing stands still in geology. These early ideas have become more refined – in a 2007 analysis by Alan Vaughan of the British Antarctic Survey, for example (Figure 10.2)[17]. Vaughan capitalised on Royer's 2004 curve of palaeotemperature and listing of multiple short-lived cool periods through time[18] to show that Frakes' Warm 1 and Warm 2 periods tended to be

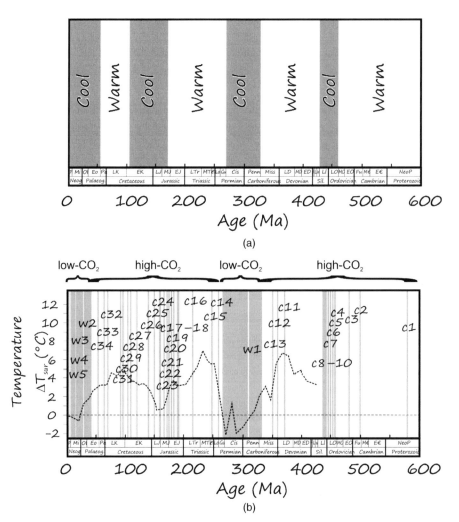

Figure 10.2 *Vaughan's view of alternating cool and warm periods through time. (a) Late Neoproterozoic and Phanerozoic climate modes of Frakes and colleagues*[15]. *(b) Dark grey = glacial or cool conditions; white = warm conditions, through late Neoproterozoic and Palaeozoic to recent time. Cool intervals are labelled (e.g. c1–c19), as are warm intervals (e.g. w1–w5). Modified from Royer et al.*[18], *and including the palaeotemperature curve from that source (dashed line). Brackets above (b) show durations of high and low CO_2 modes for Phanerozoic climate.*

interrupted by short-lived cool periods, while his Cool 2 and Cool 3 periods were interrupted by short-lived warm periods[17]. Vaughan saw Frakes' Warm 3, Cool 4 and Warm 4 periods as one long warm period containing numerous short-lived cool intervals, some of which tended to cluster close to the Cool 4 period but by no means mapped on to it. Vaughan also shortened Frakes' Cool 5 period by starting it towards the end rather than the start of the Eocene.

Summarising his view of Phanerozoic climate, Vaughan concluded that '*through the Phanerozoic, two overlapping stable climate regimes appear to have dominated: a high-CO_2 (>1000 ppmv [ppm by volume – an alternative way of expressing ppm CO_2 in the atmosphere]), largely warm climate regime, punctuated by many short-lived episodes of glaciation; and a low-CO_2 (<1000 ppmv), largely cool regime, marked by protracted episodes of superglaciation*'[17]. Vaughan's first high-CO_2 phase ran from the start of the Cambrian to the start of the Permo-Carboniferous glaciation, which was the first of his low-CO_2 climates (Figure 10.2). The second high-CO_2 phase ran from the late Permian to the late Eocene, after which Antarctic glaciation heralded the next low-CO_2 phase. We know from Royer's work that Vaughan's short-lived episodes of glaciation during the warm climate regime were associated with low CO_2[18, 19].

By Vaughan's time, geologists knew that there were areas scattered around the world characterised by the eruption of flood basalts with volumes of >100 000 km^3 magma, which must have been major sources of CO_2 from the Earth's mantle, as we shall see later. Fischer knew they existed[5] but had not built them in to his conceptual model as sources of CO_2. Vaughan's high-CO_2 climate modes tended to coincide with times of high rates of emplacement of flood basalts and of high rates of continental dispersal in the supercontinent cycle[17]. These were also periods of high rates of magmatic activity, enhanced hydrothermal activity at mid-ocean ridges, a high rate of supply of CO_2 and relatively low rates of continental weathering.

Vaughan's low-CO_2 modes, which are generally of much shorter duration, coincided with low rates of emplacement of flood basalts and times of amalgamation in the supercontinent cycles, marked by large-scale mountain building and high rates of crustal exhumation (deep rocks being brought to the surface as mountains are eroded). These are periods of low rates of magmatic activity, low rates of hydrothermal activity at mid-ocean ridges, relatively low fluxes of CO_2 and high rates of continental weathering. He thought that the short-lived cool episodes during the warm high-CO_2 modes were probably glacial, being associated

with rapid drops in sea level. They tended to be associated with the deposition of organic rich sediments and with positive excursions in $\partial^{13}C$ (due to the trapping of ^{12}C-rich organic matter in sediments) and in $\partial^{18}O$ (which is abundant when evaporated water rich in ^{16}O is trapped in ice).

Worsley continued working on global cycles at the grand scale, this time with his Ohio University colleague David Kidder. Finding a correlation in time between prolonged episodes of mountain building and icehouse climate, they saw this as '*prima facie evidence for orogenically driven CO_2 drawdown and carbon burial*'[20, 21], much as Raymo and Ruddiman had suggested (see Chapter 9).

Kidder and Worsley divided Earth's climates into icehouse, greenhouse and a kind of super greenhouse that they described as 'hothouse'[21]. In their conceptual model of Earth's climate, the average global surface ocean temperatures in the icehouse state ranged from roughly 15 °C, with 280 ppm CO_2 (expressed as 1× CO_2), to 21 °C (with 2× CO_2). Surface ocean temperatures in the cool greenhouse state ranged from 21 to 24 °C (with 4× CO_2), those in the warm greenhouse state ranged from 21 to 30 °C (with 16× CO_2) and those in the hothouse state reached almost 33 °C (with 32× CO_2). We can take these conditions to represent the natural envelope of the climate system in the Phanerozoic. The 280 ppm CO_2 chosen by Kidder and Worsley for the low end of their icehouse scale represents the present interglacial (preindustrial Earth, with 280 ppm CO_2 in the air). At peak icehouse conditions, as we shall see in Chapters 12 and 13, CO_2 fell to 180 ppm and global average temperatures fell by a further 4–5 °C.

Bill Hay, whom we met in Chapter 6, was much taken with their identification of this new hothouse state for climate and with their division of greenhouse states into cool (with small polar ice caps and Alpine glaciers, but no ice sheets capable of calving to produce icebergs) and warm (where the only ice is possible seasonal sea ice at the poles). Mapping out the alternations between icehouse, cool greenhouse, warm greenhouse and hothouse states through time over the past 750 Ma[22], Hay calculated that Earth had been in an icehouse state (with substantial ice at one or both poles) for about 25% of Phanerozoic time (the past 540 Ma), and in a greenhouse state (with little or no ice at either pole) for the remaining 75% (with 4% of the total spent in the hothouse state). How about interglacials, like the one we live in? Hay calculated that these represented only about 10% of the time spent in the icehouse state; that is, 2.5% of Phanerozoic time – a trifling amount. Evidently, we live in unusual times, under geologically rare conditions. How much would it take to tip our climate back into the greenhouse state typical of 75% of Phanerozoic time? That is one of the prime questions for this book.

Evidence for several kinds of geological and biological events following regular cycles of similar lengths through Phanerozoic time continues to accumulate, with marine organisms showing cycles of roughly 62 and 140 Ma, stratigraphic sequences showing a cycle of around 56 Ma and the strontium isotope record and the atmospheric CO_2 record both showing a 59 Ma cycle[23]. By 2013, Michael Rampino of New York University and Andreas Prokoph of Carleton University in Ottawa were ready to pose the question,[23] if these 60 and 140 Ma cycles are real, is there an underlying cause in large-scale Earth processes? Does this have something to do with mantle convection or plume activity? Mantle plumes are now thought to be responsible for the eruption of flood basalts, as we shall see later in this chapter. But first we look at the message emerging from ocean chemistry.

10.2   Ocean Chemistry

Mike Arthur figured out back in 1980 that the elemental and stable isotopic composition of marine sediments represented the chemical history of seawater[24]. Given the vast amount of CO_2 in the ocean and the small amount in the atmosphere, he realised that there had to be a link between ocean and atmospheric CO_2 that was discoverable from the ocean's chemical history. Back then, it was commonly assumed that the ocean's composition had remained constant through time, but Arthur – an integrator like his mentor Al Fischer – recognised that ocean composition reflects, on the one hand, the interplay between the composition of the atmosphere, the climate, weathering and the passage of dissolved chemical species to the ocean via rivers, and, on the other, the rates of ocean circulation, inputs of hydrothermal fluids associated with seafloor spreading and changes in biological and nonbiological processes affecting the extraction and storage of materials (e.g. plankton extracting calcium (Ca^{2+}) ions and CO_2 and using them to make $CaCO_3$ skeletons, which get stored in bottom sediment). The result of this interplay was that geochemists could use the sedimentary chemical record to deduce probable changes in climate.

Arthur also reminded us that all of the CO_2 in the atmosphere is probably cycled through plants once every 10 years or less, which is why, in order to understand climate, we have to understand the carbon cycle that controls atmospheric CO_2. This means we have to have a comprehensive understanding of the operation of the biosphere[24]; we have to know about carbon sources and sinks, the availability of nutrients, productivity, the burial of organic matter, the accumulation and dissolution of carbonates and the rates and reactions involved in the weathering of silicate and carbonate rocks. It turns out that evaporites are an important part of the equation. The massive formation of salt deposits in isolated sedimentary basins in arid environments absorbs a great deal of calcium (Ca^{2+}) ions, for example in gypsum (calcium sulphate), thus transferring Ca from the carbonate to the evaporite reservoir. As a result, there is less Ca about to form $CaCO_3$, resulting in a net transfer of CO_2 from the ocean to the atmosphere and an accompanying rise in the CCD. This scenario typifies the early Cretaceous of the narrow and slowly opening South Atlantic, where 2–3 km of evaporites accumulated across 2 Ma in the isolated Angola and Brazil Basins, preceding an abrupt rise in the CCD in the Aptian (125–112 Ma ago). Arthur went on to point out that the accumulation of 1.5 million km^3 of salt in the Mediterranean in a period of 1 Ma in the Messinian stage of the upper Miocene (7.2–5.3 Ma ago) made the Atlantic less salty. This made it easier to form sea ice in the northern North Atlantic, which may have increased albedo there sufficiently to contribute to the progressive cooling that eventually led to the formation of the Northern Hemisphere ice sheets.

Support for idea that the history of plate tectonics is represented in the chemistry of seawater comes from James Walker. Studying the global geochemical cycles of carbon, sulphur and oxygen, he concluded that there was *'a significant flux of hydrothermal sulfide to the deep sea, at least during the Cretaceous'*[25]. Assuming it came from hydrothermal vents on spreading ridges, this supports Müller's theory that there was more seafloor spreading during the Mesozoic than since[26].

We owe much of our modern understanding of the past chemistry of the atmosphere and ocean to Heinrich (Dick) Holland (1927–2012) (Box 10.3).

Holland found a close correspondence between carbonate mineralogy and sea level[27]. Comparing the distribution of aragonite-rich versus calcite-rich carbonates with the distribution of sea level through time, as mapped by Pete Vail[9], Bil Haq[28] and Tony Hallam[29], he found that

Box 10.3 Heinrich Holland.

Holland was born to Jewish parents in Germany and was sent to England to escape the Nazis as a child. He ended up in the United States with his parents in 1940. Graduating from Princeton with a degree in chemistry and acquiring a PhD in geology from Columbia University in 1952, he subsequently served on the staffs of both Princeton and Harvard. A brilliant scholar, he was made a member of the US National Academy of Sciences and received the V.M. Goldschmidt Award of the Geochemical Society in 1994, the Penrose Gold Medal of the Society of Economic Geologists in 1995 and the Leopold von Busch Medal of the Deutsche Geologische Gesellschaft in 1998.

aragonite tended to be associated with times of low sea level (early Cambrian, Carboniferous–early Jurassic and Neogene) and calcite with times of high sea level. He deduced that this correlation represented changes with time in the amount of hot hydrothermal fluid exhaled from mid-ocean ridges. This fluid was more abundant when rates of production of basaltic ocean crust and mid-ocean ridges were high, which raised sea level. Seawater circulating through these fractured rocks lost much of its dissolved magnesium (Mg) and sulphate ions[30]. The exhalation of this Mg-depleted seawater as hydrothermal fluid at ridge crests changed ocean chemistry, affecting the calcium carbonate mineralogy of skeletons formed by marine creatures. When rates of ridge production were high, the depletion of seawater in Mg favoured the deposition of calcite. When rates of ridge production were low, the enrichment of seawater in Mg favoured the deposition of aragonite. Aragonitic carbonate deposits are thus more common when continents collide and oceanic crust is being destroyed, rather than being created. Jan Zalasiewicz and Mark Williams of Leicester University call the oscillation between the two minerals the 'calcite metronome'[31]. Cool climates are associated with aragonitic carbonate deposits, warm ones with calcitic deposits[17].

Additional geochemical evidence soon arrived to support Holland's observations. In 2010, Rosalind Coggon of Imperial College London and colleagues analysed Mg/Ca and Sr/Ca ratios in carbonate veins that precipitated from circulating fluids derived from seawater in the basalts on the flanks of mid-ocean ridges[32]. Before the Neogene, and back to 170 Ma ago, the ratios of these elements were lower than they are in the modern ocean, presumably because the rate of seafloor spreading and hence the production of hydrothermal fluids at mid-ocean ridge crests has declined since the Cretaceous, increasing the Mg content and Mg/Ca ratio of seawater[30]. The calcite metronome and the Mg/Ca ratio provide further geochemical evidence, like Walker's sulphur cycle[25], that Müller was right about rates of seafloor spreading through time[26] and that Rowley was wrong[33].

As we saw in Chapter 9, the CCD also changes with climate. In warm climates, it is relatively shallow, because dissolution of CO_2 makes the ocean slightly more acid, and it is deeper in cool periods. Heiko Pälike of the United Kingdom's National Oceanography Centre, Southampton thought that the position of the CCD through time should tell us something about the changes in the balance through time between the supply of CO_2 from volcanic and metamorphic out-gassing and its removal by the weathering of silicate and carbon-bearing rocks. Along a depth transect in the equatorial Pacific, Pälike and colleagues found that the CCD tracked long-term cooling, deepening from 3.0–3.5 km at about 55 Ma ago to 4.6 km at present[34]. This pattern is consistent with an increase in weathering with time, and indicates a close correspondence between climate and the carbon cycle. Superimposed fluctuations on the CCD in the Eocene appeared to represent changes in weathering and in the mode of delivery of organic carbon to the deep ocean. The CCD deepened significantly at the Eocene–Oligocene boundary, along with growth of the Antarctic ice sheet, a fall in sea level and a shift of carbonate deposition from continental shelves to the deep sea. This is something we explore more in Chapter 11.

It took roughly 25 years to get from Plass's papers in 1956 to Arthur's geochemical paper in 1980, Walker's sulphur chemistry paper in 1981 and Fischer's papers on global climate change and stratigraphy in 1981 and 1984. In parallel with these, we benefitted from Holland's 'Chemistry of the Atmosphere and Oceans' in 1978 and 'Chemical Evolution of the Ocean and Atmosphere' in 1984. These stunning advances mean that from the early 1980s onwards, we were in an Earth system world, where everything was known to be connected, and the entire globe – including the 72% covered by ocean – was becoming sufficiently well sampled to evaluate processes operating at the global scale. Geochemists were changing the game of palaeoclimatology. The paradigm was beginning to shift.

10.3   Black Shales

At times in Earth history, sediments rich in organic matter formed widespread black shales in the deep sea and on continental margins. These deposits are of economic interest, not only as the possible source rocks for oil, but also because of their possible climatic significance. They are abundant in the Cretaceous of the deep Atlantic, where they were first drilled in 1968 on DSDP Leg 1 by Al Fischer and colleagues, who reported finding grey-black, bituminous, laminated Albian–Cenomanian radiolarian mudstones at Site 5A, just east of the Bahamas[3]. As ocean drilling progressed, organic-rich black shales stuffed with ^{12}C were seen to be common in the Aptian (125–112 Ma ago), at the Cenomanian–Turonian boundary (94 Ma ago) and in the Toarcian (early Jurassic, 183–176 Ma ago). Study of these fascinating deposits suggested that at those times the oceans may have been largely devoid of oxygen (hence anoxic), allowing for the preservation of organic remains that would otherwise have been degraded by bacteria. Occurrences were labelled 'oceanic anoxic events'[35–37]. Depositional conditions seem to have been warm, with abundant CO_2.

Sheridan thought that these deposits formed during that part of his pulsation cycle when atmospheric CO_2, temperature, sea level and the CCD were rising[13, 14]. At these times, the ocean would have been more thermally stratified, with poorly oxygenated bottom waters. Vegetation would have been lush on land, providing abundant fine-grained organic matter to the ocean, to be preserved as black shales where oxygen levels were low. He painted a convincing picture.

The Portuguese geologist João Trabucho-Alexandre and his team noted that some of these black shales formed in lakes associated with the rift phase or early stages of opening of ocean basins. Others formed as the sea flooded subsiding rift valleys, or in shallow shelf seas as continental margins subsided or sea level rose. More formed on the margins and in the deeps of the opening ocean basins, when surface waters were highly productive and bottom waters poorly oxygenated. Only a few were found in the deeps of the fully mature ocean basins, which tend to be well oxygenated[38].

In 1987, I looked into the nature and origin of the deep-water black shales of the Atlantic. I agreed with Sheridan that their formation most likely reflected internal conditions within the ocean[39]. What might those have been during the Cretaceous? In 1982, Garret Brass of the Rosensthiel School of Marine and Atmospheric Science

of the University of Miami suggested that the deeps of the Cretaceous North Atlantic would have been filled with warm, salty, dense and oxygen-poor water derived from tropical regions, where evaporation in shallow continental margin seas made the surface waters salty and sufficiently dense to sink into the deep ocean[40]. It would have contained much less oxygen than today's cold bottom water, making the development of anoxia and the accumulation of organic matter more likely[40]. Incidentally, it would also have been less able to dissolve CO_2, thus ensuring that the atmosphere contained more CO_2 than it would under cooler conditions like those of today, which helped to keep the air warm.

Today's Mediterranean Sea provides an example of Brass's 'haline' circulation, to contrast with the 'thermohaline' circulation of today's global ocean. Atlantic water makes its way at the surface through the Strait of Gibraltar to the Egyptian coast. There, evaporation in the Levantine Sea makes the surface waters sufficiently dense to sink, returning to the Atlantic as subsurface water passing over the Gibraltar sill. As they warm at the surface in the Levantine Sea, they lose dissolved oxygen to the air, warm water holding less gas than cold. Imagine that process characterising the whole ocean. Combined with the sinking and decomposition of dead organic matter, it probably created a vastly expanded oxygen minimum zone, encouraging the accumulation of organic matter on the seabed.

Oliver Friedrich of Germany's Bundesanstalt für Geowissenschaften und Rohstoffe in Hannover tested Brass's model by using $\partial^{18}O$ and Mg/Ca ratios from benthic foraminifera to reconstruct the intermediate-water characteristics of the tropical proto-Atlantic Ocean between 95 and 92 Ma ago[41]. The temperatures ranged from 20 to 25 °C, the warmest ever found for depths of 500–1000 m. Friedrich and colleagues found evidence for highly saline conditions, confirming an influx of water from surrounding epicontinental seas. The existence of these warm waters accentuated the stratification of the Atlantic basin, preconditioning it for prolonged periods of oxygen depletion.

Much the same loss of oxygen happens today in the subsurface waters of the Red Sea. The exit of its oxygen-depleted deep water contributes to the stratification and oxygen depletion of the adjacent Arabian Sea at the northeastern end of the Indian Ocean. In 1987, I suggested that, in much the same way, a subsurface current of intermediate-depth water poor in oxygen and rich in nutrients had likely entered the Atlantic from the Pacific beneath the westward-moving Atlantic surface water, thus contributing to stratification and oxygen depletion in the Cretaceous deep Atlantic[39].

Two other factors accounted for the abundant accumulation of organic matter in the Cretaceous sediments of the Atlantic. One was runoff from the surrounding land, which carried terrestrial organic material derived from lush, warm, tropical and subtropical forests. The other was wind-driven upwelling along certain continental margins. There, the upwelling of nutrient-rich subsurface waters stimulated high productivity, which enhanced oxygen depletion in bottom waters, reinforcing the already strong oxygen minimum imported from the Pacific[39].

The palaeoclimate map for the Cenomanian shown in Figures 6.9 suggests that upwelling should have been well developed at that time along the margin of northwest Africa and in the narrow gap between west Africa and Guyana, where the richest deposits of marine organic matter are found[39]. More sophisticated numerical modelling by Robin Topper of Utrecht University in 2011 confirmed that these were the areas most likely to be subject to upwelling currents in the Cenomanian[42]. Topper *et al.*'s model confirmed my 1987 prediction[39] that a subsurface current brought intermediate water into the North Atlantic basin. It was focused along the southern margin of the basin, where upwelling was best developed[42].

I thought that the unusual enrichment of Cenomanian sediments in organic matter between west Africa and Guyana might also reflect the breaking apart of Africa and South America. That would have led to an oceanic connection between the North and South Atlantic, allowing highly saline, oxygen-depleted and nutrient-rich waters from the south to enter the North Atlantic, much as Red Sea water enters and influences the Arabian Sea. Such an influx would have accentuated both productivity, through upwelling of the nutrient-rich subsurface water, and preservation of organic matter at depth in the saline, oxygen-depleted deep water[39]. The influx of this 'new' deep water would have caused significant accumulation of organic-rich sediments until the reserve of nutrients was exhausted or until continued widening of the connection to the south diminished the influx of highly saline water[39]. That nutrient limit could account for the organic enrichment in the Cenomanian and its subsequent decline.

More recently, David Kidder and Tom Worsley suggested that Brass's evaporation-driven haline circulation model may have typified ocean circulation during the extremely warm 'hothouse' intervals of climate that developed in response to the massive volcanic eruptions that produced flood basalts[20, 21]. Their hothouse climate state is an extreme version of the greenhouse state, driven ultimately by the addition of masses of CO_2 to the atmosphere from flood basalt eruptions. The '*hothouse model explains the systemic interplay among factors including warmth, rapid sea level rise, widespread ocean anoxia, ocean euxinia [oxygen depletion] that reaches the photic zone, ocean acidification, nutrient crises, latitudinal expansion of desert belts, intensification and latitudinal expansion of cyclonic storms, and more*'[21]. In their model, sinking warm, salty tropical waters would have permeated the deep ocean, eventually making their way to the surface at the poles, where they would have warmed the polar regions, eliminated polar ice and helped to reduce the Equator-to-pole thermal gradient, thus reducing the strength of major global wind systems.

Hay linked these developments to ocean productivity, pointing out that in a world of weaker winds, the supply of dust to the atmosphere would have been severely curtailed compared to what it is now, thus limiting the supply of iron (a limiting nutrient), resulting in a 'nutrient crash'[22]. At the same time, the warming of the ocean depleted it of oxygen. Waters poor in oxygen, if not actually anoxic, would have filled the subsurface waters of the ocean basins. Loss of land ice and thermal expansion of water volume would have flooded the continental margins with oxygen-depleted water. There would have been many more tropical storms, extending to higher latitudes and to deeper depths than they do today, which would have helped to maintain warm conditions in polar regions, not least by promoting the development of a warming cover of clouds. This cycle came to an end eventually, as warm, humid conditions on land encouraged chemical weathering of silicates, which brought down the CO_2 content of the atmosphere, making conditions cooler[22].

The Frakes team noticed that oceanic anoxic events tended to be associated with large $\partial^{13}C$ peaks, for example during early Cretaceous Aptian times (125–112 Ma ago)[15]. Did this indicate cooling, or was it a result of the tying up of lots of ^{12}C-rich organic matter in oceanic anoxic events? At high latitudes in the early Cretaceous, there were some indicators of cold climate in the form of dropstones, which may have derived from river or shore ice. This suggests that the early Cretaceous climate could have been cooler than was previously thought, and glaciers may have been present near the poles at that time[15].

My research back in 1984 showed that '*Deposition of sediment rich in organic matter in the Gulf [of Mexico] was not confined to a Barremian-Aptian "oceanic anoxic event" but continued at high rates throughout the Early Cretaceous, possibly because the North Atlantic (and its offshoot, the Gulf of Mexico) were separated from the rest of the world's oceans by sills'*[43]. Organic matter tends to accumulate in silled basins where the oxygen content is low, as in today's Cariaco Trench on the continental shelf off Venzuela and in today's Black Sea. This tells us that individual deep basins within the Atlantic province may preserve the history of both global events (oceanic anoxic events) and local conditions (isolation of the deep Gulf of Mexico). As Fischer pointed out, one must take care to distinguish between global and local (or regional) effects when constructing the narrative of Earth's climate history, a lesson we will return to in later chapters. Relying solely on one indicator, like $\partial^{13}C$, to tell us about past climate is likely to be unwise.

While much of the organic matter in deep North Atlantic black shales originated from the remains of marine plankton, especially along the upwelling margins of northwest Africa and Guyana, elsewhere in the basin much was terrestrial in origin[4, 39]. This land-derived material most likely reached the deep ocean in dense, rapidly moving currents of water stuffed with suspended sediment – the so-called 'turbidity currents' – which dumped their loads in 'turbidites' (deposits with distinctive layers of basal sand and later mud). More dilute suspensions of turbid water flowed slowly down the continental margin and across the basin floors to form 'hemipelagic muds': mixtures of pelagic planktonic remains and land-derived 'terrigenous' mud supplied via rivers or winds[39]. Independent confirmation for the proposal that many of the laminated black shales from the Cretaceous of the deep Atlantic were in fact thin turbidites comes from detailed sedimentological studies in both the North[44] and the South Atlantic[45]. Some terrestrial components would also have arrived as wind-blown desert dust, along with minor inputs of charcoal (fusain) blown in from forest fires.

The abundance of terrestrial plant remains, especially in the western North Atlantic and off Portugal, attests to high productivity on land at the time[46] and to a humid temperate coastal climate[47]. Terrestrial organic matter was preferentially pumped into the deep Atlantic basin at times of lowered sea level, when rivers would have discharged their loads close to the continental slope, making the sediments on the slope unstable and liable to slump, generating turbidity currents[43].

Accepting the results of Trabucho-Alexandre[44] and Stow[45], much of the mid Cretaceous deep seafloor comprises layers of thinly bedded or laminated, darkly coloured, organic-rich turbidites, which were deposited rapidly, sandwiched between layers of bioturbated, oxygenated, light-coloured, organic-poor hemipelagic sediments, which were deposited slowly. This pattern suggests that the Cretaceous deep seafloor was oxygenated, with deposition interrupted from time to time by the arrival of organic-rich turbidity currents originating on the nearby continental margin, where there must have been a strong oxygen minimum zone. The thinly bedded organic-rich turbidite layers of the deep sea are likely to tell us more about conditions on the continental margins – the source of the sediments – than about conditions on the basin floor, beneath what was most probably a rather unproductive open ocean[48]. It was once thought that the fine-scale laminations of the mid Cretaceous Atlantic black shales represented oscillations in the oxygen saturation of bottom waters, but it now seems more likely that the arrival of organic-rich material via bottom currents caused the poorly oxygenated bottom waters to become anoxic near or at the sediment–water interface, preventing bioturbation by benthic organisms and so preserving laminated structures.

There is another possible interpretation for the Aptian $\partial^{13}C$ peak. A team of Swiss scientists, led by Christina Keller, argued in 2011 that the prominent negative carbon isotope excursion that preceded the Aptian oceanic anoxic event (OAE-1) was caused by major volcanic activity on the Ontong Java Plateau in the western Pacific, which drove an increase in atmospheric CO_2[49]. Examining floral changes in Italy, they noted that at the beginning of the isotope event the climate was warm-temperate. The temperature rose across the duration of the isotope event, with the highest temperatures coinciding with arid conditions. This may have reflected a northward shift in the hot-arid northern Gondwana floral province in response to the increase in atmospheric CO_2. *'Over 200 ka after the onset of OAE-1, reduced volcanic activity and/or increased black shale deposition allowed for a drawdown of most of the excess CO_2 and a southward shift of floral belts'*, they found[49].

Isabel Montañez and Richard Norris agreed. The recent discovery of large-magnitude but short-lived $\partial^{13}C$ excursions at the onset of several Mesozoic oceanic anoxic events, they said, *'is compelling evidence for greenhouse gas forcing of these abrupt climate events, possibly by methane hydrate release from seafloor gas hydrates … methane release by magmatic intrusion into organic-rich sediments … or other greenhouse gas sources such as volcanism'*[50].

Finally, several Cretaceous black shale deposits show cyclicity reminiscent of that imposed by orbital variations in insolation (discussed in Chapter 6), demonstrating '*the sensitivity of oceanic conditions to perturbation of atmospheric circulation and continental weathering brought on by global warming*'[50].

10.4   Sea Level

These various studies suggest a close relationship between sea level and CO_2 through time. Does it exist? Gavin Foster and Eelco Rohling, then of the National Oceanography Centre, Southampton, thought so[51]. They found a well-defined relationship between CO_2 and sea level extending over the past 40 Ma that '*strongly supports the dominant role of CO_2 in determining Earth's climate on these time scales and suggests that other variables that influence long-term global climate (e.g., topography, ocean circulation) play a secondary role*'[51]. They started from the premise that sea level largely represents ice volume, and the observation that in ice cores covering the past 800 Ka, fluctuations in the level of atmospheric CO_2 closely match changes in sea level (more on that in Chapter 12). This is because CO_2 is the principal greenhouse gas that amplifies orbital forcing and so determines to a large extent the thermal state of the Earth system across glacial–interglacial cycles and thus the amount of ice stored on land. The relationship is strong despite small leads and lags. For the past 800 Ka, CO_2 did not rise above 300 ppm – not much different from the preindustrial values identified by Oeschger and Callendar. Over the last ~6000 years the lack of change in sea level shows that the ice sheets were stable, so the threshold for major ice retreat must be higher than 300 ppm CO_2.

To see what might happen to sea level if CO_2 were to increase further (as it is now doing), Foster and Rohling examined the relationship between CO_2 and sea level for periods in the Cenozoic when CO_2 was more abundant. They used past levels of CO_2 from alkenones and from boron isotopes in planktonic foraminifera and derived past sea levels from a combination of $\partial^{18}O$ data, Mg/Ca ratios and analyses of palaeowater depth on continental margins. For CO_2 levels between 200 and 400 ppm in Pliocene and Miocene times, the relationship between CO_2 and sea level was more or less the same as that for the past few hundred thousand years[51], with sea level rising in proportion to the logarithm of atmospheric CO_2. In contrast, for CO_2 levels between 400 and 650 ppm in Pliocene, Miocene and

Oligocene times, sea level estimates remained on a plateau of about +22 m (±12 m) compared to today's level. For CO_2 levels above 650 ppm, in the Eocene, rises in CO_2 were associated with rising sea level. Peter Barrett (pers. comm.) reminded me that there is a further plateau in sea level that is a consequence of the 64 m upper limit for ice-driven sea level rise once all the ice sheets have gone, no matter how high CO_2 rises.

Foster and Rohling explained the sigmoidal nature of this relationship as follows. During the Eocene, when CO_2 was above 1000 ppm, sea levels were 60–70 m higher than today, as there were no major ice sheets[51]. CO_2 declined from 1000 to 650 ppm towards the Oligocene, and sea level fell as the East Antarctic Ice Sheet grew. CO_2 continued falling, from 650 to 400 ppm, but sea level did not respond, probably because, as the oxygen isotope data suggest, very little continental ice grew or retreated during that time. This is most likely because most of the land ice was on Antarctica; the Northern Hemisphere ice sheets had not yet begun to form. The average sea level of +22 m suggests that there was neither a Greenland Ice Sheet nor a West Antarctic Ice Sheet, those being equivalent together to about +14 m, and that the East Antarctic Ice Sheet was smaller than today by the equivalent of about 10 m of sea level. According to Foster and Rohling, '*Presumably CO_2 was too high, hence the climate too warm to grow more continental ice after the 'carrying capacity' of the EAIS [East Antarctic Ice Sheet] had been reached*'[51].

Evidently, sea level does not exhibit a simple response to changing CO_2. It is modulated by the behaviour of large ice sheets, especially at levels of CO_2 between 400 and 650 ppm. The evidence '*suggests that 300–400 ppm is the approximate threshold CO_2 value for retreat and growth, respectively, of WAIS [West Antarctic Ice Sheet] and GrIS [Greenland Ice Sheet] (and possibly a more mobile portion of EAIS)*'[51]. This implies, further, that sea levels of 20–30 m above present at times during the Pliocene and Miocene, when CO_2 reached 280–400 ppm, mainly represented melting of the ice sheets of Greenland and West Antarctica, with a possible contribution from East Antarctica[51]. Sea level fell below present levels only after CO_2 dropped below 280 ppm some 2.6–2.8 Ma ago, when the Laurentide and Fennoscandian ice sheets began to grow.

Roderik Van de Wal of the University of Utrecht and colleagues took a slightly different approach to testing the notion that the gradual cooling of the climate through the Cenozoic could be attributed to a decrease in CO_2 (Figure 10.3)[52]. Collecting data on Northern Hemisphere

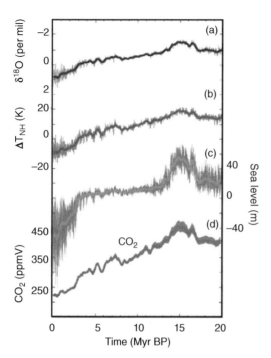

Figure 10.3 *Change in sea level and related climate variables over the past 20 Ma. Analysis of (a) the smoothed ∂¹⁸O record of Zachos leads to (b) Northern Hemisphere temperature change with respect to preindustrial conditions and (c) sea level (related to ice volume). (d) The reconstructed CO₂ record comes from inverting the relation between Northern Hemishpere temperatures and CO₂ data. Thick lines represent 400 Ka running mean. Grey error bars indicate standard deviation of model input and output.*

temperature and CO_2 for the past 20 Ma, they found that while the relationship between the two was positive and clear for some data sets, it was weak for others, notably for some of the CO_2 estimates derived from alkenones and from boron isotopes. Relying on the assumption that there was a relationship between temperature and CO_2, as demonstrated by ice core data, they excluded the apparently wayward alkenone- and boron-based CO_2 data sets, retaining the CO_2 estimates based on B/Ca ratios, combined alkenones/boron isotopes and stomatal data that were consistent with the ice core data.

Using the temperature–CO_2 relationship emerging from their data sets, Van de Wal's team found a gradual decline

of 225 ppm CO_2 from about 450 ppm in the mid Miocene warm period around 15 Ma ago to a mean level of 225 ppm during the last 1 Ma, coinciding with a fall in temperature of about 10 °C (Figure 10.3). The inception of Northern Hemisphere ice around 2.7 Ma ago took place once the long-term average concentration of CO_2 had dropped below 265 ± 20 ppm.

10.5   Biogeochemical Cycles, Gaia and Cybertectonic Earth

So far, Chapters 9 and 10 have shown that geological thinking about the evolution of Earth's climate began to change from about 1980 onwards, thanks to growing appreciation of the operation of the slow carbon cycle, which involves volcanic emission, weathering and sedimentation. This 'revolution' in thinking about the role of the carbon cycle in Earth's climate evolution owes much to the rapid expansion of geochemistry from the mid 1950s, to the comprehensive sampling of deep-ocean sediments by the Deep Sea Drilling Project (DSDP) and its successors beginning in 1968, to the discovery of gas bubbles in ice cores in 1978, to the modelling of the carbon cycle from around 1980 onwards, which was stimulated by a growing ease of access to fast numerical computers, and, to continuing efforts to improve or find new proxies for past atmospheric levels of CO_2 in the 2000s.

The rise of biogeochemistry and its holistic approach to science also played a part in the evolution of geology into Earth system science, which '*seeks to integrate various fields of academic study to understand the Earth as a system. It considers interaction between the atmosphere, hydrosphere, lithosphere (geosphere), biosphere, and heliosphere*'[53]. Some journals have even changed their names to keep up with this trend, such as *Proceedings of the Indian Academy of Sciences (Earth and Planetary Sciences)*, which in 2005 became the *Journal of Earth System Science*. Reflecting these various developments, in the past 20 years we have seen many a geology department change its name to 'department of Earth sciences'. In December 2013, the American Geophysical Union (AGU) launched a new journal focused on Earth System Science, titled *Earth's Future*. AGU's rationale for doing so was that, while we will still need disciplinary science, in order to fully understand and provide advice on '*the major challenges facing human society in the 21st century*' we need to take '*more holistic approaches that will integrate knowledge from individual disciplines'*[54]. The new journal *'deals with the state of the planet and its expected evolution. It publishes papers that emphasize the Earth as an interactive system under the influence of the human enterprise. It provides science-based knowledge on risks and opportunities related to environmental changes'*[54]. The paradigm has changed.

I have no doubt that Humboldt would have been cheered by these developments, although I know that they remain anathema to some of my more traditional contemporaries – often the same ones who resist the notion that so-called 'small' changes in CO_2 can affect our climate. It is inevitable that with change comes resistance, which often turns out to represent simply an inability to keep up. Many of the changes have been so recent, and the literature is advancing so quickly, that only the younger generation of geologists trained in Earth sciences can be expected to be fully aware of the dramatic progress that has been made and continues to be made. Even then, awareness may be only partial, not least because many of the advances are in subfields like geochemistry and biogeochemistry, which demand specialised knowledge. The general public is likely to be even less aware of these various new developments and of their significance.

One new concept has been much more in the public eye than many of the specialised elements we address in this book. It is the proposal by the English scientist and inventor James Ephraim Lovelock (1919–) that the living and nonliving parts of the Earth's surface form a complex interacting system that can be thought of as a single organism, and that the biosphere has a regulatory effect on the Earth's environment that acts to sustain life. First putting forward this proposal in a scientific paper in 1969[55], he later named it after Gaia, the Greek goddess or personification of the Earth[56]. In 1974, Lovelock fleshed out the concept with the American microbiologist Lynn Margulis (1938–2011)[57]. In her 1999 book *The Symbiotic Planet*, Margulis defined Gaia as *'the series of interacting ecosystems that compose a single huge ecosystem at the Earth's surface'* and *'an emergent property of interaction among organisms'*[58]. Lovelock popularised the concept in his 1979 book *Gaia: A New Look at Life on Earth*[59]. Harking back to both Lovelock and Vernadsky, Euan Nisbett of Royal Holloway College wrote, *'The planet shapes life, but life also shapes the planet. The maintenance of surface temperature is managed by the air: hence as life controls the composition of the air and the atmospheric greenhouse, then life sets the surface temperature'*[60]. If Lovelock and Nisbett are right then CO_2 plays an important role in keeping our climate within the natural envelopes described in this book.

The Gaia concept has been explored through a series of conferences, starting in 1985. While the concept of studying the Earth as a living whole has taken some hits, Lovelock's genius has been recognised by numerous awards, not least fellowship of the Royal Society in 1974 and the Geological Society's Wollaston Medal in 2006. His PhD student Andy Watson, also elected a fellow of the Royal Society, and Watson's PhD student Tim Lenton expanded on Lovelock's ideas, summarising the present state of play in their 2011 book *Revolutions that Made the Earth*[61]. I once met Lovelock in passing: he was on the committee that interviewed me for the post of director of the Institute of Oceanographic Sciences Deacon Laboratory in 1987. I must have pressed the right buttons on the day.

Does life shape the Earth, then, or is it shaped by plate tectonic processes, including volcanic activity and mountain building, with its attendant weathering? Mike Leeder (1947–) of the University of East Anglia, the Geological Society's Lyell medallist for 1992, was in no doubt: *'Tectonics, climate and sea level are the dominant controls on the nature and distribution of sedimentary environments … The processes of global tectonics that cause widespread mountain belt and continental plateau uplift have produced numerous "severe" events during Earth history: these often random workings-out of the plate tectonic cycle define the state of "Cybertectonic Earth" (from cyber, after the Greek* kubernan: *to steer or govern). This has worked within a usually zonally arranged series of climate belts and within the framework provided by biological evolution. At certain times in the geological past, a combination of factors arose as a response to continental uplift and have acted to cause certain Earth surface conditions and variables (notably mean global surface temperature, atmospheric pCO_2, pO_2) to have varied by very large amounts (×2–×10) compared with the present value. These large fluctuations, such as those responsible for the Neoproterozoic, Late Palaeozoic and Late Tertiary glaciations, lead one to doubt the reality of homeostatic control of surface conditions as proposed in Lovelock's Gaia hypothesis. The Gaia* kubernētēs *(steersman) had a weak hand on the helm, frequently unable to prevent vast areas of the globe experiencing rapid fluctuations in environmental conditions and inimical conditions to life for very long periods during Neoproterozoic and Phanerozoic times. At the same time, biogenic and abiogenic processes have proved capable of returning Earth to states of mean stability, although* *biogeochemical cycling models seem alarmingly ad hoc and largely untestable as scientific hypotheses in any true geological sense'*[62].

Leeder argued that Lovelock's concept of the Earth as a self-regulating entity was unsatisfactory because it '*operated without reference to, and entirely independent of, the activities of plate tectonics'*[62]. Instead, Leeder's 'cybertectonic Earth' and Lovelock's biogeochemical 'Gaia' had to work together. '*It can be argued'*, Leeder suggested, '*that this combination of tectonics and biogeochemistry is the great fulfilment of the Huttonian philosophical scheme'*, Hutton having first introduced the idea of a mobile Earth that was a '*superorganism whose proper study is physiology'*[62]. '*Modern sedimentary (and other) geologists'*, Leeder went on, '*can thrive only if they study, or are at least aware of, not only the rocks and sediment under the surface, but also the host of related disciplines that deal with the material on and above Earth'*[62]. Humboldt would have liked that. And Lyell would have seen in it an endorsement of his call for comprehensive palaeoenvironmental analysis.

Leeder's 2007 analysis was very slightly off the mark in that he relied on Dave Rowley's 2002 suggestion that rates of plate construction had not changed for the past 180 Ma[63]. But I have produced other lines of evidence (variations through time in sea level, CO_2, sulphur, and calcite versus aragonite deposition, reflecting changes in ocean chemistry) that show that rates of spreading in the Cretaceous were faster than in the Cenozoic. Müller was right and Rowley wrong. Hence, Leeder's dismissal of the Vail curve of changes in sea level through time[9], which was linked to changes in the rates of seafloor spreading, now seems suspect. Nevertheless, Leeder's notion of a cybertectonic Earth has distinct attractions to an understanding of past climate change, especially when linked to biogeochemical cycles – provided we accept in addition the occasional catastrophe imposed by asteroids of the kind that ended the Cretaceous and wiped out the dinosaurs.

10.6   Meteorite Impacts

You only have to look at the Moon through binoculars to see that its surface is pitted with craters from giant impacts with passing asteroids, and many of us will have seen the meteor showers that grace our skies from time to time. Most visitors to Arizona will have peered into 'Meteor Crater', a giant hole in the ground some

1200 m across and 170 m deep, 69 km east of Flagstaff. Many readers will recall seeing the 1998 science fiction disaster move *Armageddon*, in which NASA sent Bruce Willis into space to deflect a giant asteroid from its path towards Earth. And let's not forget Ted Nield's book *Incoming*[64]. So we should not be startled by the notion that asteroids have hit the Earth a few times in its history. Fortunately, the frequency of impact has declined substantially with time.

Massive impacts would have affected Earth's climate, at least for short periods. Georges Cuvier would have doubtless latched on to asteroid impacts as one of the missing engines for the catastrophes that he thought punctuated geological time. But they might have posed a conundrum for Charles Lyell and his doctrine of gradualism or uniformitarianism. His ignorance of their existence explains the fact that they do not disturb the pages of his *Principles*. Indeed, collisions of asteroids with the Earth do not even feature in Arthur Holmes's 1965 magnum opus, *Principles of Physical Geology*, more than a century after Lyell's *Principles* were published.

We should not be too surprised at this oversight, because it was not until 1963 that Eugene (Gene) Merle Shoemaker (1928–1997) (Box 10.4) proved conclusively that Meteor Crater was indeed an impact crater[65]. The crater was initially named Canyon Diablo Crater and was thought to be the result of a volcanic steam explosion. Daniel Barringer (1860–1929) correctly identified it as a meteorite impact structure in 1903, and it was renamed 'Barringer Crater' in his honour, although its more common name remains Meteor Crater.

Knowing what shock features to look for, geologists began searching methodically for meteor craters, discovering more than 50 by 1970. Support for their identification as meteorite strikes – where the term 'meteorite' covers everything from comets to asteroids – came from the Apollo Moon landings, starting in 1969. Because the lack of erosion on the Moon allows craters to last indefinitely, it was possible to identify the rate of cratering, which likely applied to the Earth, too[66]. Fewer craters are visible on Earth. Their traces have been obliterated by plate tectonic processes, by weathering and by burial with sediments. Those most easily found tend to be young, like Meteor Crater, which is 50 Ka old. Buried ones can only identified by geophysical survey. Large circular structures tend to be a giveaway.

One buried crater, the 180 km-wide 'Chicxulub Crater', lies beneath the northern edge of Mexico's Yucatan Peninsula and its adjacent continental shelf and slope[67]. One of the largest impact structures on Earth, it represents a collision with a bolide (the Greek for 'missile') at least 11 km in diameter – about the size of Manhattan – which took place at the Cretaceous–Tertiary (or K-T) boundary 65 Ma ago[68]. It was discovered in 1990.

Box 10.4 Eugene Merle Shoemaker.

While studying Meteor Crater for a PhD at Princeton in 1960, Shoemaker found coesite and stishovite in the ground there – rare varieties of silica formed when quartz has been severely shocked. Such 'shocked quartz' is now recognised as one of the metamorphic products of impact events. He went on to work for the US Geological Survey, where he pioneered the field of astrogeology, founding the Survey's Astrogeology Research Program in 1961. Shoemaker was in an ideal position to advise NASA about its Lunar Ranger missions to the moon, and at one point he trained as an astronaut, although he was eventually disqualified on medical grounds. Arriving at Caltech in 1969, he began a systematic search for asteroids, discovering the Apollo Asteroids. While there, he proposed that asteroid strikes on Earth had likely been 'common' on the geological time scale and would have caused sudden geological changes. Previously, impact craters were thought to be volcanic in origin – even on the Moon. In 1993, Shoemaker co-discovered a comet, named 'Shoemaker-Levy 9', which provided scientists with the first opportunity to observe a cometary impact on a planet when it slammed into Jupiter in 1994, leaving a massive 'scar'. This helped to emphasise what extraterrestrial objects might be able to do if they hit the Earth. Shoemaker was awarded the Barringer Medal in 1984 and the US National Medal of Science in 1992. In 1999, some of his ashes were taken to the Moon and buried there by the Lunar Prospector space probe.

The astonishing notion that there had been a major meteorite impact at this boundary stemmed from research undertaken by Walter Alvarez (1940–) on magnetic reversals in deep-sea limestones in Italy. There, he found a widespread clay layer right at the K-T boundary. Knowing that this was when the dinosaurs went extinct, he wondered what the clay layer meant, and discussed the matter

with his father, Luis Walter Alvarez (1911–1988). Luis was a Nobel Prize-winning physicist from the University of California at Berkeley who had worked on the Manhattan Project during the Second World War, and later used a bubble chamber to discover new fundamental particles. He persuaded colleagues from the Lawrence Berkeley Laboratory to use neutron activation to analyse the clay layer. They made one of science's greatest discoveries when in 1980 they found that the clay contained abundant iridium, a chemical element common in meteorites but not on Earth[69]. Later, the clay was found to also contain soot, glassy spherules, shocked quartz, microscopic diamonds and other materials that formed under high temperature and pressure[70]. The researchers deduced that a meteorite impact had brought the Cretaceous to a close. The crater was only found 10 years later. The immense cloud of dust and gas from the impact would have blocked sunlight, inhibited photosynthesis and cooled the atmosphere for a decade. It affected the climate sufficiently to cause a major extinction event that wiped out the dinosaurs and other creatures, including the ammonites.

The discovery upset those of a Lyellian bent[71]. Palaeontologists were especially unhappy at the intrusion of geochemists into their cosy uniformitarian world. Lyell had noted the gap in continuity of fossils across the K-T boundary, but assumed it just represented one of those annoying gaps in the geological record. Darwin, too, marvelled at the sudden disappearance of the ammonites, but agreed with Lyell's interpretation. Both Lyell and Darwin were wrong. The paradigm had shifted. Catastrophes did happen.

A review of the evidence in 2010 showed that the global ejecta layer and the extinction event coincided. Moreover, the ecological patterns in the fossil record agreed with modelled environmental perturbations (darkness and cooling). The reviewers concluded that the impact triggered the mass extinction[72]. Later, in 2013, Paul Renne of the Berkeley Geochronology Centre presented argon isotopic data which established that the impact and the extinction coincided to within 32 Ka – a very small 'error window' in geological terms. Renne and colleagues suggested that the subsequent perturbation of atmospheric carbon at the boundary likely lasted less than 5 Ka, but that recovery of the major ocean basins took much longer. The impact likely triggered a shift in the state of ecosystems that were already under stress[73].

It surprised me to learn that the Alvarezes were not the first to suggest that a bolide impact had brought the Cretaceous to an end. That honour goes to Harold Urey, who suggested in 1973 that '*it does seem possible and even probable that a comet in collision with the Earth destroyed the dinosaurs and initiated the Tertiary division of geologic time*'[74].

Not everyone agreed with the Alvarezes' claim, not least because the massive eruption of flood basalts in India's Deccan Traps occurred at about the same time and might well have had a similar catastrophic effect[75]. But it is not easy to see how the long eruptive period of the Deccan Traps – extending over about 1 Ma and spanning the K-T boundary – fits with the tightly constrained evidence for a very short extinction event. Even so, the Trap eruptions, which probably produced 10 million km^3 of lava at rates of up to or even over 1 million km^3 per year[75], may have affected the global ecosystem enough for it to have succumbed more easily to the effects of the impact (more on that later). The eruption was a response to plate tectonic processes. It immediately preceded and was possibly related to the opening of the Arabian Sea.

Were there multiple impacts at the K-T boundary – an asteroid shower, or impacts from bits of a fragmented asteroid? Two impact craters of the right age have been identified: the 24 km-diameter Boltysh Crater in the Ukraine and the 20 km-diameter Silverpit Crater in the North Sea[68]. Others, as yet unidentified, may be hidden beneath the sediments of the deep-ocean floor.

The Alvarezes were not the first to suggest that major bolide impacts might have caused biological extinctions, either. That honour goes to Digby Johns McLaren (1919–2004) (Box 10.5).

Studying the Devonian around the world, McClaren saw that the late Devonian Frasnian–Famennian boundary (374.5 Ma ago) was knife-edge sharp, synchronous globally and accompanied by extinction of 50% of the biomass. In his presidential address to the Palaeontological Society of America in 1969[76], he argued that the only explanation for such a thing was the impact of a giant meteorite. In the words of his biographer, '*the members of the Society were left in a state of shock; the general consensus was that he must have lost his marbles*'[77].

McLaren's revolutionary mechanism for explaining mass extinctions was partly inspired by the research of Robert Dietz of the Navy Electronics Laboratory, who showed in 1964 that the giant circular Sudbury structure in Ontario was an astrobleme (an impact structure)[78]. McClaren had made one of those giant intellectual leaps,

Box 10.5 Digby Johns McLaren.

Digby McClaren was born in Carrickfergus in Northern Ireland, brought up in Yorkshire in England and spent his working life with the Geological Survey of Canada (GSC), specialising in studies of the Devonian. By 1973, he was director-general of the GSC, a post he held until 1980. McLaren was honoured in many ways for his scientific contributions, not least by being made a fellow of the Royal Society of Canada in 1968 and becoming its president from 1987 to 1990. Among his several honours, he was made a fellow of London's Royal Society in 1979, and the Geological Society of London awarded him its Coke Medal in 1985 and made him an honorary fellow in 1989. The Digby McLaren Medal of the International Commission on Stratigraphy is named after him.

and the reward came when an iridium anomaly was discovered at the Frasnian–Famennian boundary in Australia's Canning Basin[79]. In his address as the retiring president of the Geological Society of America in October 1982, he reviewed progress in the search for bolides at boundaries, concluding that it was highly probable that a large-body impact had caused the extinctions of the late Devonian and end Cretaceous, possible that such a mechanism accounted for the extinctions in the late Ordovician and late Triassic and somewhat likely that it could account for the late Permian extinction[80]. Could it be that meteorite impacts were quite common? Drawing on evidence from Gene Shoemaker, McLaren said, '*several 1-km-wide objects might be expected to arrive every million years, whereas larger objects of about 10 km in diameter should arrive at an interval of between 60 and 100 m.y., or even every 50 m.y. ... [and] there is the possibility of the relatively rare arrival of a body as much as 20 km in diameter*'[80]. The main effect of such an impact would be a massive ejection of dust into the stratosphere, which would remain in place for months, if not years, reflecting sunlight and seriously cooling the planet. Given that 72% of the surface of the planet is covered by the ocean, it was likely that 70% of past meteorite strikes would have occurred in oceanic areas, and thus be difficult to detect.

In 1998, however, Tony Hallam reminded us that really good evidence for major meteorite impacts, in the form of iridium layers of global extent and shocked quartz coincident with extinction, is only available for the K-T boundary event[81]. Glassy spherules or tektites formed by meteorite strikes have been found in the geological record, but not in association with other major extinctions. That lack of a direct association between impacts on the one hand and extinctions on the other has commonly led geologists to look to other causes for extinctions, notably massive volcanic eruptions (see further on).

Before we leave the topic of extraterrestrial impacts, we should note in passing the notion that the Sun's orbit around the galactic centre every 250–300 Ma might also affect Earth's climate from time to time. One of those giving some early thought to that prospect was Herbert Friedman of the US National Research Council, in 1985[82]. Friedman noted that the Sun undulates above and below the galactic plane with a period of 27 Ma, providing the possibility of collision with dense clouds of intergalactic dust in the disc of the Milky Way. As it circles the galactic nucleus, the Sun also drifts slowly through the dust clouds in the galaxy's spiral arms. Collision of our solar system with a dust cloud would slow the solar wind, and collision of dust with the Sun's surface might raise its temperature very slightly. Both effects have the potential to change the Earth's climate a little on a regular basis. Friedman recognised that '*it might seem far-fetched … to dwell … on [such] exotic suggestions*'[82]. James Pollack, of NASA's Ames Research Center, agreed, stating that interstellar dust clouds were far too thin to affect Earth's climate[83]. Besides, present data do not show 300 or 150 Ma cycles back beyond around 650 Ma ago[17]. Furthermore, the occurrence of the short Ordovician–Silurian glaciation is an anomaly in the 300 Ma cycle mode, and is not easily explainable by the extraterrestrial argument.

As we can see from analyses of Earth history by the likes of Fischer, Worsley, Frakes, Vaughan, Hallam, Leeder and Rampino, we can find most of the answers we need by examining the behaviour of the Earth itself.

10.7 Massive Volcanic Eruptions

By the mid 1980s, Jack Sepkoski (1948–1999), a palaeontologist from the University of Chicago, had identified a number of biological extinctions through time over the past 250 Ma, and he and a colleague, David Raup, speculated that they were part of a cycle recurring with an interval of 26 Ma[84]. One of these (the K-T boundary) appeared to be related to a meteorite impact, and they speculated that all

or some of the others might be, too. I note in passing that David Raup now considers that the cycle he identified may be a sort of statistical fluke[71].

Could these extinctions be related instead to volcanic eruptions? Vincent Courtillot of the Institut de Physique du Globe in Paris thought so[85]. It was he who first drew attention to the possible linkage between the K-T boundary and the eruption of the flood basalts of the Deccan Traps in India[75]. In 1994, he reported a link between nine of Sepkoski's ten extinction events of the past 300 Ma and some of the twelve known examples of large continental flood basalt provinces from that period. He concluded that flood basalt events provided the most likely explanation for extinctions, and that the Deccan Traps eruptions had set in motion the extinction event that culminated in the meteorite impact at the K-T boundary[85]. He particularly favoured the eruption of the flood basalts of the Siberian Traps to explain the massive extinction at the Permo-Triassic (P-T) boundary. Tony Hallam agreed that the end Permian extinction correlated with the eruption of the Siberian Traps[81]. By 1994, the term 'Large Igneous Province' was widely applied to the areas of massive eruptions that produced flood basalts both on the continents and in the oceans (e.g. the submarine Ontong-Java Plateau, east of New Guinea and north of the Solomon Islands).

As the timings of Large Igneous Provinces became better established, Courtillot found more evidence for a linkage between sixteen massive igneous events and biological extinctions, naming four such events in particular as having an important causal connection: the Deccan Traps (K-T boundary), the Siberian Traps (P-T boundary), the Central Atlantic Magmatic Province (Triassic–Jurassic (T-J) boundary), and the Ethiopian and Yemen Traps (~30 Ma ago) (Figure 10.4)[86]. By 2007, he had refined this picture, noting that the four most recent large mass extinctions of the Phanerozoic associated with major flood basalt eruptions were the late Permian (Guadalupian) event at 258 Ma ago, the end Permian event at 250 Ma ago, the end Triassic event at 200 Ma ago and the end Cretaceous event at 65 Ma ago[87].

Just how big were these massive volcanic eruptions? At their maximum extent, the flood basalts of India's Deccan Traps probably covered some 1.5 million km^2 – about the size of modern India. These large volumes of magma typically erupt through fissures over geologically short periods in places well away from the boundaries of the major tectonic plates. They are not like volcanoes, which represent point sources of eruptive material. They are

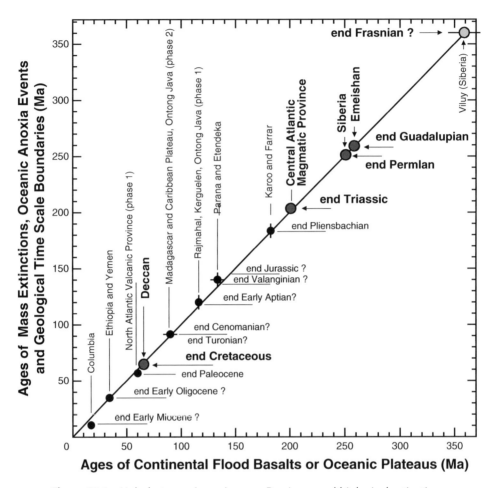

Figure 10.4   *Links between Large Igneous Provinces and biological extinctions.*

attributed to the activity of 'mantle plumes': vertical plumes of hot lava arising from point sources deep in the Earth's mantle, which enable the eruption of large volumes of material over a relatively small area. The plumes are thought to arise from periodic instabilities in the thermal boundary layer just above the Earth's core–mantle boundary. Such eruptions eject volcanic ash and sulphurous aerosols high into the air, cooling the climate for decades or even centuries. Not surprisingly, the places where they have occurred have come to be termed 'hot-spots'.

How exactly did these eruptions occur? Gerta Keller of Princeton and colleagues estimated in 2012 that the volcanic flows of the Deccan Traps took place in a number of pulses, each of which could have been as large as

$10\,000\,km^3$ and lasted up to 100 years[88]. In comparison, one of the largest historical eruptions of a single basaltic volcano, Laki, in Iceland, in 1783, produced a mere $15\,km^3$ of lava in a single year. Just one Deccan flow would have represented 667 Lakis!

The amount of carbon and sulphur dioxide emitted from one Deccan pulse is likely to have been as large as that emitted by the asteroid impact in Yucatan[88]. The SO_2 would have risen to the stratosphere, combining with water to form droplets of sulphuric acid that reflected solar energy. Cooling from any one eruptive phase would have been short-lived, because sulphuric acid droplets get rained out of the upper atmosphere within a couple of years following major eruptions. In contrast, the CO_2

would have stayed in the atmosphere for several thousands of years, contributing to long-term warming. It would also have tended to acidify the oceans, extending the range of extinction from terrestrial to oceanic.

Keller and colleagues argue that '*none of the "big five" mass extinctions was brought about by a single simultaneous event causing sudden environmental collapse. All are characterized by prolonged periods of high stress before and after mass extinctions, and three (end-Permian, end-Devonian, end-Ordovician) show multiple extinction phases, sometimes separated by hundreds of thousands of years*'[88]. Careful examination of the Cretaceous boundary extinction suggested to them that the simple impact-kill scenario was inadequate, not least because many species '*groups died out gradually or decreased in diversity and abundance well before the boundary, including dinosaurs*'[88]. She and her colleagues reiterated these conclusions following a multidisciplinary international conference in 2013 on 'Volcanism, Impacts and Mass Extinctions: Causes and Effects'[89].

We have no human memory of such enormous events. Our experience is of single volcanoes, like Laki, which may eject up to, say, 12 km^3 of lava, are active in a major way for less than a year and cause the climate to cool by up to 0.5 °C for a period of no more than a year or so. Large Igneous Provinces are not on our radar.

Tony Hallam pointed to the close association between extinctions and other possible major causes of environmental change, notably major marine regressions caused by falling sea level[81]. More important than any other factor in causing extinctions, in his view, were major transgressions caused by rising sea level associated with climatic warming and the spread of anoxic bottom waters. Intensive research cast serious doubt on the possibility that meteorite impacts had anything to do with the claim of a 26 Ma periodicity in extinction through time. Gerta Keller and colleagues agreed with Hallam that, in addition to volcanism and bolide impacts, '*sea level and climate changes (warming and cooling), ocean acidification, ocean anoxia, and atmospheric changes have to be considered in any extinction scenario to understand the causes and consequences of mass extinctions*'[89].

Was a link between Large Igneous Provinces and biodiversity justified? In 2013, statistical analyses by Rampino and Prokoph showed that both fossil diversity and the ages of Large Igneous Provinces have cycles of around 62 and 140 Ma, with an additional weaker 30–35 Ma cycle over the past 135 Ma[23]. These new data suggest a link to Fischer and Arthur's 32 Ma cycles in Earth's history.

Biological diversity was least at times of massive flood basalts. But does that matter?

In 2010, Peter Schulte of Germany's Universität Erlangen-Nürnberg argued that, while the evidence for an association between major igneous provinces and biological extinctions was reasonably sound, not all flood basalts were associated with extinctions. This led Schulte and colleagues to present the case that it was unlikely that volcanism associated with the Deccan Traps somehow destabilised the biosphere, making it more likely to collapse with the meteorite impact at the K-T boundary[90].

In 2010, David Kidder and Tom Worsley drew attention to a dozen or more examples of a tripartite link between Large Igneous Provinces, biological extinctions and geologically brief (<1 Ma) periods of exceptional warmth (their hothouse intervals)[20, 21]. Then, in 2011, David Retallack of the University of Oregon and colleagues noticed that the late Permian mass extinction was followed by an unusually prolonged recovery through the early Triassic[91]. Citing new records from Australia's Sydney Basin, they found five successive spikes of unusually high atmospheric CO_2, estimated from stomatal indices in fossil leaves, along with signs of deep chemical weathering. These 'greenhouse crises' coincided with unusually wet and warm climates at a palaeolatitude of 61° S. Between these crises were long periods of cool, dry climate with low CO_2. These patterns, they felt, '*may account for the persistence of low diversity and small size in Early Triassic plants and animals*'[91]. What might have caused these periodic events? They thought it might be '*Extraordinary atmospheric injections of isotopically light carbon, perhaps from thermal metamorphism of coal by feeder dikes to Siberian Trap lavas*'[91].

Their unexpected finding echoes those of Henrik Svensen of the University of Oslo, whose team noticed in 2008 that sills intruded into organic-rich shales and petroleum-bearing evaporites in Siberia during the eruption of the Siberian Traps at the end of the Permian baked the sediments, causing the release of '*greenhouse gases and halocarbons … in sufficient volumes to cause global warming and atmospheric ozone depletion*'[92]. The metamorphism of organic matter and petroleum could have generated >100 000 Gt CO_2 right at the Permian boundary. Emission took place through vertical pipes about 1 km across and would have included poisonous gases like methyl chloride and methyl bromide[91, 92]. Further measurements showed that the Trap magmas contained anomalously high amounts of sulphur, chlorine and fluorine. Ejection of large loads of such chemicals into the atmosphere may have led to serious deterioration in the global environment at the end of the Permian, contributing to extinction[93].

Despite that convincing evidence, the jury is still out on this question[94]. Continued searches for impact craters dated to the end Permian have found one in central Brazil. Eric Tohver of the University of Western Australia suggested that, although the 40 km-diameter crater is quite small, the bolide impact may have cracked apart underlying sediments rich in organic matter and methane across a radius of 700–3000 km. This seismic disruption could have supplied vast amounts of methane to the atmosphere in a brief period[95]. Assuming that some 1600 Gt of methane was released (equivalent to 135 000 Gt of CO_2), that would be more than enough to explain massive warming leading to an extinction. A *New Scientist* article about Tohver *et al.*'s paper makes the interesting point that the process of release of methane was in many respects equivalent to the process of fracking[94].

Recently, evidence has emerged off northwest Australia for another impact crater of approximately the right age that might help to explain the end Permian extinction[96]. We certainly do need something large to explain an extinction that wiped out 90% of terrestrial life and 70% of marine life on Earth. But is it realistic to call on bolide impacts? Investigating a claim for a massive bolide impact at the Permian–Triassic boundary, Christian Koeberl of the University of Vienna and colleagues concluded in 2002 that none of the evidence provided '*even a vague suggestion of an impact event at the P-T boundary*'[97]. The Alvarezes, too, found no evidence for an iridium spike like that of the K-T boundary in clays at the end of the Permian.

The saga continues. In March 2013, Terrence Blackburn of MIT and Washington's Carnegie Institute obtained new uranium-lead ages from zircon crystals taken from ancient lavas in North America and Morocco, which he and his colleagues used to more accurately pin down the age of the flood basalts of the Central Atlantic Magmatic Province. They found that the start of the volcanism coincided with the extinction at the end of the Triassic[98]. Over 1 million km^3 of lava poured out within less than 30 Ka, changing the ecosystem and paving the way for the dinosaurs to dominate our planet. What caused the extinction? It's difficult to say. Maybe initial cooling caused by clouds of dust and sulphuric acid droplets in the upper atmosphere, or perhaps rapid warming caused

by the emission of large volumes of CO_2 accompanied by ocean acidification.

Examining the densities of stomata in fossilised leaves from a range of species on either side of the 205 Ma Triassic–Jurassic boundary, Jenny McElwain and her colleagues at the University of Sheffield found lower densities above the boundary than below it[99]. This suggested to them that the CO_2 concentration in the atmosphere was about 600 ppm before the boundary and between 2100 and 2400 ppm after it – enough to cause a rise in mean global temperature of 4 °C, which may have interfered with the ability of large leaves to photosynthesise, leading to the extinction of 95% of land plants. The most likely origin for the rise in CO_2 was extensive volcanism as Pangaea began to break up.

Bas van der Schootbrugge of Goethe University agreed that terrestrial vegetation in Germany and Sweden had also been significantly affected by volcanic activity in the central Atlantic province at the end of the Triassic[100]. A fern-dominated association typical of disturbed ecosystems replaced Gymnosperm forests, and the associated sediments contained little charcoal but abundant polycyclic hydrocarbons, suggesting incomplete combustion of organic matter by flood basalts. This severe and abrupt shift in vegetation is unlikely to have been triggered by an increase in greenhouse gases alone. It probably also resulted from the emission of pollutants like SO_2 along with toxic aromatic hydrocarbons[100].

The research examined here confirms that there are long-lasting cycles in our planet's climate. They seem to relate mainly to processes taking place deep within the Earth. There is a growing body of evidence that mantle plumes generate hot-spots (like Iceland, for example), flood basalts in Large Igneous Provinces (like the Deccan Traps) and regional uplift and rifting. Rampino and Prokoph go so far as to suggest that '*plumes may act as a pacemaker for changes in sea level, climate and biodiversity*' and that we are looking at the exciting '*possibility of a unification of geologic processes, related in part to changes in the deep mantle*'[23]. Tectonic–volcanic cycles related to plumes and plate processes drive the slow carbon cycle, controlling the supply of CO_2 from volcanic vents and its eventual removal from the atmosphere through the weathering of newly up-thrust mountains. A fast carbon cycle uses biogeochemical processes to maintain climates suitable for life. Periodicity in the record suggests a variation on Lyell's uniformitarianism that would accommodate Cuvier's catastrophism. Steady and periodic processes were interrupted by Cuverian catastrophes in the form of massive meteorite impacts like that which ended the reign of the dinosaurs at the end of the Cretaceous, as well as occasional extended eruptions of flood basalt. There is no convincing evidence for asteroid impacts causing other major biological extinctions within the past 540 Ma.

The operations of plate tectonics and plumes evidently cause changes in the abundance of CO_2, which goes on to play a primary role as a planetary climate regulator[101]. Tectonics, volcanism, weathering, CO_2, biology and climate are inextricably linked. Their interactions kept Earth's climate cycling between fairly narrow limits of both temperature and CO_2 over the past 450 Ma. Levels of CO_2 were higher for a given temperature early in the period, when the Sun was fainter, and lower later in the period, when the Sun's output was stronger. For the most part, the climate was warm, resting in a greenhouse state, with occasional falls to glacial conditions and occasional rises to hothouse conditions. The limits of the natural envelope of Earth's climate system for the bulk of Phanerozoic time ranged from 180 ppm CO_2 and an average global temperature of around 11 °C in peak glacial conditions at the low end to somewhere between 4500 and 8500 ppm CO_2 and 30–32 °C in peak hothouse conditions at the high end.

It's now time to explore further how numerical general circulation models can aid our understanding of past climate change, and especially the role of CO_2 in the Cenozoic. With that improved understanding in mind, it is time, too, to examine some case histories of the role of CO_2 in our climate system. We will do so in Chapter 11.

References

1. Umbgrove, J.H.F. (1947) *The Pulse of the Earth*. Martinus Nijhoff, The Hague.
2. Fischer, A.G. and Arthur, M.A. (1977) Secular variations in the Pelagic realm. In: *Deep Water Carbonate Environments* (eds H.E. Cook and P. Enos). SEPM Special Publication **25**, pp. 19–50.
3. Beall, A.O. and Fischer, A.G. (1969) Sedimentology. Initial Reports of the Deep Sea Drilling Project 1. US Government Printing Office, Washington, DC, pp. 521-593.
4. Fischer, A.G. (1981) Climatic oscillations in the biosphere. In: *Biotic Crises in Ecological and Evolutionary Time* (ed. M.H. Nitecki). Academic Press, New York, pp. 103–131.
5. Fischer, A.G. (1984) The two Phanerozoic cycles. In: *Catastrophes and Earth History: The New Uniformitarianism* (eds W.A. Berggren and J.A. Van Couvering). Princeton University Press, Princeton, NJ, pp. 129–150.
6. Ager, D.V. (1993) *The Nature of the Stratigraphical Record*, 3rd edn. John Wiley and Sons, Chichester.
7. Holland, H.D. (1978) *The Chemistry of the Atmosphere*. John Wiley and Sons, New York.
8. Worsley, T.R., Moody, J.B. and Nance, R.D. (1985) Proterozoic to Recent tectonic tuning of biogeochemical cycles. In: *The Carbon Cycle and Atmospheric CO_2: Natural Variations Archean to Present* (eds E.T. Sundquist and W.S. Broecker). Geophysical Monographs 32. American Geophysical Union, Washington, DC, pp. 561–572.
9. Vail, P.R., Mitchum, R.W. and Thompson, S. (1977) Seismic stratigraphy and global changes of sea level 4. Global cycles of relative changes of sea level. *American Association of Petroleum Geologists Memoir* **26**, 82–97.
10. Wilson, J.T. (1966) Did the Atlantic close and then re-open? *Nature* **211**, 676–681.
11. Wilson, J.T. (1968) Static or mobile Earth: the current scientific revolution. *Proceedings of the American Philosophical Society* **112**, 309–320.
12. Haq, B.U. (1984) Paleoceanography: a synoptic overview of 200 million years of Earth history. In: *Marine Geology and Oceanography of Arabian Sea and Coastal Pakistan* (eds B.U. Haq and J.D. Milliman). Van Nostrand Reinhold, London, pp. 201–232.
13. Sheridan, R.E. (1986) Pulsation tectonics as the control of North Atlantic palaeoceanography. In: *North Atlantic Palaeoceanography* (eds C.P. Summerhayes and N.J. Shackleton). *Geological Society Special Publication* **21**, 255–275.
14. Sheridan, R.E. (1987) Pulsation tectonics as the control of long-term stratigraphic cycles. *Paleoceanography* **2**, 97–118.
15. Frakes, L.A., Francis, J.E. and Syktus, J.I. (1992) *Climate Modes of the Phanerozoic – The History of the Earth's Climate Over the Past 600 Million Years*. Cambridge University Press, Cambridge.
16. Budyko, M.I. and Ronov, A.B. (1979) Chemical evolution of the atmosphere in the Phanerozoic. *Geochemistry International* **16** (3), 1–9; translated from *Geokhimiya* **5**, 643–653, 1979.
17. Vaughan, A.P.M. (2007) Climate and geology – a Phanerozoic perspective. In: *Deep-Time Perspectives on Climate Change: Marrying the Signal from Computer Models and Biological Proxies* (eds M. Williams, A.M. Haywood, F.J. Gregory and D.N. Schmidt). *Micropalaeontological Society Special Publications*. The Geological Society, London, pp. 5–59.
18. Royer, D.L., Berner, R.A., Montañez, I.P., Tabor, N.J. and Beerling, D.J. (2004) CO_2 as a primary driver of Phanerozoic climate. *GSA Today* **14** (3), 4–10.
19. Royer, D.L. (2006) CO_2 – forced climate thresholds during the Phanerozoic. *Geochimica et Cosmochimica Acta* **70**, 5665–5675.
20. Kidder, D.L. and Worsley, T.R. (2010) Phanerozoic large igneous provinces (LIPs), HEATT (Haline Euxinic Acidic Thermal Transgression) episodes, and mass extinctions. *Palaeogeography, Palaeoclimatology, Palaeoecology* **295**, 162–191.

21. Kidder, D.L. and Worsley, T.R. (2012) A human-induced hot-house climate? *Geology Today* **22** (2), 4–11.

22. Hay, W.W. (2013) *Experimenting on a Small Planet – A Scholarly Entertainment*. Springer, New York.

23. Rampino, M.R. and Prokoph, A. (2013) Are mantle plumes periodic? *EOS* **94** (12), 113–114.

24. Arthur, M.A. (1982) The carbon cycle – controls on atmospheric CO_2 and climate in the geologic past. In: *Climate in Earth History: Studies in Geophysics* (eds W.H. Berger and J.C. Crowell). Report of Panel on Pre-Pleistocene Climates, for the Geophysics Study Committee of the Geophysics Research Board of the Commission on Physical Sciences, Mathematics, and Applications of the US National Research Council, based on a meeting in Toronto in 1980. National Academy Press, Washington, DC, pp. 55–67. Available from: http://www.nap.edu/catalog/11798.html (last accessed 29 January 2015).

25. Walker, J.C.G. (1986) Global geochemical cycles of carbon, sulfur and oxygen. *Marine Geology* **70**, 159–174.

26. Müller, R.D., Sdrolias, M., Gaina, C., Steinberger, B. and Heine, C. (2008) Long-term sea-level fluctuations driven by ocean basin dynamics. *Science* **319**, 1357–1362

27. Holland, H.D. (2003) The geological history of seawater. In: *The Oceans and Marine Geochemistry, Vol. 6, Treatise on Geochemistry* (eds H.D. Holland and K.K. Turekian). Elsevier-Pergamon, Oxford, pp. 583–625.

28. Haq, B.U., Hardenbol, J. and Vail, P.R. (1987) Chronology of fluctuating sea levels since the Triassic. *Science* **235**, 1156–1167.

29. Hallam, A. (1984) Pre-Quaternary sea level changes. *Annual Review of Earth and Planetary Sciences* **12**, 205–243.

30. Elderfield, H. (2010) Seawater chemistry and climate. *Science* **327**, 1092–1093.

31. Zalasiewicz, J. and Williams, M. (2012) *The Goldilocks Planet – The Four Billion Year Story of Earth's Climate*. Oxford University Press, Oxford.

32. Coggon, R.M., Teagle, D.A.H., Smith-Duque, C.E., Alt, J.C. and Cooper, M.J. (2010) Reconstructing past seawater Mg/Ca and Sr/Ca from mid-ocean ridge flank calcium carbonate veins. *Science* **327**, 1114–1117.

33. Rowley, D.B. (2008) Extrapolating oceanic age distributions: lessons from the Pacific region. *Journal of Geology* **116**, 587–598.

34. Pälike, H., Lyle, M.W., Nishi, H., Raffi, I., Ridgwell, A., Gamage, K., *et al.* (2012) A Cenozoic record of the equatorial Pacific carbonate compensation depth. *Nature* **488**, 609–615.

35. Schlanger, S.O. and Jenkyns, H.C. (1976) Cretaceous oceanic anoxic events: causes and consequences. *Geologie en Mijnbouw* **55**, 179–184.

36. Arthur, M.A., Schlanger, S.O. and Jenkyns, H.C. (1987) The Cenomanian-Turonian oceanic anoxic event, II, palaeoceanographic controls on organic-matter production and preservation. In: *Marine Petroleum Source Rocks* (eds J. Brooks and A.J. Fleet). Geological Society of London

Special Publication 26. Blackwell Science, London, pp. 401–420.

37. Jenkyns, H.C. (2010) Geochemistry of oceanic anoxic events. *Geochemistry, Geophysics, Geosystems* **11**, Q03004.

38. Trabucho-Alexandre, J., Hay, W.W. and De Boer, P.L. (2012) Phanerozoic environments of black shale deposition and the Wilson Cycle. *Solid Earth* **3**, 29–42.

39. Summerhayes, C.P. (1987) Organic-rich Cretaceous sediments from the North Atlantic. In: *Marine Petroleum Source Rocks* (eds J. Brooks and A.J. Fleet). Geological Society of London Special Publication 26. Blackwell Science, London, pp. 301–316.

40. Brass, G.W., Southam, J.R. and Peterson, W.H. (1982) Warm saline bottom waters in the ancient ocean. *Nature* **296**, 620–623.

41. Friedrich, O., Erbacher, J., Moriya, K., Wilson, P.A. and Kuhnert, H. (2008) Warm saline intermediate waters in Cretaceous tropical Atlantic Ocean. *Nature Geoscience* **1**, 453–457.

42. Topper, R.P.M., Trabucho-Alexandre, J., Tuenter, E. and Meijer, P.T. (2011) A regional ocean circulation model for the mid-Cretaceous North Atlantic basin: implications for black shale formation. *Climate of the Past* **7**, 277–297.

43. Summerhayes, C.P. and Masran, T.C. (1984) Organic facies of Cretaceous sediments from Deep Sea Drilling Project sites 535 and 540, eastern Gulf of Mexico. *Initial Reports of the Deep-Sea Drilling Project 77*. US Government Printing Office, Washington, DC.

44. Trabucho-Alexandre, J., Van Gilst, R.I., Rodriguez-Lopez, J.P. and De Boer, P.L. (2011) The sedimentary expression of oceanic anoxic event 1b in the North Atlantic. *Sedimentology* **58**, 1217–1246.

45. Stow, D.A.V. and Dean, W.E. (1984) Middle Cretaceous black shales at site 530 in the southeastern Angola Basin. *Initial Reports of the Deep-Sea Drilling Project 75*. US Government Printing Office, Washington, DC.

46. Arthur, M.A., Dean, W.E. and Schlanger, S.O. (1985) Variations in the global carbon cycle during the Cretaceous related to climate, volcanism, and changes in atmospheric CO_2. In: *The Carbon Cycle and Atmospheric CO_2: Natural Variations Archean to Present* (eds E.T. Sundquist and W.S. Broecker). Geophysical Monographs 32. American Geophysical Union, Washington, DC, pp. 504–529.

47. Summerhayes, C.P. and Masran, T.S. (1983) Organic facies of Cretaceous and Jurassic sediments from Deep Sea Drilling Project site 534 in the Blake-Bahama Basin, western North Atlantic. *Initial Reports of the Deep-Sea Drilling Project 76*. US Government Printing Office, Washington, DC.

48. Bralower, T.J. and Thierstein, H.R. (1984) Low-productivity and slow deep-water circulation in mid-Cretaceous oceans. *Geology* **12**, 614–618.

49. Keller, C.E., Hochuli, P.A., Weissert, H., Bernasconi, S., Giorgioni, M. and Garcia, T.I. (2011) Volcanically induced climate warming and floral change preceded

the onset of OAE1 (early Cretaceous). *Palaeogeography, Palaeoclimatology, Palaeoecology* **305**, 43–49.

50. Montañez, I.P., Norris, R.D., Algeo, T., Chandler, M.A., Johnson, K.R., Kennedy, M.J., *et al.* (2011) *Understanding Earth's Deep Past: Lessons For Our Climate Future.* National Academies Press, Washington, DC.

51. Foster, G.L. and Rohling, E.J. (2013) Relationship between sea level and climate forcing by CO_2 on geological timescales. *Proceedings of the National Academy of Science* **110** (4), 1209–1214.

52. Van de Wal, R.S.W., De Boer, B., Lourens, L.J., Köhler, P. and Bintanja, R. (2011) Reconstruction of a continuous high-resolution CO_2 record over the past 20 million years. *Climate of the Past* **7**, 1459–1469.

53. http://en.wikipedia.org/wiki/Earth_system_science (December 2013).

54. Brasseur, G.P. and Van der Pluijm, B. (2013) Earth's future: navigating the science of the Anthropocene. *Earth's Future* **1**, 1–2.

55. Lovelock, J.E. and Giffin, C.E. (1969) Planetary atmospheres: compositional and other changes associated with the presence of life. *Advances in the Astronautical Sciences* **25**, 179–193.

56. Lovelock, J.E. (1972) Gaia as seen through the atmosphere. *Atmospheric Environment* **6** (8), 579–580.

57. Lovelock, J.E. and Margulis, L. (1974) Homeostatic tendencies of the Earth's atmosphere. *Origins of Life* **5** (1–2), 93–103.

58. Margulis, L. (1999) *Symbiotic Planet: A New Look at Evolution.* Phoenix, London.

59. Lovelock, J. (1979) *Gaia: A New Look at Life on Earth.* Oxford University Press, Oxford.

60. Nisbett, E.G. and Fowler, C.M.R. (2005) The early history of life. In: *Biogeochemistry, Vol. 8, Treatise on Geochemistry* (eds H.D. Holland and K.K. Turekian). Elsevier-Pergamon, Oxford, pp. 1–39.

61. Lenton, T.M. and Watson, A. (2011) *Revolutions That Made the Earth.* Oxford University Press, Oxford.

62. Leeder, M. (2007) Cybertectonic Earth and Gaia's weak hand: sedimentary geology, sedimentary cycling and the Earth system. *Journal of the Geological Society of London* **164**, 277–296.

63. Rowley, D.B. (2002) Rate of plate creation and destruction: 180 Ma to present. *Bulletin of the Geological Society of America* **114** (8), 927–933.

64. Nield, T. (2011) *Incoming! or, Why We Should Stop Worrying and Learn to Love the Meteorite.* Granta Books, London.

65. http://en.wikipedia.org/wiki/Eugene_Merle_Shoemaker (December 2013).

66. http://en.wikipedia.org/wiki/Impact_crater (December 2013).

67. http://en.wikipedia.org/wiki/Chicxulub_Crater (December 2013).

68. http://en.wikipedia.org/wiki/Cretaceous–Paleogene_boundary (December 2013).

69. Alvarez, L.W., Alvarez, W., Asaro, F. and Michel, H.V. (1980) Extraterrestrial cause for the Cretaceous-Tertiary extinction: experiment and theory. *Science* **208** (4448), 1095–1108.

70. http://en.wikipedia.org/wiki/Luis_Walter_Alvarez (December 2013).

71. Kolbert, E. (2014) *The Sixth Extinction.* Bloomsbury Press, London.

72. Schulte, P., Alegret, L., Arenillas, I., Arz, J.A., Barton, P.J., Bown, P.R., *et al.* (2010) The Chicxulub asteroid impact and mass extinction at the Cretaceous-Paleogene boundary. *Science* **327** (5970), 1214–1218.

73. Renne, P.R., Deino, A.L., Hilgen, F.J., Kuiper, K.F., Mark, D.F., Mitchell, W.S. 3rd,, *et al.* (2013) Time scales of critical events around the Cretaceous-Paleogene boundary. *Science* **339** (6120), 684–687.

74. Urey, H.C. (1973) Cometary collisions and geological periods. *Nature* **242**, 32–33.

75. Courtillot, V., Besse, J., Vandamme, D., Montigny, R., Jaeger, J.-J. and Cappetta, H. (1986) Deccan flood basalts at the Cretaceous/Tertiary boundary? *Earth and Planetary Science Letters* **80**, 361–374.

76. McLaren, D.J. (1970) Time, life, and boundaries. *Journal of Paleontology* **44** (5), 801–813.

77. Hattersley-Smith, G. (2007) Digby Johns McLaren, 1 December 1919–8 December 2004: elected FRS 1979. *Biographical Memoirs* **53**, doi: 10.1098/rsbm.2007.0007. Available from: http://rsbm.royalsocietypublishing.org/content/53/237 (last accessed 29 January 2015).

78. Dietz, R.S. (1962) Sudbury structure as an Astrobleme. *Journal of Geology* **72**, 412–434.

79. McLaren, D.J. and Goodfellow, W.D. (1990) Geological and biological consequences of giant impacts. *Annual Reviews of Earth and Planetary Science* **18**, 123–171.

80. McLaren, D.J. (1983) Bolides and biostratigraphy. *Bulletin of the Geological Society of America* **94** (3), 313–324.

81. Hallam, A. (1998) Mass extinctions in Phanerozoic time. *Geological Society of London Special Publication* **140**, 259–274.

82. Friedman, H. (1985) The science of global change – an overview. In: *Global Change* (eds T.F. Malone and J.G. Roederer). Proceedings of the ICSU Symposium, Ottowa, September 25, 1984. ICSU Press and Cambridge University Press, Cambridge, pp. 20–52.

83. Pollack, J. (1982) Solar, astronomical, and atmospheric effects on climate. In: *Climate in Earth History: Studies in Geophysics* (eds W.H. Berger and J.C. Crowell). Report of Panel on Pre-Pleistocene Climates, for the Geophysics Study Committee of the Geophysics Research Board of the Commission on Physical Sciences, Mathematics, and Applications of the US National Research Council, based on a meeting in Toronto in 1980. National Academy Press, Washington, DC, pp. 68–76. Available from: http://www.nap.edu/catalog/11798.html (last accessed 29 January 2015).

84. Raup, D. and Sepkoski, J.J. (1984) Periodicity of extinctions in the geologic past. *Proceedings of the National Academy of Science* **81** (3), 801–805.

85. Courtillot, V. (1994) Mass extinctions in the last 300 million years: one impact and seven flood basalts? *Israel Journal of Earth Science* **43**, 255–266.

86. Courtillot, V.E. and Renne, P.R. (2003) On the ages of flood basalt events. *Comptes Rendus Geoscience* **335**, 113–140.

87. Courtillot, V.E. and Olsen, P. (2007) Mantle plumes link magnetic superchrons to Phanerozoic mass depletion events. *Earth and Planetary Science Letters* **260**, 495–504.

88. Keller, G., Armstrong, H., Courtillot, V., Harper, D., Joachimski, M., *et al.* (2012) Volcanism, impacts and mass extinctions. *Geoscientist* **22** (10), 10–15.

89. Keller, G., Kerr, A.C. and MacLeod, N. (2013) Exploring the causes of mass extinction events. *EOS* **94** (22), 200.

90. Schulte, P., Alegret, L., Arenillas, I., Arz, J.A., Barton, P.J., Bown, P.R., *et al.* (2010) Cretaceous extinctions: multiple causes; response to letters by Archibald *et al.*, Keller *et al.* and Courtillot and Fluteau. *Science* **328** (5981), 975–976.

91. Retallack, G.J., Sheldon, N.D., Carr, P.F., Fanning, M., Thompson, C.A., Williams, M.L., *et al.* (2011) Multiple early Triassic greenhouse crises impeded recovery from late Permian mass extinction. *Palaeogeography Palaeoclimatology Palaeoecology* **308**, 233–251.

92. Svensen, H., Planke, S., Polozov, A.G., Schmidbauer, N., Corfu, F., Podladchikov, Y.Y. and Jamtveit, B. (2008) Siberian gas venting and the end-Permian environmental crisis. *Earth and Planetary Science Letters* **277** (3–4), 490–500.

93. Black, B.A., Elkins-Tanton, L.T., Rowe, C. and Peate, I.U. (2012) Magnitude and consequences of volatile release from the Siberian Traps. *Earth and Planetary Science Letters* **317–318**, 363–373.

94. Barras, C. (2013) Fracking hell. *New Scientist* **220** (2947), 43–46.

95. Tohver, E., Cawood, P.A., Riccomini, C., Lana, C.D.C. and Trindade, R.I.F. (2013) Shaking a methane fizz: seismicity from the Araguianha impact event and the Permian-Triassic global carbon isotope record. *Palaeogeography Palaeoclimatology Palaeoecology* **387**, 66–75.

96. Becker, L., Poreda, R.J., Hunt, A.G., Bunch, T.E. and Rampino, M. (2001) Impact event at the Permian-Triassic boundary: evidence from extraterrestrial noble gases in fullerines. *Science* **291**, 1530–1533.

97. Koeberl, C., Gilmour, I., Reimold, W.U., Claeys, P. and Ivanov, B. (2002) End-Permian catastrophe by bolide impact: evidence of a gigantic release of sulfur from the mantle: comment and reply. *Geology* **30**, 855–856.

98. Blackburn, T.J., Olsen, P.E., Bowring, S.A., McLean, N.M., Kent, D.V., Puffer, J., *et al.* (2013) Zircon U-Pb geochronology links the end-Triassic extinction with the Central Atlantic Magmatic Province. *Science* **340** (6135), 941–945.

99. McElwain, J.C., Beerling, D.J. and Woodward, F.I. (1999) Fossil plants and global warming at the Triassic-Jurassic boundary. *Science* **285**, 1386–1389.

100. van de Schootbrugge, B., Quan, T.M., Lindström, S., Püttmann, W., Heunisch, C., Pross, J., *et al.* (2009) Floral changes across the Triassic/Jurassic boundary linked to flood basalt volcanism. *Nature Geoscience* **2**, 589–594.

101. Walker, J.C.G., Hays, P.B. and Kasting, J.F. (1981) A negative feedback mechanism for the long term stabilization of earth's surface temperature. *Journal of Geophysical Research* **86**, 9976–9982.

11

Numerical Climate Models and Case Histories

11.1 CO$_2$ and General Circulation Models

Having established that fluctuations in CO$_2$ are directly related to changing temperature through the operation of the slow carbon cycle, it is time to examine the effect of integrating CO$_2$ into a general circulation model (GCM) of the climate system, of the kind introduced in Chapter 6.

Realising that CO$_2$ must have played an important role in the climate story, Eric Barron and Warren Washington, whom we met in Chapter 6, modified their GCM to consider the effect of increased levels of CO$_2$ in the air. At the 1984 Tarpon Springs meeting, they reported that the addition of four times the then present-day value (in 1984) of CO$_2$ increased the globally averaged surface temperature of the Cretaceous by an additional 3.6 °C[1]. Added to the warming of 4.8 °C produced by geographical change, that gave a total global average temperature rise of 8.4 °C, which was close to that estimated from ∂^{18}O data. While this seemed convincing, the tropical temperatures in the model were a bit too high. Barron thought that might be because their GCM was primarily an atmospheric model. Lacking a dynamic ocean component, it missed the role of the oceans in poleward heat transport. This defect would take a while to correct. Meanwhile, working with Steve Schneider, Barron used his GCM for the Cretaceous to show that it was unlikely that temperatures rose above freezing in midcontinent at high latitudes[2], confirming what was known from geological data and fossils.

By 1995, advances in computing power enabled GCMs to incorporate a dynamic ocean component as well as a dynamic atmosphere. This led to the creation of GENE-SIS, which Barron used to assess the effect of ocean heat transport[3]. He ran the model with different combinations of ocean heat transport and CO$_2$ to see what best matched the observed distribution of Cretaceous temperatures. He also reviewed the latest data on Cretaceous temperatures to satisfy himself that the polar regions were free of ice on land, even if they were seasonally below freezing. The model's results confirmed that '*Higher carbon dioxide is required to achieve sufficient polar warmth, and greater oceanic heat flux is required to prevent the tropical oceans from overheating under conditions of higher carbon dioxide*'[3]. The model that best matched observations was that containing four times the 1984 value of CO$_2$ (Figure 11.1). In effect, we have come full circle, back to the picture painted in 1899 by T.C. Chamberlin and described in Chapter 4.

Was all well? Barron knew that past ocean temperatures estimated from ∂^{18}O data might be inaccurate because the ∂^{18}O ratio of seawater in an ice-free ocean might vary with salinity and the ∂^{18}O measurements were made on a wide range of organisms living at different depths and thus representing different water temperatures in the top 100 m of the water column. To adjust for those possibilities, his team compared the outputs of GENESIS with palaeotemperatures adjusted for the effects of salinity and water depth. This improved the match between the data and the model outputs[4]. Temperatures predicted by the model agreed well with isotopic temperatures for low latitudes and with the low part of the range of isotopic temperatures from high latitudes. One problem remained: the modelled temperatures from high latitudes were lower than the extremes of the range of temperatures determined from ∂^{18}O data at those latitudes. While the '*results suggest that the mid-Cretaceous climate may have been as*

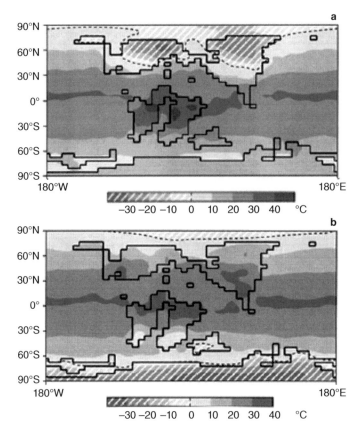

Figure 11.1   *Mid Cretaceous modelled temperatures for 4× CO₂. (a) December, January and February. (b) June, July and August. Warming is evident everywhere, especially at high latitudes. DJF Arctic warming locally exceeds 40°C. High-latitude continental interiors remain below freezing in winter. Tropical ocean surface temperatures increase by 1.5–2.0°C. Tropical continental regions experience greater warming. The polar warming approaches palaeoclimate observations, but Arctic temperatures remain too cold. Mid-latitude temperatures are close to realistic. Tropical temperatures are close to exceeding the range of observed values.*

warm or warmer than that simulated in a mid-Cretaceous 4 × CO₂ experiment', Barron recommended that additional ∂¹⁸O data from high latitudes be collected to resolve the discrepancy[4]. This is an example of how GCM outputs can identify places where new data should be collected.

Could different proxy data tell us if the 4× modern CO₂ level was a reasonable estimate for the Cretaceous climate? Bob Spicer and Judy Parrish argued that high CO₂ might have helped to maintain the polar warmth necessary to sustain vegetation in the Arctic during the late Cretaceous and Palaeocene[5]. They thought that it was *'unlikely … that levels of 4 × present CO₂ were exceeded*

because above this level many plants with conventional C₃ photosynthetic pathways close their stomata and productivity is thereby reduced'[5].

Others followed Barron's lead. For instance, in 1992, geologists from the Chevron oil company applied a GCM to the late Jurassic to see what difference might arise between an atmosphere containing the preindustrial level of 280 ppm CO₂ and one containing 1120 ppm, a level suggested for the Jurassic by Berner's models and by compilations of CO₂ proxy data[6,7]. Their GCM comprised a dynamic atmosphere coupled to a static ocean, so missed the effects of ocean heat transport.

The climate simulation with 1120 ppm best matched the distribution of palaeoclimate data. Coals occurred where rainfall was high, evaporites where rainfall was low. Warming was greatest over the high-latitude oceans and least over the tropics. Sea ice was restricted to offshore seas at high latitudes. Strong summer monsoon rains developed over South East Asia. The Trade Winds brought heavy winter rainfall to eastern Gondwana and heavy summer rains to the tropical margins of Tethys.

A comparable approach was taken in 1992 in an investigation of the climate of the Kimmeridgian (155–150 Ma), an oil-rich part of the late Jurassic period, by modeller Paul Valdes, whom we met in Chapter 6, and geologist Bruce Sellwood (1946–2007), professor of applied sedimentology at Reading University[8]. Their GCM coupled the atmosphere to a static ocean with temperatures ranging from 27 °C at the Equator to 0 °C at the poles. Like Chevron, they varied the CO_2 concentration in the air – in their case, from 350 to 1080 ppm – to see what difference this made. The most accurate reflection of the palaeoclimate data came from the simulation with high CO_2. The average global temperature was 20 °C. Their results replicated those of Chevron, but they also found the southwestern United States to be arid, southern Europe to be seasonally arid and winter temperatures over Siberia and southeastern Gondwana to have been below 0 °C, in agreement with the palaeoclimate data.

Models go on developing, and data go on accumulating. Fifteen years later, Sellwood and Valdes applied an updated GCM to palaeoreconstructions for the late Triassic, late Jurassic and early and late Cretaceous, validating the models with palaeoclimate data[9]. '*Compared to the present*', they observed, '*the Mesozoic Earth was an alien world with a greenhouse climate … dense forests grew close to both poles but experienced months-long daylight in warm summers, and months-long darkness in cold, sometimes snowy, winters. Ocean depths were warm (8 °C or more to the ocean floor) and reefs, with corals, grew 10° of latitude further north and south than at present time. The whole Earth was warmer than now by several degrees centigrade, generating high atmospheric humidity and a greatly enhanced hydrological cycle*'[9]. Much of the rainfall was focused over the oceans, leaving major expanses of desert on the continents. Polar ice sheets were absent, but local mountain glaciers could not be ruled out on Antarctica in the Cretaceous. There may have been sea ice in the Arctic, especially in the Cretaceous. During the Triassic and Jurassic, atmospheric CO_2 was at least four

times present values. Some discrepancies between data and models suggested problems with either the models or the proxy data; for instance, model simulations of continental interiors were too cold by 15 °C or more, and real ocean temperatures could have been warmer. Otherwise, the model performed well.

At this point, we might recall Hasegawa's finding, mentioned in Chapter 6, that the northern margin of the Hadley Cell advanced poleward along with global warming during the early and late Cretaceous to latitudes between about 32° and 40° N, but that during the exceptionally hot 'super greenhouse' of the middle Cretaceous the northern edge of the Hadley Cell shrank equatorward to latitudes of between 22° and 30° N[10], contrary to the expectation of Bill Hay[11]. Hasegawa and his team thought that these results '*suggest the existence of a threshold in atmospheric CO_2 level and/or global temperature, beyond which the Hadley circulation shrinks drastically*'[10]. Levels of CO_2 may have reached about 1500 ppm during the extremely warm middle Cretaceous, compared with 500–1000 ppm during the late Cretaceous.

Hasegawa supported his interpretation of events by citing recent observational and modelling studies suggesting '*that present-day Hadley circulation is expanding poleward in response to the increasing atmospheric CO_2 level and consequent global warming*'[10]. He also noted that '*Such a relationship between the width of the Hadley circulation and global temperature and/or atmospheric CO_2 levels has also been reported from paleoclimate records of glacial-interglacial transitions*'[10]. These changes in width of the Hadley circulation are related to changes in the Equator-to-pole temperature gradient, with a wider Hadley Cell during interglacial than glacial periods. Deep-ocean circulation was sluggish in the mid Cretaceous super-greenhouse period and more vigorous before and after it, indicating an atmosphere–ocean link. The reason may have been enhanced vertical mixing at high latitudes, due to enhanced storminess in the mid Cretaceous, suppressing the formation of deep water there[10], along lines also suggested by Bill Hay[11]. The causal links are not yet clear.

Models calculate global temperatures by using as boundary conditions the desired CO_2 level (say, two times present) and something called the 'radiative forcing value', calculated from experimental work on how CO_2 absorbs and re-radiates infrared radiation. For example, the radiative forcing value was 1.46 W/m^2 for the increase from 1750 to 1998, and would be 3.7 W/m^2 for a doubling of CO_2[12, 13]. The rise in temperature for a given rise in CO_2

is known as the 'climate sensitivity'. For a comprehensive view on what climate sensitivity was in the geological past, we turn to the PALEOSENS Project, comprising a large group of climatologists and palaeoclimatologists who reported in 2012[14]. Over the past 65 Ma, the climate sensitivity calculated from past temperatures and CO_2 levels implied a warming of 2.2–4.8 °C for a doubling of atmospheric CO_2[14]. That agrees reasonably well with calculations for the modern climate by the Intergovernmental Panel on Climate Change (IPCC)[15].

Calculations of climate sensitivity usually take into account only relatively fast changes in climate, including factors like snow melt, the exchange of CO_2 between the air and the ocean and the behaviour of clouds and water vapour. But, from the geological perspective, we also need to take into account factors that change slowly, such as ice sheet decay and the behaviour of the longer-term carbon cycle, including the rates of dissolution of deep-sea carbonates. This 'Earth system sensitivity' will differ from the climate sensitivity. It should govern our estimates of long-term changes in sea level and of the amount of CO_2 in the air that would be required to stabilise climate change. Studies of the Last Glacial Maximum suggest the Earth system sensitivity then ranged from 1.4 to more than 6.5 °C[14, 16].

Jim Hansen of NASA, whom we came across in Chapter 8, argued that the global temperature change of 5 °C between the peak of the last Ice Age and the Holocene implied an equilibrium climate sensitivity of 0.75 °C for each watt of forcing[17]. As a doubling of CO_2 amounts to a forcing of about 4 W/m^2, the equilibrium sensitivity of 0.75 W/m^2 implies that temperature would rise by 3 °C for such a doubling. Hansen pointed out that if the Earth were a black body without climate feedbacks, the equilibrium response to a forcing of 4 W/m^2 would be a rise in temperature of 1.2 °C. This value is more than doubled by feedbacks from the increase in water vapour caused by warming, from the decrease in albedo caused by the loss of snowfields and sea ice and from other, lesser effects. Hansen agreed that while the sensitivity of climate to fast feedbacks is about 3 °C for a doubling of CO_2, this would be amplified by the effect of slow feedbacks, including changes in ice sheet size and the long duration of CO_2 in the atmosphere[18].

This topic continues to be one of active research. For instance, one 2012 study of recent global warming suggested that the climate sensitivity to a doubling of CO_2 should be much closer to 2 °C[19]. Another, by Dana Royer, Mark Pagani and David Beerling, used Cretaceous and Eocene samples to suggest that climate sensitivity then was higher than the 3 °C suggested by several numerical climate models, and may have exceeded 6 °C in glacial times[20]. Royer and his colleagues considered that '*Climate models probably do not capture the full suite of positive climate feedbacks that amplify global temperatures during some globally warm periods, as well as other characteristic features of warm climates such as low meridional temperature gradients*'[20]; these models erred on the low side as a result. This may reflect '*the long-standing failure of GCMs in underestimating high-latitude temperatures and overestimating latitudinal temperature gradients*'[20], possibly by missing biological feedbacks. Such feedbacks, they told us, would include '*emissions of reactive biogenic gases from terrestrial ecosystems … [with] substantial impacts on atmospheric chemistry … in particular … the tropospheric concentrations of GHGs [greenhouse gases] such as ozone, methane and nitrous oxide … Climate feedbacks of these trace GHGs are missing from most pre-Pleistocene climate modeling investigations*'[20]. We will revisit the question of climate sensitivity for particular geological periods when we look at some case histories later in this chapter.

Montañez and Norris agreed in 2011 that, because current GCMs cannot reproduce the exceptional warmth of high latitudes during all past warm periods without invoking unreasonably high levels of CO_2, they must be unable '*to fully capture the processes and feedbacks governing heat transport and retention or the processes that might generate heat in the polar regions under elevated atmospheric greenhouse gases*'[21]. They are close, but not close enough. There is more work to do.

One of the gases missing from models is the dimethyl-sulphide produced by phytoplankton. This creates nuclei around which water vapour condenses, encouraging the production of low-level clouds that reflect solar energy. Did heat stress from the warm Cretaceous ocean reduce the ability of phytoplankton to produce this gas, thus reducing cloud cover and warming the atmosphere – especially at the poles? Our understanding of these 'known unknowns' is poor, reducing confidence in climate sensitivity values derived from GCMs applied to palaeoclimate problems[20]. We need more robust GCMs to address this issue. Other trace greenhouse gases may have been equally important in the past, including methane (CH_4), nitrous oxide (N_2O) and ozone (O_3)[22]. It is difficult to apply this realisation to the geological past, because we lack suitable geochemical or biological proxies for such gases back beyond the age of ice cores.

Recently, David Beerling of the University of Sheffield used an Earth systems model to evaluate how terrestrial, climatic and atmospheric processes might have interacted in the past to regulate atmospheric chemistry[23]. Beerlings *et al.*'s model includes a climate model coupled to a tropospheric chemistry–transport model and a model simulating the biogeography of terrestrial vegetation. They applied it to determine the likely concentration of greenhouse gases in the air and their influence on the planetary energy budget in the early Eocene (55 Ma ago) and late Cretaceous (90 Ma ago). The CO_2 content of the air was fixed at four times present levels, with the addition of a model with two times present levels for the early Eocene. Preindustrial levels of the trace gases were derived from studies of ice cores.

Methane concentrations calculated for these two periods were high at 2580–3614 ppb (parts per billion): double the current concentration of around 1800 ppb. This reflects the warm moist climates of those times and the increased abundance of the wetlands from which much methane comes. Nitrous oxides would also have been more abundant at these times, due to the greater extent and decomposition of plant remains. Surface ozone concentrations were calculated to be 60–70% higher than today. These increases provided additional radiative forcing at times of rapid warming and rising CO_2. Compared with the global warming produced by CO_2 alone, the extra forcing produced additional warming of 1.4 and 2.7 °C for the 2× and 4× CO_2 Eocene simulations, and of 2.2 °C for the Cretaceous[23].

How close to reality were the results? The model reproduced the mean annual land temperatures suggested by leaf fossils better than it did mean ocean temperatures. This might reflect a cool bias in proxy leaf data or a warm bias in proxy ocean data. However, the model is also missing some amplifying feedback process. It failed to reproduce accurately the Equator-to-pole temperature gradient in the ocean, and hence ocean heat transport, perhaps partly because of the relatively poor resolution of its ocean component. Nevertheless, this exercise was an interesting test of the way in which the different components of the Earth system can be represented by mathematical calculations so as to simulate the complex operation of the climate system.

Having established that modern GCMs can help us to understand how fluctuations in CO_2 and related gases may have affected past climates, we now look at the details of the Cenozoic, because its climate history is the most relevant to understanding the period in which we now live. Readers keen to explore past climate modelling further may find a 2007 review by Williams and others helpful[24].

11.2 CO_2 and Climate in the Early Cenozoic

By 2008, having absorbed all of the new information about the role of CO_2 in the climate system and the effects of mountain building on climate, Dennis Kent of Rutgers University and his colleague Giovanni Muttoni of the University of Milan felt able to explain both the increase of CO_2 and temperature from the Late Cretaceous to the Middle Eocene climatic optimum at around 50 Ma ago and the subsequent decline of both CO_2 and temperature towards the present day[25]. As a starting assumption, they accepted Dave Rowley's view that the rate of production of new seafloor (hence CO_2) was constant over the past 180 Ma[26]. Were there other ways of changing CO_2 emissions through time? Focusing on Tethys, the ocean between India and Asia, they assumed that shallow carbonate reefs would have festooned the Indian and Asian margins and that pelagic carbonate oozes would have mantled the deep-sea floor between them. As India drifted north, the floor of Tethys and its carbonate load were subducted beneath Asia, where they melted and gave off volcanic gases rich in CO_2. Eventually, the carbonate reefs were subducted too, with similar results. This output of CO_2 was augmented by CO_2 and other gases erupted from India's Deccan Traps near the Cretaceous boundary 65 Ma ago. India collided with Asia about 50 Ma ago, switching off this 'subduction factory'. By then, the Deccan Traps had drifted into the tropical rain belt, where weathering of the basalts soaked up CO_2 from the atmosphere. By the Oligocene, some 30 Ma ago, the Himalayas had started to rise as the collision of India with Tibet continued. Mountain building in a humid environment led to chemical weathering of silicates, soaking up more CO_2.

Kent and Muttoni's concept provides a convincing mechanism for the increase in CO_2 from the end of the Cretaceous to a peak some 50 Ma ago, and its subsequent decline, through increased weathering. This picture is consistent with the change in the depth of the carbonate compensation depth (CCD) with time[27]. The changing CO_2 explains the associated rise and decline in global temperature, for which there is no other plausible explanation. This interpretation is consistent with that of Froehlich and Misra, who argued that much of the rise in CO_2 in the Paleogene and its fall in the Neogene was caused by increased mountain building and associated physical and chemical weathering from the Oligocene onwards[28].

In 2013, Kent and Muttoni extended their conceptual model of CO_2 and climate back by another 30 Ma into the early Cretaceous, 120 Ma ago[29]. Once again, they accepted Rowley's assumption that seafloor spreading rates had been constant over the past 120 Ma[30]. They also assumed that rates of emission of CO_2 from spreading centres, mantle plumes and arc volcanoes were like those of the present day: about 0.26 Gt CO_2/year (see Chapter 8). Reconstructing past continental positions, they documented the times and places of the collisions of Arabia with Iran, India with Tibet and New Guinea with Indonesia, and of the southeastward migration of fragments of Asia to form South East Asia. This enabled them to see where these components had been in relation to the equatorial humid zone through time. They assumed that subduction and volcanic activity associated with India's northward march began 120 Ma ago, supplying CO_2 to the atmosphere from that time on. The rate of emission of CO_2 would have depended on the amount of carbonate-rich sediment available. As pelagic carbonates are most abundant in the tropics, the rate of CO_2 emission would have increased as plate motion brought carbonate sediment north from tropical regions of Tethys into the Asian subduction zone. They calculated that the subducted load of carbonate would have risen from 60 megatonnes (Mt)/yr at 120 Ma ago to 220 Mt/yr at 80 Ma ago and then dropped to virtually zero at collision time (50 Ma ago). For the collision of Arabia with Iran, they calculated that the subducted load of carbonate would have varied between 23 Mt/yr at 70 Ma ago and 35 Mt/yr at 20 Ma ago.

Assuming a relatively low rate of recycling of CO_2 from carbonate sediments into CO_2 emissions, Kent and Muttoni thought that their subduction mechanism would have made a modest contribution of CO_2 to the warm atmospheres of Cretaceous and Eocene times, contrary to the conclusion of their 2008 paper. Might emissions from Large Igneous Provinces have helped to increase global CO_2 levels? They thought not, since both the geological evidence and numerical models suggested that excess CO_2 supplied from such relatively short-term sources would have been adsorbed within 1 million years by negative-feedback processes[29].

So, if CO_2 emissions from seafloor spreading were more or less constant through time, and if the subduction factory added only relatively small amounts of CO_2, might variations in carbon sinks have played the key role

in controlling the concentration of atmospheric CO_2? Investigating that question, Kent and Muttoni found that the weathering of continental basalts, which today make up less than 5% of land area, accounts for between one-third and one-half of the consumption of CO_2 by the weathering of silicate minerals (Figure 11.2)[29]. Knowing that chemical weathering is strongest under warm and humid conditions typical of the tropical regions between 5° S and 5° N, they plotted the area of continental crust passing through that region between 120 Ma ago and the present, finding that it averaged about 12 million km^2. Within that area, the most weatherable silicate rocks – the basalts – were particularly abundant between 50 and 40 Ma ago, with a subsequent rise from 30 Ma ago to the present. They calculated that the rate of consumption of CO_2 in the tropics began rising 65 Ma ago to a peak 45 Ma ago, dropped back to former levels about 35 Ma ago, then rose rather steadily to the present (Figure 11.3). Significant consumption of CO_2 took place in South East Asia and New Guinea, as well as over India's Deccan Trap basalts and the Ethiopian Trap basalts. Silicate weathering under relatively tropical conditions in the rising Himalayas, as well as in the 4000 m-high mountains of Borneo and the nearly 5000 m-high mountains of New Guinea, might also have contributed to further CO_2 drawdown, they suggested.

What is to stop Kent and Muttoni's proposed tropical chemical weathering from having a runaway effect on atmospheric CO_2? Their common-sense answer was that under tropical conditions, chemical weathering leads to the development of thick soils deficient in cations, which are likely to retard further weathering of the underlying bedrock. Nevertheless, continued or periodic uplift, like that which has characterised South East Asia and New Guinea, would cause protective soil coverings to be shed from time to time, allowing CO_2 to continue to be taken up. Rains from prolonged La Niña conditions would further exacerbate chemical weathering in those areas. Continued weathering of basaltic terrains in equatorial regions might well draw down sufficient atmospheric CO_2 over time to lead to glacial conditions[29].

As I pointed out in Chapters 5 and 10, geochemical and sea level evidence strongly suggests that Rowley was wrong[26,30] and seafloor spreading rates did decline, as Müller suggested[31], from 80 to 50 Ma ago, after which they changed little. So, CO_2 supply would have been high but declining over that 30 Ma period, providing an additional source of CO_2 through the late Cretaceous and Palaeocene to add to Kent and Muttoni's calculations[25,29].

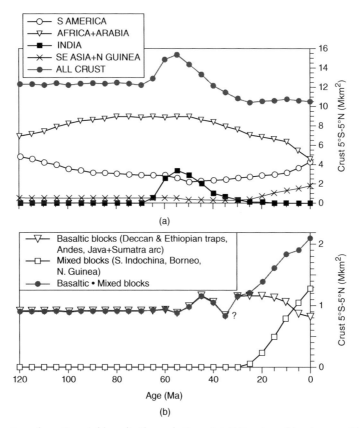

Figure 11.2 *Weathering of continental basalts through time. (a) Estimates of land area within equatorial humid belt (5° S–5° N) as a function of time since 120 Ma ago. (b) Estimates of most weatherable land areas of volcanic-arc provinces (Java, Sumatra, Andes), large basaltic provinces (Deccan Traps, Ethiopian Traps) and mixed igneous–metamorphic provinces (South Indochina, Borneo, New Guinea) in the equatorial humid belt (5° S–5° N) as a function of time from 120 Ma ago to present.*

Should we not be seeing significant emissions of CO_2 today, driven by the subduction of carbonate-rich sediments? No, because – according to geochemist John Edmond – plate motions today give rise to relatively little subduction of carbonate-rich oceanic sediments, most of which are found in the Atlantic, where subduction is uncommon[32]. It is partly this lack that has driven CO_2 levels slowly down through the Cenozoic to as low as 170–180 ppm during Pleistocene glacial periods.

Might there be other reasons for the decline in CO_2 from the mid Cretaceous to the Oligocene? Bill Hay provides two plausible explanations[11]. First, oceanic productivity fell at the Cretaceous–Tertiary boundary, when most calcareous plankton were wiped out by the asteroid impact. While this allowed the concentration of CO_2 in the air to increase in the Palaeocene, by the Eocene, *'evolution had done its work and there were new species of calcareous plankton back at work, making deep-sea carbonate ooze and lowering atmospheric CO_2 again'*[11]. Second, the warm seas of the Cretaceous and early Cenozoic were dominated by haline circulation, with warm, salty waters from tropical margins descending to abyssal depths, as in today's Mediterranean. But global cooling in the late Eocene changed ocean circulation to thermo-haline, with the deep ocean becoming filled with

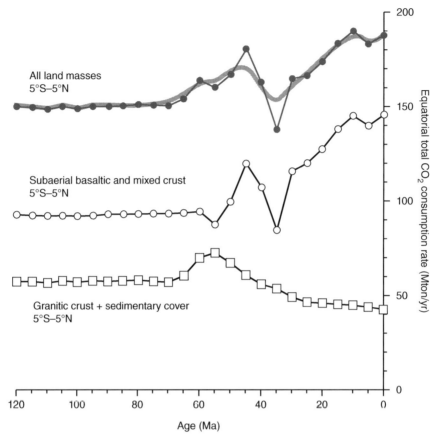

Figure 11.3 *Extraction of CO$_2$ through time by chemical weathering. Total CO$_2$ consumption rates from silicate weathering since 120 Ma ago of land areas in the equatorial humid belt (5° S–5° N), obtained by multiplying a nominal CO$_2$ consumption rate of 100 tonnes CO$_2$/yr/km^2 for basaltic provinces, 50 tonnes CO$_2$/yr/km^2 for mixed basaltic–metamorphic provinces and 5 tonnes CO$_2$/yr/km^2 for the remaining continental land areas, by the corresponding cumulative distribution curves in Figure 11.2. These total consumption rates should be divided by two for net CO$_2$ consumption, because half of the CO$_2$ consumed by silicate weathering is returned to the atmosphere–ocean during carbonate precipitation.*

cold bottom water, sinking at the poles. Cold polar water dissolves much more CO$_2$ than warm water, so this change accentuated the decline in the concentration of CO$_2$ in the air by drawing it down into the deep ocean[11].

Another angle comes from Eleanor John of the University of Cardiff, who reminded us in 2013 that the warmth of Eocene surface waters would have affected the operation of the biological pump and the cycling of carbon and related nutrients[33]. Using $\partial^{13}C$ values in planktonic foraminifera from Tanzania and Mexico, John and colleagues found that the vertical carbon isotope profile of the Eocene tropical ocean was steeper and larger than it is today. This suggests that organic matter was being decomposed and recycled at much shallower depths than today. The warm water raised metabolic rates, sped the recycling of organic matter and reduced the rate of burial of organic matter, thus minimising the loss of CO$_2$ to the deep ocean and keeping Eocene atmospheric CO$_2$ levels high. Theirs is an important conclusion, given that some 90% of Phanerozoic climate was in the warm greenhouse state. Others had also concluded that oceanic metabolism played a key role in controlling CO_2 levels in Phanerozoic climates[34, 35].

It seems highly likely that the changes in atmospheric CO_2 with time through the early to mid Cenozoic may reflect changes in CO_2 supply and demand induced by plate tectonics, as suggested by Kent and Muttoni, modified by: a decline in seafloor spreading (and CO_2 emissions) between 80 and 50 Ma ago, as suggested by Müller; changes in the population of oceanic plankton, as suggested by Hay; the changing metabolism of marine plankton in moving from warm to cool conditions, as implied by John and others; and the drawing down of CO_2 as deep-water circulation cooled, as suggested by Hay.

11.3   The First Great Ice Sheet

As we saw in Chapter 7, Jim Kennett attributed the development of the first major ice sheet on Antarctica at the Eocene–Oligocene boundary 34 Ma ago to the breaking away of Australia and South America from Antarctica, leading to the formation of the Antarctic Circumpolar Current, which isolated the continent from the warm air and warm water of the tropics. This is the 'isolation hypothesis'.

By 1994, James C. Zachos (Box 11.1) and colleagues had documented from oxygen isotopes a warming of the Southern Ocean into the early Eocene, followed by cooling into the early Oligocene[36].

Knowing that Mike Arthur and his colleagues had used carbon isotopes in organic matter from phytoplankton in 1991 to suggest that CO_2 had declined from six to three times present values between the middle Eocene

and early Oligocene[38], Zachos speculated that the cooling was probably related to the fall in CO_2[36]. But there was a problem: the early Eocene tropical temperatures were like those of today, which would seem unlikely in a high-CO_2 world. To explain this discrepancy, Zachos hypothesised that some mitigating process must have operated to take heat away from the tropics at that time, citing increased poleward transport of heat by air and ocean currents. Zachos's suggestion is consistent with the results of the modelling of Eocene climate in 2000 by Karen Bice and colleagues[39].

By 2001, as we saw in Figure 7.6, Zachos and his team had established a global pattern for bottom water temperatures[40], confirming a rise from about 8 °C at 70 Ma ago to 12 °C at about 50 Ma ago, then a decline back to about 8 °C by 42 Ma ago, to about 5 °C at the end of the Eocene and to about 0.5 °C at the Eocene–Oligocene boundary at 34 Ma ago. CO_2 showed a similar pattern, rising from a low of about 400 ppm near 70 Ma ago to a peak at about 1100 ppm between 50 and 45 Ma ago, before falling to near 400 ppm close to the Eocene–Oligocene boundary, at 28 Ma ago[41]. The relationship was close enough to suggest a causal connection.

Growing understanding of the role of CO_2 in the climate system eventually led to a reappraisal of the Antarctic 'isolation hypothesis' in 2003 by Rob DeConto of the geosciences department of the University of Massachusetts at Amherst and Dave Pollard of the Earth and Environmental Systems Institute of Penn State University[42]. DeConto and Pollard simulated '*the glacial inception and early growth of the East Antarctic Ice Sheet using a general circulation model with coupled components for atmosphere, ocean, ice sheet and sediment ... which incorporates palaeogeography, greenhouse gas, changing orbital parameters, and varying ocean heat transport*'[42]. In their model, '*declining Cenozoic CO_2 first leads to the formation of small, highly dynamic ice caps on high Antarctic plateaux ... [then] cooling due to declining pCO_2 would have gradually lowered annual snowline elevations until they intersected extensive regions of high Antarctic topography. Once some threshold was reached, feedbacks related to snow/ice-albedo and ice-sheet height/mass-balance could have initiated rapid ice-sheet growth during orbital periods favourable for the accumulation of glacial ice ... [with the ice caps] eventually coalescing into a continental-scale East Antarctic Ice Sheet*'[42]. That threshold is what we might now see as a

---

Box 11.1   James C. Zachos.

After receiving his PhD from the University of Rhode Island in 1988, Jim Zachos eventually joined the University of California Santa Cruz, where he is currently a professor in the department of Earth and planetary sciences. He specialises in studies of the biological, chemical and climatic evolution of late Cretaceous and Cenozoic oceans, using isotopic analyses to reconstruct past changes in ocean temperature and circulation, continental ice volume, productivity and carbon cycling[37].

'tipping point' – one that pushed the world into a new climate state (Figure 11.4).

According to their simulation, the transition from small ice caps on the Mühlig-Hofmann Mountains and Gamburtsev Mountains to a full ice sheet covering East Antarctica would have taken around 1 million years and resulted in a fall in global sea level of 40–50 m. They also simulated the effect of an increase in the strength of the Antarctic Circumpolar Current, but found it unlikely to have caused as rapid a change as that resulting from the decline in atmospheric CO_2, concluding that *'the*

opening of Drake Passage can only be a potential trigger for glacial inception when atmospheric CO_2 is within a relatively narrow range ... reinforcing the importance of pCO_2 as a fundamental boundary condition for Cenozoic climate change'[42]. Using a more advanced model in 2005, they found that an increase of more than eight times the preindustrial atmospheric level of CO_2, equivalent to a local temperature increase of 15–20 °C, would be needed to make the East Antarctic Ice Sheet retreat significantly – something that is not likely to happen within the next 100 years[43].

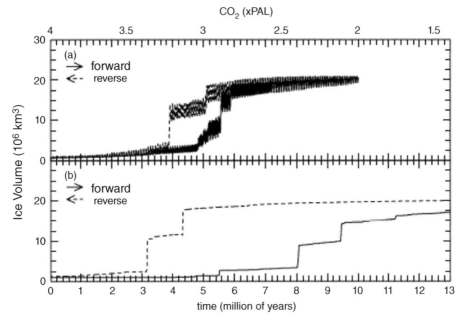

Figure 11.4   *Hysteresis in simulated development of the Antarctic ice sheet. (a) Solid curve shows ice volume from a 10 million-year simulation representing the Eocene–Oligocene transition, with imposed orbital forcing and a long-term linear decline in atmospheric CO_2 from four to two times present levels over the duration of the run. Note the relatively sudden nonlinear transition from very small ice amounts to a near continental expanse as CO_2 drops slightly below three times present levels. This stepped profile resembles the observed Eocene–Oligocene transition in $\partial^{18}O$ records. The dashed (upper) curve in (a) shows a reversed run, starting with a full continental ice sheet under a warming climate, with time running from right to left, and CO_2 increasing from two to four present values. As expected, the change displays hysteresis, with the main transition delayed compared to the forward simulation, because the pre-existing ice sheet topography is much higher and steeper than the baseline bedrock topography, so that the snowline has to be raised substantially before the overall ice sheet budget becomes negative. (b) The same pair of forward and reversed runs, but with no orbital cycles. The slow CO_2 trend is the only external forcing. Without orbital variability, the main transition and the substeps are delayed considerably, and the hysteresis between forward and reversed runs increases several-fold. Evidently, orbital cycles provide essential extra forcing on the gradual CO_2 trend, so that particular thresholds are reached earlier than in the absence of orbital forcing.*

The greenhouse–icehouse transition from a relatively deglaciated state to one characterised by a small Antarctic ice sheet is thought to have occurred over about 200–300 Ka. It seems to have taken place in a series of steps, each with an increase in ice volume and a lowering of sea level[42].

Independent support for the ideas of DeConto and Pollard came from two different sources. First, when CO_2 is abundant in the atmosphere, it is also abundant in the ocean, which thus becomes slightly more acidic. As a result, the CCD deepens, as we saw in Chapter 9. Helen Coxall of the Southampton Oceanography Centre pointed out, with colleagues, that the CCD deepened by 1 km in the latest Eocene, indicating a fall in atmospheric CO_2[44]. This deepening was more rapid than previously thought, however, taking palce in two 40 Ka-long jumps, each synchronous with the stepwise onset of the growth of the Antarctic ice sheet. Coxall thought that the glaciation was initiated, following climatic preconditioning by the fall in CO_2, by a period of cool summers favoured by orbital insolation. She considered that the changes in $\partial^{18}O$ composition across the Eocene–Oligocene boundary were too large to be explained by the growth of the ice sheet: they must also reflect contemporaneous global cooling tied to a decline in both CO_2 and insolation. Second, estimates of the levels of CO_2 across the boundary by Mark Pagani and others, based on $\partial^{13}C$ analyses of alkenones from deep-marine sediments, suggested a decline from 1000 to less than 400 ppm CO_2 between 40 and 24 Ma ago[45].

We could, of course, argue that while CO_2 clearly provides an important control on temperature through time, the development of glaciation also relies on the combination of lowered CO_2 – hence cooling – with movements of the continents that place landmasses at the Pole at the right time, echoing Lyell. But as we saw in Chapter 7, land at the poles is not enough by itself, there having been no extensive glaciation when Antarctica lay at the South Pole between 92 and 34 Ma ago. Even so, as we also learned in Chapter 7, there were probably ice caps and small glaciers on East Antarctica at that time, at least during the Eocene, providing nuclei for later ice sheets.

Did ice form rapidly on Antarctica, as Pollard and DeConto suggest? Reviewing the results of shallow offshore drilling by the SHALDRIL project east of the Antarctic Peninsula, John Anderson and Julia Wellner suggested in 2011 that the onset of glacial conditions at the Eocene–Oligocene boundary was not as rapid as they implied[46]. Anderson considered that '*This observation is more consistent with the Kennett hypothesis, which pinned glaciation on the opening of the Drake Passage … it is not one single thing, like carbon dioxide, that is driving ice sheet evolution*'[47]. Has he gone out on a shaky limb? Observations on glacial onset based on SHALDRIL are compromised by the short fragmentary nature of the cores: ~10 m from 38–36 Ma ago, and 10 m from 32–24 Ma ago. Support for the rapidity of the climate change across the Eocene–Oligocene boundary comes from Ocean Drilling Program (ODP) Leg 199 in the tropical Pacific[44]. There, as we have already seen, Helen Coxall showed that the CCD deepened abruptly across the Eocene–Oligocene boundary, consistent with rapid CO_2 fall and global cooling.

A team led by Miriam Katz of Rensselaer Polytechnic in New York later confirmed this picture[48]. Integrating $\partial^{18}O$ and Mg/Ca values from benthic foraminifera with stratigraphic data from continental margins, they constructed a record of temperature, ice volume and sea level changes to show that the transition across the Eocene–Oligocene boundary happened in three steps, with the influence of ice volume continually increasing. The Antarctic ice sheets were 25% larger than at present, while sea level fell by 67 m. They attributed the shift from greenhouse to icehouse climate as a response to a combination of factors, including the opening of seaways around the continent, the global fall in CO_2 and long, cool summers driven by Earth's orbital parameters. Large fluctuations in ice volume continued through the Oligocene[48]. Paul Pearson of Cardiff University and colleagues provided the matching CO_2 pattern, which showed a slight dip across the boundary between 34.0 and 33.5 Ma ago, associated with the fall in $\partial^{18}O$ indicative of a growth in ice volume[49]. The dip in CO_2 was followed by a brief recovery in CO_2 between 33.5 and 33.2 Ma ago, after which CO_2 continued to fall. Evidently, once the large ice sheet had formed, it was resistant to climate forcing by the brief rise in CO_2, although some reduction in ice volume was likely, due to the associated rise in temperature[49].

Comparison of the global $\partial^{18}O$ curve representing bottom water temperatures and ice volume[40] with the CO_2 curve[41] confirms that the link between temperature, ice volume and CO_2 across the Eocene–Oligocene boundary is not precise, with some high CO_2 values persisting to about 32 Ma. A more recent analysis of $\partial^{18}O$ (representing both ice volume and bottom water temperatures) and Mg/Ca (representing temperatures) through time for the past 108 Ma by Cramer and colleagues (Figure 7.7) confirms the robustness of these measures as climate proxies, especially for the past 50 Ma[50].

Cramer's reconstruction indicates '*differences between deep ocean cooling and continental ice growth in the late Cenozoic: cooling occurred gradually in the middle-late Eocene and late Miocene-Pliocene while ice growth occurred rapidly in the earliest Oligocene, middle Miocene, and Plio-Pleistocene. These differences are consistent with climate models that imply that temperatures, set by the CO_2 "steady state", should change only gradually on timescales >2 Myr, but growth of continental ice sheets may be rapid in response to climate thresholds due to feedbacks that are not yet fully understood*'[50]. This tells us that although CO_2 and plate tectonics (opening or closing ocean gateways) set the broad climate scene behind the global cooling from the mid Cretaceous to the present, from the mid Cenozoic onwards climate changes at finer scales were imposed by the vagaries of ice growth and decay. While initial ice growth was connected to the falling temperatures driven by falling CO_2, ice growth and decay at the >2 Ma scale are only crudely connected to the fall in CO_2 from the Eocene–Oligocene boundary onwards, presumably because they also respond to changing insolation at the prevailing lowish levels of CO_2[50].

Katz and Coxall both agree that the effects of orbital variations were superimposed on the effects on climate of plate tectonics and CO_2 during the Cenozoic. In the late 1990s, studies of the variability of $\partial^{18}O$ in deep-sea cores showed that between 34 and 15 Ma ago orbital cycles caused fluctuations in global temperatures and the volumes of ice at high latitudes. Off Cape Roberts in the western Ross Sea in 2001, Tim Naish and others found numerous cyclic variations linking the extent of the East Antarctic Ice Sheet to orbital cycles during the Oligocene–Miocene transition (24.4–23.7 Ma ago)[51]. By 2007, that picture was extended to fluctuations between 34 and 17 Ma ago, with the 40 Ka obliquity cycle and 100 Ka eccentricity cycle being prominent[52].

Heiko Pälike and his colleagues from the United Kingdom's National Oceanography Centre tell us that orbital cycles formed 'The Heartbeat of the Oligocene Climate System'[53]. Their research was based on a 13 Ma-long record of variations in $\partial^{13}C$, representing the carbon cycle, and $\partial^{18}O$, representing ocean temperature and ice volume, from a deep-sea drill core from the equatorial Pacific. Results confirmed that the climate system responds to intricate orbital variations, which induce a fundamental relationship between solar forcing, the carbon cycle and glacial events. Periodically recurring glacial and carbon cycle events showed that the heartbeat comprised cycles of orbital eccentricity of 405, 127 and 96 Ka in length, and a 1.2 Ma cycle in obliquity. The orbitally modulated variations in the carbon cycle induced changes in deep-ocean acidity, manifest as changes in the amount of $CaCO_3$ in bottom sediments. Pushing their analogy to the human body further, Pälike concluded that '*Earth seems to "breathe" on time scales ranging from the annual to the orbital. We hypothesize that in all cases these cycles are driven by the expansion and contraction of biosphere productivity in response to changes in solar insolation*'[53].

Investigating the onset of glaciation at the Eocene–Oligocene boundary, Pälike's team found from numerical models that '*by imposing a gradual decrease in atmospheric CO_2 levels some time before the E-O [Eocene–Oligocene] transition…we obtain a rapid onset of glaciation at the time that it is observed in our records. We find that onset of glaciation is independent of the exact timing of CO_2 reduction and is triggered by astronomical forcing as soon as atmospheric CO_2 levels are close to a threshold value. Our model results therefore confirm the view [of DeConto and Pollard] that a decrease in atmospheric CO_2 is a possible mechanism to explain the record across the E-O transition*'[53].

The formation of the ice sheet undoubtedly influenced local circulation, facilitating the formation of Antarctic Bottom Water and Antarctic Intermediate Water, which expanded through the world ocean to further influence circulation and climate. Ice, too, had its role to play.

11.4   Hyperthermal Events

Palaeocene and Eocene times hold one more fascinating indication of climate change. The warm climate of the early Eocene was punctuated by two substantial short-lived warming events linked to the release of CO_2, known as the early Eocene hyperthermals[54]. These brief warm events each lasted no more than about 200 Ka and were marked by the release of abundant ^{12}C-rich carbon, ocean acidification and the dissolution of deep-sea carbonates. They are also identified in sediments on land, for example in the Bighorn Basin of Wyoming. One of these hyperthermal events, Eocene Thermal Maximum 2 (about 53.5 Ma ago) can be seen even in the Arctic, where sea surface temperatures rose by 3–5 °C, accompanied by eutrophicaton of the surface waters, which led to oxygen-poor conditions in the photic zone[55]. Temperatures during the coldest months remained above 8 °C in the Arctic, possibly due to cloudy conditions reducing winter cooling.

Philip Sexton, from both Scripps and the United Kingdom's National Oceanography Centre, identified, along with colleagues, six hyperthermal events in the Palaeocene and Eocene epochs at 65.2, 58.2, 53.7, 53.2, 52.5 and 41.8 Ma ago[56]. Generally lasting about 40 Ka, they involved rapid redistributions of carbon between Earth's surface reservoirs. Carbon was supplied to the air by periodic rapid release of dissolved organic carbon, due to oxidation of deep waters, then rapidly reabsorbed back by the ocean.

The six new events were not alone. They were the more extreme variants of a large number of events with a spacing of about 100 Ka, which were probably paced by the cycles in the eccentricity of the Earth's orbit[56]. During these transient events, the deep ocean warmed by 2–4 °C. They were associated with a reduction in the deposition of sedimentary carbonate, indicating increases in deep-ocean acidity. And their duration, of about 41 Ka, suggests that they were forced by the Earth's obliquity cycle, which has its greatest effects at high latitudes. Because the dissolution of $CaCO_3$ during these events, as seen in deep-sea cores, increases towards the high-latitude South Atlantic, the Southern Ocean is implicated as a possible source region. Sexton suggested that the warmth of the Eocene ocean led to depletion in oxygen of the deep ocean, which led to a build-up of dissolved organic carbon from the decay of sinking organic matter. Periodic cooling at high latitudes would have supplied the deep ocean with more dissolved oxygen, resulting in the oxidation of the carbon in the abyssal ocean reservoir and its release to the atmosphere to cause warming. These oceanic processes would have been rapid, given the relatively short overturning time of the ocean (300–1000 years) in relation to the long cycles in eccentricity (100 Ka) and tilt (40 Ka). These events were associated with abrupt and extreme climate change, an accelerated hydrological cycle and ocean acidification. The rates of negative-feedback processes like weathering were too slow to restore the global carbon cycle to a steady state in a short space of time[21].

Among these assorted hyperthermal events in the early part of the Cenozoic was one that was much larger than the rest: the Palaeocene–Eocene Thermal Maximum (PETM) at the Palaeocene–Eocene boundary. As with the Pliocene, the data from the PETM carry an important message for us. While the 55 Ma PETM event is not an exact analogue of

what is happening to Earth's climate today, because the rate of growth in CO_2 then was slower than it is today, we can use it, like the Pliocene, as a **case history** to suggest what might happen to our climate if we keep emitting greenhouse gases at an increasing rate.

11.5 Case History: The Palaeocene–Eocene Boundary

Convincing evidence of the role of CO_2 in raising temperature comes from the end of the Palaeocene, about 55 Ma ago, when there was a sudden short-lived warming event in which temperatures rose by about 5–6 °C globally and by as much as 8 °C at the poles. This event, the PETM, caused one of the largest extinctions of deep-sea benthic organisms in the last 90 Ma. Discovered as recently as 1991[57], it has been analysed in detail by Jim Zachos and colleagues[40], who interpret their $\partial^{13}C$ data to suggest that it was accompanied by a major release of more than 2000 Gt of ^{12}C-rich carbon as CO_2 into the ocean and air (Figure 11.5). Its signal appears everywhere[58,59], including the Arctic, where ocean temperatures near the North Pole rose from 18 to >23 °C, at the same time as nearby lands warmed from 17 to 25 °C[60]. Appy Sluijs of Utrecht University and colleagues found evidence for a rise in temperature of 7 °C as far south as the Tasman Sea at a palaeolatitude of 65° S[61]. Rises of this magnitude seem typical of continental-margin rather than open-ocean settings, where it was less. The Tasman site is far enough south to suggest that the warming there may represent polar amplification[61]. Maximum temperatures at that location ranged between 29–34 °C, indicating a low Equator-to-pole thermal gradient. Field evidence in the form of transgressive sedimentary deposits shows that the warming led to a rise in sea level, with thermal expansion of the ocean contributing around 5 m, and the melting of all or part of the relatively small Antarctic mountain ice caps that may have existed at that time contributing some additional amount, likely less than 10 m[62].

How fast was the carbon emitted? The distribution of fossil remains across the Palaeocene–Eocene boundary suggests that the onset of the event took place in a geological instant – less than 500 years[63]. Ying Cui of Penn State and colleagues found that the peak rate of addition of carbon to the Earth system during the PETM in Svalbard was probably in the range of 0.3–1.7 Gt C/year[64]. Both the 500-year onset and rate of emission of CO_2 were much

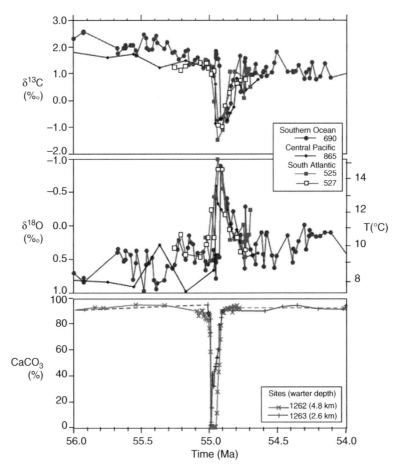

Figure 11.5 *The Palaeocene–Eocene Thermal Maximum (PETM), as recorded in benthic foraminiferal isotopic records. A rapid decrease in carbon isotope ratios (top panel) indicates a large increase in atmospheric greenhouse gases (CO₂ and CH₄), coincident with 5°C global warming (centre panel). Much of the added CO₂ would have been absorbed by the ocean, thereby lowering seawater pH and causing widespread dissolution of seafloor carbonates (lower panel), manifest as a transient reduction in the carbonate (CaCO₃) content of sediments. The ocean's carbonate saturation horizon rapidly shoaled more than 2 km, then gradually recovered as buffering processes slowly restored the chemical balance of the ocean.*

less than the duration of onset and rate of increase in CO_2 today, at about 1%/year[21,a].

Where did the CO_2 come from? Gerald Dickens of Rice University suggested that a massive disturbance of

the seabed led to the rapid release of methane (CH_4), quickly oxidised to CO_2[66]. But evidence is emerging that the warming associated with the $\partial^{13}C$ excursion may have begun a short while before the excursion itself took place, and might have triggered the event[67]. Good evidence for such a precursor recently came from studies of the PETM in the Bighorn Basin in Wyoming, where Ross Secord of the University of Nebraska found continental warming of

---

[a] Note added in press: A new high-resolution carbon isotope record from the Bighorn Basin in Wyoming shows that rates of carbon release during the PETM were in fact similar to those associated with modern carbon emissions, making the PETM a likely analogue for modern conditions[65].

about 5 °C prior to the event[68]. Given that the precursor warming is unrelated to changes in marine carbon ($\partial^{13}C$), the results of Secord and colleagues suggest that there were at least two sources of warming, the earlier of which was unlikely to be marine methane. One possible source is mantle-derived CO_2 released during North Atlantic volcanic activity[68]. Nevertheless, the size of the shift in the $\partial^{13}C$ ratio measured in the marine carbonates would have required the transfer into the atmosphere of an amount of terrestrial carbon equivalent to most of today's terrestrial biosphere, including soils[69]. Rather than a volcanic source, it seems more likely that we are looking at the release of a massive amount of seafloor methane, which is significantly depleted in ^{13}C. As Dickens suggested[66], it most probably came from the ice methane hydrates known as clathrates, which grow in the sediments of the continental slope. These form where molecules of water under high pressure in the cold pores of the sediment form cages of ice around molecules of methane gas released into the sediment by the decomposition of organic matter. One possibility is that warming destabilised these clathrates, leading to a burst of methane that rapidly converted to CO_2 in the atmosphere[67].

Not everyone agreed with Dickens. John Higgins and Daniel Schrag of Harvard argued that factors such as the magnitude of the warming, the rise in tropical sea surface temperature, the abrupt warming of the deep sea, the ocean acidification and the extent of benthic extinctions at the time meant that the size of the addition of carbon was larger than could be accounted for by his methane hypothesis[70]. They concluded that oxidation of 5000 Gt of organic carbon was required. Logical sources included contact metamorphism associated with intrusion of a large igneous province into organic-rich sediments, a peatland fire of global dimensions and desiccation of a large epicontinental sea.

In his 2010 book, *Challenged by Carbon*, Bryan Lovell of the Earth sciences department of Cambridge University suggested a different cause. 55 Ma ago, the hot plume of magma beneath the volcanic island of Iceland produced a lateral pulse of magmatic fluid that pushed eastward beneath Scotland, lifting the terrain. This igneous-tectonic process is likely to have destabilised the adjacent continental slopes of the northeast Atlantic, making the soft sediments slump and releasing methane gas from frozen methane hydrates, which rapidly oxidised to CO_2. CO_2 levels in the atmosphere were already abundant, but the additional CO_2 warmed it even more.

Lovell's imaginative concept draws on research by his Cambridge colleagues, who used three-dimensional seismic reflection data to build a picture of a buried late Palaeocene landscape beneath the seabed north of Scotland[71]. This landscape, they contended, was uplifted by between 600 and 1200 m by the passage of the subterranean pulse of magma from Iceland, before being reburied after about 1 Ma ago. The uplift took place at the time of the PETM, which is why Lovell connected it to the possible release of methane from gas hydrates. The igneous activity to which he refers is likely to have been associated with the million years or so of massive flood-basalt volcanism in East Greenland that began just before and ended just after the PETM, which marked a key stage in the separation of Greenland from western Europe[72]. Magma interacting with sedimentary basins full of Greenland's carbon-rich sedimentary rocks could have provided an excess of CO_2 at this time[73]. This linkage may perhaps explain Secord's observation that a CO_2-induced warming event, perhaps related to Atlantic volcanic activity, immediately preceded the main PETM event[68,b].

As we know from Chapter 9, increasing the amount of CO_2 in the ocean makes it slightly more acid, dissolving deep-sea carbonates and raising the CCD to shallower depths. Not surprisingly, then, Zachos found that calcium carbonate disappeared from the deep sea during the warming event, which lasted 170 Ka[75]. Whereas the CCD occurs in the Atlantic today at water depths of around 5 km (see Chapter 9), during the PETM it shoaled to 2.5 km. It took at least 100 Ka for the CCD to fall back to its original level, consistent with the residence time the average CO_2 molecule spends in the air[75]. There is a good summary of Zachos's findings in the October 2011 issue of *National Geographic*[76].

A side effect of making the ocean more acid and raising the CCD was to kill off the benthic foraminiferal species of the deep-sea floor, as well as to cause most shallow-water coral reefs to vanish. Coral reefs did not become widespread again until the middle Eocene, about 49 Ma ago[21]. For more information on ocean acidification, see Box 9.1 and the growing literature on the geological record of ocean acidification[77] and on its cause of crises in ancient coral reefs[78].

---

[b] Note added in press: It was recently proposed that both the PETM and the Eocene hyperthermal events can be explained by the warming driven by orbital forcing that triggered decomposition of organic carbon in polar permafrost[74].

Do we know everything we need to about the PETM event? Not yet. Richard Zeebe of the University of Hawaii, along with Jim Zachos and Gerald Dickens, used the record of dissolution of deep-sea carbonates through the PETM to suggest that the release of carbon must have been pulsed rather than instantaneous[79]. Atmospheric CO_2 would have increased to 1700 ppm from a base level of 1000 ppm. Given the IPCC's accepted range of climate sensitivity as being a rise of 1.5–4.5 °C for a doubling of CO_2[15], this 700 ppm increase should have caused at most 3.5 °C of warming – less than that observed. Zeebe and his team concluded that '*our results imply a fundamental gap in our understanding of the amplitude of global warming associated with large and abrupt climate perturbations. This gap needs to be filled to confidently predict future climate change*'[79]. This begs the question, did they give enough consideration to slow feedback elements of the climate system, such as changes in terrestrial ecosystems?

Why did it take around 100 Ka for conditions to return to normal? Isn't the residence time of CO_2 in the ocean short? This is a popular misconception, as we saw in Chapter 8. As David Archer of the University of Chicago pointed out, '*The carbon cycle of the biosphere will take a long time to completely neutralise and sequester anthropogenic CO_2*'[80]. He calculated '*that 17–33% of the fossil fuel carbon will still reside in the atmosphere 1 kyr from now, decreasing to 10–15% at 10 kyr, and 7% at 100 kyr. The mean lifetime of fossil fuel CO_2 is about 30–35 kyr*'[80].

The only way to take the CO_2 out of the ocean naturally is by the gradual accumulation in sediment of the organic carbon and calcium carbonate remains of marine organisms and the remains of terrestrial vegetable matter brought down to the coast by rivers, which is extremely slow. In effect, then, the PETM gives us a realistic illustration of what happens naturally when a pulse of carbon is added to the atmosphere. If the PETM is anything to go by, we should expect our current, even more rapid additions to lead to significant warming, a slightly more acidic ocean, destruction of the carbonate-based organisms of deep water, a profound decrease in coral reefs, a substantial rise in sea level and a long recovery time.

What was the effect of the PETM on terrestrial vegetation? An international team led by Gabriel Bowen and David Beerling suggested in 2004 that the release of carbon would have had distinct regional effects[81]. Additional warmth, along with increased rainfall and enhanced CO_2, would have encouraged vigorous plant growth and an increase in the turnover of soil organic matter, doubling the rate of fast carbon cycling. Higher rainfall and temperature would have expanded wetlands, leading to an increase in atmospheric methane – another greenhouse gas.

The PETM event is not the only one of its kind. Anthony Cohen and colleagues found another one in the early Jurassic Toarcian period (183–178 Ma ago)[82]. Like the PETM, this was associated with a major $\partial^{13}C$ excursion, significant biotic extinctions, severe global warming and an enhanced hydrological cycle. It was also accompanied by widespread seawater anoxia, forming a prominent oceanic anoxic event. In the United Kingdom, the rich deposit of organic matter formed by this event crops out in the cliffs of Whitby, Yorkshire, as the Jet Rock – a black layer of fossil wood from the monkey puzzle tree. It was once popular as jewellery. In mourning for Prince Albert, Queen Victoria declared that only jet jewellery could be worn at court for a year[83]. Like the PETM, the Toarcian event is most readily explained '*by the abrupt, large-scale dissociation of methane hydrate that followed a period of more gradual environmental change linked to the emplacement of a large igneous province*'[82].

Can we model the PETM? The simple answer is, not yet. As Paul Valdes explains, the background climate state of the period was characterised by an extremely flat temperature gradient between the Equator and the poles[84]. Climate models have so far been unable to simulate the intense warmth of the polar regions of the time. According to Valdes, '*Not being able to start from a realistic global temperature distribution for the late Palaeocene makes it unrealistic to simulate further abrupt warming associated with the Palaeocene-Eocene Thermal Maximum*'[84]. More worryingly, '*similarly flat latitudinal temperature gradients are a common feature of extreme warm climates of the past, suggesting that IPCC-type, complex climate models may not be well suited to simulating climate dynamics during the past, extremely warm periods*'[84].

11.6 CO_2 and Climate in the Late Cenozoic

What were the respective roles of plate tectonics, orbital changes and CO_2 in modifying the climate of the later Cenozoic? Plate tectonics must have played some role, as it did in the early Cenozoic[29]. For example, Australia and New Guinea began colliding with the island arcs of South East Asia, constricting the link between the Indian and Pacific Oceans. Numerical modelling suggests that this constriction would have affected the location of deep-water formation, oceanic heat transport and sea surface temperatures at both high and low latitudes. Those changes would likely also have affected the marine carbon cycle and atmospheric CO_2[85]. That collision may help to explain the origin of the Oligocene–Miocene boundary around 23 Ma ago. The northward movement of Australia also widened the Southern Ocean between Tasmania and Antarctica at roughly the same time as the Drake Passage opened between Tierra del Fuego and the Antarctic Peninsula. This widening strengthened the Antarctic Circumpolar Current, further isolating Antarctica thermally and helping to keep it cool, which may be why Southern Hemisphere ice sheets came to be more or less permanent features of the climate system from 34 Ma onwards.

What about the role of CO_2, which declined quite rapidly from levels near 750 ppm at the Eocene–Oligocene boundary to about 400 ppm by the end of the Oligocene[41, 45, 86]? Cramer and colleagues tell us that Oligocene bottom water temperatures stayed flat at about 6 °C, while there was a persistent but dynamic Antarctic ice sheet (see Figure 7.7)[50]. Bottom waters warmed abruptly to about 7 °C 26 Ma ago, and stayed warm across the Oligocene–Miocene boundary until about 16 Ma ago[50]. The existence of some high CO_2 values at about 26 Ma ago[41] suggests that the warming then may have had something to do with increasing CO_2. But CO_2 then fell to lowish values across the Oligocene–Miocene boundary[41], while bottom temperatures stayed on the warm side[50]. High-resolution $\partial^{18}O$ data at the boundary have revealed a 2 Ma-long cold event coinciding with a fall in sea level, interpreted as a period of growth in the Antarctic ice sheet[85]. The boundary event correlated with minima in the amplitudes of both the low-frequency (400 Ka) eccentricity cycle and the high-frequency (41 Ka) obliquity cycle, which sustained unusually cold summers. While this orbital 'event' was transient, it may have been enough to expand the Antarctic ice sheet to dimensions like those of the Last Glacial Maximum for some 200 Ka[87]. Orbital variations are readily identifiable for the Oligocene[53] and the Miocene[51].

During the middle Miocene, there was a warm event lasting from 18 to 14 Ma ago, when bottom water temperatures reached 7 °C[50]. This so-called 'Mid Miocene Climatic Optimum' was associated with high values of CO_2, reaching 500–600 ppm according to data from leaf stomata[41]. Mark Pagani's original alkenone data suggested that CO_2 levels were close to 200–300 ppm during the Miocene[45], but a later compilation of alkenone data from a deep-ocean drill site showed that they were closer

to 400 ppm[88], in line with data from a variety of other sources[41]. Deriving surface CO_2 values from B/Ca ratios in planktonic foraminifera and surface ocean temperatures from Mg/Ca ratios, Aradhna Tripati (from Cambridge University and the University of California, Los Angeles) and colleagues found that the mid Miocene warm period was characterised by CO_2 values averaging about 400 ppm (ranging between 350 and 450 ppm)[89].

There is widespread evidence for this warming event on land, as we saw in Chapters 6 and 7. Antarctica warmed significantly, with temperatures in the Ross Sea region averaging 10 °C in January, a reduction in sea ice and a proliferation of woody plants on land[90]. Mg/Ca ratios from planktonic and benthic foraminifera show that warm waters prevailed around Antarctica[91]. ANDRILL data suggest that the East Antarctic Ice Sheet may have retreated into the Transantarctic Mountains. Drill cores confirm that the ice sheet fluctuated dynamically during this period[85]. The warming coincided with the eruption of the Columbia River Plateau basalts – a Large Igneous Province that was at its most vigorous between 17 and 14 Ma ago. These eruptions may have contributed to the elevated CO_2 of the time[92]. Nevertheless, it is fair to say that we do not yet understand quite what drove the warming of this period, at a time of what seemed to be relatively low values of atmospheric CO_2. As is so often the case, further research is needed to resolve the uncertainties.

The climate cooled sharply at about 14 Ma ago, with bottom waters beginning a long temperature decline[50, 86]. This pronounced cooling led to strengthening of the Antarctic Circumpolar Current and enlargement of the Antarctic ice sheet[89]. The cooling went along with a slow decline of CO_2 levels to between 200 and 350 ppm between 12 and 5 Ma ago. In contrast, mid-latitude Pacific Ocean surface waters stayed warm[93]. Sea level also rose at the time[89]. Stable isotope measurements from the western equatorial Pacific show that over this period the thermocline gradually became shallower, which eventually led to the stronger coupling between CO_2, sea surface temperature and climate typical of the Pliocene and Pleistocene[93]. Between 8 and 2.5 Ma ago, bottom water temperatures fell from around 6 to less than 2 °C[50]. CO_2 levels fell in parallel[94].

Evidently, while there were still strong connections between CO_2 and temperature, as in the early Cenozoic, they do not explain all the changes that we see, most likely for two main reasons. First, plate tectonics caused gateways to open or close, changing ocean circulation and thus changing the distribution of heat around the globe. Second, as Foster and Rohling pointed out[95], the relationship between CO_2 and temperature is disrupted at times when there are large ice sheets, which alter the thermal state of the Earth system, as we saw in Chapter 10.

This analysis of the relation of CO_2 to the picture of global change in deep-ocean bottom water derives from stacking together a number of oxygen isotopic and Mg/Ca records from deep-sea cores[40, 50]. But this does not tell us what was going on region-by-region at the surface. Sam Savin, whom we met in Chapter 7, found that most of the cooling was confined to high latitudes and deep bottom waters. In contrast, the equatorial region warmed, and mid-latitude waters showed no change[96]. This constitutes something of a paradox. Savin found no evidence for global refrigeration at the time of the onset of new Antarctic ice growth 14 Ma ago. We know from the CO_2 data of Pagani, Zhang and Beerling and Royer that this ice growth took place without much change in CO_2. What, then, made the ice grow?

Much of Savin's picture of the cooling of high southern latitudes with time may have to do with the stabilising effect on global climate of the existence of an Antarctic ice sheet. Undoubtedly, the ice sheet helped to keep the surrounding air and water cool. But as the Southern Ocean widened, the Antarctic Circumpolar Current migrated north, expanding the area of cold water south of the polar front. The circulation of the Southern Ocean of the time was probably much like it is today, with seasonal sea ice and seasonal production of Antarctic Bottom Water around the Antarctic coast, northward movement of cold surface water driven by the westerly wind, replacement of surface water by the upwelling of Circum-Polar Deep Water and the sinking of cool intermediate water at the polar front. These processes are balanced today by the sinking of cold North Atlantic Deep Water in the Nordic Seas. But in the Oligocene and Miocene there was no major north polar ice sheet, and in any case the Fram Strait between Svalbard and Greenland was closed until 12 Ma ago, which suggests that the water from the north balancing the outward flow of Antarctic Bottom Water and Intermediate Water from the south would have been fairly warm compared with what it is today. The ocean and atmosphere processes associated with Antarctica and its growing Southern Ocean would have kept southern high latitudes cool and stable, but growing colder as the Southern Ocean widened. Changes might be expected from time to time as a result of changes in orbital forcing, but global temperature and CO_2 may not have changed much due to the absence of Northern Hemisphere ice

sheets and the corresponding weakness of the supply of North Atlantic Deep Water.

The key change driven by the growing Southern Ocean would have been, as Savin saw, gradual intensification of the Equator-to-pole thermal gradient, leading to a more vigorous thermohaline circulation, stronger winds and enhanced upwelling – hence cooling – on continental margins[96], including around Antarctica. But Savin missed another factor – a change in the source of deep water – as we shall see.

As in all of the developments we have examined, ideas have changed with time. Let's look at the late Miocene. At the Tarpon Springs meeting in 1984, the palaeontologist Edith Vincent and geochemist Wolfgang Berger, of Scripps, pointed out that this cooling step occurred within a substantial 'excursion' of carbon isotopes towards heavier values (higher $\partial^{13}C$), signifying the excess trapping of ^{12}C-rich organic matter in sediments[97]. The timing coincided with the deposition of widespread diatomaceous deposits on the Pacific margin, including the shales rich in diatomaceous remains and organic matter of California's Monterey Formation. That association led them to propose that global cooling had increased the Equator-to-pole thermal gradient, strengthening coastal winds, which in turn enhanced upwelling, productivity, the development of the oxygen minimum zone and the sedimentation and entrapment of organic matter, so trapping ^{12}C-rich material. The massive production, deposition and entrapment of organic carbon in sediments combined to pull CO_2 out of the atmosphere, they thought, which in turn contributed to a further lowering of global temperature through the greenhouse mechanism. This cycle would have been broken when the nutrients in the upwelling waters, especially phosphorus, became exhausted, causing production to decline[97].

While it sounded eminently reasonable at the time, some 15 years later, and with the benefit of access to a global set of deep-ocean drill cores, this 'Monterey hypothesis' was found wanting by a team led by Liselotte Diester-Haas of the University of Saarland in Saarbrücken, Germany[98]. Diester-Haas used multiple proxies of climate change to analyse the sediment records of middle Miocene age in deep-sea drill cores from the Atlantic, which crossed the prominent positive mid Miocene excursion in $\partial^{13}C$ between 17.5 and 13.5 Ma ago, at the heart of the Mid Miocene Climatic Optimum. Her team made two significant discoveries. First, marine productivity in the Atlantic was not related to the $\partial^{13}C$ excursion, so the excursion could not represent a global marine productivity event of the kind envisaged by Vincent and Berger. Second, the $\partial^{13}C$ values of bulk organic matter in deep-marine sediments derived from the ocean surface paralleled those in benthic foraminifera that drew their carbon from dissolved organic carbon in bottom waters. The absence of a surface-to-seabed gradient in carbon isotopes made it highly unlikely that some marine productivity event had led to a significant change in the CO_2 content of the air during this period. How, then, did the ocean come to be enriched in ^{13}C at this time? The answer, thought Diester-Haas, must lie on land. The ^{13}C excursion in the ocean coincided with the Mid Miocene Climate Optimum, when warm conditions prevailed and ^{12}C-rich vegetation spread to high latitudes. Evergreen forests extended north to 45° N in North America and 52° N in Europe, leading to major deposits of brown coal, or lignite, worldwide. CO_2 was relatively abundant at the time, although it is not entirely clear why[88]. The trapping of ^{12}C in terrestrial organic matter enriched the ocean and its sediments in ^{13}C[98].

Global cooling and the growth of the Antarctic ice sheet at 14 Ma ago followed the Mid Miocene Climatic Optimum. Grasslands replaced the forests, and the deposition of brown coal ceased[99]. The Monterey hypothesis failed to explain the time lag between the onset of the $\partial^{13}C$ excursion at 17.5 Ma ago and the global cooling that followed at around 14 Ma ago. Not only that, but the hypothesis also failed to explain significant changes in atmospheric CO_2, which it was required to do[98, 99].

Like Savin, Diester-Haass called on a major reorganisation of ocean circulation to explain the late Miocene cooling[98]. We can imagine such a reorganisation resulting from the continued reorganisation of land masses and ocean gateways as the continents continued to move. One of the key changes has been suggested by differences in the $\partial^{13}C$ ratios of benthic foraminifera from different ocean basins. These indicate that, in the early Miocene, the Southern Ocean received warm saline deep water from a tropical source, probably in the Indonesian region, and that the flow of this water ceased in the mid Miocene, perhaps due to the collision of New Guinea with the Indonesian islands, thus reducing meridional heat transport and possibly triggering expansion of the Antarctic ice sheet[94].

Vincent and Berger may still be right about the Monterey Formation arising from an increase in the Equator-to-pole thermal gradient, accompanied by stronger winds and more coastal upwelling. My own research on deep-sea drilling sites from the continental margin west of California shows that organic enrichment extended into deeper water at that time, likely representing an increase in the thickness of the oxygen minimum zone resulting from increased productivity in upwelling centres in the California Current[100]. These data suggest that Savin was wrong to imply that the global climate did not cool in the late Miocene. Indeed, as we saw in Chapters 6 and 7, there is ample evidence on land for a global cooling at that time.

Most of the increase in ice volume in the late Miocene occurred on East Antarctica[94]. While seismic studies also pointed to an expansion of ice caps on West Antarctica, several lines of evidence show that there was no major ice sheet on West Antarctic or on the Antarctic Peninsula from the mid Miocene to the early part of the late Miocene. A large and thick West Antarctic Ice Sheet advanced up to the edge of the continental shelf during the latest Miocene and early Pliocene, and the Antarctic Peninsula Ice Sheet expanded at the same time. While those changes were taking place, the ice sheets were expanding and contracting periodically in response to orbital changes in insolation[94].

An intriguing resolution of the mid Miocene climate paradox came in 2014 from new research by the Alfred Wegener Institute for Polar and Marine Research (AWI) in Bremerhaven, Germany. Gregor Knorr and Gerrit Lohmann used numerical model simulations of the Miocene climate to find out why, when the Antarctic ice sheet grew to its present size around 14 Ma ago, some parts of the world, including parts of the Southern Ocean, became warmer[101]. The expansion of the ice sheet increased the elevation of Antarctica. The resulting lapse rate (a fall of 6–10 °C/1000 m) cooled the surface by up to 22 °C in the high central part of the ice sheet. Katabatic winds spilling down to the coast from the cold, high interior favoured the development and northward export of sea ice, cooling the surface waters in the Ross Sea region. In contrast, the coastal wind patterns for the Weddell Sea region were primarily onshore, leading to the *import* of warmer waters from the north, accompanied by the retreat of the sea ice. These changes in the wind pattern served to increase the rate of deep-water formation in the Weddell Sea and to decrease it in the Ross Sea region. Superimposed on these regional changes, which included local warming of the Southern Ocean, was a fall in temperature associated with the continued decline in CO_2. High-latitude air cooled by around 6 °C in both hemispheres as sea ice, and thus albedo, increased. The decline in CO_2 and associated fall in global temperature had less effect on the wind field of the Southern Ocean than did the growth of the ice sheet, which, in effect, caused pronounced regional change. Hence, the mid-Miocene $\partial^{18}O$ shift reflected the growth of the ice sheet, a regional event that encouraged regional warming of the Southern Ocean in the Atlantic sector during an overall global cooling trend in sea surface temperature and bottom water temperature driven by the global decline in CO_2.

Evidently, understanding climate change in the Miocene requires paying careful attention to the respective roles of: plate tectonics and their effects on ocean circulation; the changing volume of Antarctic ice and its effects on the wind field; and the overall decline in CO_2. Changing geographic gateways forced changes on global ocean and atmospheric circulation. The climate was generally stable, probably due to the influence of stable oceanographic processes driven by the Antarctic thermostat, which kept southern high-latitude temperatures cool, along the lines suggested by Ralph Keeling and Martin Visbeck[102]. The thermostat works by strengthening the flow of Antarctic Bottom Water into the interior and cooling the deep ocean when other sources of deep water are too warm, and by turning off or reducing the supply of Antarctic Bottom Water when deep bottom waters are too cold, thus allowing the deep ocean to warm again through the input of warm deep water from elsewhere, eventually achieving a steady state. In due course, plate tectonic processes switched off the supply of warm, tropically sourced deep water, allowing cold deep waters to predominate and making oceanic and atmospheric circulation more vigorous. This led temperatures at high latitudes to cool enough to enable a bigger ice sheet to grow. The widening Southern Ocean played a key role in these developments. With the opening of Fram Strait at 12 Ma ago and onwards, the ocean became preconditioned for the development of ice in the Northern Hemisphere.

We will now examine a case history for the Pliocene, to see what it can tell us about the kinds of change that increases in CO_2 may lead to in the future.

11.7   Case History: The Pliocene

As we saw in Chapter 7, the cooling trend of the late Cenozoic was interrupted during the Pliocene (5.33–2.58 Ma ago) by conditions warm enough to melt Antarctica's Ross Ice Shelf and deposit diatomaceous ooze there. In their 2001 compilation of $\partial^{18}O$ data for the past 65 Ma, Zachos and colleagues reported that '*The early Pliocene is marked by a subtle warming trend until 3.2 Ma*'[40]. The available

geological and geochemical data were comprehensively reviewed in 2011 by a team led by Ulrich Salzmann of Northumbria University in Newcastle upon Tyne, who found that the global mean surface temperatures of the Pliocene were 2–3 °C warmer than they are today, and that they were accentuated at high latitudes, reducing the Equator-to-pole thermal gradient[103]. '*The Pliocene world*', they said, '*was not only warmer but on most continents also wetter than today, leading to an expansion of tropical savannas and forests in Australia and Africa at the expense of deserts*'[103].

Arctic temperatures were estimated from Pliocene tree rings in fossil wood from Ellesmere Island in Arctic Canada[104]. For each tree ring, oxygen isotopes provided mean annual temperatures, carbon isotopes provided growing season temperatures and hydrogen isotopes provided relative humidity. Growing season temperatures averaged 15.8 ± 5 °C, ranging by up to 4 °C from year to year. They were about 12 °C warmer than today's summer temperatures. Mean annual temperatures averaged -1.4 ± 4 °C, varying by up to 2 °C from year to year. They were around 18 °C warmer than today's average. These 'fossil temperatures' are typical of boreal forests growing today at 15–20° N latitude *south* of Ellesmere Island. There would have been almost none of the tundra that today marks the Canadian and Siberian Arctic; '*instead the polar sun rose across ... [a] well-nigh endless green Pliocene forest*'[81].

Coring in Lake El'gygytgyn in northeast Arctic Russia recently showed that Pliocene summer temperatures there were about 8 °C warmer 3.6–3.4 Ma ago than today, at a time when atmospheric pCO_2 was about 400 ppm[105]. Arctic summers stayed warm until 2.2 Ma ago, following the onset of Northern Hemisphere glaciation, confirming that Arctic cooling was not sufficient to support large ice sheets until the early Pleistocene[105]. Evidently, mid Pliocene temperatures were significantly warmer at high latitudes and in the Arctic than globally – a sign of polar amplification of the global warming signal of the times[106].

The Antarctic Peninsula, in contrast, was covered by a permanent ice sheet at the time[107]. Even so, the fossil seashells of Cockburn Island in the northwestern Weddell Sea east of the Antarctic Peninsula tell us that '*during warm intervals of the Pliocene there was little or no seasonal ice in the northern part of the Weddell Sea*', where the large pectinid scallop *Chlamys* flourished[83]. Close relatives of this species now live further north, on the coast of Patagonia. An increase in biological productivity suggested a reduction in sea ice in the Southern Ocean during the Pliocene, and silicoflagellate fossil assemblages suggested that at times the Pliocene Southern Ocean was about 5 °C warmer than it is today[94].

Conditions began to cool in the late Pliocene as the Antarctic ice sheets expanded towards their present state between about 3.0 and 2.5 Ma ago. This is when the West Antarctic ice streams advanced into the Ross Sea. Ice shelves began to develop, along with offshore floating extensions of glacier termini, replacing relatively warm and often land-based ice margins[108]. Even so, the picture was not simple. Ice streams along the Peninsula repeatedly advanced to the shelf break throughout the late Pliocene. '*Enhanced progradation of the continental shelf and slope started all around Antarctica at ca. 3 Ma*'[94], which is when the Antarctic ice sheets became primarily of the cold-mode type, more or less at the same time as the build-up of major ice sheets in the Northern Hemisphere.

Given the warmth of the mid Pliocene event, we should not be surprised at the evidence for sea level having been much higher then than it is today. Ken Miller of Rutgers University, whom we met in Chapter 5, examined a global spread of data with his team to suggest that peak sea level between 3.2 and 2.7 Ma ago was 22 ± 10 m higher that it is today[109]. It may even have reached as high as 36 m above today's level[21]. These large rises help to explain why the Pliocene coastal cliff line in the eastern United States sits tens of kilometres inland as the Orangeburg Scarp between Florida and North Carolina.

As there was no large Northern Hemisphere ice sheet in the Pliocene, Miller's estimate of a peak Pliocene sea level of +22 m above today's level requires the melting of the West Antarctic Ice Sheet (5 m of sea level) and a significant part of the East Antarctic Ice Sheet (15+ m of sea level). Pliocene palaeoclimate data show that warm and cool periods alternated with one another, and that while sea level was higher than at present in the warm periods, it was close to the present level in the cool periods. The ~20 m range varied with orbitally induced ice volume changes, which took place mainly in Antarctica[109, 110].

Maureen Raymo of Boston University was concerned that these estimates of sea level rise failed to take into account the kinds of isostatic adjustment made as ice was removed from land masses, so she and her colleagues used a numerical model to investigate such effects[111]. Their results suggested that the actual eustatic change in the mid Pliocene warm period may have been significantly less than measurements of fossil sea level features implied. They concluded that '*much needed constraints on Pliocene SL [sea level], and therefore ice volume, can only be achieved with a large global matrix of palaeoshoreline data, which does not yet exist, evaluated within the context of model predictions ... Indeed, results described here can be used to target diagnostic regions where evidence for palaeoshorelines should be sought*'[111].

Miller agreed that estimates of past sea level at each site are complicated by the effects of glacial isostatic adjustment (GIA – land rising when ice is removed), which is partly why he cited a potential error range of ± 10 m[109]. But GIA should not affect the average sea level calculated for multiple global sites.

Eelco Rohling of the University of Southampton estimated with colleagues that an equilibrium concentration of 387 ppm CO_2 in the atmosphere (the level in 2009) would have led to a sea level of 25 ± 5 m above present during interglacials of the Pliocene warm period[112], which is well within the range of Miller's estimates. Of course, the global average hides the range of different estimates of Pliocene sea levels from individual coastal sites as a consequence of GIA, along with the effect of mass redistribution on the planet's gravity field (e.g. when land rises as Northern Hemisphere ice is removed, the seabed is lowered up to a few thousand kilometres to the south in compensation)[111, 113].

Sea level rise implies ice sheet collapse. What direct evidence is there for that on or around Antarctica? Drilling through the 80 m-thick McMurdo Ice Shelf and into the underlying seafloor, as described in Chapter 7, confirmed that in the mid Pliocene, when global temperatures were 2–3 °C warmer than today and atmospheric CO_2 was around 400 ppm, there was no Ross Ice Shelf and most likely no West Antarctic Ice Sheet[114]. Dave Pollard and Rob Deconto simulated the ice sheet cycles recorded in the drill core and estimated a rise of up to +7 m in sea level, most of it from the West Antarctic Ice Sheet[115]. Their subsequent modelling of the mid Pliocene retreats of the Antarctic ice sheets indicates an even greater loss of ice, with up to about 20 m of sea level rise coming from additional melting of the East Antarctic Ice Sheet during the warmest intervals of the Pliocene[116]. Further drilling, by the International Phase of Ocean Drilling (IPOD) some 300 km off the coast of Adélie Land in 2009, provides evidence for partial melting of the massive East Antarctic Ice Sheet during the Pliocene. Carys Cook of Imperial College London and colleagues used ratios of neodymium to strontium isotopes to show that rocks from the Wilkes Sub-Basin, a large, low-lying part of East Antarctica, had been eroded in the mid Pliocene in the absence of ice[117, 118].

My take on the complementary data emerging from the studies of Miller, Rohling, De Conto, Pollard and Cook is that we are converging on the idea that during Pliocene interglacial warm periods, average global sea level may well have reached on the order of 20 m above today's level for periods of, say, 1000–2000 years during each 40 Ka cycle. Raymo may be right to assume that sea levels varied widely from place to place, due to GIAs[111], but it is difficult to argue with the observed convergence. The different data sets imply that neither the West nor the East Antarctic Ice Sheet is stable under warm conditions like those of today.

Knowing what we now do about the relation of carbon dioxide to temperature, it is hardly surprising that a 2011 compilation of proxy data for CO_2 by David Beerling and Dana Royer showed that the Pliocene was characterised by values reaching almost 450 ppm[41]. '*Elevated atmospheric CO_2-concentrations, ranging from 330–425 ppmv during warm interglacials have been quoted as one of the main reasons for higher global temperatures during the Pliocene*', according to Salzmann's team[103]. Tripati recorded values reaching almost 350 ppm from a much more limited data set[89].

Marcus Badger of the University of Bristol confirmed with colleagues in 2013 that Pliocene CO_2 values also followed cycles of about 41 Ka, indicative of orbital control[119]. These CO_2 variations were small, about ±40 ppm, suggesting that the climate was quite stable but that there were fluctuations in ice volume, temperature and CO_2 driven by or linked in some tight way to orbital changes.

The Pliocene warm period has engaged the attention of the climate modelling community because, as Alan Haywood and Mark Williams of the British Antarctic Survey reminded us in 2005, '*The mid-Pliocene is the last time in geological history when our planet's climate was significantly warmer, for a prolonged period, than it is today*'[120]. Not only that but, as Salzmann's team pointed out, the Pliocene could almost form an analogue for today's climate, given that the land geography, ocean bathymetry and marine and terrestrial biology were all quite similar to today's[103]. According to Zalasiewicz and Williams, it '*looks more and more like a potential late twenty-first century climate scenario on Earth*'[83], much as Mikhail Budyko had warned. As a result, a concerted effort is being made both to assemble Pliocene temperature data and to model the Pliocene climate. Even so, the

Pliocene cannot be a direct analogue for our present and future climate because, first, the rate of rise of temperature today is faster than during the Pliocene, and, second, our trajectory of CO_2 increase will take it far above anything measured in Pliocene samples.

Among the leading figures in modelling the Pliocene climate is palaeontologist Harry Dowsett of the US Geological Survey (USGS) in Reston, Virginia, who is addressing the question of Pliocene warmth through the USGS Pliocene Research Interpretation and Synoptic Mapping (PRISM) project, which started in the early 1990s[121]. He and his international team fill the PRISM database with data for the warm Pliocene world of 3 Ma ago. They have used their data to map Pliocene sea surface temperatures globally, and found a good fit between the data and the output of an atmosphere–ocean GCM[121]. The model outputs confirm that equatorial temperatures were higher than today, polar temperatures were much higher than today and polar ice sheets were smaller, especially in Greenland. The agreement between the 'real' data estimated from proxies for temperature and the 'simulated' data in outputs from the model shows that palaeoclimate modellers are getting better at their job. There are some discrepancies, for example in the North Atlantic and Arctic, where the use of modern bathymetry for the Pliocene might not be appropriate. To test this possibility, the model was rerun with a slight deepening of the Greenland–Scotland Ridge, a barrier to the northward movement of warm water from the Atlantic to the Arctic[122]. The modelled Pliocene ocean currents increased the poleward heat transport, thus increasing Arctic sea surface temperatures. This provides a possible mechanism for warming of the Arctic and may explain the discrepancy between 'real' and 'simulated' data.

The subject is advancing apace. In 2007, Dowsett's PRISM database indicated significant Pliocene warming in the polar regions but not in the tropics[123]. Applying these data in a numerical climate model, Alan Haywood and his colleagues suggested that the warming of the polar regions must reflect an increase in the flow of currents taking heat from the Equator to the poles, and not an increase in CO_2 in the atmosphere, which would have warmed the tropics as well as the poles[124].

Were Dowsett's tropical data accurate? Kira Trillium Lawrence of Lafayette College in Easton, Pennsylvania and her team used alkenones from marine phytoplankton to show that tropical temperatures in the mid Pliocene had in fact warmed considerably relative to modern conditions[125]. This finding confirmed that there had been global warming at that time, most likely related to the increase in CO_2. Lawrence found that the pronounced cooling between the mid Pliocene and the Pleistocene preceded the onset of significant Northern Hemisphere glaciation. The onset of the glaciation seems to have been the culmination of more gradual changes in the Earth's climate, possibly related to changes taking place in the Southern Hemisphere. The global cooling accentuated the regional cooling in areas of coastal upwelling, such as the Benguela Current off Namibia, where the sea surface temperature dropped 10 °C over this period. In the eastern equatorial Pacific, surface temperatures were as warm as they are today (around 28 °C) in the western Pacific warm pool off New Guinea, suggesting that El Niño-like conditions dominated Pacific equatorial circulation. We will consider the implications of expanded El Niño conditions in a moment.

These data-to-model comparisons for past time slices enable modellers to test the GCMs used to predict how the climate may change in the future, by adding certain likely forcings such as CO_2 to the atmosphere. Carrying out these tests on the Pliocene is especially important because its temperatures and CO_2 levels were so similar to those forecast for the future Earth. Geological data, then, have become a critical part of the effort to improve the numerical models used to suggest what range of future climate conditions we may face with continued greenhouse gas emissions.

Recent studies of Pliocene temperatures in relation to the content of CO_2 in the atmosphere have raised questions about climate sensitivity: the response of mean global temperature to a doubling of CO_2. As we saw earlier in this chapter, the sensitivity proposed by the IPCC in 2007 was a 1.5–4.5 °C warming for a doubling of CO_2[15]. But as geochemist Mark Pagani and colleagues pointed out in *Nature* in 2010, '*this value incorporates only relatively rapid feedbacks such as changes in atmospheric water vapour concentrations, and the distributions of sea ice, clouds and aerosols … [it excludes] the effects of long-term feedbacks such as changes in continental ice-sheet extent, terrestrial ecosystems and the production of greenhouse gases other than CO_2*'[126]. Examining the data from the Pliocene, they found that '*only a relatively small rise in atmospheric CO_2 levels was associated with substantial global warming*'[126]. In other words, the climate system appears to have been more sensitive than the IPCC supposed to an increase in atmospheric CO_2.

Pagani and his team concluded that the climate sensitivity of Earth's ice–ocean–atmosphere system was '*significantly higher over the past five million years than estimated from fast feedbacks alone*'[126]. Their concern was echoed by a team led by Daniel Lunt of the University of Bristol, who used a coupled atmosphere–ocean GCM to simulate the climate of the mid Pliocene warm period and compared the outputs with proxy records of mid Pliocene sea surface temperature[127]. They estimated '*that the response of the Earth system to elevated atmospheric carbon dioxide concentrations is 30–50% greater than the response based on those fast-adjusting components of the climate system that are used traditionally to estimate climate sensitivity*'[127]. They concluded, '*targets for the long-term stabilization of atmospheric greenhouse-gas concentrations aimed at preventing a dangerous human interference with the climate system should take into account this higher sensitivity of the Earth system*'[127].

The possible change in climate sensitivity with time was one of the topics reviewed by Isabel Montañez and Richard Norris[21]. For the most recent period of global warming, the middle Pliocene, they point to evidence suggesting this sensitivity may have been as high as 7.0–9.6 ± 1.4 °C per doubling of CO_2. Why so high? '*Long-term feedbacks operating at accelerated timescales (decadal to centennial) promoted by global warming can substantially magnify an initial temperature increase*'[21].

Warming is not the only factor of interest. More evaporation in a warmer world will both change the hydrological cycle and provide a positive feedback to warming, as water vapour is a greenhouse gas. Not surprisingly, then, climate models of global warming predict an intensified hydrological cycle and, on a global scale, enhanced precipitation[21]. Increased atmospheric water vapour will enhanced the transfer of latent heat from the tropics to the poles, thus sustaining polar warmth, melting polar ice (especially sea ice), decreasing albedo and reinforcing greenhouse conditions[21].

Both global warming and the hydrological cycle will have been affected by the presence or absence of El Niño-like conditions in the Pacific. During the Pliocene, the Pacific Ocean was characterised by a mean state resembling El Niño-like conditions, which contributed to overall warming. Palaeoclimate proxies like Mg/Ca ratios and the alkenone ($U^{K'}_{37}$) index tell us that there was only a very slight or negligible temperature gradient between the eastern and western equatorial Pacific from 5 to 2 Ma ago, with surface waters averaging about 28 °C, after which the western Pacific became warmer (averaging 30 °C) and the eastern Pacific became cooler (averaging 23 °C)[21]. The emission of heat during El Niño events raises global temperatures above the mean temperature today, and is likely to have done so in the past. El Niño events bring rains to the western coasts of North, Central and South America and aridity to Australia, southern Africa and northeastern Brazil. Abundant warm surface waters also tend to keep CO_2 in the air, rather than dissolved in the ocean.

Weak Trade Winds are a key feature of El Niño events, so, as pointed out by Ana Ravelo of the University of California, Santa Cruz, the low (1.5 °C) temperature difference from east to west across the Pacific, compared with the 5 °C difference there today, implies that the Trade Winds were weaker in the mid Pliocene, and therefore that upwelling may have been suppressed along the Peruvian and Californian margins[128]. The palaeoclimate records are not of high enough resolution to distinguish individual El Niño events. All they can tell us is that the mean state of the Pacific at this time resembled warm El Niño conditions. Ravelo and colleagues wrote: '*The oceanic processes that cause change in the long-term mean surface temperature pattern through the Pliocene are thought to be different from the rapid air-sea processes that generate interannual variability and El Niño events in today's climate*'[128]. Whatever the conditions that kept the Pliocene warm, they ended by 3 Ma ago and were replaced by conditions like those of the present by 2 Ma ago[128].

Alexey Fedorov of Yale thought that expanded hurricane activity might have played a role in sustaining the 'permanent' El Niño state typical of the warm Pliocene, which was about 2–3 °C warmer than today[129]. Fedorov and his team pointed out that tropical cyclones (hurricanes) increase vertical mixing by deepening the ocean mixed layer to depths of 120–200 m in their wake, in effect pumping heat from the surface down into the interior. Applying a GCM with the boundary conditions for the Pliocene, they found that tropical cyclones would have been much more common and longer-lasting globally then than they are today. They felt confident that their model was correct, since with modern boundary conditions it correctly simulated the present distribution of tropical cyclones. They concluded that the expanded warm pool enhanced hurricane activity in the subtropical Pacific, which led to stronger vertical mixing, which warmed deep parcels of water moving east in the equatorial undercurrent, which then warmed the eastern tropical Pacific and deepened the tropical thermocline. This positive feedback sustained permanent El Niño conditions and strong hurricane activity across the equatorial Pacific[129]. The implication of Fedorov's findings is that a warmer

ocean will generate more tropical cyclones, increase the frequency of El Niño events and warm the ocean, much as happened in the Pliocene tropics and subtropics[130].

The transition from warm Pliocene to cool Ice Age conditions began at least 4 Ma ago, with marked cooling from 3.5 Ma ago[128]. But the timing of the establishment of Ice Age conditions was different in high compared with low latitudes. Mean ice volume has been greater than today for the past 2.5 Ma, but coastal upwelling systems have been cooler than today for only the past 1.6 Ma, implying that the cause of cooling at low latitudes must be somewhat independent of the cause of changes in the size of ice sheets at high latitudes[128]. Ravelo explained the difference by calling on changes in the conditions of subsurface water. For example, where the thermocline is deep today, as in the eastern equatorial Indian Ocean, upwelled water is warm, while where the thermocline is shallow, as off Namibia or California, upwelled water is cold. Similarly, the thermocline in the eastern equatorial Pacific today may be either deep, during warm El Niño events, or shallow, during cold La Niña events. Changes in the mean state of the Pacific in the Pliocene could thus reflect a shift from El Niño-dominated to La Niña-dominated conditions. Thus, the cooling of upwelling regions with time during the Pliocene could represent shoaling of the thermocline rather than an increase in wind strength; indeed, the palaeoenvironmental evidence favours this explanation[128]. Possibly the same applied to the late Miocene.

Pliocene warming is a front-line research topic of some urgency, and new results are emerging every year. In 2012, Dan Lunt and his colleagues refined their picture[131]. Evidence had accumulated for large fluctuations in ice cover on Greenland and West Antarctica, which were likely free of ice during the warmest periods, as were parts of East Antarctic over the Aurora and Wilkes subglacial basins. In addition, the Arctic Ocean may have been seasonally free of sea ice. Melting ice and snow would have exposed land and ocean; thus, less solar energy would have been reflected, heating the Earth. Melting sea ice would have exposed more ocean as a source of CO_2 for the atmosphere. Rising sea levels flooded the continental margins, providing an even greater oceanic surface area for exchange of CO_2 with the atmosphere. These changes help to explain why atmospheric concentrations of CO_2 rose to perhaps 450 ppm. The warming ocean would have encouraged evaporation, increasing water vapour – another greenhouse gas – in the air. At this time, East African highlands were higher by about 500 m, while the mountains along the western side of the Americas were lower; the effect was to decrease the amount of cool land. Lowering the mountains also changed the location of the Northern Hemisphere's jet stream.

Is there a smoking gun – some one change that accounts for the Pliocene warming? Lunt and his colleagues calculated that 48% of the global average Pliocene temperature rise of 3.3 °C came from CO_2, 21% was caused by the changes in mountain height, 10% came from the loss of sea and land ice, lowering albedo, and 21% was caused by vegetation changes – increased 'greening' of the Arctic (meaning an increase in Arctic vegetation as snow decreased) – further lowering the albedo there[131]. Warming was highest in the polar regions – the phenomenon known as polar amplification – due to the loss of ice, the gain of vegetation and the consequent decrease in albedo. The combined increases in CO_2 and water vapour contributed 61% of the total surface temperature change. Partitioning the responsibility for aspects of the warming still does not provide us with a cause, but we do need to explain the rise in CO_2.

Evidently, the Pliocene warming was not like that at the PETM. We do not need to call on methane clathrates or volcanic activity to explain it. Perhaps the answer ultimately comes down to changes in global tectonics. The gradual emergence of the Isthmus of Panama closed the Central American Seaway, severely limiting the equatorial exchange of seawater between the Atlantic and the Pacific Oceans at around 4.7 Ma ago, and cutting it off completely by around 2.5 Ma ago[132]. Restricting the east–west movement of equatorial surface waters forced them to move north and south in the Atlantic, taking more heat towards the poles. This gradual process may have primed the Earth system to warm. An increase in the amount of solar radiation due to subtle changes in orbital eccentricity and axial tilt may have further done so. Gradual warming would have stimulated a rise in CO_2 and water vapour, which fed back into further temperature rises[131].

Michael Sarnthein of the University of Kiel favoured the plate tectonic solution[133]. He and his colleagues noted that the severe deterioration of climate occurred in three steps between 3.2 and 2.7 Ma ago (Figure 11.6). Data from deep-ocean drill cores suggested clear linkages between the onset of Northern Hemisphere glaciation and three steps in the final closure of the Central American Seaway, which they deduced from rising salinity differences between the Caribbean Sea and the East Pacific. They concluded that each closing event strengthened the poleward transport of salt and heat in the Atlantic, which warmed the temperate regions. At the same time, there was an increase

in the northward transport of moisture in the air, which increased precipitation and runoff in northern Eurasia. This lowered salinity in the Arctic, which increased sea ice and thus albedo, cooling the Arctic. The combination of more salty water in the Norwegian-Greenland Sea and Arctic cooling enhanced the sinking of North Atlantic Deep Water and thus strengthened the Atlantic Meridional Overturning Circulation, which in turn drew more warm salty water north via the Gulf Stream to sustain warming around the North Atlantic.

Their new evidence also showed that closing the Central American Seaway increased sea level in the North Pacific, which doubled the through-flow of cool water from the Bering Strait through the Arctic to the East Greenland Current, ultimately cooling the Labrador Sea by 6 °C. Once the East Greenland Current had been established, it *'led to robust thermal isolation of East Greenland from the poleward heat transport further east through the North Atlantic and Norwegian Currents. Since this time the EGC [East Greenland Current] formed a barrier important to promote the growth of continental ice [sic]'*[133]. As a result, while most of the Northern Hemisphere warmed after the closure of the seaway, Greenland experienced cooling and accelerated snowfall. Formation of the Greenland Ice Sheet enhanced the polar high-pressure cell, which accelerated the northerly winds along its eastern margin, further strengthening the current. Once the ice sheet was established, about 3.18–3.12 Ma ago, and especially after 2.9 Ma ago, it formed a nucleus for the development of Quaternary glaciations in the Northern Hemisphere[133]. The increased runoff to the Arctic and the resulting increase in sea ice further contributed to the eventual development of ice sheets from about 2.8 Ma ago and on. While Sarnthein may be right about the intensification of glaciation on Greenland, actual development of glacial conditions there may be as old as 18 Ma, as we saw in Chapter 7.

Sarnthein's argument does not require a significant rise in CO_2 to explain the warming of the mid Pliocene. Instead, it regards the warming of the North Atlantic region as sufficient to influence the global average temperature. The warming caused CO_2 to be released from surface waters, providing positive feedback and further warming. In due course, the increased heat transport to the north was overcompensated for by the flow of low-salinity cold water through the Arctic from the North Pacific, inducing dramatic cooling and freshening of the East Greenland Current and the onset of major Northern Hemisphere glaciation[133].

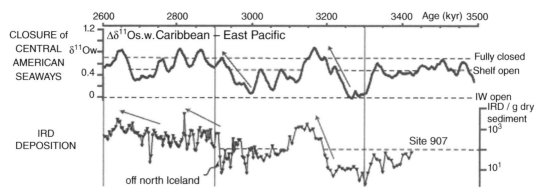

Figure 11.6 *Closing of the Central American Seaway linked to episodes of ice-rafting in the North Atlantic. The closing of the seaway is estimated from the difference in salinity between Caribbean ODP site 999 and East Pacific ODP site 1241, deduced from variations in $\partial^{18}O$ and Mg/Ca-based sea surface temperatures in Globigerinoides sacculifer, a species growing at 50–100 m water depth. A $\partial^{18}O$ gradient of 0.6–0.7 equates to a sea surface salinity gradient of 1.2–1.8 salinity units, which corresponds to full closure of the seaway. Salinity increases in the Caribbean and decreases in the East Pacific with closure. IRD = ice-rafted debris abundance at site 907, north of Iceland. IW = Intermediate Water connection. Arrows indicate closures centred on (i) 3.2 Ma, (ii) 2.9 Ma and (iii) 2.65 Ma, after which connection was essentially closed. Sea surface temperatures in the Irminger Current and North Atlantic Current increased with closures, by 2–3°C, while deep-water temperatures decreased by 1.5–2.0°C. Warming mainly affected temperature at temperate sites (609 and 610, west of Ireland) and subpolar sites (984, south of Iceland). At the same time, flow increased from the Pacific to the Arctic and thermally isolated Greenland, leading to the onset of major Northern Hemisphere glaciation near 2.8 Ma ago.*

So, Sarnthein's model explains both the warming of the Pliocene and the ensuing cooling that led to the initiation of the Pleistocene Ice Age[c]. Cronin agreed that closing the Isthmus of Panama intensified the global thermohaline circulation system and strengthened production of North Atlantic Deep Water, preconditioning the high latitudes of the North Atlantic for a later build-up of large ice sheets[135]. Closure induced greater transport of moisture from the equator to the pole in the Atlantic, enhancing the runoff of rivers into the Arctic and thereby encouraging the formation of sea ice. The Trade Winds intensified, strengthening upwelling along the African margins and thus enhancing productivity and further draw-down of CO_2, driving further cooling. Nonetheless, as Cronin reminds us, we still have no specific causal connection between the closure of the isthmus and the development of the Northern Hemisphere ice sheets. Central America was not the only seaway to close: there was also the closing of the connection between the Mediterranean and the Atlantic, when Spain met Africa around 5–6 Ma ago and shut off supplies of warm water from the Med.

Ana Ravelo and her colleagues disagreed that closure of the Central American Seaway triggered the development of the Ice Age 2.6 Ma ago[128]. But she accepted that the closure did occur just before a steep decline in deep-ocean temperature. North Atlantic bottom water like that of today first formed at around 3.5 Ma ago, and signs of major ice-rafting developed over the Rockall Plateau west of the United Kingdom around 2.5 Ma ago[136, 137]. Shackleton and Kennett had already deduced from Southern Ocean studies in 1975 that the onset of the Northern Hemisphere glaciation must have taken place 2.6 Ma ago[138].

Montañez and colleagues thought that it was the decline in CO_2 from around 450 to the ~200 ppm typical of the Pleistocene glacials that '*most probably accounted for the initiation and growth of Northern Hemisphere ice sheets at around 3 Ma*'[21]. While that agrees with the analysis of Barry Saltzman, as we shall see in Chapter 13, it begs the question: What was driving the CO_2, if not ocean

---

[c] Note added in press: The processes leading to the end of the Pliocene warm period and the start of the Quaternary glaciation have recently been comprehensively reviewed by Sarnthein[134].

temperatures? Ravelo and colleagues argued that because mid Pliocene CO_2 levels were only slightly higher than they are today, while temperatures were 2–3 °C warmer, global cooling from the Pliocene into the Ice Age could only be explained by decreasing albedo from the formation of polar ice and/or decreases in water vapour[128]. The change from 'permanent' El Niño to more modern conditions might explain about 1 °C of cooling, while the shallowing of the thermocline and cooling of the sea surface temperatures in upwelling currents along continental margins might explain another 0.6 °C, leaving unidentified processes or factors to explain the remaining ~2 °C[128]. Ravelo and her team considered as one possibility the closing of the Central American Seaway. While it seemed unlikely to them that the resulting northward transport of heat would have contributed to generating an Ice Age, they thought it possible that the closure would have stimulated shallowing of the thermocline, making upwelling currents much colder, decreasing water vapour and so cooling the climate. Demise of 'permanent' El Niño conditions in the Pacific would have cooled North America, providing conditions under which ice sheets could grow there. This, in turn, would have increased albedo, enhanced the Equator-to-pole thermal gradient, strengthened winds and further enhanced upwelling and hence further cooling in a positive-feedback loop. Increased circulation and ocean stratification would have contributed to more CO_2 being sequestered in the ocean, adding to cooling.

Further evidence that plate tectonics, rather than CO_2, may have played a role in initiating this final cooling comes from Cyrus Karas of the Leibnitz Institute of Marine Sciences. Karas and his colleagues suggested in 2009 that the continued northward motion of New Guinea restricted flow through the Indonesian 'gateway' between the Pacific and Indian Oceans. This cut off the supply of warm water from the central and South Pacific, replacing it with cool water from the North Pacific between 4 and 3 Ma ago[139]. Karas *et al.* used '$\partial^{18}O$ and Mg/Ca ratios of planktonic foraminifera to reconstruct the thermal structure of the eastern tropical Indian Ocean' to show that '*subsurface waters freshened and cooled by about 4 °C between 3.5 and 2.95 Myr ago*'[139]. This would have caused the thermocline in the Indian Ocean to shallow and cool, eventually cooling the Benguela Current on the warm-water route from the Indian Ocean to the North Atlantic, and possibly also leading to the development of the eastern equatorial Pacific cold tongue[139]. These changes would have stimulated or acted as further feedbacks to global cooling.

On a final note, regarding Hutton's dictum on the use of the past to suggest what might happen in the future, sea level expert Ken Miller and his team were quick to point out that when our future climate reaches Pliocene CO_2 levels (we are almost there), and the climate stays warm long enough for equilibrium conditions to be reached, we could in the long term be facing a sea level rise of a similar magnitude. Even if global sea level only rose by 12 m, to the low end of Miller's estimated rise, it would still be uncomfortable for coastal communities.

While these Pliocene changes are broadly consistent with computer climate model outputs for that period[140], they are much larger than we have experienced since 1900, despite the similarity in CO_2 levels. This is because the high concentrations of CO_2 in the Pliocene were maintained for millennia, allowing ice sheets and oceans to come to equilibrium. Our present climate is not in equilibrium with the current rapid rise in CO_2. Hence, assuming that CO_2 levels like those of today persist, it may still take a long time – perhaps between 500 and 2500 years – for the equilibrium level of sea level to be reached[95].

Looking at all the evidence in this section, it seems to me that in the Pliocene, as in the Miocene, there is ample room for slowly changing plate tectonic processes to have set the scene for climate change. I find the arguments that the closure of the Panama seaway first warmed the polar regions through the north–south deflection of formerly east–west tropical currents to be convincing. The resulting warming of the global water surface would have decreased the ocean's ability to take up CO_2, leaving more of it in the atmosphere, reinforcing warming via positive feedback. Eventually, the change in sea level of the North Pacific would have cooled the East Greenland Current, isolating Greenland. While warm water was pushing north in the North Atlantic Current, it was being countered by cold water moving south in the East Greenland Current. Eventually, with the cooling of the Labrador Sea, a source of cool deep water, conditions would have approached those that we have today, with the sinking of North Atlantic Deep Water fed from the Labrador and Norwegian-Greenland Seas, further cooling the north and leading to the development of the other Northern Hemisphere ice sheets. The process would have been accentuated by the cooling in the south driven by the shrinkage of the Indonesian through-flow from the Pacific into the Indian Ocean.

Clearly, there is ample evidence for the primacy of plate tectonic process as the ultimate drivers of climate change through modulations in the supply of CO_2 via volcanoes, the uplift of mountains stimulating increases in weathering and the opening or closing of ocean gateways. Rises in CO_2 may precede temperature change, as in the Eocene, or follow it, as in the Pliocene. In both cases, the two move in lock step, because CO_2 works through positive feedback. CO_2 may be either chicken or egg, depending on the plate-tectonic circumstances. That does not mean that the Pliocene is useless as a case history when looking at today's climate change. It provides a concrete example of what happens when CO_2 rises and reinforces an ongoing rise in temperature triggered by some other mechanism (such as the closing of the Panama seaway).

In this chapter, we have examined a few aspects of the climate of the past 450 Ma in broad-brush terms and the history of the Cenozoic in some detail, along with two case histories, from the PETM and the Pliocene, in order to illustrate the role of CO_2 in changing the climate. It is now time to turn our attention to the Pleistocene Ice Age and its aftermath: the Holocene times in which we now live.

References

1. Barron, E.J. and Washington, W.M. (1985) Warm Cretaceous climates: high atmospheric CO_2 as a plausible mechanism. In: *Carbon Cycle and Atmospheric CO_2: Natural Variations Archean to Present* (eds E.T. Sundquist and W.S. Broecker). Geophysical Monographs Series 32. American Geophysical Union, Washington, DC, pp. 546–553.

2. Schneider, S.H., Thompson, S.L. and Barron, E.J. (1985) Mid-Cretaceous continental surface temperatures: are high CO_2 concentrations needed to simulate above-freezing winter conditions. In: *Carbon Cycle and Atmospheric CO_2: Natural Variations Archean to Present* (eds E.T. Sundquist and W.S. Broecker). Geophysical Monographs Series 32. American Geophysical Union, Washington, DC, pp. 554–559.

3. Barron, E.J., Fawcett, P.J., Peterson, W.H., Pollard, D. and Thompson S.L. (1995) A 'simulation' of mid-Cretaceous climate. *Paleoceanography* **10** (2), 953–962.

4. Poulson, C.J., Barron, E.J., Peterson, W.H. and Wilson, P.A. (1999) A reinterpretation of mid-Cretaceous shallow marine temperatures through model-data comparison. *Paleoceanography* **14** (6), 679–697.

5. Spicer, R.A. and Parrish, J.T. (1990) Late Cretaceous-early Tertiary palaeoclimates of northern high latitudes: a quantitative view. *Journal of the Geological Society of London* **147**, 329–341.

6. Moore, G.T., Hayashida, D.N., Ross, C.A. and Jacobson, S.R. (1992) Paleoclimate of the Kimmeridgian/Tithonian

7. (Late Jurassic) world. I. Results using a general circulation model. *Palaeogeography, Palaeoclimatology, Palaeoecology* **93**, 113–150.

7. Moore, G.T., Sloan, L.C., Hayashida, D.N. and Umrigar, N.P. (1992) Paleoclimate of the Kimmeridgian/Tithonian (Late Jurassic) world. II. Sensitivity tests comparing three different paleotopographic settings. *Palaeogeography, Palaeoclimatology, Palaeoecology* **95**, 229–252.

8. Valdes, P.J. and Sellwood, B.W. (1992) A palaeoclimate model for the Kimmeridgian. *Palaeogeography, Palaeoclimatology, Palaeoecology* **95**, 47–72.

9. Sellwood, B.W. and Valdes, P.J. (2007) Mesozoic climates. In: *Deep-Time Perspectives on Climate Change: Marrying the Signal from Computer Models and Biological Proxies* (eds M. Williams, A.M. Haywood, F.J. Gregory and D.N. Schmidt). Micropalaeolgoical Society Special Publication. The Geological Society, London, pp. 201–224.

10. Hasegawa, H., Tada, R., Jiang, X., Suganuma, Y., Imsamut, S., Charusiri, P., *et al.* (2012) Drastic shrinking of the Hadley circulation during the mid-Cretaceous Supergreenhouse. *Climate of the Past* **8**, 1323–1337.

11. Hay, W.W. (2013) *Experimenting on a Small Planet: A Scholarly Entertainment*. Springer, New York.

12. Ramaswamy, V., Boucher, O., Haigh, J., et al. (2001) Radiative forcing of climate change. IPCC Third Assessment Report. Available from http://www.grida.no/climate/ipcc _tar/wg1/pdf/tar-06.pdf (last accessed 29 January 2015).

13. Myhre, G., Highwood, E.J., Shine, K.P. and Stordal, F. (1998) New estimates of radiative forcing due to well mixed greenhouse gases. *Geophysical Research Letters* **25** (14), 2715–2718.

14. PALEOSENS Project Members (2012) Making sense of palaeoclimate sensitivity. *Nature* **491**, 683–691.

15. Solomon, S., Qin, D., Manning, M., Chen, Z., Marquis, M., Averyt, K.B., *et al.* (2007) *Climate Change 2007: The Physical Science Basis*. Cambridge University Press, Cambridge.

16. Schmittner, A., Urban, N.M., Shakun, J.D., Mahowald, N.M., Clark, P.U., Bartlein, P.J., *et al.* (2011) Climate sensitivity estimated from temperature reconstructions of the Last Glacial Maximum. *Science* **334**, 1385–1388.

17. Hansen, J.E. and Sato, M. (2012) Paleoclimate implications for human-made climate change. In: *Climate Change: Inferences from Paleoclimate and Regional Aspects* (eds A. Berger, F. Mesinger and D. Šijački). Springer, Berlin, pp. 21–48.

18. Hansen, J., Sato, M., Russell, G. and Kharecha, P. (2013) Climate sensitivity, sea level and atmospheric carbon dioxide. In: *Warm Climates of the Past – A Lesson for the Future* (eds D.J. Lunt, H. Elderfield, R. Pancost and A. Ridgwell). Philosophical Transactions of The Royal Society A **371** (2001), 20120294.

19. Ring, M.R., Lindner, D., Cross, E.F. and Schlesinger, M.E. (2012) Causes of the global warming observed since the 19th century. *Atmospheric and Climate Sciences* **2**, 401–415.

20. Royer, D.L., Pagani, M. and Beerling, D.J. (2012) Geobiological constraints on Earth system sensitivity to CO_2 during the Cretaceous and Cenozoic. *Geobiology* **10** (4), 298–310.

21. Montañez, I.P., Norris, R.D., Algeo, T., Chandler, M.A., Johnson, K.R., Kennedy, M.J., *et al.* (2011) *Understanding Earth's Deep Past: Lessons for Our Climate Future.* National Academies Press, Washington, DC.

22. Ramanathan, V. (1998) Trace-gas greenhouse effect and global warming: underlying principles and outstanding issues: Volvo Environmental Prize Lecture 1997. *Ambio* **27** (3), 187–197.

23. Beerling, D.J., Fox, A., Stevenson, D.S. and Valdes, P.J. (2011) Enhanced chemistry-climate feedbacks in past greenhouse worlds. *Proceedings of the National Academies of Science* **108** (24), 9770–9775.

24. Williams, M., Haywood, A.M., Gregory, F.J. and Schmidt, D.N., (eds.) (2007) Deep-time perspectives on climate change: marrying the signal from computer models and biological proxies. Micropalaeolgoical Society Special Publication. The Geological Society, London.

25. Kent, D.V. and Muttoni, G. (2008) Equatorial convergence of India and early Cenozoic climate trends. *Proceedings of the National Academies of Science* **105** (42), 16065–16070.

26. Rowley, D.B. (2002) Rate of plate creation and destruction: 180 Ma to present. *Bulletin of the Geological Society of America Bulletin* **114**, 927–933.

27. Pälike, H., Lyle, M.W., Nishi, H., Raffi, I., Ridgwell, A., Gamage, K., *et al.* (2012) A Cenozoic record of the equatorial Pacific carbonate compensation depth. *Nature* **488**, 609–615.

28. Froehlich, P. and Misra, S. (2014) Was the late Paleocene-early Eocene hot because Earth was flat? *Oceanography* **27** (1), 36–49.

29. Kent, D.V. and Muttoni, G. (2013) Modulation of late Cretaceous and Cenozoic climate by variable drawdown of atmospheric pCO_2 from weathering of basaltic provinces on continents drifting through the equatorial humid belt. *Climate of the Past* **9**, 525–546.

30. Rowley, D.B. (2008) Extrapolating oceanic age distributions: lessons from the Pacific region. *Journal of Geology* **116**, 587–598.

31. Müller, R.D., Sdrolias, M., Gaina, C., Steinberger, B. and Heine, C. (2008) Long-term sea-level fluctuations driven by ocean basin dynamics. *Science* **319**, 1357–1362.

32. Edmond, J.M. and Huh, Y. (2003) Non-steady state carbonate recycling and implications for the evolution of atmospheric pCO_2. *Earth and Planetary Science Letters* **216**, 125–139.

33. John, E.H., Pearson, P.N., Coxall, H.K., Birch, H., Wade, B.S. and Foster, G.L. (2013) Warm ocean processes and carbon cycling in the Eocene. *Philosophical Transactions of The Royal Society A* **371** (2001), 20130099.

34. Olivarez, L.A. and Lyle, M.W. (2006) Missing organic carbon in Eocene marine sediments: is metabolism the

35. biological feedback that maintains end-member climates? *Paleoceanography* **21**, PA2007.

36. Stanley, S.M. (2010) Relation of Phanerozoic stable isotope excursions to climate, bacterial metabolism, and major extinctions. *Proceedings of the National Academies of Science* **107**, 19185–19189.

37. Zachos, J.C., Stott, L.D. and Lohmann, K.C. (1994) Evolution of early Cenozoic marine temperatures. *Paleoceanography* **9** (2), 353–387.

38. eps.ucsc.edu/faculty/Profiles/singleton.php?&singleton =true&cruz_id=jzachos (last accessed 29 January 2015).

39. Arthur, M.A., Hinga, K.R., Pilson, M.E.Q., Whitaker, E. and Allard, D. (1991) Estimates of pCO_2 for the last 120 Ma based on the $\partial^{13}C$ of marine phytoplanktonic organic matter. *EOS* **72** (17), 166.

40. Bice, K.L., Scotese, C.R., Seidov, D. and Barron, E.J. (2000) Quantifying the role of geographic change in Cenozoic ocean heat transport using uncoupled atmosphere and ocean models. *Palaeogeography, Palaeoclimatology, Palaeoecology* **161**, 295–310.

41. Zachos, J., Pagani, M., Sloan, L., Thomas, E. and Billups, K. (2001) Trends, rhythms, and aberrations in global climate 65 Ma to present. *Science* **292**, 686–693.

42. Beerling, D.J. and Royer, D.L. (2011) Convergent Cenozoic CO_2 history. *Nature Geoscience* **4**, 418–420.

43. DeConto, R.M. and Pollard, D. (2003) Rapid Cenozoic glaciation of Antarctica induced by declining atmospheric CO_2. *Nature* **421**, 245–249.

44. Pollard, D. and DeConto, R.M. (2005) Hysteresis in Cenozoic Antarctic ice-sheet variations. *Global and Planetary Change* **45**, 9–21.

45. Coxall, H., Wilson, P.A., Pälike, H., Lear, C.H. and Backman, J. (2005) Rapid stepwise onset of Antarctic glaciation and deeper calcite compensation in the Pacific Ocean. *Nature* **433**, 53–57.

46. Pagani, M., Zachos, J.C., Freeman, K.H., Tipple, B. and Bohaty, S. (2005) Marked decline in atmospheric carbon dioxide concentrations during the Paleogene. *Science* **309**, 600–603.

47. Anderson, J.B. and Wellner, J.S. (2011) *Tectonic, Climate, and Cryospheric Evolution of the Antarctic Peninsula.* Special Publications 63. American Geophysical Union, Washington, DC.

48. Schultz, C. (2013) Tectonic, climate, and cryospheric evolution of the Antarctic Peninsula. *EOS* **94** (23), 210.

49. Katz, M.E., Miller, K.G., Wright, J.D., Wade, B.S., Browning, J.V., Cramer, B.S. and Rosenthal, Y. (2008) Stepwise yransition from the Eocene greenhouse to the Oligocene icehouse. *Nature Geoscience* **1**, 329–334.

50. Pearson, P.N., Foster, G.L. and Wade, B.S. (2009) Atmospheric carbon dioxide through the Eocene-Oligocene climate transition. *Nature* **461**, 1110–1114.

51. Cramer, B.S., Miller, K.G., Barrett, P.J. and Wright, J.D. (2011) Late Cretaceous–Neogene trends in deep ocean temperature and continental ice volume: reconciling records

of benthic foraminiferal geochemistry ($\partial^{18}O$ and Mg/Ca) with sea level history. *Journal of Geophysical Research* **116**, C12023.

51. Naish, T.R., Woolfe, K.J., Barrett, P.J., Wilson, G.S., Atkins, C., Bohaty, S.M., *et al.* (2001) Orbitally induced oscillations in the East Antarctic ice sheet at the Oligocene/Miocene boundary. *Nature* **423**, 719–723.

52. Barrett, P.J. (2007) Cenozoic climate and sea level history from glacimarine strata off the Victoria Land Coast, Cape Roberts Project, Antarctica. In: *Glacial Processes and Products* (eds M.J. Hambrey, P. Christoffersen, N.F. Glasser and B. Hubbart). *International Association of Sedimentology Special Publication* **39**, 259–287.

53. Pälike, H., Norris, R.D., Herrie, J.O., Wilson, P.A., Coxall, H.K., Lear, C.H., *et al.* (2006) The heartbeat of the Oligocene climate system. *Science* **314**, 1894–1898.

54. Secord, R. (2012) Hot spells on land. *Nature Geoscience* **5**, 306–307.

55. Sluijs, A., Schuten, S., Donders, T.H., Schoon, P.L., Röhl, U., Reichart, G.-J., *et al.* (2009) Warm and wet conditions in the Arctic region during Eocene Thermal Maximum 2. *Nature Geoscience* **2**, 777–780.

56. Sexton, P.F., Norris, R.D., Wilson, P.A., Pälike, H., Westerhold, T., Röhl, U., *et al.* (2011) Eocene global warming events driven by ventilation of oceanic dissolved organic carbon. *Nature* **471**, 349–352.

57. Kennett, J.P. and Stott, L.D. (1991) Abrupt deep-sea warming, palaeoceanographic changes and benthic extinctions at the end of the Palaeocene. *Nature* **353**, 319–322.

58. Dypvik H., Riber L., Burca F., Rüther, D., Jargvoll, D., Nagy, J. and Jochmann, M. (2011) The Paleocene–Eocene thermal maximum (PETM) in Svalbard – clay mineral and geochemical signals. *Palaeogeography, Palaeoclimatology, Palaeoecology* **302**, 156–169.

59. Handley, L., Crouch, E.M. and Pancost (2011) A New Zealand record of sea level rise and environmental change during the Paleocene–Eocene Thermal Maximum. *Palaeogeography, Palaeoclimatology, Palaeoecology* **305** (2011), 185–200.

60. Miller, G.H., Brigham-Grette, J., Alley, R.B., et al. (2010) Temperature and precipitation history of the Arctic. *Quaternary Science Reviews* **29**, 1679–1715.

61. Sluijs, A., Bijl, P.K., Schouten, S., Röhl, U., Reichart, G.-J. and Brinkhuis, H. (2011) Southern ocean warming, sea level and hydrological change during the Paleocene-Eocene thermal maximum. *Climate of the Past* **7**, 47–61.

62. Sluijs, A., Brinkhuis, H., Crouch, E.M., John, C.M., Handley, L., Munsterman, D., *et al.* (2008) Eustatic variations during the Paleocene–Eocene greenhouse world. *Paleoceanography* **23**, doi:10.1029/2008PA001615.

63. Zachos, J.C., Bohaty, S.M., John, C.M., McCarren, H., Kelly, D.C. and Nielsen, T. (2010) The Palaeocene-Eocene carbon isotope excursion: constraints from individual planktonic foraminifer record. *Philosophical Transactions of The Royal Society A* **365**, 1829–1842.

64. Cui, Y., Kump, L.R., Ridgwell, A.J., Charles, A.J., Junium, C.K., Diefendorf, A.F., *et al.* (2011) Slow release of fossil carbon during the Palaeocene-Eocene Thermal Maximum. *Nature Geoscience* **4**, doi: 10.1038/NGEO1179.

65. Bowen, G.J., Maibauer, B.J., Kraus, M.J., Röhl, U., Westerhold, T., Steimke, A., *et al.* (2015) Two massive, rapid releases of carbon during the onset of the Palaeocene-Eocene thermal maximum. *Nature Geoscience* **8**, 44–47.

66. Dickens, G.R. (2011) Down the rabbit hole: toward appropriate discussion of methane release from gas hydrate systems during the Paleocene-Eocene thermal maximum and other past hyperthermal events. *Climate of the Past* **7**, 831–846.

67. Sluijs A., Brinkhuis H., Schouten S., Bohaty, S.M., John, C.D., Zachos, J.C., *et al.* (2007) Environmental precursors to rapid light carbon injection at the Palaeocene/Eocene boundary. *Nature* **450**, 1218–1221.

68. Secord, R., Gingerich, P.D., Lohmann, K.C. and McLeod, K.G. (2010) Continental warming preceding the Palaeocene-Eocene Thermal Maximum. *Nature* **467**, 955–958.

69. Sundquist, E.T. and Visser, K. (2005) The geologic history of the carbon cycle. In: *Biogeochemistry, Vol. 8, Treatise on Geochemistry* (eds H.D. Holland and K.K. Turekian). Elsevier-Pergamon, Oxford, pp. 425–513.

70. Higgins, J.A. and Schrag, D.P. (2006) Beyond methane: towards a theory for the Paleocene-Eocene Thermal Maximum. *Earth and Planetary Science Letters* **245**, 523–537.

71. Hartley, R.A., Roberts, G.G., White, N. and Richardson, C. (2011) Transient convective uplift of an ancient buried landscape. *Nature Geoscience* **4**, 563–565.

72. Storey, M., Duncan, R.A. and Swisher, C.C. (2007) Paleocene-Eocene Thermal Maximum and the opening of the Northeast Atlantic. *Science* **316**, 587–589.

73. Svensen, H., Planke, S., Melthe-Sørenssen, A., Jamtveit, B., Myklebust, R., Eidem, T.R. and Rey, S.S. (2004) Release of methane from a volcanic basin as a mechanism for initial Eocene global warming. *Nature* **429**, 542–545.

74. DeConto, R., Geleotti, S., Pagani, M., Tracy, D., Schaefer, K., Zhang, T., *et al.* (2012) Past extreme warming events linked to massive carbon release from thawing permafrost. *Nature* **484**, 87–91.

75. Zachos, J.C., Dickens, G.R. and Zeebe, R.E. (2008) An early Cenozoic perspective on greenhouse warming and carbon-cycle dynamics. *Nature* **451**, 279–283.

76. Kunzig, R. (2011) Hothouse Earth: world without ice. *National Geographic* **220**, 90.

77. Honisch, B., Ridgwell, A., Schmidt, D.N., Thomas, E., Gibbs, S.J., Sluijs, A., *et al.* (2012) The geological record of ocean acidification. *Science* **335** (6072), 1058–1063.

78. Kiessling, W. and Simpson, C. (2011) On the potential for ocean acidification to be a general cause of ancient reef crises. *Global Change Biology* **17**, 56–67.

79. Zeebe, R., Zachos, J.C. and Dickens, G.R. (2009) Carbon dioxide forcing alone insufficient to explain Palaeocene–Eocene Thermal Maximum warming. *Nature Geoscience* **2**, 576–580.

80. Archer, D. (2005) Fate of fossil fuel CO_2 in geologic time. *Journal of Geophysical Research* **110**, C09S05.

81. Bowen, G.J., Beerling, D.J., Koch, P.L., Zachos, J.C. and Quattlebaum, T. (2004) A humic climate state during the Paleocene/Eocene Thermal Maximum. *Nature* **432** (7016), 495–499.

82. Cohen, A.S., Coe, A.L. and Kemp, D.B. (2007) The Late Palaeocene–Early Eocene and Toarcian (Early Jurassic) carbon isotope excursions: a comparison of their time scales, associated environmental changes, causes and consequences. *Journal of the Geological Society of London* **164** (6), 1092–1108.

83. Zalasiewicz, J. and Williams, M. (2012) *The Goldilocks Planet – the Four Billion Year Story of Earth's Climate.* Oxford University Press, Oxford.

84. Valdes, P. (2011) Built for stability. *Nature Geoscience* **4**, 414–416.

85. Wilson, G.S., Pekar, S.F., Nash, T.R., Passchier, S. and Deconto, R.M. (2009) The Oligocene-Miocene boundary – Antarctic climate response to orbital forcing. In: *Antarctic Climate Evolution* (eds F. Florindo and M. Siegert). Developments in Earth and Environmental Sciences 8. Elsevier, Amsterdam, pp. 369–400.

86. Pagani, M., Arthur, M.A. and Freeman, K.H. (1999) Miocene evolution of atmospheric carbon dioxide. *Paleoceanography* **14**, 273–292.

87. Sluijs, A., Bowen, G.J., Brinkhuis, H., Lourens, L.J. and Thomas, E. (2007) The Palaeocene-Eocene Thermal Maximum super greenhouse: biotic and geochemical signatures, age models and mechanisms of global change. In: *Deep-Time Perspectives on Climate Change: Marrying the Signal from Computer Models and Biological Proxies* (eds M. Williams, A.M. Haywood, F.J. Gregory, and D.N. and Schmidt). Micropalaeolgoical Society Special Publication. The Geological Society, London, pp. 23–349.

88. Zhang, Y.G., Pagani, M., Liu, A., Bohaty, S.M. and Deconto, R. (2013) A 40-million-year history of atmospheric CO_2. *Philosophical Transactions of The Royal Society A* **371** (2001), 20130096.

89. Tripati, A.K., Roberts, C.D. and Eagle, R.A. (2009) Coupling of CO_2 and ice sheet stability over major climate transitions of the last 20 million years. *Science* **326**, 1394–1397.

90. Warny, S., Askin, R.A., Hannah, M.J., Mohr, B.A.R., Raine, J.I., Harwood, D.M. and Florindo, F. (2009) Palynomorphs from a sediment core reveal a sudden remarkably warm Antarctica during the middle Miocene. *Geology* **37** (10), 865–960.

91. Shevenell, A.E. and Bohaty, S.M. (2012) Southern exposure: new paleoclimate insights from Southern Ocean and Antarctic margin sediments. *Oceanography* **25** (3), 106–117.

92. Foster, G.L., Lear, C.H. and Rae, J.W.B. (2012) The evolution of pCO_2, ice volume and climate during the middle Miocene. *Earth and Planetary Science Letters* **341–344**, 243–254.

93. LaRiviere, J.P., Ravelo, A.C., Vrimmins, A., Dekens, P.S., Ford, H.L., Lyle, M. and Wara, M.W. (2012) Late Miocene decoupling of oceanic warmth and atmospheric carbon dioxide forcing. *Nature* **486**, 97–100.

94. Haywood, A.M., Smellie, J.L., Ashworth, A.C., Cantrill, D.J., Florindo, F., Hambrey, M.J., *et al.* (2009) Middle Miocene to Pliocene history of Antarctica and the Southern Ocean. In: *Antarctic Climate Change* (eds F. Florindo and M. Siegert). Developments in Earth and Environmental Sciences 8. Elsevier, Amsterdam, pp. 401–463.

95. Foster, G.L. and Rohling, E.J. (2013) Relationship between sea level and climate forcing by CO_2 on geological timescales. *Proceedings of the National Academies of Science* **110** (4), 1209–1214.

96. Savin, S.M. and Woodruff, F. (1990) Isotopic evidence for temperature and productivity in the Tethyan Ocean. In: *Phosphate Deposits of the World, Vol. 3, Neogene to Modern Phosphorites* (eds W.C. Burnett and S.R. Riggs). Cambridge University Press, Cambridge, pp. 241–259.

97. Vincent, E. and Berger, W.H. (1985) Carbon dioxide and polar cooling in the Miocene: the Monterey Hypothesis. In: *The Carbon Cycle and Atmospheric CO_2: Natural Variations Archean to Present* (eds E.T. Sundquist and W.S. Broecker). Geophysical Monographs 32. American Geophysical Union, Washington, DC, pp. 455–468.

98. Diester-Haass, L., Billups, K., Gröcke, D.R., Francois, L., Lefebvre, V. and Emeis, K.C. (2009) Mid-Miocene paleoproductivity in the Atlantic Ocean and implications for the global carbon cycle. *Paleoceanography* **24** (1), PA1209.

99. Bertler, N.A.N. and Barrett, P.J. (2010) Vanishing polar ice sheets. In: *Changing Climates, Earth Systems and Society, International Year of Planet Earth* (ed. J. Dodson). Springer, New York, pp. 49–83.

100. Summerhayes, C.P. (1981) Oceanographic controls on organic matter in the Miocene Monterey Formation, offshore California. In: *The Monterey Formation and Related Siliceous Rocks of California* (eds R.E. Garrison and R.G. Douglas). Society for Economic Paleontologists and Mineralogists Pacific Section, Upland, CA, pp. 213–219.

101. Knorr, G. and Lohmann, G. (2014) Climate warming during Antarctic ice sheet expansion at the Middle Miocene transition. *Nature Geoscience* **7**, 376–381.

102. Keeling, R.F. and Visbeck, M. (2011) On the linkage between Antarctic surface water stratification and global deep-water temperature. *Journal of Climate* **24**, 3545–3557.

103. Salzmann U., Williams M., Haywood A.M., Johnson, A.L.A., Kender, S. and Zalasiewicz, J. (2011) Climate and environment of a Pliocene warm world. *Palaeogeography, Palaeoclimatology, Palaeoecology* **309** (1–2), 1–8.

104. Csank A.Z., Patterson W.P., Eglington B.M., Rybczynski, N. and Basinger, J.F. (2011) Climate variability in the early

Pliocene Arctic: annually resolved evidence from stable isotope values of sub-fossil wood, Ellesmere Island, Canada. *Palaeogeography, Palaeoclimatology, Palaeoecology* **308**, 339–349.

105. Brigham-Grette, J., Melles, M. Minyuk, P., Andreev, A., Tarasov, P., Deconto, R., *et al.* (2013) Pliocene warmth, polar amplification, and stepped Pleistocene cooling recorded in NE Arctic Russia. *Science* **340** (6139), 1421–1427.

106. Dowsett, H.J., Robinson, M.M., Haywood, A.M., Hill, D.J., Dolan, A.M., Stoll, D.K., *et al.* (2012) Assessing confidence in Pliocene sea surface temperatures to evaluate predictive models. *Nature Climate Change* **2**, 365–371.

107. Salzmann U., Riding J.B., Nelson A.E. and Smellie, J.L. (2011) How likely was a green Antarctic Peninsula during warm Pliocene interglacials? A critical reassessment based on new palynofloras from James Ross Island. *Palaeogeography, Palaeoclimatology, Palaeoecology* **309** (2011), 73–82.

108. Naish, T., Carter, L., Wolff, E., Pollard, D. and Powell, R. (2009) Late Pliocene–Pleistocene Antarctic climate variability at orbital and suborbital scale: ice sheet, ocean and atmospheric interactions. In: *Antarctic Climate Change* (eds F. Florindo and M. Siegert). Developments in Earth and Environmental Sciences 8. Elsevier, Amsterdam, pp. 465–529.

109. Miller, K.G., Wright J.D., Browning J.V., Kulpecz, A., Kominz, M., Naish, T.R., *et al.* (2008) High tide of the warm Pliocene: implications of global sea level for Antarctic deglaciation. *Geology* **40** (5), 407–410.

110. Dwyer, G.S. and Chandler, M.A. (2009) Mid-Pliocene sea level and continental ice volume based on coupled benthic Mg/Ca palaeotemperatures and oxygen isotopes. *Philosophical Transactions of the Royal Society A* **367**, 157–168.

111. Raymo, M.E., Mitrovica, J.X., O'Leary, M.J., DeConto, R.M. and Hearty, P.J. (2011) Departures from eustacy in Pliocene sea-level records. *Nature Geoscience* **4**, 328–332.

112. Rohling, E.J., Grant, K., Bolshaw, M., Roberts, A.P., Siddall, M., Hemleben, C. and Kucera, M. (2009) Antarctic temperature and global sea level closely coupled over the past five glacial cycles. *Nature Geoscience* **2**, 500–504.

113. Stocchi, P., Escutia, C., Houben, A.J.P., Vermeersen, B.L.A., Bijl, P.K., Brinkhuis, H., *et al.* (2013) Relative sea-level rise around East Antarctica during Oligocene glaciation. *Nature Geoscience* **6**, 380–384.

114. Naish, T.R., Powell, R., Levy, R., Wilson, G., Scherer, R., Talarico, F., *et al.* (2009) Obliquity-paced Pliocene West Antarctic Ice Sheet oscillations. *Nature* **458**, 322–329.

115. Pollard, D. and DeConto, R.M. (2009) Modelling West Antarctic Ice Sheet growth and collapse through the past five million years. *Nature* **458**, 329–333.

116. DeConto, R.M. and Pollard, D. (2013) Pliocene retreat of Greenland and Antarctic ice sheet margins. Abstract, AGU Fall Meeting, San Francisco, 9–13 December 2013. Available from http://adsabs.harvard.edu/abs /2013AGUFMPP52A..08D (last accessed 29 January 2015).

117. Cook, C.P., van de Flierdt, T., Williams, T., Hemming, S.R., Iwai, M., Kobayashi, M., *et al.* (2013) Dynamic behaviour of the East Antarctic Ice Sheet during Pliocene warmth. *Nature Geoscience* **6**, 765–759.

118. Gramling, C. (2013) East Antarctica's ice sheet not as stable as thought. *Science NOW*. Available from http://news.sciencemag.org/2013/07/east-antarcticas-ice-sheet-not-stable-thought (last accessed 29 January 2015).

119. Badger, M.P.S., Schmidt, D.N., Mackensen, A. and Dancost, R.D. (2013) High-resolution alkenone palaeo-barometry indicates relatively stable pCO_2 during the Pliocene (3.3–2.8 Ma). *Philosophical Transactions of The Royal Society A* **371** (2001), 20130094.

120. Haywood, A. and Williams, M. (2005) The climate of the future: clues from three million years ago. *Geology Today* **21** (4), 138–143.

121. Dowsett, H.J., Haywood, A.M., Valdes P.J., Robinson, M.M., Lunt, D.J., Hill, D.J., *et al.* (2011) Sea surface temperatures of the mid-Piacenzian warm period: a comparison of PRISM3 and HadCM3. *Palaeogeography, Palaeoclimatology, Palaeoecology* **309** (2011), 83–91.

122. Robinson, M.M., Valdes, P.J., Haywood, A.M., Dowsett, H.J., Hill, D.J. and Jones, S.M. (2011) Bathymetric controls on Pliocene North Atlantic and Arctic sea surface temperature and deepwater production. *Palaeogeography, Palaeoclimatology, Palaeoecology* **309**, 92–97.

123. Dowsett, H.J. (2007) The PRISM palaeoclimate reconstruction and Pliocene sea-surface temperature. In: *Deep-Time Perspectives on Climate Change: Marrying the Signal from Computer Models and Biological Proxies* (eds M. Williams, A.M. Haywood, F.J. Gregory and D.N. Schmidt). Micropalaeolgoical Society Special Publication. The Geological Society, London, pp. 459–480.

124. Haywood, A.M., Valdes, P.J., Hill, D.J. and Williams, M. (2007) The Mid-Pliocene warm period: a test-bed for integrating data and models. In: *Deep-Time Perspectives on Climate Change: Marrying the Signal from Computer Models and Biological Proxies* (eds M. Williams, A.M. Haywood, F.J. Gregory and D.N. Schmidt). Micropalaeolgoical Society Special Publication. The Geological Society, London, pp. 43–457.

125. Lawrence, K.T., Herbert, T.D., Dekens, P.S. and Ravelo, A.C. (2007) The application of the alkenone organic proxy to the study of Plio-Pleistocene climate. In: *Deep-Time Perspectives on Climate Change: Marrying the Signal from Computer Models and Biological Proxies* (eds M. Williams, A.M. Haywood, F.J. Gregory and D.N. Schmidt). Micropalaeolgoical Society Special Publication. The Geological Society, London, pp. 539–562.

126. Pagani, M., Liu Z., LaRiviere, J. and Ravelo, A.C. (2010) High Earth-system climate sensitivity determined from Pliocene carbon dioxide concentrations. *Nature Geoscience* **3**, 27–30.

127. Lunt, D.J., Haywood A.M., Schmidt G.A., Salzmann, U., Valdes, P.J. and Dowsett, H.J. (2010) Earth system sensitivity inferred from Pliocene modelling and data. *Nature Geoscience* **3**, 60–64.

128. Ravelo, A.C., Billups, K., Dekens, P.S., Herbert, T.D. and Lawrence, K.T. (2007) Onto the ice ages: proxy evidence for the onset of Northern Hemisphere glaciation. In: *Deep-Time Perspectives on Climate Change: Marrying the Signal from Computer Models and Biological Proxies* (eds M. Williams, A.M. Haywood, F.J. Gregory and D.N. Schmidt). Micropalaeolgoical Society Special Publication. The Geological Society, London, pp. 563–573.

129. Fedorov, A.V., Brierley, C.M. and Emanuel, K. (2010) Tropical cyclones and permanent El Niño in the Early Pliocene Epoch. *Nature* **463**, 1066–1070.

130. Sriver, R.L. (2010) Tropical cyclones in the mix. *Nature* **463**, 1032–1033.

131. Lunt, D.J., Haywood A.M., Schmidt G.A., Salzmann, U., Valdes, P.J., Dowsett, H.J. and Lopston, C.A. (2012) On the causes of mid-Pliocene warmth and polar amplification. *Earth and Planetary Science Letters* **321–322**, 128–138.

132. Schmidt, D.N. (2007) The closure history of the Central American Seaway: evidence from isotopes and fossils to models and molecules. In: *Deep-Time Perspectives on Climate Change: Marrying the Signal from Computer Models and Biological Proxies* (eds M. Williams, A.M. Haywood, F.J. Gregory and D.N. Schmidt). Micropalaeolgoical Society Special Publication. The Geological Society, London, pp. 427–442.

133. Sarnthein, M., Bartoli, G., Prange, M., Schmittner, A., Schneider, B., Weinelt, M., *et al.* (2009) Mid-Pliocene shifts in ocean overturning circulation and the onset of Quaternary-style climates. *Climate of the Past* **5**, 269–283.

134. Sarnthein, M. (2013) Transition from late Neogene to early Quaternary environments. In: Elias S.A. (ed.) *The Encyclopedia of Quaternary Science*, Vol. 2 (ed. S.A. Elias). Elsevier, Amsterdam, pp. 151–166.

135. Cronin, T.M. (2010) *Paleoclimates: Understanding Climate Change Past and Present*. Columbia University Press, New York.

136. Shackleton, N.J. and Hall, M.A. (1984) Oxygen and carbon isotope stratigraphy of Deep Sea Drilling Project hole 552A: Plio-Pleistocene glacial history. Initial Reports Of the DSDP 81. US Government Print Office, Washington, DC, pp. 599–609.

137. Shackleton, N.J., Backman, J., Zimmerman, H., Kent, D.V., Hall, M.A., Roberts, D.G., *et al.* (1984) Oxygen isotope calibration of the onset of ice-rafting and history of glaciation in the North Atlantic region. *Nature* **307**, 620–623.

138. Shackleton, N.J. and Kennett, J.P. (1975) Late Cenozoic oxygen and carbon isotopic changes at DSDP Site 284: implications for glacial history of the Northern Hemisphere and Antarctica. Initial Report of the Deep-Sea Drilling Project 24. US Government Printing Office, Washington, DC, pp. 801–807.

139. Karas, C., Nürnberg, D., Gupta, A.K., Tiedemann, R., Mohan, K. and Bickert, T. (2009) Mid-Pliocene climate change amplified by a switch in Indonesian subsurface throughflow. *Nature Geoscience* **2**, 434–438.

140. Haywood, A.M., Hill, D.J., Dolan, A.M., Otto-Bliesner, B.L., Bragg, F., Chan, W.-L., *et al.* (2013) Large-scale features of Pliocene climate: results from Pliocene Model Intercomparison Project. *Climate of the Past* **9**, 191–209.

12

Solving the Ice Age Mystery: The Deep-Ocean Solution

12.1   Astronomical Drivers

By the time of the Pleistocene, the Earth had cooled to its lowest point since the Carboniferous glaciation 300 Ma ago. During the Pleistocene, much of the variation in the Earth's climate was due not to the plate tectonic processes that were important in older times, but to celestial mechanics. This led to cycles of 100 Ka in the eccentricity of the Earth's orbit, of 41 Ka in the Earth's axial tilt (which dominates radiation at high latitudes) and of 22 Ka in the precession of the equinoxes (which dominates radiation at low latitudes)[1]. As we saw in Chapter 6, Milutin Milankovitch set out the basis for our understanding of this process between 1920 and 1941.

In 1945, Frederick Zeuner (1905–1963) of London's Institute of Archaeology tested Milankovitch's theory by examining how it applied to what was known of the Pleistocene period on land[2]. Finding a close match between the sequence of Ice Age strata and the variations in insolation, he concluded, '*no objection can be raised against the astronomical theory of the glacial and interglacial phases of the Pleistocene*'[2]. Among those intrigued by what controlled the changes between glacial and interglacial periods was Richard Foster Flint (1901–1976) of Yale University, who, along with other honours, would be awarded the Prestwich Medal by the Geological Society of London in 1972 for his contributions to our understanding of the Ice Age. Flint was one of the most influential figures in Quaternary science in the 20th century, much admired for his seminal 1957 text *Glacial and Pleistocene Geology*[3]. This book concluded, '*the geometric scheme of distribution of insolation heating must be considered inadequate in itself to explain the Pleistocene climatic changes*'[3].

Things have changed since then. As Mike Walker of the University of Wales and John Lowe of Royal Holloway College London pointed out in 2007, Flint's approach was rooted in glacial geology. Since his day, those investigating Quaternary science have moved '*away from Flint's somewhat narrow glacial-geological paradigm towards the multi- and inter-disciplinary approach to the study of recent Earth history that is practiced today*'[3]. With this new approach, we can now analyse the '*rich and often readily accessible Quaternary record ... at a level of detail not normally possible for older geological periods*'[3]. Milankovitch's theory has come to stay[4].

Milankovitch lacked computers. André Léon Georges Chevalier Berger (1942–) (Box 12.1) used them to refine his calculations[5–8]. Figure 12.1 provides an introduction to Berger's findings, which we explore in more detail in Chapter 13.

Full comprehension of Ice Age climate change hinges on novel studies of deep-ocean sediments collected by piston cores and deep-ocean drilling, as we see in this chapter, and of ice cores, which we examine in Chapter 13. These studies fall into the science of the Quaternary, which comprises the Pleistocene, starting at 2.6 Ma ago, and the Holocene – the last 11 700 years[3]. To help move the field forward, scientists formed the International Quaternary Union (INQUA) in 1928[3]. Several

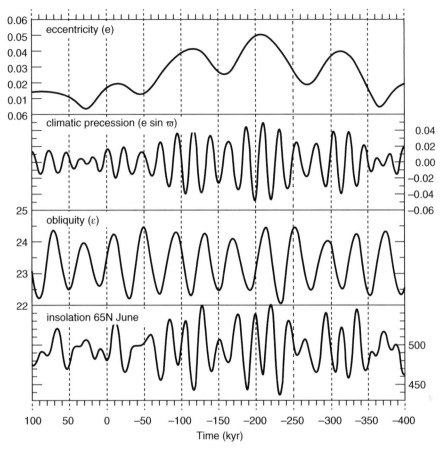

Figure 12.1   *The Berger Astronomical Model of Orbital Variability Present and Future. These curves have been produced in numerous formats in several publications by André Berger and Marie-France Loutre. Obliquity is expressed in degrees of tilt of the Earth's axis. Insolation at the summer solstice at 65° N is expressed in W/m².*

Box 12.1 André Léon Georges Chevalier Berger.

André Berger has a master's degree in meteorology from MIT (1971) and a doctorate from the Catholic University of Louvain, Belgium (1973). He is renowned for contributing to the renaissance of Milankovitch's theory of climate change, for making major contributions to simulating future climate change and for working on the first Earth model of intermediate complexity. He was professor of meteorology and climatology at Louvain, and then director of the Institute of Astronomy and Geophysics Georges Lemaître from 1978 to 2001, where he now has emeritus status. He has served as president or chairman of several national and international scientific organisations and committees and was honorary president of the European Geosciences Union. He was on the steering committee for the International Geosphere-Biosphere Programme (IGBP) and initiated the Palaeoclimate Modeling Intercomparison Project (PMIP). He has received many honours for his discoveries, including the Milutin Milankovitch Medal of the European Geophysical Society, and in 1996 he was made a knight of the realm by King Albert II.

scientific journals emerged to meet their needs: *Quaternary Research* (1971), *Boreas* (1972), *Quaternary Science Reviews* (1982), *Journal of Quaternary Science* (1985), *Quaternary International* (1989) and *The Holocene* (1991). Researchers also make good use of journals like *Paleoceanography* and *Palaeogeography, Palaeoecology, Palaeoclimatology*, as well as the four-volume *Encyclopaedia of Quaternary Science* (2006)[9].

12.2   An Ice Age Climate Signal Emerges from the Deep Ocean

As we saw in Chapter 7, the first to exploit the new technology of piston coring in order to examine the history of climate recorded in deep-sea sediments was Gustaf Arrhenius. In 1952, Arrhenius attributed alternations between carbonate-rich and carbonate-poor sediments in east Pacific cores to changes in the 'aggressiveness' of polar bottom waters. During the Ice Age, he thought, large volumes of bottom water were derived from the polar regions. Being cold, they carried large amounts of dissolved CO_2, which enabled them to dissolve, or to prevent the deposition of, deep-sea carbonates. Bottom waters of intervening warm periods carried less dissolved CO_2, so were less 'aggressive'. At the time, Arrhenius, like Flint, dismissed Milankovitch's ideas[10], although he later adopted them.

At about the same time, in the early 1950s, Lamont began its routine collection of long piston cores from the world's oceans. David Ericson, who was in charge of the new Lamont core store, found that although the distribution of the planktonic foraminiferan *Globorotalia menardii* indicated warm conditions, it was also influenced by ocean currents. Another species, *Globigerina pachyderma*, was a cold-water indicator, as were *Globigerina inflata* and *Globigerina bulloides*. Changes from warm to cold were also indicated by changes in the coiling direction of *Globorotalia truncatulinoides*. From the distribution of these species down-core, Ericson built a Quaternary stratigraphy incorporating the Holocene, the last glaciation and the previous interglacial[11].

Ericson and Arrhenius's interests were shared by Cesar Emiliani, whom we met in Chapters 6 and 7. In 1955, Emiliani analysed the oxygen isotopes in planktonic foraminifera collected from eight piston cores from the Swedish Deep-Sea Expedition and four from the Lamont core store. Finding fluctuations in the $\partial^{18}O$ ratio with time, he interpreted them as representing the variations

in climate between glacial and interglacial periods[12]. By analysing the same species of surface-dwelling foraminiferan, he eliminated the effect of metabolic differences between different species. The resulting variations were due to differences in either the temperature or the isotopic composition of seawater, the latter representing the amount of water tied up as ice on land. Emiliani '*guessed that 60% of the signal was due to the temperature effect, 40% to the ice effect*'[13]. Following Zeuner's reasoning[2], he thought that the variations in $\partial^{18}O$ with time represented changing insolation[12]. He invented the nomenclature that is still in use today, in which each warm or cold period is identified as a marine isotope stage (MIS). MISs with even numbers are cold stages, those with odd numbers are warm stages. Some can be subdivided into substages (e.g. 5a, 5b, 5c).

In March 1961, Emiliani teamed up with Flint to link the Pleistocene record in continental and deep-sea sediments[14]. They considered that MIS 1 was the Holocene, MIS 2 was the main Würm glacial stage on land, MIS 3 was the early to main Würm glacial interval on land, MIS 4 was the early Würm on land and MIS 5 was the last interglacial. Thinking about the possible drivers for the Ice Age, they favoured a model of glaciation based on Milankovitch's concept that ice accumulated during cool summers, driven by insolation in the Northern Hemisphere. Evidently, Flint had changed his mind about Milankovitch since 1957. They discounted Plass's idea that ice ages were in some way controlled by atmospheric CO_2, because they thought, wrongly, that Revelle and Suess[15] had concluded, based on studies of ^{14}C, that atmospheric CO_2 would be rapidly taken up by the ocean. It would be a while before the role of CO_2 in the Ice Age climate would be fully understood.

Before Emiliani's isotopic analyses of deep-sea cores, it was thought that there were only four major glacial periods during the Ice Age[2]. Emiliani found more than twice that! While this shocked classical Quaternary geologists, physicist Nicholas J. Shackleton (Box 7.3) agreed with Emiliani[16]. The fly in the ointment was the inadequacy of methods for accurately dating Emiliani's cores. Emiliani guessed that his main $\partial^{18}O$ cycles were 41 Ka long, but they were later found to last about 100 Ka.

Subtle differences in the ways in which Ericson and Emiliani interpreted their data meant that, for a while, they disagreed about the precise sequence of cold and warm events. When they did agree, we had two independent means of mapping changes between glacial and interglacial periods in deep-sea cores – one palaeontological,

Figure 12.2 *John Imbrie.*

from microscopic fossils of marine plankton (microfossils), and the other geochemical, from oxygen isotopes in those same fossils.

By the mid 1960s, two new figures occupied centre stage: Nick Shackleton and the micropalaeontologist and stratigrapher John Imbrie (1925–), of Brown University (Figure 12.2, Box 12.2). Our understanding of Ice Age climate owes a great deal to these two men.

Box 12.2 John Imbrie.

Imbrie obtained a BA from Princeton in 1948, after serving with the US 10th Mountain Division in Italy during the Second World War. Having obtained a PhD from Yale in 1951, he taught at Columbia University until 1967. Joining Brown University, he held the Henry L. Doherty Chair in Oceanography, and he now holds emeritus status there. He pioneered the use of computers to demonstrate the relation of assemblages of plankton to the temperature of surface waters, thus providing palaeoceanography with one of its key tools. His book *Ice Ages: Solving the Mystery*[18], written with his daughter Katherine, won the 1976 Phi Beta Kappa Prize. Imbrie was co-author with Hays and Shackleton of the 1976 paper in *Science* that linked Milankovitch variations to the sediment record. He was elected to the US National Academy of Sciences in 1978 and received the Maurice Ewing Medal of the American Geophysical Union in 1986,

the Twenhofel Medal of the Society of Sedimentary Geology, the Lyell Medal of the Geological Society of London in 1991 and the Vetlesen Prize in 1996.

Using his refined mass-spectrometric technique (see Chapter 7), Shackleton showed by 1967 that a distinctive $\partial^{18}O$ pattern is evident in the benthic foraminifera that live on the seabed, which are bathed in cold oxygen-rich bottom water sinking from the surface in the polar regions[17]. Most of this water is colder than 4 °C, so short-term variations in the $\partial^{18}O$ ratio in these organisms tell us more about changing seawater temperature than changing ice volume. Shackleton found that planktonic foraminifera growing in surface waters displayed much the same signal. He calculated that about 66% of the $\partial^{18}O$ shift between glacial and interglacial periods was due to changes in ice volume, not to the influence of temperature, which was the opposite of what Emiliani had concluded. This revelation caused a paradigm shift in our understanding of $\partial^{18}O$ ratios in the service of palaeothermometry. Urey had assumed that the oxygen isotopic composition of seawater would be invariant[19]. Clearly it was not. Today, the $\partial^{18}O$ ratios of benthic foraminifera are taken as representing the temperatures of polar surface waters.

By 1969, Imbrie realised that because the total assemblage of planktonic foraminifera in surface waters should reflect the environment in which they lived, and multivariate statistical analyses could quantify that relationship, statistical analyses of faunal assemblages down-core could be used to ascertain past climate change[20]. Reanalysing the Caribbean cores analysed by Ericson and Emiliani, he and Nilva Kipp showed that the temperatures of surface waters of glacial periods there fell by just 2 °C, not 6 °C. Because the fluctuations in their data agreed with those determined by Emiliani from isotopes, it was clear that *Globorotalia menardii*, which Ericson used to identify warm periods, fluctuated in ways unrelated to temperature, explaining the discrepancy between Ericson and Emiliani's data[21, 22]. This was a breakthrough.

By the late 1960s, as the Deep Sea Drilling Project (DSDP) got underway, analyses of $\partial^{18}O$ changes down piston cores from different parts of the ocean showed that sediments could be routinely subdivided into Emiliani's MISs representing glacial and interglacial periods. These stages coincided with intervals defined by Imbrie's assemblages of microfossils and could be correlated from one core to another over vast oceanic distances, suggesting planetary control.

Was that planetary control the same as Milankovitch's astronomically controlled insolation? To find out, geologists needed closely spaced dates down-core. In the late 1960s to early 1970s, microfossils could tell us about environmental change from cold to warm and back, but not about the ages of cold and warm stages. As we saw in Chapter 6, radiocarbon dating was useful, but only in sediments less than 50 Ka old. Layers of volcanic ash older than 100 Ka could be dated by the potassium–argon (K-Ar) method. Changes in the Earth's magnetic field could be used to date specific sedimentary horizons, but only in sediments older than 780 Ka. Together, these independent techniques provided a crude means of dating sediment layers in cores. Over time, more techniques would become available. A major breakthrough in radiocarbon (^{14}C) dating came about in 1977, when a new technique – accelerator mass spectrometry (AMS) – enabled us to count ^{14}C atoms, as opposed to measuring ^{14}C decay. AMS can date samples as small as a pinhead-sized microfossil. The technique is fast and cheap, but it is still limited to sediments less than about 50 Ka old.

Determining the ages of the glacial and interglacial sedimentary stages identified by ∂^{18}O and microfossil analyses became a major objective of the international Climate Long Range Investigation, Mapping and Prediction (CLIMAP) project. Founded by Imbrie, Shackleton and others, CLIMAP began in spring 1971 as part of the International Decade of Ocean Exploration[23]. The project aimed to establish average boundary conditions for the Last Glacial Maximum at 18 Ka ago. Those conditions included the geography of the continents, the albedo of land and ice surfaces, the extent and elevation of permanent ice and the sea surface temperature. Modellers would use those conditions in atmospheric General Circulation Models (GCMs) to map the climate of the Last Glacial Maximum. CLIMAP would then test model outputs against palaeoclimate data. The first simulation was for August 18 Ka ago[23]. Sea surface temperature values were derived from ∂^{18}O data and from Imbrie's statistical analyses of planktonic faunal assemblages. The extent of sea ice in the polar regions was estimated from the presence or absence of diatomaceous sediments, with absence indicating ice.

The CLIMAP data showed that extensive cooling at the poles and an expanded area of land and sea ice steepened the thermal gradient between the Equator and the poles, strengthening the winds. In the Southern Ocean, the Antarctic polar front moved north, along with Antarctic

sea ice. The Subtropical Front moved far enough north to limit the passage of warm Indian Ocean water around South Africa and into the South Atlantic. This cooled the South Atlantic and created a closed anticlockwise gyre in the Indian Ocean. Upwelling increased where it is found today, along continental margins and along the Equator, as the winds that drove it increased in strength. On land, grasslands, steppes, deserts and ice spread at the expense of forests, increasing the Earth's albedo.

Palaeoclimatologists noticed that the wiggles in the ∂^{18}O curves down sediment cores seemed to mimic the wiggles in the patterns of Earth's insolation through time. If the match was real then it offered an opportunity to date the age of the sediments from the pattern of wiggles in the oxygen isotope data. Starting with just a few radiometric dates as tie points, the CLIMAP scientists assumed that the rates of sedimentation in different MISs were constant down-core. This enabled them to estimate the age of each wiggle on the curve of variation in ∂^{18}O back through time. Applying spectral analysis to the ∂^{18}O curve dated in this way, James Hays (1938–) of Lamont, together with Imbrie and Shackleton, demonstrated in a landmark paper in 1976 that in cores where sedimentation was undisturbed, the variations in ∂^{18}O over the past 450 Ka accurately mimicked the orbital signals calculated by André Berger[24]. Furthermore, climate changes in the Northern Hemisphere were essentially synchronous with those observed in the Southern Hemisphere. The correlation between the astronomical variables and ∂^{18}O told them that '*changes in the earth's orbital geometry are the fundamental cause of the succession of Quaternary Ice Ages*'[24]. The three Milankovitch mechanisms – eccentricity, tilt and precession – worked in unison to provide the 'pacemaker of the ice ages'.

Not only that, but the fact that the same pattern of wiggles occurred everywhere meant that even cores without precise radiometric dates could be dated by reference to a standard ∂^{18}O curve derived by merging data from several well-dated cores. Wiggle matching offered an incredible opportunity to date intervals of time as small as about 1000 years long. Patterns of tree rings back through time offer much the same possibility, provided they come from much the same area and so experienced more or less the same climate changes through time. While this is very helpful for the past 11 Ka, tree rings do not provide us with a lengthy and globally distributed data base of the kind provided by deep-sea cores. Wiggle matching of oxygen isotope curves from core to core does contain the assumption that the section has not been disturbed by either erosion or the lateral introduction of material by turbidity currents, but these possibilities are identified or eliminated by comparing individual cores with the global standard. Indeed, wiggle matching can identify how much section has been removed! Nowadays, wiggle matching has enabled palaeoclimatologists to push the $\partial^{18}O$-calibrated time scale back into the Oligocene, about 30 Ma ago[13]. Amazing!

This incredible breakthrough refined the resolution of the geological time scale beyond anything previously imaginable, except where annual layering was preserved in tree rings, corals, stalagmites or lake sediments, most of which did not allow dating back beyond about 2000 years ago. Palaeoclimatologists who started their careers in the 1980s and later take these advances for granted, but they were astonishing developments to the geologists of my age group. As Mike Leeder pointed out in 2011, this new understanding was '*arguably as big an earth sciences discovery as that of plate tectonics*'[25]. Mike Walker and John Lowe agreed, citing the 1976 Hays paper as '*perhaps the most important Quaternary paper of the past 50 years*'[3]. It definitively overturned the long-held perception that there had been four major Quaternary glacial periods[26], by showing that there had been many more cycles of climate, ice volume and sea level in the late Cenozoic, and that these cycles formed in response to variations in Earth's orbital parameters. As Nick McCave and Harry Elderfield point out in Shackleton's obituary, '*This clear recognition of orbital control is also now revolutionizing the whole of stratigraphy (the study of geological strata) because it provides in principle a means of correlating beds at separated parts of the Earth to a precision of 20 000 years at a time of hundreds of millions of years ago, and of determining precise "orbitally tuned" age-calibrated stratigraphies back to about 250 Ma ago*'[27].

This was not all. Given that the climate was governed by celestial mechanics, and that Berger's data projected Earth's orbital properties and insolation far into the future, Hays, Imbrie and Shackleton deduced that '*the long term trend over the next 20,000 years is towards extensive Northern Hemisphere glaciation and a cooler climate*'[24]. That's not quite the picture we have today, but it's close. Berger's data show that because the Earth's orbit is at present close to circular, the present warm Holocene interglacial should last some 30–50 Ka, of which we have already experienced 10 Ka[28, 29]. So, we should have ~20 Ka more relatively warm climate before the next glacial period. For today, Berger's data show that Earth should be experiencing

a slight cooling trend, which started around 10 Ka ago (Figure 12.1). We look at that more closely in later chapters. These various dramatic developments were ably summarised by John and Katherine Imbrie in their 1979 book *Ice Ages: Solving the Mystery*[18]. It remains a classic.

In the 1980s, the CLIMAP project was succeeded by the SPECMAP (Spectral Mapping) project, designed by Imbrie, Shackleton and others to produce continuous time series of Ice Age climate change from deep-sea sediments and to facilitate studying their spectral properties. A key SPECMAP achievement was publication of a time scale for the last 780 Ka, based on a $\partial^{18}O$ reference curve compiled by stacking together planktonic foraminiferal $\partial^{18}O$ records from five low- and middle-latitude sites. Stacking avoids local 'noise' interfering with the underlying signals[30]. Known as the SPECMAP stack, this curve, which was tuned (phase-locked) to the oscillations of precession and obliquity, provided a continuous geological time scale for the late Pleistocene, the divisions of which were accurate to within ±5000 years, an astoundingly high resolution for geological records. The SPECMAP $\partial^{18}O$ stack was improved and extended over time[31, 32]. In 2005, Lorraine Lisiecki, then at Brown University, and Maureen Raymo, then at Boston University, replaced it with a stack made from combinations of benthic foraminiferal data (Figure 12.3)[33, 34], which show less variability than planktonic data. Over 100 MISs have been identified, going back some 6.6 Ma. The stack is a 'type section' against which new core measurements are compared.

The Last Glacial Maximum, or MIS 2, extended from 30 to 15 Ka ago. On average, compared with today, it was thought to be 5–6 °C colder globally. It was much colder in the Arctic, with temperatures in central Greenland depressed by as much as 20 °C[35]. At that time, much Arctic land lay beneath continental ice sheets, and the Arctic Ocean was mantled by continuous sea ice and entrapped icebergs. The lack of northward transport of warm, salty water during the winters made them exceptionally cold. Polar desert replaced tundra. Ice volume peaked about 21 Ka ago, after which rising insolation caused ice sheets and glaciers to melt back, with most coastlines becoming ice-free before 13 Ka ago[35]. We will discuss deglaciation later.

The climate signal in cores was not smoothly varying, unlike the variation in orbital and axial properties (Figure 12.1). As we can see from Figure 12.3, it was saw-toothed – something first pointed out by Broecker

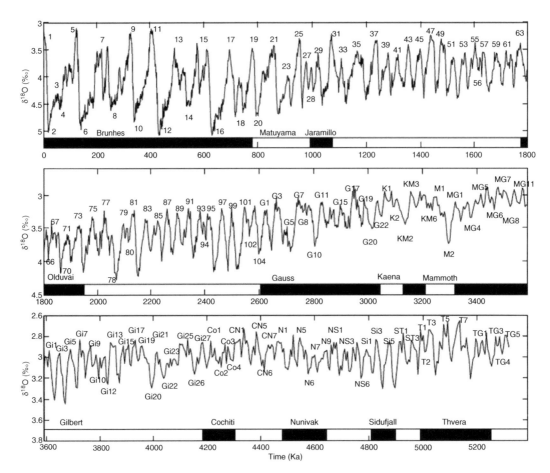

Figure 12.3 *Benthic oxygen isotope stack, constructed by the graphic correlation of 57 globally distributed benthic $\partial^{18}O$ records covering 5.3 Ma. Note that the scale of the vertical axis changes from panel to panel. From this stack, a number of new MISs were identified in the early Pliocene. MISs are identified by number back to 2.6 Ma ago; before that, the lettering refers to the name of the magnetic chron in which the isotope peaks appear (e.g. Si = Sidufjall, Co = Cochiti etc).*

and Van Donk[36]. The saw-tooth shape represents the slow growth of ice followed by rapid deglaciation, calling for strong positive-feedback mechanisms to accelerate melting. We examine this in more detail in Chapter 13.

CLIMAP's maps of sea surface temperature for the winter and summer of the Last Glacial Maximum (21.5–18.0 Ka ago) were later updated by the Glacial Atlantic Mapping Project (GLAMAP)[37], which reconstructed the glacial Atlantic Ocean from 275 deep-sea sediment cores[38]. During the northern winter, sea ice extended south as far as about 50° N, close to the latitude

of Cork, on the south coast of Ireland. During the northern summer, warm surface waters moved northwards into the Norwegian-Greenland Sea much as they do today, but they were between ice sheets on both sides, meeting sea ice at about latitude 70° N. As today, there was a return flow from the Arctic down the east side of Greenland. A proto-Gulf Stream marched eastward from Labrador to Lisbon, where the sea surface temperature averaged about 16 °C.

In parallel with GLAMAP, a comparable project began to examine Environmental Processes of the Ice Age:

Land, Oceans, Glaciers (EPILOG). It focused on the 21 Ka interval, this being the age of minimal summer insolation at 65 °N, and the time when ice sheets reached their maximal volume, as represented by sea level fall[39]. GLAMAP and EPILOG were followed by the Multiproxy Approach for the Reconstruction of the Glacial Ocean project (MARGO)[40]. These various studies and others concluded that the last deglaciation began between 22 and 18 Ka ago. Two schools of thought emerged, one suggesting that deglaciation began in the Southern Hemisphere, with surface and deep-ocean warming followed by tropical sea surface temperatures and by atmospheric CO_2, and the other suggesting that it began in the Northern Hemisphere, where summer insolation at high northern latitudes was the trigger for ice-sheet decay and sea level rise[41]. We explore these issues in Chapter 13.

The MARGO project team constructed maps of sea surface temperature to provide constraints on ocean cooling at the Last Glacial Maximum[42]. Like CLIMAP, they found that the strongest mean annual cooling (-10 °C) occurred in the mid-latitude North Atlantic, extending into the western Mediterranean, but unlike CLIMAP they found that this cooling was most pronounced in the east. Indeed, most ocean basins had cooler eastern than western sides. This eastern cooling was probably due to increased upwelling of cold water forced by stronger coastal winds. It was not replicated by existing GCMs for the Last Glacial Maximum, for reasons not then fully understood (in 2009). In contrast with CLIMAP, the MARGO team found that conditions were ice-free in the summer in the Nordic seas, and that the tropics were on average 1.7 °C cooler than CLIMAP had thought. In the Southern Ocean, the polar front shifted north from near 60 to 45° S, associated with a cooling of 2–6 °C in the austral winter.

One of the key features evident from the stack of $\partial^{18}O$ data (Figure 12.3), noted in 1976 by Shackleton and Opdyke[43], is that climate variability has grown with time[19]. In the earlier part of the Pleistocene, the signal comprised relatively small glacial–interglacial changes, in which signals of precession (22 Ka cycles) and obliquity (41 Ka cycles) predominated. The signals became much larger at about 900 Ka ago, after which a signal with a spacing of 100 Ka intervals predominated. The 100 Ka signal itself became larger with time, especially from around 430 Ka ago (MIS 11) onwards. The change at 900 Ka ago formed the Mid Pleistocene Transition (MPT). What did it represent?

Harry Elderfield and colleagues used Mg/Ca ratios to establish what part of the $\partial^{18}O$ signal at the transition was due to ice volume rather than water temperature[44]. Changes in ice volume from glacial to interglacial were much smaller before the transition than since, presumably because the older ice sheets were smaller in area and/or thickness. The transition was a sudden jump, not the result of a long-term trend towards increased ice volume and colder temperatures. Elderfield's team concluded that it represented '*an abrupt reorganization of the climate system*'[44]. The trigger seemed to be a brief period of anomalously low summer insolation in the Southern Hemisphere during the warm MIS 23. This suppressed the melting of ice formed previously in cold MIS 24, allowing unusually extended ice growth in the following cold MIS 22, at 900 Ka ago, to yield a very large ice sheet, associated with a lowering of sea level of about 120 m[44].

Investigating the behaviours of ice volume and temperature, Elderfield's team confirmed that ice volume followed a saw-toothed pattern, growing steadily from low amounts during interglacials to high amounts during glacials, then suddenly retreating. In contrast, bottom water temperatures followed a square wave pattern, falling to a certain level as ice volume grew, then staying more or less constant, before rising again as ice volume decreased. The temperatures of bottom waters during glacial periods remained constant at -1.5–2.0 °C, because, once the temperature of surface water in the source region fell to about the freezing point of salt water, it would fall no further. Bottom water temperatures warmed to 3 °C during interglacial periods.

Prior to the transition, sea level fell to 70 m below present levels. The drop by a further 50 m at the transition exposed continental shelf sediments to erosion, transferring marine organic matter rich in ^{12}C to the deep sea and lowering the $\partial^{13}C$ ratio of bottom water and benthic organisms. Elderfield's team calculated that about half of the fall in $\partial^{13}C$ at 900 Ka was due to this change in carbon reservoirs, with the other half coming from a reduction in the influence of North Atlantic Deep Water[44].

As the volume of ice increased across the transition, the supply of aeolian dust, represented by the rates of accumulation of sedimentary iron and terrestrial leaf waxes, doubled in the Southern Ocean[45]. The increase in the dust supply tells us that the surrounding lands dried out as the globe cooled. The increase in iron helped to cool the globe further, via positive feedback, because an increase in iron as a key nutrient stimulates productivity, drawing CO_2 from the atmosphere, as we see in more detail later[45]. This interpretation is supported by an increase in the sedimentation of opal, representing diatom productivity, at the same time. The rise in productivity drove a 30 ppm reduction in CO_2 across the transition. By driving a descent into deep, cold, glacial periods, the insolation/dust/CO_2 feedback may have initiated the strong 100 Ka periodicity that characterised subsequent climate change. On the global scale, an increase in the supply of dust also coincided with the start of the major Northern Hemisphere glaciation at about 2.6 Ma ago, which drew down CO_2 in much the same way.

The notion of using a single stack of $\partial^{18}O$ values to represent Earth's recent glacial history would have seemed odd to Joseph Adhémar and James Croll, whom we met in Chapter 3, because they thought that cooling related to precession would alternate between the two hemispheres. As we now know, glaciers and ice sheets in Patagonia and Antarctica actually advance and retreat at more or less the same times as those in the Northern Hemisphere. Why? The answer lies in those feedbacks to which Croll first introduced us. Antarctica is an ice-covered continent surrounded by ocean – there is nowhere for its land ice to expand into. When Antarctic ice is at a maximum, global ice can only increase by growing on the Northern Hemisphere continents. This growth lowers sea level, exposing Antarctica's continental shelf and so providing space for yet more ice growth. Sea level links ice growth on Antarctica to that on the northern continents. Thus, glaciation tends to become more or less synchronous in both hemispheres, even while the insolation is opposite[13]. Besides that, insolation has certain seasonal characteristics that help to align glaciation in the two hemispheres even though their annual insolation signal is opposed, as we see in Chapter 13.

The latest view of the temperature of the Last Glacial Maximum is that it was probably $4.0 \pm 0.8\,°C$ cooler than the modern preindustrial climate[46].

The growing literature on orbital variations and their record in cores from ice, sediments, corals and stalactites fuelled intensive discussion about the precise mechanisms underlying the climate changes of the Ice Age, which we review in Chapter 13.

12.3   The Ice Age CO_2 Signal Hidden on the Deep-Sea Floor

Could carbon isotopes from marine sediments tell us about the abundance of CO_2 in the Ice Age atmosphere? Wally Broecker suggested that the atmospheric CO_2 signal could be represented by the difference in $\partial^{13}C$ ratios between surface planktonic foraminifera and bottom-dwelling benthic foraminifera, an idea followed up by Nick Shackleton and colleagues in 1983[47]. The variations they detected in CO_2 by this means matched those found in ice cores by Oeschger, as mentioned in Chapter 9. Clearly, CO_2 rose and fell with rises and falls in temperature during the Ice Age. How reliable was the association between estimated CO_2 and the $\partial^{18}O$ values used to estimate temperature? In 1985, working with Nick Pisias of Oregon State University to analyse the $\partial^{13}C$ and $\partial^{18}O$ profiles through the past 340 Ka in a core from 3091 m depth in the Pacific, Shackleton found that CO_2 closely followed temperature[48]. It slightly lagged orbital insolation in June at 65° N, led the response in ice volume (documented by variations in $\partial^{18}O$) by about 2500 years and was closely linked to variations in axial tilt, which dominates the insolation signal at middle to high latitudes. Shackleton and Pisias concluded that the CO_2 signal was forced by high-latitude orbital insolation through '*a mechanism at present not fully understood*'[48] – probably the effect of that insolation on ocean circulation. As changes in CO_2 led changes in ice volume in the North Atlantic, the CO_2 must have contributed to the forcing of changes in ice volume there. It was a forcing factor. Insolation warmed the ocean, which released CO_2, which enhanced temperature, stimulating an eventual decrease in ice volume.

These conclusions were consistent with the proposal by Wally Broecker and Tsung-Hung Peng in their 1982 classic *Tracers in the Sea* that, since the ocean contains about 60 times more carbon than the atmosphere, the glacial–interglacial change in atmospheric CO_2 content must have been driven by changes in ocean chemistry[49]. Broecker seems to have been the first to suggest that a glacial increase in the strength of the biological pump drove down CO_2 levels. His initial ideas about the biological pump revolutionised the field of chemical oceanography[50].

Broecker's ideas got John Martin (1935–1993) thinking. Director of the Moss Landing Marine Laboratories in California, and crippled by polio when he was 19, Martin set himself the task of figuring out the role of phytoplankton – the grass of the sea – in the global climate system[51]. Phytoplankton use CO_2 for photosynthesis. When their remains sink to the seafloor and decompose, the CO_2 is returned to deep-ocean waters or trapped in sediment and so can no longer contribute to warming the planet. In order to determine how much plankton sank to the seafloor in a given time, Martin organised the Vertical Transport and Exchange of Oceanic Particulate Program (VERTEX) in 1981, placing sediment traps across the North Pacific to sample the flux and composition of settling particulates. Among other things, he discovered that the parts of the ocean that are high in nutrients but low in chlorophyll were depleted in iron (Fe)[52]. Joseph Hart, an English scientist, had speculated in the 1930s that this might be the case, but was unable to prove it. Martin proposed in 1990 that Fe was a limiting nutrient and that production of phytoplankton could be negatively affected by its supply, for example in airborne dust[53].

During the Last Glacial Maximum, the supply of dust was 50 times greater than today, enhancing productivity enough to draw CO_2 out of the air. Lack of Fe-rich dust during interglacials slowed productivity, leaving CO_2 in the air. Martin suggested testing his hypothesis through Fe-enrichment experiments even at the scale of the whole Southern Ocean. Several such experiments were carried out, the first in 1993, although none at the scale of an entire ocean. The early experiments found that the excess organic matter created by the addition of Fe was recycled in the water column; it did not settle to the seabed as Martin had imagined[54]. That changed in 2012, when Victor Smetacek of Germany's Alfred Wegener Institute for Polar and Marine Research performed a 5-week-long Fe-fertilisation experiment in the Antarctic Circumpolar Current with colleagues and discovered that at least half of the diatom bloom caused by the fertilisation sank far below 1000 m, with much being likely to have reached the deep-sea floor[55]. This confirmed the view that '*iron-fertilized diatom blooms may sequester carbon for timescales of centuries in ocean bottom water and for longer in the sediments*'[55]. Here was support for the geoengineering notion that iron fertilisation of the ocean could transport CO_2 out of the surface waters and into the deeps, thus drawing down atmospheric CO_2.

The abundance of CO_2 in the air between glacial and interglacial times was also governed by the presence or absence of sea ice[56]. Growing sea ice placed a lid on polar surface waters, preventing them from absorbing CO_2 from the air. As a result, CO_2 would gradually accumulate in the air, causing it to warm. Melting of sea ice exposed the cold ocean, enabling it to absorb CO_2 from the air, contributing to eventual cooling.

12.4   Flip-Flops in the Conveyor

Palaeoceanographic studies radically changed our understanding of the variability of Ice Age climate and the role of the ocean in climate change. One key result was the realisation that the circulation of the ocean had different stable states for glacials and for interglacials. During interglacials like the Holocene, in which we live now, ocean circulation was much as it is today (Figure 12.4).

Warm, salty surface water is drawn towards the Arctic through the northern branch of the Gulf Stream, losing heat to the atmosphere en route. This heat warms northwest Europe. By the time the salty water reaches the Norwegian-Greenland Sea, it has cooled to the point of becoming dense enough to sink and form North Atlantic Deep Water, which moves south towards Antarctica and fills the mid-water depths of the Atlantic, Indian and Pacific Oceans. The strong westerly winds blowing around Antarctica towards the east force these northern-sourced deep waters to the surface through the process of upwelling. The newly upwelled surface waters then return to the North Atlantic to close the cycle through two pathways. First, under the influence of Antarctica's coastal easterly winds, some water moves south on to the Antarctic continental shelf, where the excretion of salt from sea ice forming at the ocean's surface makes it dense enough to sink to the deep-ocean floor. This deep, cold Antarctic Bottom Water moves back to the north through the Atlantic, Indian and Pacific Oceans. Because cold water dissolves larger amounts of oxygen from the atmosphere than does warm water, these deep waters rich in oxygen aerate the bottom of the world's oceans. Second, much of the rest of the Circumpolar Deep Water eventually wells up to the surface in the Pacific. There, it becomes entrained in the major surface currents that move west from the North Pacific through the Indonesian archipelago, across the Indian Ocean, down the East African coast in the Agulhas Current and across the South Atlantic to the Equator in the Benguela Current, gaining salt and heat along the way. This water ends up feeding in to the southern end of the Gulf Stream, to repeat the cycle. This global pattern of southward-moving North Atlantic Deep Water and northward-moving warm salty surface water forms the so-called Thermohaline Conveyor Belt (from '*thermo*', meaning heat, and '*haline*', meaning salt), which moves heat and salt around the globe. Wally Broecker is credited with devising this cartoon of ocean circulation[57]. As we'll see later, there is a third pathway not shown in Figure 12.4. Northward-moving surface water in the Southern Ocean eventually sinks at the polar front near 60° S to form Antarctic Intermediate Water, which circulates through the world ocean at depths of 600–1000 m.

The Thermohaline Conveyor, these days referred to by physical oceanographers as the Meridional Overturning Circulation (MOC), is geologically young. It did not exist before the opening of the Drake Passage. Robbie Toggweiler from Princeton and H. Bjornsson from Iceland used experiments with an ocean model in 2001 to show that, prior to the opening of the passage, ocean temperature should have been symmetric about the Equator, with meridional overturning being driven by deep-water formation at the poles in both hemispheres[58]. With the passage open, the overturning took the form of an interhemispheric conveyor, with deep-water formation primarily in the Northern Hemisphere. The conveyor made temperatures rise in the Northern Hemisphere and fall in the Southern Hemisphere, as the ocean transported heat north across the Equator, especially in the Atlantic. The high salt content of the warm surface water allowed northern waters to become dense when cooled, thus driving the return flow at depth as North Atlantic Deep Water. While salinity differences are obviously important in driving the conveyor, Toggweiler's model showed that the conveyor could not be entirely driven by buoyancy. The westerly winds funnelled through Drake Passage do more than 'set the stage' for the work of the buoyancy forces in the North Atlantic: they are an indispensible part of the conveyor circulation, because they drive the upwelling of deep water around Antarctica to bring North Atlantic Deep Water to the surface[58].

In glacial times, the power of the Thermohaline Conveyor was much reduced, as we can see from the geochemistry of microfossils from cores collected at different water depths. In 1982, Ed Boyle from MIT and Lloyd Keigwin from Woods Hole Oceanographic Institution (WHOI) found that the shells of benthic foraminifera contained variations down-core in the ratio of cadmium (Cd) to calcium (Ca)[59]. The ocean distribution of Cd follows that of nutrients, so the Cd/Ca ratio in these shells records variations in the nutrient content of bottom waters. North Atlantic Deep Water turned out to be relatively poor in Cd compared with deep water of southern origin. The Cd/Ca evidence told Boyle and Keigwin that the intensity of the northern source relative to the southern one diminished by a factor of two during severe glaciations. $\partial^{13}C$ values also document the distribution of nutrients, being low in old waters containing abundant dissolved organic carbon rich in ^{12}C. Both Cd/Ca and $\partial^{13}C$ distributions suggest a strong stratification in the North Atlantic at the Last Glacial Maximum, with a low-nutrient, high-$\partial^{13}C$ water

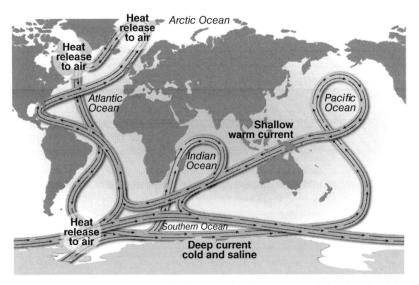

Figure 12.4 *Ocean thermohaline conveyor belt, showing the directions and depths of cold, salty, oxygen-rich deep currents, warm surface currents and vertical connections from deep to shallow and vice versa.*

mass (Glacial North Atlantic Intermediate Water) occupying depths down to around 2000 m, and a high-nutrient, low-$\partial^{13}C$ water mass of southern origin below that[60]. Laurent Labeyrie of the Centre National de la Recherche Scientifique (CNRS) laboratory in Gif-sur-Yvette, southwest of Paris, a winner of the European Geophysical Union's Hans Oeschger Medal in 2005, confirmed that the overturning circulation of the glacial Atlantic was shallower and weaker than today's[61]. Production of North Atlantic Deep Water was much reduced during the Last Glacial Maximum.

With the development of vast ice sheets on western Europe and North America, the sea's surface froze during the winter as far south, at times, as 40° N – the latitude of Boston in the west and Lisbon in the east. With the freezing over of the Norwegian-Greenland Sea, there was no longer a significant source for North Atlantic Deep Water[62, 63]. Boyle realised that this icing up prevented deep-water formation[64]. The Cd/Ca signal showed strong periodicity at 41 Ka, synchronous with changes in the tilt of the Earth's axis, confirming that Northern Hemisphere ice cover (at least in the Norwegian-Greenland Sea) was controlled by insolation at high latitudes. Increasing tilt produced higher summer insolation and less ice cover.

The icing up of the North Atlantic north of 40° N at the Last Glacial Maximum switched off the branch of the Gulf Stream that extended into the Nordic Seas and deflected the Gulf Stream east between New York and Lisbon. Sea ice also extended 10° closer to the Equator in the Southern Hemisphere, putting a lid on Southern Ocean processes. With the growth of ice on land, sea level fell 120–130 m. The growth of land ice and sea ice increased Earth's albedo significantly, helping to cool the planet.

12.5 A Surprise Millennial Signal Emerges

In the 1970s, marine geologists dredged large angular boulders of continental rocks such as granite from the Mid-Atlantic Ridge in the North Atlantic, and some people unwisely took this to mean that the ridge was made of continental rock[65]. By 1977, Bill Ruddiman of Lamont-Geological Observatory had mapped wide swathes of ice-rafted glacial debris over much of the North Atlantic seafloor north of a line connecting Boston with Lisbon, and it was realised that the angular boulders were also ice-rafted[66–68]. More detailed studies of piston cores from the North Atlantic by the German geologist Hartmut Heinrich (1952–), of the Deutsches Hydrographisches Institut in Hamburg, found ice-rafted debris concentrated in six layers deposited half a precession unit (11 Ka) apart[69]. Heinrich's discovery, in 1988, aroused intense interest, and an international team formed under the leadership of Gerard Clark Bond (1940–2005) of Lamont in order to investigate these sediments, which Bond's group named 'Heinrich layers'[70].

The layers formed between 10 and 60 Ka ago, apparently in response to surging within the Canadian ice sheet, which led to it breaking up into myriads of icebergs – an armada carrying rocks across the ocean. Charles Lyell would have been pleased, since this was the mechanism he had proposed back in the 1830s to explain the widespread distribution of glacial debris across western Europe. However, as we saw in Chapter 2, it is not reasonable to call on drifting icebergs to cover Europe – there, we must call on a grounded ice sheet to explain the origin of the boulders and associated boulder clay. Along with the melting icebergs came vast volumes of cold, fresh water, cooling the North Atlantic. Each event lasted around 750 years and began suddenly – within about a decade. These outbreaks reached the continental margin off Portugal, as I discovered in 1995, when I went to sea with Nick Shackleton on the maiden voyage of the new French research vessel *Marion Dufresne*, with Yves Lancelot as chief scientist. My goal was to collect samples for a project on 'Northeast Atlantic Palaeoceanography and Climate Change' that I had cooked up with geochemist John Thomson, from my institute at Wormley in Surrey. We cored the Portuguese continental margin en route from the Azores to Marseilles. I can heartily recommend French research cruises to those with a taste for cordon bleu cuisine and fine wines.

Much to our delight, Thomson and I found in the *Marion Dufresne*'s 40 m-long giant piston cores certain layers rich in magnesium derived from dolomite rock deposited as ice-rafted debris by the armadas of melting icebergs from Canada[71, 72]. Later, I spent a short sabbatical at Lancelot's laboratory at Aix-en-Provence, working up some of the data from our long cores. I remember his kindness in loaning me a car to help me get around during my month there. It was also a pleasure to meet other prominent French palaeoclimatologists during my stay, among them Edith Vincent and Edouard Bard.

The periodic irruption of ice-rafting in Heinrich events showed that the climate of the Ice Age was variable at the millennial scale, as well as at Milankovitch's orbital frequencies. Bond was keen to find out more about these millennial-scale processes, and in 1995, with Rusty Lotti, carried out close-spaced analyses down two cores collected west of Ireland and spanning the past 9–38 Ka[73]. Between each of the Heinrich layers they found yet more layers of ice-rafted debris, but of lesser magnitude. They deduced that iceberg calving had recurred at intervals of 2000–3000 years. Examination of the rock grains in these layers showed that, while the carbonate-rich ones came from Canada and were concentrated in Heinrich layers, red (hematite)-stained rock grains from multiple sources were common, along with grains of volcanic glass from Iceland or Jan Mayen Island, in the other millennial-scale layers. By 1997, yet more detailed studies by Bond and his team suggested that the red (hematite)-stained rock grains in these layers came from East Greenland or Svalbard[74]. The ice-rafting took place on average every 1536 ± 563 years. This cyclicity persisted into the Holocene, where the frequency was 1374 ± 502 years – statistically indistinguishable from the cyclicity of the glacial period. Averaging the Holocene and glacial signals gave a frequency of 1470 ± 532 years, which, they thought, '*reflects the presence of a pervasive, at least quasi-periodic, climate cycle occurring independently of the glacial-interglacial climate state*'[74]. Furthermore, '*ocean circulation … [is implicated] as a major factor in forcing the climate signal and in amplifying it during the last glaciation … with a single mechanism – an oscillating ocean surface circulation, we can explain at once the synchronous ocean surface coolings, changes in IRD [ice-rafted debris] … and foraminiferal concentrations, and changes in petrologic tracers*'[74]. Laurent Labeyrie's French team distilled this jargon-rich waffle down to the conclusion that a 'climate oscillator' caused the millennial oscillations[75].

As ever, the science marched on. By 1999, Bond's team had confirmed the persistence of the 1470-year cycle back to 80 Ka ago[76]. Why weren't each of these ice-rafting events associated with massive iceberg outbreaks of the kind forming Heinrich events? Bond thought that after a massive purge of ice during a Heinrich event, it took a few thousand years for the ice at source to grow back and reach the unstable conditions needed for another massive discharge[76]. Noting that Heinrich events got closer together between MIS 4 and the Last Glacial Maximum, he thought that this might reflect deterioration of global climate after the last interglacial, with recovery of the ice in Hudson Strait taking progressively less time

as the climate cooled. Heinrich events were not strictly periodic, and the oscillations were internal to the system.

Bond's team found signs of oceanic cooling leading up to Heinrich events[76], including a southward extension of polar surface waters and disruption of the North Atlantic Current, the branch of the Gulf Stream that transports warm waters north[75]. Mark Maslin of the Environment Change Research Centre at University College London suggested that some of Bond's 1500-year cooling events may have produced ice surges from Iceland and East Greenland that were large enough to raise sea level to the extent that it undercut and destabilised the edges of the Laurentide Ice Sheet, precipitating a full-scale Heinrich event[77].

Following up Bond's work, Bill Curry and his team from the WHOI, on Cape Cod, used $\partial^{18}O$, $\partial^{13}C$, and Cd/Ca ratios from benthic foraminifera from a core near Iceland to show that these cooling events occurred at more or less the same time as decreases in the production of North Atlantic Deep Water and cooling of surface waters in the western equatorial Atlantic[78]. After each Heinrich event, warm, saline surface water re-entered the area, and thermohaline circulation resumed[75, 78]. Labeyrie and his French team agreed that millennial ice-rafting events were associated with cooling of the surface waters and southward extension of cold and low-salinity Arctic waters, but found that the widespread low-salinity meltwater accompanying these events delayed resumption of the production of deep water by several hundred years by forming a lid on the ocean[75].

Changes at the millennial scale, such as Heinrich events, Bond cycles and the Younger Dryas cold event at the end of the last glaciation (discussed later), are visible in the reconstructions of past sea surface temperatures made for the Mediterranean and North Atlantic from alkenone palaeothermometry[79]. The Canary Current transports these millennial signals south into the tropics along the coast of northwest Africa, where, as a result, the coldest time at the sea surface in the past 80 Ka ($-12\,^{\circ}C$) was not at the Last Glacial Maximum, at around 20 Ka ago, but during Heinrich event 2, just before the Last Glacial Maximum, and Heinrich event 1, just after the Last Glacial Maximum[80]. Surface currents also transported these signals, including all those seen in Greenland ice cores for the past 50 Ka, into the Mediterranean, affecting the climate there, too[81]. As off northwest Africa, the temperatures of these events were colder in the Mediterranean than were those of the Last Glacial Maximum.

Some of the millennial changes were quite fast, and coincided with large, rapid climate changes recorded in ice cores from Greenland. The discovery of these large and sudden changes in piston cores and ice cores was a revelation, proving that the glacial period was far from being as stable as was once supposed. It now appeared that slow, steady changes in the climate of that time led eventually to 'tipping points' at which the climate changed to a different state, before eventually tipping back to its previous one[77]. Some of the most pronounced changes, especially around the northern North Atlantic, occurred within a few decades, or even just a few years. These sudden step-like transitions would have had significant effects on human life at the time, making it prudent for us to reflect on what caused them in the past and what might do so in the future as global warming continues. Was this 'flickering' between one state and another typical just of glacial periods, or might it occur also in interglacials like the one we are now living through? We revisit this question in Chapters 13 and 14, when we explore these rapid changes in some detail.

Before concluding this section, we should note the conclusion of Julian Dowdeswell, of Scott Polar Research Institute, concerning the behaviour of ice sheet margins. The response of an ice sheet to a climatic event or a rise in sea level is not necessarily uniform from one ice margin to another[82]. Random processes could have led to surges in Northern Hemisphere ice sheets during the last glacial period, which means that even if some external forcing agent is involved, there may be a random component to Heinrich events. Readers who want more detail on the competing models explaining periodic surging by ice sheets may find it useful to consult Cronin's *Paleoclimates*[41].

12.6   Ice Age Productivity

Prior to my *Marion Dufresne* cruise, I had been trying to test the hypothesis that the increase in the steepness of the thermal gradient between the Equator and the poles during glacial times increased the strength of the Trade Winds and so enhanced upwelling on continental margins. I did so in 1995, using a piston core that I had collected back in 1973 from the continental slope off Namibia when I was chief scientist on one leg of RV *Chain* cruise 115 between Dakar and Cape Town[83]. I wanted to know whether the upwelling associated with the Benguela Current changed with time,

and, if so, how and when. The answer required assembling a multidisciplinary team.

To assess the temperature history, I needed alkenone data. Fortunately, I knew Geoff Eglinton well, having first met him in the late 1970s, when we were both members of the Organic Geochemistry Panel advising the DSDP. Geoff agreed to provide the alkenone data we needed to determine sea surface temperatures over the past 70 Ka[83]. To our surprise, we found that they were coldest and most productive during MIS 3 (60–24 Ka ago), a warm interstadial during the last glacial period. We deduced that the alongshore Trade Winds had been strongest during stage 3, thus driving more upwelling of cold, nutrient-rich and highly productive water. In the colder isotopic stages above and below (MIS 2 and MIS 4), waters were slightly warmer and slightly less productive. While this could indicate reduced wind strength, the evidence suggested that wind directions might have changed, there being more winds rich in desert dust blowing directly offshore, and fewer of the alongshore Trades that drove upwelling currents. In contrast, today's sea surface along that margin is very much warmer and less productive than it was in glacial times, although upwelling still prevails there, and surface waters are still highly productive – this is one of the world's great fishing grounds. Our organic carbon signal fluctuated through time on a cycle of about 22 Ka, evidently driven by variations in the precession of the Earth's orbit.

Several other researchers were extracting climate signals from piston cores from close by in the southeastern Atlantic at the same time, and we pooled our resources to show that the Heinrich events, when icebergs were most abundant in the North Atlantic, were represented by warming oceanographic signals in the South Atlantic[84]. We explore the reasons for this unexpected hemispheric climatic connection in Chapter 13.

Looking at the Benguela Current system in rather more detail, a group of German researchers used the alkenone method to show that the warming characteristic of the Last Glacial Maximum began before it, probably in response to a change in the winds that allowed subtropical surface waters to move south down the coast from Angola[85]. This coincided with conditions less favourable for upwelling, which helps to explain the decrease in organic carbon accumulation we found on the continental slope off Walvis Bay. Timothy Herbert of Brown University found much the same thing off southern California – just a slight cooling at the Last Glacial Maximum, close to the coast. In both of these environments, the cores from the open ocean farther offshore contained temperature profiles typical of those seen in the global SPECMAP stack, with the coldest sea surface temperatures at the time of maximum ice volume – the Last Glacial Maximum[79]. Along the Benguela and California coast, then, upwelling was stronger and more productive than during the Holocene during glacials, including at the Last Glacial Maximum, but was not as strong or productive at the Last Glacial Maximum as it was in the interstadial period (MIS 3).

Was our finding that upwelling had decreased during peak glacial times (MIS 2) typical, I wondered? Yes. As Sigman and Haug explained, the coastal upwelling zones off California and Mexico in the north and off Peru in the south were less productive during the recent glacial period[50]. Sigman and Haug attribute this to the effect of continental cooling (and a large North American ice sheet, in the case of the California Current) on the winds that currently drive coastal upwelling. Upwelling associated with monsoonal circulation in the Somali Current of the western Indian Ocean also decreased, because the cooling of the Tibetan Plateau weakened the southwest monsoonal winds. In contrast, upwelling was strengthened in the equatorial Indian Ocean, where the northeast monsoonal winds remained strong; the same applied in the South China Sea. It was also increased in the eastern equatorial Pacific[86] and the equatorial Atlantic[87].

Yet another geochemical technique helped to ascertain the history of productivity in the Southern Ocean during the Ice Age. Because the element thorium (Th) is rapidly adsorbed from the ocean on to sinking particles, and there is little lateral transport of dissolved Th from its site of production to its site of deposition, an isotope of thorium (^{230}Th) can be used as a proxy for the vertical downward flux of sediment. Use of this technique helped to determine the vertical fluxes of opal, barium, organic carbon and other proxies for palaeoproductivity in the Southern Ocean[88]. Compared with the Holocene, productivity was lower south of the polar front and higher north of the polar front during the Last Glacial Maximum in the Atlantic and Indian Ocean sectors[51,90]. The main planktonic organisms in these cold waters are siliceous diatoms. Diatom production shifted north as temperatures cooled. While this applied in the Atlantic and Indian Ocean sectors, it did not apply in the Pacific sector, where productivity was lower in the Last Glacial Maximum than in the Holocene.

The northward shift in productivity reflects northward migration of the oceanic fronts and their accompanying sea ice during glacial times. The absence of high productivity in the Pacific sector was probably due to its excessive distance from the westerly sources of dust that transported Fe to fertilise the ocean. The lower productivity of Antarctic waters during the recent glacial period was most likely due to decreasing supply of deep water to the surface, resulting from the diminished supply of North Atlantic Deep Water, which would have driven a relative fall in atmospheric CO_2. In addition, the prevailing westerly winds shifted northwards as the Hadley Cell shrank, reducing upwelling in the Antarctic coastal sector. Besides that, more extensive cover of sea ice in the Southern Ocean limited the exposure of the ocean to the air, contributing to a fall in atmospheric CO_2[89–91]. At the Last Glacial Maximum, sea ice was double its present extent, both in winter and in summer[91]. Sigman and Haug suggest that salinity stratification associated with sea ice was a major limiting factor on CO_2 exchange with the air during glacial times[51]. Whether there was a net change in total productivity of the Southern Ocean from the Last Glacial Maximum to the Holocene remains a topic for debate[88]. It seems more likely that marine productivity stayed the same but underwent a lateral shift from south to north in the Last Glacial Maximum. Regardless of what happened in the polar regions, studies of ^{230}Th in the equatorial Pacific show little change in productivity from glacial to interglacial[88].

12.7 Observations on Deglaciation and Past Interglacials

The last deglaciation was the most massive change in Earth's climate in the past 25 Ka. The Northern Hemisphere ice sheets began to melt back around 21 Ka ago, as insolation and CO_2 began to rise. Rising seas contributed to the rapid decay of those ice sheets, encouraging an increase in the rate of flow of ice streams draining the interior, thus thinning the ice sheets and facilitating their collapse[92]. Increased melting formed large meltwater lakes on the southern fringes of the ice sheets, especially in North America, where Lake Agassiz covered an area about the size of the Black Sea. Its remnants today form Lakes Winnipeg and Manitoba. The sudden drainage of the lake put a freshwater cap on the North Atlantic, shutting down the northern arm of the Thermohaline Conveyor. This cap was probably responsible for the Younger Dryas cold period or stadial interrupting the deglaciation between 12 800 and 11 500 years ago, which caused temperatures to drop 5 °C in the United Kingdom, for instance. A further drainage from the lake gave rise to a brief cooling 8200 years ago, which we examine in Chapter 14. Other meltwater pulses occurred at around 14 200 and 11 000 years ago[93]. In many respects, the cold Younger Dryas period represented a temporary return to the glacial circulation pattern of reduced North Atlantic Deep Water. Surprising though it may seem, it was primarily a Northern Hemisphere phenomenon, although the alkenone data show that sea surface temperatures fell by some 12 °C off western North America, and demonstrate cooling of the same age in the South China Sea, the Indian Ocean and the South Atlantic[79]. The puzzle of how Lake Agassiz drained into the ocean was eventually solved. From gravels and a regional erosion plain in northern Canada, Julian Murton and colleagues at the University of Sussex showed in 2010 that it discharged along the path of the Mackenzie River[94].

Temperatures derived from alkenone data can also be used to check the temperatures obtained for the Last Glacial Maximum by CLIMAP researchers. The alkenone data show that the surface ocean was cooler then than CLIMAP researchers thought, but that the tropics cooled much less than the high latitudes, perhaps by only 1 °C[79].

Are past interglacials analogues for the Holocene – the interglacial we are now living in? The simple answer is: no. Interglacials are not all alike. The modulating effect of the roughly 400 Ka cycle of eccentricity means that the interglacial most similar to our own is that from roughly 400 Ka ago, during MIS 11[95]. Analyses of the $\partial^{18}O$ ratios in samples of the right-coiling planktonic foraminifera *Neogloboquadrina pachyderma* from stage 11 in deep-sea drill cores in the northeast Atlantic show that sea surface temperatures varied by less than ±1 °C from the long-term mean for at least 30 Ka[96]. The near-circular orbit of the Earth at the time prevented the 20 Ka precession signal from having much effect within this isotope stage. In effect, the Milankovitch cycle 'missed a beat', prolonging the interglacial to close on 50 Ka. MIS 11 was in effect about two precession cycles long, instead of one.

As André Droxler of Rice University in Houston pointed out, MIS 11 and the present interglacial are similar because their orbital variables are almost identical. According to Droxler and colleagues, '*both interglacials correspond to times when the eccentricity of the Earth orbit was at its minimum, so that the amplitude of the precessional cycle was damped*'[97]. The strongest and longest Pleistocene interglacial, stage 11, had prolonged intense warmth, sea level stands up to perhaps 13–20 m above present levels[98] and significant poleward penetration of warm waters. It lasted twice as long as more recent interglacial stages. The

Holocene is likely to be just as long. In Chapter 13 we will look at possible explanations for these patterns.

The warming in stage 11 was important for the establishment of coral reefs. Wolf Berger of Scripps and Gerold Wefer of the University of Bremen used deep-sea drill core data to show that the western Pacific warm pool of surface water expanded dramatically some 400 Ka ago, helping to explain the growth of Australia's Great Barrier Reef[99]. At that time, shallow carbonate platforms grew to the point where they clogged the flow of surface water through the Indonesian islands between the Pacific and Indian Oceans. The warming also triggered the establishment of other barrier reefs, like that off Belize[97]. These reefs grew when the large rise in sea level at the end of the previous glacial maximum extensively flooded fluvial plains, preventing the former supply of riverborne silt and sand from reaching offshore reef sites.

'*Will such warm conditions be replicated as the Holocene continues?*' asked Droxler and colleagues[97]. Yes, they concluded, '*we can expect another ~20,000 years of interglacial conditions, independent of any anthropogenic forcing*'[97]. Is MIS 11 an exact analogue for the Holocene? Not according to David Hodell of the University of Florida at Gainsville, who found maxima in the $\partial^{13}C$ of the planktonic foraminifera *Globigerina bulloides* and in fragmented foraminiferal remains in stage 11 sediments from a deep-sea drill core from the Cape Basin off South Africa[100]. The same indicators in sediments from the last 100 Ka and the Holocene have much lower values, however, and the carbonate compensation depth was 600 m shallower in stage 11 than it is now. These patterns suggest a lowering in the concentration of carbonate ions in the ocean at that time, possibly related to the massive building of barrier reefs in shallow waters. Atmospheric and oceanic feedbacks are not operating in exactly the same way today as they were then. Even so, a comparison of sea surface temperatures and of $\partial^{18}O$ ratios from benthic foraminifera in the southeast Atlantic showed that the 11.7 Ka of the Holocene are indeed comparable to the first 12 Ka of MIS 11[101].

What about the last interglacial, the Eemian, during MIS 5, which began at around 135 Ka ago and lasted until around 110 Ka ago? Alkenone and Mg/Ca data from marine sediments suggest that it was warmer than the late Holocene by up to 3 °C – consistent with stage 5 experiencing significantly higher orbital insolation[79]. These new data improve on the CLIMAP data, which suggested that there was little difference between stage 5 and today. It now seems likely that stage 5 was as warm as the early

Holocene climatic optimum, when insolation was much higher than it is today. We can extract more evidence of Eemian climate change from pollen and lake records in central Europe, loess sediments from central China and marine sediment cores from the eastern subtropical Atlantic. These data show evidence for a single, sudden cool event in the middle Eemian at about 122–120 Ka ago, showing that short, sharp cold periods can occur in interglacials.

Compared with today, global ice volumes were smaller and solar radiation was 13% stronger over the Arctic in summer during the Eemian interglacial. According to Gifford Miller and his team, sea ice and permafrost were vastly reduced, boreal forest expanded to the Arctic shore and most Northern Hemisphere glaciers melted[35]. Summer temperature anomalies over Arctic lands were 4–5 °C above present values, especially in the Atlantic sector. Northern Canada and parts of Greenland were 5 °C warmer than today in summer, but Alaska and Siberia were only about 2 °C warmer. Interpretation of marine data is complicated by the stratification of the Arctic Ocean, which commonly has a cool relatively fresh cap (< −1 °C) overlying warmer subsurface waters (> 1 °C).

A final point to bear in mind is that interglacials as warm as the present one occurred for only about 10% of the time in the late Quaternary[102]. The climate of the past 800 Ka was predominantly cold.

12.8   Sea Level

Rising sea level is one of the most highly visible results of a warming world, driven by the melting of ice on land and the expansion of warming seawater. These changes are termed 'eustatic'[102]. Although changes in sea level provide us with yet another proxy for past climate change – especially for ice volume – the relation between climate and sea level is not simple, as we saw in Chapter 11. As R. Lawrence Edwards of the University of Minnesota and his colleagues remind us[103], sea level can also change as the result of tectonic uplift or sinking of the Earth's surface, or of isostatic changes, through which land sinks beneath ice sheets but rises around their periphery to form a 'fore-bulge' – a process that reverses when the ice sheets melt. Only eustatic change is truly global. Tectonic and isostatic adjustments cause local or regional changes that complicate the extraction of a global sea level signal. Lyell knew all about tectonic effects, having observed that the so-called Temple at Serapis in Italy had first been

partly drowned and then uplifted. These competing signals must be unravelled to separate the local from the global signal, as we see in some detail in Chapter 14.

There is also the question of rates. For example, the Scandinavian ice sheet melted by about 6000 years ago, but Scandinavia is still slowly rising. So too is Scotland, which lost its ice long ago. In contrast, southern England, the southern edge of the Baltic and the west coasts of Germany and the Netherlands, which were on the fore-bulge around the European ice sheets, are still slowly subsiding. Similarly, the parts of the northernmost United States and Canada that lay beneath the Laurentide Ice Sheet are now rising, while the southern United States, which formed the fore-bulge area outside that ice sheet, is slowly sinking. Within the area of the former Laurentide Ice Sheet, its core region – Hudson's Bay – is still depressed below sea level, although it is rising slowly.

Past sea levels can be determined directly by using the ^{14}C or other radiometric techniques to date carbonates such as reefs and other features that formed at or very close to sea level[103]. These techniques include U/Th dating, which involves calculating ages from radioactive decay relationships between ^{238}U, ^{234}U and ^{230}Th; this is also known as ^{230}Th dating. A further check on accuracy can be obtained from U/Pa dating, in which ages are calculated from the relationship between Uranium-235 (^{235}U) and its daughter, Protoactinium-231 (^{231}Pa). U/Th and U/Pa dating extend the range of ^{14}C dating (maximum 50 Ka) to 250 Ka (^{230}Pa) and 600 Ka (^{230}Th). These techniques, like AMS ^{14}C dating, came into their own after the mid 1980s, with the development of mass spectrometric measurements that reduced sample size and increased the speed and precision of analysis. Even so, despite the accuracy of the dates, all estimates of past sea level come with some uncertainty.

In this section, we focus on how high sea level might have been during past warm interglacials. The data available to Edwards in 2003 suggested that sea levels were up to 20 m above today's level in MIS 11 (400 Ka ago); up to 29 m above in MIS 9 (330 Ka ago); up to 9 m above in MIS 7 (240 Ka ago); and around 5 ± 3 m above in MIS 5, the last interglacial (100 Ka ago)[103]. The rise in sea level from a low point of about −130 m following the Last Glacial Maximum is known in some detail, thanks to comparable data from the New Guinea's Huon Peninsula, Tahiti, South East Asia's Sunda Shelf and northwest Australia's Bonaparte Gulf. These are far-field sites remote from polar ice sheets. Isostatic adjustments are unimportant, and the data reflect a true global signal.

These estimates were refined for the last interglacial (MIS 5) by a team led by Robert Kopp of Princeton, who in 2009 compiled a large number of indicators of local sea level change and applied a statistical approach to estimating global sea level[104]. They found a 95% probability that global sea level peaked at least 6.6 m higher than today, and a 67% probability that it exceeded 8 m, but only a 33% likelihood that it exceeded 9.4 m. Rates of sea level rise could have varied between about 56 and 92 cm per century. For comparison, the present rate of sea level rise is around 33 cm per century. The last interglacial was only slightly warmer than the present – by about 2 °C. Achieving a sea level rise in excess of 6.6 m higher than present '*is likely to have required major melting of both the Greenland and West Antarctic ice sheets*', they concluded[104].

Eelco Rohling of the University of Southampton used data from the Red Sea to suggest, like Kopp, that during the last interglacial, sea level reached a mean position of +6 m, with individual short-term peak positions up to about +9 m compared with today's level. The rate of rise of sea level, according to Rohling and colleagues, was about 1.6 m per century, which '*would correspond to disappearance of an ice sheet the size of Greenland in roughly four centuries*'[105]. This rate occurred when global mean temperature was 2 °C higher than today. Kurt Lambeck (1941–) (Box 12.3) of the Australian National University evaluated sea level for the last interglacial by using tectonically stable sites in the 'far field', estimating that it was 5.5–9.0 m above today's[106], which is consistent with the findings of both Kopp and Rohling.

Maureen Raymo of Boston University looked with a member of Kopp's team, Jerry Mitrovica of Harvard, into the contentious suggestion that Pleistocene shoreline features on the tectonically stable islands of Bermuda and

Box 12.3 Kurt Lambeck.

Lambeck, professor of geophysics at the Australian National University in Canberra, was born in Utrecht in the Netherlands. From 2006 to 2010 he was president of the Australian Academy of Science. He has been honoured with several awards, among them fellowship in the French and US Academies of Science and the United Kingdom's Royal Society, the international Balzan Price (2012) and the Wollaston Medal of the Geological Society of London (2013).

the Bahamas were more than 20 m higher than today in MIS 11, some 400 Ka ago[107]. They found both sites to be located on the outer edge of the peripheral bulge of the Laurentide Ice Sheet. To account for postglacial crustal subsidence at these sites, their elevations were adjusted downwards by about 10 m. That reduced eustatic sea level rise to ~6–13 m above the today's level in the second half of MIS 11. This rise was caused by prolonged warmth leading to the collapse of both the Greenland and West Antarctic Ice Sheets. Given that the likely maximum rises in sea level for the melting of the Greenland and West Antarctic Ice Sheets are 7 and 5 m respectively, the change of 6–13 m suggests that changes in the volume of the East Antarctic Ice Sheet were minor.

Roland Gehrels of the University of Plymouth drew attention to other flaws in the analysis of sea level change, focusing on the rise of sea level since the Last Glacial Maximum[108]. Fairbanks' classic paper on sea level rise, published in 1989, and based largely on data from Barbados, suggested that sea level was 120 m below present at the Last Glacial Maximum[109]. Gehrels cited three possible sources of error in Fairbanks' data. First, Barbados lies in an active tectonic setting on the edge of the Caribbean Plate. Even slight tectonic changes could have affected the absolute amount of sea level change registered on the island. Second, Barbados lay on the trailing edge of the glacial fore-bulge pushed up around the margins of the Laurentide Ice Sheet, and the collapse of that feature would have created further vertical change. Third, Fairbanks' curve included data from other islands (Martinique, Bahamas, Puerto Rico and St Croix), '*thereby introducing errors resulting from differential isostatic movements and regional sea-level variations*'[108].

Focusing on far-field sites, Yokoyama and colleagues calculated in 2000 that global sea level was as low as 130–135 m below present levels at the Last Glacial Maximum[110]. Claire Waelbroeck and colleagues provided much the same picture, with a maximal lowering to −135 m at the Last Glacial Maximum[111]. They also calculated that sea level fell to about −125 m in MIS 6 (140 Ka ago) and MIS 10 (345 Ka ago), and to about −110 m during MIS 8 (250 Ka ago). Their values differ significantly from those derived by Nick Shackleton[112] and provide good reasons for discounting Shackleton's data. Lambeck recently estimated sea level lowering at the Last Glacial Maximum as −134 m[113], explaining that this was a measure of grounded ice volume, including ice grounded on shelves. Along far-field continental margins, the Last Glacial Maximum sea levels would generally be less than this due to isostatic/gravitational effects, while in mid oceans they would exceed this (James Scourse, pers. comm.).

When it was published in 1989, the Barbados sea level curve[109] gained a great deal of attention because it showed evidence for episodes of very rapid sea level rise, most notably the event known as meltwater pulse 1a, dated to about 14 Ka ago, when sea levels rose by 15–25 m at rates of over 40 mm/yr[108]. The jury is still out with respect to the source of the meltwater pulse, which could have originated in surges of the Laurentide or the Antarctic Ice Sheets, or from the discharge of large glacial lakes.

Clearly, sea level has changed through time in response to the waxing and waning of ice sheets, and measurements of past sea level can be used as a proxy for ice volume change. In order to refine these calculations further, there is much still to learn about regional variations in sea level, which depend on local tectonics and glacial isostatic adjustments of the Earth's surface to the addition or removal of large masses of ice. Knowing that '*regional sea-level changes resulting from polar ice melt can depart by up to 30% from the global mean*', Gehrels concluded that '*regional sea-level variability precludes the use of the term "eustasy" in the traditional sense (i.e. global average sea-level change). The recognition that "eustasy" is only a concept should … lead to [improved] regional sea level predictions*'[108].

As we saw in Chapter 11, one way to avoid problems created by the use of past shorelines to establish past sea levels is to use the $\partial^{18}O$ composition of seawater, which is related to both ocean temperature and ice volume, both of which are, in effect, global. Subtracting the temperature signal enables us to determine ice volume and hence sea level[114,115].

Using stable oxygen isotope analyses of planktonic foraminifera and bulk sediments from the Red Sea, Eelco Rohling and his team developed a relative sea level record for the past 520 Ka[98]. It shows a striking similarity to the record of Antarctic temperature, a relationship that remains the same regardless of whether the climate system is shifting towards glaciation or deglaciation, and which does not drift back through time. As this is a robust relationship within the climate system, it could be applied in estimating the effects of future climate change (see Chapter 16).

Jacqueline Austermann of Harvard agreed with the revisions to the Fairbanks model of sea level change[116]. Austermann and colleagues' model confirmed that at the Last Glacial Maximum, sea level should have been lowered

to about −130 m, not −120 m as Fairbanks thought. That left a significant volume of ice in the Northern Hemisphere unaccounted for: it appeared from sea level data that more ice must have melted than had been available in the ice sheets of Laurentia (North America) and Fennoscandia. A joint team from Germany's Alfred Wegener Institute and the Korean Polar Research Institute discovered in 2013 that the furrows that arise when large ice sheets become grounded on the seabed are widespread on the seabed off the coast of northeast Siberia. The team estimated that the furrows represented the former existence of an Arctic ice sheet that covered an area at least as large as Scandinavia and was up to 1200 m thick. This previously missing ice may well explain the accounting discrepancy[117].

The study of sea level is worth an entire book. Readers wishing to probe further might like to start with a 2010 compilation of global data entitled *Understanding Sea-Level Rise and Variability*[118].

Returning to the astronomical calculations, it is now abundantly clear that regular changes in the Earth's orbit and axial tilt cause the amount of insolation we receive to vary within narrow limits, a discovery as influential in its own way as plate tectonic theory. The limits define a 'natural envelope' in which the maxima and minima are seldom if ever exceeded. These limits apply in turn to global temperature, which varied over the narrow global range of 4–5 °C between glacial and interglacial times. Back in 1982, Wolf Berger realised that this was '*a striking phenomenon, important especially for the survival of higher organisms*'[119]. This Ice Age natural envelope was superimposed on a background climate whose extremes varied within another natural envelope, in which, as we saw at the end of Chapter 9, the variation in CO_2 was driven by plate tectonic processes including the emission of CO_2 from volcanoes and its extraction by weathering, especially in mountainous areas, and by sedimentation in growing ocean basins. The natural envelope of CO_2 from 200 to 1000 ppm (Chapter 9) was only occasionally exceeded, as we saw in Chapter 10, when hothouse conditions prevailed.

To summarise, our view of Pleistocene climate changed dramatically from the mid 1960s onwards, when piston corers and deep-ocean drilling enabled us to study for the first time the climate history recorded over the 66% of the Earth's surface covered by water depths of more than 200 m. Application of novel palaeontological and geochemical techniques showed that Earth had experienced many more substantial variations in climate than was apparent from studies of glaciation on land, where the advances of later glaciers and ice sheets removed the records of earlier ones. The realisation that changes in insolation were intimately linked to changes in temperature and ice volume enabled palaeoclimatologists to tune their signals of climate change to orbital changes, thus deriving a novel method for dating core horizons to an unheard of accuracy of ± 2000 years, over periods of more than 1 million years. Furthermore, as clockwork variations in insolation could be projected into the future, it became possible to estimate the extent, duration and timing of the next glaciation.

The growing global array of deep-ocean cores enabled comparisons to be made between glacial and interglacial conditions caused by changes in the extent of sea ice and in the Thermohaline Conveyor Belt. This confirmed the validity of orbital cycles and highlighted the saw-toothed pattern of actual climate change, reflecting the slow build-up of ice sheets and their rapid eventual demise, a pattern suggesting that once warming caused melting to reach some critical rate, land ice reservoirs collapsed to produce a glacial termination[119]. Carbon isotopes could be used to estimate the amount of CO_2 in the air, showing that CO_2 varied with temperature, presumably because carbon reservoirs slowly built up in peat beds, rain forest debris, fine-grained organic-rich sediments and deep-ocean waters as ice accumulated, before decaying rapidly and releasing CO_2 as ice melted and the climate warmed[119].

CO_2 provided one positive feedback, affecting temperature. Sea ice provided another, first through its affect on albedo, and second through governing the exchange of CO_2 between ocean and atmosphere[119]. Dust provided a third, increasing fertilisation of the ocean with iron in glacial times, thereby enhancing CO_2 draw-down; its absence in interglacials had the opposite effect. Water vapour provided a fourth, following CO_2, influencing temperature and governing change in the water cycle. Sea level provided a fifth, increasing or decreasing the area of ocean available for the exchange of CO_2 and water vapour between ocean and atmosphere. Sea levels in past interglacials may have been as high as 9 m above today's.

A millennial level of natural variability became apparent from concentrations of ice-rafted debris. Large glacial outbreaks formed Heinrich events; small ones formed 1500-year Bond cycles. They corresponded with cold periods in the North Atlantic and warm periods in the South Atlantic. Outbreaks from massive glacial lakes flooded the northern oceans from time to time, causing the northern freeze known as the Younger Dryas.

In the next chapter, we will look at the exciting discoveries that the ice core drillers were making on land, and compare them with those emerging from studies of the ocean floor. The history of fossil CO_2 in ice cores enables us to further explore the relationship between CO_2 and the curious 100 Ka climate cycle. We will also examine possible mechanisms for glacial–interglacial climate change.

References

1. Milankovitch, M. (1941) Kanon der erdbestrahlung and seine andwendung auf das eiszeitproblem. Special Publication 133. Royal Serbian Academy, Belgrade; English translation published by Israel Program for Scientific Translations, US Dept of Commerce, 1969.
2. Zeuner, F.E. (1959) *The Pleistocene Period – Its Climate, Chronology and Faunal Successions*, 2nd edn. Hutchinson Scientific and Technical, London.
3. Walker, M. and Lowe, J. (2007) Quaternary science 2007: a 50-year retrospective. *Journal of the Geological Society of London* **164**, 1073–1092.
4. Weedon, G. (2003) *Time-Series Analysis and Cyclostratigraphy*. Cambridge University Press, Cambridge.
5. Berger, A. (1976) Long-term variations of daily and monthly insolation during the Last Ice Age. *EOS* **57** (4), 254.
6. Berger, A. (1976) Obliquity and general precession for the last 5,000,000 years. *Astronomy and Astrophysics* **51**, 127–135.
7. Berger, A. (1977) Support for the astronomical theory of climatic change. *Nature* **268**, 44–45.
8. Berger, A. (1978) Long-term variations of caloric insolation resulting from the Earth's orbital elements. *Quaternary Research* **9**, 139–167.
9. Elias, S. (2006) *Encyclopedia of Quaternary Science*. Elsevier Science, Amsterdam.
10. Arrhenius, G. (1952) *Sediment Cores from the East Pacific*. Reports of the Swedish Deep Sea Expedition 5. Elanders boktryckeri aktiebolag, Sweden.
11. Ericson, D.B. (1953) *Sediments of the Atlantic Ocean*. Technical Report on Submarine Geology 1, Lamont Geological Observatory, Palisades, New York. Columbia University Press, New York.
12. Emiliani, C. (1955) Pleistocene temperatures. *Journal of Geology* **63**, 538–578.
13. Hay, W.W. (2013) *Experimenting on a Small Planet: A Scholarly Entertainment*. Springer, New York.
14. Emiliani, C. and Flint, R.F. (1963) The Pleistocene record. In: *The Sea: Ideas and Observations on progress in the Study of the Seas, vol. 3, The Earth Beneath the Sea – History* (ed. M.N. Hill). John Wiley & Sons, London, pp. 888–927.
15. Revelle, R. and Suess, H.E. (1957) Carbon dioxide exchange between atmosphere and ocean and the question of an

increase of atmospheric CO_2 during the past decades. *Tellus* **9** (1), 18–27.

16. Shackleton, N.J. and Turner, C. (1967) Correlation between marine and terrestrial Pleistocene successions. *Nature* **216**, 1079–1082.

17. Shackleton, N.J. (1967) Oxygen isotope analyses and Pleistocene temperatures re-assessed. *Nature* **215**, 15–17.

18. Imbrie, J. and Imbrie, K. (1979) *Ice Ages: Solving the Mystery*. Enslow, Berkeley Heights, NJ.

19. Lea, D.W. (2003) Elemental and isotopic proxies of past ocean temperatures. In: *The Oceans and Marine Geochemistry, Vol. 6, Treatise on Geochemistry* (eds H.D. Holland and K.K. Turekian). Elsevier-Pergamon, Oxford, pp. 365–390.

20. Imbrie, J. and Kipp, N.G. (1969) Quantitative interpretation of late Pleistocene climate based on planktonic foraminiferal assemblages in Atlantic cores (abstract). Geological Society of America Meeting Program for 1969, part 7, p. 113.

21. Imbrie, J. and Kipp, N.G. (1971) A new micropaleontological method for quantitative paleoclimatology: application to a late Pleistocene Caribbean core. In: *Cenozoic Glacial Ages* (ed. K.K. Turekian). Yale University Press, New Haven, CT, pp. 71–181.

22. Imbrie, J., Van Donk, J. and Kipp, N.G. (1973) Paleoclimatic investigation of a Caribbean core: comparison of isotopic and faunal methods. *Quaternary Research* **3**, 10–38.

23. CLIMAP Project Members (1976) The surface of the ice age Earth. *Science* **191** (4232), 1131–1137.

24. Hays, J.D., Imbrie, J. and Shackleton, N.J. (1976) Variations in the earth's orbit: pacemaker of the ice ages. *Science* **194** (4270), 1121–1132.

25. Leeder, M. (2011) *Sedimentology and Sedimentary Basins – From Turbulence to Tectonics*, 2nd edn. Wiley-Blackwell, Chichester.

26. Flint, R.F. (1971) *Glacial and Quaternary Geology*. John Wiley & Sons, New York.

27. McCave, I.N. and Elderfield, H. (2011) Sir Nicholas John Shackleton, 23 June 1937–24 January 2006. Biographical Memoirs of Fellows of the Royal Society **57**, pp. 435–462.

28. Berger, A. and Loutre, M.F. (1994) Astronomical forcing through geological time. Special Publications of the International Association of Sedimentology **19**, pp. 15–24.

29. Berger, A. and Loutre, M.F. (2002) An exceptionally long interglacial ahead? *Science* **297**, 1287–1288.

30. Imbrie, J., Hays, J.D., Martinson, D.G., McIntyre, A., Morley, J.J., Pisias, N.G., *et al.* (1984) The orbital theory of Pleistocene climate: support from a revised chronology of the marine $\partial^{18}O$ record. In: *Milankovitch and Climate, Part I* (eds A.L. Berger, J. Imbrie, J. Hays, G. Kukla, B. Saltzman). D. Reidel, Dordrecht, pp. 269–305.

31. Imbrie, J., Boyle, E.A., Clemens, S.C., Duffy, A., Howard, W.R., Kukla, G., *et al.* (1992) On the structure and origin of major glaciation cycles. *I. Linear responses to Milankovitch forcing*. Paleoceanography **7**, 701–738.

32. Imbrie, J., Berger, A., Boyle, E.A., Clemens, S.C., Duffy, A., Howard, W.R., *et al.* (1993) On the structure and origin of major glaciation cycles. II. The 100,000-year cycle. *Paleoceanography* **8**, 699–735.

33. Lisiecki, L.E. and Raymo, M.E. (2005) A Pliocene-Pleistocene stack of 57 globally distributed benthic $\delta^{18}O$ records. *Paleoceanography* **20**, PA1003.

34. http://lorraine-lisiecki.com/stack.html (last accessed 29 January 2015).

35. Miller, G.H., Brigham-Grette, J., Alley, R.B., Anderson, L., Bauch, H.A., Douglas, M.S.V., *et al.* (2010) Temperature and precipitation history of the Arctic. *Quaternary Science Reviews* **29**, 1679–1715.

36. Broecker, W.S. and Van Donk, J. (1970) Insolation changes, ice volumes, and the O-18 record in deep-sea cores. *Reviews in Geophysics and Space Physics* **8**, 169–197.

37. Pflaumann, U., Sarnthein, M., Chapman, M., d'Abreu, L., Funnell, B., Huels, M., *et al.* (2003) Glacial North Atlantic: sea-surface conditions reconstructed by GLAMAP 2000. *Paleoceanography* **18** (1065), PA000774.

38. Sarnthein, M., Gersonde, R., Niebler, S., Pflaumann, U., Spielhagen, R., Thiede, J., *et al.* (2003) Overview of glacial Atlantic Ocean mapping (GLAMAP 2000). *Paleoceanography* **18** (1030), PA000769.

39. Mix, A.C., Bard, E. and Schneider, R. (2001) Environmental processes of the last ice age: land, oceans, and glaciers (EPILOG). *Quaternary Science Reviews* **20**, 627–657.

40. Kucera, M., Rosell-Melé, A., Schneider, R., Waelbroeck, C. and Weinelt, M. (2005) Multiproxy approach for the reconstruction of the glacial ocean surface (MARGO). *Quaternary Science Reviews* **24**, 813–819.

41. Cronin, T.M. (2010) *Paleoclimates: Understanding Climate Change Past and Present*. Columbia University Press, New York.

42. MARGO Project Members (2009) Constraints on the magnitude and patterns of ocean cooling at the Last Glacial Maximum. *Nature Geoscience* **2**, 127–132.

43. Shackleton, N.J. and Opdyke, N. (1976) Oxygen isotope and paleomagnetic stratigraphy of Pacific core V28-239 late Pliocene to latest Pleistocene. In: *Investigation of Late Quaternary Paleoceanography and Paleoclimatology* (eds R.M. Cline and J.D. Hays). Geological Society of America Memoirs **145**, 449–464.

44. Elderfield, H., Perretti, P., Greaves, M., Crowhurst, S., McCave, I.N., Hodell, D. and Piotrowski, A.M. (2012) Evolution of ocean temperature and ice volume through the mid-Pleistocene climate transition. *Science* **337**, 704–709.

45. Martinez-Garcia, A., Rosell-Melé, A., Jaccard, S.L., Geibert, W., Sigman, D.M. and Haug, G.H. (2011) Southern Ocean dust-climate coupling over the past four million years. *Nature* **476**, 312–315.

46. Annan, J.D. and Hargreaves, J.C. (2013) A new global reconstruction of temperature changes at the Last Glacial Maximum. *Climate of the Past* **9**, 367–376.

47. Shackleton, N.J., Hall, M.A., Line, J. and Cang, S. (1983) Carbon isotope data in core V19-30 confirm reduced carbon dioxide concentration of the ice age atmosphere. *Nature* **306**, 319–322.

48. Shackleton, N.J. and Pisias, N.G. (1985) Atmospheric carbon dioxide, orbital forcing, and climate. In: *The Carbon Cycle and Atmospheric CO₂: Natural Variations Archean to Present* (eds E.T. Sundquist and W.S. Broecker). Geophysical Monographs 32. American Geophysical Union, Washington, DC, pp. 303–317.

49. Broecker, W.S. and Peng, T.H. (1982) *Tracers in the Sea.* LDGO, Columbia University Press, New York.

50. Sigman, D.M. and Haug, G.H. (2003) Biological pump in the past. In: *The Oceans and Marine Geochemistry, Vol. 6, Treatise on Geochemistry* (eds H.D. Holland and K.K. Turekian). Elsevier-Pergamon, Oxford, pp. 491–528.

51. http://earthobservatory.nasa.gov/Features/Martin/martin_4.php (last accessed 29 January 2015).

52. Martin, J.H. and Fitzwater, S.E. (1988) Iron deficiency limits phytoplankton growth in the North-East Pacific Subarctic. *Nature* **331** (6154), 341.

53. Martin, J.J. (1990) Glacial-interglacial CO₂ change: the iron hypothesis. *Paleocanography* **5** (1), 1–13.

54. De La Rocha, C.L. (2003) The biological pump. In: *The Oceans and Marine Geochemistry, Vol. 6, Treatise on Geochemistry* (eds H.D. Holland and K.K. Turekian). Elsevier-Pergamon, Oxford, pp. 83–111.

55. Smetacek, V., Klaas, C., Strass, V.H., Assmy, P., Montresor, M., Cisewski, B., *et al.* (2012) Deep carbon export from a Southern Ocean iron-fertilized diatom bloom. *Nature* **487**, 313–319.

56. Frakes, L.A., Francis, J.E. and Syktus, J.I. (1992) *Climate Modes of the Phanerozoic – The History of the Earth's Climate over the Past 600 Million Years.* Cambridge University Press, Cambridge.

57. Broecker, W.S. (1987) The great ocean conveyor. *Natural History Magazine* **97**, 74–82.

58. Bjornsson, H. and Toggweiler, J.R. (2001) The climatic influence of Drake Passage. In: *The Oceans and Rapid Climate Change: Past, Present and Future* (eds D. Seidov, B.J. Haupt and M. Maslin). Geophysical Monographs 126. American Geophysical Union, Washington, DC, pp. 243–259.

59. Boyle, E.A. and Keigwin, L.D. (1982) Deep circulation of the North Atlantic over the last 200,000 years: geochemical evidence. *Science* **218** (4574), 784–787.

60. Lynch-Stieglitz, J. (2003) Tracers of past ocean circulation. In: *The Oceans and Marine Geochemistry, Vol. 6, Treatise on Geochemistry* (eds H.D. Holland and K.K. Turekian). Elsevier-Pergamon, Oxford, pp. 433–451.

61. Labeyrie, L.D. (1992) Changes in the vertical structure of the North Atlantic Ocean between glacial and modern times. *Quaternary Science Reviews* **11**, 401–413.

62. McIntyre, A., Kipp, N.G., Bé, A.W.H., Crowley, T., Kellogg, T., Gardner, J.V., *et al.* (1976) Glacial North Atlantic 18,000 years ago: a CLIMAP reconstruction. *Geological Society of America Memoirs* **145**, 43–76.

63. CLIMAP (1981) Seasonal reconstructions of the Earth's surface at the last glacial maximum. Map Series, Technical Report MC-36. Geological Society of America, Boulder, Colorado.

64. Boyle, E.A. (1984) *Cadmium in Benthic Foraminifera and Abyssal Hydrography: Evidence for a 41 kyr Obliquity Cycle.* Geophysical Monographs 29. American Geophysical Union, Washington, DC, pp. 360–368.

65. Meyerhoff, A.A. and Meyerhoff, H.A. (1972) The new global tectonics: major inconsistencies. *Bulletin of the American Association of Petroleum Geologists* **56** (2), 269–336.

66. Ruddiman, W.F. (1977) Late Quaternary deposition of ice-rafted sand in the subpolar North Atlantic (Lat. 40° to 65°N). *Bulletin of the Geological Society of America* **88**, 1813–1827.

67. Ruddiman, W.F. and McIntyre, A. (1977) Late Quaternary surface ocean kinematics and climate change in the high-latitude North Atlantic. *Journal of Geophysical Research* **82** (27), 3877–3887.

68. Ruddiman, W.F. (1977) North Atlantic ice rafting: a major change at 75,000 yr B.P. *Science* **196**, 1208–1211.

69. Heinrich, H. (1988) Origin and consequences of cyclic ice rafting in the northeast Atlantic Ocean during the past 130,000 years. *Quaternary Research* **29**, 142–152.

70. Bond, G.C., Heinrich, H., Broecker, W.S., Labeyrie, L., McManus, J., Andrews, J., *et al.* (1992) Evidence for massive discharges of icebergs into the North Atlantic Ocean during the last glacial period. *Nature* **360**, 245–249.

71. Thomson, J., Nixon, S., Summerhayes, C.P., Schönfeld, J., Zahn, R. and Grootes, P. (1999) Implications for sediment changes on the Iberian margin over the last two glacial/interglacial transitions from (²³⁰Th-excess) systematics. *Earth and Planetary Science Letters* **165**, 255–270.

72. Thomson, J., Nixon, S., Summerhayes, C.P., Rohling, E.J., Schönfeld, J., Zahn, R., *et al.* (2000) Enhanced productivity on the Iberian margin during glacial/interglacial transitions revealed by barium and diatoms. *Journal of the Geological Society of London* **157**, 667–677.

73. Bond, G.C. and Lotti, R. (1995) Iceberg discharges into the North Atlantic on millennial time scales during the last glaciation. *Science* **267**, 1005–1010.

74. Bond, G., Showers, W., Cheseby, M., Lotti, R., Almasi, P., DeMonocal, P., *et al.* (1997) A pervasive millennial-scale cycle in North Atlantic Holocene and glacial climates. *Science* **278**, 1257–1266.

75. Labeyrie, L., Leclaire, H., Waelbroeck, C., Cortijo, E., Duplessy, J.C., Vidal, L., *et al.* (1999) Temporal variability of the surface and deep waters of the north west Atlantic Ocean at orbital and millennial scales. In: *Mechanisms of Global Climate Change at Millennial Time Scales* (eds P.U. Clark, R.S. Webb and L.D. Keigwin). Geophysical Monographs

112. American Geophysical Union, Washington, DC, pp. 77–98.

76. Bond, G.C., Showers, W., Elliot, M., Evans, M., Lotti, R., Hajdas, I., *et al.* (1999) The North Atlantic's 1-2kyr climate rhythm: relation to Heinrich Events, Dansgaard/Oeschger Cycles and the Little Ice Age. In: *Mechanisms of Global Climate Change at Millennial Time Scales* (eds P.U. Clark, R.S. Webb, L.D. and Keigwin), *Geophysical Monographs* **112**, American Geophysical Union, Washington, DC, pp. 35–58.

77. Maslin, M., Seidov, D. and Lowe, J. (2001) Synthesis of the nature and causes of rapid climate transitions during the Quaternary. In: *The Oceans and Rapid Climate Change: Past, Present and Future* (eds D. Seidov, B.J. Haupt and M. Maslin). Geophysical Monographs 126. American Geophysical Union, Washington, DC, pp. 9–52.

78. Curry, W.B., Marchitto, T.M., McManus, J.F., Oppo, D.W. and Laarkamp, K.L. (1999) Millennial-scale changes in ventilation of the thermocline, intermediate, and deep waters of the glacial North Atlantic. In: *Mechanisms of Global Climate Change at Millennial Time Scales* (eds P.U. Clark, R.S. Webb and L.D. Keigwin). Geophysical Monographs 112. American Geophysical Union, Washington, DC, pp. 59–76.

79. Herbert, T.D. (2003) Alkenone paleotemperature determinations. In: *The Oceans and Marine Geochemistry, Vol. 6, Treatise on Geochemistry* (eds H.D. Holland and K.K. Turekian). Elsevier-Pergamon, Oxford, pp. 391–432.

80. Zhao, M., Beveridge, N.A.S., Shackleton, N.J. and Sarnthein, M, (1995) Molecular stratigraphy of cores off northwest Africa: sea surface temperature history over the last 80ka. *Paleoceanography* **10**, 661–675.

81. Cacho, I., Grimalt, J.O. and Canals, M. (2002) Response of the western Mediterranean Sea to rapid climatic variability during the last 50,000 years: a molecular biomarker approach. *Journal of Marine Systems* **33–34** (C), 253–272.

82. Dowdeswell, J.A., Elverhoi, A., Andrews, J.T. and Hebbeln, D. (1999) Asynchronous deposition of ice-rafted layers in the Nordic seas and North Atlantic Ocean. *Nature* **400**, 348–351.

83. Summerhayes, C.P., Kroon, D., Rosell-Mele, A., Jordan, R.W., Schrader, H.J., Hearn, R., *et al.* (1995) Variability in the Benguela Current upwelling system over the past 70,000 years. *Progress in Oceanography* **35**, 207–251.

84. Little, M.G., Schneider, R.R., Kroon, D., Price, B., Summerhayes, C.P. and Segl, M. (1997) Trade Wind forcing of upwelling, seasonality, and Heinrich Events as a response to sub-Milankovitch climate variability. *Paleoceanography* **12** (4), 568–576.

85. Kirst, G., Schneider, R.R., Müller, P.J., von Storch, I. and Wefer, G. (1999) Late Quaternary temperature variability in the Benguela Current system derived from alkenones. *Quaternary Research* **52**, 92–103.

86. Pedersen, T.F. (1983) Increased productivity in the eastern equatorial Pacific during the Last Glacial Maximum (19,000 to 14,000 yr B.P.). *Geology* **11**, 16–19.

87. Lyle, M. (1988) Climatically forced organic carbon burial in equatorial Atlantic and Pacific Oceans. *Nature* **335**, 529–532.

88. Anderson, R.F. (2003) Chemical tracers of particle transport. In: *The Oceans and Marine Geochemistry, Vol. 6, Treatise on Geochemistry* (eds H.D. Holland and K.K. Turekian). Elsevier-Pergamon, Oxford, pp. 247–291.

89. Crosta, X., Pichon, J.J. and Burckle, L.H. (1998) Application of modern analogue technique to marine antarctic diatoms: reconstruction of maximum sea-ice extent at the Last Glacial Maximum. *Paleoceanography* **13** (3), 284–297.

90. Crosta, X., Sturm, A., Armand, L. and Pichon, J.J. (2004) Late Qaternary sea ice history in the Indian sector of the Southern Ocean as recorded by diatom assemblages. *Marine Micropalaeology* **50**, 209–223.

91. Gersonde, R., Crosta, X., Abelmann, A. and Armand, L. (2005) Sea-surface temperature and sea ice distribution of the Southern Ocean at the EPILOG Last Glacial Maximum – a Circum-Antarctic view based on siliceous microfossil records. *Quaternary Science Reviews* **24**, 869–896.

92. Denton, G.H. and Hughes, T.J. (1986) *The Last Great Ice Sheets*. John Wiley and Sons, New York.

93. Blanchon, P. (2011) Meltwater pulses. In: *Encyclopedia of Modern Coral Reefs* (ed. D. Hopley). Springer, New York, pp. 683–690.

94. Murton, J.B., Bateman, M.D., Dallimore, S.R., Teller, J.T. and Yang, Z. (2010) Identification of Younger Dryas outburst flood path from Lake Agassiz to the Arctic Ocean. *Nature* **464**, 740–743.

95. Berger, A. (2009) Astronomical theory of climate change. In: *Encyclopedia of Paleoclimatology and Ancient Environments* (ed. V. Gornitz). Springer, Dordecht, pp. 51–57.

96. McManus, J., Oppo, D., Cullen, J. and Healey, S. (2003) Marine isotope stage 11 (MIS 11): analog for Holocene and future climate? In: *Earth's Climate and Orbital Eccentricity: The Marine Isotope Stage 11 Question* (eds A. Droxler, R.Z. Poore and L.H. Burckle). Geophysical Monographs 137. American Geophysical Union, Washington, DC, pp. 69–85.

97. Droxler, A., Alley, R.B., Howard, W.R., Poore, R.Z. and Burckle, L.H. (2003) Introduction: unique and exceptionally long interglacial marine isotope stage 11: window into earth warm future climate. In: *Earth's Climate and Orbital Eccentricity: The Marine Isotope Stage 11 Question* (eds A. Droxler, R.Z. Poore and L.H. Burckle). Geophysical Monographs 137. American Geophysical Union, Washington, DC, pp. 1–14.

98. Rohling, E.J., Grant, K., Bolshaw, M., Roberts, A.P., Siddall, M., Hemleben, C. and Kucera, M. (2009) Antarctic temperature and global sea level closely coupled over the past five glacial cycles. *Nature Geoscience* **2**, 500–504.

99. Berger, W.H. and Wefer, G. (2003) On the dynamics of the ice ages: stage 11 paradox, mid-Brunhes climate shift, and 100-ky cycle. In: *Earth's Climate and Orbital Eccentricity:*

The Marine Isotope Stage 11 Question (eds A. Droxler, R.Z. Poore and L.H. Burckle). Geophysical Monographs 137. American Geophysical Union, Washington, DC, pp. 41–59.

100. Hodell, D.A., Kanfoush, S.L., Venz, K.A., Charles, C.D. and Sierro, F.J. (2003) The mid-Brunhes transition in ODP sites 1089 and 1090 (subantarctic South Atlantic). In: *Earth's Climate and Orbital Eccentricity: The Marine Isotope Stage 11 Question* (eds A. Droxler, R.Z. Poore and L.H. Burckle). Geophysical Monographs 137. American Geophysical Union, Washington, DC, pp. 113–129.

101. Dickson, A.J., Beer, C.J., Dempsey, C., Maslin, M.A., Bendle, J.A., McClymont, E.L. and Pancost, R.D. (2009) Oceanic forcing of the Marine Isotope Stage 11 interglacial. *Nature Geoscience* **2**, 428–433.

102. Montañez, I.P., Norris, R.D., Algeo, T., Chandler, M.A., Johnson, K.R., Kennedy, M.J., *et al.* (2011) *Understanding Earth's Deep Past: Lessons for Our Climate Future.* National Academies Press, Washington, DC.

103. Edwards, R.L., Cutler, K.B., Cheng, H. and Gallup, C.D. (2003) Geochemical evidence for Quaternary sea-level changes. In: *The Oceans and Marine Geochemistry, Vol. 6, Treatise on Geochemistry* (eds H.D. Holland and K.K. Turekian). Elsevier-Pergamon, Oxford, pp. 343–364.

104. Kopp, R.E., Simons, F.J., Mitrovica, J.X., Maloof, A.C. and Oppenheimer, M. (2009) Probabilistic assessment of sea level during the last interglacial stage. *Nature* **462**, 863–868.

105. Rohling, E.J., Grant, K., Hemleben, C., Siddall, M., Hoogakker, B.A.A., Bolshaw, M. and Kucera, M. (2008) High rates of sea-level rise during the last interglacial period. *Nature Geoscience* **1**, 38–42.

106. Dutton, A. and Lambeck, K. (2012) Ice volume and sea level during the last interglacial. *Science* **337** (6091), 216–219.

107. Raymo, M.E. and Mitrovica, J.X. (2012) Collapse of polar ice sheets during the stage 11 interglacial. *Nature* **483**, 453–456.

108. Gehrels, R. (2010) Sea-level changes since the Last Glacial Maximum: an appraisal of the IPCC Fourth Assessment Report. *Journal of Quaternary Science* **25** (1), 26–38.

109. Fairbanks, R.G. (1989) A 17,000-year glacio-eustatic sea level record: influence of glacial melting rates on the Younger Dryas event and deep-ocean circulation. *Nature* **342**, 637–642.

110. Yokoyama Y., Lambeck K., de Deckker P., Johnston, P. and Fifield, L.K. (2000) Timing of the Last Glacial Maximum from observed sea-level minima. *Nature* **406**, 713–716.

111. Waelbroeck, C., Labeyrie, L., Michel, E., Duplessy, J.C., McManus, J.F., Lambeck, K., *et al.* (2002) Sea level and deep water temperature changes derived from benthic foraminifera isotopic records. *Quaternary Science Reviews* **21**, 295–305.

112. Shackleton, N.J. (2000) The 100,000-year ice-age cycle identified and found to lag temperature, carbon dioxide, and orbital eccentricity. *Science* **289**, 1897–1902.

113. Lambeck, K., Yokoyama, Y. and Purcell, A. (2002) Into and out of the Last glacial Maximum: sea level change during oxygen isotope stages 3-2. *Quaternary Science Reviews* **21**, 343–360.

114. Chappell, J. and Shackleton, N.J. (1986) Oxygen isotopes and sea level. *Nature* **324**, 137–140.

115. Elderfield, H. and Ganssen, G. (2000) Past temperature and $\partial^{18}O$ of surface ocean waters inferred from foraminiferal Mg/Ca ratios. *Nature* **405**, 442–445.

116. Austermann, J., Mitrovica, J., Ltychev, K. and Milne, G.A. (2013) Barbados-based estimate of ice volume at Last Glacial Maximum affected by subducted plate. *Nature Geoscience* **6**, 553–557.

117. Niessen, F., Hong, J.K., Hegewald, A., Matthiessen, J., Stein, R., Kim, H., *et al.* (2013) Repeated Pleistocene glaciation of the east Siberian continental margin. *Nature Geoscience* **6**, 842–846.

118. Church, J.A., Woodworth, P.L., Aarup, T. and Wilson, W.S. (2010) *Understanding Sea-Level Rise and Variability.* Wiley-Blackwell, Chichester and Oxford.

119. Berger, W.H. (1982) Climate steps in ocean history – lessons from the Pleistocene. In: *Climate In Earth History: Studies in Geophysics* (ed W.H. Berger and J.C. Crowell). Report of Panel on Pre-Pleistocene Climates, for the Geophysics Study Committee of the Geophysics Research Board of the Commission on Physical Sciences, Mathematics, and Applications of the US National Research Council, based on a meeting in Toronto in 1980. National Academies Press, Washington, DC, pp. 43–54. (Available from http://www.nap.edu/catalog/11798.html, last accessed 29 January 2015).

13

Solving the Ice Age Mystery:
The Ice Core Tale

13.1 The Great Ice Sheets

This book is not the place for a detailed review of the Northern Hemisphere ice sheets, but it is useful background to know where they were. The vast Laurentide Ice Sheet covered all of Canada and a bordering strip of the United States (Figure 13.1). Large lateral moraines, including those forming Long Island and Cape Cod, mark its extreme southern edge. Ice covered about half of Alaska, all of the Canadian islands and Greenland. Aside from the ice sheet in Greenland, a tiny remnant remains in the Barnes Ice Cap on Baffin Island. The North American Ice Sheet was much like Antarctica is today, covering a similar area (almost 13 million km^2) to about the same maximum thickness (about 4 km). Its height and surface albedo kept the region cold, and it formed a significant barrier to atmospheric circulation. A massive and equally thick Fennoscandian Ice Sheet of about 8.5 million km^2 covered Scandinavia and most of northern Siberia, extending south to a line running from just north of London in the west to just south of Berlin and Moscow in the east. The ice sheets of both poles extended to the edges of the continental shelf when sea level was low. The map excludes the northeast Siberian Ice Sheet, which was newly identified in 2013 (see Chapter 12). Floating ice shelves probably bordered the Arctic ice sheets, as in Antarctica today. During the winters, at least, they would have been surrounded by sea ice. The Antarctic ice sheet is described in Chapter 7.

The weight of the great ice sheets depressed the crust in the Arctic by as much as 700–800 m^1. Temperatures close to the edges of the ice sheets were probably about 10 °C lower than today on average, and 15–20 °C lower in winter. Conditions became much drier in the polar regions and may have been about 50% drier near the ice sheets, resulting in a significant increase in the atmospheric transport of dust, recorded in ice cores from Greenland and Antarctica.

13.2 The Greenland Story

Palaeoclimatologists were keen to drill through the Greenland and Antarctic ice sheets, which – like ocean or lake sediments – were expected to operate like gigantic tape recorders, with successive snowfalls capturing details of climate change through time. Confirmation that the palaeoceanographers were right in their identifications of Ice Age climate changes began to emerge in the very early 1980s from studies of the oxygen isotopes down ice cores from Greenland. At the 4th Maurice Ewing Symposium in Palisades, New York in 1982, the Danish and Swiss teams working on those cores presented their findings, led by Willi Dansgaard (1922–2011) (Figure 13.2). Dansgaard was professor of geophysics at the University of Copenhagen, a member of the Academies of Science of Denmark, Sweden and Iceland and soon to be a winner of the 1995 Swedish Crafoord Prize.

Earth's Climate Evolution, First Edition. Colin P. Summerhayes.
© 2015 John Wiley & Sons, Ltd. Published 2015 by John Wiley & Sons, Ltd.

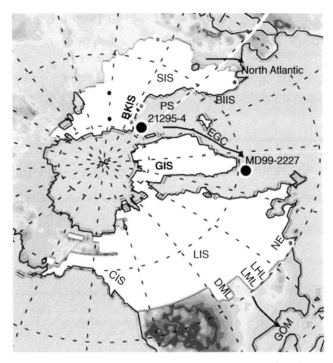

Figure 13.1   *The large ice sheets of the Northern Hemisphere at the Last Glacial Maximum. Cordilleran (CIS), Laurentide (LIS), British Isles (BIIS), Scandinavian (SIS) and Barent-Kara (BKIS) ice sheets are labelled, as are the Lake Michigan (LML), Lake Huron (LHL) and Des Moines (DML) ice lobes and the New England margin (NE). Small dots = dated points. Large dots = specific ocean cores. Black arrows = directions taken by meltwater flow from BKIS via East Greenland Current (EGC) and from LIS to the Gulf of Mexico (GOM).*

Dansgaard worked on the 1390 m-long Camp Century core, drilled in the 1960s by the US Army Cold Regions Research and Engineering Laboratory, and led the drilling party for the DYE-3 ice core, drilled 1400 km away in South Greenland in the 1970s. The DYE-3 core was a product of the Greenland Ice Sheet Project (GISP), run by the Danes, Americans and Swiss from 1971 to 1981. In 1979, it reached bedrock at 2038 m and retrieved ice possibly as old as the last interglacial, around 130 Ka ago[2].

Much to their amazement, Dansgaard and the Swiss ice scientist Hans Oeschger (see Chapter 9) found in their Greenland ice cores evidence for large, abrupt variations in $\partial^{18}O$ ratios within the last glaciation. They interpreted these changes as being caused by fast changes in the latitude of the polar front. Abrupt rises in temperature of up to 6 °C took place in as little as 2.5 years,

remained stable for 500–1000 years, then plunged[3]. The fluctuations seemed to follow cycles of about 2550 years, now named 'Dansgaard–Oeschger cycles', and there was some slight indication of one with a frequency near that of the 1470-year cycle that Bond found in North Atlantic sediment cores[4,5]. The two men thought that these cycles followed some feedback mechanism within the climate system, rather than some as yet unknown solar cycle[2]. The cycles became colder with each cool event, the coldest being associated with ice-rafting in a Heinrich event, after which conditions warmed and the process repeated itself. Oeschger and Dansgaard also showed that the $\partial^{18}O$ temperature record from the last 60 Ka of the DYE-3 core paralleled the $\partial^{18}O$ record from Lake Gerzensee in Switzerland[6]. This convinced them that advances and retreats of cold Atlantic surface water were influencing the climate of the region.

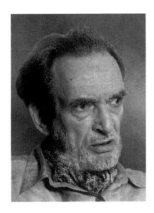

Figure 13.2   *Willi Dansgaard.*

GISP was followed by GRIP, the international Green-land Ice Core Project, which drilled a 3038 m core to basement at Summit in central Greenland between 1989 and 1992. As with the GISP core, the ice layers at the bottom of the hole were contorted due to squeezing by the weight of the ice sheet. To avoid distortions created by ice flow, another hole was drilled in North Greenland at a site where the basal topography was flat. The North GRIP team obtained a core 3085 m long between 1999 and 2003. At its base were 5000 undisturbed years of the last interglacial – the Eemian. Dansgaard–Oeschger events were well represented in the GRIP Summit core[7], reinforcing the conclusion from ocean core studies that the Ice Age climate oscillated between two stable states. The temperature increases during these events ranged between 5 and 15 °C in and around Greenland[8].

By 1993, it was obvious to Gerard Bond and his team that millennial variations in ice-rafted debris from North Atlantic cores were synchronous with Dansgaard–Oeschger events in Greenland ice, and that this relationship held up over the past 90 Ka (Figure 13.3)[9]. Making progressively more detailed studies in 1997 and 1999, they found that the millennial cycles were bundled into cooling cycles lasting around 10–15 Ka, which had asymmetric saw-tooth shapes due to their fast warm beginnings and slow cooling endings. The most massive discharges of icebergs into the North Atlantic (Heinrich events) happened at the end of each such bundle, followed by warm Dansgaard–Oeschger events (Figure 13.3). As we saw in Chapter 12, it seemed likely that a climate oscillator within the ocean–atmosphere system periodically injected cold fresh water into the Nordic Seas[4]. Those injections weakened or shut down the thermohaline circulation that imported warm surface waters from the south. Iceberg outbreaks occurred once a threshold was crossed[5].

The longest and warmest Dansgaard–Oeschger events were associated with elevated amounts of methane (Figure 13.3), because warmth encouraged the formation of wetlands, from which CH_4 is emitted. Initially, it appeared from Greenland ice cores that these events were also associated with high levels of CO_2, but those are now known to be artefacts caused by acids deposited from volcanic eruptions interacting with carbonate-rich dust blown on to the ice from Canada[10]. Such reactions do not affect Antarctic ice, which contains less dust and virtually no carbonate.

Enter the well-known palaeoclimatologist Richard Alley (1957–) of Pennsylvania State University, a mem-ber of the US National Academy of Sciences and a recipient of the Louis Agassiz Medal of the European Geosciences Union and of the Seligman Crystal of the International Glaciological Society. Summarising the results of an AGU Chapman Conference on 'Mechanisms of Millennial-Scale Global Climate Change' that took place in Snowbird, Utah in June 1998, Alley reminded us that although the Dansgaard–Oeschger oscillation was primarily an oceanic process centred on the North Atlantic, winds transported this climate signal rapidly to surrounding areas, including the monsoonal areas of Africa and Asia. The signal is also identified in cores from the Cariaco Basin off Venezuela, the Arabian Sea and the Santa Barbara Basin off California[11]. Although the oscil-lation seemed to be periodic, there was much variability in the spacing of its signal. Variability of the kind represented by Dansgaard–Oeschger events has subsequently been detected in older glacial periods (marine isotope stages (MISs) 6 and 8), suggesting that the climate of glacial periods is predisposed to unstable behaviour[12, 13]. More on that later.

13.3   Antarctic Ice

Another set of stunning discoveries emerged from drilling into the Antarctic ice sheet. The first results were not inspiring – a 2164 m-long core drilled to bedrock by the Americans in Marie Byrd Land in 1968 reached ice almost 90 Ka old. Major success came from a 3623 m-long core from the USSR's Vostok Station, 3490 m up on the Polar Plateau, where the Russians began drilling in the 1970s. Cores from depths greater than 1000 m were not drilled until 1984, and the oldest ice was not cored until 1996. The

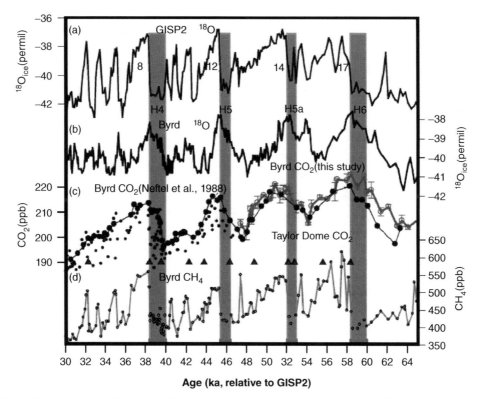

Figure 13.3 *Links between Dansgaard–Oeschger and Heinrich events and CO_2 (65–30 Ka ago). (a) $\delta^{18}O_{ice}$ from GRIP 2 as a proxy for surface temperature. Bold numbers = Dansgaard–Oeschger warm events. (b) $\delta^{18}O_{ice}$ from Byrd station and Antarctic warm events A1–A4. (c) Atmospheric CO_2. Small solid circles back to 47 Ka are existing data from Byrd; open circles prior to 47 Ka are new data from Byrd; large solid dots are data from Taylor Dome; triangles are control points for the synchronisation of gas ages between Byrd and Taylor Dome records. (d) CH_4 from Byrd ice core. Ages are adjusted to synchronise abrupt increases of CH_4 and Greenland temperature. Vertical shaded bars represent cold Heinrich events. Note that atmospheric CO_2 rose several thousand years before abrupt warming in Greenland associated with Dansgaard–Oeschger events 8, 12, 14 and 17, the four large warm events that followed Heinrich events. The CO_2 rise predated Heinrich events associated with these Dansgaard–Oeschger events and terminated at the onset of Greenland warming for each of them. Atmospheric CO_2 is strongly correlated with the Antarctic isotopic temperature proxy, with an average time lag of 720 ± 370 years (mean $\pm 1\sigma$) during the time interval studied.*

Vostok ice core reached ice 420 Ka old and revealed four glacial cycles. Many of the geochemical analyses of this core were carried out by French scientists led by Claude Lorius (1932–) (Figure 13.4, Box 13.1) at the Laboratoire de Glaciologie et de Géophysique de l'Environnement in St Martin d'Hères, near Grenoble, and by Jean Jouzel (1947–) of the Laboratoire de Géochimie Isotopique of the Laboratoire d'Océanographie Dynamique

et de Climatologie (LODYC) at Gif-sur-Yvette, near Paris.

Landmark papers announced the French and Russian results. In August 1985, Lorius and others presented the temperature record for the past 150 Ka, deduced from analyses of $\partial^{18}O$ of the earliest (hence, shallowest and youngest) cores retrieved[15]. This was the first ice core record extending through both the last glacial and the

Figure 13.4   *Claude Lorius.*

Box 13.1 Claude Lorius.

Lorius was one of three young Frenchmen who were the first scientists to winter over on the Polar Plateau during the International Geophysical Year of 1957–58. He was one of the first to demonstrate a relationship between hydrogen isotopes and temperature in Antarctic ice, thus discovering a new palaeothermometer, ∂D or $\partial^2 D$ (the ratio in parts per thousand (‰) between hydrogen and its isotope deuterium)[14]. For his work on ice cores, Lorius has been showered with medals and prizes, including the Blue Planet Prize in 2008. He and Jean Jouzel were jointly awarded France's highest scientific award, the Gold Medal of the Centre National de la Recherche Scientifique (CNRS), in 2002. Among other awards, France named him a Commandeur of the Légion d'Honneur in 2009, and he is a member of the French Academy of Sciences. Along with Dansgaard and Oeschger, he received the Tyler Prize for Environmental Achievement in 1996.

last interglacial, something not achieved in Greenland. In October 1987, Jouzel and others presented the temperature record for the past 160 Ka, deduced from analyses of hydrogen isotopes[16]. The isotopic data showed that the Last Glacial Maximum in Antarctica was about 9 °C colder than the early Holocene, while the last interglacial around 140 Ka ago was around 2–3 °C warmer. The last glacial period contained several warm and cool intervals,

repeating on a cycle of 40 Ka, typical of the behaviour of Earth's axial tilt. In due course, the core provided a record of four major interglacials before the present, centred at 125, 240, 320 and 420 Ka ago (Figure 13.5)[17, 18].

Vostok's astounding success was followed by even greater penetration back through time by the European Project for Ice Coring in Antarctica (EPICA). EPICA drilled two holes on opposite sides of the Polar Plateau. One, with a low rate of accumulation, was drilled at the Franco-Italian Concordia base on Dome C, 560 km from Vostok and at a slightly lower altitude of 3233 m. It reached a depth of 3270 m in December 2004 and recovered ice as old as 800 Ka, containing eight glacial cycles (Figure 13.6)[19]. The other, with a faster rate of accumulation, was drilled at Germany's Kohnen summer station in Dronning Maud Land, at a height of 2892 m. It reached a depth of 2774 m in January 2006, and ice at a depth of 2416 m was 150 Ka old[20]. This is the EPICA Dronning Maud Land (EDML) core. The Japanese also drilled a long core, on Dome F, the site of Japan's Dome Fuji station and one of the highest points in Dronning Maud Land, at 3810 m. Drilling began in August 1995, achieving a depth of 3035 m in the southern winter of 2006–07 and reaching back 720 Ka. The Japanese results confirmed findings made in the Vostok and EPICA cores[21].

I visited the Kohnen station on the Polar Plateau in November 2004, as a guest of the Alfred Wegener Institute for Marine and Polar Research (AWI). We landed in a twin-engined Dornier and wandered around outside the orange-coloured laboratories and workshops made from shipping containers and placed on stilts to prevent the snow piling up against them. Nearby was the drilling pit, but the drilling season had not started – the place was deserted and the drill rig was dismantled. The vast, uninterrupted white expanse of the Polar Plateau swept away in all directions with deceptive smoothness. Seen close up, it was roughened by 30 cm-high ridges of hard ice – 'sastrugi' – carved by the wind and set about half a metre apart. They made the surface look like an old-fashioned washboard. Taking off to return to our base, the German Neumayer station down on the coast, was a nightmare. The plane had so much fuel aboard, and the air was so thin at that altitude, that it took the pilot 13 attempts at full throttle, including taxiing back to the start, all the while bouncing over the sastrugi, before we finally lifted off. What it must have been like for Scott and his men towing their own sledges over such terrain, I cannot imagine. But I can empathise with his remark on reaching

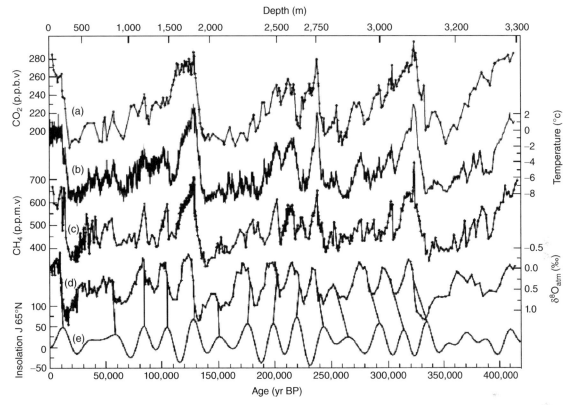

Figure 13.5 *Vostok: the climate of the past 400000 years. (a) CO_2. (b) Isotopic temperature of the atmosphere. (c) CH_4. (d) $\partial^{18}O_{atm}$. (e) Mid-June insolation at 65° N (in W/m^2).*

the pole in January 1912: '*Great God, this is an awful place*'.

The long cores from Vostok, Fuji and Dome C confirmed the picture from deep-ocean drilling. There have been five major interglacial periods spaced around 100 Ka apart over the past 400 Ka[22]. We are living in the last of them: the Holocene. Each of the four prior interglacials was about 2–3 °C warmer than the present, and sea level was higher than it is today by between 2 and 9 m[23]. In the 1990s, the mean global difference between the Ice Age and the present was thought to be about 5 °C. Recent research shows that it was closer to 4 °C[24]. Prior to 420 Ka ago, the Dome C and Fuji cores showed that major interglacial periods occurred at intervals of around 40 Ka, rather than 100 Ka (Figure 13.6), just as we see in deep-sea sediment cores (Figure 12.3).

The French geochemists found that there was more airborne dust and sea salt during cold periods. Nearby

dry areas either expanded at the time or were subject to stronger winds that took more of their dust offshore, while Southern Ocean winds strengthened, whipping up waves and enhancing the supply of sea salt to the air. Analyses of sulphate showed no long-term link between volcanism and climate[25,26].

By 2009, Tim Naish of Victoria University of Wellington had accumulated a great deal of information about Ice Age climate change in and around Antarctica from ice cores, continental margin drill cores and deep-ocean drilling cores from the Southern Ocean[27]. He and his team found an interlude of anomalous continental-scale warmth dated to about 1 Ma ago, correlating with MIS 31. It was represented by foraminiferal ooze and coccolith-bearing assemblages in cores from the Weddell Sea and Prydz Bay, along with bioclastic limestone in the Ross Sea. Sea surface temperatures locally warmed by 4–6 °C, apparently due to an extended interval of unusually warm or

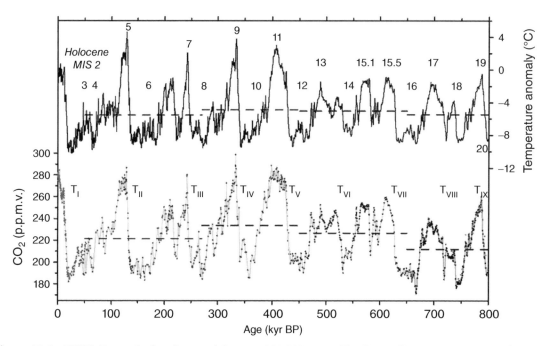

Figure 13.6 *EPICA Dome C: the climate of the past 800 000 years. The Dome C temperature anomaly record with respect to the mean temperature of the last millennium (based on deuterium isotope data). Data for CO_2 are from Taylor Dome (75–25 Ka ago), Vostok (425–25 Ka ago) and Dome C. Horizontal lines are the mean values of temperature and CO_2 for the time periods 799–650, 650–450, 450–270 and 270–50 Ka ago. T_I: glacial termination. MISs are numbered.*

long summers forced by orbital variations. Following it, as seen in marine sediment cores (Chapter 12) and ice cores, was the Mid Pleistocene Transition (MPT) between 900 and 700 Ka ago, when orbital cycles changed in frequency from an older set dominated by 41 Ka cycles to a younger set dominated by 100 Ka cycles. After the MPT, evidence for substantial discharge of meltwater disappears, suggesting that the ice sheet had become more dry-based: a sign of cooling. Naish's team described the transition as '*one of the most poorly understood events in palaeoclimatology*'[27]. While that was true back in 2009, by 2012 Elderfield had come up with an explanation for it[28], as we saw in Chapter 12.

Naish's team found that during glacial periods, Antarctic sea ice advanced further north, sea surface temperatures fell by about 6 °C, oceanic frontal zones migrated north by 5–10° latitude and zonal winds intensified, strengthening the Antarctic Circumpolar Current. These processes intensified between 3000 and 7000 years before maximum ice volume was reached and the ice sheet covered

the continental shelf. Expansion and contraction of the ice sheet across the full width of the continental shelf contributed 15–20 m to the global changes in sea level of the past 700 Ka. During the Last Glacial Maximum, ice built up around the continental margins by up to several hundred metres due to increased snowfall, but diminished in the interior by 100–150 m due to lower rates of snowfall there. The dust blown into the region by high winds during glacials supplied nutrients (especially iron, Fe) to the Southern Ocean, increasing its productivity and helping to draw down the content of CO_2 in the atmosphere.

13.4   Seesaws

The fine-scale signals of climate change within the last glaciation that Dansgaard and Oeschger found in Greenland ice cores are also identified in Antarctic ice cores, where they are much less abrupt (Figure 13.3)[27]. To compare these signals, the cores from the two poles had to be matched in time. What clever means could be used to do that? The answer: methane! Methane gas is rapidly mixed throughout the atmosphere, so a prominent methane signal will be trapped in ice at both poles at the same time. Superimposing the patterns of wiggles in the methane signal through time in the cores from both hemispheres enabled the relationships in time between the temperature signals from the two regions to be fixed[29–31]. Slow warming in Antarctica preceded each sudden warming in Greenland (Figure 13.3). Greenland warming rose rapidly as Antarctic warming reached a peak and began to cool. Greenland then cooled rapidly as the Antarctic slowly started to warm again. What was going on? The answer is what Wally Broecker and Thomas Stocker called the 'bipolar seesaw'[32, 33].

This out-of-phase relationship between the climates of the two hemispheres is one of the most unexpected finds in the study of millennial-scale climate change[34], and one of the most exciting additions to the Thermohaline Conveyor Belt paradigm[35]. Initial research suggested that the bipolar seesaw works as follows. Warming in the north leads to iceberg outbreaks in the North Atlantic. As the bergs melt, they cool the surface ocean and flood it with a freshwater lid, which stops or reduces the production of North Atlantic Deep Water. Warm Gulf Stream waters are no longer pulled north to replace the subsiding North Atlantic Deep Water; instead, they return south, building up heat in the South Atlantic, which spreads through the Southern Ocean. Meanwhile, deep northward-moving cold Antarctic Bottom Water displaces slightly warmer southward-moving North Atlantic Deep Water as the main source of cross-equatorial flow at depth (this does not necessarily imply an increase in Antarctic Bottom Water production – merely a change in the balance between Antarctic Bottom Water and North Atlantic Deep Water at depth). Once the northern iceberg armada shrinks, the freshwater lid disappears, and warm, salty surface waters return north. North Atlantic Deep Water production resumes, pulling yet more warm, salty Gulf Stream water north, enabling the cycle to repeat itself. Meanwhile, in the south, the warming of the Southern Ocean eventually melts sea ice and ice shelves, making the Antarctic ice sheet surge, which cools and freshens surface waters, slowing Antarctic Bottom Water production. The net result is a continual off-and-on reversal of cross-equatorial flow – an 'internal oscillator'. No external forcing is required.

Wally Broecker thought that the seesaw was all down to salt, which changes the density of ocean water[36]. The North Atlantic Deep Water exports salt from the North Atlantic via the sinking of water with a relatively high salt content, made denser by cooling in the Norwegian-Greenland and Labrador Seas. When little or no North Atlantic Deep Water is produced, salt transported north by the Gulf Stream builds up in the northernmost Atlantic to the point where eventually the North Atlantic Deep Water has to switch back on to remove it. In due course, this export lowers the salt content of the North Atlantic again to the point where the North Atlantic Deep Water once more switches off, because surface waters are no longer dense enough to sink. Hence, deep-water processes are at the root of the bipolar seesaw. Other amplifiers also come into play. As the North Atlantic cools, the sea ice front moves south, increasing the Earth's albedo and thus causing further cooling. Expansion of sea ice limits the formation of North Atlantic Deep Water. Cooling steepens the thermal gradient between the Equator and the Pole, strengthening winds and lofting dust and sea salt into the air to reflect solar energy and further amplify cooling. In due course, salt build-up turns North Atlantic Deep Water production back on.

Dan Seidov and his team from Penn State used an advanced ocean model to explore the chicken-and-egg implications of these findings. They concluded that while the North Atlantic Deep Water is the main driver of the Thermohaline Conveyor, in some circumstances the Southern Ocean can overpower that Northern Hemisphere driver and become a major player in long-term climate change[37]. They agreed that deep-ocean circulation drives the seesaw.

Mark Maslin and colleagues explained the seesaw in terms of 'heat piracy'[34]. During warm periods, the North Atlantic 'steals' heat from the Southern Hemisphere, sustaining a strong Gulf Stream and preventing ice build-up in the north. This is northern heat piracy. During a Heinrich event, when meltwater covers the North Atlantic, heat is transported south across the Equator, warming the southern oceans. The South Atlantic 'steals' heat from the north. This is southern heat piracy. Under 'normal' glacial conditions, the production of Antarctic Bottom Water and North Atlantic Deep Water is in balance and there is no heat piracy.

The abrupt nature of each warming in Greenland suggests that the climate there moved from cool to warm within a decade. This implies that the arrival of heat from the south pushed local temperature over a threshold, a 'tipping point', beyond which it rose suddenly. The incoming heat maintained the new warm stable state around Greenland for a time, during which the south cooled as it lost heat to the north. The cool waters from the south eventually arrived in the north and tipped the northern system back below the threshold, returning it to its original cold state. The threshold for stopping North Atlantic Deep Water formation is much higher than the threshold for restarting it, because of a phenomenon termed 'hysteresis', a Greek word meaning 'a coming short' or 'deficiency' – we use it in palaeoclimate studies to indicate where an effect is lagging behind its cause. A larger response than would be expected from gradual linear forcing is referred to as an 'overshoot'[34]. Overshoots are difficult to reverse even when the forcing subsequently declines below the original threshold. Hysteresis typically gives rise to loop-like behaviour. For instance, a gradually increasing freshwater lid eventually gets to the point where North Atlantic overturning, represented by the production of North Atlantic Deep Water, abruptly stops: a high threshold value, in this case abundant freshwater at the surface, has been passed – we have an overshoot. Before North Atlantic Deep Water production can restart, freshwater has to decline far below the point at which it stopped that production[38].

Two researchers from Germany's Potsdam Institute for Climate Impact Research, Andrey Ganopolski and Stefan Rahmstorf, used a coupled ocean–atmosphere climate model to see how stable the Thermohaline Conveyor was, and to explore possible mechanisms for rapid change[39]. While for the modern climate they found two stable modes for Atlantic circulation – the 'warm' and the 'off' mode – they found only one – 'cold' – for glacial times. The 'cold' mode has deep-water formation south of Iceland and relatively shallow overturning, with weak outflow to the South Atlantic. The 'warm' mode, with deep-water formation north of Iceland, has deeper overturning and strong outflow to the South Atlantic. It is unstable under glacial conditions, where it can occur for short periods when the flux of freshwater to the North Atlantic is low. Temporary transitions from the 'warm' to the 'cold' mode and back explain the observed Dansgaard–Oeschger cycles. The model outputs confirmed that changes in freshwater supply in the North Atlantic could drive warming in the south. Given the nonlinear threshold response of the North Atlantic circulation to forcing, a weak climate cycle can trigger large-amplitude episodic warm events. Rapid transitions can occur because the sites for deep-water formation between the 'cold' mode (south of Iceland) and the 'warm' mode (north of Iceland) are geographically far apart.

As mentioned in Chapter 12, I was part of a group that identified the bipolar seesaw effect in sediment cores from the South Atlantic in 1997. We found that cold Heinrich events in the North Atlantic were represented as warm events in the South Atlantic[40]. This is consistent with what one would expect if the thermohaline circulation collapsed at those times[38], as explained earlier.

Is Bond's internal oscillator, which appears to be an element of Broecker's bipolar seesaw, truly periodic, or does it just seem that way? Richard Alley was one of those to puzzle over that question. In 1999, he and some others thought that the extent to which Bond's cycle was truly regular remained to be seen[11]. Some thought that it was a free oscillation, possibly related to the El Niño–Southern Oscillation characteristic of the Pacific Ocean[41]. The length of the cycle suggested a major role for the ocean, rather than control by changes in the ice or the atmosphere[11]. By 2001, Alley and his team had an answer[42]. Analysing temperature records from Greenland ice cores, they found peaks in periodicity at 1500, 3000 and 4500 years, diminishing in amplitude from 1500 to 4500 years. This pattern is typical of 'stochastic resonance', meaning that it is largely random or nondeterministic. Financial markets use stochastic models to represent the seemingly random behaviour of the stock market. Bond's sediment cores showed the same pattern. Alley's team suggested that weak periodicity in combination with statistical noise in an internal oceanic oscillator made the climate switch from one mode to another with an apparent periodicity of about 1500 years, driven by random fluctuations in freshwater supply. We will revisit this issue in Chapters 14 and 15. For the moment, it is worth recalling that Bond's 1470-year cycle is roughly equivalent to the overturning age of the ocean. Vikings sailed their longships on the 1000-year-old fossil surface water now welling up in the North Pacific.

In 2011, Stephen Barker of Cardiff University set out with colleagues to test the notion that Dansgaard–Oeschger fluctuations in climate persisted back through time beyond the limit of the Greenland ice cores, and that the events in Greenland and Antarctica were connected through the Atlantic Meridional Overturning Circulation, the part of the Thermohaline Conveyor that links the two poles[43]. Using the bipolar seesaw model, they constructed an 800 Ka-long synthetic record of climate variability. It agreed with what was known of climate change in Greenland for the past 100 Ka and with what was known of climate change from a well-dated Chinese speleothem record covering the past 400 Ka, providing confidence that this same climatic behaviour probably extended back through time for the full 800 Ka record seen in Antarctic ice cores. Mark Siddall of the University of Bristol carried out a similar exercise[44], probing the Dome C record to see if older intervals than the last glacial period contained similar millennial-scale events. Over the past 500 Ka, Siddall and colleagues found clusters of millennial events recurring with a period of about 21 Ka, like that of orbital precession, especially during periods of intermediate ice volume, when sea level was lowered by 40–80 m. Millennial variability was absent during periods of high ice volume, probably because the southward extension of large ice sheets and widening of the belt of sea ice during glacial maxima affect the zonality of atmospheric circulation, impeding the development of millennial variability. Intermediate amounts of land ice would not have had the same effect, allowing changes in precession to force changes in sea ice through zonal changes in atmospheric circulation and surface air temperature. The lack of a consistent correlation between millennial variability and ice-rafted debris of the kind one might expect from Bond's theory suggested to Siddall that a different cause must be sought for Bond cycles than iceberg release into the North Atlantic.

As ever, science marches on. Examining the Dansgaard–Oeschger events in MISs 3 (at 60–28 Ka) and 5 (at 123.0–73.5 Ka) in Greenland and their counterparts in Antarctica, Emilie Caprion of the Institut Pierre-Simon Laplace of the Laboratoire des Sciences du Climat et de l'Environnement at Gif-sur-Yvette, France found in 2010 that these sudden warm events were preceded by a short warm period lasting less than 100 years, and were followed by a brief and low-amplitude 'rebound' event[45]. Moreover, the duration of individual Dansgaard–Oeschger events varied in relation to sea level: the longer ones were associated with the greatest sea level rise, and hence the largest amount of ice melt. Caprion and colleagues' exceptionally high-resolution studies confirm that more research is needed to quantify the influence of insolation, ice sheet volume, sea ice and the hydrological cycle on variability at the submillennial scale – the precursor and rebound events[45]. While Dansgaard–Oeschger events and their Antarctic counterparts must be linked through changes in the Atlantic Meridional Overturning Circulation, the link is not as simple as was first supposed.

The existence of Dansgaard–Oeschger and Heinrich events tells us that slow, gradual changes in climate forcing can lead to abrupt climate shifts. Now, there's a good reason to study the processes and interactions within our climate system!

13.5 CO_2 in the Ice Age Atmosphere

So much for temperature; now, what about CO_2 in the Ice Age air? As we saw in Chapter 9, early work by Oeschger and Dansgaard on CO_2 in ice cores showed that the air carried 180–200 ppm CO_2 at the Last Glacial Maximum and 260–300 ppm in the preindustrial era[6, 46]. By 1984, they showed that there was a good correlation between the CO_2 in bubbles of fossil air and the $\partial^{18}O$ composition of the ice, an indicator of temperature. As warming increased, so did atmospheric CO_2. They concluded that the rise in CO_2 at the end of the last glaciation contributed to the rise in temperature[46].

With their results from Vostok, the French reached much the same conclusion by 1987. Jean-Marc Barnola of the Laboratoire de Glaciologie et Géophysique de l'Environnement at St Martin d'Heres showed with colleagues that CO_2 shifted from 190 to 280 ppm as conditions warmed at the ends of glacial periods, confirming a strong link between the climate cycle and the carbon cycle[47]. Cristophe Genthon of the Laboratoire de Géochimie Isotopique within the Laboratoire d'Océanographie Dynamique et Climatologie at Gif sur Yvette concluded that CO_2 was the dominant factor mediating between the two hemispheres, with their different insolation forcings[48]. If Genthon and colleagues were right then the 100 Ka glacial–interglacial cycle might originate from variations in CO_2 rather than from processes associated with the growth and decay of ice sheets suggested by Imbrie's SPECMAP project (indeed, Genthon's suggestion would cause Imbrie to modify the SPECMAP concept).

Claude Lorius summarised the Vostok data at a symposium in Vancouver in August 1987[25, 26]. CO_2 in Greenland fluctuated between 190–200 ppm in cold periods and 260–280 ppm in warm periods. The results, he said '*provide the first direct evidence of a close association between atmospheric CO_2 and climatic (temperature) changes in a glacial-interglacial time scale*'[25, 26]. He went on to say that '*a simple linear multivariate analysis suggests that CO_2 changes may have accounted for more than 50% of the Vostok temperature variability, the remaining part being associated with orbital forcing …The close correlation between CO_2 and temperature records and their spectral characteristics supports the idea that climatic changes* *could be triggered by an insolation input, with the relatively weak orbital forcing strongly amplified by possibly orbitally induced CO_2 changes'[25,26]*. So, fluctuations in ice volume might not, after all, reflect just orbital forcing; CO_2 acting as a greenhouse gas must play some role.

Aside from the close similarity between the trends of CO_2, CH_4 and local temperature calculated from hydrogen isotopes ($\partial^2 H$) in the Vostok core (Figure 13.5), a conspicuous feature of these trends was their 'saw-toothed' pattern, like that seen in deep-ocean sediment cores (Figure 12.3). Since insolation, the main heat source, has a sinusoidal character, the saw-toothed pattern had to reflect slow ice growth and rapid decay[10]. By 2008, analyses of CO_2 had been extended to the bottom of the Dome C core (Figure 13.6)[49], as had analyses of methane[50]. These confirmed the stability of the saw-toothed CO_2 and temperature patterns back through time. This begged the question, which parts of the saw-tooth represented mostly the insolation signal, and which the ice signal?

Both at Vostok and at Dome C, maximum values of CO_2 and CH_4 reached 280–300 ppm and 650–700 ppb, respectively, in interglacials, and fell to 180 ppm and 320–350 ppb in glacials. These limits describe what I call the 'natural envelope' for these gases (Figures 13.5 and 13.6). That natural envelope coincides in turn with the natural envelope for Ice Age temperatures (Figures 12.3 and 13.6), described in Chapter 12, providing compelling evidence that the carbon cycle was intimately connected to the climate system throughout the 2.6 Ma of the Pleistocene period, most probably through connecting feedback processes. Paul Falkowski of Rutgers University and his coworkers agreed, noting that '*The remarkable consistency of the upper and lower limits of the glacial-interglacial atmospheric CO_2 concentrations, and the apparent fine control over periods of many thousands of years around those limits, suggest strong feedbacks that constrain the sink strengths in both the oceans and terrestrial ecosystem*'[51] How close is the relationship between CO_2 and temperature with time? Not surprisingly, the parallel oscillations between the two (Figure 13.6)[52] are reflected in statistics showing a moderately good correlation (correlation coefficient ~0.7)[18,49].

Is the CO_2 signal reliable back through time down-core? Yes. First, over the past 800 Ka the values of CO_2 in the Dome C ice core have remained within the 'natural envelope' of 180–280 ppm (Figure 13.6)[52]. Second, the consistency of the correlation between CO_2 and temperature, regardless of age down-core, rules out the prospect

of gas migration after compaction of the firn[49,53]. Third, this pattern is consistent from one site to another (Vostok, Dronning Maud Land, Fuji and Dome C). Fourth, measurements of CO_2 covering the past 1000 years in Antarctic ice cores from Law Dome, Siple Dome, South Pole and Kohnen station show the same CO_2 patterns despite being very different in location, altitude and rate of accumulation. That rules out anything that might affect the concentration of CO_2 after compaction of the firn has stopped gas migration[53]. For example, Law Dome is a coastal site with a mean annual temperature of −19 °C, a relatively high concentration of some impurities and a high rate of accumulation, while Dome C is a high-altitude inland site with a mean annual temperature of −44.6 °C, low accumulation and low concentrations of impurities. All of these sites show more or less no change from 280 ppm until the late 1700s or early 1800s, after which their CO_2 abundances rise exponentially to merge imperceptibly with those from Keeling's air measurements begun in the late 1950s (Figure 9.2).

It is of particular interest for studies of future climate change, as well as for understanding of past climate change, that the modest 0.7 correlation coefficient between CO_2 and temperature shows that this is not a simple one-to-one relationship. That piqued the interest of Barry Saltzman (1931–2001) of Yale (Box 13.2).

Box 13.2 Barry Saltzman.

Saltzman's work in 1962 on chaos theory led to the development of the 'Saltzman–Lorenz attractor', but he turned his attention to climate change in 1980, and spent 2 decades at Yale developing models and theories of how ice sheets, winds, ocean currents, CO_2 concentrations and other factors work together to make the climate oscillate in a 100 Ka cycle[54,55]. Perhaps not surprisingly given his earlier involvement in the development of chaos theory, with its emphasis on attractors, he became widely known for developing a numerical approach to modelling past climate change. Among other things, Saltzman's models predicted the possibility of 1–2 Ka oscillations of the climate system many years before widespread interest in this topic developed in the palaeoclimate community[54]. He thought that a long-term tectonically forced decrease in atmospheric CO_2 could lead to a bifurcation of the system from steady state to a near-100 Ka auto-oscillation, a subject he explored with Kirk Allen Maasch[56–58]. For his research on how the climate system works, the American Meteorological Society awarded him its highest honour, the Carl Gustaf Rossby Research Medal, in 1998.

Working with Mikhail Verbitsky in 1994, Saltzman had the innovative idea of plotting CO_2 data from Jouzel's Vostok ice core against ice volume data from Imbrie's SPECMAP $\partial^{18}O$ data set and against sea surface temperature data from a North Atlantic core[59, 60]. Plotting these data in 2000-year time steps, Saltzman and Jouzel expected to find that CO_2 decreased as temperatures cooled and ice volume grew, and increased as temperatures warmed and ice volume shrank. While that was broadly true, the data deviated from the expected pattern. Starting from a fully glacial cold mode, CO_2 increased while the temperature stayed constant, then the temperature jumped while CO_2 stayed constant until a fully warm interglacial mode was reached. CO_2 then decreased while temperature stayed constant, then temperature decreased while CO_2 stayed constant, until the cool mode was reached again. This pattern forms a 'hysteresis loop' of the kind discussed earlier. As both temperature and ice volume lagged behind CO_2, Saltzman assumed that the change in CO_2 caused the changes in the other two properties. He thought he had discovered an oscillation within the climate system between two stable states (glacial and interglacial) terminated by internal feedbacks, and that CO_2 was the key forcing agent in this system. His novel findings are summarised in his 2002 book *Dynamical Paleoclimatology*[61]. As we shall see, his analysis may be distorted by the assumption that there was a significant lag between rising temperature and rising CO_2. We explore his ideas in some depth later.

Stimulated by Saltzman's research, in 1999 Hubertus Fischer of Scripps carried out a more detailed study of the relationship between CO_2 and temperature in ice cores[62]. In the Vostok and Taylor Dome cores, Fischer and colleagues found that while CO_2 and temperature rose more or less together going into an interglacial, the temperature dropped more sharply than did the CO_2 at its end. The same was true for changes in temperature within glacials – compare the $\partial^{18}O$ signal, equivalent to temperature, with the CO_2 signal in Figure 13.3. As Fischer put it, '*high carbon dioxide concentrations can be sustained for thousands of years during glaciations*'[62]. He recommended that '*the carbon cycle–climate relation*

should be separated into (at least) a deglaciation and a glaciation mode'[62]. In the deglaciation mode, when orbital insolation increased and led to warming, '*a net transfer of carbon from the ocean to the atmosphere*' directly connected temperature and CO_2 as warming exuded CO_2 from the ocean[62]. But when insolation decreased and led to cooling, several processes maintained abundant CO_2 in the air until temperatures reached their lowest point. The growth of sea ice limited the uptake of CO_2 by surface waters. A decline in terrestrial vegetation released CO_2 into the air, as did the decomposing marine organic matter when falling sea level exposed continental shelves. Once temperatures were at their lowest, land sources of CO_2 ceased supplying CO_2 to the air and the cooler ocean away from sea ice areas took up more CO_2, reducing the amount in the air.

In 1961, Plass had predicted a further disconnect between temperature and CO_2 that should be evident during deglaciation in the Arctic. While increasing insolation warmed mid latitudes and tropics, releasing CO_2 from a warming ocean, in the Arctic the increased insolation energy went into melting ice sheets, keeping the surroundings cold despite the increase of CO_2 in the globally circulating air[63].

These various observations confirm that the relationship between temperature and CO_2 during the Ice Age is not a simple one. In effect, the climate sensitivity differs between glaciation and deglaciation modes. Paul Falkowski and his team reached much the same conclusions[51]. They called for further investigation, because understanding the carbon cycle is essential to deciding whether or not we can '*distinguish between anthropogenic perturbations and natural variability in biogeochemical cycles and climate*', and, further, to determining '*the sensitivity of Earth's climate to changes in atmospheric CO_2*'[51].

This raised a question: was there evidence that CO_2 and temperature changed in lock step as the world warmed? Or were there leads and lags in the system? If you have ever walked on fresh snow, you will recall the crunching sound your footsteps make. That's the sound of the snow being compressed and causing pockets of trapped air to collapse. Without human intervention, the deposition of more and more snow eventually compresses earlier layers into dense snow, or firn. The air between the snow particles in the firn can still mix with air at the surface down to depths of 50 or 100 m, at which point the firn is converted to glacial ice and the air is locked in as bubbles. Comparison of the properties of the air bubbles with those of the surrounding ice requires an adjustment to the time scale for the air bubbles that takes into account the younger age of the trapped air than of the surrounding ice. From such calculations, it was thought that the rise in CO_2 lagged behind the rise in temperature by 600–1900 years[49]. This lag was a topic of controversy, first because its calculation was based on assumptions that might or might not be accurate about the 'lock-in depth' at which air was finally trapped, and second because basic physics showed that temperature and CO_2 should rise together in a warming ocean. As Ed Brook said, we have to know, '*Does CO_2 drive climate cycles or is it a feedback in the system that contributes to warming?*'[64]. If the model of firn compaction used to determine the difference between the gas age and the age of its surrounding ice were wrong, the lag could be more, less or even non-existent.

The controversy has been resolved recently by the latest exciting development in studies of ice core CO_2 published in March 2013 by an international team led by Frédéric Parrenin of the Laboratoire de Glaciologie et Géophysique de l'Environnement in Grenoble[65]. Parrenin and colleagues used nitrogen isotopes to establish the difference between the gas age and the age of the surrounding ice in several Antarctic ice cores. The ratio of $^{15}N/^{14}N$ (expressed as $\partial^{15}N$) in the air bubbles is enriched in firn due to gravitational settling, and depends on the firn's thickness[64]. Given this ratio, the offset in depth between the gas and the ice of the same age can be determined along with the amount of time represented. Using this new technique, Parrenin's team found that CO_2 concentrations and Antarctic temperatures were tightly coupled throughout the last deglaciation, within a quoted uncertainty of less than ±200 years[64, 65]. The correlation between CO_2 and Antarctic temperature using the new gas-age chronology is very high, at 0.993, suggesting that the rise in CO_2 did contribute to much of the rise in temperature in Antarctica during the last deglaciation, even at its onset, more or less as Lorius had initially suggested. The authors theorised that insolation warmed the ocean, which simultaneously released CO_2 to the air, providing an immediate positive feedback that enhanced further warming. The same warming released water vapour, another greenhouse gas, which acted in concert with the CO_2. There was no delay between warming and CO_2 emission. The link between CO_2 and temperature is not a coincidence – it is causal, and both are tied to insolation. Physics rules!

Independent support for these conclusions came from Joel Pedro of the Antarctic Climate and Ecosystems Cooperative Research Centre at the University of Tasmania, Hobart (Figure 13.7)[66]. Using multiple Antarctic ice cores with both high and low rates of accumulation to refine estimates of the lock-in depth of air bubbles in the firn, Pedro and colleagues found in 2012 that during the last deglaciation '*the increase in CO_2 likely lagged the increase in regional Antarctic temperature by less than 400 yr and that even a short lead of CO_2 over temperature cannot be excluded. This result … implies a faster coupling between temperature and CO_2 than previous estimates, which had permitted up to millennial-scale lags*'[66]. Pedro agreed with Parrenin: '*Mounting evidence attributes a large component of the deglacial CO_2 increase to release of old CO_2 from the deep Southern Ocean through changes in its biogeochemistry and physical circulation*'[66].

Yet another international team, led by Jeremy Shakun of Harvard, agreed in 2012 that '*the covariation of carbon dioxide (CO_2) concentration and temperature in Antarctic ice-core records suggests a close link between CO_2 and climate during the Pleistocene ice ages*'[24]. Their analysis, which was not as refined as Parrenin's, suggested that '*The role and relative importance of CO_2 in producing these climate changes remains unclear … in part because the ice-core deuterium record [used as a proxy for temperature] reflects local rather than global temperature*'[24]. Using a record of *global* temperature constructed from 80 proxy records, they showed '*that temperature is correlated with and generally lags CO_2 [globally] during the last … deglaciation*'[24]. This is '*consistent with CO_2 acting as a primary driver of global warming, although its continuing increase is presumably a feedback from changes in other aspects of the climate system*'[24].

The message is that we need to differentiate local from global signals, much as Al Fischer recommended (Chapter 10). Examining Shakun's results, Pedro explained that CO_2 led the Northern Hemisphere temperature reconstruction by 720 ± 330 years, led the global temperature reconstruction by 460 ± 340 years and lagged the Southern Hemisphere temperature reconstruction by 620 ± 660 years[66]. Given the uncertainties in dating, they thought that the 620 ± 660-year lag reported by Shakun's team for the Southern Hemisphere was '*not inconsistent*' with their own conclusion that the lag in the Antarctic was less than 400 years and that CO_2 might even lead temperature change[66]. That being so, Shakun's result for the Southern Hemisphere is not inconsistent with Parrenin's finding that during the last deglaciation, CO_2 and temperature changed synchronously in the Southern Hemisphere. This begs the question, **what delayed warming in the north?**

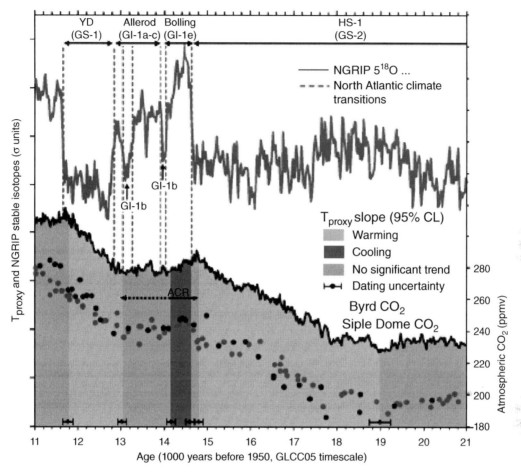

Figure 13.7 *Synchronous proxy temperature (T_{proxy}) and atmospheric CO_2 signals in the last deglaciation at Byrd and Siple coring sites, displaying the bipolar seesaw. Significant warming and cooling trends in T_{proxy} are represented by shaded vertical bands. Climate in the North Atlantic region is represented by the NorthGRIP ice core $\partial^{18}O$ record at top. Changes in the slope of Antarctic T_{proxy} are synchronous with climate transitions in the North Atlantic (vertical dashed lines), within relative dating uncertainties (horizontal error bars). The deglacial increase in CO_2 occurs in two steps, starting at 19 and 13 Ka and corresponding to significant warming trends in T_{proxy}. A pause in the CO_2 rise is aligned with a break in the Antarctic warming trend during the Antarctic Cold Reversal (ACR). Within the core of the Antarctic Cold Reversal, significant cooling in T_{proxy} (dark shaded band) coincides with an apparent decrease in CO_2. Note that the North Atlantic cools while the Antarctic warms from 19.0 to 14.8 Ka ago, then the North Atlantic warms into the Bølling–Allerød warm stage as Antarctic Cold Reversal begins. The Antarctic then warms as Younger Dryas cooling takes place. The Younger Dryas ends as Antarctic warming stops, at 11.8 Ka ago. Fast-acting interhemispheric coupling mechanisms linking Antarctica, Greenland and the Southern Ocean are required to satisfy these timing constraints.*

Examined in detail, the parallel curves of CO_2 and temperature in Antarctic ice cores through the last deglaciation display two warming trends interrupted by the cooling of the Antarctic Cold Reversal between 14.7 and 13.1 Ka ago (Figure 13.7)[66]. Pedro's team concluded that *'Evidence from Southern Ocean marine sediment cores directly links each of these two warming steps with release of CO_2 accumulated in the deep Southern Ocean during the last glacial period … [including] pulses in upwelling (represented by opal fluxes) … and coincident with negative excursions in atmospheric $\partial^{13}C$ … while cores … identify a source of old (^{14}C-depleted) carbon-rich water that dissipated over … corresponding intervals'*[66] Pedro proposed *'that the increases in [wind-driven] upwelling were responsible for the simultaneous delivery of both sequestered heat and CO_2 to the atmosphere around Antarctica'*[66].

Figure 13.3 shows how CO_2 behaves over small time steps, such as individual Dansgaard–Oeschger events during the bipolar seesaw. In 2000, analysing an ice core from Antarctica's Taylor Dome, Andreas Indermühle and colleagues from the University of Bern's Climate and Environmental Physics group showed that CO_2 rose by about 20 ppm as the environment warmed and fell by the same amount as it cooled during climate changes between 60 and 20 Ka ago[67]. In 2007, comparing Greenland's GISP 2 core with Antarctica's Byrd core, Jinho Ahn and Ed Brook of Oregon State University confirmed that CO_2 rose along with temperature in the Antarctic predecessor to a Greenland Dansgaard–Oeschger event (Figure 13.3)[68]. As temperature and CO_2 reached their peak in the south and started to decline, a Dansgaard–Oeschger event suddenly began in the north. The decline in CO_2 was less rapid than the decline in temperature in the south, much as was found by Saltzman and by Fischer (see earlier). Ahn and Brook argued that the close relationship between CO_2 and Antarctic temperatures at both the coarse scale of orbital change (Figure 13.5) and the fine scale of Dansgaard–Oeschger change (Figure 13.3), and the fact that the pattern is saw-toothed at both scales, reinforces the argument that atmospheric CO_2 must be governed by physical processes that control surface air and ocean temperatures in the Southern Ocean. These processes likely included an increase in warming (driven by some forcing factor, as explored later), which caused a decline in sea ice cover, allowing winds to interact with the ocean surface, causing upwelling and thus overturning southern deep water, decreasing the stratification of the Southern Ocean and bringing CO_2 to the surface in parallel with ocean warming.

Comparing the climate records from Greenland with those from Antarctica (Figure 13.7), Pedro's team found that *'there is little or no time lag separating the major millennial and sub millennial climate transitions in Greenland … from the onsets and ends of the warming and cooling trends in Antarctica'*[66], suggesting that there must be rapid coupling between Antarctic temperatures, Greenland temperatures and atmospheric CO_2. They concluded that *'The ice core observations point to a tightly coupled system operating with little or no time delay between the onsets/terminations of North Atlantic climate stages and near simultaneous trend changes in both Antarctic temperature and atmospheric CO_2 … [which] lends support to the current concept of an atmospheric teleconnection between the northern and southern high latitudes which forces wind-driven CO_2 release from the Southern Ocean [via increased upwelling]'*[66]. To put it another way, CO_2 and temperature vary in the same way at the glacial–interglacial scale as at the millennial scale: they first rise together in the south.

While these studies showed that the distribution of CO_2, like temperature, responded to the bipolar seesaw, they did not identify what warmed the Southern Ocean to set the seesaw off. As we saw earlier, initial arguments suggested the warming of the Southern Ocean was influenced by fluctuations in the production of North Atlantic Deep Water. To test that possibility, Andreas Schmittner of Oregon State University and Eric Galbraith of Princeton applied a coupled ocean-atmosphere–sea ice–biosphere model, forcing it by varying freshwater inputs to the North Atlantic[69]. Much as expected, adding freshwater stopped the circulation, reduced the northward transport of heat, cooled the North Atlantic and warmed the Southern Ocean. It also reduced the transport of salt from the North Atlantic to the Southern Ocean via North Atlantic Deep Water, reducing the stratification of the Southern Ocean and allowing more Circum-Polar Deep Water to reach the surface to release CO_2, thus increasing atmospheric CO_2 concentrations in concert with Southern Ocean (and Antarctic) temperatures. These changes overrode the ability of the biological pump to store CO_2 in deep water.

Did the Southern Ocean contain enough CO_2 during the last glaciation to cause the rapid rises in atmospheric CO_2 seen during the subsequent deglaciation? Yes. Luke Skinner of the Godwin Laboratory of Cambridge University showed in 2010 that the deep waters of the Southern Ocean during the Last Glacial Maximum were depleted in ^{14}C, which increased as CO_2 was degassed from the ocean during the deglaciation, and diluted the CO_2 containing excess ^{14}C that had accumulated in the air during the glacial period[70]. Skinner and colleagues' data confirmed that the Atlantic sector of the Southern Ocean contained a poorly ventilated deep carbon pool during the Last Glacial Maximum. That reservoir sequestered CO_2, reducing its concentration in the air. The increase in CO_2 and decrease in ^{14}C in the air at deglaciation went along with a decrease in sea ice, estimated from the increased supply of sea salt to coastal Antarctic sites. Shrinkage of sea ice from its maximal extent in the Last Glacial Maximum allowed westerly winds to act directly on the ocean, causing Circum-Polar Deep Water to well up from below and provide old CO_2 to the air[71]. Models and data agree with this interpretation. While Schmittner and Galbraith's model seems to confirm that this process was driven by what happened in the north (the freshwater lid model), we must not neglect the possibility that the ultimate trigger for the seesaw lay in the south. More on that later.

13.6 The Ultimate Climate Flicker: The Younger Dryas Event

As we saw in Chapter 10, one of the best-known features of the warming from the last glacial maximum to the Holocene was its interruption, in the Northern Hemisphere, by the 1300-year-long cold period known as the Younger Dryas, named after the white Arctic flower *Dryas octopetala*, which spread southwards at the time (Figure 13.7). It was preceded by an initial warm period – the Bølling–Allerød interstadial – before temperatures plunged almost back to the levels of the Last Glacial Maximum, and it was followed by rapid warming to the start of the Holocene. Mapping these changes in detail has only been possible through the advent of ice cores, although marine sediment cores, despite their lower resolution in time, have also been useful.

One of the biggest surprises to emerge from a comparison of the Greenland and Antarctic ice cores was that the warm Bølling–Allerød interstadial and cold Younger Dryas stadial of Greenland ice cores were missing from Antarctica (Figure 13.7)[72]. Even so, both of these events are visible in Antarctic ice cores, for example at Dome C, where they are represented not by temperature but by methane (Figure 13.5). Jeff Severinghaus of Scripps and colleagues showed that the CH_4 began to rise within 0–30 years of the beginning of the temperature rise at the end of the Younger Dryas event[73].

Plants decaying under low-oxygen (anaerobic) conditions emit methane, mostly from wetlands. These were thought to lie mostly in the Northern Hemisphere, where there is more land, although some scientists suggested that the rapidity of the CH_4 rise reflected a sudden release from large volumes of methane stored in marine and continental sediments[10]. Although the Arctic was commonly cited as a possible source, Bill Ruddiman and Maureen Raymo showed from the gradients of CH_4 between the hemispheres that most of the methane came from wetlands in the northern tropics, controlled by the monsoons and their response to changes in insolation at 30° N[74]. Regardless of its origin, once emitted, CH_4 spread rapidly everywhere.

In Antarctica, a slight cooling – the Antarctic Cold Reversal – coincided with Greenland's warm Bølling–Allerød interstadial, and was followed by warming throughout the period represented by Greenland's cold Younger Dryas (Figure 13.7)[72]. In addition, the rise in temperature in Antarctica from 18 Ka ago coincided with flat or declining temperature in Greenland (Figure 13.7)[75]. These opposing trends mimic the bipolar seesaw, which likely reflected changes within the Atlantic part of the Thermohaline Conveyor.

How might we account for the cooling in the Younger Dryas? As we saw earlier in this chapter and in Chapter 12, the prevailing theory, proposed by Wally Broecker and colleagues in 1989, was that the Younger Dryas came about because of a sudden shutdown of the global ocean conveyor, when melting ice in North America allowed an enormous glacial lake, Lake Agassiz, to drain to the North Atlantic via the St Lawrence seaway, providing a temporary freshwater lid that stopped the formation of deep water in the Norwegian-Greenland Sea[76]. While Lev Tarasov and Richard Peltier of the University of Toronto agreed with that notion in principle, they showed in 2005 that the discharge most likely ran down the Mackenzie River into the Arctic[77]. Peltier is the holder of several awards for his contributions to climate science, including the Milutin Milankovic Medal of the European Geosciences Union, which he received in 2008, and he was elected a fellow of the Royal Society of Canada in 1986. This is not the place to go into detail about the history of Lake Agassiz, but, as we saw in Chapter 12, outwash gravels have been found that confirm Tarasov and Peltier's interpretation of the route. It now seems likely that there were also many smaller discharges from the lake between 17 and 8 Ka ago, several of which can be correlated with $\partial^{18}O$ events in Greenland ice cores[13]. The most recent such event occurred 8.2 Ka ago (see Chapter 14). I discount the recently proposed notion that the cooling of the Younger Dryas was the result of a major bolide impact around 12.9 Ka ago[78], since the evidence for it is currently controversial.

Analyses of the GRIP and GISP 2 ice core records show that the warm Bølling–Allerød interstadial arose remarkably quickly at 14.7 Ka ago, and that the transition from the cold of the Younger Dryas to the warmth of the Holocene was also extremely quick, with about half the warming concentrated into about 15 years (Figure 13.7)[34].

It was once thought that the Younger Dryas was present in sedimentary sequences in the Southern Hemisphere[13]. But in 2007, the Waiho Loop, a large moraine associated with New Zealand's Franz Josef Glacier, was re-dated using beryllium (^{10}Be) and chlorine (^{36}Cl) isotopes produced in the upper atmosphere, and shown to be older than the Younger Dryas period[79]. Further examination of glacial moraines in New Zealand's Southern Alps in 2010 showed that the glaciers there advanced during the Antarctic Cold Reversal. They were older than the Younger Dryas[80]. Michael Kaplan of Lamont and colleagues used maps of landforms, high-precision ^{10}Be dating of the period of exposure of boulder surfaces and reconstructions of former snowlines and ice extents to show that New Zealand glaciers melted back during the Younger Dryas[81]. Glacier resurgence in New Zealand peaked 13 Ka ago, coincident with the Antarctic Cold Reversal, probably in response to northward migration of the southern subtropical front, bringing cold Southern Ocean water close to New Zealand[80]. Like others, Kaplan argued that sea ice forming in the Northern Hemisphere in the Younger Dryas curtailed Atlantic meridional overturning, leading to stronger winds over the Southern Ocean, which increased upwelling, releasing CO_2 from the Southern Ocean to make the Southern Hemisphere warm and the Northern Hemisphere cool[81]. Other signs of the disconnect between the Younger Dryas and the Southern Hemisphere include warming of sea surface temperatures in the Tasman Sea right through the Younger Dryas into the Holocene[79] and the advance of glaciers in Patagonia after the Younger Dryas[82].

To conclude this section, two observations spring to mind. First, as Cronin noted, Broecker's 1989 paper on the Younger Dryas '*generated a worldwide search for evidence for the Younger Dryas and other abrupt climate reversals. More generally, it shifted emphasis from research on orbital-scale climate dynamics to suborbital timescales, especially the abrupt onset and termination of millennial events, which remain directly relevant today as a reminder about the vulnerability of the climate system to abrupt changes*'[13]. Second, as Eric Sundquist reminded us, the melting of the continental ice sheets of the Northern Hemisphere took place well after the rise of CO_2 and associated warming had begun in the south. Thus, the role of those ice sheets as amplifiers of cooling via the albedo effect must have been less important than the role of CO_2 as an amplifier of global warming at the time[10]. We explore that in the next section.

13.7   Problems in the Milankovitch Garden

Palaeoclimate indicators in ice cores replicate the cyclicity seen in marine sediment cores and expected from the orbital periodicities calculated by Milankovitch: 100 Ka cycles of eccentricity; 41 Ka cycles of axial tilt; and 23 and 19 Ka cycles in the precession of the equinoxes. While many saw this as indicating direct control of Earth's climate by orbital forcing, which dictates the amount of insolation received from the Sun at the top of the atmosphere, the change in eccentricity changes insolation by less than 0.2%, which is insufficient to explain the magnitude of the 100 Ka cycles seen in the proxy climate data. How could this tiny change lead to major cyclic glaciations[83]?

The puzzle is even more enigmatic than that simple question might imply. André Berger found that the most important theoretical period of orbital eccentricity, 400 Ka, was weak before 1 Ma ago, but strengthened towards the present, while the strength of the theoretically observed 100 Ka signal decreased after 900 Ka ago. Paradoxically, the weakening of this 100 Ka period in the astronomical calculations coincided with the strengthening of the same period in the palaeoclimate records. '*This implies that the 100 ka period found in paleoclimatic records cannot, by any means, be considered to be linearly related to the eccentricity*', according to Berger[84]. Feedbacks must be at work to give us the strong 100 Ka signal that we see in the record – something Croll had noticed.

This paradox applies also to the long interglacial of MIS 11, about 400 Ka ago. Simple 100 Ka orbital forcing cannot explain this long-lived event, as pointed out in 2003 by Wolf Berger of Scripps and Gerold Wefer of the University of Bremen, among others[85]. Its existence suggests that the climate system must be responding, in addition, to internal fluctuations, not just orbital change. Along with Marie-France Loutre, André Berger argued from an analysis of orbital parameters in 2003 that the low amplitude of insolation change during both MIS 11 and the Holocene made the climate more sensitive to changes in the concentration of CO_2 in the atmosphere at those times than at times of higher amplitude of the insolation signal[86]. Considering that possibility for the record of MIS 11 in the Vostok ice core, Dominic Reynaud and colleagues interpreted the results of a climate model in 2003 to suggest that insolation alone could not have been responsible for the duration of this stage. Atmospheric CO_2 must have made an important contribution to the sustained warming[87]. In 2012, an Indian research team led by Das Sharma agreed[88]. Using advanced statistical techniques, they found that atmospheric CO_2 was the driving signal for change in MIS 11, while all the other climate proxies (like sea surface temperature and the carbon isotopic composition of organic carbon) were responses.

There's another complication. As Thomas Cronin reminded us[13], the interglacial peaks of the late Pleistocene do not repeat with a strict 100 Ka beat – they can be anywhere from 82 to 123 Ka apart. Some palaeoclimatologists have argued that forcing by precession influenced climate more than did eccentricity, and that the 100 Ka peaks occurred at times of low precession, recurring every fourth or fifth precession cycle[89]. Others suggested that variations in obliquity drove the 100 Ka cycles[90, 91].

Aside from the thorny 100 Ka issue, we were faced with a puzzling chicken-and-egg problem. Milankovitch realised that in order for us to compare insolation with climate, we needed to know how much insolation reached particular latitudes, especially polar latitudes, since those have the greatest effect on the preservation or melting of snow, and hence the presence or absence of ice sheets. His calculations, and later improvements upon them by André Berger and Marie-France Loutre[92], confirmed that, if insolation is high in the Northern Hemisphere, it will generally be low in the Southern Hemisphere. As Peter Huybers of Harvard explained, the climates of the two hemispheres should be disconnected: '*One implication of orbital geometry is that at the time when precession aligns Earth's closest approach to the Sun (perihelion) with Northern Hemisphere summer, Earth is furthest away from the Sun (at aphelion) during the Southern Hemisphere summer*'[93]. Obviously, this difference between the hemispheres has a greater effect when the orbit is elliptical than when it is circular. Why does this matter? Paradoxically, despite the implication that the climates of the two hemispheres should not follow each other exactly,

Southern Hemisphere climate proxies do tend to closely follow the changing intensity of Northern Hemisphere insolation. That led to the notion that northern insolation controlled southern climate, probably through the Thermohaline Conveyor – something we have seen arguments for already in this chapter. Is this indeed the case, or did the two hemispheres operate independently, yet more or less synchronously, for some other reason?

As we saw earlier, a third key question concerned the role of CO_2 during the Ice Age. Did it lead or lag temperature? Did it drive climate change, causing temperatures to rise as we saw in past ages, or did it just act as a feedback to climate change, enhancing the effect of an already rising temperature? Perhaps insolation caused temperature and CO_2 to rise simultaneously globally by causing a warming ocean to carry less CO_2. Teasing out the answer was complicated by the fact that CO_2 seemed to change with temperature change in the Antarctic but ahead of temperature change in the Arctic.

Then there was the question of the role of ice. Were changes in the great ice sheets solely driven by changing temperature (under the influence of insolation), or did their own internal dynamics play a key role in driving change?

Back in 2002, Barry Saltzman was not surprised that nobody yet had a definitive answer to the question of how the Pleistocene Ice Age worked. He reminded us that '*Earth's climate system is delicately poised near the freezing point of water, allowing relatively slow but major fluctuations in the proportion of surficial ice to liquid water that involve only small perturbations to the global energy cycle. The low levels of energy flux involved, occurring in a complex, heterogeneous, nonlinear, nonequilibrium system, pose a class of problems that is as difficult and important as those more commonly treated in modern physics*'[61]. Putting it another way, he went on: '*The paleoclimatological (e.g., ice-age) problem … is, in short, a problem in ultraslow, complex evolution in which the rates of change, and the fluxes of mass, momentum, and energy that accompany and drive them, are too small to be calculable directly or in some cases even measurable, though we are sure they occurred and are still occurring*'[61] A thorny problem, indeed.

Peter Huybers saw that a remarkable 33 out of 35 deglaciation features occurred when axial tilt (obliquity) was anomalously large[94]. During the early Pleistocene, deglaciations occurred at 40 Ka intervals with nearly every obliquity cycle, while in the late Pleistocene they skipped one or two obliquity beats, providing interglacials with a spacing of 80–120 Ka, averaging about 100 Ka. Huybers discovered that the character of the $\partial^{18}O$ record in sediment cores changed regularly and progressively following an increasing trend throughout the Ice Age, the signals being smaller 2 Ma ago and growing steadily with time. He deduced that the cyclicity at each point in the record must have been derived from the same underlying mechanism, and that eccentricity did not pace the glacial cycles. The origin of the gradual increase with time was unclear – perhaps it reflected a gradual cooling through slow progressive loss of CO_2, ultimately reflecting control by plate tectonic processes. We do not as yet have an ice core long enough to check this possibility.

Despite appearances to the contrary, then, for some considerable time everything has not been rosy in the Milankovitch garden, and much time and effort has been spent by many palaeoclimatologists on attempts to resolve these complex issues[95]. In this book, we are only going to look at a few of these attempts, in order to give you some idea of where current thinking is headed. For more comprehensive reviews, you might like to consult recent palaeoclimate textbooks by Cronin[13], Ruddiman[96] or Hay[97].

13.8   The Mechanics of Change

To address these various questions, we'll start our tour back in the CLIMAP era, in the mid-1970s. Inevitably, the rapid accumulation of new data from seabed cores spawned efforts to see whether the new generation of atmospheric general circulation models (GCMs) could be used to simulate the climate of the Ice Age[71]. In 1976, W. Lawrence Gates (1928–), then at Oregon State University, used the CLIMAP data set and an atmospheric GCM to simulate the July climate in the Last Glacial Maximum, at 18 Ka ago[98]. Compared with the present, he found the Last Glacial Maximum was cooler and drier, especially over the Northern Hemisphere, due to enhanced anticyclonic circulation over the major ice sheets and reduced summer monsoonal circulation. The mid-latitude westerlies were stronger. There was not much change in global cloudiness or relative humidity, but summer precipitation was 20% below that for today's July. Syukuro Manabe and Douglas Hahn carried out a similar exercise in 1977. Comparing modern and Ice Age conditions, they found that the tropics were much drier in the Ice Age. This was because the continents cooled more than the oceans,

reducing the flow of air between them. Increased continental albedo weakened the Asian monsoon in the Ice Age simulation[99].

At that time, we still lacked data on atmospheric CO_2. The arrival of CO_2 data at the end of the 1970s changed the picture. Jim Hansen and his space physicists from NASA were the first to apply climate models to assess the effects of changing CO_2 on the climate of the Last Glacial Maximum as mapped by the CLIMAP team. In a paper for the 1982 Maurice Ewing Symposium in New York, Hansen explained, '*records of past climate provide a valuable means to test our understanding of climate feedback mechanisms, even in the absence of a complete understanding of what caused the climate change*'[100]. His team simulated the climate of the Last Glacial Maximum at 18 Ka ago by incorporating climate boundary conditions, including continental ice, sea ice and sea surface temperature, from the CLIMAP project. They assessed the effects of different feedbacks on the planetary radiation balance, essentially testing the sensitivity of the model to forcing by different parameters. Global surface temperature was 3.6 °C cooler than today, and cooler in the polar regions than elsewhere. A fall in CO_2 to 200 ppm at 18 Ka ago (from an assumed interglacial level of 300 ppm) caused about 0.6 °C of the estimated global cooling, the rest coming from reductions in water vapour and clouds, a growth in land and sea ice (hence albedo) and a decrease in vegetation cover (also affecting albedo)[100].

Hansen's results suggested an overall climate sensitivity ranging from 2.5 to 5.0 °C, averaging about 4 °C, for a doubling of atmospheric CO_2 (implicit was the accompanying rise in water vapour and fall in albedo). In recent years, with more data available, that estimate dropped to about 3 °C, as we saw in Chapter 11. Given the slow nature of ocean circulation, Hansen calculated that '*the response time of surface temperature to a change of climate forcing is of the order of 100 years*'[100]. Thus, any change would be around for a long time. This limit is now thought to be ~100 000 years, as we saw in Chapter 11.

One of Saltzman's early numerical modelling exercises on this topic, in 1984, which used a dynamical model, not a GCM, showed that the 100 Ka cycles could be explained by dynamical variations in land ice, marine ice and the mean climate (represented by ocean temperature), and feedbacks between them, paced by orbital variations in insolation[101]. In a nutshell, orbital cycles could modulate and pace a self-oscillating climate system. Way back in 1980, he was one of the early proponents of the notion that CO_2 played a key role in such a self-oscillating system[102].

At the Tarpon Springs meeting in 1984, two ocean modellers from Princeton, Robbie Toggweiler and Jorge Louis Sarmiento, tested the idea that the proposed glacial–interglacial changes in CO_2 were related to changes in the nutrient content of high-latitude surface waters[103]. During glacial periods, when the Thermohaline Conveyor was switched off or reduced, the deep ocean accumulated nutrients and dissolved CO_2 through the action of the solubility pump (cold water holds more CO_2 gas) and the biological pump (sinking organic matter dissolves in the deep ocean, enriching deep waters in CO_2), as we saw in Chapter 9. The deep-ocean waters were unable to reach the atmosphere because the surface of the Southern Ocean was capped by a stable thermocline topped with sea ice. Much as we saw from later studies, Toggweiler and Sarmiento's model showed that in the deglaciation, the supply of North Atlantic Deep Water would be enhanced, the Southern Ocean thermocline – which had inhibited deep convection – would be disrupted, sea ice would melt and CO_2 would be released from the deep ocean. They concluded that '*climate forcing of atmospheric CO_2 probably played a pivotal role in amplifying orbital forcing which paced the climate change*'[103], thereby providing independent support to the Shackleton and Pisias model of glacial–interglacial climate change, and to Saltzman's concept. These ideas have a 30-year pedigree.

By 1987, Syukuro Manabe of the Geophysical Fluid Dynamics Laboratory at Princeton was ready to use a GCM combining the atmosphere with a mixed-layer ocean in order to examine the effects of expanded continental ice, reduced atmospheric CO_2 and changes in land albedo on the climate of the Last Glacial Maximum[104]. Expanded continental ice and reduced CO_2, he found, had a substantial impact on global mean temperature. Increasing albedo from growing Northern Hemisphere ice sheets cooled that region. The Antarctic ice sheet hardly affected the picture, because it already occupied most of the available land space, so could not grow much. Reduced CO_2 cooled both hemispheres. The presence of the northern ice sheets caused substantial change to atmospheric circulation in winter, amplifying the westerlies near the ice sheets. The sea surface temperature in the model was reduced in much the same way as shown in the CLIMAP data, with greater cooling in the Northern than in the Southern Hemisphere. Half of the global 1.9 °C cooling in the Last Glacial Maximum sea surface temperatures was attributable to the CO_2 effect, which was larger in the Southern Hemisphere. Reduction in sea surface temperature was greatest near the margin of the sea ice. It did

not decrease further poleward because temperatures there were already close to freezing. Air temperatures were especially cold over both land and sea ice, and there was a larger area of cold air in the north than in the south, reflecting the larger area of Northern Hemisphere land and ice. Cooling due to a decline in CO_2 was less than 1 °C in the tropics, and much greater at high latitudes, mainly during the cold season. This study '*supports the hypothesis that glacial-interglacial variations in CO_2 concentration may provide a linkage between the two hemispheres*', Manabe said[104]. His model suggested that the presence of large ice sheets alone was insufficient to explain the glacial climate of the Southern Hemisphere, where the cooling was most likely driven by the fall in atmospheric CO_2.

By 1989, Wally Broecker of Lamont and George Denton, a glaciologist from the University of Maine at Orono, had entered the ring[105]. In addition to the puzzle that forcing by precession and obliquity seemed to produce bigger signals than eccentricity, they wanted to know '*What might explain the asymmetric shape of the 100 ky cycle, with its rapid start and slow decay?*'[105]. Long intervals of gradual increase in $\partial^{18}O$ in marine cores (signifying cooling) ended abruptly with a rapid decrease in $\partial^{18}O$ (signifying warming) at what they called 'glacial terminations' as the world changed from a glacial to an interglacial climate. Explaining how those rapid terminations were produced at 100 Ka intervals posed a difficult challenge.

Broecker and Denton thought that the answers to their questions lay in the ocean, and that glacial–interglacial transitions involved major reorganisations of the ocean–atmosphere system. Basically, the climate oscillated between two stable states of operation, which caused changes in the greenhouse gas content of the atmosphere and the albedo of the planet's surface. Only in this way could they account for the rapidity of the glacial terminations, the apparent synchroneity between the two hemispheres and the large variations in air temperatures and dust concentrations. They thought that the connection between insolation and climate was driven by orbital change, leading to changing freshwater input and its impact on ocean salinity, and hence density and ocean transport. Their arguments for the role of the Thermohaline Conveyor have been rehearsed in detail earlier in this chapter, having been adopted by later researchers.

Their research suggested that no one element changed enough by itself to cause a glacial termination: changes in insolation were too small, as were decreases in albedo caused by the reducing area of sea ice, the 80 ppm increase in atmospheric CO_2 emitted from areas of the ocean formerly covered by sea ice and the decrease in dust caused by weakening westerly winds. Whether there were more or fewer clouds may have been a factor, but not a major one. All these things together provided positive feedback, making glacial–interglacial change much faster and more extensive than it would otherwise have been, so accounting, in their minds, for the asymmetry of the 100 Ka peaks in the record. They saw no basis for rejecting Saltzman's idea that the changes in mode from glacial to interglacial were part of a self-sustaining internal oscillation that operated even in the absence of orbital changes, but which, in the presence of orbital changes (especially at low temperatures), was paced by those changes.

Over the next decade, palaeoclimatologist Maureen Raymo of Lamont noticed that the excess ice typical of 100 Ka climate cycles tended to accumulate when July insolation at 65° N was unusually low for more than a full precessional cycle (21 Ka), and that once established it did not last beyond the next precessional maximum in summer insolation: '*Thus, the timing of the growth and decay of large 100-kyr ice sheets, as depicted in the deep sea delta (18)O record, is strongly (and semipredictably) influenced by eccentricity through its modulation of the orbital precession component of Northern Hemisphere summer insolation*'[106]. Referring to studies of ice sheet decay by Oerlemans[1] and Pollard[107], she deduced that the rapid collapse of the ice sheet at the precessional high point might be triggered as some threshold was passed, such as a critical level of isostatic adjustment of the bedrock beneath the growing ice[89]. Nevertheless, she also thought that '*the global carbon cycle is likely to be a critical component of the mechanism(s) controlling the 100 kyr cycles*'[89]. We will explore her argument for isostatic adjustment later.

While the behaviour of the global Thermohaline Conveyor got a lot of attention for providing a means of connecting the two hemispheres via the ocean, by 1999 Mark Cane and Amy Clement at Lamont were ready to draw attention '*to a part of the system that is known from the modern climate record to be capable of organising global scale climate events: the tropics*'[108]. Behind that goal was the growing realisation that El Niño events were driven by an internal climate oscillator and not by external (solar) forcing. During an El Niño event, the Pacific Ocean warms. It then cools during the following La Niña event, in a pattern that repeats on a 2–7-year cycle and has a global reach, as we saw in Chapter 11[109]. El Niños warm the globe by spreading warm water across the surface of

the Pacific, while La Niñas cool the globe by making those same waters cold enough to absorb heat. A key feature in this system is the oscillation of the 'warm pool' in the western Pacific, near New Guinea. The strong easterly Trade Winds of a La Niña pile warm water up in the west to feed the warm pool; the weak Trades of an El Niño collapse the warm pool, allowing warm surface water to slosh back to the east.

Cane and Clement wanted to know whether the tropical Pacific climate could change *on its own*, with no influence from higher latitudes[108]. Their model showed that nonlinear interactions in the tropical Pacific could generate variations in sea surface temperature on both orbital and millennial time scales through essentially the same physics as involved in El Niño–La Niña events. With that understanding, they went on to suggest that '*global scale millennial and glacial cycles may be initiated from the tropical Pacific*'[41]. Low orbital insolation cooled the tropics by between 2 and 5 °C, creating a cooler, more La Niña-like system in the tropical Pacific. The changes in sea surface temperature altered the character and location of atmospheric convection, which altered global climate through long-range connections (teleconnections). Through the natural oscillations in the system, the cold (La Niña-like) phase would tend to cool the Earth further by increasing low cloud cover, so reducing planetary albedo, and by reducing atmospheric water vapour – a greenhouse gas. This would increase glaciation over North America. The warm (El Niño-like) phase would do the opposite. As nonlinear processes drove this millennial variability, peaks would tend to cluster in broad time bands rather than at particular time intervals[108].

They argued that there had been too much emphasis on fluctuations in North Atlantic Deep Water as a primary control of glacial–interglacial climate change[41]. Centres of tropical convection tend to lie over the warmest water. Moving the locus of that water alters the convection, which changes the impacts on distant locations. This works for the fast 2–7-year scale El Niño–La Niña changes. They also thought that there was no intrinsic reason why it should not work on longer time scales, with similar global effects, and be accentuated by global changes triggered by orbital variations. Given the nonlinear nature of the system, they argued that, while their model runs showed peaks near 1500 years, like those found by Bond, such peaks might well be artefacts of the short record. The longer the record in both models and palaeoclimate data, the more one expects a broad band of peaks originating from nonlinear processes. They went on to point out that '*in common with other non-linear systems ... the tropical Pacific ocean-atmosphere may exhibit regime-like behaviours which persist far longer than any obvious intrinsic physical timescale*'[41]. It may vary on millennial time scales, independent of influences from elsewhere.

The Pacific climate does, in fact, vary on longer time scales than the short-wavelength El Niño–La Niña cycle. This cycle is superimposed on a longer-wavelength oscillation, the Pacific Decadal Oscillation. In the positive phase of the Pacific Decadal Oscillation, the equatorial region is warm (like an El Niño) and the Gulf of Alaska is cool, while in the negative phase, the equatorial region is cool (like a La Niña) and the Gulf of Alaska is warm. Each of these phases can last from 10 to 25 years[110]. Cane and Clement were suggesting that oscillations like these, if sufficiently extended in time, could have a substantial and possibly lasting effect on the global climate system.

Imbrie, Ruddiman and Shackleton, three of the CLIMAP scientists, also had views on the mechanisms of climate change. Initially, in devising the SPECMAP model of 1984, Imbrie and his coworkers followed Milankovitch in accepting that summer insolation in the Northern Hemisphere forced the growth and decay of ice sheets directly. This was consistent with the original CLIMAP view published by Hays, Imbrie and Shackleton in 1976[111], which we examined in Chapter 12. Their conclusions were published before we knew about the history of CO_2 in ice cores. Recognising what ice cores were telling us, Imbrie changed his mind in the final SPECMAP models of 1992–93. He proposed that Northern Hemisphere summer insolation triggered a train of climatic responses that were transmitted via deep water to the Southern Hemisphere, where they caused changes in CO_2 and other feedbacks that then affected Northern Hemisphere ice sheets. Direct forcing of the Northern Hemisphere ice sheets had changed to indirect.

Shackleton supported this interpretation. In 2000, he used the $\partial^{18}O$ record from Vostok to separate the ice volume component from the ocean temperature component of the $\partial^{18}O$ signal of marine sediments. His results showed that atmospheric CO_2, Vostok air temperature and deep-water temperature were all in phase with orbital eccentricity, while ice volume lagged behind these three variables. '*The coherences and phases in the 100-ky band strongly suggest that atmospheric CO_2 has a direct and immediate control on deep water temperature (presumably with high latitude air temperature as an intermediary)*', he wrote[112]. '*Hence, the 100,000-year cycle does not*

arise from ice sheet dynamics; instead, it is probably the response of the global carbon cycle that generates the eccentricity signal by causing changes in atmospheric carbon dioxide concentrations*'[112]. Ice volume then responds to these changes. Finally, '*The effect of orbital eccentricity probably enters the paleoclimatic record through an influence on the concentration of atmospheric CO_2*'[112]. This was independent support for Saltzman's model.

Bill Ruddiman thought this was unnecessarily complicated[113]. In a nutshell, he felt that although the 100 Ka cycle of eccentricity was too small to achieve much change by itself, it grew in importance by accentuating the patterns of insolation caused by variations in precession and obliquity (tilt). It accentuated the summer (mid-July) insolation that forced Northern Hemisphere ice sheets at the 41 Ka obliquity period, helping to change ice volume, sea surface temperature, dust supply and the production of North Atlantic Deep Water, which produced a strong positive CO_2 feedback that further amplified ice volume changes. It also accentuated insolation at the roughly 21 Ka precession period, focused in the tropics, which influenced wetlands through monsoonal changes that drove fast feedbacks in methane. Clearly, Ruddiman thought that CO_2 was not a primary driver of ice sheet change in the Northern Hemisphere. Instead, he saw it as providing positive feedback to accentuate the climate signal. These effects were stronger at the MPT, where '*gradual global cooling allowed ice sheets to survive during weak precession insolation maxima and grow large enough during 41,000-year ice-volume maxima to generate strong positive CO_2 feedback*'[113].

Ruddiman highlighted the importance of combining Milankovitch's three orbital signals in order to generate the 100 Ka cyclicity: '*each broad eccentricity maximum at the 100,000-year cycle spans 2 or 3 individual insolation maxima at the precessional cycle. During the last several deglaciations, the climate system response has latched onto one or other of these precession maxima in creating the observed termination. As a result, all terminations occur at or very near even multiples of 4 or 5 precession cycles (90,000–115,000 years). The specific precession maximum chosen by the climate system depends on ... close alignment with a nearby obliquity maximum. Modulation of precession by the longer-term 400,000-year eccentricity cycle also plays a role: it makes all precession maxima in a particular 100,000-year cycle either weaker or stronger, thereby affecting which peak is chosen for the termination*'[113]. As a result, some interglacials started almost in phase with eccentricity (e.g. Termination I, at

20 Ka), while others might lead eccentricity by 17 000 years or more (e.g. Termination II, at 135 Ka).

There is no doubt that the ocean did play a key role in storing and releasing CO_2 during the Ice Age. Thanks to the discovery that the ratio of boron to calcium (B/Ca) in benthic foraminifera is directly related to the concentration of carbonate ions (CO_3^{2-}) in bottom waters[114], we can now use this ratio to show changes in the concentration of carbonate ions (CO_3^{2-}) in oceanic deep waters over the past 25 Ka[115]. The data show that the biological pump caused large amounts of CO_2 to be stored in the deep glacial ocean, thus helping to reduce the levels of CO_2 in the air during the Last Glacial Maximum. The reduced CO_3^{2-} in bottom waters made them more corrosive, dissolving $CaCO_3$ and so raising the level of the carbonate compensation depth (CCD). On deglaciation, CO_2 was released back into the air from the rising deep water in the Southern Ocean as sea ice declined, exposing the ocean to the atmosphere, and as the declining supply of iron-rich dust reduced ocean productivity, thus lessening the extraction of CO_2 from the air. The resulting increase in CO_3^{2-} concentration lowered the CCD, allowing more $CaCO_3$-rich sediments to be preserved.

Was it reasonable to consider the ocean as the main source of change in CO_2? In 1996, Guy Munhoven and Louise François of the University of Liege in Belgium suggested that not enough attention had been paid to the possibility that variations in rock weathering might have altered CO_2 levels between glacial and interglacial periods[116]. They used the ratio of germanium to silicon (Ge/Si) in marine sediments as an indicator of the consumption of CO_2 by the weathering of silicate rocks, noting that the Ge/Si ratio was lowest where the rate of weathering was highest, at the Last Glacial Maximum. That is because increased cycles of freezing and thawing enhance mechanical weathering, increasing the surface area of exposed rock that can be subjected to chemical weathering and thus supplying more dissolved silicon to the ocean. This kind of weathering is most efficient in mountains, and might have been enough to significantly lower atmospheric CO_2. They calculated that CO_2 consumption by rock weathering in glacials could have reduced atmospheric CO_2 by 50–60 ppm. Previously, rock weathering was dismissed as too slow to influence glacial–interglacial variations in CO_2, but the existence of the Ge/Si signal in marine sediments suggests that this perception was misguided.

Criticising palaeoclimatologists for focusing on a search for changes in external forcing to explain palaeoclimatic

variability, Barry Saltzman thought that much of it could have resulted from internal instability within a system forced steadily by the thermal gradient between Equator and Pole[61]. As a scientist steeped in chaos theory, he was puzzled by the fact that although the character of orbital forcing was unchanged over the full Pleistocene period and before, there were signs of significant instability, including the rather sudden onset of the Pleistocene glacial epoch at about 2.5 Ma ago, the MPT at about 900 Ka ago and the dominance of 100 Ka cyclicity in the past 400 Ka. He concluded that '*the main variations of planetary ice mass do not represent a linear response to the known orbitally induced radiative forcing, having a temporal spectrum that is much different than that of the forcing ... Although it is possible that the orbital external forcings may be a necessary condition for the observed ice variation, they cannot be a sufficient condition*'[61]. This greatly interested him, because other modellers assumed that external forcing controlled the 100 Ka oscillations, with the implication that if those forcings were removed then the 100 Ka oscillations would vanish. Those other models, he emphasised, could not account for the transitions at 2.5 Ma or 900 Ka ago. He declared that '*If ... the tectonic decrease in CO_2 over the Late Cenozoic provided the threshold state for the initiation of the major ice build ups, it would then deserve recognition as the "cause" of the ice epoch and its oscillations*'[61]. These were profound and counterintuitive insights.

Saltzman thought that Ice Age variability resulted from a mix of both external forced and internal free effects, and that the broad spectral peak centred on 100 Ka ago resulted from an internally driven fluctuation caused by instability, rather than the response to an externally driven forcing like that at the 41 Ka ago period. He interpreted Ice Age changes with time as showing that there was strong free variability within the climate system, part of which might be described as 'climatic turbulence' stimulated when the tectonically forced value of CO_2 achieved a critically low range of values Recognising that the system was complicated and the solution to the problem would not be easy, the intrepid Saltzman set out '*to develop [a] quantitative theory of climate in which all relative forcings, feedbacks, and competitive physical factors are taken into account simultaneously. That is, we consider the explanations of variations in the climate system as a problem in mathematical physics, in which the basic conservation laws for mass, momentum, and energy are expressed in symbolic forms so that the power of mathematical deductive logic can be used to extract quantitative relationships ... It is the purpose of theory to provide a predictive connection between the known external forcing … and the observed internal behaviour*[61].

His dynamical system model of the Northern Hemisphere assumed that atmospheric CO_2 linearly decreased from 350 ppm at 5 Ma ago to 250 ppm in preindustrial times in response to tectonic forcing, and that the only other forcing came from orbitally induced variations in summer insolation at high northern latitudes. '*In essence*', he explained, his model unified '*the two major theories of the ice ages: the CO_2 theory (in which longwave radiation is altered by the greenhouse effect) and the Milankovitch theory (in which the distribution of shortwave radiation is altered by Earth-orbital changes), supplemented by a new third major theory resting on the possible role of internal instability*'[61] His numerical model showed that the global ice mass over the past 5 Ma responded with a significant jump at 2.5 Ma ago, coincident with the $\partial^{18}O$ record from ocean drilling cores. Hence, '*the imposition of the slow tectonic forcing of CO_2 transforms what would otherwise be the chaotic, intermittent … [distribution of ice volume through time] into an organized sequence of clearly defined regimes separated by well-developed transitions*'[61]. One of these regimes started at about 900 Ka ago with the emergence of 100 Ka variability.

Saltzman's model of the increase of Pleistocene ice volume at 2.5 Ma ago is an emergent property of the systematic decline in CO_2 with time He thought that the 100 Ka fluctuations were associated with and probably driven by internally generated CO_2 fluctuations, while the 20–40 Ka fluctuations were driven by externally imposed orbital variations and the associated instability of the ice sheet, leading to calving[56]. Calving instabilities contributed to the rapid deglaciation of the Laurentide and Fennoscandian Ice Sheets by causing an earlier collapse of ice mass than predicted by the insolation and CO_2 variations alone. His model's predictions have been validated by testing against the data from Vostok and Dome C. André Berger, too, used a numerical model to demonstrate that, given a linear decline in CO_2, a shift from glacial cycles dominated by periodicity of 41 Ka to cycles dominated by periodicity of 100 Ka took place about 1 Ma ago at the MPT[117].

Elderfield was not convinced that they were right. Having found that the MPT was associated with a particular pattern of insolation, he concluded that '*Data of CO_2 … are as yet too sparse to determine the respective roles of temperature and the carbon system*' in causing the MPT[28]. Bärbel Hönisch of Lamont agreed[118]. Using

boron isotopes in planktonic foraminiferal shells as a proxy for the partial pressure of CO_2 in ocean surface waters from the past 2.1 Ma, and comparing glacial and interglacial values before and after the MPT, Hönisch and colleagues found that while CO_2 was slightly higher in the *glacial* periods from before the MPT, **it was not in the interglacial periods**. While this confirmed a close linkage between atmospheric CO_2 concentration and global climate, it militated against long-term drawdown of atmospheric CO_2 as being the ultimate cause of the MPT.

By 2010, André Berger was inclined to agree. Using an Earth system model of intermediate complexity, he and his colleague Yin found that the later interglacials were all warmer than those before the MPT, mainly because global mean temperatures increased during Northern Hemisphere winters, due to increased insolation during the winter season. Changes in CO_2, water vapour, sea ice and land vegetation had the secondary effect of amplifying the astronomically induced boreal wintertime warming and counteracting the astronomically induced boreal summertime cooling[119]. During boreal winter (austral summer), about 60% of the warming was due to greenhouse gases and 30% to insolation. The Southern Hemisphere had a more important role than the Northern Hemisphere, as it warmed significantly in both winter and summer.

Huybers, too, investigated the MPT[94]. The pacing of deglaciations by obliquity throughout the Pleistocene, mentioned earlier, plus the regular and progressive increase in the properties of glacial cyclicity, which followed a trend from smaller to larger signals over the past 2 Ma, suggested that there was no sudden onset of 100 Ka cyclicity. Hence, the MPT must be an artefact.

Despite Saltzman's valiant attempts to find out how the Ice Ages worked, there was still much to consider. For instance, his model did not examine the role of sea ice. Yet, the changing distribution of sea ice evidently played a key role in governing the exchange of CO_2 between air and ocean, and in the switching on and off of the supply of North Atlantic Deep Water[120], as we saw earlier. These arguments usually apply to the interactions of Arctic sea ice and the production of North Atlantic Deep Water. But Ralph Keeling from Scripps and Britton Stephens from the University of Colorado argued that the ocean Thermohaline Conveyor could also be destabilised by the influence of Antarctic sea ice[121]. They supposed '*that changes in the freshwater budget of high southern latitudes may provide the link between Antarctic warming and sudden Greenland climate changes associated with long-lived D/O events*' and that '*the duration of the short-lived interstadial events would be linked to the timescale for NADW [North Atlantic Deep Water] to propagate from the North Atlantic to high southern latitudes*[121].

Trond Dokken of the Bjerknes Centre for Climate Research in Bergen, Norway agreed that northern sea ice played an important role in the development of Dansgaard–Oeschger events[122]. New sediment core data showed Dokken and colleagues that warm subsurface Atlantic water had flowed into the Nordic seas beneath the sea ice and its associated freshwater cap. Eventually, that warm water destabilised the cool surface system and its associated sea ice, venting heat to the air and warming the region abruptly by as much as 10 °C. This warmth then gradually melted the Fennoscandian Ice Sheet, recreating the freshwater cap on the Nordic seas, allowing sea ice to reform. Dokken *et al.*'s hypothesis avoids *ad hoc* proposals for the periodic arrival of a freshwater cap.

Confirmation that variations in atmospheric CO_2 were most likely related to variations in the ocean came from an Earth system model of intermediate complexity (CLIMBER-2) run by Brovkin and others in 2007[123]. This model showed that the change from interglacial to glacial reduced Atlantic thermohaline circulation, caused the thermocline to shallow, allowed southern deep waters to penetrate further north and drew down atmospheric CO_2 by 43 ppm. Upwelling and dust fertilised the Southern Ocean in the Atlantic and Indian Ocean sector north of the polar front, drawing down a further 37 ppm CO_2. There was an accompanying decrease in the cooled terrestrial biosphere, thus diminishing photosynthesis, which increased CO_2 in the atmosphere by 15 ppm, as well as an increase in ocean salinity resulting from the conversion of water to ice, which led to a further rise of 12 ppm. A decrease in deposition of $CaCO_3$ in shallow water following the fall of sea level drew down atmospheric CO_2 by 12 ppm. These various mechanisms explained 65 ppm (more than two-thirds) of the fall in CO_2 during glacial times, suggesting that the model captured reasonably well the effects of reorganisation of biogeochemistry in the Atlantic Ocean. The rest of the fall might be explained by less well-known processes, including changes in terrestrial weathering and iron fertilisation of the sub-Antarctic Pacific Ocean.

Despite the intriguing outputs of Saltzman's model, Michael Crucifix pointed to its *'failure to reproduce the steadily increasing trend in CO_2 concentration during marine isotope stage (MIS) 11 … [suggesting that] some stabilizing mechanisms may have been ignored in this model'*[124]. He reminded us that the correct length for

MIS 11 (two precession cycles) had been predicted by a model devised in 2001 by Didier Paillard of the French Laboratoire des Sciences du Climat et l'Environnement[125]. Paillard's model featured three possible climate regimes (glacial, mild glacial, interglacial) to which the climate system was successively attracted, depending on insolation and ice volume. For his pioneering ideas on the response of Quaternary climate system dynamics and the carbon cycle to Milankovitch forcing, Paillard was awarded the Milutin Milankovitch Medal of the European Geosciences Union.

Reviewing progress up to the mid-2000s in the development of our understanding of ice age variability, I am reminded of Crowley and North's summary of the position back in 1991: *'A diversity of ideas exists as to the origin of the ice-age CO_2 fluctuations … Working on the carbon cycle on this time-scale is like trying to piece together a giant puzzle for which some of the pieces are missing and some of the rules not thoroughly understood. But the progress that has been made … is impressive, and we are optimistic that a revisitation of this subject in a few years will indicate considerable advances over what has been presented'*[71].

Part of the problem, as Michael Crucifix explained in 2009, was that *'There is presently no comprehensive model … capable of representing the interactions between the slow components of the climate system satisfactorily enough to predict the evolution of ice volume and greenhouse gas concentrations over several glacial-interglacial cycles'*[124].

Significant further progress has been made since the mid-2000s, as we see further on. This stems in particular from results emerging from the completed analyses of the long Antarctic ice cores collected from Dome Fuji in 2007, Dome C in 2004 and Dronning Maud Land in 2007. These cores have been available for only 7 years at the time of writing, so it is not surprising that our understanding of the operation of the Ice Age climate system has evolved recently. Ice cores from Greenland have been available for longer, but they do not go past the last interglacial and are unreliable for CO_2.

In 2010, Lorraine Lisiecki of the University of California at Santa Barbara carried out a statistical analysis of the links between eccentricity and the 100 Ka glacial cycle[126]. She showed that the cycle is indeed paced by eccentricity, but that, paradoxically, strong eccentricity is associated with weak power. She argued that strong forcing by precession disrupted the internal climate feedbacks that drive the

100 Ka glacial cycle. Her findings '*support the hypothesis that internally driven climate feedbacks are the source of the 100,000-year climate variations*'[126]. These internal feedbacks must be phase-locked to eccentricity, and vary slowly over long periods, as do the carbon cycle and the ice sheets. Glacial terminations are driven by precession and obliquity after large ice sheets develop, and are paced by eccentricity (which affects the amplitude of precession). Lisiecki agreed that the 100 Ka cycle was likely to originate from processes associated with the carbon cycle, as Saltzman had suggested.

What might make the carbon cycle vary in that way? In 2008, Robbie Toggweiler of the National Oceanic and Atmospheric Administration (NOAA)'s Geophysical Research Laboratory in Princeton thought that the answer might lie in the Southern Ocean[127]. There, the effect of salinity on the overturning and mixing of the Southern Ocean grows as polar waters approach the freezing point, with overturning being particularly strong at temperatures between 1 and 3 °C, and weak at temperatures below 0 °C, when sea ice tends to form. Given this constraint, he saw that, as CO_2 and temperatures fell during the late Cenozoic, temperatures would eventually reach the critical point where polar waters became prone to overturning and mixing. Where those temperatures were relatively warm (1–3 °C), strong overturning and mixing would supply deep-water CO_2 to the air; where they cooled to less than 0 °C, reduction of overturning and mixing would encourage CO_2 to dissolve in the ocean and be carried to the depths. In warm times, when CO_2 was released to the air, the deficit of carbonate (CO_3^-) ions in the deep ocean would switch to an excess, enhancing the burial of $CaCO_3$. This would release more CO_2, leading to more warming and more overturning, which would lead to more CO_2 release and more $CaCO_3$ burial and so on, converting a relatively minor overturning fluctuation into a major transition. Over a period of 50 Ka, the excess of CO_3^- ions in the deep ocean was erased by $CaCO_3$ deposition, lowering the partial pressure of CO_2 in the ocean and making more CO_2 from the air dissolve in the ocean to begin the cycle again. In Toggweiler's model, the booms and busts in atmospheric CO_2 took 50 Ka each, together making up a 100 Ka cycle. In the 'on' state, when the CO_2 was high, the Southern Ocean and Antarctica were warm. This process could have triggered the global seesaw. Toggweiler concluded that '*most of the 100,000-year temporal variability in the ocean is a greenhouse response to CO_2 cycles from the south, as suggested by Shackleton*'[127]. Thus, '*taken together the Northern and Southern Hemispheres would*

seem to have dominant influences and dominant periods of variability that are basically independent: precession and tilt make the ice sheets grow and shrink in the north; the internal mechanism warms and cools the south. The greenhouse effect from the internal mechanism in the south transmits some of the 100,000-year southern variability to the northern ocean and the northern ice sheets'[127].

How independent were the Northern and Southern Hemispheres? In 2007, Kenji Kawamura of Japan's Tohoku University decided with colleagues to use the ratio of oxygen to nitrogen molecules in fossil air from the Dome Fuji and Vostok ice cores as a proxy for local summer insolation (stronger insolation diminishes the O_2 concentration), allowing them to examine the phase relationships between climate records from the ice cores and changes in insolation[128]. They found that southern summer insolation was out of phase with Antarctic climate change, and interpreted this to mean that Antarctic climate change on orbital time scales must be paced by northern summer insolation and its effects on northern ice sheets and the northern oceans They concluded that '*Northern Hemisphere summer insolation triggered the last four deglaciations*'[128]. Like Saltzman and Fischer, they thought that Antarctic cooling into past glacial periods began earlier by several millennia than the corresponding CO_2 falls. That led them to suggest '*that post-interglacial cooling began in the Northern Hemisphere with ice area growth and was transferred to Antarctica quickly through modulation of poleward heat transport and methane concentration decrease – before the reduced CO_2 forcing, or the sea level drop caused by northern ice volume growth, became significant*'[128] By their reckoning, then, CO_2 was an amplifier of orbital input, not a primary driver of change. However, as we learned earlier, at least one of their assumptions – that CO_2 preceded warming – is now known to be wrong, as shown by Parrenin[65] and Pedro[66].

Were they right to focus on southern summer insolation? Examining the month-by-month changes in Southern Hemisphere insolation, Peter Huybers, together with George Denton of the University of Maine, found that some aspects of that insolation did in fact co-vary with the pattern of insolation in the Northern Hemisphere, contrary to expectation[129]. As usual, the devil is in the detail. In principle, when summer insolation was highest at 65° N due to Earth being at perihelion (closest to the Sun), summer insolation at 77° S should have been weak due to it being at aphelion (furthest from the Sun). But Huybers and Denton realised from Kepler's second law that although summers have less intense insolation when

Earth is at aphelion, they are longer, and the associated winters shorter, than the average. Calculating for the Southern Hemisphere the number of summer days experiencing daily average insolation of more than $250 \, \text{W/m}^2$ and the number of winter days experiencing daily average insolation of less than $250 \, \text{W/m}^2$, they found that Northern Hemisphere insolation co-varied positively with the duration of southern summers and negatively with the duration of southern winters over the past 350 Ka. Spring insolation intensity at high southern latitudes also varied closely with the duration of the southern summer[129]. Hence, orbitally driven changes in the south could happen simultaneously with different orbitally driven changes in the north. Evidently, when trying to assess the relationship between insolation and climate in different hemispheres, the insolation of all seasons has to be considered, along with their duration. Milankovitch knew that, but dismissed Antarctica as too cold to allow changes in southern insolation to influence its volume of ice[93]. Evidently, he was wrong. By focusing on annual insolation, Kawamura's team had missed the fact that the duration of southern summer, and the pattern of spring insolation in the south, might explain what they saw[128]. Their focus was too narrow.

Huybers and Denton speculated '*that the increasing summer and decreasing winter durations caused by the alignment of aphelion with southern summer solstice coordinates the effects of summer radiation balance, winter sea ice and atmospheric CO_2 so as to increase Antarctic temperature. Variations in sea ice and CO_2 may also explain why climate variations similar to those in Antarctica are observed in mid-latitude southern marine and continental environments*'[129]. This freed southern data from northern forcing at precession and obliquity time scales. Northern climate would respond to summer insolation intensity à la Milankovitch, while southern climate would respond to the duration of summer and winter seasons, perhaps also reflecting the contrasting distributions of land and sea and ice sheets in the two hemispheres[129]. The net result was that the two hemispheres operated in sync. And as mentioned in Chapter 12, the two hemispheres were also linked by changes in sea level.

One key difference between the two hemispheres, apart from the larger amount of land in the north, was the fact that the Antarctic ice sheet covered an entire continent. Aside from the narrow continental shelf, there was no room for expansion in glacials. As a result, the area and volume of its ice was relatively stable compared with the Northern Hemisphere ice sheets, which waxed and waned

considerably with changing insolation, thus affording significant opportunities for changes in both elevation and albedo to affect northern temperatures. Independent hemispheric responses to insolation at orbital time scales would imply that there was no need to invoke causality in explaining lead–lag relationships between the hemispheres. Nevertheless, Huybers and Denton went on to say that '*An Antarctic response to local changes in insolation is consistent with hypotheses calling on terminations to be triggered by changes in southern insolation … If a long summer and a short winter lead to a decrease in production and extent of Antarctic sea ice, they may also increase the outgassing of CO_2 from the Southern Ocean by decreasing near-surface stratification … Once the northern ice sheets are sufficiently large to become unstable, the combination of a long southern summer and an intense northern summer may be the one-two punch that leads to the collapse of northern ice sheets*'[129].

Nevertheless, Huybers agreed that the beat of Northern Hemisphere insolation could influence Antarctic climate through the transfer of heat across the Equator either in the atmosphere or in the ocean[89]. Regardless of that possibility, there was no doubt that CO_2 amplified temperatures measured in Antarctic ice cores by about half, was sourced mainly from the Southern Ocean and was a good candidate for orchestrating global climate change, since it was well mixed through the atmosphere. Deciding between the various options would require more research on precisely how the annual climate signal became fixed in Antarctic ice[93].

By 2009, Eric Wolff of the British Antarctic Survey agreed that the Southern Hemisphere was in the driving seat for glacial terminations. Wolff and colleagues noticed that '*the initial stages of glacial terminations are indistinguishable from the warming stage of events in Antarctica known as Antarctic Isotopic Maxima, which occur frequently during glacial periods*'[130]. Those warmings, which are associated with increasing CO_2, are directly associated with Dansgaard–Oeschger events in the north (Figure 13.3). The Antarctic warmings begin to reverse with the onset of the warm Dansgaard–Oeschger events. Wolff and his team argued that glacial terminations were in effect an extreme variety of this relationship, in which there was no reversal of the Antarctic warming. As that warming continued, a full deglaciation became inevitable. In these findings, there was both an implication and a question. The implication was that if the millennial Antarctic isotopic maxima were identical with Antarctic warmings at the start of deglaciations, then the timings of the deglaciations were probably not orbitally controlled.

The question was, why did some Antarctic warming events not reverse?

Wolff noticed that before each termination, the global climate reached a cold maximum, when ice sheets and sea ice had their largest extent. This excessive cold may have stopped the system from producing a Dansgaard–Oeschger event following an Antarctic warming. Eventually, a Dansgaard–Oeschger event did occur, but too late to prevent the Antarctic warming from continuing virtually unchecked. In effect, '*terminations are caused by southern warming that runs away because the north cannot produce a DO event*'[130]. The seesaw was temporarily switched off for long enough to allow Antarctic warming to swamp the system.

In an independent review in 2010, Daniel Sigman and Gerald Haug confirmed the importance of the Southern Ocean in controlling Ice Age CO_2[131]. Knowing that the modern Southern Ocean releases old CO_2 from upwelling deep water into the air, they surmised that this 'leak' was suppressed during glacial periods, thus increasing the storage of CO_2 within the deep ocean. This made the deep ocean more acid, causing deep-ocean carbonates to dissolve, which in turn made the global ocean more alkaline, increasing the solubility of CO_2 in seawater and so driving more uptake of CO_2, hence driving further cooling through positive feedback, much as Toggweiler had suggested. The ocean must drive these changes, because it is by far the largest reservoir of CO_2 on the planet. Sea ice played a key role by limiting the 'leak' of CO_2 to the atmosphere in cold periods. A decrease in the flow of North Atlantic Deep Water served to reduce oxygenation of carbon-rich southern-sourced deep water. Thus, both northerly *and* southerly processes exacerbated the build-up of CO_2 at abyssal depths. Surface water productivity also played a role, especially in the sub-Antarctic, where high productivity and efficient grazing of phytoplankton by zooplankton led to massive increases in siliceous ooze composed mostly of diatoms in sediments from glacial periods. The biological pump was efficiently transferring abundant CO_2 directly to the deep sea. This excessive production may reflect an increase in iron fertilisation, stemming from an enhanced influx of wind-blown dust in glacial times. What we end up with is robust coupling of atmospheric CO_2 to climate cycles driven largely by changes in the ocean[131].

In 2013, Feng He of the University of Wisconsin-Madison set out to test the Huybers and Denton model of independent hemispheric response to forcing. He and colleagues used a coupled atmosphere–ocean GCM to identify the impacts of forcing on air temperature from changes in orbits, CO_2, ice sheets and the Atlantic Meridional Overturning Circulation connecting the two hemispheres via the ocean[132]. They interpreted their results to suggest that rising insolation in spring and summer in the Northern Hemisphere initiated the last deglaciation and controlled the timing and the magnitude of the evolution of surface temperature in the Southern Hemisphere, which would appear to support the Milankovitch model. They concluded that the orbitally induced retreat of the Northern Hemisphere ice sheets stimulated changes in the Atlantic Meridional Overturning Circulation that prompted deglacial warming in the Southern Hemisphere and its subsequent lead over Northern Hemisphere temperature. CO_2 rising with the Southern Ocean warming provided a critical feedback encouraging global warming and deglaciation. Other researchers disagreed, as we see further on.

In 2012, Joel Pedro and his team suggested two possible mechanisms for the last deglaciation, one involving fast connection through the atmosphere, the other a slower connection through the ocean[66]. Their atmospheric model starts with an orbitally induced increase in northern summer insolation initiating local warming and retreat of the Northern Hemisphere ice sheets. Melting supplies freshwater to the North Atlantic, weakening the Atlantic Meridional Overturning Circulation and leading to warming of North Atlantic subsurface waters. These warm waters destabilise ice shelves, driving further ice retreat and releasing more freshwater. The surface cooling stimulated by freshwater makes sea ice expand, cooling the air and pushing the Inter-Tropical Convergence Zone south, which displaces southward and strengthens the Southern Hemisphere's mid-latitude westerly winds. These winds generate more upwelling, which draws up warm deep waters to release heat and old CO_2, further warming the air. In this atmospheric scenario, North Atlantic cooling and the release of CO_2 from the high-latitude Southern Ocean are almost simultaneous.

In contrast, Pedro's ocean pathway invokes the bipolar seesaw along the lines suggested by Broecker, as discussed earlier[32]. Weakening the Atlantic Meridional Overturning Circulation reduces northward heat transport, allowing heat to accumulate in the south.

Whatever the solution turns out to be in terms of atmospheric versus oceanic connections, Pedro noted that '*the ice core observations point to a tightly-coupled system operating with little or no time delay between the onset/terminations of North Atlantic climate stages* *and a near-simultaneous trend change in both Antarctic temperature and atmospheric CO_2'*[66].

Thomas Crowley agreed that the ocean route played a significant role, introducing a novel additional idea. Switching off the thermohaline conveyor by reducing the formation of North Atlantic Deep Water meant that heat entering the South Atlantic from the Indian Ocean around South Africa would not be able to escape to the north via the Gulf Stream, as it did in warm times. Instead, it would turn south along the South American coast in the Brazil Current, thus pumping heat into the Southern Ocean[133].

Jeremy Shakun and his team found in 2012 that their observed temperature variations in Antarctic ice cores closely matched variations in the strength of the Atlantic Meridional Overturning Circulation, interpreting this to '*suggest that ocean circulation changes driven primarily by freshwater flux, rather than by direct forcing from greenhouse gases or orbits, are plausible causes of the hemispheric differences in temperature change seen in the proxy records*'[24]. The fact that their Southern Hemisphere temperature stack led Northern Hemisphere temperatures during the deglaciation supported their '*inference that AMOC [Atlantic Meridional Overturning Circulation]-driven internal heat redistributions explain the Antarctic temperature lead and global temperature lag relative to CO_2*'[24]. What then triggered deglacial warming? They found that '*substantial temperature change at all latitudes ... as well as a net global warming of about 0.3 °C ... precedes the initial increase in CO_2 concentration at 17.5 kyr ago, suggesting that CO_2 did not initiate deglacial warming*'[24]. They went on to '*suggest that these spatiotemporal patterns of temperature change are consistent with warming at northern mid to high latitudes, leading to a reduction in the AMOC at ~19 kyr ago, being the trigger for the global deglacial warming that followed*'[24]. The trigger may have been '*rising boreal insolation driving northern warming*'[24]. Then the Northern Hemisphere ice sheets retreated, and the resulting influx of freshwater reduced the Atlantic Meridional Overturning Circulation and thus warmed the Southern Hemisphere through the bipolar seesaw, leading to the release of old CO_2. The observed pattern '*is difficult to reconcile with hypotheses invoking a southern high latitude trigger for deglaciation*'[24]. Even so, they concluded, '*CO_2 [was] a key mechanism of global warming during the last deglaciation*'[24]. However, Parrenin's results now suggest that the core of Shakun's model – the assumption that CO_2 lagged Antarctic rise in temperature – is suspect.

The correspondence between bipolar seesaw oscillations and changes in atmospheric CO_2 during glacial terminations back through time suggested to Stephen Barker and his team in 2011 that the bipolar seesaw played a key role in the mechanism of deglaciation through positive feedbacks associated with increasing CO_2[43]. '*With the supercritical size of continental ice sheets as a possible precondition, and in combination with the right insolation forcing and ice albedo feedbacks, the CO_2 rise associated with an oscillation of the bipolar seesaw could provide the necessary additional forcing to promote deglaciation*', they felt[43]. Thus, the mechanism of glacial termination might result from the timely and necessary interaction between millennial variations (Dansgaard–Oeschger events) and orbital time-scale variations, much as Wolff had suggested.

Clearly, Frédéric Parrenin and his team disagreed with Shakun and other Meridional Overturning enthusiasts. The tight correlation they discovered between CO_2 and temperature in Antarctica meant that '*invoking changes in the strength of the Antarctic meridional overturning circulation is no longer required to explain the [previously apparent] lead of [Antarctic temperature] over [atmospheric] CO_2*'[65]. They went on to suggest that '*Given the importance of the Southern Ocean in carbon cycle processes ... one should not exclude the possibility that [atmospheric] CO_2 and [Antarctic temperature] are interconnected through another common mechanism such as a relationship between sea ice cover and ocean stratification*'[65]. This latest finding lends support to the Huybers and Denton model, in which southern spring insolation and the length of southern summers control climate change in the south at the same time that summer insolation at 65° N controls what happens in the north[129]. Toggweiler and Wolff would no doubt agree.

A recent twist in the complex tale of the evolution of Ice Age climate theory came in 2013 from a team led by Ayako Abe-Ouchi of the University of Tokyo[134] and including Maureen Raymo, whom we met earlier arguing for a role for glacial rebound. They used numerical physical models in combination with a GCM to assess the relative importance of internal mechanisms that might drive the 100 Ka glacial cycles. These included delayed bedrock rebound (i.e. glacial isostatic adjustment), the calving of icebergs from ice sheet margins, variations in CO_2 and feedback from aeolian dust and the oceans. The advantage of their method was that '*The ice-sheet model with the climate parameterization ... can represent fast feedbacks, such as water vapour, cloud and sea-ice feedbacks ... [as well as] slow feedbacks, such as albedo/temperature/ice-sheet and lapse rate/temperature/ice sheet feedbacks'*[134]. Lapse rate was included because it is about 6.5 °C per 1000 m in moist air, making ice sheets cooler as they grow upwards and warmer as they shrink downwards. Their rising and shrinking may be functions not only of precipitation, but also of isostatic adjustment, when the ice mass depresses the underlying crust. Abe-Ouchi's approach thus took into consideration variables that had been missing in previous theoretical analyses, including extent of sea ice, processes governing growth and decay of ice sheets and isostatic adjustments to ice sheets.

The team calculated ice sheet variation for the past 400 Ka forced by insolation and atmospheric CO_2 content, validated the results using proxy palaeoclimate data and conducted sensitivity experiments to investigate the possible mechanisms controlling the 100 Ka glacial cycles. Their model took into account the fact that as ice sheets thicken, they depress the land beneath them, which may lower their tops into areas of warmer air, increasing the amount of melt and the area exposed to melting. Ice sheets also flow, which may widen their area and lower their height, thus exposing them to melting. These various processes become effective late in the glacial cycle, because it takes a long time for ice sheets to grow to the point where they will both lose height and spread. By incorporating these and other feedbacks, the model *'realistically simulates the sawtooth characteristics of glacial cycles, the timing of the terminations and the amplitude of the Northern Hemisphere ice-volume variations ... as well as their geographical patterns at the Last Glacial Maximum and the subsequent deglaciation'*[134]. Among other things, the team found that *'The ~100-kyr periodicity, the sawtooth pattern and the timing of the terminations are reproduced [in sensitivity experiments] ... with constant CO_2 levels ... [and that] the crucial mechanism for the ~100-kyr cycles is the delayed glacial isostatic rebound, which keeps the ice elevation low, and, therefore, the ice ablation high, when the ice retreats'*[134].

Abe-Ouchi's team found that the relationship between ice volume and temperature for the North American (Laurentide) ice sheet followed a hysteresis loop, with ice sheet volume first declining gradually as temperature anomalies increased from −5 to 0 °C, then declining rapidly as temperatures increased to +2 °C. Regrowth happened rapidly only after temperature anomalies fell to zero and declined below it. When the ice sheet was large enough in extent, its rapid disintegration was triggered by just a modest increase in insolation, enhanced by low

elevation due to the delayed isostatic response, calving into proglacial lakes, increasing CO_2 concentrations (amplifying warming), dust feedback and basal sliding as water made its way to the base of the ice sheet. Thus, *'insolation and internal feedbacks between the climate, the ice sheets and the lithosphere–asthenosphere system explain the 100,000-year periodicity ... The larger the ice sheet grows and extends towards lower latitudes, the smaller is the insolation required to make the mass balance negative. Therefore, once a large ice sheet is established, a moderate increase in insolation is sufficient to trigger a negative mass balance, leading to an almost complete retreat of the ice sheet within several thousand years. This fast retreat is governed mainly by rapid ablation due to the lowered surface elevation resulting from delayed isostatic rebound, which is the lithosphere–asthenosphere response. Carbon dioxide is involved, but is not determinative, in the evolution of the 100,000-year glacial cycles'*[134] The observation about CO_2 is fair enough, in that, as we now know, it started growing long before the warming and collapse of the Northern Hemisphere ice sheets. The possibility emerges, then, that CO_2 output is dominantly a Southern Hemisphere phenomenon and has its greatest effect on climate change there, while the events in the north are much more closely controlled by the dynamics of ice sheet response to northern (boreal) insolation forcing. I am reminded of Plass's hypothesis from the 1950s that Arctic temperatures would remain low while ice sheets melted, even though CO_2 was rising, simply because melting ice absorbs huge amounts of energy. Hence, we should not expect a direct link between Northern Hemisphere ice and CO_2.

The European ice sheet behaved differently from the North American one because it was thinner, less extensive and located in a warmer climate[135]. It also followed a hysteresis loop, with ice sheet volume declining rapidly between −2 and 0 °C compared with modern conditions, and regrowing rapidly only after temperature anomalies fell to −1 °C and further declined. It responded mainly to insolation following the obliquity cycle of ~40 Ka. The difference from its North American equivalent was caused in part because summers in Europe are warmer than in North America. Under these conditions, the European ice sheet could not sustain large ice volumes for long. Obliquity was much less important for the North American ice sheet, where eccentricity modulated the amplitude of the precession signal by causing critical changes in summer insolation to create the 100 Ka cycle. The team reached the remarkable conclusion that *'the 100-kyr glacial cycle exists only because of the unique geographic and climatological setting of the North American ice sheet with respect to received insolation*[134].

Others were working along similar lines. In 2011, Andrey Ganopolski and Reinhard Calov of the Potsdam Institute for Climate Impact Research in Germany used an Earth system model of intermediate complexity to show that 100 Ka variations in ice volume and in the timing of the terminations of glacial periods could be simulated as nonlinear responses of the climate–cryosphere system to orbital forcing, provided that CO_2 levels were below those typical of interglacial periods[136]. Like Abe-Ouchi, they attributed the existence of long glacial cycles mainly to the behaviour of the Laurentide Ice Sheet of North America. Although this behaviour could be simulated without any variation in CO_2, they found that variations in CO_2 with time amplified the 100 Ka cycles. They also found that the development of the 100 Ka pattern depended on former ice sheets having cleared northern North America of sediments that might have enhanced sliding at the base of the ice sheet. It was important in the model for the ice sheet to be sitting on rock. Their model showed that ice formed when summer insolation fell below a certain threshold, allowing ice to remain all year. Rapid terminations of the ice sheets were strongly related to their coverage by dust, which reduced albedo, enhanced melting (ablation) and thus amplified the response of the ice sheet to rising insolation.

The solid Earth also seems to have been important in another way, through CO_2 provided by volcanic eruptions. In 2004, comparing volcanic activity with the deuterium (∂^2D) proxy for temperature in the EPICA Dome C ice core, a team led by Jean Jouzel of Gif-sur-Yvette found no clear evidence for a close relationship between climate change and volcanism over the past 45 Ka[137]. But, a later, more detailed examination by Peter Huybers and Charlie Langmuir of Harvard, in 2009, showed that subaerial volcanism increased globally by two to six times above background levels during the last deglaciation 12–7 Ka ago[138]. That rise was consistent with an increase of 40 ppm in atmospheric CO_2 during the second half of the last deglaciation. Huybers and Langmuir suggested that the glacial isostatic adjustments associated with shrinkage of ice caps decompressed the mantle in deglaciating regions. That increased the number and intensity of volcanic eruptions, which raised atmospheric CO_2, which warmed the atmosphere and caused more deglaciation, and so on. According to Huybers and Langmuir, '*Such a positive feedback may contribute to the rapid passage*

from glacial to interglacial periods'[138]. Their hypothesis provides us with a further modification to the commonly accepted notion that glacial–interglacial variations in CO_2 are primarily attributable to oceanic processes[a].

The story is almost complete. The latest data from West Antarctica provide us with yet another angle to consider. In 2011, the members of the West Antarctic Ice Sheet (WAIS) Divide team drilled the 3405 m-long WAIS Divide Ice Core (WDC) in central West Antarctica at an altitude of 1766 m, recovering ice as old as 68 Ka[139]. The data appeared in 2013. The key thing to bear in mind is that the East Antarctic coring sites (Vostok, Fuji, Dome C and Dronning Maud Land) lie high up on the Polar Plateau, where they are isolated from the influence of changes of circulation and sea ice within the Southern Ocean, whereas the West Antarctic site is lower, hence warmer, and much closer to the sea, so is subject to the influence of marine air. It thus preserves a clearer record of changes in ocean circulation and sea ice. Warming began at the WDC site 20 Ka ago, at least 2 Ka before significant warming in East Antarctica. At the same time, sea-salt sodium (Na) increased, suggesting a decline in the cover of sea ice. This agrees with data from a marine core from the southwest Atlantic showing that sea ice began to retreat shortly before 22 Ka ago. The warming began *before* the decrease in the Atlantic Meridional Overturning Circulation that had been called upon to explain Southern Ocean warming. Its likely cause was local orbital forcing, with annual insolation at 65° S increasing 1% between 22 and 18 Ka ago. Longer summers and shorter winters likely melted sea ice and warmed the ocean. Decreasing sea ice would have decreased albedo, causing further warming.

Can we tie the Northern and Southern Hemisphere pictures together now? The WAIS Divide team summarised the picture as follows: '*While the abrupt onset of East Antarctic warming, increasing CO_2 and decreasing AMOC 18 kyr ago has supported the view that deglaciation in the Southern Hemisphere is primarily a response to changes in the Northern Hemisphere … the evidence of warming in West Antarctica and corresponding evidence for sea-ice decline in the [southeast] Atlantic show that climate changes were ongoing in the Southern Ocean before 18 kyr ago, supporting an important role for local*

---

[a] When the pull of other planets made Earth's orbit most eccentric, Earth's shape would have been sufficiently distorted to enhance submarine volcanism and the exhalation of CO_2 especially at mid-ocean ridge crests. That may have helped to enhance glacial-interglacial cycles and may help to explain the dominance of the 100 kyr cycle (Tolstoy, M., (2015) Mid-ocean ridge eruptions as a climate valve. Geophysical Research Letters, in press).

orbital forcing. Warming in the high latitudes of both hemispheres before 18 kyr ago implies little change in the interhemispheric temperature gradient that largely determines the position of the intertropical convergence zone and the position and intensity of the mid-latitude westerlies[139]. They proposed that '*when Northern Hemisphere cooling occurred ~18 kyr ago, coupled with an already-warming Southern Hemisphere, the intertropical convergence zone and mid-latitude westerlies shifted southwards in response. The increased wind stress in the Southern Ocean drove upwelling, venting of CO_2 from the deep ocean, and warming in both West Antarctica and East Antarctica*'[139]. This new West Antarctic Ice Sheet core confirms an active role for the Southern Hemisphere in initiating global deglaciation. It looks like Huybers, Denton, Toggweiler and Wolff were right.

So, by the beginning of the 21st century, much as hoped by Crowley and North[69], palaeoclimatologists had brought us close to a definitive explanation of the variability of the Pleistocene Ice Age. Access to very recently obtained long cores from the Antarctic ice sheet was critical to this growth in our understanding, and much of the data emerging from these cores has yet to make it into the literature available to the wider public. These data have been obtained at a considerable cost, much like those obtained from deep-sea drilling. Without the results from drill cores through the ice sheets and the ocean floor, our understanding of the Earth's climate evolution would be dim indeed.

Looking at these various developments, we can now see that CO_2 and the circulation of the Southern Ocean play key roles within the glacial–interglacial climate system. Rising spring insolation and the long duration of the summer in the Southern Hemisphere caused the Southern Ocean to lose its sea ice, to warm and to simultaneously release CO_2, which further accentuated the warming. The ocean began to carry less CO_2 as it warmed and as the rate of production of deep water decreased. Atmospheric CO_2 mixed rapidly to the Northern Hemisphere, whose rise in temperature lagged that in the south because of the thermal inertia of the Northern Hemisphere ice sheets. The Atlantic Meridional Ocean Circulation connected the two hemispheres, feeding 'old' CO_2 south, where it emerged at the surface of the Southern Ocean. The strengthening of the Atlantic Meridional Overturning Circulation, previously called upon to explain Southern Ocean warming, began *after* warming started in the south, indicating the importance of local southern factors. Tropical warming increased wetlands, which provided

methane, and enhanced oceanic evaporation, providing water vapour, both of which stimulated further warming. Rising seas provide a larger surface area for the production of water vapour. Volcanic eruptions associated with glacial isostatic adjustment as ice melted in the north provided yet more CO_2 during the deglaciation. The *three* increasing greenhouse gases in the atmosphere (CO_2, H_2O, CH_4) added to warming, which was further accentuated by rising insolation in the Northern Hemisphere and by the eventual disappearance of Northern Hemisphere ice, which reduced albedo, and in due course we arrived at a warm interglacial. The disappearance of Northern Hemisphere ice was largely controlled by the response of ice to local insolation, combined with the decay of the growing ice sheet as it sank into warmer air and spread widely across its surroundings.

In contrast, during cold conditions, CO_2 was drawn down from the atmosphere into the ocean by dissolution in cold surface waters and was taken into deep water by enhanced deep-water production. Intensified weathering in mountainous areas further drew down atmospheric CO_2. A low equilibrium level of atmospheric CO_2 was reached at glacial maxima, when sea level was at its lowest, reducing ocean areas, and sea ice was most abundant, further reducing the area of ocean exposed to the atmosphere. At that time, expanded ice sheets reduced the area covered by vegetation, a potential source of CO_2 and CH_4. The area of tropical wetlands, a major source of CH_4, was also reduced. Nevertheless, exchange of CO_2 between ocean and atmosphere continued, especially where strengthening winds intensified upwelling along the Equator, beneath the westerly winds of the Southern Ocean and along continental margins, bringing 'old' CO_2 to the surface. Falling seas reduced the area for the production of water vapour. The various factors combined to cool the planet. Eventually insolation rose, and the warming cycle began again, leading from glaciation to deglaciation.

We have learned one further important lesson. Temperature *per se* is not absolutely dependent on CO_2. It is also governed by insolation, by albedo (e.g. the extent of ice sheets and sea ice), by the local melting of land ice and by the movement of warm currents. In contrast, while CO_2 is dependent on factors like the warmth of seawater (the solution pump), it is also affected by productivity (the biological pump), the distribution of carbonate (the CCD), the presence or absence of sea ice, the area of the ocean (which changes with sea level) and the amount and type of vegetation on land. Under 'normal' interglacial conditions, CO_2 reaches an equilibrium level (about 280 ppm), driven by the various feedbacks within the climate system, before declining insolation begins to drive the system back towards cooler conditions. That decline decreases the conditions that lead to high concentrations of atmospheric CO_2, which in turn aids cooling. The rate of decline in temperature in encroaching glaciation exceeds the rate of decline in CO_2, being driven more by falling insolation and increasing albedo due to the growing extent of land and sea ice, the last of which prevents CO_2 decline in polar regions. Eventually, CO_2 reaches a new equilibrium level (about 180 ppm), driven by the feedbacks within the climate system. These processes account for the fluctuation of CO_2 during the Ice Age within a 'natural envelope' of 180–280 ppm.

Neither insolation nor CO_2 is the sole key to understanding the 100 Ka cycles. Both hemispheres are connected through the ocean and the atmosphere, which means that each can influence the other, for example through the Atlantic Meridional Ocean Circulation. Aspects of the insolation characteristics of each hemisphere mean that both can warm or cool at more or less the same time, despite their separate drivers. With the latest data in hand from Frédéric Parrenin[65] and Joel Pedro[66], CO_2 can no longer be seen as merely a follower of temperature during the Ice Age. At least around Antarctica, the two vary in lock-step during deglaciations and the Antarctic equivalents of Dansgaard–Oeschger events. Instead of millennial variations in the bipolar seesaw being kicked off in the north, as was formerly thought, they now appear to be kicked off in the south, where temperature and CO_2 rise first, including during deglaciations. It now seems that deglaciations occur when Northern Hemisphere ice sheets (especially the Laurentide) grow too large to respond quickly to warming from the south, so fail to produce the necessary Dansgaard–Oeschger event to counter it. Deglaciations seem to occur in a 100 Ka cycle following the behaviour of the Laurentide Ice Sheet. Once it gets too big, it can easily be removed by relatively small changes in northern insolation and warming from the south, independently of global CO_2 levels.

We have also learned that isostatic adjustments during deglaciation in the north stimulate volcanic activity that provides new CO_2 unrelated to plate tectonic processes. And we now know that weathering can proceed fast enough to draw down CO_2 in glacial times. Hence, CO_2 can have both a primary role (via volcanic activity and weathering) and a secondary role (via emergence from the ocean as it warms, then enhancement of warming through positive feedback), with water vapour and methane adding

to the feedback. We understand the complexities much better than we used to.

CO_2 and temperature follow the same saw-toothed pattern at orbital scales (Figure 3.5) and at Dansgaard–Oeschger scales (Figure 3.3). CO_2 does not operate independently of temperature. Both respond at the same time to some common external forcing, which drives the upwelling that releases old CO_2 from the Southern Ocean. Contrary to initial ideas, this forcing kicks off the process in the south, possibly during long southern summers. What would account for the same patterns at orbital and millennial scales? We can expect there to be harmonics in the orbital signal that lead to millennial-scale variation between the main orbital periods. Alternatively, once the oscillations are set up, as Broecker initially suggested, they may simply persist with no external driver.

Having established how the Ice Age climate system works, we turn in Chapter 14 to a detailed examination of climate change in the latest interglacial period, the Holocene, starting 11 700 years ago.

References

1. Oerlemans, J. (1982) Glacial cycles and ice-sheet modelling. *Climate Change* **4**, 353–374.
2. Dansgaard, W., Johnsen, S.J., Clausen, Dahl-Jensen, D., Gundestrup, N., Hammer, C.U. and Oeschger, H. (1984) North Atlantic climatic oscillations revealed by deep Greenland ice cores. In: *Climate Processes and Climate Sensitivity* (eds J.E. Hansen and T. Takahashi). Geophysical Monographs 29. American Geophysical Union, Washington, DC, pp. 288–298.
3. Steffensen, J.P., Andersen, K.K., Bigler, M., Clausen, H.B., Dahl-Jensen, D., Fischer, H., *et al.* (2008) High-resolution Greenland ice core data show abrupt climate change happens in few years. *Science* **321**, 680–684.
4. Bond, G., Showers, W., Cheseby, M., Lotti, R., Almasi, P., deMenocal, P., *et al.* (1997) A pervasive millennial-scale cycle in North Atlantic Holocene and glacial climates. *Science* **278**, 1257–1266.
5. Bond, G.C., Showers, W., Elliot, M., Evans, M., Lotti, R., Hajdas, I., *et al.* (1999) The North Atlantic's 1-2kyr climate rhythm: relation to Heinrich Events, Dansgaard/Oeschger Cycles and the Little Ice Age. In: *Mechanisms of Global Climate Change at Millennial Time Scales* (eds P.U. Clark, R.S. Webb and L.D. Keigwin). Geophysical Monographs 112. American Geophysical Union, Washington, DC, pp. 35–58.
6. Oeschger, H., Beer, J., Siegenthaler, U., Stauffer, B., Dansgaard, W. and Langway, C.C. (1984) Late glacial climate history from ice cores. In: *Climate Processes and Climate Sensitivity* (eds J.E. Hansen and T. Takahashi). Geophysical

Monographs 29. American Geophysical Union, Washington, DC, pp. 299–306.

7. Dansgaard, W., Johnsen, S.J., Clausen, H.B., Dahl-Jensen, D., Gundestrup, N.S., Hammer, C.U., *et al.* (1993) Evidence for general instability of past climate from a 250 kyr ice core record. *Nature* **364**, 218–220.

8. Jouzel, J. (2006) Water stable isotopes: atmospheric composition and applications in polar ice core studies. In: *The Atmosphere, Vol. 4, Treatise on Geochemistry* (eds H.D. Holland and K.K. Turekian). Elsevier-Pergamon, Oxford, pp. 213–243.

9. Bond, G., Broecker, W.S., Johnsen, S., McManus, J., Labeyrie, L., Jouzel, J. and Bonani, G. (1993) Correlations between climate records from North Atlantic sediments and Greenland ice. *Nature* **365**, 143–147.

10. Sundquist, E.T. and Visser, K. (2005) The geologic history of the carbon cycle. In: *Biogeochemistry, Vol. 8, Treatise on Geochemistry* (eds H.D. Holland and K.K. Turekian). Elsevier-Pergamon, Oxford, pp. 425–513.

11. Alley, R.B., Clark, P.U., Keigwin, L.D. and Webb, R.S. (1999) Making sense of millennial-scale climate change. In: *Mechanisms of Global Climate Change at Millennial Time Scales* (eds P.U. Clark, R.S. Webb and L.D. Keigwin). Geophysical Monographs 112. American Geophysical Union, Washington, DC, pp. 385–394.

12. Siddall, M., Stocker, T.F., Blunier, T., Spahni, R., Schwander, J., Barnola, J.-M. and Chappellaz, J. (2007) Marine Isotope Stage (MIS) 8 millennial variability stratigraphically identical to MIS 3. *Paleoceanography* **22**, PA1208.

13. Cronin, T.M. (2010) *Paleoclimates: Understanding Climate Change Past and Present* Columbia University Press, New York.

14. Lorius, C. and Merlivat, L. (1977) Distribution of mean surface stable isotope values in East Antarctica; observed change with depth in a coastal area. In: *Isotopes and Impurities in Snow and Ice*. Proceedings of the Grenoble Symposium 1975. International Association of Hydrological Sciences, Wallingford, pp. 125–137.

15. Lorius, C., Jouzel, J., Ritz, C., Merlivat, L., Barkov, N.I., Korotkevich, Y.S. and Kotlyakov, V.M. (1985) A 150,000-year climatic record from Antarctic ice. *Nature* **316**, 591–596.

16. Jouzel, J., Genthon, C., Lorius, C., Petit, J.R. and Barkov, N.I. (1987) Vostok ice core: a continuous isotope temperature record over the last climatic cycle (160,000 years). *Nature* **329**, 403–408.

17. Petit, J.R., Basile, I., Leruyuet, A., Raynaud, D., Lorius, C., Jouzel, J., *et al.* (1997) Four climate cycles in Vostok ice core. *Nature* **387**, 359–360.

18. Petit, J.R., Jouzel, J., Raynaud, D., Barkov, N.I., Barnola, J.-M., Basile, I., *et al.* (1999) Climate and atmospheric history of the past 420,000 years from the Vostok ice core, Antarctica. *Nature* **399**, 429–436.

19. Augustin, L., Barbante, C., Barnes, P.R.F., Barnola, J.M., Bigler, M., Castellano, E., *et al.* (2004) Eight glacial cycles from an Antarctic ice core. *Nature* **429**, 623–628.

20. Oerter, H., Drücker, C., Kipfstuhl, S. and Wilhelms, F. (2008) Kohnen Station – the drilling camp for the EPICA deep ice core in Dronning Maud Land. *Polarforschung* **78** (1–2), 1–23.

21. Watanabe, O., Kamiyama, K., Motoyama, H., Fujii, Y., Shoji, H. and Satow, K. (1999) The paleoclimate record in the ice core at Dome Fuji Station, East Antarctica. *Annals of Glaciology* **29** (1), 176–178.

22. Jouzel, J., Masson-Delmotte, V., Cattani, O., Dreyfus, G., Falourd, S., Hoffman, G., *et al.* (2007) Orbital and millennial Antarctic climate variability over the past 800,000 years. *Science* **317** (5839), 793–796.

23. Kopp, R.E., Simons, F.J., Mitrovica, J.X., Maloof, A.C. and Oppenheimer, M. (2009) Probabilistic assessment of sea level during the last interglacial stage. *Nature* **462**, 863–868.

24. Shakun, J. D., Clark, P. U., He, F., Marcott, S.A., Mix, A.C., Liu, Z., *et al.* (2012) Global warming preceded by increasing carbon dioxide concentrations during the last deglaciation. *Nature* **484**, 49–54.

25. Lorius, C., Barnola, J-M., Legrand, M., Petit, J.R., Raynaud, D., Ritz, C., *et al.* (1989) Long-term climatic and environmental records from Antarctic ice. In: *Understanding Climate Change* (eds A. Berger, R.E. Dickinson and J.W. Kidson). Geophysical Monographs 52. American Geophysical Union, Washington, DC.

26. Lorius, C., Barkov, N.I., Jouzel, J., Korotkevich, Y.S., Kotlyakov, V.M. and Raynaud, D. (1988) Antarctic ice core: CO_2 and climatic change over the last climatic cycle. *EOS* **69** (26), 681, 683–684.

27. Naish, T., Carter, L., Wolff, E., Pollard, D. and Powell, R. (2009) Late Pliocene–Pleistocene Antarctic climate variability at orbital and suborbital scale: ice sheet, ocean and atmospheric interactions. In: *Antarctic Climate Change* (eds F. Florindo and M. Siegert). Developments in Earth and Environmental Sciences 8. Elsevier, Amsterdam, pp. 465–529.

28. Elderfield, H., Perretti, P., Greaves, M., Crowhurst, S., McCave, I.N., Hodell, D. and Piotrowski, A.M. (2012) Evolution of ocean temperature and ice volume through the mid-Pleistocene climate transition. *Science* **337**, 704–709.

29. Blunier, T., Chappellaz, J., Schwander, J., Dällenbach, A., Stauffer, B., Stocker, T.F., *et al.* (1998) Asynchrony of Antarctic and Greenland climate change during the last glacial period. *Nature* **394**, 739–743.

30. Barbante, C., Barnola, J.-M., Becagli, S., Beer, J., Bigler, M., Boutron, C., *et al.* (2006) One-to-one coupling of glacial climate variability in Greenland and Antarctica. *Nature* **444**, 195–198.

31. Blunier, T., Spahni, R., Barnola, J.-M., Chappellaz, J., Loulergue, L. and Schwander, J. (2007) Synchronization of ice core records via atmospheric gases. *Climate of the Past* **3**, 325–330.

32. Broecker, W.S. (1998) Paleocean circulation during the last deglaciation: a bipolar seesaw? *Paleoceanography* **13**, 119–121.

33. Stocker, T.F. (1998) The seesaw effect. *Science* **282**, 61–62.

34. Maslin, M., Seidov, D. and Lowe, J. (2001) Synthesis of the nature and causes of rapid climate transitions during the Quaternary. In: *The Oceans and Rapid Climate Change: Past, Present and Future* (eds D. Seidov, B.J. Haupt and M. Maslin). Geophysical Monographs 126. American Geophysical Union, Washington, DC, pp. 9–52.

35. Barron, E.J. and Seidov, D. (2001) Ocean currents of change: introduction. In: *The Oceans and Rapid Climate Change: Past, Present and Future* (eds D. Seidov, B.J. Haupt and M. Maslin). Geophysical Monographs 126. American Geophysical Union, Washington, DC, pp. 1–5.

36. Broecker, W.S. (2001) The big climate amplifier: ocean circulation-sea ice-storminess-dustiness-albedo. In: *The Oceans and Rapid Climate Change: Past, Present and Future* (eds D. Seidov, B.J. Haupt and M. Maslin). Geophysical Monographs 126. American Geophysical Union, Washington, DC, pp. 53–56.

37. Seidov, D., Haupt, B.J., Barron, E.J. and Maslin, M. (2001) Ocean bi-polar seesaw and climate: southern versus northern meltwater impacts. In: *The Oceans and Rapid Climate Change: Past, Present and Future* (eds D. Seidov, B.J. Haupt and M. Maslin). Geophysical Monographs 126. American Geophysical Union, Washington, DC, pp. 147–167.

38. Stocker, T.F., Knutti, R. and Plattner, G.-K. (2001) The future of the thermohaline circulation – a perspective. In: *The Oceans and Rapid Climate Change: Past, Present and Future* (eds D. Seidov, B.J. Haupt and M. Maslin). Geophysical Monographs 126. American Geophysical Union, Washington, DC, pp. 277–293.

39. Ganopolski, A. and Rahmstorf, S. (2001) Stability and variability of the thermohaline circulation of the past and future: a study with a coupled model of intermediate complexity. In: *The Oceans and Rapid Climate Change: Past, Present and Future* (eds D. Seidov, B.J. Haupt and M. Maslin). Geophysical Monographs 126. American Geophysical Union, Washington, DC, pp. 261–275.

40. Little, M.G., Schneider, R.R., Kroon, D., Price, B., Summerhayes, C.P. and Segl, M. (1997) Trade Wind forcing of upwelling, seasonality, and Heinrich Events as a response to sub-Milankovitch climate variability. *Paleoceanography* **12** (4), 568–576.

41. Cane, M. and Clement, A.C. (1999) A role for the tropical Pacific coupled ocean-atmosphere system on Milankovitch and millennial timescales: Part II: global impacts. In: *Mechanisms of Global Climate Change at Millennial Time Scales* (eds P.U. Clark, R.S. Webb and L.D. Keigwin). Geophysical Monographs 112. American Geophysical Union, Washington, DC, pp. 373–383.

42. Alley, R.B., Anandakrishnan, S., Jung, P. and Clough, A. (2001) Stochastic resonance in the North Atlantic: further insights. In: *The Oceans and Rapid Climate Change:*

Past, Present and Future (eds D. Seidov, B.J. Haupt and M. Maslin). Geophysical Monographs 126. American Geophysical Union, Washington, DC, pp. 57–68.

43. Barker, S., Knorr, G., Edwards, R.L., Parrenin, F., Putnam, A.E., Skinner, L.C., *et al.* (2011) 800,000 years of abrupt climate variability. *Science* **334**, 347–351.

44. Siddall, M., Rohling, E.J., Blunier, T. and Spahni, R. (2010) Patterns of millennial variability over the last 500 ka. *Climate of the Past* **6**, 295–303.

45. Capron, E., Landais, A., Chappellaz, J., Schilt, A., Buiron, D., Dahl-Jensen, D., *et al.* (2010) Millennial and sub-millennial scale climatic variations recorded in polar ice cores over the last glacial period. *Climate of the Past* **6**, 345–365.

46. Oeschger, H., Stauffer, B., Finkel, R. and Langway, C.C. (1985) Variations of the CO_2 concentration of occluded air and of anions and dust in polar ice cores. In: *The Carbon Cycle and Atmospheric CO_2: Natural Variations Archean to Present* (eds E.T. Sundquist and W.S. Broecker). Geophysical Monographs 32. American Geophysical Union, Washington, DC, pp. 132–142.

47. Barnola, J.M., Raynaud, D., Korotkevitch, Y.S. and Lorius, C. (1987) Vostok ice core: a 160,000 year record of atmospheric CO_2. *Nature* **329**, 408–414.

48. Genthon, C., Jouzel, J., Barnola, J.M., Raynaud, D. and Lorius, C. (1987) Vostok ice core: the climate response to CO_2 and orbital forcing changes over the last climate cycle (160,000 years). *Nature* **329**, 414–418.

49. Siegenthaler, U., Stocker, T.F., Monnin, E., Lüthi, D., Schwander, J., Stauffer, B., *et al.* (2005) Stable carbon cycle-climate relationship during the late Pleistocene. *Science* **310**, 1313–1317.

50. Loulergue, L., Schilt, A., Spahni, R., Masson-Delmotte, V., Blunier, T., Lemieux, B., *et al.* (2008) Orbital and millennial-scale features of atmospheric CH_4 over the past 800,000 years. *Nature* **453**, 383–386.

51. Falkowski, P., Scholes, R.J., Boyle, E., Canadell, J., Canfield, D., Elser, J., *et al.* (2000) The global carbon cycle: a test of our knowledge of Earth as a system. *Science* **290**, 291–296.

52. Lüthi, D., Le Floch, M., Bereiter, B., Blunier, T., Barnola, J.-M., Siegenthaler, U., *et al.* (2008) High resolution carbon dioxide concentration record 650,000-800,000 years before present. *Nature* **453**, 379–382.

53. Wolff, E.W. (2011) Greenhouse gases in the Earth System: a palaeoclimate perspective. *Philosophical Transactions of the Royal Society A* **369**, 2133–2147.

54. Anon . (2001) Professor Barry Saltzman, a pioneer in the study of the atmosphere and climate, dies. *Yale Bulletin and Calendar* **29** (18). Available from http://www.yale.edu/opa/arc-ybc/v29.n18/story18.html (last accessed 29 January 2015).

55. Maasch, K.A., Oglesby, R.J. and Fournier, A. (2005) Barry Saltzman and the theory of climate. *Journal of Climate* **18**, 2142–2150.

56. Maasch, K.A. and Saltzman, B. (1990) A low-order dynamical model of global climatic variability over the full Pleistocene. *Journal of Geophysical Research* **95**, 1955–1963.

57. Saltzman, B. and Maasch, K.A. (1990) A first-order global model of late Cenozoic climate change. *Transactions of the Royal Society of Edinburgh* **81**, 315–325.

58. Saltzman, B. and Maasch, K.A. (1991) A first-order global model of late Cenozoic climate change II: a simplification of CO_2 dynamics. *Climate Dynamics* **5**, 201–210.

59. Saltzman, B. and Verbitsky, M. (1994) Late Pleistocene climate trajectory in the phase space of global ice, ocean state, and CO_2: observations and theory. *Paleoceanography* **9**, 767–779.

60. Saltzman, B. and Verbitsky, M. (1994) CO_2 and the glacial cycles. *Nature* **367**, 418.

61. Saltzman, B. (2002) *Dynamical Paleoclimatology: Generalized Theory of Global Climate Change* Academic Press, London.

62. Fischer, H., Wahlen, M., Smith, J., Mastroianni, D. and Deck, B. (1999) Ice core records of atmospheric CO_2 around the last three glacial terminations. *Science* **283**, 1712–1714.

63. Plass, G.N. (1961) The influence of infrared absorptive molecules on the climate. *Annals of the New York Academy of Sciences* **95**, 61–71.

64. Brook, E.J. (2013) Leads and lags at the end of the last ice age. *Science* **339**, 1042–1043.

65. Parrenin, F., Masson-Delmotte, V., Köhler, P., Raynaud, D., Paillard, D., Schwander, J., *et al.* (2013) Synchronous change of atmospheric CO_2 and Antarctic temperature during the last deglacial warming. *Science* **339**, 1060–1063.

66. Pedro, J.B., Rasmussen, S.O. and Van Ommen, T.D. (2012) Tightened constraints on the time-lag between Antarctic temperature and CO_2 during the last deglaciation. *Climate of the Past* **8**, 1213–1221.

67. Indermühle, A., Monnin, E., Stauffer, B., Stocker, T.F. and Wahlen, M. (2000) Atmospheric CO_2 concentration from 60 to 20 kyr BP from the Taylor Dome ice core, Antarctica. *Geophysical Research Letters* **27** (5), 735–738.

68. Ahn, J. and Brook, E.J. (2007) Atmospheric CO_2 and climate from 65 to 30 ka B.P. *Geophysical Research Letters* **34**, L10703, GL029551.

69. Schmittner, A. and Galbraith, E.D. (2008) Glacial greenhouse-gas fluctuations controlled by ocean circulation changes. *Nature* **456**, 373–376.

70. Skinner, L.C., Fallon, S., Waelbroeck, C., Michel, E. and Barker, S. (2010) Ventilation of the deep Southern Ocean and deglacial CO_2 rise. *Science* **328**, 1147–1151.

71. Crowley, T.J. and North, G.R. (2011) *Paleoclimatology* Oxford Monographs on Geology and Geophysics 18. Oxford University Press, Oxford.

72. Monnin, E., Indermühle, A., Dällenbach, A., Flückiger, J., Stauffer, B., Stocker, T.F., *et al.* (2001) Atmospheric CO_2 concentrations over the last glacial termination. *Science* **291**, 112–114.

73. Severinghaus, J.P., Sowers, T., Brook, E.J., Alley, R.B. and Bender, M.L. (1998) Timing of abrupt climate change at the end of the Younger Dryas interval from thermally fractionated gases in polar ice. *Nature* **391**, 141–146.

74. Ruddiman, W.F. and Raymo, M.E. (2003) A methane-based time scale for Vostok ice. *Quaternary Science Reviews* **22** (2–4), 141–155.

75. Severinghaus, J. (2009) Southern see-saw seen. *Nature* **457**, 1093–1094.

76. Broecker, W.S., Kennett, J.P., Flower, B.P., Teller, J.T., Trumbore, S., Bonani, G. and Wolfli, W. (1989) Routing of meltwater from the Laurentide ice sheet during the Younger Dryas cold episode. *Nature* **341**, 318–321.

77. Tarasov, L. and Peltier, W.R. (2005) Arctic freshwater forcing of the Younger Dryas cold reversal. *Nature* **435**, 662–665.

78. Firestone, R.B., West, A. Kennett, J.P., Becker, L., Bunch, T.E., Revay, Z.S., *et al.* (2007) Evidence for extraterrestrial impact 12,900 years ago that contributed to the megafaunal extinctions and Younger Dryas cooling. *Proceedings of the National Academies of Science* **104**, 16016–16021.

79. Barrows, T.T., Lehman, S.J., Fifield, L.K. and De Deckker, P. (2007) Absence of cooling in New Zealand and the adjacent ocean during the Younger Dryas Chronozone. *Science* **318**, 86–89.

80. Putnam, A.E., Denton, G.H., Schaefer, J.M., Barrell, D.J.A., Andersen, B.G., Finkel, R.C., *et al.* (2010) Glacier advance in southern middle-latitudes during the Antarctic Cold Reversal. *Nature Geoscience* **3**, 700–704.

81. Kaplan, M.R., Schaefer, J.M., Denton, G.H., Barrell, D.J., Chinn, T.J., Putnam, A.E., *et al.* (2010) Glacier retreat in New Zealand during the Younger Dryas Stadial. *Nature* **467**, 194–197.

82. Ackert, R.P., Becker, R.A., Singer, B.S., Kurz, M.D., Caffee, M.W. and Mickelson, D.M. (2008) Patagonian glacier response during the late Glacial-Holocene transition. *Science* **321**, 392–395.

83. Frakes, L.A., Francis, J.E. and Syktus, J.I. (1992) *Climate Modes of the Phanerozoic* Cambridge University Press, Cambridge.

84. Berger, A. (2009) Astronomical theory of climate change. In: *Encyclopedia of Paleoclimatology and Ancient Environments* (ed V. Gornitz). Springer, Dordrecht, pp. 51–57.

85. Berger, W.H. and Wefer, G. (2003) On the dynamics of the ice ages: stage 11 paradox, mid-Brunhes climate shift, and 100-ky cycle. In: *Earth's Climate and Orbital Eccentricity: The Marine Isotope Stage 11 Question* (eds A. Droxler, R.Z. Poore and L.H. Burckle). Geophysical Monographs 137. American Geophysical Union, Washington, DC, pp. 41–59.

86. Berger, A. and Loutre, M.-F. (2003) Climate 400,000 years ago, a key to the future? In: *Earth's Climate and Orbital Eccentricity: The Marine Isotope Stage 11 Question* (eds A. Droxler, R.Z. Poore and L.H. Burckle). Geophysical Monographs 137. American Geophysical Union, Washington, DC, pp. 17–26.

87. Raynaud, D., Loutre, M.F., Ritz, C., Chappellaz, J., Barnola, J.-M., Jouzel, J., *et al.* (2003) Marine isotope stage (MIS) 11 in the Vostok ice core: CO_2 forcing and stability of East Antarctica. In: *Earth's Climate and Orbital Eccentricity: The Marine Isotope Stage 11 Question* (eds A. Droxler, R.Z. Poore and L.H. Burckle). Geophysical Monographs 137. American Geophysical Union, Washington, DC, pp. 27–40.

88. Das Sharma, S., Ramesh, D.S., Bapanayya, C. and Raiju, P.A. (2012) Sea surface temperatures in cooler climate stages bear more similarity with atmospheric CO_2 forcing. *Journal of Geophysical Research: Atmospheres* **117** (D13), doi: 10.1029/2012JD017725.

89. Ridgwell, A.J., Watson, A.J. and Raymo, M.E. (1999) Is the spectral signature of the 100 kyr glacial cycle consistent with a Milankovitch origin? *Paleoceanography* **14** (4), 437–440.

90. Masson-Delmotte, V., Kageyama, M., Baconnot, P., Charbit, S., Krinner, G., Ritz, C., *et al.* (2006) Past and future polar amplification of climate change: climate model intercomparisons and ice-core constraints. *Climate Dynamics* **27**, 437–440.

91. Huybers, P. and Wunsch, C. (2005) Obliquity pacing of the late Pleistocene glacial terminations. *Nature* **434**, 491–494.

92. Berger, A. and Loutre, M.-F. (1991) Insolation values for the climate of the last 10 million years. *Quaternary Science Reviews* **10**, 297–317.

93. Huybers, P. (2009) Antarctica's orbital beat. *Science* **325**, 1085–1086.

94. Huybers, H. (2007) Glacial variability over the last two million years: an extended depth-derived age model, continuous obliquity pacing, and the Pleistocene progression. *Quaternary Science Reviews* **26**, 37–55.

95. Elkibbi, M. and Rial, J.A. (2001) An outsider's review of the astronomical theory of the climate: is the eccentricity-driven insolation the main driver of the ice ages? *Earth Science Reviews* **56**, 161–177.

96. Ruddiman, W.F. (2008) *Earth's Climate, Past and Future*, 2nd edn. W.H. Freeman and Co., New York.

97. Hay, W.W. (2013) *Experimenting on a Small Planet: A Scholarly Entertainment* Springer, New York.

98. Gates, W.L. (1976) The numerical simulation of ice-age climate with a global general circulation model. *Journal of Atmosphere Science* **33**, 1844–1873.

99. Manabe, S. and Hahn, D. (1977) Simulation of the tropical climate of an ice age. *Journal of Geophysical Research* **82** (C27), 3889–3911.

100. Hansen, J.E., Lacis, A., Rind, D., Russell, G., Stone, P., Fung, I., *et al.* (1984) Climate sensitivity: analysis of feedback mechanisms. In: *Climate Processes and Climate Sensitivity* (eds J.E. Hansen and T. Takahashi). Geophysical Monographs 29. American Geophysical Union, Washington, DC, pp. 130–163.

101. Saltzman, B., Hansen, A.G. and Maasch, K.A. (1984) The late Quaternary glaciations as the response of a three-component feedback system to earth orbital forcing. *Journal of Atmosphere Science* **41**, 3380–3389.

102. Saltzman, B. and Moritz, R. (1980) A time-dependent climatic feedback system involving sea-ice extent, ocean temperature, and CO_2. *Tellus* **32**, 93–118.

103. Toggweiler, J.R. and Sarmiento, J.L. (1985) Glacial to interglacial changes in atmospheric carbon dioxide: the critical role of ocean surface water in high latitudes. In: *The Carbon Cycle and Atmospheric CO_2: Natural Variations Archaean to Present* (eds E.T. Sundquist and W.S. Broecker). Geophysical Monographs 32. Proceedings of the Chapman Conference, Tarpon Springs, Florida, 9–13 January 1984, pp. 163–184.

104. Broccoli, A.J. and Manabe, S. (1987) The influence of continental ice, atmospheric CO_2, and land albedo on the climate of the Last Glacial Maximum. *Climate Dynamics* **1**, 87–99.

105. Broecker, W.S. and Denton, G.H. (1989) The role of ocean-atmosphere reorganization in glacial cycles. *Geochimica et Cosmochimica Acta* **53**, 2465–2501.

106. Raymo, M.E. (1997) The timing of major climate terminations. *Paleoceanography* **12** (4), 577–585.

107. Pollard, D. (1982) A simple ice sheet model yields realistic 100 kyr glacial cycles. *Nature* **296**, 334–338.

108. Clement, A.C. and Cane, M. (1999) A role for the tropical Pacific coupled ocean-atmosphere system on Milankovitch and millennial timescales. Part I: A modelling study of tropical Pacific variability. In: *Mechanisms of Global Climate Change at Millennial Time Scales* (eds P.U. Clark, R.S. Webb and L.D. Keigwin). Geophysical Monographs 112. American Geophysical Union, Washington, DC, pp. 363–371.

109. Kenyon, J. and Hegerl, G.C. (2010) Influence of modes of climate variability on global precipitation extremes. *Journal of Climate* **23**, 6248–6262.

110. Chavez, F.P., Ryan, J., Lluch-Cota, S.E. and Ñiquen, M. (2003) From anchovies to sardines and back: multidecadal change in the Pacific Ocean. *Science* **299**, 217–221.

111. Hays, J.D., Imbrie, J. and Shackleton, N.J. (1976) Variations in the Earth's orbit: pacemaker of the ice ages. *Science* **194** (4270), 1121–1132.

112. Shackleton, N.J. (2000) The 100,000-year ice-age cycle identified and found to lag temperature, carbon dioxide, and orbital eccentricity. *Science* **289**, 1897–1902.

113. Ruddiman, W.F. (2003) Orbital insolation, ice volume, and greenhouse gases. *Quaternary Science Reviews* **22**, 1597–1629.

114. Yu, J.M. and Elderfield, H. (2007) Benthic foraminiferal B/Ca ratios reflect deep water carbonate saturation state. *Earth and Planetary Science Letters* **258**, 73–86.

115. Yu, J.M., Anderson, R.F. and Rohling, E.J. (2013) Deep ocean carbonate chemistry and glacial-interglacial atmospheric CO_2. *Oceanography* **27**(1), 16–25.

116. Munhoven, G. and François, L.M. (1996) Glacial and inter-glacial variability of atmospheric CO_2 due to changing continental silicate rock weathering: a model study. *Journal of Geophysical Research* **101** (D16), 21 423–21 437.

117. Berger, A., Li, X.S. and Loutre, M.F. (1999) Modelling Northern Hemisphere ice volume over the last 3 Ma. *Quaternary Science Reviews* **18**, 1–11.

118. Hönisch, B., Hemming, N.G., Archer, D., Siddall, M. and McManus, J.F. (2009) Atmospheric carbon dioxide concentration across the mid-Pleistocene transition. *Science* **324**, 1551–1554.

119. Yin, Q.Z. and Berger, A. (2010) Insolation and CO_2 contribution to the interglacial climate before and after the mid-Brunhes event. *Nature Geoscience* **3**, 243–246.

120. Sigman, D.M. and Haug, G.H. (2003) Biological pump in the past. In: *The Oceans and Marine Geochemistry, Vol. 6, Treatise on Geochemistry* (eds H.D. Holland and K.K. Turekian). Elsevier-Pergamon, Oxford, pp. 491–528.

121. Keeling, R. and Stephens, B.B. (2001) Antarctic sea ice and the control of Pleistocene climate instability. *Paleoceanography* **16**, 112–131 and *Paleoceanography* **16** (3), 330–334.

122. Dokken, T., Nisancioglu, K.H., Li, C., Battisti, D.S. and Kissel, C. (2013) Dansgaard-Oeschger cycles: interactions between ocean and sea ice intrinsic to the Nordic Seas. *Paleoceanography* **28** (3), 491–502.

123. Brovkin, V., Ganopolski, A., Archer, D. and Rahmstorf, S. (2007) Lowering of glacial atmospheric CO_2 in response to changes in oceanic circulation and marine biogeochemistry. *Paleoceanography* **22**, PA4202.

124. Crucifix, M. (2009) Modeling the climate of the Holocene. In: *Natural Climate Variability and Global Warming: A Holocene Perspective* (eds R.W. Battarbee and H.A. Binney). Wiley-Blackwell, Chichester and Oxford, pp. 98–122.

125. Paillard, D. (2001) Glacial cycles: towards a new paradigm. *Reviews of Geophysics* **39**, 325–346.

126. Lisiecki, L.E. (2010) Links between eccentricity forcing and the 100,000-year glacial cycle. *Nature Geoscience* **4**, 349–352.

127. Toggweiler, J.R. (2008) Origin of the 100,000-year timescale in Antarctic temperatures and atmospheric CO_2. *Paleoceanography* **23**, PA2211.

128. Kawamura, K., Parrenin, F., Lisiecki, L., Uemura, R., Vimeux, F., Severinghaus, J.P., *et al.* (2007) Northern Hemisphere forcing of climatic cycles in Antarctica over the past 360,000 years. *Nature* **448**, 972–977.

129. Huybers, P. and Denton, G. (2008) Antarctic temperature at orbital timescales controlled by local summer duration. *Nature Geoscience* **1**, 787–792.

130. Wolff, E.W., Fischer, H. and Röthlisberger, R. (2009) Glacial terminations as southern warmings without northern control. *Nature Geoscience* **2**, 206–209.

131. Sigman, D.M., Hain, M.P. and Haug, G.H. (2010) The polar ocean and glacial cycles in atmospheric CO_2 concentration. *Nature* **466**, 47–55.

132. He, F., Shakun, J.D., Clark, P.U., Carlson, A.E., Liu, Z., Otto-Bliesner, B.L. and Kutzbach, J.E. (2013) Northern Hemisphere forcing of Southern Hemisphere climate during the last deglaciation. *Nature* **494**, 81–85.

133. Crowley, T.J. (2011) Proximal trigger for late glacial Antarctic circulation and CO_2 changes. *PAGES News* **19** (2), 70–71.

134. Abe-Ouchi, A., Saito, F., Kawamura, K., Raymo, M.E., Okuno, J., Takahashi, K. and Blatter, H. (2013) Insolation-driven 100,000-year glacial cycles and hysteresis of ice-sheet volume. *Nature* **500**, 190–194.

135. Marshall, S.J. (2013) Solution proposed for ice-age mystery. *Nature* **500**, 159–160.

136. Ganopolski, A. and Calov, R. (2011) The role of orbital forcing, carbon dioxide and regolith in 100 kyr glacial cycles. *Climate of the Past* **7**, 1415–1425.

137. Castellano, E., Becagli, S., Jouzel, J., Migliori, A., Severi, M., Steffensen, J.P., *et al.* (2004) Volcanic eruption frequency in the last 45 ky as recorded in EPICA-Dome-C ice core (East Antarctica) and its relationship to climate changes. *Global and Planetary Change* **42**, 195–205.

138. Huybers, P. and Langmuir, C. (2009) Feedback between deglaciation, volcanism and atmospheric CO_2. *Earth and Planetary Science Letters* **286**, 479–491.

139. Fudge, T.J., Steig, E.J., Markle, B.R., Schoenemann, S.W., Ding, Q., Taylor, K.C., *et al.* (2013) Onset of deglacial warming in West Antarctica driven by local orbital forcing. *Nature* **500**, 440–446.

14

The Holocene Interglacial

14.1 Holocene Climate Change

We live in the Holocene. If we want to get some idea of how its (and our) climate may evolve into the future, we need to see how it developed over the past 11 700 years, since the end of the Younger Dryas cold event. A recent review by John Birks of the Bjerknes Centre for Climate Research of the University of Bergen explores the early stirrings of inquisitiveness about Holocene climate change[1]. Natural historians noted back in the late 18th century *'the impressive occurrence of large fossil trunks and stumps (megafossils) of pine trees buried in peat bogs in northwest Europe'*[1]. A Mr H. Maxwell observed in 1815 that *'one of the greatest enigmas of natural science is presented in the remains of pine forest buried under a dismal treeless expanse on the Moor of Rannoch, and on the Highland hills up to and beyond 2000 feet altitude'*[1]. Similar changes in Denmark were attributed in the late 19th century to changes in moisture and to cooling in the postglacial period. The Norwegian botanist Axel Blytt (1843–1898) interpreted tree layers in peat bogs and changes from dark, humified peat to pale, fresh peat as evidence for alternations between dry and wet periods. The Swedish botanist Rutger Sernander (1866–1944) added ideas about summer temperature changes to propose the famous four Blytt-Sernander periods of postglacial time: Boreal (warm, dry), from 10 000 to 8000 years ago; Atlantic (warmest, wet), from 8000 to 5000 years ago; Sub-Boreal (warm, dry), from 5000 to 2500 years ago; and Sub-Atlantic (cool, wet), from 2500 years ago to the present. These were preceded by a Pre-Boreal period (cool, sub-Arctic) prior to 10 000 years ago. This Eurocentric scheme *'became the dominant paradigm for Holocene climate history'* during the early part of the 20th century[1].

Birks also reminded us that another late-19th-century scientist, the Swedish botanist Gunnar Andersson (1865–1928), while searching peat bogs, had discovered plant fossils, including hazel nuts, well north of their current range, suggesting that the climate had once been warmer than today. This led Anderson to present the idea of a gradually rising temperature curve reaching a long early to mid Holocene period of temperature warmer than today (a thermal maximum), followed by a subsequent decline. The notion of a Holocene thermal maximum seems to have arisen independently in the minds of other natural historians, including the Scottish palaeontologist Thomas F. Jamieson (1829–1913), based on his study of the molluscan fauna of mid Holocene estuarine clays in Scotland[1].

Early in the 20th century, a Swedish geologist, Lennart Von Post (1884–1951) proposed that *'pollen analysis … [be used] as a technique for relative dating and for reconstructing past vegetation and past climate'*[1]. In contrast to megafossils and macrofossils, it could provide a continuous record of changing vegetation and climate. Von Post integrated the Blytt and Salander phases of climate with Andersson's notion of gradual climate change and the findings of his own pollen analyses. This led to pollen analysis being used as a key means of determining Holocene geological time and climate change prior to about 1960. Quoting Ed Deevey, Birks told us that *'Von Post's simple idea that a series of changes in pollen proportions in accumulating peat was a four-dimensional look at vegetation, must rank with the double-helix as one of the most productive suggestions of modern times*[1]. So it should be no surprise that, for his contributions, Von Post was awarded the Vega Medal of the Swedish Anthropological and Geographical Society in 1944.

Holocene climate change was the stuff of particular fascination for H.H. Lamb, whom we met in Chapter 8. In his 1966 book *The Changing Climate* (written in 1964), Lamb produced a chronicle of climate change for the Holocene[2]. The first version of this chronicle, dated 1959, began with the disappearance of the last major ice sheet from Scandinavia between 10 and 9 Ka ago, followed by warming to a postglacial 'climatic optimum' of 6–4 Ka ago, when he thought that world temperatures were 2–3 °C warmer than now. Later, we will see just how 'global' that so-called 'optimum' was. The warming roughly coincided with a 'subpluvial' period in North Africa between 7.0 and 4.4 Ka ago, when there were settlements in the Sahara. As well as archaeological evidence for the presence of humans, the development of major river systems draining from the interior down to the Atlantic coast of the Sahara, now almost completely dry, attests to a formerly much more humid climate. These rivers include the Oued Souss (reaching the coast at about 30° 30′ N) and the Oued Dra (reaching the coast at about 28° 30′ N), which are largely dry, and the Seguia del Hamra (reaching the coast at 27° 30′ N), which is almost always dry[3]. The 'climatic optimum' was well known to students of the Holocene, and was recognised in deep-sea cores by Emiliani in 1955, for instance[4].

Lamb found that subsequent cooling and increased rainfall in Europe was followed by a drier and warmer climate during the Roman Era, and by a second climatic optimum between 400 and 1200 AD, with a peak around 800–1000 AD. During this medieval peak, the Vikings colonised the southern coast of Greenland and reached the shores of North America, and wine was produced in England as far north as York (almost 54° N). Summer temperatures in England were about 1–2 °C higher than today. Cooling then set in towards the period 1550–1850 AD, which Lamb referred to as the Little Ice Age (a term first introduced by Matthes in 1939, used to describe an '*epoch of renewed but moderate glaciation which followed the warmest part of the Holocene*'[5,6]) – a period whose cold winters were immortalised by Pieter Brueghel the Elder in his February 1565 painting 'Hunters in the Snow'. Lamb's research suggested that the decline from the 1300s onwards was not uniform and that there was a partial recovery in the period 1440–1500 AD. A warming at this time seems slightly odd in retrospect, because it coincides

with the Spörer sunspot minimum (1450–1550 AD). We will look at the effects of sunspots in more detail later. Although Lamb was attempting to build a global picture, his focus at the time was Eurocentric, not least because that's where he got most of his data.

By 1961, in a paper delivered in Rome to the WMO/UNESCO Symposium on Changes of Climate, Lamb expanded on his chronicle by incorporating climatic data from other parts of the world[2]. By this time, he had concluded that – compared with present values (meaning 1960) – temperatures were raised 2–3 °C in the warm epochs of the Holocene and lowered 1–2 °C in the Little Ice Age. According to Lamb, the Little Ice Age ranged from 1430 to 1850 AD and was coldest from 1550 to 1700 AD. He associated the cooling with weakening of both atmospheric circulation and incident radiation, and thought that the latter might have been caused by a veil of dust from volcanic activity.

Following the Holocene Climatic Optimum, glaciers began to readvance. George Denton of the University of Maine referred to this new period of glacial advance (the Little Ice Age of Matthes) as the 'Neoglaciation'[7,8]. It incorporates the narrower range of Lamb's Little Ice Age.

Lamb referred to his secondary Holocene climatic optimum (1000–1200 AD) as the 'Early Middle Ages Warm Epoch'. Elsewhere in *The Changing Climate*, he called it the 'Little Optimum'. Later, in 1965, he referred to it as '*The early medieval warm epoch*'[9]. It is now commonly termed the Medieval Warm Period, Medieval Climatic Optimum or Medieval Climate Anomaly. We will look at the Little Ice Age and Medieval Warm Period in more detail in Chapter 15. In passing, we should note that, aside from suffering from a paucity of records compared with what was available later, Lamb also had a lower-resolution chronology and fewer climate proxies to work with.

Moving forward in time, the success of CLIMAP in mapping the climate of the Last Glacial Maximum (see Chapter 12) stimulated the formation of a successor project, COHMAP, to look at the climate of the Holocene. Starting out as 'Climates of the Holocene – Mapping Based on Pollen Data', COHMAP quickly became global both in its scope and in the climate proxies it considered. Renamed the 'Cooperative Holocene Mapping Project', it ran from 1977 to 1995. COHMAP was a turning point for the palaeoclimate community studying the Holocene. It tied Holocene models and proxy data together for the first time, and stimulated international collaboration in Holocene research[1]. Key organisers were John Kutzbach, whom we met in Chapter 6, and his colleagues Tom Webb and Herb Wright. They aimed to use the Community Climate Model, an atmospheric general circulation model (GCM) of the US National Center for Atmospheric Research (NCAR), to simulate past climates in 3000-year-long time slices centred on 18, 15, 12, 9, 6, 3 and 0 Ka ago and to compare the outputs with palaeoclimate maps based on data from pollen, lake levels, pack-rat middens and marine plankton[10, 11]. These simulations ignored events of short duration, like Bond cycles.

André Berger's astronomical calculations showed that at the start of the Holocene around 10 Ka ago, Northern Hemisphere summers (in June at 65° N) were warm, while Southern Hemisphere summers (in December at 65° S) were cool[12]. Over time, summer insolation in the Northern Hemisphere fell gradually by 45 W/m^2 to present levels, while that in the Southern Hemisphere rose by 30 W/m^2. Winters in both hemispheres were cool at the start of the Holocene (December at 65° N and June at 65° S), then gradually warmed very slightly, by 5 W/m^2 in the Northern Hemisphere and by 3 W/m^2 in the Southern (Figure 14.1)[13].

The Northern Hemisphere now has cooler summers and warmer winters than the Southern Hemisphere. Matters are a little more complicated than these data might suggest, however. We can draw on the work of Huybers and Denton[14] (see Chapter 13) to infer that, although insolation increased during the Holocene in the Southern Hemisphere summer, it decreased during the Southern Hemisphere spring. Also, when summer insolation was high in the Southern Hemisphere, the length of summer was shorter and the length of winter was longer than average, because the Earth was then at perihelion (close to the sun). These factors conspired to cool the Earth throughout the Holocene. If insolation was cooling the Earth, how do we explain the mid Holocene Climatic Optimum? As we shall see in more detail further on, it was a Northern Hemisphere phenomenon that arose because the melting of the great ice sheets at the beginning of the Holocene used up all the available solar energy, keeping the region cool until the ice sheets had gone.

Berger and Loutre's astronomical calculations of the influence of the other planets on the Earth's orbit and axial tilt enable us to determine insolation well into the future[15]. Their calculations show that the Earth should have cooled from 10 000 years ago to the present, and will remain cool for the next 5000 years. We can confidently predict that the present low level of summer insolation at 65° N, which has persisted for the past 2000 years, will remain more or less unchanged for at least another 1000 years.

Berger and Loutre's solar radiation parameters were entered into the COHMAP model as an external forcing variable. Other boundary conditions within the model provided internal climate forcing through changes in aerosol loading (dust), ice volume, sea surface temperature and atmospheric CO_2. These were not allowed to change between 18 and 15 Ka ago. After 15 Ka ago, dust was decreased to around present levels by about 12 Ka ago, and ice volume to around present levels by 9 Ka ago (there was still some ice on North America and Scandinavia at that time). Sea surface temperature increased to present levels and CO_2 to preindustrial levels between 15 and 9 Ka ago, with CO_2 increasing again in modern times[10].

COHMAP's results, which started appearing in 1988, were stunning[10]. The combination of climate data and models enabled palaeoclimatologists to understand for the first time what caused global climate changes to occur at different times between the last glacial and the present, and how these changes were distributed across different regions. Previously, the role of the Earth's orbital variations had been discounted, the full impact of changes in ice sheets had not been understood and climate models had not been well enough developed to test the influence of these and other factors.

Changes in the precession and tilt cycles made seasonality increase in the Northern Hemisphere and decrease in the Southern Hemisphere between the 15 and 9 Ka-ago time slices. By 9 Ka ago, the Northern Hemisphere received 8% more solar radiation in July and 8% less in January than today. After this, these extremes diminished towards modern values. These changes increased the thermal contrast between land and ocean, causing strong monsoons between 12 and 6 Ka ago in the northern tropics and subtropics, particularly in Africa and Asia, and raising lake levels in regions that are arid today, such as the Sahara. Crocodiles and hippos expanded into the Saharan region, and people settled on the lake shores. At the same time, summers became warm and dry in the northern interiors.

After 6 Ka ago, as July insolation continued to decrease, temperatures fell over the land, monsoon rains weakened, deserts expanded and the present climate regime developed. In the southern tropics, the changed insolation (more in July, less in January) had the opposite effect, producing decreased seasonality and less intense rains in tropical South America, southern Africa and Australia.

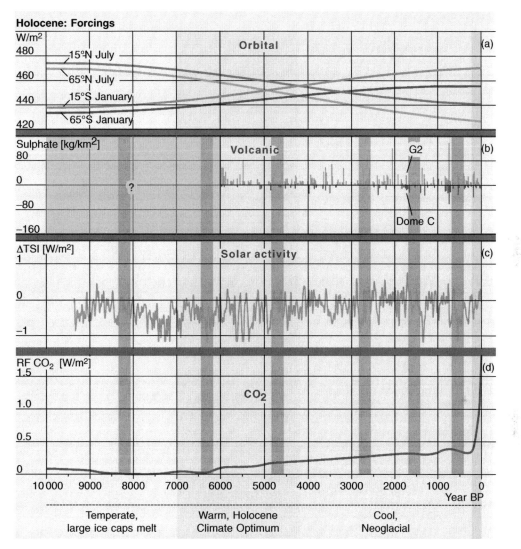

Figure 14.1 *Main forcings during the Holocene. (a) Solar insolation due to orbital changes from two specific sites in the Northern and Southern Hemispheres during the corresponding summer. (b) Volcanic forcing during the past 6 Ka, depicted by the sulphate concentration of two ice cores form Greenland (above the line) and Antarctica (below the line). (c) Solar activity fluctuations based on ^{10}Be measurements in polar ice. (d) Forcing due to rising CO_2 concentrations. The six vertical bars show the timing but not the duration of six cold periods.*

    The initial COHMAP model suggested that, during the glacial period (the 18 and 15 Ka-ago time slices), the large North American ice sheet split the westerly winter jet stream over North America into a northerly branch, running along the Arctic shore then down the Labrador

Sea between Canada and Greenland, and a southerly branch, running over the American southwest and joining the northerly branch over the mid Atlantic, before heading for Europe. A later, more advanced version of the model, along with a mixed-layer ocean model and an ice sheet reconstruction that gave a lower height to the Laurentide Ice Sheet, found that the ice sheet did not split the jet stream at 18 Ka ago[16]. Apart from that, the results seemed robust, with good coherence between the outputs from the earlier and later versions of the model and between both models and the palaeoclimate data.

The COHMAP data and models showed that there is no simple description of Holocene climate. Orbital change forced climate changes that varied across the globe, not only in magnitude, but also in space and time. Patterns of variation in both precipitation and seasonality changed along with temperature. Lamb's Europe was not globally representative.

As a result of the increase in Northern Hemisphere insolation, which peaked around 10 Ka ago (Figure 14.1), and the subsequent absorption of heat to melt the Northern Hemisphere ice sheets, temperatures remained cool in Europe until about 9 Ka ago, when the combination of moderately high insolation and much reduced ice sheets created the Holocene Climatic Optimum between 9 and 5 Ka ago; this is also known as the 'Holocene Thermal Maximum' and the 'hypsithermal'. In the Arctic, at 70° N, June insolation 11 Ka ago was about 45 W/m² greater than today. By 4 Ka ago, it had dropped to about 15 W/m² greater than today[17]. During the optimum, temperatures were up to 4 °C above later Holocene levels there, but while northwest Europe warmed, southern Europe cooled, and there was little or no change in the tropics. This so-called 'event' was not globally uniform in either magnitude or timing.

Overall, the decrease in insolation from 10 Ka ago to the present cooled the Northern Hemisphere significantly. For example, sea surface temperatures off Cap Blanc, Mauretania, declined by between 4 and 6 °C. There, the input of dust rose abruptly at around 5.5 Ka ago, when the African Humid Period came to an abrupt end[18, 19]. This is an example of a 'tipping point': an abrupt response emerging from a gradual change. The former existence around 6 Ka ago of a wide belt of enlarged lakes occupying the tropics between 32° N and 18° S implied a northward shift of the equatorial rain belt, as well as greatly enhanced monsoonal transport of moisture into the tropical continents at that time[20].

As more data poured in from tree rings, corals, ice cores, stalactites and sediment cores from lakes and oceans, the Holocene climate record was further refined. Richard Alley was among the first to assess Holocene climate change from Greenland ice cores[21]. His calculations of past temperature change there were based on '*site-specific*

calibrations using ice-isotopic ratios, borehole temperatures, and gas-isotopic ratios' and were modified from earlier data provided by other researchers[22]. They show some warm periods alternating with cold periods at the start of the Holocene between 10 and 7 Ka ago, rather lower values between 7 and 5 Ka ago, moderately high values between 3.5 and 2.0 Ka ago, a steep decline to around 1850 AD and then a gradual increase leading into the modern warm period. It was not obvious from Alley's data[21, 22] that Greenland had experienced the gradual decline in temperature throughout the Holocene expected from the decline in Northern Hemisphere insolation[15].

This begs the question, just how valid was Alley's Greenland temperature record? The question was addressed by Bo Vinther of the Centre for Ice and Climate at the Niels Bohr Institute of the University of Copenhagen[23]. Examining ice cores from Greenland and Arctic Canada in 2009, Vinther's team saw that both altitude and past thinning caused by warming shaped the $\partial^{18}O$ record that Alley had used for his temperature reconstruction. '*Contrary to the earlier interpretation of $\partial^{18}O$ evidence from ice cores*', Vinther said, '*our new temperature history reveals a pronounced Holocene climatic optimum in Greenland coinciding with maximum thinning near the GIS [Greenland Ice Sheet] margins*'[23]. His new record of Greenland's temperature history (Figure 14.2) shows a steep warming of around 6 °C between 12 and 10 Ka ago, a slight rise of a further 0.5 °C to a peak around 8 Ka ago, then a steady fall of around 2.5 °C to between 1600 and 1850 AD, followed by the rise into the modern era that Alley had also seen. Evidently, the broad pattern of rise and fall in temperature that Vinther found in Greenland ice[23], which Alley had missed in examining a more limited Greenland data set[21], did broadly follow the pattern of Northern Hemisphere summer insolation.

Incidentally, Vinther's data show that the Greenland Ice Sheet responds much more rapidly to warming than had formerly been supposed, making it '*entirely possible that a future temperature increase of a few degrees Celsius in Greenland will result in GIS mass loss and contribution to sea level change larger than previously projected*'[23].

The existence of an early to mid Holocene thermal maximum is well documented in proxy records of sea surface temperature from the high-latitude North Atlantic and the Nordic Seas, where polar amplification is clear[24]. The sea surface temperature maximum there was forced by the summer insolation maximum, and is not simply a result of advection of sea surface temperature anomalies from further south. Superimposed on the sea surface

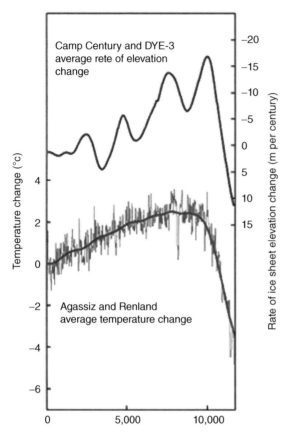

Figure 14.2 Greenland temperatures adjusted for changing elevation. Temperature change in Greenland derived from Agassiz and Renland ^{18}O records and average rate of elevation change of ice sheets at the DYE-3 and Camp Century drill sites. Elevation changes distort the temperature records for these drill sites.

temperature trends is variability at the century or millennial scale, the amplitude of which increased after the mid Holocene. There is very little persistence in the main frequencies of variability through the Holocene. The increased climatic variability affected primarily wintertime conditions, such as sea ice, and was consistent with the onset of Neoglacial conditions in Europe. According to Eystein Jansen and his team, 'Increased sea ice cover following the reduced summer insolation may have put in place amplification mechanisms leading to stronger ocean temperature variability ... In the absence

of a clear attribution of this variability to external forcings (e.g. solar, volcanic ...), it appears most likely that the century to millennial scale variability is primarily caused by the long time-scale internal dynamics of the climate system'[24].

Polar amplification can come about for a variety of reasons, not least of which is reduction in sea ice, land ice and snow cover, which reduces albedo and allows the exposed ocean to absorb heat, which is then given up to the atmosphere, causing more sea ice to be lost and so on, through positive feedback. Gifford Miller of the Institute of Arctic and Alpine Research of the University of Colorado, Boulder explored this phenomenon in some detail in the Arctic in 2010[25]. Miller and his team found that Arctic temperature change exceeded the Northern Hemisphere average by a factor of 3 to 4. For example, warming compared to today during the Holocene Thermal Maximum was $+1.7 \pm 0.8\,°C$ (Arctic), $+0.5 \pm 0.3\,°C$ (Northern Hemisphere) and $0 \pm 0.5\,°C$ (global). During the last interglacial, it was $+5 \pm 1\,°C$ (Arctic) and $+1 \pm 1\,°C$ (global and Northern Hemisphere). And during the mid Pliocene it was $+12 \pm 3\,°C$ (Arctic) and $+4 \pm 2\,°C$ (global and Northern Hemisphere)[25].

A comprehensive survey of Holocene climate variability, based on some 50 globally distributed palaeoclimate records from a wide range of environments extending from Greenland to Antarctica, was carried out in 2004 by a team led by Paul Mayewski of the Climate Change Institute of the University of Maine at Orono[26]. In recognition of his extensive research on climate change in Greenland and Antarctica, Mayewski was awarded the Seligman Crystal of the International Glaciological Society in 2009 and the science medal of the Scientific Committee on Antarctic Research in 2006. He has an Antarctic peak named after him.

Mayewski's team found clear signs of cooling in line with the decline in summer insolation in the Northern Hemisphere from about 7400 years ago to the present[26]. Norwegian glaciers advanced, the Swedish tree line moved downwards and temperatures (from $\partial^{18}O$) fell in Soreq Cave, Israel. Africa's Lake Victoria and Ethiopia's Lake Ahhe began shrinking, and the Trade Winds began to weaken over Venezuela's Cariaco Basin. Peru's Huascaran ice cap and Antarctica's Taylor Dome began cooling, as did sea surface temperatures in Namibia's Benguela Current.

In 2006, Stephan Lorenz, then at the Max-Planck-Institut für Meteorologie, Hamburg, used alkenone data and a coupled ocean–atmosphere circulation model forced by orbital changes to examine, with colleagues, global patterns of sea surface temperature for the past 7000 years[27]. As in COHMAP, the patterns proved to be heterogeneous. While the higher latitudes cooled over time, the tropics warmed slightly. In the North Atlantic region, many aspects of the regional climate are dictated by the behaviour of the North Atlantic Oscillation, measured from the difference in air pressure between the high-latitude Iceland low-pressure centre and the low-latitude Azores high-pressure centre. When the pressure difference is high, the North Atlantic Oscillation is positive, westerly winds are strong, Europe has cool summers and mild, wet winters and the Mediterranean area is dry and cool. This difference has decreased towards the present, making the North Atlantic Oscillation more negative, which has weakened the westerlies, making European winters colder and dryer and the Mediterranean warmer and wetter. This trend is consistent with the observation that the Little Ice Age late in the Holocene was characterised in Europe by very cold winters (more on this in Chapter 15). The trend in the North Atlantic Oscillation operated *against* the trend in orbital insolation, which *increased* in winter in the Northern Hemisphere – a good example of how local effects can override Milankovitch variations. Overall, summer insolation in the Northern Hemisphere over the past 7000 years decreased by more than 30 W/m² at middle and northern latitudes, while winter insolation increased by about 25 W/m² at low latitudes. Lorenz's study noted that '**Northern Hemisphere summer cooling during the Holocene** [meaning the last 7000 years] **is of the same order of magnitude as the warming trend over the last 100 years**'[27] (my emphasis).

Several more extensive studies of Holocene climate change were later carried out by Heinz Wanner (1945–), the former director of the Oeschger Centre for Climate Change Research at the University of Bern in Switzerland, which was named after the ice core expert Hans Oeschger. Among other things, in 2011 Wanner and his colleagues used high-resolution records of Holocene climate change to produce a Holocene Climate Atlas containing 100 anomaly maps representing 100-year averages of climatic conditions for the last 10 Ka[28]. In recognition of his contributions to climate change research, he was awarded the Vautrin Lud Prize – regarded unofficially as the Nobel Prize in Geography – in 2006.

In 2008, Wanner and his colleagues published a comprehensive review of mid to late Holocene climate change, spanning the last 6000 years[29]. They chose this period because '*the boundary conditions of the climate system did not change dramatically*' during it[29]. The large continental Northern Hemisphere ice sheets had melted, and there were no large outflows of freshwater from melting ice sheets, nor any major rises in sea level. Plus, there were abundant, detailed regional palaeoclimatic proxy records. The team mapped the distribution of temperature and precipitation through time, supplementing the data with results from GCMs and Earth system models of intermediate complexity (EMICs) fed with data on the agents forcing climate change: orbital variations, solar variations, large volcanic eruptions and changes in land cover and greenhouse gases. The goal was '*to establish a comprehensive explanatory framework for climate changes from the Mid-Holocene (MH) to pre-industrial time*'[29]. One of their sources of information was the international Paleoclimate Modeling Intercomparison Project (PMIP), which started in the early 1990s in an effort to improve palaeoclimate models. Another source was the IGBP's Paleovegetation Mapping Project (BIOME 6000), which provides a global data set derived from pollen and plant fossils for use as a benchmark against which to test the outputs from palaeoclimate models.

In 2013, Wanner was part of another team, which produced a detailed analysis of the global climate of the past 2000 years[30] as a contribution to the IGBP's Past Global Changes (PAGES) programme, a successor to COHMAP. One of the PAGES subgroups, the '2k Network', aims to produce a global array of regional climate reconstructions for the past 2000 years. It coordinates with the NOAA World Data Center for Paleoclimatology to maintain a benchmark database of proxy climate records for that period[30]. By 2013, the PAGES 2k data set included 511 time series of tree rings, pollen, corals, lake and marine sediments, glacier ice, speleothems and historical documents recording changes in processes sensitive to variations in temperature. Resolution is annual, enabling the team to examine multidecadal variability by focusing on 30-year mean temperatures.

Wanner's various studies, made at much higher resolution than the COHMAP ones, confirmed that decreasing solar insolation in the Northern Hemisphere summer led not only to Northern Hemisphere cooling but also to a southward shift of the summer position of the Inter-Tropical Convergence Zone (ITCZ) and a weakening of the Northern Hemisphere summer monsoon systems in Africa and Asia, associated with increasing dryness and desertification[13, 29]. The southward shift in the ITCZ and the weakening of the monsoons came about as a result of the interplay between decreasing insolation in the

Northern Hemisphere summer and increasing insolation in the Southern Hemisphere summer. Insolation in the Northern Hemisphere declined by more than insolation in the Southern Hemisphere increased (Figure 14.1), so the cooling associated with the former had a wider effect on tropical systems than did the warming associated with the latter.

The cooling of the Northern Hemisphere summer with time increased the activity of the El Niño–La Niña system in the Pacific up to around 1300 AD, since when that activity has fluctuated significantly. The cooling also led to development of an increasingly negative North Atlantic Oscillation between 6 and 2 Ka ago, followed by a weak reversal[29]. As mentioned earlier, the negative phase of the North Atlantic Oscillation is associated with colder winters over Europe and a warmer, drier climate over the Mediterranean. Sea surface temperatures declined in the North Atlantic and the Norwegian Sea, along with southward retreat of the Arctic tree line, implying declining summer temperatures[29]. Glaciers advanced across the Arctic, in the Alps and in the Western Cordillera of North America, and decreased in the Western Cordillera of South America. In the Southern Hemisphere, lake levels were low from 6.0 to 4.5 Ka ago, but increased towards the present – the opposite of the trend in the Northern Hemisphere subtropics. This seems to reflect an intensification of northward-migrating westerlies, consistent with an increase in upwelling along the Pacific coast of South America[29].

The spectacular decrease of vegetation in the Sahara between 6 and 4 Ka ago was '*related to a positive atmosphere-vegetation feedback, triggered by comparatively slow changes in orbital forcing*'[29]. As Wanner explained, '*Due to a decrease in the intensity of the African monsoon, related to the decrease in summer insolation, precipitation decreases in the Sahara during the Holocene. This induces a decrease in the vegetation cover, and thus an additional cooling [caused by an increase in albedo, or reflectivity] and reduction of precipitation that amplifies the initial decrease in vegetation cover. The amplification is particularly strong when a threshold is crossed, leading to a rapid desertification and ... fast changes*'[29].

In marine records, the coherent long-term cooling of between 1 and 2 °C over the past 9–10 Ka was not confined to the North Atlantic Ocean. It also extended to the Mediterranean[31–33], which is consistent with glacial readvance in Iceland, with Greenland ice cores and with pollen data in Europe and North America that indicate

southward migration of cool spruce forest[34]. Tim Herbert pointed out[34] that this appeared to be a regional rather than global pattern, because alkenone data from the Indian Ocean, South China Sea and western tropical Atlantic showed very slight warming from the early Holocene to the present[35], while data from the western margin of North America showed no trend at all during last 9 Ka, apart from minor millennial oscillations of about 1 °C.

The PAGES 2k team showed that the long-term continental cooling driven by the fall in insolation continued right through the past 2000 years, during which '*all regions experienced a long-term cooling trend followed by recent warming during the 20th century*'[30]. I noticed one important caveat: while Antarctica showed the same cooling as the other continents, it did not – in the data available to the PAGES group – show the recent warming of the 20th century[30]. This perception is clearly wrong, as subsequent data show, because some parts of Antarctica – in particular, the Antarctic Peninsula, but also, to a lesser extent, West Antarctica – do show this recent warming[36]. It is evident in an ice core from the ice cap of James Ross Island in the western Weddell Sea, for example (Figure 14.3)[37, 38], and at an Ocean Drilling Programme site (1089) in the Palmer Deep just south of Anvers Island[39, 40]. More on that in Chapter 15.

Studying the Antarctic changes, Amelia Shevenell and her team found that the insolation in the southern *spring* at 65° S (September–November) *decreased* in parallel with the decrease in temperature seen in cores from the Ross Sea and the Palmer Deep[39]. This suggests local orbital control. As we saw in Chapter 13, there are ample reasons for invoking local orbital control on Antarctic climate rather than forcing by the orbital patterns of the Northern Hemisphere, where summer insolation at 65° N followed that same decrease[41]. This local insolation influenced the westerlies near the Antarctic Peninsula, which play a key role in warming and cooling locally, as well as in regulating the emission of CO_2 from the Southern Ocean and its subduction in Antarctic Bottom Water or Intermediate Water[39].

Based on the declining insolation in the north and its effects, Wanner subdivided the Holocene into three periods[13]. First was an early deglaciation phase, between 11.7 and 7.0 Ka ago, characterised by high summer insolation in the Northern Hemisphere, a cool, temperate climate near the melting ice sheets in North America and Eurasia and strong monsoonal activity in Africa and Asia. Second came the Holocene Thermal Optimum, between 7.0 and 4.2 Ka ago, which had high summer temperatures in

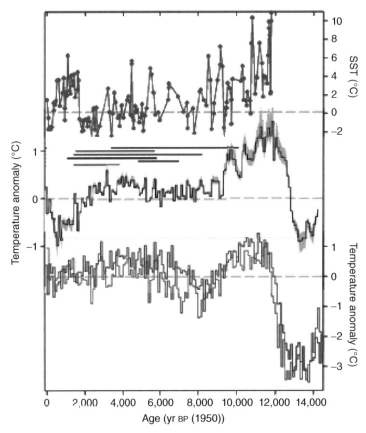

Figure 14.3   *Holocene temperature history of the Antarctic Peninsula. Top panel: Sea surface temperature (SST) reconstruction from off the shore of the western Antarctic Peninsula. Middle panel: James Ross Island ice-core temperature reconstruction relative to the 1961–1990 mean, in 100-year averages, with grey shading indicating the standard error of the calibration. Lower panel: Temperature reconstructions from the Dome C (uppermost) and Dronning Maud Land (lowermost) ice cores from East Antarctica. Horizontal bars show intervals when open water was present in the area of the Prince Gustav Channel.*

mid- and high-latitude parts of the Northern Hemisphere and active but weakening monsoonal systems at low latitudes. Third, there was a neoglacial period, with falling summer temperatures in the Northern Hemisphere, terminating with the sharp rise in global temperature as we entered the modern era.

In 2013, Shaun Marcott of Oregon State University and colleagues presented a slightly different reconstruction of global temperature change for the Holocene[42]. Most of their data came from marine cores, but they tell basically the same story as that of Wanner and his colleagues. Combining their data into a single global temperature stack, Marcott and colleagues found a rise in temperature

of around 0.6 °C between 11.3 and 10.0 Ka, with Early Holocene warmth remaining more or less stable between 10 and 7 Ka ago. This stable period was followed by 0.7 °C of cooling, largely biased by a 2 °C cooling of the North Atlantic, which culminated in the coldest temperatures of the Holocene in the Little Ice Age about 200 years ago. After that there was a sudden rise of 1 °C to the warm temperatures of the modern era. The global stack was most similar to the data from the Northern Hemisphere between 30 and 90° N, suggesting a primary control by Northern Hemisphere summer insolation. The tropics (30° N to 30° S) warmed by 0.3 °C from 11.3 to 5.0 Ka ago, then cooled by a similar amount up to around 250 years ago, before warming sharply to modern values. Southern Hemisphere temperatures varied more. The narrowly defined mid Holocene thermal optimum between 6 to 7 and 4 Ka ago in the Northern Hemisphere identified by Lamb and Wanner was not readily evident in Marcott's global data.

Marcott's data show that the Northern Hemisphere signal of climate change tends to dominate the Holocene global signal, as it does today[49]. In part, the predominance of this signal is due to the greater area of land in the Northern Hemisphere. Land tends to heat and cool much faster than water, and water dominates the Southern Hemisphere. The global cooling had an impact on the North Atlantic Current, which takes heat to western Europe. Less heat was transported northwards, helping Europe to cool[43].

In the Arctic, Gifford Miller's team found that sea ice first decreased during the early Holocene as insolation increased, then increased in the late Holocene as insolation declined[17]. Likewise, the tree line expanded northwards to as much as 200 km beyond its current position, before beginning to retreat southward starting 3–4 Ka ago. Permafrost followed much the same pattern, melting south of the Arctic Circle, then refreezing after 3 Ka ago. Summer temperature anomalies along the northern margins of Eurasia in the thermal optimum ranged from 1 to 3 °C above today's, and sea surface temperatures were up to 5 °C warmer than today's. As elsewhere, cooling began between 6 and 3 Ka ago. Most Arctic mountain glaciers and ice caps expanded during the 'neoglaciation' of the late Holocene, as did the Greenland Ice Sheet. This cooling trend ended in most places in the mid 19th century[17, 44, 45].

Much the same broad-brush picture of sea ice was true of the Antarctic (Figure 14.4). There, Xavier Crosta found evidence for minimal distribution of sea ice as Holocene temperatures rose in the summer months between 9 and 4 Ka ago, followed by a substantial expansion as temperatures fell over the next 3000 years[46].

The results of Mayweski[26], Wanner[13] and Marcott[42] confirm that the continued existence of ice sheets in the Northern Hemisphere until around 7 Ka ago prevented temperatures from rising far despite high insolation between 10 and 7 Ka ago. The disappearance of the great ice sheets, combined with moderately high insolation, created the Holocene Thermal Optimum between 7.0 and 4.2 Ka ago. Then neoglacial conditions set in, with glacier advances across the Northern Hemisphere and the reformation of ice shelves along the northern coast of Ellesmere Island in northern Canada[47]. The most recent

advance of glaciers, in the Little Ice Age, was the most extensive in all areas, making it the coldest episode of the Holocene[47]. The cooling of the North Atlantic after the thermal optimum may represent weakening of the Thermohaline Conveyor system, associated with the increase in Northern Hemisphere winter insolation[43]. Volcanic activity continued at variable levels throughout the Holocene, with no significant trend, making forcing from this source an unlikely driver of the overall cooling in the climate system (Figure 14.1)[13, 26, 42].

The pattern of climate change leading into the Late Holocene neoglacial and the Little Ice Age reflects the pattern of insolation calculated by André Berger and Marie-France Loutre[15, 48]. Insolation peaked in the Northern Hemisphere 11 Ka ago, then declined to the present, with a substantial flattening of the rate of decline over the past 1000 years or so. Insolation is calculated to stay just as low for at least another 1000 years[15, 48]. Taken together, the astronomical and the palaeoclimate data suggest that the world should have continued to cool, so it is more than a little surprising that '*in the brief interval of less than two centuries, the Northern Hemisphere (at least) has experienced the warmest and the coldest extremes of the late Holocene*'[47]. Recent warming bucks the trend imposed by orbital forcing, which should have kept our climate in a cool neoglacial state. We will explore this issue further in Chapter 15.

14.2 The Role of Greenhouse Gases: Carbon Dioxide and Methane

Did changes in CO_2 have a significant role to play in driving the cooling seen in the Holocene? Eric Sundquist of the US Geological Survey at Woods Hole reminded us that measurements of fossil atmospheric CO_2 from the Taylor Dome ice core, from near McMurdo Sound in the Ross Sea, showed that CO_2 declined from around 270 ppm at 10.5 Ka ago to 260 ppm at 8 Ka ago, and then increased to values near 285 ppm by 1000 AD (see Figure 14.1)[49]. Much the same picture emerged from ice cores at Law Dome in the Australian sector of Antarctica and at the Russian Vostok station on the Polar Plateau[50]. Andreas Indermühle of the University of Bern and colleagues explained the initial decrease in CO_2 as being due to the expansion of vegetation into formerly glaciated areas in the Northern Hemisphere, and the subsequent increase as due to a gradual release of carbon caused by global cooling and drying, the cooling being a response to the

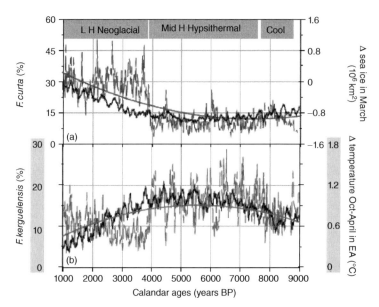

Figure 14.4 *Sea ice area around Antarctica. Relative abundances of (a) the* Fragilariopsis curta *group (ragged grey line) and (b) the* Fragilariopsis kerguelensis *group (ragged grey line) versus calendar ages in core MD03-2601 close to the coast of Adelie Land, Antarctica. F. curta is a sea ice diatom, whose abundance represents denser and longer-lasting coverage of the ocean by sea ice. F. kerguelensis, in contrast, is most abundant along the polar front, where there is less sea ice and temperatures are warmer. Polynomial regressions indicate the first-order evolution for the F. curta group (solid smooth line in (a)) and the F. kerguelensis group (solid smooth line in (b)), compared to simulated time series of the March sea ice area (black zigzag line in (a)) and the October–April temperature for East Antarctica over the last 9000 years (black zigzag line in (b)), plotted as deviation from the preindustrial mean. The data show warming from a cool early Holocene into a warm mid Holocene hypsithermal with low sea ice, then into a cool late Holocene neoglacial with abundant sea ice.*

orbitally controlled decline in insolation[50]. Glen Macdonald of the University of California at Los Angeles and colleagues found in 2006 that circum-Arctic peatlands began to develop rapidly as ice and snow began to disappear 16.6 Ka ago and expanded between 12 and 8 Ka ago, when insolation was at its highest, drawing down CO_2 during the early Holocene[51]. In a moment we'll see what Bill Ruddiman thought of this CO_2 picture.

Methane in ice cores from Taylor Dome and from Dome C changed over the same period, but in a different way, as shown by Jérôme Chappelaz of the Laboratoire de Glaciologie et Géophysique de l'Environnement of the University of Grenoble and colleagues in 1997. CH_4 declined from values near 700 ppb at 10.5 Ka ago to values between 550 and 600 ppb at 5 Ka ago, before increasing to near 700 ppb again by 1000 AD[52]. Many scientists attributed the late Holocene increase in CH_4 values to

an expansion of boreal wetlands. Not surprisingly, as most wetlands are found in the Northern Hemisphere, the Holocene data showed a gradient in CH_4 from north to south, with between 30 and 50 ppb more in air from the north. The addition of recent anthropogenic sources of CH_4 in the north increased this gradient by a factor of three[49]. The CH_4 data showed a sudden short-term drop at 8 Ka ago, which corresponded to the widespread climatic event linked to a major flood of freshwater into the North Atlantic from a melting event in the Laurentide Ice Sheet.

The rise in atmospheric CO_2 values from 260 ppm around 8 Ka ago to 285 ppm 200–400 years ago (Figure 14.1) might be expected to have led to a small warming. Instead, cooling continued, showing that the forcing by declining Northern Hemisphere insolation had a much greater effect on climate than did forcing by the 25 ppm addition of CO_2. Much the same argument applies to the rise in methane, which increased by about 160 ppb towards the preindustrial era[26] – mostly over the past 3 Ka, despite continued cooling[29]. It is worth recalling at this point the divergence of temperature and CO_2 typical of the re-entry into glacial conditions from the peaks of past interglacials that we saw in Chapter 13, with temperature falling faster than CO_2. As in those cases, the lack of correlation between CO_2 or CH_4 and temperature between the early and late Holocene is likely to have much to do not only with the decrease in insolation *per se*, which drives temperature, but also with its effect on albedo, through the growth in sea ice, land ice and snow cover into the neoglacial. This temporary lack of correlation changed as we moved into the modern era, when massive increases in the two gases caused a sharp increase in temperatures unprecedented in the previous millennium[47], as we see in more detail in Chapter 15.

This seems like a good point to review the ideas of Bill Ruddiman concerning the effect humans may have had on climate since the beginning of the Holocene[53]. But we should first recall Berger's calculations suggesting that the Holocene interglacial may last 30 Ka or more – like marine isotope stage (MIS) 11, which occurred some 400 Ka ago (see Chapter 12)[54,55]. Furthermore, Michael Crucifix reminds us that EMICs[a] show that when the CO_2 was maintained above 240 ppm in the Holocene, the system remained in an interglacial state[56]. This, he says, '*further shows that even if CO_2 concentration decreased in the future down to glacial levels, glaciation will not occur before 50,000 years*'[56], because insolation does not get low enough. In contrast, Bill Ruddiman argued that the fall in insolation during the Holocene should have led to the formation of new ice sheets by now[57]. He suggested that '*CO_2 and CH_4 concentrations should [also] have fallen steadily from 11,000 years ago until now*'[57] Indeed, they did start to fall from the beginning of the Holocene, but then '*CO_2 and CH_4 began anomalous increases at 8000 and 5000 years ago, respectively*'[57]. He thought that these rises were most probably due to human activities, and, further, that those increases had prevented the occurrence of a new glaciation. He attributed the slow, small rise of CO_2 over the past 8 Ka or so to forest clearance and the development of agriculture[57,58].

Birks reminded us that there is some independent support for this idea, in that the distribution of charcoal, an index of the existence of fires, parallels the increase in CO_2 concentrations, suggesting that biomass burning may have

been a cause for the rise in CO_2[1]. Nevertheless, 'mainstream' thinking suggests that the rise most probably had several causes, including changes in calcite compensation in the ocean, changes in sea surface temperature and the postglacial build-up of coral reefs, and was not caused by major changes in the storage of carbon on land. More data are needed before we have a definitive solution[29]. For example, Broecker argued in 2006 that the latest measurements of $\partial^{13}C$ in CO_2 from Antarctic ice suggest that the main changes in atmospheric CO_2 over this period were the result of changes in the world's oceans[59].

Reviewing the various arguments for and against Ruddiman's hypothesis in 2009, Michael Crucifix concluded, '*the early anthropogenic theory implies – if it is correct – that there was a bifurcation point during the past 6000 years during which the climate system hesitated before opting for a glacial inception or staying interglacial. The anthropogenic perturbation gave it the necessary kick to opt for a long interglacial*'[56]. While that may sound satisfying, he went on to remind us that '*This hypothesis cannot be easily proved or disproved*'[56].

This did not stop people from trying. As we saw in Chapter 13, the discovery in 2007 that the ratio of boron to calcium (B/Ca) in species of benthic foraminifera is directly related to the concentration of carbonate ions (CO_3^{2-}) in bottom water has greatly improved our understanding of the behaviour of the carbonate system in the ocean[60,61]. Jimin Yu and colleagues used the B/Ca ratios in deep-sea cores to show that the concentration of CO_3^{2-} in the deep waters of the Pacific and Indian Oceans declined during the Holocene[61]. They thought it likely that the build-up of coral reefs during the Holocene caused the whole ocean concentration of CO_3^{2-} to decline. This decline reduced the ocean's alkalinity, causing the solubility of CO_2 in the ocean to decline, so contributing to the 20 ppm rise in atmospheric CO_2 over the past 8 Ka. Much the same conclusion is reached by numerical modelling[62]. Ruddiman may thus be wrong about early humans having increased CO_2 in the air.

Having carefully examined Ruddiman's hypothesis, Wally Broecker and Thomas Stocker reasoned that, as the amount of CO_2 in the air had increased, so too must the amount of CO_2 in the ocean. In other words, much more CO_2 had been emitted than could be accounted for by deforestation alone[63]. Instead, they proposed that the CO_2 rise was triggered by the ocean's response to the extraction of CO_2 from the ocean to create the early Holocene increase in forest cover. Removing that CO_2 from the ocean lowered early Holocene CO_2 in the air

---

[a] EMICs = Earth Models of Intermediate Complexity

and increased the carbonate ion concentration of ocean water, hence deepening the carbonate compensation depth (CCD) and causing calcium carbonate to accumulate. That drew down the carbonate ion concentration, making the CCD rise and increasing the CO_2 content of ocean water, which fed back to increase the CO_2 in the air. This is a story of natural ocean chemistry, not anthropogenic deforestation.

Ruddiman attributed the small rise in CH_4 that began about 5 Ka ago to the increasing culture of rice in Asian wetlands[57,58]. Although Wanner thought in 2008 that the cause still *'eludes a simple explanation'*[29], a numerical simulation of the climate system by Joy Singarayer, Paul Valdes and colleagues in 2011 found that the rise in methane was most likely the result of a natural expansion in wetlands[64]. That, in turn, was related to changes in orbital precession and its control over insolation, especially at low latitudes. As Eric Wolff explained, *'Insolation reaches its maximum during the part of the precession cycle when the elliptical orbit of the Earth takes the planet closest to the Sun during northern summer. The result is a stronger monsoon in Asia and other regions, with more summer precipitation, and consequently greater wetland areas and methane production by soil-dwelling microorganisms'*[65]. The methane increase of the past 5 Ka departed from that pattern, increasing when northern summer insolation was on the wane. The reason seemed to be that during the late Holocene there was an increase in wetland sources in South America that outweighed the expected decreases in Eurasia and East Asia. The South American source was reacting to Southern Hemisphere insolation with a different phase from that in the Northern Hemisphere. Hence, there was no need to call upon human intervention as Ruddiman had.

More recent data provide Ruddiman with some support. Celia Sapart of Utrecht University and colleagues recently reviewed variations in the abundance of atmospheric methane and its carbon isotopic composition for the past 2000 years[66]. Their new high-resolution data came from the recent North Greenland Eemian Ice Drilling programme (NEEM) and EUROCORE ice cores from the Summit site in central Greenland. These data confirmed that methane had increased by about 70 ppb from 100 BC to 1800 AD, and that there was an accompanying fall in its $\partial^{13}C$ composition. These trends appeared mainly to be driven by biogenic emissions related to a growing increase in agriculture. After 1800 AD, CH_4 rose abruptly, with an accompanying rapid rise in its $\partial^{13}C$ composition, consistent with increased emissions from fossil fuel burning associated with the onset of industrialisation. The measurements also revealed three centennial-scale falls in $\partial^{13}C$ concentration between 100 BC and 1600 AD that were not previously recognised, and which were superimposed on the declining long-term trend in $\partial^{13}C$. These were attributed to biomass burning associated with the clearance or maintenance of cropland. Clearly, human activities must have played an important role in governing the long-term increase in CH_4 over this period[66].

Nevertheless, the analysts of the Vostok core, led by Jean Jouzel, were keen to point out that in MIS 11 – the 420 Ka interglacial sampled in the ice cores from Vostok and EPICA Dome C, – *'the highest levels [of methane] were observed at the very beginning of the interglacial period, then, rather similarly for the Holocene, decreased for 5,000 years, and then began to increase. This increase could obviously not be attributed to human activity ... and it weakened Ruddiman's argument, which was also not obviously supported by the records of concentrations of carbon dioxide recorded at Vostok and at Dome C throughout that interglacial period'*[67]. They regarded this as crucially important information, because the long interglacial of MIS 11 is likely to be the best analogue for the Holocene.

Much has been made in the popular press of the possibility that continued warming may melt the Arctic permafrost, releasing large amounts of methane trapped within or beneath it. Eric Wolff[65] pointed out that the modelling by Singarayer and others[64] showed that, during the last interglacial, atmospheric methane decreased slightly at a time when temperatures globally and in the Arctic were a few degrees warmer than at present. This suggests that a large influx of additional emissions from methane hydrates currently trapped in permafrost may be unlikely with continued warming in the immediate future[65]. The existence of a natural envelope for methane, like that for CO_2, over the past 800 Ka of the Dome C core (Figure 13.5) suggests that continued warming to the levels typical of past interglacials will not release large extra amounts of CH_4.

Before we leave this topic, we should return to the subject of volcanoes and their emission of CO_2. As we saw in Chapter 13, volcanism increased two to six times above background levels during the last deglaciation between 12 and 7 Ka ago, probably due to decompression of the Earth's mantle in deglaciating regions[68]. The increase in CO_2 from these increased volcanic emissions provided a positive feedback, helping to warm the deglaciating world, and may have accounted for the addition of 40 ppm

CO_2 to the atmosphere. Huybers and Langmuir pointed out that, '*Conversely, waning volcanic activity during the Holocene would contribute to cooling and reglaciation, thus tending to suppress [further] volcanic activity and promote the onset of an ice age*'[68]. However, there is no evidence to suggest that volcanic activity contributed to the cooling of the late Holocene (Figure 14.1).

Having a strong interest in predicting what the effect on our climate and sea level of further emissions of CO_2 might be, Jim Hansen of NASA was keen to see how much warmer than the Holocene were some recent interglacial periods[69]. He found that the temperatures of the major interglacials of the past 450 Ka tended to be close to or above those of the Holocene, while the smaller interglacials of the preceding 400 Ka tended to be cooler than the Holocene. The large warm interglacials, he thought, had '*moved into a regime in which there was less summer ice around the Antarctic and Greenland land masses, there was summer melting on the lowest elevation of the ice sheets, and there was summer melting on the ice shelves, which thus largely disappeared. In this regime, we expect warming on the top of the ice sheet to be more than twice global mean warming*'[69]. Why? Because melting of the sea ice and ice shelves created large areas of warm open ocean that reduced albedo and thus affected temperature year round, something unlikely to have happened in the smaller interglacials prior to 450 Ka ago.

Hansen went on to suggest that the relative stability of Holocene climate came about because orbital controls kept global temperature just below the level required to melt enough ice to decrease albedo in the same way. His calculations suggested that during the peak warm interglacials at 420 and 120 Ka ago, the global climate was only about 1 °C warmer than the Holocene Climatic Optimum, or about 2 °C warmer than the preindustrial Holocene. For that reason, Hansen and his colleague Sato considered that, with the modern temperature rise having returned Earth's mean temperature close to the Holocene maximum, the planet was poised to experience the strong amplifying polar feedbacks that likely led to the warmest interglacials of the past 450 Ka. The continued decline in extent of September sea ice in the Arctic is one sign that Earth's climate is already on that path. Melting of all or parts of the Greenland and West Antarctic Ice Sheets would be expected to follow, albeit much more slowly.

Undoubtedly, there have been human influences on climate through the Holocene, not least via fire, forest clearance, agriculture, the killing of large mammals and changes to the hydrological cycle through the damming of rivers. Even so, the vast majority of the Holocene changes we examine in this chapter came about naturally, rather than being affected by human activities.

Sceptics of the influence of CO_2 on climate point to the fact that the decline in temperature over the latter part of the Holocene accompanied a 25 ppm rise in CO_2, taking this as evidence against CO_2 as a driver of modern climate change. Looking at just temperature and CO_2 is not the most appropriate way to check the influence of CO_2 on climate, however, since global temperature is driven by other factors, too, such as change in albedo, which may be caused by both changes in vegetation and changes in ice cover, along with changes in insolation, and the balance between them all. All forcings have to be considered simultaneously. Climate change is not a single-issue game.

14.3   Climate Variability

As we saw in Chapter 12, Gerard Bond discovered that the deposition of ice-rafted debris during the last glacial period seemed to follow a climate cycle of roughly 1500 years, which continued into the Holocene[70]. Although Bond's cycles were very much more prominent during the last glacial period, dating back to 80 Ka ago, than during the Holocene, the amount of variability in any one grain type (Icelandic glass or red-stained grains) was about the same in both periods; only the absolute abundance decreased[70]. That suggested that the cycles were independent of orbital forcing and carried on regardless of glacial state. Some periodic or quasi-periodic forcing agent must cause ice streams to grow, thereby increasing the rates of discharge of icebergs and thus freshening surface waters in the North Atlantic. During the glacial periods of the Ice Age, some threshold (or tipping point) was eventually passed, beyond which there was a major iceberg discharge and cooling event, from which the system eventually recovered, passing back across the threshold as it warmed up. Such tipping points were not exceeded during the Holocene interglacial, because, with the melting of the major ice sheets on North America and Scandinavia, there was no longer a sufficient mass of cold ice to respond in such a way in the Northern Hemisphere. Consequently, the cycles in the Holocene had a very much lower amplitude than those in the last glacial period.

Looked at more closely, the length of Bond's cycles varied from a low of 1328 ± 539 years between 43 and 31 Ka ago to a high of 1795 ± 425 years between 79 and 64 Ka ago, with the Holocene cycle being 1374 ± 502 years[71].

Although there is a difference of 400 years between the extremes of these averages, the standard deviations about the means all overlap to some degree, making it difficult to differentiate between them statistically. Spectral analysis of the record of red (hematite)-stained grains revealed a slightly different picture, with peaks at around 4.7 and 1.8 Ka ago[70, 71]. Ignoring the apparent cycle at 4.7 Ka ago, Bond felt confident that his team had identified a natural millennial-scale cycle of 1–2 Ka, and he thought that the Little Ice Age, which we'll come to later, was the most recent cold phase of that cycle[71].

Bond was not alone. In 1999, Giancarlo Bianchi and Nick McCave of Cambridge University confirmed that sediments from a Holocene core taken south of Iceland also followed a quasi-periodic 1500-year cycle[72]. Peter De Menocal found much the same signal in marine sediments off west Africa[18]. The supposed 1500-year climate signal was also identified in 2011 in Holocene cave formations known as speleothems in Israel[73]. These are precipitates of calcium carbonate, which most people know as stalactites (growing down from the cave ceiling, like icicles) and stalagmites (growing up from the cave floor). Their internal layering is a reflection of the history of local rainfall, and the individual layers may be analysed for their $\partial^{18}O$ characteristics, which provide climate signals through time.

More recently, a team led by Philippe Sorrel found similar variability in its examination of Holocene records of high-energy estuarine and coastal sediments from the south coast of the English Channel[74]. '*High storm activity occurred periodically with a frequency of about 1,500 years, closely related to cold and windy periods diagnosed earlier*', Sorrel's team concluded[74]. These oscillations, linked to Bond cycles, appeared within an array of different spectral signatures, ranging from a 2500-year cycle to a 1000-year one. There was no consistent correlation between spectral maxima in records of storminess and solar irradiation, making solar activity seem an unlikely cause of millennial-scale variability. Rather, to Sorrel, the storminess reflected a natural periodic cooling of the North Atlantic.

Also in 2012, Dennis Darby of Old Dominion University, Virginia, found Bond's 1500-year cycle in an 8000-year record of ice-rafted iron-rich grains in sediments from the Arctic[75]. These grains were rafted to the Alaskan coast from Russia's Kara Sea during strongly positive phases of the Arctic Oscillation. Darby and colleagues used the sediment record to document an 8000-year history of the Arctic Oscillation. Recognising

that there was no 1500-year solar cycle (see Chapter 15 for a detailed discussion of solar cycles), they attributed the forcing to internal variability within the climate system, or to an indirect response to solar forcing at low latitudes[75]. Low-latitude palaeoclimate records do show significant linear solar forcing, suggesting that the El Niño–Southern Oscillation system in the Pacific acts as a mediator of the solar influence on the climate system's low-latitude heat engine, rather than on the high-latitude one[76]. That may explain why Bond's ice-rafting events correlate well with sea surface temperatures from the low-latitude Atlantic[74]. Solar heating of the tropics would create a stronger Equator-to-pole thermal gradient, strengthening winds and storms, and in due course leading to southwardly-directed outbreaks of ice-rafting[75]. The link would be indirect. As yet, the precise mechanism for propagating low-latitude forcing through the climate system to high latitudes is unknown[76].

Other researchers also realised that, superimposed on the significant changes of temperature that were driven by orbital changes in insolation through the Holocene, there were slight variations on the millennial scale of the kind – although not the magnitude – identified by Dansgaard and Oeschger in ice records and by Bond in marine sediment records from the last glacial period. For instance, Vinther's data showed that temperatures on the Greenland Ice Sheet varied on the millennial time scale by up to 1 °C both above and below the background Holocene temperature trend[23]. Among these variations was a warm period centred near 1000 AD that might represent Lamb's Medieval Warm Epoch, followed by a subsequent cold period more or less coinciding with Lamb's Little Ice Age. Alley, too, identified in Greenland ice a warm period centred on about 1030 AD, followed by a cold one reaching maximum lows between around 1650 and 1850 AD[21].

Working with Mayewski and others, Alley also identified a significant short-lived cooling event at 8.2 Ka ago[77], most likely caused by a North American glacial lake draining into the adjacent North Atlantic, as mentioned earlier. Vinther calculated the amount of cooling in that event to be around 2 °C[23]. Careful scrutiny of palaeoclimate records from the early Holocene by Eeelco Rohling and Heiko Pälike of the UK's National Oceanography Centre in Southampton showed that the 8.2 Ka event was the peak of a cooling event that started at 8.6 Ka ago and lasted 400–600 years[78]. Rohling and Pälike concluded that, while the peak event may well have been caused by outflow from glacial Lake Ojibway-Agassiz, its regional and even global extent was difficult to estimate because of its occurrence within a longish cool period of the Holocene. Nevertheless there was good evidence to suggest that the freshwater lid formed over the North Atlantic by discharge from the lake did cause the Meridional Overturning Circulation to slow at about 8.2 Ka ago, with global effects[79].

Mayewski's team's analysis of a global data set provided an opportunity to obtain a global picture of millennial events in the Holocene[26]. They identified six globally distributed events they called 'rapid climate change events' at 9–8 Ka, 6–5 Ka, 4.2–3.8 Ka, 3.5–2.5 Ka, 1.2–1.0 Ka and 600–150 years ago. Polar cooling, tropical aridity and major changes in atmospheric circulation characterised most of these events, while the 6–5 Ka-ago event marked the end of the Holocene humid period in tropical Africa, beginning a trend towards tropical aridity. Several of these events coincided with major disruptions of civilisation, suggesting the impact of climatic variability on past civilisations.

Like Alley and Vinther, Mayewski documented the brief cooling event at 8.2 Ka ago, which lay within his 9–8 Ka-ago period. At that time, there were still remnants of the ice sheets on North America and Scandinavia. Conditions were cool over much of the Northern Hemisphere. There were major episodes of ice-rafting, stronger winds over the North Atlantic and Siberia, outbreaks of polar air over the Aegean Sea and glacier advances in Scandinavia and North America. In contrast, glaciers retreated in the Alps, as the air became drier there. In the tropics, the monsoon weakened and there were widespread droughts. In the Southern Hemisphere, sea surface temperatures warmed around southern Africa, and grounded ice retreated in the Ross Sea.

Most of the other global events reviewed by Mayewski shared the twinning of cool poles with dry topics that also characterised cool periods during the Ice Age. Evidence for the events at 4.2–3.8 and 1.2–1.0 Ka ago (the latter equivalent to 800–1000 AD) appeared in fewer of the records, but their widespread distribution and synchrony suggested the operation of global connections. At these latter times, winds weakened over the North Atlantic and Siberia, and temperatures fell in North America and Eurasia.

The most recent of these events covered the end of the Holocene, from around 600 years ago (or 1400 AD) to the beginning of the modern era[26]. It differed from previous events in being characterised by cool poles and wet tropics. There were rather fewer records to draw upon than one might expect for this relatively recent period, leading Mayewski to complain: '*Unfortunately, determining the nature and duration of later stages of this interval is difficult because high-resolution records for this time are relatively scarce and because several records are missing recent sections as an artifact of sampling. Moreover, interpretation is complicated by potential anthropogenic influences*'[26]. As a consequence, his team investigated this event only from 600 to 150 years ago (1400–1850 AD). In the Northern Hemisphere, this cool event had the fastest and strongest onset of any of the Holocene events, with glaciers advancing and westerlies strengthening. While Venezuela, Haiti and Florida became more arid, tropical Africa became more humid and monsoon rains increased in India. While parts of the Antarctic Peninsula warmed, East Antarctic cooled. Glaciers advanced in New Zealand, rainfall increased in Chile and southern Africa became cool and dry. This event corresponds to the Little Ice Age.

Heinz Wanner and his group also identified six 'cold relapses' in their collection of global samples from the Holocene, with peaks at 8.2 Ka, 6.3 Ka, 4.7 Ka, 2.7 Ka, 1.55 Ka and 550 years ago[13]. Each of these peaks lay amid a period spanning about 500 years. One might expect these cool peaks to correlate reasonably well with the cool events identified by Mayewski and his team and with the cold peaks of Bond cycles. Table 14.1 compares the ranges of Wanner's and Bond's cold peaks and Mayewski's cold events.

The comparison looks close for all three for Bond cycles 5a and 5b, 2 and 0. It is less good for Bond cycle 4, where Wanner's cool period does not closely match the other two. Moreover, Mayewski's 6–5 Ka-ago event is not, strictly speaking, a cold event, but rather the end of the Saharan pluvial period. The comparison is also less good for Bond cycle 3, which does not fit Wanner's cool period, although it does match Wanner's glacial advance. Bond cycle 1 fits with Wanner's glacial advance, but only overlaps slightly with Mayewski's and Wanner's cool periods, which themselves fail to overlap. Neither Wanner nor Mayewski identified Bond cycle 6 as a cool period.

Discrepancies like these may arise because of inadequacies in the chronology of the core samples used. Can we rely on Mayewski and Wanner's data? They all come from published sources, but one of these was later found to be suspect: Alley's 2000 AD Greenland data profile[21], which Vinther showed to be misleading as a guide to Greenland temperature[23]. Nevertheless, the discrepancies are likely to be real and to reflect the geographical and temporal heterogeneity of the climate signal globally. For instance, Mayewski observed advances in glaciers over North America and Scandinavia during his 9–8 Ka-ago event, at the same time that glaciers were retreating in the

Table 14.1 Comparison of the ranges of Heinz Wanner's and Gerard Bond's cold peaks (both measured from Figure 3 in Wanner, H., Solomina, O., Grosjean, M., Ritz, S.P. and Jetel, M. (2011) Structure and origin of Holocene cold events. *Quaternary Science Reviews* **30**, 3109–3123) and Paul Mayewski's cold events, in 1000 years (Ka) before present.

Bond cycles (Ka)	Mayewski events (Ka)	Wanner peaks (Ka)	Wanner glacial advances (Ka)
6. 9.6–9.2			
5a. 8.6–8.25	9–8	8.55–8.0	8.6–8.1
5b. 7.8–7.2			7.8–7.5
4. 5.95–5.1	6–5	6.45–5.9	
3. 4.6–3.8	4.2–3.8	4.8–4.6	4.5–4.0
2. 3.4–2.65	3.5–2.5	3.35–2.45	3.6–3.1
1. 1.6–1.0	1.2–1.0	1.8–1.5	1.8–1.0
0. 700–100 years	600–150 years	850–150 years	600–150 years

Alps, and each of Wanner's events showed a high degree of spatial and temporal variability. Selecting the boundaries of a global cool period is likely to be a more subjective process than identifying a Bond cycle or a glacial advance. Discrepancies between the North American and European climate are not unexpected, because the North American ice sheet took much longer to melt away than did the Scandinavian and British ice sheets.

What about the rapidity of change? Contrary to Mayewski's conclusions, with regard to the past 6 Ka, Wanner's PAGES team did not '*find any time period for which a rapid or dramatic climate transition appears even in a majority of … time series*', although rapid shifts were recognised at certain times and in particular regions[30]. However, Wanner did agree that there had been a rapid and short-term climate change event at 8.2 Ka ago[13].

What are we to make of the various warm and cool periods that are superimposed on the gradual cooling driven by the orbital decline in insolation throughout the Holocene? If we are to extrapolate into the future what we know of the past, we need to be sure the cycles are real and to understand what caused them. The MIT oceanographer Carl Wunsch (1941–) called Bond's 1500-year cycle into question on the grounds that it might be a simple alias of inadequately sampled seasonal cycles[80]. When Wunsch removed this signal from ice-core and marine-core data, climate variability appeared as a continuous process, suggesting that the finding of a narrowly defined 1500-year cyclicity in the data set might not represent actual millennial events. His analysis supported the idea that Heinrich events and Dansgaard-Oeschger events might be quasi-periodic and driven by several different influences.

Richard Alley disagreed. Since the periodicity was evident in a wide variety of analyses, regardless of sampling interval and other details, Alley thought it could not be an alias of any shorter periodicity, such as the annual cycle, as Wunsch had suggested[81]. Wanner, on the other hand, was inclined to agree with Wunsch, noting, '*There is thus scant evidence for consistent periodicities and it seems likely that much of the higher frequency variability observed is due to internal variability or complex feedback processes that would not be expected to show strict spectral coherence*'[29]. Wanner's view of Bond cycles was that '*the origin of these cycles remains unknown*'[29], and he went on to question the relevance of Bond cycles for the Southern Hemisphere.

Bianchi and McCave agreed with Bond that the 1500-year cycles probably represented some manifestation of the internal circulation of the ocean of unknown cause, but they did not rule out the possibility of some modulation of climate behaviour by Earth's orbital properties[72]. Bond and his team dismissed that possibility. Millennial-scale climate change could arise from harmonics and combinations of the three main orbital periodicities, but cycles originating in those ways were mostly longer than the observed 1500-year cycle[70].

How might the internal circulation of the ocean have changed? The most likely culprit was the Atlantic Meridional Overturning Circulation, which transports warm salty water to high latitudes, where they cool, sink and return southwards at depth. David Thornally, Harry Elderfield and Nick McCave investigated this possibility by using Mg/Ca and $\partial^{18}O$ ratios measured in foraminifera from a sediment core taken in 1938 m of water close to Iceland[82]. The temperatures of near-surface waters oscillated between about 10 and 11 °C over the past 10 Ka, while their salinity slowly increased. Subsurface waters from below the thermocline showed much greater variability in both salinity and temperature, depending on whether they were drawn from the cold, fresh subpolar gyre or the warm, saline subtropical gyre. From

12.0 to 8.4 Ka ago, the North Atlantic was well strati-fied, with fresh surface water – probably from melting ice – overlying warm, saline tropical gyre water. Then there was a switch to well-mixed waters, followed by an oscillation between stratified and well-mixed waters roughly every 1500 years, attributable to changes in ocean dynamics, much as suggested by Debret and colleagues in 2007[83]. This appeared to be very much a North Atlantic phenomenon. A strong link is likely between the behaviour of the Atlantic Meridional Overturning Circulation and the North Atlantic Oscillation, which is in turn linked to the behaviour of the Arctic Oscillation. And as we saw earlier, the 1500-year cycle also shows up in the Arctic, where ice-rafting occurred during positive phases of the Arctic Oscillation[75]. These linkages help to explain why Bond cycles tend to be focussed around the North Atlantic, rather than elsewhere.

Greenhouse gases were an equally unlikely cause of mil-lennial change. Neither CO_2 nor CH_4 showed sufficient variability at the millennial scale during the Holocene to have caused the observed millennial cooling events, nor did volcanic eruptions[13, 26].

Can we blame multicentennial internal oscillations in the ocean for these millennial events? There is ample evidence for their operation on the decadal scale, for instance in the shape of the Pacific Decadal Oscillation or the Atlantic Multi-Decadal Oscillation[29]. When the Pacific Decadal Oscillation is positive, we get warm conditions in the central tropical Pacific and up the American west coast to Alaska, along with cold conditions in the northwest Pacific. The opposite occurs during the negative phase. Analyses of tree-ring data from the region show that the Pacific Decadal Oscillation signal is recognisable at least back to 1470 AD, but is not a persistently dominant feature. The Atlantic Multi-Decadal Oscillation is recog-nised from sea surface temperature patterns in the North Atlantic, which warm by around 0.2 °C in the positive phase, and decline by the same amount in the negative phase, at intervals of around 20 years – much like the time scale of the Pacific Decadal Oscillation. Part of the global warming in the 1930s is likely to have been caused by the positive mode of the Atlantic Multi-Decadal Oscilla-tion, as is the American dustbowl of the same era[56]. These quasi-periodic oscillations supply much of the background high-frequency 'noise' in the climate spectrum. Similar variability, on time scales of between 35 and 120 years, has been detected in the Atlantic Meridional Overturning Circulation – the Atlantic branch of the global ocean conveyor belt. Shorter-term variations attributable to El Niño are superimposed on these larger-scale cycles.

Long-period oscillations may arise within the ocean system through nonlinear processes affecting either **advection** – the transport of heat and salt – or **convec-tion** – the vigorous vertical mixing of water that occurs when denser water lies atop lighter water[56]. Much of what we know about advection started with Henry (Hank) Stommel (1920–1992), who was appointed to the US National Academy of Science in 1962 and awarded the US National Medal of Science in 1989. His 1961 box model of the thermohaline circulation of the oceans showed that it could feature a sharp decrease in advective transport, which we might think of as a shutdown of the Thermohaline Conveyor[84]. While such shutdowns did occur in glacial times, they have not taken place during the Holocene, unless temporarily during the 8.2 Ka-ago cool event. According to Crucifix, '*A refined version of Stommel's model incorporating realistic propagation times suggests that advective processes may also cause sustained oscillations … associated with the propaga-tion of temperature and salinity anomalies through the conveyor belt, and their periods range between two and four millennia*'[84]. Such advective oscillations may explain Bond's millennial cycles[56].

The role of oceanic convection is to restore gravitational stability. As Crucifix explained, it is a self-maintained process – the exchange of heat between the ocean and the atmosphere can make surface waters denser, promoting further convection. Deep convection occurs today in the Norwegian-Greenland Sea, the Labrador Sea and the Weddell and Ross Seas. Modelling shows that convec-tive instability may induce repeated stops and starts of convection, which may explain the abrupt warming and cooling observed by Bond in the Norwegian-Greenland Sea during the Holocene. During a convective shutdown, sea ice may advance southwards, maintaining the shut-down and pushing the system into a cold state. Such a state may become persistent in the presence of external forcing, for example in the form of a decrease in solar radiation, or cooling caused by a volcanic eruption. The probability of such events occurring will tend to increase through the Holocene, with the long-term cooling brought about by the persistent decrease in orbital insolation during summer in the Northern Hemisphere. Convective instability and the inception of a temporary cold state may have pro-longed the cooling 8.2 Ka ago caused by the discharge of freshwater from Lake Agassiz. That discharge would have taken place within a few years at most, while the cooling event lasted at least 100 years[56].

Pulling these various bits of information together, the basic theme of Holocene climate is one of initially high insolation leading to the demise of the Northern Hemisphere ice sheets, apart from Greenland, followed by a mid Holocene Climatic Optimum in the Northern Hemisphere as the influence of ice diminished, followed by declining insolation leading to the development of a neoglacial period culminating in the Little Ice Age. Most of these changes were typical of the Northern Hemisphere. The tropics tended to warm slightly.

Superimposed on this overall decline were climate variations on the millennial or centennial scale[19]. Some distinct oscillations or climate steps appear to be widespread, for example across the North Atlantic region, notably at 8.2, 5.5–5.3 and 2.5 Ka ago, where they punctuated the gradual Holocene decline in temperature. In that region, the coldest points of each of these millennial steps in sea surface temperature were 2–4 °C colder than the warmest part[19]. The event at 8.2 Ka ago was the most sudden and striking of the Holocene, bringing cool, dry conditions to the North Atlantic region. It is picked up in ice cores from as far away as Greenland and Antarctica. That particular event represents a massive flood of glacial lakewater into the Arctic and North Atlantic. There was a broader and less intense shift to colder and drier conditions at 5.5 Ka ago, and a similar shift at 4 Ka ago that coincided with the collapse of a number of early civilisations[19]. While it is possible that solar forcing played a role in causing these cycles, it may have merely exacerbated the effects of oscillations caused by natural variability in oceanic advection and convection. That being said, it is more likely that the effects of advection and convection will be focused in specific regions, like the North Atlantic and Arctic provinces, rather than globally.

Now we turn to Chapter 15, to see what role the Sun may have played in driving millennial or centennial climate change, and to explore in some detail climate change over the last 2000 years.

References

1. Birks, H.J.B. (2009) Holocene climate research – progress, paradigms, and problems. In: *Natural Climate Variability and Global Warming: A Holocene Perspective* (eds R.W. Battarbee and H.A. Binney). Wiley-Blackwell, Chichester and Oxford, pp. 7–57.
2. Lamb, H.H. (1966) *The Changing Climate* Methuen, London.
3. Summerhayes, C.P., Milliman, J.D., Briggs, S.R., Bee, A.G. and Hogan, C. (1976) Northwest African shelf sediments: influence of climate and sedimentary processes. *Journal of Geology* **84**, 277–300.
4. Emiliani, C. (1955) Pleistocene temperatures. *Journal of Geology* **63** (6), 538–578.
5. Matthes, F.E. (1939) Report of committee on glaciation, April 1939. *Transactions of the American Geophysical Union* **20**, 518–523.
6. Crowley, T.J. and North, G.R. (2011) *Paleoclimatology*. Oxford Monographs on Geology and Geophysics 18. Oxford University Press, Oxford.
7. Porter, S.C. and Denton, G.H. (1967) Chronology of neoglaciation in the North American Cordillera. *American Journal of Science* **265**, 177–210.
8. Denton, G.H. and Karlen, W. (1973) Holocene climatic variations – their pattern and possible causes. *Quaternary Research* **3**, 155–205.
9. Lamb, H.H. (1965) The Early Medieval Warm Epoch and its sequel. *Palaeogeography, Palaeoclimatology, Palaeoecology* **1**, 13–37.
10. COHMAP Members (1988) Climatic changes of the last 18,000 years: observations and model simulations. *Science* **241**, 1043–1052.
11. Wright, H.E., Kutzbach, J.E., Webb, T. III, Ruddiman, W.F. and Street-Perrott, F.A. (1993) *Global Climates Since the Last Glacial Maximum*. University of Minnesota Press, Minneapolis, MN.
12. Berger, A. (1978) Long-term variations of daily insolation and Quaternary climate changes. *Journal of Atmosphere Science* **35**, 2362–2367.
13. Wanner, H., Solomina, O., Grosjean, M., Ritz, S.P. and Jetel, M. (2011) Structure and origin of Holocene cold events. *Quaternary Science Reviews* **30**, 3109–3123.
14. Huybers, P. and Denton, G. (2008) Antarctic temperature at orbital timescales controlled by local summer duration. *Nature Geoscience* **1**, 787–792.
15. Loutre, M.-F. and Berger, A. (2000) Future climate changes: are we entering an exceptionally long interglacial? *Climate Change* **46**, 61–90.
16. Kutzbach, J.E., Gallimore, R., Harrison, S.P., Behling, P., Selin, R. and Laarif, F. (1998) Climate and biome simulations for the past 21,000 years. *Quaternary Science Reviews* **17**, 473–506.
17. Miller, G.H., Geiersdottir, A., Zhong, Y., Larsen, D.J., Otto-Bliesner, B.L., Holland, M.M., *et al.* (2012) Abrupt onset of the Little Ice Age triggered by volcanism and sustained by sea-ice/ocean feedbacks. *Geophysical Research Letters* **39** (2), doi:10.1029/2011GL050168.
18. De Menocal, P., Ortiz, J., Guilderson, T. and Sarnthein, M. (2000) Coherent high- and low- latitude climate variability during the Holocene warm period. *Science* **288**, 2198–2202.
19. Maslin, M., Seidov, D. and Lowe, J. (2001) Synthesis of the nature and causes of rapid climate transitions during the

Quaternary. In: *The Oceans and Rapid Climate Change: Past, Present and Future* (eds D. Seidov, B.J. Haupt and M. Maslin). Geophysical Monographs 126. American Geophysical Union, Washington, DC, pp. 9–52.

20. Frakes, L.A., Francis, J.E. and Syktus, J.I. (1992) *Climate Modes of the Phanerozoic – The History of the Earth's Climate Over the Past 600 Million Years.* Cambridge University Press, Cambridge.

21. Alley, R.B. (2000) The Younger Dryas cold interval as viewed from central Greenland. *Quaternary Science Reviews* **19**, 213–226.

22. Cuffey, K.M. and Clow. G.D. (1997) Temperature, accumulation, and ice sheet elevation in central Greenland through the last deglacial transition. *Journal of Geophysical Research* **102**, 26 383–26 396.

23. Vinther, B.M., Buchardt, S.L., Clausen, H.B., Dahl-Jensen, D., Johnsen, S.J., Fisher, D.A., *et al.* (2009) Holocene thinning of the Greenland Ice Sheet. *Nature* **461**, 385–388.

24. Jansen, E., Andersson, C., Moros, M., Nisancioglu, K.H., Nyland, B.F. and Telford, R.J. (2009) The early to mid-Holocene thermal optimum in the North Atlantic. In: *Natural Climate Variability and Global Warming: A Holocene Perspective* (eds R.W. Battarbee and H.A. Binney). Wiley-Blackwell, Chichester and Oxford, pp. 123–137.

25. Miller, G.H., Alley, R.B., Brigham-Grette, J., Fitzpatrick, J.J., Polyak, L., Serreze, M.C. and White, J.W.C. (2010) Arctic amplification: can the past constrain the future? *Quaternary Science Reviews* **29** (15–16), 1779–1790.

26. Mayewski, P.A., Rohling, E., Stager, J.C., Karlen, W., Maasch, K.A., Meeker, L.D., *et al.* (2004) Holocene climatic variability. *Quaternary Research* **62**, 243–255.

27. Lorenz, S.J., Kim, J.-H., Rimbu, N., Schneider, R.R. and Lohmann, G. (2006) Orbitally driven insolation forcing on Holocene climate trends: evidence from alkenone data and climate modeling. *Paleoceanography* **21**, PA1002.

28. HOCLAT: A Web-based Holocene Climate Atlas. http://www.oeschger.unibe.ch/research/projects/holocene_atlas/ (last accessed 29 January 2015).

29. Wanner, H., Beer, J., Bütikofer , Crowley, T.J., Cubasch, U., Flückiger, J., *et al.* (2008) Mid- to late Holocene climate change: an overview. *Quaternary Science Reviews* **27**, 1791–1828.

30. PAGES 2k Consortium (2013) Continental-scale temperature variability during the last two millennia. *Nature Geoscience* **6**, 339–346.

31. Zhao, M., Beveridge, N.A.S., Shackleton, N.J. and Sarnthein, M. (1995) Molecular stratigraphy of cores off Northwest Africa: sea surface temperature history over the last 80 ka. *Paleoceanography* **10**, 661–675.

32. Calvo, E., Grimalt, J. and Jansen, E. (2002) High resolution U^k_{37} sea surface temperature reconstruction in the Norwegian Sea during the Holocene. *Quaternary Science Reviews* **21**, 1385–1394.

33. Marchal, O., Cacho, I., Stocker, T.F., Grimalt, J.O., Calvo, E., Martrat, B., *et al.* (2000) Apparent long-term cooling of the sea surface in the Northeast Atlantic and Mediterranean during the Holocene. *Quaternary Science Reviews* **21**, 455–483.

34. Herbert, T.D. (2003) Alkenone paleotemperature determinations. In: *The Oceans and Marine Geochemistry, Vol. 6, Treatise on Geochemistry* (eds H.D. Holland and K.K. Turekian). Elsevier-Pergamon, Oxford, pp. 391–432.

35. Bard, E., Rostek, F. and Sonzogni, C. (1997) Interhemispheric synchrony of the last deglaciation inferred from alkenone paleothermometry. *Nature* **385**, 707–710.

36. Turner, J., Bindschadler, R.A., Convey, P., Di Prisco, G., Fahrbach, E., Gutt, J., *et al.* (2009) *Antarctic Climate Change and the Environment.* Scientific Committee on Antarctic Research, Cambridge.

37. Mulvaney, R., Abram, N.J., Hindmarsh, R.C.A., Arrowsmith, C., Fleet, L., Triest, J., *et al.* (2012) Recent Antarctic Peninsula warming relative to Holocene climate and ice-shelf history. *Nature* **489**, 141–145.

38. Abrams, N.J., Mulvaney, R., Wolff, E.J., Triest, J., Kipfstuhl, S., Trusel, L.D., *et al.* (2013) Acceleration of snow melt in an Antarctic Peninsula ice core during the twentieth century. *Nature Geoscience* **6**, 404–411.

39. Shevenell, A.E., Ingalls, A.E., Domack, E.W. and Kelly, C. (2011) Holocene Southern Ocean surface temperature variability west of the Antarctic Peninsula. *Nature* **470**, 250–254.

40. Etourneau, J., Collins, L.G., Willmott, V., Kim, J.H., Barbara, L., Schouten, J.S., *et al.* (2013) Holocene climate variations in the western Antarctic Peninsula: evidence for sea ice extent predominantly controlled by changes in insolation and ENSO variability. *Climate of the Past* **9**, 1431–1446.

41. Renssen, H., Goosse, H., Fichefet, T., Masson-Delmotte, V. and Koç, N. (2005) Holocene climate evolution in the high-latitude Southern Hemisphere simulated by a coupled atmosphere-sea ice-ocean-vegetation model. *The Holocene* **15** (7), 951–984.

42. Marcott, S.A., Shakun, J.D., Clark, P.U. and Mix, A.C. (2013) A reconstruction of regional and global temperature for the past 11,300 years. *Science* **339**, 1198–1201.

43. Hoogakker, B.A.A., Chapman, M.R., McCave, I.N., Hillaire-Marcel, C., Ellison, C.R.W., Hall, I.R. and Telford, R.J. (2011) Dynamics of North Atlantic deep water masses during the Holocene. *Paleoceanography* **26** (4), 4214.

44. Alley, R.B., Andrews, J.T., Brigham-Grette, J., Clarke, G.K.C., Cuffey, K.M., Fitzpatrick, J.J., *et al.* (2013) History of the Greenland Ice Sheet: paleoclimatic insights. *Quaternary Science Reviews* **29**, 1728–1756.

45. Öberg, L. and Kullman, L. (2011) Recent glacier recession – a new source of postglacial treeline and climate history in the Swedish Scandes. *Landscape Online* **26**, 1–38.

46. Crosta, X., Debret, M., Denis, D., Courty, M.A. and Ther, O. (2007) Holocene long- and short-term climate changes off Adelie Land, East Antarctica. *Geochemistry, Geophysics, Geosystems* **8** (11), doi:10.1029/2007GC001718.

47. Bradley, R.S. (2009) Holocene perspectives on future climate change. In: *Natural Climate Variability and Global Warming:*

A *Holocene Perspective* (eds R.W. Battabee and H.A. Binney). Wiley-Blackwell, Chichester and Oxford, pp. 254–268.

48. Berger, A. and Loutre, M.-F. (2002) An exceptionally long interglacial ahead. *Science* **297**, 1287–1288.

49. Sundquist, E.T. and Visser, K. (2005) The geologic history of the carbon cycle. In: *Biogeochemistry, Vol. 8, Treatise on Geochemistry* (eds H.D. Holland and K.K. Turekian). Elsevier-Pergamon, Oxford, pp. 425–513.

50. Indermühle, A., Stocker, T.F., Joos, F., Fischer, H., Smith, H.J., Wahlen, M., *et al*. (1999) Holocene carbon-cycle dynamics based on CO_2 trapped in ice at Taylor Dome, Antarctica. *Nature* **398**, 121–126.

51. MacDonald, G.M., Beilman, D.W., Kremenetski, K.V., Sheng, Y., Smith, L.C. and Velichko, A.A. (2006) Rapid early development of circumarctic peatlands and atmospheric CH_4 and CO_2 variations. *Science* **314**, 285–288

52. Chappelaz, J., Blunier, T., Kints, S., Dällenbach, A., Barnola, J.-M., Schwander, J., *et al*. (1997) Changes in the atmospheric CH_4 gradient between Greenland and Antarctica during the Holocene. *Journal of Geophysical Research* **102**, 15 987–15 997.

53. Ruddiman, W.F. and Thomson, J.S. (2001) The case for human causes of increased atmospheric CH_4. *Quaternary Science Reviews* **20**, 1769–1777.

54. Berger, A. (2009) Astronomical theory of climate change. In: *Encyclopedia of Paleoclimatology and Ancient Environments* (ed V. Gornitz). Springer, Dordecht, pp. 51–57.

55. Berger, A. and Loutre, M.-F. (2003) Climate 400,000 years ago, a key to the future? In: *Earth's Climate and Orbital Eccentricity: The Marine Isotope Stage 11 Question* (eds A. Droxler, R.Z. Poore and L.H. Burckle). Geophysical Monographs 137. American Geophysical Union, Washington, DC, pp. 17–26.

56. Crucifix, M. (2009) Modeling the climate of the Holocene. In: *Natural Climate Variability and Global Warming: A Holocene Perspective* (eds R.W. Battabee and H.A. Binney). Wiley-Blackwell, Chichester and Oxford, pp. 98–122.

57. Ruddiman, W.F. (2003) Orbital insolation, ice volume, and greenhouse gases. *Quaternary Science Reviews* **22**,1597–1629.

58. Ruddiman, W.F. (2005) *Plows, Plagues and Petroleum: How Humans Took Control of Climate* Princeton University Press, Princeton, NJ.

59. Broecker, W.S. (2006) The Holocene CO_2 rise: anthropogenic or natural? *EOS* **87**, 27.

60. Yu, J.M. and Elderfield, H. (2007) Benthic foraminiferal B/Ca ratios reflect deep water carbonate saturation state. *Earth and Planetary Science Letters* **258**, 73–86.

61. Yu, J.M., Anderson, R.F. and Rohling, E.J. (2013) Deep ocean carbonate chemistry and glacial-interglacial atmospheric CO_2. *Oceanography* **27** (1), 16–25.

62. Menviel, L. and Joos, F. (2012) Toward explaining the Holocene carbon dioxide and carbon isotope records: results from transient ocean carbon cycle-climate simulations. *Paleoceanography* **27** (1), doi:10.1029/2011PA002224.

63. Broecker, W.S. and Stocker, T.F. (2006) The Holocene CO_2 rise: anthropogenic or natural? *EOS* **87** (3), 27.

64. Singarayer, J.S., Valdes, P.J., Friedlingstein, P., Nelson, S. and Beerling, D.J. (2011) Late Holocene methane caused by orbitally controlled increase in tropical sources. *Nature* **470**, 82–85.

65. Wolff, E.W. (2010) Methane and monsoons. *Nature* **470**, 49–50.

66. Sapart, C.J., Monteil, G., Prokopiou, M., van de Wal, R.S.W., Kaplan, J.O., Sperlich, P., *et al*. (2012) Natural and anthropogenic variations in methane sources during the past two millennia. *Nature* **490**, 85–88.

67. Jouzel, J., Lorius, C. and Raynaud, D. (2013) *The White Planet: The Evolution and Future of our Frozen World.* Princeton University Press, Princeton, NJ.

68. Huybers, P. and Langmuir, C. (2009) Feedback between deglaciation, volcanism and atmospheric CO_2. *Earth and Planetary Science Letters* **286**, 479–491.

69. Hansen, J.E. and Sato, M. (2012) Paleoclimate implications for human-made climate change. In: *Climate Change: Inferences from Paleoclimate and Regional Aspects* (eds A. Berger, F. Mesinger and D. Šijački). Springer, Berlin, pp. 21–48.

70. Bond, G., Showers, W., Cheseby, M., Lotti, R., Almasi, P., deMenocal, P., *et al*. (1997) A pervasive millennial-scale cycle in North Atlantic Holocene and glacial climates. *Science* **278**, 1257–1266.

71. Bond, G.C., Showers, W., Elliot, M., Evans, M., Lotti, R., Hajdas, I., *et al*. (1999) The North Atlantic's 1-2kyr climate rhythm: relation to Heinrich Events, Dansgaard/Oeschger Cycles and the Little Ice Age. In: *Mechanisms of Global Climate Change at Millennial Time Scales* (eds P.U. Clark, R.S. Webb and L.D. Keigwin). Geophysical Monographs 112. American Geophysical Union, Washington, DC, pp. 35–58.

72. Bianchi, C. and McCave, I.N. (1999) Holocene periodicity in North Atlantic climate and deep-ocean flow south of Iceland. *Nature* **397**, 515–517.

73. Bar-Matthews, M. and Ayalon, A. (2011) Mid-Holocene climate variations revealed by high-resolution speleothem records from Soreq Cave, Israel and their correlation with cultural changes. *The Holocene* **21** (1), 163–171.

74. Sorrel, P., Debret, M., Billeaud, I., Jaccard, S.L., McManus, J.F. and Tessier, B. (2012) Persistent non-solar forcing of Holocene storm dynamics in coastal sedimentary archives. *Nature Geoscience* **5**, 892–896.

75. Darby, D.A., Ortiz, J.D., Grosch, C.E. and Lund, S.P. (2012) 1,500-year cycle in the Arctic oscillation identified in Holocene Arctic sea-ice drift. *Nature Geoscience* **5**, 897–900.

76. Marchitto, T.M., Muschelere, R., Ortiz, J.D., Carriquiry, J.D. and van Geen, A. (2010) Dynamical response of the tropical Pacific Ocean to solar forcing during the early Holocene. *Science* **330**, 1378–1381.

77. Alley, R.B., Mayewski, P.A., Sowers, T., Stuiver, M. Taylor, K.C. and Clark, P.U. (1997) Holocene climatic instability: a prominent widespread event 8200 years ago. *Geology* **25**, 483–486

78. Rohling, E.J. and Pälike, H. (2005) Centennial-scale climate cooling with a sudden cold event around 8200 years ago. *Nature* **434**, 975–979.

79. Cronin, T.M. (2010) *Paleoclimates – Understanding Climate Change Past and Present* Columbia University Press, Cambridge.

80. Wunsch, C. (2000) On sharp spectral lines in the climate record and the millennial peak. *Paleoceanography* **15**, 417–424.

81. Alley, R.B., Anandakrishnan, S., Jung, P. and Clough, A. (2001) Stochastic resonance in the North Atlantic: further insights. In: *The Oceans and Rapid Climate Change: Past, Present and Future* (eds D. Seidov, B.J. Haupt and M. Maslin). Geophysical Monographs 126. American Geophysical Union, Washington, DC, pp. 57–68.

82. Thornally, D.J.R., Elderfield, H. and McCave, I.N. (2009) Holocene oscillations in temperature and salinity of the surface subpolar North Atlantic. *Nature* **457**, 711–714.

83. Debret, M., Bout-Roumazeilles, V., Grousset, F., Desmet, M., McManus, J.F., Massei, N., *et al.* (2007) The origin of the 1500-year climate cycles in Holocene North Atlantic records. *Climate of the Past* **3**, 569–575.

84. Stommel, H. (1961) Thermohaline convection with two stable regimes of flow. *Tellus* **13**, 224–230.

15

Medieval Warming, the Little Ice Age and the Sun

15.1 Solar Activity and Cosmic Rays

Given the ongoing fall in orbital insolation that cooled the world from the mid Holocene Climatic Optimum into the neoglacial climate of the Little Ice Age[1,2], what climate forcing explains the Medieval Warm Period? Could the same forcing be responsible for the warming of the modern era? To address these questions, we must first examine variations in the output of the Sun over the past 2000 years.

The sun's face suffers from spots. They come and go in a roughly 11-year cycle, the Schwabe cycle, named after the German astronomer Heinrich Schwabe (1789–1875), in which insolation varies by a tiny 0.1%, or 0.25 W/m² [3]. When sunspots are abundant, our climate is very slightly warmer; when they are fewer, it is very slightly cooler. Back in 1801, the astronomer William Herschel noticed that the price of wheat rose when the number of sunspots fell[4]. The French historian Emmanuel Le Roy Ladurie (1929–) found a similar association between sunspots and the French wine harvest[5]. He also found that the dates of French grape harvests between 1370 and 1879 were inversely proportional to April–September temperatures in Paris[6–8]. The Little Ice Age cooling was not manifest in his Parisian summer records[9], confirming that in Europe, at least, the Little Ice Age was most effective in winter[10].

The sunspot cycle varies in both length and amplitude. Its length ranges from 9.7 to 11.8 years. The longer cycles are cooler, while shorter ones are warmer[11]. The amplitude of the cycle is modulated by three other cycles.

The Gleissberg cycle, named after the German astronomer Wolfgang Gleissberg (1903–1986), lasts 88 years and ranges from 70 to 100 years. The Suess–De Vries cycle, named after Hans Suess and the Dutch physicist Hessel De Vries (1916–1959), lasts 208 years and ranges from 170 to 260 years. The Hallstatt cycle, named after a cool and wet period in Europe when glaciers advanced, lasts 2300 years. One of the first to appreciate the modulation of the sunspot cycle was the American astronomer John (Jack) Eddy (1931–2009), who pointed out that '*the long-term envelope of sunspot activity carries the indelible signature of slow changes in solar radiation which surely affect our climate*'[12]. In 1976, he drew attention to the period of zero sunspots between 1645 and 1715 known as the Maunder Minimum, after the English astronomer Edward Walter Maunder (1851–1928)[13]. In 1987, the US National Academy of Sciences awarded Eddy the Arctowski Medal for his work on long solar cycles.

The variability of solar radiation depends on wavelength and can reach 100% in the ultraviolet (UV) part of the spectrum. UV radiation is most effective in the upper atmosphere. A 1% increase in UV radiation at the peak of a solar cycle generates 1–2% more ozone in the stratosphere. This warms the lower stratosphere, strengthens stratospheric winds and displaces poleward the westerly jet streams that steer the mid-latitude storm tracks[3]. These displacements are quite small.

Observations of the sunspot cycle through telescopes began in 1601. Back beyond that, we rely on proxy measurements of solar activity in the form of the

Earth's Climate Evolution, First Edition. Colin P. Summerhayes.
© 2015 John Wiley & Sons, Ltd. Published 2015 by John Wiley & Sons, Ltd.

radioactive nuclides of carbon (^{14}C) and beryllium (^{10}Be), with half-lives of 5730 years and 1.39 million years, respectively. Bombardment by cosmic rays creates these radioisotopes in the upper atmosphere. They are least abundant when there are many sunspots, the Sun's output is high and the solar wind deflects galactic cosmic rays away from our outer atmosphere. Cosmic rays are also deflected when the Earth's internal magnetic field is strong. Subtracting the effect of the magnetic field on the abundance of ^{14}C and ^{10}Be gives us residual ^{14}C and ^{10}Be signals, representing solar output[14]. These signals are trapped in Earth materials, like tree rings. Single living trees can provide annual rings extending back 1000 or more years. Ancient trees preserved in sediments and overlapping in age can provide a continuous annual record dating back several thousand years. A treasure trove of these trees in the flood deposits of rivers like the Rhine and the Danube gets us back to 11 919 years ago[14]. It shows that ^{14}C declined steadily through the Holocene until 1500 years ago. The decline represents the strengthening of the Earth's magnetic field that began ~40 Ka ago, when the field almost disappeared for a few thousand years during the Laschamp geomagnetic event[15]. Because this field deflects cosmic rays, its growing strength led to the production of progressively less ^{14}C in the upper atmosphere. Superimposed on that trend are minor excursions lasting about 100 years, mostly caused by solar activity[16,17]. The excursions in ^{14}C are not a precise guide to solar change, because carbon is influenced by exchange with organisms and the ocean, which damps the solar signal[14,15]. ^{10}Be is not influenced by the carbon cycle, so with both ^{14}C and ^{10}Be from ice cores or tree rings, we can reconstruct the past intensity of solar activity.

Back in the early 1980s, Minze Stuiver found that the ^{14}C in the fossil CO_2 from ice cores increased during the Maunder Minimum, confirming that solar output was weak at that time[16,17]. The solar connection was confirmed when Dansgaard and Oeschger found that ^{10}Be was also abundant then[19,20]. Stuiver found similar increases in ^{14}C at the times of the Wolf Sunspot Minimum (1280–1345 AD) and the Spörer Sunspot Minimum (1420–1540 AD). Later analyses of ^{14}C and ^{10}Be by Manfred Schüssler and Dieter Schmitt of the Max Planck Institute for Solar System Research at Katlenburg-Lindau found maximum values coinciding with the Oort Sunspot Minimum (1040–1080 AD) and the Dalton Sunspot Minimum (1790–1820 AD)[21]. Another sunspot minimum, the Gleissberg Minimum, covered the period 1890–1910 AD[22]. Low ^{14}C and ^{10}Be values, suggesting high sunspot activity,

occurred in medieval times (1100–1250 AD) and the modern era (1960–1990 AD)[16,17].

By 1998, Stuiver and his colleagues had extended the variation of ^{14}C and the solar signal back to 24 Ka ago[18]. Eduard Bard of the Université d'Aix-Marseille and colleagues confirmed Stuiver's findings for solar activity for the past 1200 years, finding radionuclide highs associated with the Maunder, Dalton and Gleissberg sunspot minima[23]. The radionuclide high associated with the Maunder Minimum lay within a rather long period of low solar irradiance between 1450 and 1750 AD. In between the solar minima were solar maxima representing warm periods, including one with values slightly higher than today centred on 1200 AD. Bard concluded that the radionuclide data supported the idea that variations in solar output contributed to the Medieval Warm Period and the Little Ice Age[23]. His findings were refined in 2007 by Raimund Muscheler of Lund University in Sweden[24]. Muscheler and colleagues identified solar maxima tied to warming at 1100–1200 AD (the Medieval Warm Period), at 1750–1800 AD, with a peak at 1790, and at 1960 AD, with somewhat lesser ones at 1370, 1550–1630 and 1850–1870 AD. Intervening solar minima were associated with cool periods typical of the Little Ice Age.

Friedhelm Steinhilber of the Swiss Federal Institute of Aquatic Science and Technology further refined the picture for most of the Holocene in 2012, producing graphs of 'cosmic intensity' through time, where cosmic intensity equals cosmic radiation and is inversely proportional to the number of sunspots (Figure 15.1)[22]. Steinhilber and colleagues confirmed that solar minima and maxima alternated over the past 2000 years. Their data are confined within the limits of what I call the 'natural envelope' of solar variation through time (Figure 15.1). Assuming that this natural envelope does represent solar variability, then, in the absence of any change in the underlying orbital insolation, we would not expect solar variability to increase global temperatures above levels reached in about 1100, 1790 and 1960 AD[a].

Analysing the spectrum of ^{14}C and ^{10}Be data, Stuiver identified the 2300-year-long Hallstatt solar cycle in 1998[18]. By 2004, Raimund Muscheler and colleagues had showed that the 208-year Suess–De Vries cycle, which had persisted for at least the past 50 Ka, caused the Maunder, Spörer, Wolf and Oort solar minima[15].

---

[a] Note added in press: Corrections to the record of sunspot numbers[25] show that between 1749 and the present there was hardly any difference between the peaks of solar activity in 1780–90, 1840–70 and 1960–90. There was no trend in solar cycle amplitude.

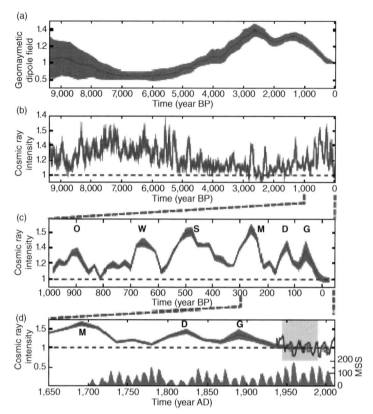

Figure 15.1 *Variations in solar energy with time. (a) Geomagnetic dipole field strength relative to today, with variance. (b) Cosmic radiation based on the first principal component of several radionuclide records, 22-year averages, over the last 8000 years. Time is given as year before present (BP). Grey band represents the standard deviation of the individual radionuclide records. Black dashed line represents the average cosmic ray intensity for 1944–1988 AD. (c) Same as (b), but just for the past millennium. Capital letters mark grand solar minima: O, Oort; W, Wolf; S, Spörer; M, Maunder; D, Dalton; G, Gleissberg. (d) Same as (c), but just for the past 350 years. Time is given as year AD. Circles and zigzag curve for post-1940 are 22-year averages and yearly averages of cosmic ray intensity calculated using the solar modulation potential, obtained from neutron monitor and ionisation chamber data. At bottom are the annual sunspot numbers.*

Paul Damon and Alexie Peristykh of the University of Arizona then identified the 208-year Suess–De Vries cycle and the 88-year Gleissberg cycle and their respective 'overtones' at 104 and 44 years in Greenland ice cores[26]. Overtones recur at half the interval of major peaks. They also found 'combination tones', where signals of different wavelength interfere with one another to accentuate or diminish the underlying signal. Modification of the 88-year Gleissberg cycle by the 208-year Suess cycle produced combination tones at 152 and 61.8 years. They also found the 2300-year Hallstatt cycle in the ice core data.

Heinz Wanner and his team carried out similar spectral analyses, finding peaks in solar variability at 208, 150, 104 and 88 years[27]. Steinhilber's team also confirmed the existence of the Suess–De Vries cycle, which they found to vary with a period of about 2200 years (the Hallstatt cycle). Its highest amplitudes occurred when the Hallstatt cycle was at a low – at 8200, 5500, 2500 and 500 years ago – and its lowest occurred during grand solar minima. They also found the Eddy cycle (~1000 years) and an unnamed cycle (~350 years), among other, less significant cycles[22].

In Steinhilber's data (Figure 15.1)[22], we can sense the warmth of the Roman period (0–600 AD) and the Medieval Warm Period (1050–1250 AD), and the cold of the Dark Ages (600–1050 AD) and the main phase of the Little Ice Age (1250–1750 AD), which then continues with lesser intensity through the Dalton Minimum (1820 AD) and the Gleissberg Minimum (1900 AD), where the fall in cosmic ray intensity and rise in sunspot activity towards the present begins.

15.2   Solar Cycles in the Geological Record

We can see these various solar cycles in the palaeoclimate record. By 1993, Michel Magny of the Laboratoire de Chrono-Environnement in Besançon had found that lake levels in Europe were high, indicating cool and wet conditions, when ^{14}C increased by 5% (= low solar activity), and that glaciers advanced and the tree line descended when ^{14}C increased above 10% (= even lower solar activity) (Figure 15.2)[28]. Cooling and high lake levels occurred during the Maunder and Spörer sunspot minima. There was an earlier 'little ice age' of high lake levels with high ^{14}C at around 750 BC. This was the cool and wet Hallstattzeit, or Hubert Lamb's 'early Iron Age cold epoch'. The cooling of the Little Ice Age, with its ^{14}C peaks at 1500 and 1700 AD, and the cooling of the Hallstattzeit, with its ^{14}C peak at 750 BC, were about 2300 years apart – the period of the Hallstatt solar cycle.

Between 1999 and 2007, Magny and his colleagues used Stuiver's ^{14}C data to extend their analysis of the relationship between ^{14}C peaks and mid-European lake levels back to the start of the Holocene (Figure 15.2)[29–32]. High lake levels[29] indicative of cooling and high precipitation corresponded with peaks in ^{14}C and [10]Be, indicating low solar output. The largest positive ^{14}C excursions and lake high stands of the past 3000 years were those at 700 BC (the Hallstattzeit) and 1500 AD (the Little Ice Age). The larger of the lake high stands corresponded to peaks in Bond cycles of ice-rafting. Although I found Magny's evidence for a correlation between ^{14}C and European lake levels convincing, Wanner thought it was 'very weak'[27].

Complementing Magny's work, in 2009, Dirk Verschuren of Ghent University and Dan Charman, then at the University of Plymouth, found evidence for a decrease in monsoonal activity associated with these cool events, suggesting that the Inter-Tropical Convergence Zone (ITCZ) extended less far north in cool periods (solar minima) than in warm ones[33]. That agreed with the results of Steinhilber's examination of data from the Dongge Cave in southern China, where low $\partial^{18}O$ corresponded to strong Asian monsoons, high solar irradiance and warming[22].

Could quasi-regular variations in the Sun's output explain Bond's cycles? In 1997, Bond and his team thought not – because there was no known solar cycle with such a period. But, working with specialists in solar signals – Raimund Muscheler and Jürg Beer – Bond found by 2001 that the ratio of red-coated (iron-stained) rock grains to total ice-rafted grains in the Holocene correlated with fluctuations in ^{14}C and [10]Be, which were abundant when the Sun cooled[34]. Red ice-rafted grains had a more northerly source, suggesting that cooling forced ice south. The correlation coefficients were not high, at 0.44 for ^{14}C and 0.56 for [10]Be. While the linkage convinced Bond that small cyclic temperature changes, with a period of around 1500 years, might indeed be paced by changes in solar luminosity in the Holocene, they recognised that the large errors in the ages estimated from their marine sediment cores made the correlations with solar output fuzzy. Nevertheless, they considered that the signal was likely to be global, because there was a clear link between strong solar minima, cooler ice-bearing waters in the North Atlantic and reduced monsoonal activity and rainfall in the tropics. Evidently, what they called '*the enigmatic, at best quasi-periodic, "1500-year" cycle*' had to be '*a pervasive feature of the climate system ... linked to variations in solar irradiance*'[34]. They knew that general circulation models (GCMs) implied that '*at times of reduced solar irradiance, the downward-propagating effects triggered by changes in stratospheric ozone lead to cooling of the high northern latitude atmosphere, a slight southward shift of the northern subtropical jet, and a decrease in the Northern Hadley circulation*', which would increase North Atlantic drift ice, cool the ocean surface and the atmosphere above Greenland and reduce precipitation at low latitudes[34]. This cooling and freshening of the northern seas would have temporarily reduced the production of North Atlantic Deep Water. In conclusion, they said, '*If forcing of North Atlantic drift ice and surface hydrography is fundamentally linked to the Sun and begins in the stratosphere, then atmospheric dynamics and their link to the ocean's circulation are much more important for interpreting centennial and millennial time sales of climate variability than has been assumed*'[34].

In 2004, Damon and Peristykh set out to test that interpretation. Among a number of small peaks of millennial

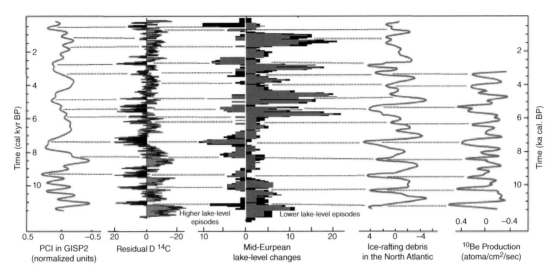

Figure 15.2 *Solar variation and European lake levels. Comparison between the Polar Circulation Index (PCI) at GISP2, the atmospheric residual ^{14}C variations, the Greenland ^{10}Be record, the mid-European phases of higher lake level and the ice-rafting debris (IRD) events in the North Atlantic Ocean.*

duration in the ^{14}C record in Holocene tree rings, they found one at 1433 years – about the same interval as the 1470-year peak in Bond cycles[26]. But this peak was '*too insignificant compared to the noise level to be counted as real*'[26]. Their analysis of the ^{14}C spectrum identified peaks spaced roughly 800 years apart over the past 9000 years. Half of these were larger than the others and spaced around 1600 years apart. These peaks were emergent properties of the system, created by a variety of combination tones and overtones rather than primary solar forcing signals. Nevertheless, they do suggest that emergent millennial variation in the centennial solar cycle at approximately 1600-year intervals might possibly account for the ice-rafting events, even if this millennial variation did not reflect specific independent solar signals with a regular beat.

Holger Braun of the Heidelberg Academy of Sciences agreed. In 2005, Braun and colleagues found that the pronounced solar cycles of the 88-year Gleissberg and 208-year Suess–De Vries cycles could have led to periodic inputs of freshwater to the North Atlantic, such that '*superposition of the two … cycles, together with strongly nonlinear dynamics and the long characteristic timescale of the thermohaline circulation*' could have combined to provide the robust 1470-year climate cycle underpinning Dansgaard–Oeschger events during the last glaciation[35]. Thus, despite the absence of a 1470-year

solar cycle, solar forcing could have triggered those events through nonlinear connections within the climate system. Although that was the case for the last glaciation, they could not get it to apply to the Holocene, possibly because of the large differences in the thermohaline circulation between glacial and Holocene times.

A subsequent, more sophisticated statistical analyses of Bond's Holocene data by a French group in 2007 suggested that his millennial cycles were – as Bond had first thought – most probably caused by internal oceanic oscillations[36]. They did find evidence in Bond's data for solar cycles, but with periodicities of 1000 years (the Eddy cycle) and 2500 years (the Hallstatt cycle), not 1500 years[36].

Nevertheless, in 2014, analysing $\partial^{18}O$ and Mg/Ca ratios from a deep-water sediment core off Iceland, Paola Moffa-Sánchez of Cardiff University found a correlation between solar output, climate and variations in ocean circulation for the period from 818 to 1780 AD[37]. She and her colleagues inferred '*that the hydrographic changes probably reflect variability in the strength of the subpolar gyre associated with changes in atmospheric circulation. Specifically … low solar irradiance promotes the development of frequent and persistent atmospheric blocking events, in which a quasi-stationary high-pressure system in the eastern North Atlantic modifies the flow of the westerly winds'*[37]. They concluded *'that this process could have contributed to the consecutive cold winters documented in Europe during the Little Ice Age'*[37]. The blocking events are mid-latitude weather systems in which a quasi-stationary high-pressure system over the North Atlantic modifies the flow of westerly winds by blocking or diverting their path. Blocking events derive from the meanders in the jet stream, mostly in winter, in association with a negative phase of the North Atlantic Oscillation. They keep western Europe cold by blocking the northeastward transport of warm maritime air from the southwest and replacing it with cold southerly transport of air from Scandinavia or Russia. This pattern caused the cold winters of 1963, 2009, 2010 and 2013. Lamb had already called on blocking to explain the severe winters of the Maunder Minimum[38].

In 2005, a group of palaeoclimatologists comparing their climate data with ^{14}C and ^{10}Be records confirmed that the major cooling event of the past 2000 years occurred during the period of low solar activity that embraces the Little Ice Age[39]. Much like Magny, they concluded that there was *'a first-order relationship between a variable Sun and climate. The relationship is seen on a global scale'*[39].

Given these relationships, it is not entirely clear why the cool periods identified by Wanner and Mayewski and examined in Chapter 14[40, 41] do not correlate more closely with the ^{14}C, ^{10}Be and lake high-stand data and Bond cycles. Three of Mayewski's long cool periods (600–150, 3.5–2.5 Ka and 6–5 Ka ago) correlate with major positive ^{14}C excursions, but one (9–8 Ka ago) corresponds to a major depletion in ^{14}C[41]. All four of his events seem to have been selected based on the abundance of potassium (K^{+}) ions in the GISP 2 ice core. The event at 9–8 Ka ago occurred at a time of abundant volcanic activity, which may explain its high K^{+} signal.

Considering these various analyses, it seems highly likely that most of the short cool periods of the Holocene represent periods of low amplitude in century-scale solar cycles. These cool periods occurred irregularly within a cycle with a broad envelope of 1–2 Ka, much as Bond had suggested in moving away from a precisely defined 1500-year signal[34, 42]. They led to high European lake levels, North Atlantic ice-rafting and associated freshwater outbreaks during Bond cycles, which slowed the Thermohaline Conveyor system temporarily. That, in turn, transmitted Northern Hemisphere cooling signals slowly to the Southern Ocean, thereby putting Southern Hemisphere responses to the same forcing events out of phase with their Northern Hemisphere counterparts,

and thus making the identification of originally global signals of solar output more difficult. Wanner reminded us that, given the different hemispheric responses to solar changes, we should not expect uniform global warming or cooling – there is, and ought to be, significant regional heterogeneity[27]. The incidence of large volcanic eruptions complicates matters by contributing additional cooling at times.

Should we be confident of the results emerging from examinations of the increasing flood of proxy solar data? By the late 1980s, Lamb knew that variations in solar activity might have controlled the climate of the past 2000 years, but he was not convinced that it had, because the signal of climate change was too heterogeneous[43]. In his day, not only were the signals of climate change geographically heterogeneous, but also the timings of so-called climatic events did not coincide precisely with the extent of known solar events. Even as recently as 2012, Gerald North of Texas A & M University agreed that current palaeoclimatic evidence for solar forcing is limited – it is difficult to find clear indications of the impact of solar variations on climate[44]. Most records do not provide the necessary resolution or signal strength to detect a weak solar signal even if it is present. Other forcing factors, like volcanic eruptions or oceanic oscillations, may be operating on similar time scales, thus complicating signal detection at certain frequencies. For instance, more than 90% of the variance in temperatures in the central England temperature record since about 1650 AD can be accounted for by volcanic activity or by fluctuations in the Atlantic Multidecadal Oscillation. That begs the question, to what extent were these oscillations linked to solar activity? In the Pacific, there may be a link between peak sunspot years and cool La Niña events. North and his team called for more interdisciplinary research and better linkages between data and models as the basis for recognising solar signals in the palaeoclimate record[44]. Looking at the distribution of key radionuclides (^{14}C and ^{10}Be), Cronin observed, in 2010, that *'geomagnetic reconstructions vary greatly depending on the region and source of measurements, that equally complex obstacles surround reconstruction of long-term modulation of solar energy flux by solar wind and other processes, and that these factors preclude a simple quantification of millennial and short-term irradiance changes from nuclear records ... [thus,] a great deal of caution is needed when interpreting paleoclimate records before A.D. 1600 because reconstructed radionuclide records are not direct measures of solar activity'*[45].

Despite these conclusions, it seems to me that we now have, in Steinhilber's 2012 data[22], an up-to-date view of solar variability with time, along with stronger evidence in Moffa-Sánchez's 2014 report[37] for a clear relationship between local climate and solar forcing in the North Atlantic and good evidence in Magny's data[29] for a direct link between solar activity and European lake levels. If the association between declining solar activity and cooling proves to be robust then the present decline in solar forcing suggests a return to more severe European winters – unless it is countered by anthropogenic global warming.

Before moving on to examine the Medieval Warm Period and the Little Ice Age in detail, I should comment on the theory that an increase in cosmic rays causes an increase in cloud cover that cools climate. The relationship between the solar cycle and the Earth's magnetic field described in this section offers a unique opportunity to test that notion[46]. If the hypothesis is correct, temperatures should have cooled during the Laschamp geomagnetic reversal. Comparing ^{10}Be and ∂^{18}O in Greenland ice cores shows that Greenland did not cool during the reversal[15]. The latest evaluation of the effect of cosmic rays on climate, by Mike Lockwood (1954–) of the Meteorology Department of the University of Reading in 2012, concluded that galactic cosmic rays provide an '*increasingly inadequate explanation of observations*'[47]. A recent review of solar influences on climate agreed that current data do not support the postulated link between cloud cover and cosmic rays[48].

15.3   The Medieval Warm Period and the Little Ice Age

In the context of global warming, the best known, or the most talked about, climatic events during the Holocene are the Medieval Warm Period and the Little Ice Age. These events fascinated Hubert Lamb[49]. During the Little Ice Age, he reported, the Thames froze over four times in the 1500s, eight times in the 1600s and six times in the 1700s, encouraging people to hold 'frost fairs' on the ice. The average temperature for January in central England in the 1780s was 2.5 °C lower than it was in the 1920s and 1930s, although summers in southern England were not much different in the 1780s from what they were in the 1950s. The Gulf Stream was further south in 1780–1820 than it is now, and ice was more extensive in northern seas. Nevertheless, there were warm periods

within the Little Ice Age, in the 1630s, 1730s, 1770s and 1840s. Superimposed on these broad patterns were large oscillations lasting on the order of 20–60 years, affecting ice extent and winter temperatures, especially in Europe – where, for instance, there were cold winters with much ice in the 1880s and early 1890s. I have seen photographs of the Thames frozen over in London in February 1896, but there were hardly any cold winters between 1896/97 and 1939/40. Lamb thought that global temperatures fell 1–2 °C in the Little Ice Age, which he dated as ranging from 1430 to 1850, with its coldest period being from 1550 to 1700. His observation that the Little Ice Age was primarily a wintertime phenomenon has been confirmed by more modern data[10]. In considering his estimates of temperature, we have to bear in mind that much of his data was from western Europe.

In September 1964, Lamb presented to the Southampton meeting of the British Association for the Advancement of Science an illustration of temperature change through time that would continue to cause confusion about climate change in the minds of the general public almost 50 years later (Figure 15.3)[50]. Entitled 'Temperatures (°C) Prevailing in Central England, 50 Year Averages', his illustration included graphs for average annual, summer and winter temperatures since 900 AD in central England. He grouped the data into 50-year averages to eliminate high-frequency noise caused by decadal variability. The graph of annual averages shows a smooth rise from 900 AD to a broad peak between 1140 and 1270 AD, representing his 'Early Middle Ages Warm Epoch'). The following Little Ice Age was represented by low values in ~1470 AD and between 1570 and 1670 AD, slight rises in ~1520 AD and between 1720 and 1870 AD, and a final rise to 1950 AD. Much the same figure appeared in 1988 in Lamb's book, *Climate, History and the Modern World*[43].

Lamb's graph of annual average temperatures for the past 1000 years *for central England* (Figure 15.3) became iconic following its reproduction more or less unchanged in Chapter 7 of the First Assessment Report of the Intergovernmental Panel on Climate Change (IPCC) in 1990 with the misleading title 'Schematic Diagrams of **Global** Temperature Variations Since the Last Thousand Years'[51] (my emphasis) (Figure 15.4). Although the caption cited no specific source for the figure, the text of the report referred to a 1988 paper by Lamb as the probable source[52]. Not surprisingly, it is remarkably similar to the one in his 1988 book[43] and his 1964 Southampton paper[50]. Given that caption, readers could be forgiven for taking it as a

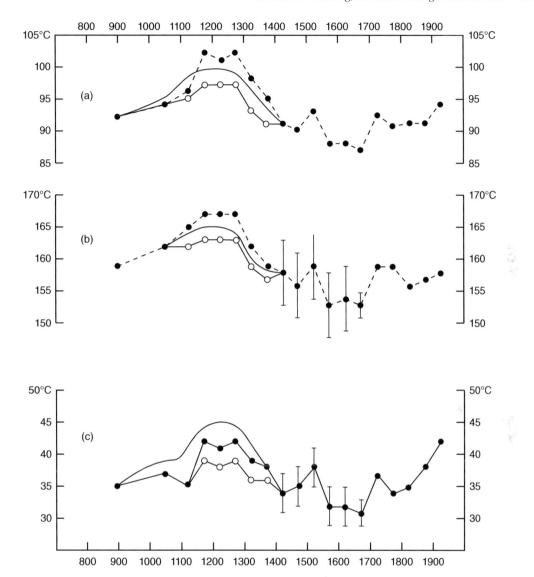

Figure 15.3 *Lamb's graphs of temperature in central England, showing 50-year averages: (a) annual, (b) summer (July and August), (c) winter (December, January and February). The ranges indicated by vertical bars are three times the standard error of the estimates. Heavy solid line (from 1680) indicates observed values. Fine dotted line indicates unadjusted values based on purely meteorological evidence. Heavy dashed line indicates preferred values, including temperatures adjusted to fit botanical indications. Heavy dotted line (pre 1150 AD) connects points corresponding to 100–200-year means indicated by sparse data. Thin solid line (back from 1400 AD) is Lamb's preferred option.*

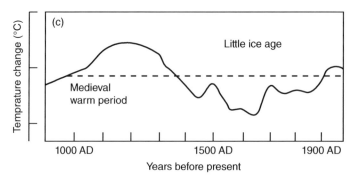

Figure 15.4   *The figure the IPCC got wrong – Lamb's graph of English temperature taken as global. Captioned 'Schematic Diagrams of Global Temperature Variations Since the Last Thousand Years. The Dashed Line Nominally Represents Conditions Near the Beginning of the Twentieth Century'.*

representation of *global temperature*, rather than what it was was in fact: the *temperature of central England*[b].

How widespread was Lamb's 'Early Middle Ages Warm Epoch' (now the Medieval Warm Period) and what was its magnitude? This is a question that has come to bedevil discussions about global warming. Many of those who think the Medieval Warm Period was warm everywhere seem to have forgotten, if they ever knew, that Lamb saw it as asymmetric over the Northern Hemisphere[43]. While medieval warming was widespread across Europe, North America and the European Arctic, it was cold in China and Japan, where there was warming from 650 to 850 AD, when it was cold in Europe[43] Even within the Atlantic sector, there were variations from one part to another. The warm phase passed its peak in Greenland, Iceland and eastern Europe in the 12th century, but continued in western Europe, reaching a maximum about 1300 AD[43]. Clearly, the medieval warming could not be considered global even based just on Northern Hemisphere data. This does not mean it was not triggered by a global solar event. It means only that processes internal to the Earth's climate system interfere with such signals, so they don't have a uniform global impact.

By 1995, Lamb's view that the medieval warm epoch might not be global had gained some traction amongst the climate cognoscenti. By then, a great many proxy data for temperature over the past 1000 years had been obtained

from isotopic analyses and assessments of changes in tree rings[53]. In Chapter 3 of the IPCC's Second Assessment Report, published in 1996, lead author Neville Nicholls from the Australian Bureau of Meteorology Research Centre concluded that the evidence for a Medieval Warm Period between the 9th and 14th centuries was geographically limited and equivocal[53]. Evidence for it was clear for parts of Europe, but elsewhere there was no such evidence, or else warmer conditions prevailed, but at different times. More and better-calibrated proxy records were needed in order to obtain a clearer picture. Bearing that in mind, Nicholls concluded: *'it is not possible to say whether, on a hemispheric scale, temperatures declined from the 11-12th to the 16-17th century. Nor, therefore, is it possible to conclude that global temperatures in the Medieval Warm Period were comparable to the warm decades of the late 20th century'*[53]. The emphasis in the last sentence should lie on the word *global*, as it is clear that *European* temperatures in the Medieval Warm Period were warmer than they were in the cool 1960s, when Lamb produced his graphs.

With regard to the Little Ice Age, Nicholls concluded that *'the climate of the last few centuries was more spatially and temporally complex than this simple concept implies ... It was a period of both warm and cold climatic anomalies that varied in importance geographically'*[53]. Nevertheless, *'despite the spatial and temporal complexity, it does appear that much of the world was cooler in the few centuries prior to the present century'*[53].

Nicholls and his team drew heavily on three papers by the climatologist Raymond S. Bradley (1948–), of the University of Massachusetts at Amherst, and Lamb's

---

[b] Only after noticing this error did I discover that this had earlier been pointed out (rather obscurely) in Appendix A to a paper by Jones, P.D., Briffa, K.R., Osborn, T.J., Lough, J.M., et al. (2009) High-resolution palaeoclimatology of the last millennium: a review of current status and future prospects. The Holocene 19(1), 3–49.

colleague Phil Jones, of East Anglia's Climate Research Unit (CRU)[54-56]. Jones was listed as a key contributor to Nicholls' IPCC report, and Bradley also contributed to it[53]. In due course, Bradley was to rise to the position of research director of UMass Amherst's Climate System Research Centre, and Jones to the position of director of the CRU. Jones would later star in the 'Climategate' saga of the winter of 2009/10.

Lamb, too, documented the climate of the Little Ice Age, finding that, unlike the period around 1000 AD, temperatures in China closely paralleled those in Europe at that time. Most of North America was also cold[49]. He found almost no cooling in the tropics, and rather mild conditions in Antarctica.

By the early 2000s, the Little Ice Age was seen as divisible into two main parts: an early Little Ice Age phase, or Medieval Cold Period, between about 1250 and 1550 AD, and the main Little Ice Age, lasting from 1550 to 1850 AD, with peak cooling between 1750 and 1850 AD[33,57]. Crowley and North agreed that the main Little Ice Age '*consisted of two main cold stages of about a century's length … in the seventeenth and nineteenth centuries … [and that] the coldest decades occurred in the mid-late 1600s, the early 1800s, and the late 1800s*'[9]. The start of the early Little Ice Age phase coincided with pronounced glacial advance in the European Alps, an expansion of sea ice around Iceland and a return to pluvial conditions following medieval drought in tropical Africa[33]. Crowley and North pointed out that '*The Little Ice Age is considered to end around 1890, although different authors might choose slightly different dates*'[9]. For instance, in eastern Africa it ended with an abrupt switch to wet conditions in the early 1800s[33]. Later, we will assess evidence for the end to the Little Ice Age.

Lamb associated the Little Ice Age with a weakening of both atmospheric circulation and incident radiation[49]. As we saw in Chapter 14, he thought that a veil of dust from volcanic activity might have caused this weakening. Gerard Bond disagreed, because his team had identified the Little Ice Age as one of his ice-rafting cycles. Linking the Little Ice Age to the cold phase of Bond's persistent climate cycle ruled out volcanism as a cause, since volcanism would not be expected to vary in a continuous and regular manner[42]. Wanner agreed with Bond, because the distribution of large climate-changing eruptions over the past 6 Ka was '*highly inhomogeneous*'[27]. Nevertheless, Wanner reminded us, '*more large tropical volcanic eruptions [of the kind likely to cool the climate for a year or two] have occurred during certain intervals of the last millennium, i.e. between AD 1200 and 1350 or*

around AD 1700 and 1800, than at other times during the Holocene. These maxima of volcanic activity happen to coincide with both low orbitally induced insolation in the [Northern Hemisphere] and an unusual concentration of solar activity minima. Therefore, it seems plausible that the cold intervals of the past millennium, including the LIA [Little Ice Age], might be attributed to a combination of orbital, volcanic and solar forcing'[27]. To be more explicit, Wanner attributed the cooling of the Little Ice Age to three factors: the continued decrease in insolation attributable to orbital forcing, occasional declines in solar activity (which we now know were driven by the 208-year Suess–De Vries cycle) and the output of some large tropical volcanoes[27]. The PAGES team identified five particular short periods of cooling attributable to either volcanic activity or a combination of volcanic and solar activity across the past 2000 years, at: 1251–1310 AD (initial volcanic activity followed by the solar low of the Wolf Minimum); 1431–1520 AD (initial volcanic activity within the solar low of the Spörer Minimum); 1581–1610 AD (volcanic activity during a sunspot high); 1641–1700 AD (combined volcanic activity within the Maunder Minimum); and 1791–1820 AD (volcanic activity followed by and overlapping with the Dalton Minimum)[58].

Can the relative effects of volcanic and solar activity be disentangled? Andrew Schurer and colleagues at the University of Edinburgh attempted to do so across the Northern Hemisphere for the past millennium[59]. They concluded that '*Although solar forcing may be relatively unimportant for large-scale climate change, it could still play a significant role in regional and seasonal variability, owing to its influence on climate dynamics, an influence that is strongly diminished when averaging annually and over the whole Northern Hemisphere*'[59]. Instead, they thought that the changes of the past millennium were influenced more by volcanic eruptions and changes in greenhouse gas concentrations than by changes in solar output. When I examined their data, I found 12 major volcanic eruptions within the solar minima between 800 and 2000 AD, and 15 within the intervening solar maxima. I therefore remain unconvinced that volcanic activity was a significant control on decadal to centennial change through the last millennium. Recognising that volcanic activity generally leads to cooling lasting no more than a few months to 2 years, it seems far more likely to me that Schurer's team underestimated the solar effect.

Undoubtedly, eruptive events did influence the climate of the past 1000 years. For instance, Rosanne D'Arrigo of Lamont and colleagues found some connections between low-latitude volcanic eruptions and tropical climate over the past 400 years[60]. Among the eruptions causing temporary global cooling were those of Tambora in 1815 and Krakatoa in 1883. The stratospheric veil of dust and sulphuric acid droplets caused by these eruptions scattered incoming solar radiation and led to fine red sunsets, which may have influenced the paintings of J.M.W. Turner (following the eruption of Tambora) and of Edvard Munch (following the eruption of Krakatoa). Tambora exploded during the Dalton Sunspot Minimum, and temporarily contributed to its cooling effects. This combination led to the 'year without a summer', when cooling and excessive rains made the harvests fail in western Europe in 1816. Similar effects may have resulted from the eruptions in 1902 and 1911/12 that coincided with the Gleissberg Minimum. Even so, my calculations suggest that it is unwise to assume that this association weakens the argument for the importance of solar activity. After all, large eruptions are short-lived, while solar minima are not.

Until 2013, the source of the volcanic activity of the period 1251–1310, identified from ash in polar ice, was a mystery. We now know that in 1257 there was a colossal eruption of the Salamas volcano on Lombok Island in Indonesia. It was among the largest eruptions of the Holocene, with a dust column reaching an altitude of 43 km and depositing 40 km[3] of volcanic ash[61]. This eruption may explain the rapidity with which the Little Ice Age set in. A team headed by Gifford Miller of the University of Boulder, Colorado found in 2012 that *'precisely dated records of ice-cap growth from Arctic Canada and Iceland [show] that LIA summer cold and ice growth began abruptly between 1275 and 1300 AD, followed by a substantial intensification 1430-1455 AD. Intervals of sudden ice growth coincide with two of the most volcanically perturbed half centuries of the past millennium'*[62]. Sea ice cover also expanded north of Iceland at these times: *'The persistence of cold summers is best explained by consequent sea-ice/ocean feedbacks during a hemispheric summer insolation minimum – large changes in solar irradiance are not required'*[62]. Miller viewed the underlying low orbital insolation of the late Holocene as the main contributor to the coolness of the Little Ice Age. Volcanism and sunspot minima exacerbated what would have been a cold period in any case. Once the sea ice formed, and after the volcanic activity ceased, low insolation and brief sunspot minima helped to maintain a large area of sea ice that suppressed regional summer temperatures for centuries. According to Miller, *'an explanation of the LIA does not require a solar trigger'*[62]. Nor does it require a volcanic one, although the recovery of

the climate system after a very large eruption may take a decade, due to the persistence of local cold anomalies in the ocean. Orbital insolation is quite enough by itself to make cold summers persist.

The PAGES team agreed that declining insolation was an important forcing factor. The cooling of the continents over the past 2 Ka was *'consistent with the cooling of global sea surface temperatures from year 1 to 1800 CE [Common Era, or AD] exhibited in the PAGES Ocean2k synthesis'*[58]. The continents cooled at about 0.2 °C per 1000 years over that period. In mid to high latitudes of the Northern Hemisphere, this was due to a decrease in orbitally driven local summer insolation, while in the Southern Hemisphere it was a delayed response to the decrease in spring insolation modulated by the Southern Ocean's thermal inertia.

What temperature changes could be discerned in the Medieval Warm Period and the Little Ice Age? Lamb concluded in the mid 1960s that – compared with then present values – temperatures were raised 2–3 °C in the warm epochs of the Holocene and lowered 1–2 °C in the Little Ice Age, which was a time of glacial advance in virtually all of the world's mountain regions, and of advancing sea ice around Iceland[49]. Were his temperature suggestions reasonable? No. His data base was skewed towards the United Kingdom in particular and western Europe in general. By 2000, Bard thought that the world had cooled by 0.5–1.0 °C during the Little Ice Age, and that the Medieval Warm Period had been as warm as the mid 20th century[23].

In 2001, researchers used $\partial^{18}O$ analyses on fossil shells from lake sediments in Alaska's Alaska Range to extract a record of growing-season temperatures for the past 2000 years. They found three periods of comparable warmth: 0–300 AD (Roman Period), 850–1200 AD (Medieval Warm Period) and post 1800 AD[63]. The Little Ice Age peaked in 1700 AD with a climate 1.7 °C colder than in 2000 AD. Another prominent cooling event at 600 AD corresponded to the European Dark Ages. Both cold periods were times of significant glacial advance in Alaska and were wetter than the warmer periods. Later work by Michael Loso of Alaska Pacific University in Anchorage on a varved glacial lake from southern Alaska showed that temperatures over the past 1500 years varied by about 1.1 °C, with the maximum in the late 20th century[64]. The Medieval Warm Period there was cooler than the climate of recent decades.

Jan Esper of the Swiss Federal Research Institute was keen to test the idea that tree rings could provide an accurate picture of climate change over the past 1000 years.

Analysing tree rings from high elevation and middle- to high-latitude sites in the Northern Hemisphere, Esper and colleagues showed in 2012 that temperatures were above average in the Medieval Warm Period (900–1300 AD), although there were local differences in the timing of peak warmth then, indicating a high degree of spatial variability in the Medieval Warm Period signal[65]. Low temperatures characterised the Little Ice Age between 1200 and 1850 AD, followed by a warming trend. Esper wondered how reliable tree-ring data were, and in a later paper showed that differences of up to 0.5 °C could be derived by different research teams from much the same data, due to differences in the methods used to calculate temperature. Great care was required in interpreting results derived by different teams. Ideally, all researchers should follow the same methodology[66].

By 2005, a team led by Anders Moberg from the department of meteorology of Stockholm University found that high temperatures like those observed in the 20th century before 1990 had occurred in the Northern Hemisphere around 1000–1100 AD in the Medieval Warm Period, and that minimum temperatures of about 0.7 °C below the average of 1961–90 had occurred around 1600 AD, during the Little Ice Age[67]. Jasper Kirby of the European Organization for Nuclear Research (CERN) in Switzerland confirmed that temperatures in the Medieval Warm Period in Europe were about 1.7 °C warmer than at their minimum in the Little Ice Age[68]. As we saw in Figure 15.1, the cosmic intensity values typical of the Medieval Warm Period were about the same as those in the first part of the 20th century[22], suggesting that the Medieval Warm Period was not warmer than the early part of the 20th century, except, as Lamb showed, in western Europe.

More recently, in 2012, a team from Stockholm University led by Fredrik Ljungqvist used a large set of temperature-sensitive proxies and historical records to assess the temperature history of the Northern Hemisphere back to about 800 AD (Figure 15.5)[69]. They found widespread positive temperature anomalies from the 9th to the 11th centuries comparable to the 20th-century mean and widespread negative anomalies from the 16th to the 18th centuries. Most of their data were from Europe, Greenland, North America and China. More data are needed from interior Asia, North Africa and the Middle East. Temperatures varied regionally: *'almost all of North America, western Europe and much of central and eastern Asia warmed from the 17th to the 18th century but not Greenland, eastern Europe and northwestern Asia. Notable cooling occurred from the 18th to 19th century in* *northern Europe and much of Asia except in the south to southwest. This cooling caused the 19th century to be the coldest over much of northwestern Eurasia'*[69].

In 2011, a team led by Kristin Werner of the Leibnitz Institute for Marine Science used climate proxies from a marine sediment core from the Svalbard continental margin on the east side of Fram Strait to show that warm water inflow from the Atlantic led to seasonally ice-free conditions at about 80° N during the Medieval Warm Period[70]. Summer sea surface temperatures reached 4.4 °C between 650 and 1400 AD. The site was at or close to the sea ice margin in the Little Ice Age after 1400 AD and was covered by extensive sea ice and icebergs after 1730 AD. The coldest part of the Little Ice Age on Svalbard occurred between 1760 and 1900 AD. Warm Atlantic water inflow increased after about 1860 AD. Trond Dokken of the Bjerknes Centre for Climate Research, Bergen, Norway noted that warm subsurface Atlantic water periodically flows into Nordic seas beneath the sea ice, destabilising it and rapidly warming the region, leading to formation of a freshwater cap, associated initially with ice-rafting. Dokken and colleagues suggested that the interplay between warm inflow and local sea ice played an important role in the development of Dansgaard–Oeschger events, as we saw in Chapter 13[71]. It may have helped control Bond Cycles, and it may be operating now, promoting development of an ice-free Arctic[71].

On the west side of Fram Strait, analyses of winter-season $\partial^{18}O$ ratios and borehole temperatures in Greenland ice cores suggested to Bo Vinther and colleagues, in 2010, that *'temperatures during the warmest intervals of the Medieval Warm Period were as warm as or slightly warmer than present day Greenland temperatures'*[72]. The following year, Takuro Kobashi of the National Institute of Polar Research in Tokyo, working with Vinther and colleagues, found that the average modern snow temperature at the GISP 2 site near the summit of the Greenland Ice Sheet was −29.9 °C (for the period 2001–10) and had not yet risen above the natural variability of the past 4 Ka, which averaged −30.7 ± 1.0 °C[73]. Over the past 2 Ka, the warmest temperatures, rising some 2 °C above the average, were at 750 AD, after which they fell to about −33.4 °C at the heart of the Little Ice Age, close to 1700 AD. Brief peaks of warmth, about the same level as today's, occurred during the Medieval Warm Period in the 1140s and in the 1930s–40s. Temperatures were generally higher early in the past 4 Ka, which is consistent with the decline in insolation since then. The authors recognised that their temperature reconstruction differed

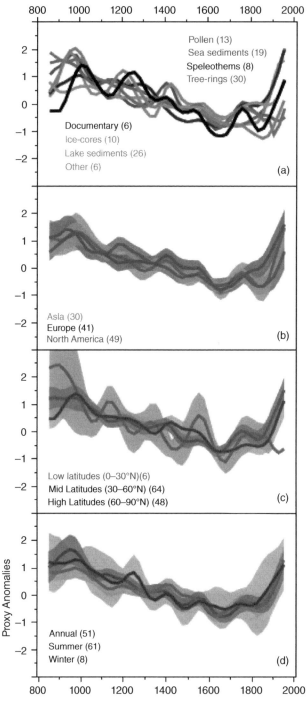

Figure 15.5 *Mean time series of centennial proxy anomalies separated by (a) data type, (b) continents, (c) latitude and (d) seasonality of signal. The curves in (b–d) show the mean confidence intervals (±2σ). The numbers in parentheses indicates the number of proxies in each category.*

from the reconstruction of Arctic summer air temperature over the past 2 Ka, which showed a long cooling trend, ending with pronounced warming[73]. They attributed this to three factors: first, the Greenland record is a mean annual temperature, rather than a summer temperature, whereas Arctic temperatures may represent summer maxima; second, Greenland temperature may be affected by cloud cover and wind speed; and, third, Greenland is not representative of the whole Arctic[73]. Then again, Greenland is a large block of ice and is likely to have its own microclimate.

Reviewing Arctic temperature records, Miller and his team found that the most consistent records of medieval warming came from the North Atlantic sector, and that the evidence for medieval warmth elsewhere in the Arctic was less clear. This regional variability meant that *'the Arctic as a whole was not anomalously warm throughout Medieval time'*[62].

Studies of the Great Barrier Reef and coral islands in the tropical southwest Pacific led Erica Hendry of the Australian National University, Canberra to conclude that, during the Little Ice Age, local surface waters were relatively warm and saline, and that *'Cooling and abrupt freshening of the tropical southwestern Pacific coincided with the weakening of atmospheric circulation at the end of the LIA, when glaciers worldwide began to retreat'*[74]. She suggested that the Equator-to-pole temperature gradient steepened during the Little Ice Age, strengthening the easterly Trade Winds, which pushed warm surface waters west towards the reefs. These warm waters would also have been more subject to evaporation, possibly providing a source for the moisture that helped polar glaciers to grow between 1600 and 1860 AD.

Analysis of a coastal sediment core from San Francisco Bay showed Mary McGann of the US Geological Survey that the Medieval Warm Period was warm and dry on the California coast, while the Little Ice Age was humid, with an influx of freshwater[75]. A 500-year record of alkenone temperatures ($U^{k'}_{37}$) from deep-sea cores along the Californian margin showed no dip in sea surface temperature during the Little Ice Age, probably because the entire core came from within the Little Ice Age[76]. The tops of the cores carried evidence of recent warming[76].

In 2006, Paul Mayewski and his colleague Kirk Allen Maasch used the abundances of sodium (Na^+) and calcium (Ca^{2+}) ions in ice cores to map polar climate change over the past 2000 years[77]. Stronger winds blow more dust and sea salt into the interiors of ice sheets, so changes in those components seen in a network of cores over an ice sheet provide information about changes in wind speed and direction, and hence changes in atmospheric circulation, for comparison with changes in temperature deduced from $\partial^{18}O$ or ∂^2D measurements. The warm temperatures typical of the Medieval Warm Period (800–1400 AD) were associated with lower wind strengths, while the cool temperatures typical of the Little Ice Age (1401–1930 AD) were associated with higher wind strengths (Figure 15.6).

Comparing the behaviour of the two polar regions, Mayewski and Maasch found that the warming associated with the Medieval Warm Period and the cooling associated with the Little Ice Age seemed to begin 300–400 years earlier in the south than in the north and were much stronger in the north[77]. Furthermore, while the peak of the Little Ice Age appeared to coincide with the Maunder Minimum in solar activity in the north, it did not in the south. For those reasons, they advised against applying the same terminology to the Antarctic. In contrast with this lag in time between Northern and Southern Hemisphere events, they observed that present-day warming was unusual in occurring simultaneously in both hemispheres – there was no lag. Their observation that the cooling between the Medieval Warm Period and the Little Ice Age began earlier in the south than in the north begs the question, is this another example of the climate seesaw, with warming starting first in the Southern Ocean and being transmitted to the north through the Atlantic Meridional Overturning Circulation?

In 2009, I worked with Mayewski and others on a review of the state of the Antarctic and Southern Ocean ecosystem[78]. Among other things, we found that the climate of West Antarctica, represented by an ice core from Siple Dome in the Ross Sea area, began to warm around 6 Ka ago, in parallel with the rise in both summer and winter insolation at 60° S and the gentle global rise in CO_2. This warming steepened after around 1750 AD, following the curve of rising CO_2. Over the past 1200 years, the concentrations of calcium and sea salt increased at Siple Dome, telling us that the westerly winds were getting stronger there, bringing in more dust from surrounding continents and warm air from the open ocean. In contrast, in East Antarctica – represented by an ice core from Law Dome – we saw a cooling trend continuing from around 1300 AD to the present. The opposing climatic trends in East and West Antarctica tell us about the development of the wind system around the continent. In a perfect world, the westerly winds would run in a circle around the South Pole. But the Antarctic is far from perfect. The biggest ice cube in the world, the enormous upstanding mass of East Antarctica, which reaches a height of 4092 m, is offset to the east of the Pole, distorting the wind pattern and

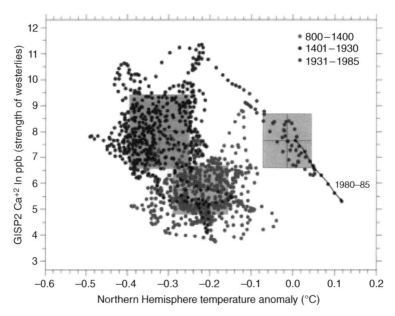

Figure 15.6 *Phase diagram for Northern Hemisphere temperature versus an ice core proxy for Northern Hemisphere westerlies (calcium ions in the GISP ice core). Cluster over box at lower right centre = 800–1400 AD; cluster over box at upper left centre = 1401–1930 AD; cluster over box at far right and along the line of dots stretching from mid top centre to lower right = 1931–85 AD. Shaded boxes represent the mean ± 1σ for each period. Black line = 1980–85, the last 5 years of the data record. Clearly, climatic conditions since 1930 diverge from those of both the Medieval Warm Period and the Little Ice Age.*

allowing the formation of a giant eddy – the Amundsen Sea low-pressure centre – over West Antarctica in the lee of the ice cube[79]. This 'Amundsen Sea Low' pulls warm air from the north down the Antarctic Peninsula and carries warm, salt-laden marine air over West Antarctica's Siple Dome. Intensification of the Amundsen Sea Low with time accounts for the difference in climate regimes between East and West Antarctica.

In a follow-up study in 2013, Mayewski, Maasch and colleagues examined West Antarctica's sensitivity to natural and human-forced climate change over the Holocene[80]. Comparing ice cores from Siple Dome in West Antarctic and Taylor Dome in East Antarctica, they suggested that global warming over the past 150 years strengthened the Southern Ocean westerlies and made them migrate towards Antarctica. This change in zonal wind flow warmed West Antarctic and the Antarctic Peninsula, while cooling East Antarctica, much as described by SCAR scientists in their reports on *Antarctic Climate Change and the Environment* in 2009[78, 79]. They also observed that the wind system intensified abruptly both in Greenland

and at Siple Dome around 1400 AD, close to the beginning of the Little Ice Age. While this could suggest the global imprint of a solar signal, there was no abrupt solar event at the time[22]. Instead, this intensification could reflect a global response to the atmospheric signal of the major volcanic episodes identified by Miller and his team as taking place in the Arctic between 1430 and 1455 AD[62].

Antarctic studies highlight the difference in climate between the high-standing block of East Antarctica (averaging 3000 m high) and the much lower block of West Antarctica (averaging 1800 m high)[78, 79]. Warm, wet air makes precipitation high in West Antarctica, while precipitation is as low as that in a desert over the high and extremely cold interior of East Antarctica.

There is no sign either at James Ross Island east of the Antarctic Peninsula or in the Bellingshausen Sea west of the Peninsula of the Little Ice Age or the Medieval Warm Period (Figure 15.7). Indeed, between 1671 and 1777 AD, at the heart of the Little Ice Age in the Northern Hemisphere, temperatures on James Ross Island rose as rapidly as 1.5 °C per century[81]. Sea ice expanded around

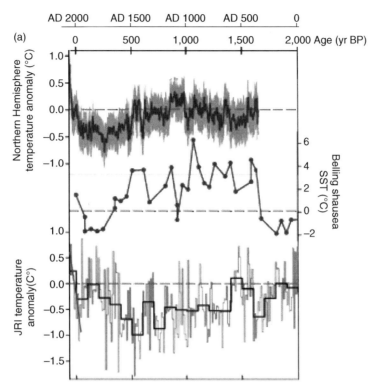

Figure 15.7   *Temperature history of the Antarctic Peninsula over 2000 years. Lower panel: the James Ross Island (JRI) ice core temperature reconstruction with 100-year averaging (heavy line) and 10-year averaging (grey ranges) relative to the 1961–1990 mean (dashed line). Warming by 1.56°C over the past 100 years is highly unusual in the context of natural variability. Middle panel: sea surface temperature (SST) record from Ocean Drilling Program site 1098 in Bellingshausen Sea west of the Antarctic Peninsula. Upper panel: reconstructed Northern Hemisphere temperature anomaly relative to the 1961–1990 mean, with envelope showing 95% confidence interval.*

the Antarctic Peninsula as the temperatures fell from the Holocene thermal optimum towards the late Holocene[82].

Jennifer Pike of Cardiff University and colleagues concluded from their study of Palmer Deep, west of the Antarctic Peninsula, that warming there from about 2200 years ago was driven by peak summer insolation at 60° S and an increase in the intensity of La Niña events, which drove increasing discharge of glacial ice from the land[83]. What was not clear was the cause of the increase in La Niña events in the equatorial Pacific, although, as we shall see later, such increases are consistent with the development of the Medieval Warm Period and may have a solar link.

In a 2008 review of global climate change in the mid to late Holocene, Wanner confirmed that the time of

the medieval warm peak was not simultaneous around the globe[27]. Among other things, that convinced him that the transition between the Medieval Warm Period and Little Ice Age was not caused by external forcing. In contrast, the glacial maxima at the time of the Little Ice Age were reasonably synchronous around the world[27]. By 2013, as a member of the PAGES team, Wanner had firmed up his opinions[58]. The PAGES team's 2013 compilation of 511 data sets for the past 2000 years proved that '*at multi-decadal to centennial scales temperature variability shows distinctly different regional patterns, with more similarity within each hemisphere than between them. There were no globally synchronous multi-decadal warm or cold intervals that define a worldwide Medieval Warm Period or Little Ice Age, but all reconstructions show generally cold conditions between 1580 and 1880 CE, punctuated in some regions by warm decades during the 18th century. The transition to these colder conditions occurred earlier in the Arctic, Europe and Asia than in North America or the Southern Hemisphere regions*'[58]. In reaching those conclusions, the PAGES team averaged data for each continent. By doing so, as pointed out earlier, they missed the differentiation between East and West Antarctica noted by Mayewski[80].

The evidently complex global distribution of climatic patterns during both the Medieval Warm Period and the Little Ice Age underscores the perceptive observation of Damon and Peristykh that, because '*climate is complex involving the immense thermohaline currents of the oceans, global circulation of air masses, cyclical phenomena like the El Niño-La Niña, quasi-decadal circulation of the oceans, aerosols injected into the stratosphere by volcanoes, etc ... even during a climatic event like the Little Ice Age only regionally sensitive areas will be observably cooler*' (my emphasis)[26]. Jonathan Cowie also reminded us that we would be unwise to expect a uniformly global effect, given that the Northern Hemisphere is predominantly land, while the Southern Hemisphere is dominantly ocean, and both react differently to incoming solar energy[84,c].

Bard and his colleagues thought that climate models would not necessarily show the Medieval Warm Period and Little Ice Age as global, because advective processes that moved heat around in the atmosphere meant that not every location would experience cooling in the Little Ice Age, for instance[23]. Advective changes that moved heat around within the ocean would also cause regional variability in the signal of the Little Ice Age. There is some evidence that, during the Little Ice Age, the Gulf Stream's normal flow was reduced by 10%. That would have reduced the flow of heat to northwest Europe, and may explain why the Little Ice Age was so pronounced there[85]. Michael Schlesinger and Natalia Andronova of the University of Illinois at Urbana explain that reduction. They calculated that the average global temperature was around 0.34 °C lower in the Maunder Minimum than immediately before it, resulting in a shift towards the negative phase of the Arctic Oscillation and North Atlantic Oscillation that led to higher pressure over the

Arctic and lower pressure over the mid-latitude North Atlantic, weakening the polar vortex and reducing the transport of warm air from the oceans to the continents, thus cooling Europe and eastern North America by around 1–2 °C[86]. During the negative phase of the North Atlantic Oscillation, Europe experiences hot, dry summers and very cold, dry winters, while the Mediterranean region is wet. This would help to explain why the Little Ice Age seemed mostly to be a winter phenomenon[10].

As we saw earlier, Greenland temperatures are something of a puzzle, due to their divergence from other Arctic temperature records over the past 2000 years. Kobashi and colleagues recently compared Greenland temperatures from the GISP 2 Summit ice core with global temperatures over the past 800 years to seek the origin of Greenland's climate changes[87]. Changes in Greenland's temperature generally followed those of the Northern Hemisphere, but were more strongly influenced than were Arctic temperatures by the North Atlantic Oscillation, which has its origins further south. Greenland Summit temperatures correlate significantly with North Atlantic temperatures, which were warmer than average Northern Hemisphere temperatures in the 1400s and cooler during the past few centuries. During periods of lower solar activity, the Atlantic Meridional Overturning Circulation strengthened, warming the North Atlantic and Greenland; the opposite was the case at times of higher solar activity. Presumably, when solar activity was low, the Norwegian-Greenland Sea cooled, making surface waters denser and more prone to sink, and thus enhancing the overturning circulation and increasing the supply of warm Gulf Stream water to Greenland. With more solar activity, the surface waters of the Norwegian-Greenland Sea would have become less dense, reducing the overturning circulation and thus bringing less Gulf Stream water northwards. That may explain why Greenland showed less pronounced warming than did the Northern Hemisphere as a whole in the latter half of the 20th century[87].

These various modern analyses help to confirm that, while Lamb's suggested temperatures for medieval warming and the Little Ice Age seem reasonable for western Europe, they are not reasonable for the globe as a whole.

---

[c] Recent data show that the Medieval Warm Period (MWP) was delayed by 200 years in the Southern Hemisphere, while the Little Ice Age (LIA) and the modern warming were more or less synchronous in both hemispheres. That suggests the MWP was controlled more by regional ocean variability, while the LIA and modern warming were more strongly externally forced (Neukom *et al.*, 2014, Nature Climate change 4, 362–367).

15.4   The End of the Little Ice Age

We can use Steinhilber's cosmic intensity (Figure 15.1)[22] to define the most likely boundaries of solar-driven warm and cool periods. Accepting that the Medieval Warm

Period lay between the Oort and Wolf sunspot minima, it lasted from 1070 to 1270 AD, when cosmic intensity was <1.25. Using that same cut-off, we can define the Little Ice Age as lasting from 1270 to 1920 AD and comprising the Wolf, Spörer, Maunder, Dalton and Gleissberg sunspot minima and their intervening warm periods, of which there were four (at 1350–1390, 1560–1640, 1730–1790 and 1840–1870 AD) when cosmic intensity was <1.25 and the climate was likely to have been nearly as warm as it was in the Medieval Warm Period. While many scientists (e.g. Folland, Karl and Vinnikov[51]) seem to regard 1850 as a rough guide to the end of the Little Ice Age (end of the Dalton sunspot minimum, with its peak low at about 1820), we might reasonably extend that to about 1920 (end of the Gleissberg sunspot minimum), following which there was a continued rise in sunspot activity to the peak solar maximum of the modern era between 1960 and 1990[88]. As pointed out in footnote 'a', that maximum now seems little different from those of the 1780s and 1850s. **The key question is: has the Little Ice Age come to a natural end, or have we simply been living in a post-Gleissberg sunspot maximum, which was merely a temporary warm period of the Little Ice Age and is now in decline?**

Many studies confirm that both modern temperatures and sunspot activity are the highest they have been since about 1900. For example, a 2007 review by Jasper Kirby of CERN confirmed that sunspots, as represented by ^{14}C and ^{10}Be, were about as abundant in the Medieval Warm Period as they were in the late 20th century[68], much as found by Steinhilber and his team in 2012[22]. Nevertheless, the devil as usual is in the detail, and as we shall see further on, the detailed pattern of temperature change for the late 20th century does not match the sunspot record.

While global temperature data fell from 1880 to a low in 1910, during the Gleissberg solar minimum, then rose to a high in 1940 following the solar data[89], solar data and temperature data then diverged (Figure 15.8)[90]. In effect, temperature fell when it should have been rising, according to solar activity (1950 and 1960), and rose when it should have been falling (1990 through 2010)[90, 91]. Meanwhile, CO_2 continued to rise exponentially – something that began in the late 1700s with the increased burning of coal at the start of the Industrial Revolution (Figure 9.2). It is these convergences (CO_2 and temperature) and divergences (temperature and solar data) that lead the vast majority of climate scientists to suggest that we are now living in an era in which CO_2 plays a more important role than solar activity in driving global temperature.

It seems most likely that the shape of the global temperature curve since 1900 reflects a combination of the solar activity signal plus irregular volcanic activity plus the temporary effects of short-term (2–6-year) internal oscillations like El Niño-La Niña events plus longer-term (10–30-year) fluctuations in the North Atlantic Oscillation and the Pacific Decadal Oscillation[92], superimposed on a trend driven by greenhouse gas forcing that increased with time[89]. As we saw in Chapter 8, aerosols from human activities also affected global temperature, reflecting solar energy and keeping temperatures low, especially in the late 1950s and 1960s, as demonstrated by Budyko[93]. The increase in aerosols was driven by the massive rise in industrial output that accompanied and followed the Second World War and (see footnote on page 139) was most effective before the introduction of increasingly comprehensive clean air acts by several industrial nations between 1950 and 1970[89].

The observation that temperature and solar signals diverged after 1940, most likely due to anthropogenic activities, is reinforced by results emerging from global numerical modelling. In 2009, Caspar Amman of the National Center for Climate Research in Boulder, Colorado (NCAR) used a coupled ocean–atmosphere GCM to see whether solar irradiance patterns determined from ^{10}Be, along with data on past explosive volcanic emissions (represented by volcanic aerosols in ice cores), might explain the variations seen in climate data from the Northern Hemisphere since 850 AD[94]. Amman and colleagues concluded that the climate of the Northern Hemisphere for the past 1150 years was directly attributable to small fluctuations in solar output and explosive volcanic output up until about 1870[94]. Those same climatic conditions should have continued with little change – a slight warming of perhaps 0.2 °C to the 1940s and 1950s, driven by solar factors, followed by a decrease of about 0.1 °C to the present (meaning 2009), driven by small increases in explosive volcanism from eruptions like that of Mount Pinatubo in 1991. Superimposed on this natural background was a progressive warming from about 1900, driven by increases in human-added greenhouse gases. They deduced, '*the 20th century warming is not a reflection of a rebound from the Little Ice Age cool period, but it is largely caused by anthropogenic forcing*'[94].

In a 2010 review of solar influences on climate, Lesley Gray of Reading University's meteorology department supported Amman's conclusions[48]. Gray and colleagues found that, although the surface responses of greenhouse gases and solar forcing might be similar, the greenhouse gas response in the stratosphere was the opposite (cooling)

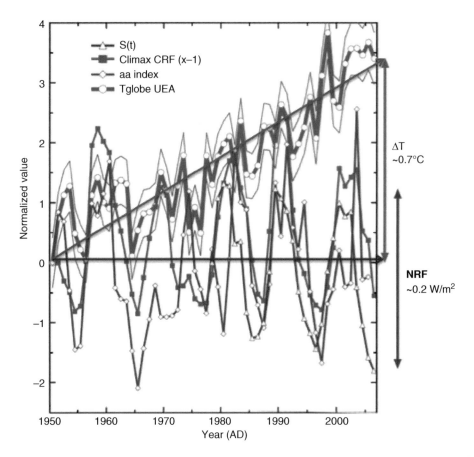

Figure 15.8 *Divergence of temperature and solar data, showing annual averages of solar activity indices and global surface temperature since the year 1950. S(t) is total solar irradiance measured by satellites (only since 1978). Climax CRF is the cosmic ray flux measured at Climax in Colorado. The aa index is a geomagnetic index prepared by the International Service of Geomagnetic Indices. Tglobe UEA is global surface temperature anomalies from HadCRUT3, with total uncertainty at the 95% level. NRF is the net radiative forcing of the Sun (i.e. S(t) divided by 4 and multiplied by 0.7). Only the Tglobe curve has an upward trend (0.11 °C per decade, r = 0.87, since 1950). All other curves oscillate about a trend that is basically flat or, in the case of S(t), slightly declining. Large dips of the Tglobe curve occurred just after major volcanic eruptions (e.g. 1963, 1982 and 1991).*

of what would be expected from solar forcing (warming). Model simulations of all known forcings showed that, while much of the global warming in the first half of the 20th century was natural and much was due to solar activity (a slow rise in sunspot activity), this did not apply in the second half of the 20th century or the beginning of the 21st, when solar forcing played at most a weak role in temperature trends[48]. A similar review by a panel of the US National Academy of Sciences meeting in 2012

had no doubt that changes in total solar irradiance were unrelated to the increase in global temperature over the past 50 years[44]. France's Edouard Bard agreed[90].

Evidence compiled from the Arctic by Gifford Miller and his team showed that warming began in the late 19th century in most palaeoclimate records[62]. During the last millennium, orbital forcing caused midsummer insolation to fall about 1 W/m² at 75° N and about 2 W/m² at 90° N. Because insolation was weakening, '*additional forcing was needed in the 20th century to give the same sum-mertime temperatures as achieved in the Medieval Warm Period*[62]. That *additional forcing* must have been large, as illustrated, for example, by the emergence of dead vegetation from beneath ice caps on Baffin Island, which grew 1600 years ago and lasted through the Medieval Warm Period, before melting early in the 20th century. Similarly, the percentage of summer melting of the Agassiz Ice Cap in the Canadian High Arctic decreased along with the insolation trend through the Holocene, but increased significantly during the past century, to the point where current rates of melting are greater than at any time in the past 1700 years. Tree ring data show that summer temperatures now are the most favourable for tree growth within the past 4 Ka. Across the Arctic, lakes were dominated by a first-order cooling trend for most of the past 2 Ka, due to declining insolation, but this trend reversed in the 20th century *despite* continued reduction of summer insolation. The warmest 50-year interval in the 2 Ka composite record occurred between 1959 and 2000 AD[62]. Miller concluded, '*The strong warming trend of the past century across the Arctic, and of the past 50 years in particular, stands in stark contrast to the first-order Holocene cooling trend, and is very likely a result of increased greenhouse gases that are a direct consequence of anthropogenic activities*'[95]. Given a consistent polar amplification of global temperatures in the Arctic by a factor of 3–4 in times past (see Chapter 14), Miller suggested that '*Arctic warming will continue to greatly exceed the global average over the coming century, with concomitant reductions in terrestrial ice masses and, consequently, an increasing rate of sea level rise*'[95].

A prior review of Arctic temperature data in 2003 by Igor Polyakov and colleagues suggested that the rise from 1900 to 1940 had been followed, after a temperature dip, by a second rise to similar levels since about 1970[96]. They interpreted these changes to be responses to a natural cycle unrelated to greenhouse gas emissions. The idea that Arctic climate change is subject to a natural 60-year cycle has been popular with some Russian authors[97]. In contrast, American investigators Kevin Wood of the University of Washington and James Overland of NOAA, writing in 2010, attributed the early 20th-century warming to increasing southerly winds bringing warm air north from the Atlantic in the form of '*a random climate excursion imposed on top of the steadily rising global mean temperature*'[98]. Those winds were associated with the arrival of anomalous high pressure over Europe and a deepening of the Iceland low, implying strengthening of the North Atlantic Oscillation. The

following year, Overland and Wood, with Muyin Wang, pointed out that the pattern of sea ice during the early 20th-century warm event was quite different from that of the late 20th-century warming[99]. Surface air temperatures during the earlier event were warmest in the Atlantic sector of the Arctic. The distribution of sea ice in August 1938 reflected that pattern, disappearing all along the European and Russian coasts but remaining along the coast of Canada, the United States and Greenland, where the Northwest Passage remained closed[99]. The modern distribution of sea ice is quite different. Sea ice is being lost from all Arctic margins except the northern edges of Greenland and the Canadian Arctic islands, leaving the Northwest Passage open. These differences confirm that different forcing factors were responsible for the two warming periods. Polyakov's most recent contribution, in 2012, showed remarkable warming of the Arctic from 1990 to 2008, associated with a dramatic loss of sea ice since 1979, which he attributed to polar amplification of warming[100].

The variation in Arctic sea ice through time provides a useful backdrop to these investigations. In 2008, Christophe Kinnard of the University of Ottawa published time series of maximum and minimum Arctic sea ice extent from 1870 to 2003[101]. The area of winter ice was essentially constant from 1870 until 1950, after which it steadily declined. The area of summer ice fell slightly from 1900 to 1950, then began a steep decline. These data confirm that different climatic forcing was in effect in the early part of the 20th century compared with the latter part.

In 2011, Kinnard and colleagues used a range of climate proxies to extend the study of Arctic sea ice back to 1450 years before present[102]. Remarkably, they found that '*both the duration and magnitude of the current decline in sea ice seem to be unprecedented for the past 1450 years*'[102]. Not only has sea ice shrunk in area, it has also thinned, making the volume of sea ice decline radically in recent years. The area covered by thin 1-year ice expanded greatly at the expense of older and thicker ice. Because the 1-year ice tends to melt each season, this has meant a great expansion in the area covered by seasonal ice. The extremely low sea ice extent observed since the mid 1990s is well below the range of natural variability inferred by the team's reconstructions. Nevertheless, the extent of sea ice was also low before 1200 AD, in association with the warm conditions of the Medieval Warm Period, although not as low as it is today. As far as they could tell, the driver for declining sea ice seemed to be enhanced advection of warm Atlantic water to the

Arctic. The warming in medieval times and at the end of the 20th century was related to an increase in the positive index of the North Atlantic Oscillation, which pushed the westerly winds north between Scotland and Iceland, funnelling warm air and ocean water towards the Arctic. There was also a period of warming and reduced sea ice in the middle of the Little Ice Age at about 1470–1520, likely driven by the same mechanism. Given the present state of knowledge, Kinnard and his team thought that anthropogenic 'greenhouse gas' warming explained the record warming and sea ice loss of recent decades[102]. The trend towards an increasing area of seasonal ice shows no sign of diminishing, although the area fluctuates from year to year.

A 2013 study by Martin Tingley and Peter Huybers of Harvard confirmed that the magnitude and frequency of recent warm extreme temperatures of high northern latitudes were the highest for 600 years and that the high-latitude summers of 2005, 2007, 2010 and 2011 were warmer than those of all prior years back to 1400 AD[103]. These extremes greatly exceeded those expected about a stationary climate, but were consistent with variability about an increasing mean temperature.

A synthesis of proxy temperature records for the past 2000 years from north of latitude 60° N, by Darrell Kauffman of Northern Arizona University and colleagues, concluded in 2009 that, excluding short-term, low-amplitude events like the Medieval Warm Period and the Little Ice Age, the long-term trend comprised gradual cooling up to the end of the 19th century[104]. That cooling then reversed, with four of the five warmest decades of the 2000-year-long reconstruction occurring between 1950 and 2000[105]. The 20th-century warming took temperatures above anything that their proxy data set revealed in the previous 19 centuries. This is consistent with the fact that, over the past 60 years, the Arctic has warmed by more than 2 °C – more than double the global average warming for the same period – due to polar amplification driven by positive feedback from melting sea ice[105].

Anne Bjune of the Bjerknes Centre for Climate Research at the University of Bergen observed a similar pattern[106]. From studies of pollen from lakes in Fennoscandia and on the Kola Peninsula, she and her colleagues found in 2009 that mean July temperatures were about 0.2 °C above present between 0 and 1100 AD and fell to about 0.2 °C below present in the Little Ice Age. Abrupt warming occurred at about 1900 AD, and the 20th century was the warmest century since 1100 AD. They were unable to detect a Medieval Warm Period.

Bo Vinther and colleagues reported from an analysis of winter-season $\partial^{18}O$ and borehole temperatures in three Greenland ice cores that '*the warming that commenced in the early 20th century has brought Greenland temperatures to a level matching the warmest periods of the Medieval Warm Period some 900–1300 years ago. After a cold spell in the 1970s and 1980s Greenland temperatures have increased rapidly and present day temperatures have just about reached the same level as during the warm period in 20th century. This result implies that further warming of present day Greenland climate will result in temperature conditions that are warmer than anything seen in the past 1400 years*'[72]. As we saw earlier, the surrounding Arctic has warmed more than Greenland.

In Fram Strait, between Svalbard and Greenland at about 80° N, summer sea surface temperatures increased significantly after about 1800 AD, reaching a maximum of 6 °C in recent years, which is warmer than conditions in the Medieval Warm Period[70]. Even so, surface water stayed cool and sea ice and ice-rafted debris continued to be abundant well into the 20th century, probably because of an increase in glacial meltwater, accompanied by icebergs. Evidence from planktonic foraminifera suggests the persistence of a thick, cold and fresh mixed-water surface layer overlying warm, saline Atlantic water that penetrates the Arctic in the subsurface, much as suggested by Dokken and colleagues[71].

Exciting new finds confirm the dramatic warming of the Arctic. Miller's 2013 examination of rooted tundra plants exposed by receding cold-based ice caps in eastern Canada shows that 5000 years of regional summertime cooling has been reversed, taking the average summer temperatures of the past 100 years to levels higher than in any summer period for more than 44 Ka[107]. That includes the peak warmth of the early Holocene, when Arctic summer insolation was 9% above modern levels. Summers cooled by some 2.7 °C over the past 5000 years, until the reversal of the modern era. Miller and colleagues' findings show that, in the Arctic, '*anthropogenic emissions of greenhouse gases have now resulted in unprecedented recent summer warmth that is well outside the range of that attributable to natural climate variability*'[107].

Independent confirmation of the unprecedented nature of modern Arctic warming comes from a study in northern Scandinavia by Jan Esper and colleagues (Figure 15.9)[108]. Like others, Esper observed that the climatic trend for the past 2000 years was one of cooling driven by the Holocene decline in insolation. His study of maximum latewood

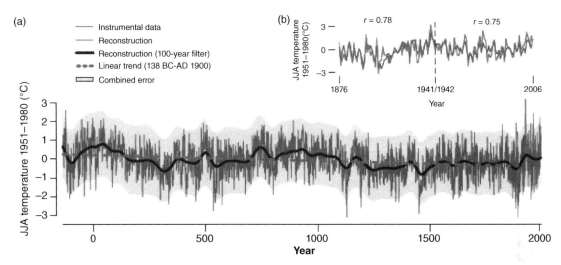

Figure 15.9 *Northern Hemisphere temperature reconstruction based on maximum density (MXD) tree ring data from northern Scandinavia. Summer (June, July, August; JJA) temperature reconstruction and fit with regional instrumental data. (a) Reconstruction extending back to 138 BC, highlighting cool and warm periods on decadal to centennial scales (black curve, 100-year spline filter) amid variation between extremely cool and warm summers (jagged dark data) and a long-term cooling trend of −0.31 °C per 1000 years from 138 BC to 1900 AD (dashed line), within an overall uncertainty area (grey) representing 2σ. (b) MXD trend (darker line) compared with measured JJA temperatures (light line) from 1876 to 2006 AD. Correlation between MXD and instrumental data is 0.77. Lighter jagged values in (a) post-1876 indicate measured air temperatures from (b).*

density in trees indicated a decline of ∼0.31 °C per 1000 years from 138 BC to 1900 AD – a signal that is missing from published tree ring proxy records of climate change. Peak warmth occurred in Roman and medieval times, alternating with severe cold conditions in the 4th and 14th centuries. A decline in orbital summer insolation by ∼6 W/m^2 in 2000 years made the Roman period (21–50 AD) 1.05 °C warmer than the 1951–1980 mean and ∼2 °C warmer than the coldest period (1451–1480 AD), which was 1.19 °C colder than the mean and ∼0.5 °C warmer than the maximum 20th-century warmth of 1921–50. Temperature rose sharply into the modern era from 1900 AD, cutting across the orbitally driven decline[108].

Similar evidence is emerging from other regions. For instance, Lonnie Thompson of the Byrd Polar and Climate Research Center of Ohio State University found that glaciers at Quelccaya, 5670 m up in the Peruvian Andes, declined rapidly in ice cover between 1983 and 2003, just like those of other low-latitude ice caps and glaciers[109]. These retreats reversed advances that began at the end of the Holocene Climatic Optimum, which ended some 5 Ka ago. This is an abrupt climatic transformation.

Mayewski and Maasch found that the climate conditions for the modern era (1931–1980) in the Northern Hemisphere differ from those of both the Medieval Warm Period and the Little Ice Age, having variable wind strengths associated with significantly higher temperatures (Figure 15.6)[77]. Much the same is true for the Southern Hemisphere, except that today's winds tend to be stronger there than in the Medieval or Little Ice Age climate eras, while the temperatures are not much different. Evidently, the modern climate in both polar regions differs from what it was in the past. But the climates in the two regions also differ from one another, with stronger winds and less warming in the south. To a polar expert, this is not surprise. After all, the Arctic has an ocean surrounded by land, while the Antarctic has a continent surrounded by ocean. The wall of winds around Antarctica keeps warm air out; there is no comparable wall around the Arctic.

In Europe, studies of tree rings, published in 2011 and 2013, confirm that the recent warming at the end of the 20th century is unprecedented in the eastern part of the continent over the past 1000 years and in the central part over the past 2500 years[110, 111].

Ljungqvist's team from the University of Stockholm found that the warmth of the Northern Hemisphere in the 9th to 11th centuries was comparable to that of the 20th-century mean[69]. But the rate of warming from the 19th century to the 20th was by far the fastest between any two centuries in the past 1200 years, and was '*unprecedented in the context of the last 1200 yr*'[69]. They also noted that analyses of instrumental data[112] showed that the last decade of the 20th century was much warmer than the 20th-century mean nearly everywhere over Northern Hemisphere land areas, '*thus providing evidence that the long-term, large-scale, NH [Northern Hemisphere] warming that began in the 17th century and accelerated in the 20th century has continued unabated*'[69].

Looking at the global picture, the PAGES group, analysing the climate of the past 2000 years, reported in 2013 that '*Recent warming reversed the long term cooling [of the Holocene]; during the last 30-year period (1971-2000 CE), the area-weighted average reconstructed temperature was likely higher than anytime in nearly 1400 years*'[58]. Evidently, they concluded, '*The global warming that has occurred since the end of the 19th century reversed a persistent long-term cooling trend*'[58].

Like the PAGES team, Sean Marcott and colleagues found that global temperatures for 2000–09 were warmer than 72% of the stacked Holocene record, but had not yet exceeded the highs of the early Holocene between 10 and 5 Ka ago. In contrast, the decadal mean global temperature for 1900–09 (the Gleissberg Minimum) was cooler than 95% of the Holocene temperature distribution. '*Global temperature, therefore, has risen from near the coldest to the warmest levels of the Holocene within the past century, reversing the cooling trend that began ~5000 yr B.P.*', Marcott concluded[113].

The present climate is also an anomaly as far as CO_2 is concerned. Hans Oeschger's team in Switzerland showed that the preindustrial values of about 280 ppm in fossil air from shallow ice cores began to rise exponentially following the mid 1700s, driven by the Industrial Revolution (Chapter 9). These rising values mesh with the values measured in background air (Figure 9.2)[114]. As far as CO_2 is concerned, there is a seamless transition between modern air and fossil air[115]. There is no analogue deeper in the ice core record for the changes in CO_2 seen since the 19th century. The fastest rate of increase in CO_2 seen during the last glacial termination was 20 ppm per 1000 years; CO_2 rose by the same amount in the 11 years before 2010, or 100 times as fast[115]. Changes like that led Nobel chemist Paul Crutzen to describe the period of time since

the late 18th century when humans began to impact the Earth's climate as the 'Anthropocene'[116]. This anomalous rise in CO_2 is closely connected to the rise in temperature since 1900.

According to Raimund Muscheler and his team, the most recent minimum of the 208-year Suess–De Vries sunspot cycle was combined with a minimum in the Gleissberg sunspot cycle at about 1900[15]. The corresponding maximum in the Suess–De Vries cycle should have occurred between 1985 and 2030, which may account for the peak in sunspot activity at about 1990, and its subsequent decline[91]. Assuming that Muscheler and his team are right, the next cold minimum in the Suess–De Vries cycle should occur between about 2070 and 2160. Similarly, assuming that the 2300-year Hallstatt solar cycle is real, and that it was manifest as a solar minimum and cold period in 750–500 BC, then the minimum should have recurred at about 1500 AD in the Little Ice Age, which may explain the rather low temperatures of that period, and should recur again at about 3800–4000 AD. If Bond was right, on the other hand, and there is a 1500-year cycle, of which the last major cooling peak was ~500 years ago during the Maunder Minimum, then the next cooling peak should occur about 1000 years from now, with a warm peak in between at about 2350 AD. Time will tell.

To conclude this section, we should examine the flattening of the warming signal since the year 2000, which has attracted much attention. As we have seen already in this book, one should not expect there to be one-to-one relationship between CO_2 and temperature at all times. CO_2 is not the only thing to affect temperature, and vice versa. Global temperature can also be affected, for instance, by internal oceanic oscillations like El Niño–La Niña and the Pacific Decadal Oscillation. As mentioned in Chapter 13, during the positive phase of the Pacific Decadal Oscillation, the equatorial region is warm (like an El Niño) and the Gulf of Alaska is cool, while in the negative phase, the equatorial region is cool (like a La Niña) and the Gulf of Alaska is warm. Each of these phases can last from 10 to 25 years. During a La Niña, and during the negative phase of the Pacific Decadal Oscillation, cold water spreads out over the central and eastern Pacific Ocean. This unusually large area of cold water cools the air, bringing down the global average temperature[92, 117–119]. But, as is apparent from studies of temperature change elsewhere, this cooling effect is highly localised. Many other areas, notably the Arctic, have gone on warming. If it weren't for the *temporary* behaviour of the Pacific, the global warming signal would likely have gone on climbing. When the Pacific Decadal Oscillation reverses, it seems likely that the global warming signal will jump up.

The standstill in temperature began in 2000, when the Pacific Decadal Oscillation reversed[92]. It did not begin in 1998, when a massive El Niño event temporarily increased global warming. That was followed by an equally large La Niña and associated global cooling. Given the eventual reversal of the Pacific Decadal Oscillation, and the continued rise in CO_2 beyond the 400 ppm level reached in May 2014, we can expect to see the warming pause end within about a decade[120]. This will make it just as temporary as the 'hiatus' observed between 1950 and 1970, when the Pacific Decadal Oscillation was last negative[92]. The flattening of the curve of temperature rise since 2000 may also reflect a slight increase in volcanic aerosols emitted into the upper atmosphere[121], for example from the eruption of Soufrière Hills in Montserrat and Tavurvur in Papua New Guinea, both in 2006[122]. There are also growing emissions of aerosols from the rapid increase in coal burning accompanying industrial expansion in India and China[123]. China's coal consumption more than doubled from 2002 to 2007, accounting for 77% of the rise in global coal use and increasing global sulphur emissions by 26%[123]. Volcanoes and coal-burning power stations emit sulphur dioxide. Combined with water, sulphur dioxide forms sulphuric acid, whose droplets reflect solar radiation. These increases in aerosols may help to maintain the present standstill in average global temperature. Note, again, that this standstill is not seen in the Arctic, only in the average of global temperature[d] .

15.5   The Hockey Stick Controversy

By this time, some readers may be wondering at the lack of any mention of the person the climate contrarians love to hate: palaeoclimatologist Michael Mann (1965–). I left him out because I wanted to be able to show, without his contribution, that there is abundant palaeoclimatic evidence that temperature declined over the past 2000

years and then rose rapidly to high levels after about 1900. That pattern should now be blatantly obvious to you.

In 1996, after completing his PhD at Yale, Michael Mann became a post-doctoral fellow in the laboratory of Raymond Bradley at the University of Massachusetts in Amhurst. There, the National Science Foundation and the US Department of Energy sponsored him to '*take a new statistical approach to reconstructing global patterns of annual temperature … based on the calibration of multiproxy data networks by the dominant patterns of temperature variability in the instrumental record*'[126], as a contribution to the Analysis of Rapid and Recent Climate Change project sponsored by the National Science Foundation (NSF) and the National Oceanic and Atmospheric Administration (NOAA). The thinking behind this study went as follows: '*If a faithful empirical description of climate variability could be obtained for the past several centuries, a more confident estimation could be made of the role of different external forcings [such as volcanic activity, solar irradiance and greenhouse gases] and internal sources of variability on past and recent climate*'[126]. Much the same rationale underpinned the work of Vinther, Wanner, Mayewski, Marcott, Moberg, Esper, Lungqvist, Kauffman and others that we examined earlier.

Mann's initial results, published in 1998 with Raymond Bradley of UMass and Malcolm Hughes of the Laboratory of Tree Ring Research at the University of Arizona, reconstructed surface temperature patterns for the Northern Hemisphere over the past 600 years[126]. They resembled those published by Bradley and Jones in the early 1990s[54-56]. There were '*pronounced cold periods during the mid-seventeenth and nineteenth centuries, and somewhat warmer intervals during the mid-sixteenth and late eighteenth centuries, with almost all years before the twentieth century well below the twentieth-century climatological mean*'[126]. Furthermore, '*the years 1990, 1995 and 1997 … each show anomalies that are greater than any other year back to 1400*'[126]. Solar forcing proved to be especially important in and around the Maunder Minimum, and in association with the warming through the 19th century. Forcing by greenhouse gases worked jointly with solar forcing for most of the past 200 years: while both were increasing, then greenhouse gas forcing increased further to become more significant as solar forcing levelled off after the mid 20th century. Explosive volcanism had only a spasmodic influence on climate, for example with the widespread cooling in 1816 caused by the eruption of Tambora. The authors hoped that '*It may soon be possible to faithfully reconstruct mean global*

---

[d] Note added in press: A comprehensive analysis of ocean warming has subsequently demonstrated that global surface warming slowed at the end of the 20th century as more heat moved into the deeps of the Atlantic and Southern Oceans. Cooling associated with this deep heat sequestration usually lasts 20–35 years, after which surface warming should accelerate[124]. Equally, new research has demonstrated that the 'apparent' recent slowdown in global warming was not as slow as had been thought[124]; the more recent data are biased on the cool side.

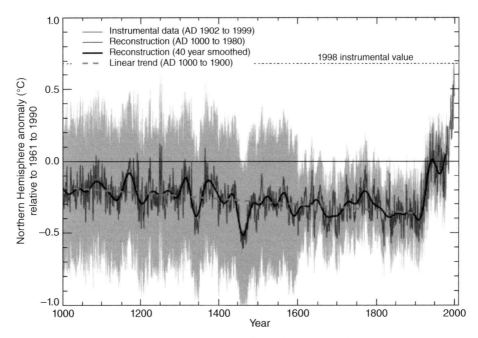

Figure 15.10 *Mann's Hockey Stick graph, 2001. Millennial Northern Hemisphere temperature reconstruction for 1000 to 1999 AD (jagged dark profile), with smoothed version (heavy dark line) and trend line to 1850 (dashed line), within 2σ error limits (grey envelope). From 1902, lighter jagged line represents addition of instrumental data.*

temperatures back over the entire millennium, resolving, for example the enigmatic medieval period'[126].

Enigmatic? Yes, because they were referring to a 1994 study by Malcolm Hughes of the University of Arizona and Henry Diaz of NOAA, which found that, while some areas of the globe were warmer in summer during the medieval era than they are today (Scandinavia, China, the Sierra Nevada, the Rockies and Tasmania), others experienced similar temperatures at later times (the southeastern United States, the Mediterranean and parts of South America)[127]. '*Taken together*', they concluded, '*the available evidence does not support a global Medieval Warm Period, although more support for such a phenomenon could be drawn*'[127], perhaps from high elevation records. This was not a new problem: it had been identified by Lamb in 1988[43], investigated in the early 1990s by Mann's co-author Raymond Bradley[54–56] and highlighted in the 1995 IPCC report by Nicholls[53]. Mann was the new kid on the block, within a team that had been carrying out this kind of research for several years.

By 1999, the trio had extended their temperature reconstruction for the Northern Hemisphere back a full 1000

years[128], producing a graph that became known as the 'Hockey Stick' (Figure 15.10). It showed minor variability about a trend of steadily declining temperature from 1000 through to about 1900 (the handle of the stick), after which there was a sharp upward kick (the blade) to the high temperature of 1998. Reappearing as Figure 2.20 on page 134 in Chapter 2 of the IPCC's third assessment report in 2001[129], this graph became just as iconic as Lamb's diagram in the IPCC report of 1990 (Figure 15.3)[51].

Had Mann demolished the Medieval Warm Period? No. On the contrary, he said, '*Our reconstruction thus supports the notion of relatively warm hemispheric conditions earlier in the millennium*'[128]. But, in comparison with Lamb's graph, his results reduced the amplitude of that medieval warming for the Northern Hemisphere. This reversal of received wisdom by a young upstart created considerable fury in the climate contrarian community, at least in part because **they failed to appreciate that Lamb's graph was not global**, even though Mann and his team carefully noted that, with respect to the Medieval Warm Period, '*Lamb, examining evidence mostly from western Europe, never suggested this was a global phenomenon*'[128].

Context is everything, and many critics missed the point that, in a very real sense, Mann was not saying anything new. He was working with two experienced researchers, Bradley and Hughes, using additional data and additional ways of processing them to reach basically the same conclusions as they had already reached some years before. His 1999 reconstruction[128] was done in greater detail, with higher resolution in time and with more sites than Lamb's original diagram (Figure 15.3), and provided abundant evidence for the natural variability of the climate system above and below the mean – the mean being the cooling trend line after 1000 AD. In contrast, Lamb had smoothed his data to eliminate the noise of natural variability. We should expect a difference between the Lamb and Mann curves because Mann's reconstruction was for the entire Northern Hemisphere, whereas Lamb's was primarily for the UK, representing western Europe. Like others referred to earlier, Mann saw his Northern Hemisphere temperature curve as a response to declining orbitally driven insolation, pointing out that '*the recent warming is especially striking if viewed as defying a long-term cooling trend associated with astronomical forcing*'[128].

A key question for his critics was: How good were his data and his techniques for processing them? Mann's statistical methods were challenged by Stephen McIntyre (1947–), a Canadian mathematician and semiretired mining consultant, and Ross McKitrick, a Canadian economist and policy analyst from the University of Guelph[130, 131]. They helped to trigger debates about the validity of Mann's graph, with the contrarian viewpoint being comprehensively covered by the science journalist Andrew Montford in 2010[132]. Mann responded by summarising his own position in 2012[133]. Flames were fanned further for a different reason, when it became clear that Mann and his co-authors had contributed to the chapter of the 2001 IPCC report that incorporated their own diagram[129]. This suggested a conflict of interest[132].

In response to the furore, the US National Research Council (NRC), the working arm of the US National Academies, decided to review the topic of *Surface Temperature Reconstructions for the Last 2000 Years*, appointing a committee for the purpose chaired by Gerald North of Texas A & M University[134]. In the foreword to the 2006 North report, the president of the National Academy of Sciences and chair of the NRC said it was '*a report that … provides policy makers and the scientific community with a critical view of surface temperature reconstructions and how they are evolving over time, as well as a good sense of how important our*

understanding of the paleoclimate temperature record is within the overall state of knowledge on global climate change'[134].

North prefaced his report by explaining where Mann's research fitted in to investigations of climate change. '*Science*', he reminded his readers, '*is a process of exploration of ideas – hypotheses are proposed and research is conducted to investigate. Other scientists work on the issue, producing supporting or negating evidence, and each hypothesis either survives for another round, evolves into other ideas, or is proven false and rejected. In the case of the hockey stick, the scientific process has proceeded for the last few years with many researchers testing and debating the results. Critics of the original papers have argued that the statistical methods were flawed, that the choice of data was biased, and that the data and procedures used were not shared so others could verify the work. This report is an opportunity to examine the strengths and limitations of surface temperature reconstructions and the role that they play in improving our understanding of climate. The reconstruction produced by Dr. Mann and his colleagues was just one step in a long process of research, and it is not (as sometimes presented) a clinching argument for anthropogenic global warming, but rather one of many independent lines of research on global climate change. Using multiple types of proxy data to infer temperature time series over large geographic regions is a relatively new area of scientific research, although it builds upon the considerable progress that has been made in deducing past temperature variations at single sites and local regions. Surface temperature reconstructions often combine data from a number of specialized disciplines, and few individuals have expertise in all aspects of the work. The procedures for dealing with these data are evolving – there is no one "right" way to proceed. It is my opinion that this field is progressing in a healthy manner. As in all scientific endeavors, research reported in the scientific literature is often "work in progress" aimed at other investigators, not always to be taken as individual calls for action in the policy community*'[134]. Wise words, indeed.

The 2006 North report is a useful primer on palaeoclimate analysis. As one test of Mann's hockey stick, presented in updated form in 2003 by Mann and Phil Jones from East Anglia's CRU[135], the report provided a comparison with two other independent climate reconstructions back to 900 AD, one from Anders Moberg of Stockholm University[67, 136], the other from Jan Esper of the Swiss Federal Institute of Technology (ETH)[65].

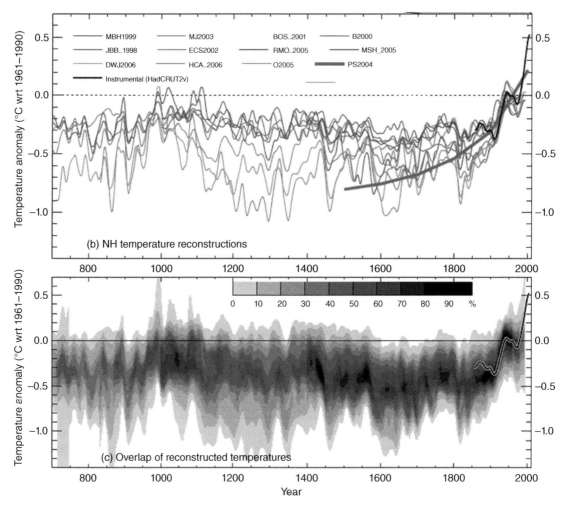

Figure 15.11 *IPCC temperature reconstruction 2007 for Northern Hemisphere temperature variation over the last 1.3 Ka. (a) Annual mean instrumental temperature records. (b) Reconstructions using multiple climate proxy records and the HadCRUT2v instrumental temperature record (heavy black line from c.1850 onward). (c) Overlap of the published multidecadal time scale uncertainty ranges of all temperature records, with temperatures within ±1 standard error (SE) of a reconstruction 'scoring' 10% and regions within the 5–95% range 'scoring' 5% (the maximum 100% is obtained only for temperatures that fall within ±1 SE of all 10 reconstructions). The HadCRUT2v instrumental temperature record is shown in black. All series have been smoothed to remove fluctuations on time scales less than 30 years. All temperatures represent anomalies (°C) from the 1961–90 mean.*

Despite their independent construction, all showed much the same thing, with the medieval warming of the Northern Hemisphere just about reaching temperatures typical of the 1950s, although the Mann and Jones curve peaked around 0.2 °C lower in the medieval period than did the other two. By 2007, Eystein Jansen of the University of Bergen in Norway and Jonathan Overpeck of the University of Arizona had expanded this comparison[137]. They used 12 independent reconstructions, mostly back to 800 AD, to show that peak temperatures for the medieval period lay 0.1–0.2 °C below those typical of 1950 (Figure 15.11).

North's report concluded, like the IPCC in 1996[53], that relatively warm conditions were centred near 1000 AD in the Medieval Warm Period, and relatively cold conditions were centred around 1700 AD in the Little Ice Age, especially in the Northern Hemisphere. Evidence for a Little Ice Age between 1500 and 1850 AD was widespread, but evidence for a Medieval Warm Period was limited. The timing and duration of warm periods varied from region to region, and the magnitude and geographic extent of the warmth were uncertain[134]. Jansen and Overpeck reached much the same conclusion in 2007[137], and the 2013 PAGES report agreed with them[58].

Examining not only surface temperature reconstructions but also data on melting ice caps and retreating glaciers, North's panel agreed with Mann that the late 20th-century warmth of the Northern Hemisphere was unprecedented for at least the last 1000 years. Given the effort made on this issue since the early 1990s, and the wide-ranging support from the likes of Vinther, Wanner, Mayewski, Moberg, Esper, Kauffman, Amman and Marcott, North's conclusions seem beyond reproach.

One of the criticisms associated with Mann's work concerned the fact that some tree ring data seemed to show that temperatures had fallen in recent decades. Rosanne D'Arrigo of Lamont and colleagues explained that this anomalous pattern, known as the 'divergence problem', is attributable to several causes including, *'temperature-induced drought stress, nonlinear thresholds or time-dependent responses to recent warming, delayed snowmelt and related changes in seasonality, and differential growth/climate relationships inferred for maximum, minimum and mean temperatures'*[138]. The key is the changing balance between precipitation and temperature, both of which affect ring patterns. Another possibility is 'global dimming', the decrease in solar radiation attributable to aerosols. This would have its greatest effect at high northern latitudes, consistent with tree ring observations.

At the International Polar Year Conference in Montreal in April 2012, Trevor Porter reported on his observations of 'divergence' in tree rings from white spruce trees in the Yukon Territory[139]. In this one relatively small area of boreal forest, the trees of about half the sites responded negatively to summer temperature increase in the period 1930–2007, while those from the other half responded positively. Both shared a common response prior to 1930. A wider geographic examination confirmed that most white spruce sites reacted positively to temperature increases over the period between 1300 and 1930 AD.

Replying to a question that I posed as the moderator of his oral session, Porter explained that the leading hypothesis for the decline seen in some of his studied trees was localised drought stress, and that the region had become significantly drier in recent decades, with precipitation down to 200–400 mm/year. However, not all tree sites were affected. Both positive and negative responses were found even within a few kilometres of one another. The reason may lie in the soil moisture levels. Local changes in permafrost, with areas of melting sited close to areas of permanently frozen ground, depending on the slope of the ground in relation to the Sun, may account for extreme local variation in tree response, but factors such as the density of stands of trees and the amounts of organic material supplying nutrients to the surface layer may also play a part. More research is needed. Clearly, great care needs to be taken, especially in boreal areas, in selecting the trees that most appropriately reflect the actual temperature conditions through time, rather than those responding to water stress.

Undeterred by contrarian sceptics, and in response to the suggestions of North's NRC report, in 2008 Mann and colleagues reconstructed surface temperature at both hemispheric and global scales for much of the last 2000 years, using an expanded set of proxy data[140]. They concluded that *'the hemispheric-scale warmth of the past decade for the NH is likely anomalous in the context of not just the past 1,000 years ... but longer ... [and] appears to hold for at least the past 1,300 years'*[140]. Moreover, *'this conclusion can be extended back to at least the past 1,700 years if tree ring data are used'*[140]. The picture was *'less definitive for the SH [Southern Hemisphere] and globe, which we attribute to large uncertainties arising from the sparser available proxy data in the SH'*[140]. Much the same picture has emerged from the studies of Moberg, Marcott and PAGES, as we saw earlier.

The following year, Mann and his team extended their analysis[141]. Using a network of 1000 climate proxy records from various sites around the world, they concluded, *'The Medieval period is found to display warmth that matches or exceeds that of the past decade in some regions, but which falls well below recent levels globally'*[141]. Similarly, *'The coldest temperatures of the Little Ice Age are observed ... over the extratropical Northern Hemisphere continents'*[141]. Wanner and his PAGES team reached much the same conclusions[27, 58]. How could this variation be explained?

Mann noticed that La Niña-like conditions, making the tropics cool, tended to predominate during the medieval period in the tropical Pacific, most noticeably between 950 and 1100 AD. At the same time, sea surface temperatures warmed significantly in the North Atlantic and the North Pacific. This pattern is typical of the negative phase of the Pacific Decadal Oscillation. Mann thought that the effects of orbital, solar, volcanic and greenhouse forcings were being modulated regionally through oceanic responses[141]. The medieval period showed enhanced warmth over the interior of North America and the Eurasian Arctic, and cooling over central Eurasia '*suggestive of the positive phase of the North Atlantic Oscillation (NAO) and the closely related Arctic Oscillation (AO) sea-level pressure (SLP) pattern*'[141]. It would seem that medieval warmth was associated with the positive phase of the North Atlantic Oscillation, which was in turn associated with relatively strong solar forcing. The positive phase of the North Atlantic Oscillation enhanced the westerlies, creating cool summers and mild winters with frequent rain in northwest Europe, snow in Scandinavia and dry conditions in the Mediterranean. In contrast, the Little Ice Age was associated with a predominantly negative phase of the North Atlantic Oscillation, driven by weak solar forcing. This suppressed the strength of the westerlies, bringing hot, dry summers and very cold winters to northwest Europe and rain to the Mediterranean. This helps to explain why the Little Ice Age was mainly a winter phenomenon[10].

Rather like Mann, both the PAGES group[58] and Marcott[113] found *global* medieval temperatures around 1000 AD to be around 0.1 °C lower than the 1961–91 reference period. These recent studies vindicate Mann's earlier conclusion that the Medieval Warm Period, although real, was not a major global climatic phenomenon. In support of this conclusion, Steinhilber's solar data (Figure 15.1)[22] show that cosmic intensity (the inverse of the intensity of sunspots) was not much stronger during the Medieval Warm Period than it was during warm periods within the Little Ice Age. The main difference from those warm intervals, which lasted ~50 years, was that the Medieval Warm Period lasted longer, at ~200 years. Thus, in terms of absolute *global* warming, the Medieval Warm Period is no more than a small upward blip on the orbitally driven downward curve of Holocene cooling. This picture emerges time and time again, for example in 2014, as shown in Figure 15.12 [142].

In retrospect, then, Mann is just one of a number of palaeoclimatologists confirming that, over the past 2000 years, there was a brief and modestly warm period in the Middle Ages, followed by a long, cold Little Ice Age punctuated by warm periods and terminated by

rapid warming into the middle of the 20th century. These changes were driven primarily by fluctuations in solar output superimposed on the slowly declining base of orbitally imposed insolation that formed a distinctly cool end to the Holocene. Volcanoes contributed additional minor short-term cooling during this time, locally accentuating the depth of cooling events within the Little Ice Age. The regional redistribution of heat by the ocean, particularly through the operations of the North Atlantic Oscillation, probably underpinned by the Atlantic Meridional Overturning Circulation, smeared the impact of the relatively weak solar signals, making it difficult to identify global signals and enhancing the effects of the Medieval Warm Period and Little Ice Age in Europe. In effect, the Little Ice Age has not ended. It should be continuing, following orbital insolation, modulated by variations in solar output and the occasional volcano.

As we saw earlier, this conclusion was tested in 2009 using a numerical climate model that considered all the possible interactions between different forcing factors over the past 1000 years or so. Amman's results confirmed that modern warming was attributable to the addition of greenhouse gases[94].

That same year, another modeller, Hugues Goosse, a colleague of André Berger at the Catholic University of Louvain in Belgium, together with Hans Renssen of the Free University of Amsterdam, carried out a similar modelling exercise, this time in association with Michael Mann[143]. Considering the effects of changes in land use, solar irradiance, aerosols, volcanoes and greenhouse gases, their model confirmed that solar forcing explained the high temperatures of the 1100s, 1200s and 1900s and the low temperatures of the 1500s and late 1700s to early 1800s. Large volcanic forcing explained short but particularly cold periods in 1258, 1452, 1600, 1641 and 1815–16. Land use changes were associated with a long-term cooling trend, while cooling associated with increasing (industrial) aerosols was confined to the late 20th century. Greenhouse gas forcing induced warming during the past 150 years. Responses to forcing were faster and larger in the Northern Hemisphere and slower and lower in the Southern Hemisphere, due to its large area of ocean. The high heat capacity of the Southern Ocean damped the signals of global warming or cooling. In addition, upwelling brought to the surface of the Southern Ocean old cold ocean water that had acquired its characteristics decades to centuries earlier, which resulted in a more complex interaction between forcing and response than we see in the north[143].

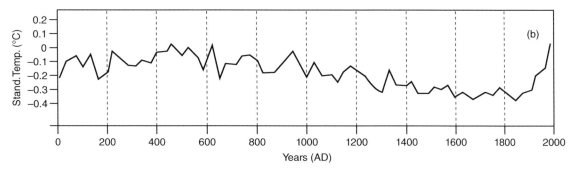

Figure 15.12 *Global temperature for the last 2000 years, represented by 30-year means. The zero on the temperature scale represents the average from 1950 to 2000, which is also the last data point.*

Goose and his team also used model simulations to investigate the response of particular parts of the climate system to external forcing[143]. Their results '*indicate a counterintuitive tendency towards El Niño (warm eastern and central tropical Pacific) conditions in response to negative radiative forcing (past explosive tropical volcanic eruptions or decreases in solar irradiance) and a tendency for La Niña-like conditions [cool tropical Pacific] in response to positive radiative forcing (i.e., increases in solar irradiance). This prediction matches available evidence from tropical Pacific coral records*'[143]. This counterintuitive finding explains '*why the tropical Pacific appears to have been in a cold La Niña-like state during the so-called "Medieval Warm Period" and a warm El Niño-like state during the "Little Ice Age"*'[143]

It will be important in the future to expand the ways in which models and data can be used together, as in that study, to establish the nature and origins of climate variability with progressively higher degrees of confidence. Goose concluded that '*Improvement in the representation of past changes by models or in the number and quality of proxy records would certainly be very beneficial*'[143].

Given that independent data from many different groups of researchers confirm what Mann and his colleagues found, it looks very much as though the Hockey Stick controversy was largely manufactured.

15.6   Sea Level

Taking into account tectonic changes and glacial isostatic adjustments, sea level has been rising throughout the past

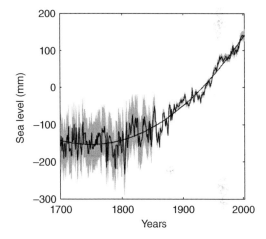

Figure 15.13 *Sea level reconstruction since 1700. Shading represents the error window of the reconstruction. Fitted curve is a second-order polynomial.*

200 years alongside the rise in temperature typical of the modern era (Figure 15.13)[144]. It has risen in response to both ice melt (increase in water mass) and thermal expansion (increase in water volume), which are themselves responses to global warming. As John Church of the Centre for Australian Weather and Climate Research pointed out, a 1000 m column of seawater expands by about 1 or 2 cm for every 0.1 °C of warming[145]. According to Konrad Steffen of the University of Colorado at Boulder and his team, thermal expansion of ocean water accounted for about 0.4 mm/year of sea level rise for the past 4 or 5 decades, rising to about 1.5 mm/year during the 2000s, while melting glaciers and ice caps contributed

a rise of 0.30–0.45 mm/year over the past 50 years, rising to 0.8 mm/year during the 2000s[146]. While that leaves an unexplained contribution of about 1 mm/year, which probably came from the large ice sheets, John Church explained that this apparent 'enigma' in resolving the sea-level budget (the relationship between actual measurement and likely contribution from different sources) has now been largely resolved '*by combining … revised estimates for upper ocean thermal expansion and glacier and ice-cap contributions with reasonable but more poorly known estimates of contribution from deep-ocean thermal expansion and the Greenland and Antarctic ice sheet contributions*'[147]. The Intergovernmental Oceanographic Commission awarded Church the 2006 Roger Revelle Medal.

Salt marshes are useful for estimating past sea levels. John Church and his team noted in 2008 that '*changes in the local sea level estimated from sediment cores collected in salt marshes reveal an increase in the rate of sea-level rise in the western and eastern Atlantic Ocean during the 19th and early 20th centuries … consistent with the few long tide-gauge records from Europe and North America*'[148]. In 2010, Roland Gehrels of the University of Plymouth, and president of the INQUA Commission on Coastal and Marine Processes for 2011–15, recalled that '*The AR4 [the IPCC's 4th Assessment Report] states with "high confidence" that the onset of modern rates of sea-level rise occurred between 1850 and 1950*'[149]. He went on to observe that '*The timing of the change in trend has been more precisely determined in proxy records from salt marshes in eastern North America … and New Zealand … and occurred between 1880 and 1920 … This acceleration is also visible in the long tide-gauge record of Brest … In Iceland, evidence from salt marsh sediments shows that sea level started rising rapidly between 1800 and 1840, possibly as a result of regional ocean warming, and continued to the present, without showing an inflexion at the beginning of the 20th century … An acceleration of sea-level rise starting in the early 1800s has been proposed on the basis of the oldest sea-level measurements in the world at Amsterdam … The first sea-level acceleration within the 20th century occurred in the 1930s … and is demonstrated by model-dependent global reconstructions of tide-gauge data*'[149].

Church observed that '*sea-level reconstructions based on salt-marsh sediments from eastern North America and Europe show that centennial variability during the late Holocene was probably of the order of 0.2 m or less … [in agreement with] micro-atoll studies from Australia [which] also limit rates of Holocene sea-level change to 0.1–0.2 m per century*'[147]. Kurt Lambeck (see Box 12.3) of the Australian National University concurred, noting that '*The absence of pre-20th-century sea-level fluctuations in proxy records from North America suggests that the 20th-century sea-level acceleration was unprecedented in (at least) the past millennium*'[150]. Seeing that there was some evidence from the Baltic for small changes of less than 1 m during the late Holocene, he went on to observe that '*coral evidence from the Great Barrier Reef … or from the Abrolhos Islands of Western Australia … does not require oscillations during the Late Holocene and suggests that what is seen in the Baltic is a regional rather than a global signal*'[150]. He reminded us that '*Baltic levels are strongly correlated with meteorological conditions … and it is possible that what is recorded … is atmospherically forced*'[150].

Phil Woodworth of the Proudman Oceanographic Laboratory in Liverpool, winner of the 2010 Vening Meinesz Medal of the European Geosciences Union, explained in 2011 that records of sea-level change from salt marshes overlap with tide gauge records but have lower resolution, preventing examination of interannual variability[151]. Woodworth and colleagues concluded that '*The most we can say with regard to findings from this relatively new field is that salt marsh … and tide gauge … data appear to reflect similar features of acceleration between the 19th and 20th centuries, with important details to be resolved by further research*'[151]. Svetlana Jevrejeva, of the same laboratory, concluded with her colleagues that sea levels started to rise at about the end of the 18th century and rose by 6 cm during the 19th century and by 19 cm in the 20th (Figure 15.13)[144, 152].

Andrew Kemp of the University of Pennsylvania and colleagues used data from salt-marsh sediments on the Atlantic coast of North Carolina and similar environments, corrected for glacial isostatic adjustment, to show that '*North Carolina sea level was stable from at least BC 100 until AD 950. Sea-level rose at a rate of 0.6 mm/y from about AD 950 to 1400 as a consequence of Medieval warmth*'[153]. This was '*followed by a further period of stable, or slightly falling, sea-level that persisted until the late 19th century … due to cooler temperatures associated with the Little Ice Age. A second increase in the rate of sea-level rise occurred between around AD 1880-1920; in North Carolina the mean rate was 2.1 mm/y, in response to 20th century warming. This historical rate of rise was greater than any other persistent century-scale trend during the past 2100 y*'[153].

Kemp's team included Stefan Rahmstorf of the Potsdam Institute for Climate Impact Research at Potsdam University, near Berlin, and Martin Vermeer of Helsinki University of Technology. In 2007, Rahmstorf found a linear relationship between global temperature and the rate of sea level rise, which suggested that sea level could rise by up to about 1.4 m by 2100, given greenhouse gas emissions continuing 'business as usual'[154]. By 2009, Vermeer and Rahmstorf refined this calculation to suggest that the rise might be up to 1.9 m[155]. The Kemp team used these semi-empirical relationships to convert temperatures estimated from proxies by Michael Mann and colleagues in 2008[140] into past sea levels for comparison with the observations from North Carolina[153]. They found that the estimated sea levels diverged from those measured in salt marsh sediments prior to 1000 AD, possibly because global temperatures estimated from proxies for the period from 500 to 1100 AD were too high, by about 0.2 °C.

Using a comparable model, Aslak Grinsted of the Arctic Centre at Finland's University of Lapland calculated in 2010 that sea levels in about 600 AD were some 10 cm below today's (1980–99) average level, reached 12–20 cm above present average levels in the medieval warm period and fell to about 19–26 cm below present levels in the Little Ice Age[156]. The levels Grinsted and colleagues estimated for the 200-year-long medieval warm period reflect the long time it takes to melt substantial amounts of land ice, and imply that it takes 100 years or more for sea level to reach an equilibrium position with respect to temperature rise. They used two independent estimates of global temperature as inputs to their model, and found that the 2005 palaeotemperature estimates of Moberg[67] gave a more realistic result than did the 2004 estimates of Jones and Mann[157].

Much the same results emerged from another modelling experiment, carried out by Jonathan Gregory of the meteorology department at the University of Reading and colleagues from the Hadley Centre for Climate Prediction and Research of the United Kingdom's Meteorological Office[158]. They found that sea level fell during the 16th and 17th centuries due to ocean cooling associated with the Little Ice Age. With increased warming at the end of the Maunder Minimum, and fewer volcanoes, the model showed sea level rising in the 18th century, before falling in the early 19th century, with cooling associated with the Dalton (sunspot) Minimum (1790–1830) and the gigantic Tambora volcanic eruption of 1815–16. Sea level rose back to former levels after 1820, as sunspots increased following the Dalton Minimum and volcanic activity

decreased. Gregory concluded '*that sea level did not go outside natural variability until the twentieth century*'[158].

Nevertheless, we do have to be cautious about model outputs. Gehrels reminded us not to interpret them too literally, not only for past times, but also for the present. They cannot yet replicate accurately either observed trends or the magnitude of the interannual to decadal variability of modern times[149]. More research is needed to reconcile observations with models.

One would certainly expect increasing emissions of greenhouse gases to have caused a continuous acceleration of the rate of sea level rise in response to the global warming documented in this section. And we do observe that. From 1870 to 2001, the rate of rise averaged 1.7 mm/year, with an increase in the rate of rise over this period from 1 mm/year prior to the 1930s to about 2.5 mm/year in the late 1950s[146]. Large volcanic eruptions between 1963 and 1991 caused temporary cooling, which briefly lowered the rate of rise to less than 2 mm/year. Against this backdrop, the rate of sea level rise since the mid 1980s is unusual. Church noted that data from satellite altimeters '*indicate that global average sea level has been rising at 3.1 ± 0.4 mm [per year] … [which] is faster (by almost a factor of two) than the average rate of rise during the 20th century, which, in turn, was an order of magnitude larger than the rate of rise over the two millennia prior to the 18th century*'[148]. Roland Gehrels confirmed that '*Rapid rates of sea-level rise during the 20th century, as recorded in many places around the global, represent a significant departure from late Holocene trends of sea-level change*'[149]. Kurt Lambeck agreed, as we have already seen.

These various lines of evidence show that, after the Last Glacial Maximum, sea level rose quite rapidly until about 7 Ka ago, by which time the major Northern Hemisphere ice sheets had mostly melted. The rise then slowed to extremely low rates over the past 7 Ka[148]. Comparison of recent sea level with that of 2 Ka ago, deduced from the positions of Roman fish tanks, suggests that there was little further change until the start of the 19th century. This picture was influenced by Fairbanks' development of a global sea level curve based on data from Barbados[159], as discussed in Chapter 12. Careful examination of the rationale behind the construction of this curve led Roland Gehrels to conclude that '*The Holocene part of the Barbados curve [produced by Fairbanks 1989] is … without much merit*'[149]. A key part of the story, as we saw in Chapter 12, is the change to the Earth's crust when the ice load is removed. Land that was formerly depressed by ice sheets begins to rise, while land that was forced to bulge up around the ice-laden crust begins to sink. This siphons water from the tropics towards the poles to fill the space left by the former peripheral bulge. Similarly, the rise in sea level after the Last Glacial Maximum drowned the continental shelves, which sank under the extra weight of water. Water was siphoned from the open ocean into the space created by this sinking. This siphoning lowered sea level in tropical and mid latitudes, which explains why Holocene shorelines in those areas are commonly found several metres above present sea level. In response to these various effects, models suggest that Holocene sea level should have fallen by 0.28–0.36 mm/year along with the fall in insolation and accompanying global cooling. It did not do so. There is some evidence that the Greenland Ice Sheet and small Northern Hemisphere glaciers were growing in the late Holocene (the neoglacial), which implies – if sea level was more or less constant – that ice must have been melting in Antarctica. There is some evidence for that, for example from the Ross Sea[149]. Getting precise data on late Holocene sea level change, according to Gehrels, '*requires an interdisciplinary approach by glaciologists, GIA [glacial isostatic adjustment] modelers, sea-level scientists and space geodesists*'[149]. Care is needed in measuring recent regional changes in sea level where coasts are subsiding, as they are in many estuarine and deltaic environments due to the weight of sediments brought down by rivers like the Rhine. Even so, the vast majority of scientists making sea level measurements agree that sea level began to rise in the late 1800s and that the rate has increased since then[144, 152].

In 2009, Eelco Rohling and colleagues[160] used their analysis of the relationship between Antarctic temperature and sea level, and the relationship between CO_2 and temperature in Antarctic ice cores[161], to conclude that a high CO_2 value like that of today's world could lead over a few hundred years to an equilibrium rise in temperature of up to 16 °C and an equilibrium rise in sea level of around 25 ± 5 m, like that observed for the Pliocene (see Chapter 11). In contrast, climate models suggest that the relationship between CO_2 and temperature may diverge from the '*natural and linear pattern*'[160] that they observed for the Pliocene, and that a modern CO_2 value would equate to an Antarctic temperature of 5 °C, suggesting a sea level rise of up to only 5 m. The discrepancy between the end points of 5 and 25 m may arise either because the natural linear relationship remains so with increasing CO_2 or because the high-side model projections for the relationship between temperature and CO_2 do not use a

high enough climate sensitivity (i.e. the rise in temperature for a doubling of CO_2 may be much greater than 3 °C, as some have suggested). In any case, the upper limit of the projections (25 m) is based not on models but on actual observations (see Chapter 11). What we need to know is how long such a change might take.

Rohling later extended his study of the relationship between CO_2 and sea level back through time for 40 Ma, as we saw in Chapter 10[162]. Given modern atmospheric concentrations of CO_2, and knowing the long-term equilibrium relationship between CO_2 and sea level, he estimated that, without any further rise in CO_2, the long-term equilibrium rise in sea level compared with the present would reach between +9 and +31 m, averaging +24 m, over some 5–25 centuries. Rises of CO_2 above the modern level will cause larger long-term change.

The estimates of Rahmstorf[154] and his colleague Vermeer[155] are more than double the maximal sea level rise of about 80 cm suggested for 2100 by the IPCC in its 4th and 5th Assessment Reports. Jim Hansen of NASA believed that Rahmstorf's various projections failed to consider nonlinearities in the climate system, and that the ongoing rise in temperature could destabilise ice sheets and lead to a cumulative rise in sea level of 5 m by 2100[163]. Tad Pfeffer of the Institute of Arctic and Alpine Research of the University of Colorado at Boulder disagreed, suggesting that such a rapid rate of rise was physically untenable, given what was known about the behaviour of ice sheets. Considering all the available data, he and his team opted instead for a maximum rise of 2 m and a likely rise of up to 0.8 m by 2100[164].

Hansen thought that Pfeffer and his team were too conservative in their estimation of the amount of forcing likely from growing emissions of greenhouse gases, which by mid-century was expected to lead to extensive summer melting in a long melt season in Greenland and to the loss of the ice shelves that prevented inland ice streams from discharging land ice rapidly at the coast of Antarctica[165]. Ice loss from the Pine Island Glacier (really a 50 km-wide ice stream – as wide as the English Channel between Folkestone and Boulogne) and its sister glaciers draining West Antarctica is significant and increasing. It is attributed not to surface warming but to the penetration on to the continental shelf beneath the glacier of Circum-Polar Deep Water with a temperature of about 1 °C, which carries sufficient warmth to thin the glacier from beneath, making it discharge faster[165]. The ice streams represented by the glaciers draining from West Antarctica into the Amundsen Sea Embayment are all moving faster, posing the risk of eventual collapse of the West Antarctic Ice Sheet and a possible sea level rise of close to 3 m[78, 79]. In addition, recent analyses suggest that the ice of the Wilkes Basin in East Antarctica is also at risk of collapse, providing the prospect of an additional rise of sea level of 3–4 m[166]. The risk increases markedly where erosion by warm seawater beneath the seaward edge of the ice shelf there removes a volume of ice equivalent to a sea level rise of 8 cm. That is enough to melt the ice plug that currently stops incursion of warm water into Wilkes Basin.

The current global rise in sea level is attributed in roughly equal measure to thermal expansion of the ocean, the melting of mountain glaciers and the melting of ice caps and ice sheets on land. A group led by Alex Gardner of the University of Alberta showed in 2011 that the ice caps of the Canadian Archipelago on Baffin Island and Ellesmere Island had lost 61 ± 7 Gt/year between 2004–06 and 2007–09 in response to summer warming[167]. This was the single largest contributor to sea level rise (0.17 ± 0.02 mm/year) outside Greenland and Antarctica.

Returning to Robert Kopp of Princeton, whom we met in Chapter 12, we are reminded that '*Incorporating a large database of palaeoclimatic constraints … highlights the vulnerability of ice sheets to even relatively low levels of sustained global warming*'[168]. Kopp's team found that with just 2–3 °C of warming above present levels in the last interglacial, sea levels peaked at least 6.6 m higher than today.

In 2010, an IPCC workshop on sea level change included the observation that '*ice sheets are capable of highly nonlinear dynamical behaviour that could contribute significantly to short-term sea level rise (to 2100), and may also produce a long-term commitment (e.g. centuries-long) to substantial (many metres) of sea level also*'[169]. John Church agreed, noting in 2011 that while '*major deficiencies in our understanding remain … perhaps the major challenge is the response of ice sheets, particularly those parts grounded below sea-level*'[170]. We do know that rates of loss of ice from Greenland and Antarctica are increasing with time[171]. But more research on the range of rates of ice sheet decay under different thermal regimes is needed before we can answer the question, 'How much and how fast?' with confidence. Meanwhile, engineers responsible for London's Thames Barrier plan to redesign it to withstand a rise of 2 m.

This chapter has confirmed that the climate continued to cool over the past 2000 years in response to the decrease in Northern Hemisphere summer insolation. Superimposed on that cooling trend were several short-term warming and cooling events lasting 50–200 years and attributable to variations in solar output, modified by shorter-term coolings lasting up to a decade or so and caused by large volcanic eruptions with short-term effects. This pattern was modified by natural internal variations (oscillations) within the ocean–atmosphere system, such as the Pacific Decadal Oscillation, the North Atlantic Oscillation and the El Niño–Southern Oscillation. As the trend in insolation is expected to continue more or less flat for at least another 1000 years, Earth's climate should remain in a cool state rather like that of the Little Ice Age, with occasional warmings and coolings.

We are now in the 'modern warm period'. Warming from 1900 up to around 1945 is attributable in part to an increase in solar energy caused by development of the latest peak in the Suess–De Vries solar cycle and in part to a gradually growing contribution from climbing emissions of CO_2. A disconnect between solar output and global warming began in about 1950. Solar energy increased towards the double peak of the solar cycle between 1960 and 1990[47, 88], while global temperatures flattened, probably due to cooling induced by a growing load of industrial aerosols between 1945 and 1970[92, 93]. Thereafter, temperatures rose while solar energy remained constant[90]. Temperatures continued to rise after the cycle came to an end in about 1990, when solar output began its present decline[88].

Greenhouse gases started to increase exponentially from about 1776, when the Industrial Revolution took off[172]. The rise in greenhouse gases began to contribute significantly to the rise in temperature from about 1900 onwards, increasing eventually to the point where they became the main drivers of warming after 1950, accentuated by water vapour evaporated from the warming ocean. The global temperatures of the mid 20th century were not much different from those of the Medieval Warm Period. But present temperatures are significantly higher, especially in the Arctic, where the shrinkage of sea ice makes the ocean and lower atmosphere warm much faster than elsewhere. These warming trends cut across both the long-term trend of cooling driven by declining insolation and the recent decline in sunspot activity representative of solar output.

In a nutshell, then, orbital insolation tells us we should still be in the neoglacial Little Ice Age of the Holocene and experiencing periodic minor warming and cooling, due mainly to predictable fluctuations in solar energy modified by occasional volcanic outpourings. In that context, the warming since 1950 is unprecedented in the past 2000 years, and nobody has yet come up with a better explanation for it than that it is caused by our rising outputs of greenhouse gases.

References

1. Loutre, M.F. and Berger, A. (2000) Future climatic changes: are we entering an exceptionally long interglacial? *Climatic Change* **46**, 61–90.
2. Berger, A. and Loutre, M.-F. (2002) An exceptionally long interglacial ahead. *Science* **297**, 1287–1288.
3. Beer, J. and Van Geel, B. (2009) Holocene climate change and the evidence for solar and other forcings. In: *Natural Climate Variability and Global Warming: A Holocene Perspective* (eds R.W. Battarbee and H.A. Binney). Wiley-Blackwell, Chichester and Oxford, pp. 138–162.
4. Herschel., W. (1801) Observations tending to investigate the nature of the sun in order to find the causes or symptoms of its variable emission of light and heat. *Philosophical Transactions of the Royal Society of London* **91**, 265–318.
5. Le Roy Ladurie, E. (1988) *Times of Feast, Times of Famine: A History of Climate Since the Year 1000*. Farrar Straus and Giroux, New York.
6. Le Roy Ladurie, E. and Baulant, M. (1980) Grape Harvests from the fifteenth through the nineteenth centuries. *Journal of Interdisciplinary History* **10**, 839–849.
7. Chuine, I., Yiou, P., Viovy, N., Seguin, B., Daux, V. and Le Roy Ladurie, E. (2004) Historical phenology: grape ripening as a past climate indicator. *Nature* **432** (7015), 289–290.
8. Le Roy Ladurie, E. (2005) [Dog-days, cold periods, grape-harvests (France, 15–19th centuries)]. *Comptes Rendus Biologies* **328** (3), 213–222.
9. Crowley, T.J. and North, G.R. (2011) *Paleoclimatology*. Oxford Monographs on Geology and Geophysics 18. Oxford University Press, Oxford.
10. Pfister, C. (1985) *CLIMHIST: A Weather Data Bank for Central Europe*. Meteotest, Bern.
11. Friis-Christensen, F. and Lassen, L. (1991) Length of the solar cycle: an indicator of solar activity closely associated with climate. *Science* **254**, 698–700.
12. Eddy, J.A. (1977) Climate and the changing sun. *Climatic Change* **1**, 173–190.
13. Eddy, J.A. (1976) The Maunder Minimum. *Science* **192** (4245), 1189–1202.
14. Broecker, W. (2006) Radiocarbon. In: *The Atmosphere, Vol. 4, Treatise on Geochemistry* (eds H.D. Holland and K.K. Turekian). Elsevier-Pergamon, Oxford, pp. 245–260.
15. Muscheler, R., Beer, J. and Kubik, P.W. (2004) Long-term solar variability and climate change based on radionuclide data from ice cores. In: *Solar Variability and its Effects on Climate* (eds J.M. Pap and P. Fox). Geophysical Monographs 114. American Geophysical Union, Washington, DC, pp. 221–235.
16. Stuiver, M. and Quay, P.D. (1980) Changes in atmospheric carbon-14 attributed to a variable sun. *Science* **207**, 11–19.
17. Stuiver, M. and Quay, P.D. (1981) Atmospheric ^{14}C changes resulting from fossil fuel CO_2 release and cosmic ray flux variability. *Earth and Planetary Science Letters* **53**, 349–362.
18. Stuiver, M., Reimer, P.J., Bard, E., Beck, J.W., Burr, G.S., Hughen, K.A., *et al.* (1998) INTCAL98 radiocarbon age calibration, 24,000-0 cal BP. *Radiocarbon* **40** (3), 1041–1083.
19. Dansgaard, W., Johnsen, S.J., Clausen, H.B., Dahl-Jensen, D., Gundestrup, N., Hammer, C.U. and Oeschger, H. (1984) North Atlantic climatic oscillations revealed by deep Greenland ice cores. In: *Climate Processes and Climate Sensitivity* (eds J.E. Hansen and T. Takahashi). Geophysical Monographs 29. American Geophysical Union, Washington, DC, pp. 288–298.
20. Oeschger, H., Beer, J., Siegenthaler, U., Stauffer, B., Dansgaard, W. and Langway, C.C. (1984) Late glacial climate history from ice cores. In: *Climate Processes and Climate Sensitivity* (eds J.E. Hansen and T. Takahashi). Geophysical Monographs 29. American Geophysical Union, Washington, DC, pp. 299–306.
21. Schüssler, M. and Schmitt, D. (2004) Theoretical models of solar magnetic variability. In: *Solar Variability and its Effects on Climate* (eds J.M. Pap and P. Fox). Geophysical Monographs 114. American Geophysical Union, Washington, DC, pp. 33–49.
22. Steinhilber, F., Abreu, J.A., Beer, J., Brunner, I., Christl, M., Fischer, H., *et al.* (2012) 9,400 years of cosmic radiation and solar activity from ice cores and tree rings. *Proceedings of the National Academy of Sciences* **109** (16), 5967–5971.
23. Bard, E., Raisbeck, G., Yiou, F. and Jouzel, J. (2000) Solar irradiance during the last 1200 years based on cosmogenic nuclides. *Tellus* **52B**, 985–992.
24. Muscheler, R., Joos, F., Beer, J., Müller, S.A., Vonmoos, M. and Snowball, I. (2007) Reply to the comment by Bard et al. on 'Solar activity during the last 1000 yr inferred from radionuclide records'. *Quaternary Science Reviews* **26**, 2301–2308.
25. Clette, F., Svalgaard, L., Vaquero, J.M. and Cliver, E.W. (2014) Revisiting the Sunspot number. *Space Science Review* **186**, 35–103.
26. Damon, P. and Peristykh, A.N. (2004) Solar and climatic implications of the centennial and millennial periodicities in atmospheric Δ14C variations. In: *Solar Variability and its Effects on Climate* (eds J.M. Pap and P. Fox). Geophysical Monographs 114. American Geophysical Union, Washington, DC, pp. 237–249.
27. Wanner, H., Beer, J., Bütikofer, J., Crowley, T.J., Cubasch, U., Flückiger, J., *et al.* (2008) Mid- to late Holocene climate change; an overview. *Quaternary Science Reviews* **27**, 1791–1828.
28. Magny, M. (1993) Solar influences on Holocene climatic changes illustrated by correlations between past lake level

fluctuations and the atmospheric ^{14}C record. *Quaternary Research* **40**, 1–9.

29. Magny, M. (2007) West-Central Europe. In: *Lake Level Studies, Encyclopedia of Quaternary Science*. Elsevier, Berlin, pp. 1389–1399.

30. Magny, M., De Beaulieu, J.-L., Drescher-Schneider, R., Vannière, B., Walter-Simonnet, A.-V., Miras, Y., *et al.* (2007) Holocene climate changes in the central Mediterranean as recorded by lake-level fluctuations at Lake Accesa (Tuscany, Italy). *Quaternary Science Reviews* **26** (13–14), 1736–1758.

31. Magny, M. (2004) Holocene climatic variability as reflected by mid-European lake-level fluctuations, and its probable impact on prehistoric human settlements. *Quaternary International* **113**, 65–79.

32. Magny, M. (1999) Lake-level fluctuations in the Jura and French Subalpine Ranges associated with ice-rafting events in the North Atlantic and variations in the polar atmospheric circulation. *Quaternaire* **10**, 61–64.

33. Verschuren, D. and Charman, D.J. (2009) Latitudinal linkages in late Holocene moisture-balance variation. In: *Natural Climate Variability and Global Warming: A Holocene Perspective* (eds R.W. Battarbee and H.A. Binney). Wiley-Blackwell, Chichester and Oxford, pp. 189–231.

34. Bond, G.C., Kromer, B., Beer, J., Muscheler, R., Evans, M.N., Showers, W., *et al.* (2001) Persistent solar influence on North Atlantic climate during the Holocene. *Science* **294**, 2130–2152.

35. Braun, H., Christl, M., Rahmstorf, S., Ganopolski, A., Mangini, A., Kubatzki, C., *et al.* (2005) Possible solar origin of the 1,470-year glacial climate cycle demonstrated in a coupled model. *Nature* **438**, 208–211.

36. Debret, M., Bout-Roumazeilles, V., Grousset, F., Desmet, M., McManus, J.F., Massei, N., *et al.* (2007) The origin of the 1500-year climate cycles in Holocene North Atlantic records. *Climate of the Past* **3**, 569–575.

37. Moffa-Sánchez, P., Born, A., Hall, I.R., Thornalley, D.J.R. and Barker, S. (2014) Solar forcing of North Atlantic surface temperature and salinity over the past millennium. *Nature Geoscience* **7**, 275–278.

38. Lamb, H. (1979) Climatic variation and changes in the wind and ocean circulation: the Little Ice Age in the northeast Atlantic. *Quaternary Research* **11**, 1–20.

39. Maasch, K.A., Mayewski, P.A., Rohling, E.J., Stager, J.C., Karlen, W., Meeker, L.D. and Meyerson, E.A. (2005) A 2000-year context for modern climate change. *Geografiska Annaler* **87A**, 7–15.

40. Wanner, H., Solomina, O., Grosjean, M., Ritz, S.P. and Jetel, M. (2011) Structure and origin of Holocene cold events. *Quaternary Science Reviews* **30**, 3109–3123.

41. Mayewski, P.A., Rohling, E., Stager, J.C., Karlen, W., Maasch, K.A., Meeker, L.D., *et al.* (2004) Holocene climatic variability. *Quaternary Research* **62**, 243–255.

42. Bond, G.C., Showers, W., Elliot, M., Evans, M., Lotti, R., Hajdas, I., *et al.* (1999) The North Atlantic's 1-2kyr climate rhythm: relation to Heinrich Events, Dansgaard/Oeschger Cycles and the Little Ice Age. In: *Mechanisms of Global Climate Change at Millennial Time Scales* (eds P.U. Clark, R.S. Webb and L.D. Keigwin). Geophysical Monographs 112. American Geophysical Union, Washington, DC, pp. 35–58.

43. Lamb, H.H. (1995) *Climate, History and the Modern World*, 2nd edn. Routledge, London.

44. North, G.R., Baker, D.N., Bradley, R.S., Foukal, P., Haigh, J.D., Held, I.M., *et al.* (2012) *The Effects of Solar Variability on Earth's Climate*. National Academies Press, Washington, DC.

45. Cronin, T.M. (2010) *Paleoclimates: Understanding Climate Change Past and Present*. Columbia University Press.

46. Svensmark, H. and Friis-Christensen, E. (1997) Variation of cosmic ray flux and global cloud coverage – a missing link in solar-climate relationships. *Journal of Atmospheric and Solar-Terrestrial Physics* **59** (11), 1225–1232.

47. Lockwood, M. (2012) Solar influence on global and regional climates. *Surveys in Geophysics* **33** (3–4), 505–534.

48. Gray, L.J., Beer, J., Geller, M., Haigh, J.D., Lockwood, M., Matthes, K., *et al.* (2010) Solar influences on climate. *Reviews of Geophysics* **48**, RG4001.

49. Lamb, H.H. (1966) *The Changing Climate: Selected Papers by H.H. Lamb*. Methuen, London.

50. Lamb, H.H. (1966) Britain's climate in the past. In: *The Changing Climate: Selected Papers by H.H. Lamb* (H.H. Lamb). Methuen, London, pp. 170–195.

51. Folland, C.K., Karl, T.R. and Vinnikov, K.Y. (1990) Observed climate variations and change. Chapter 7 in: *Report of Working Group 1, Climate Change* (eds J.T. Houghton, G.J. Jenkins and J.J. Ephraums). The IPCC First Assessment Report. Cambridge University Press, Cambridge, pp. 195–238.

52. Lamb, H.H. (1988) Climate and life during the Middle Ages, studied especially in the mountains of Europe. In: *Weather, Climate and Human Affairs*. Routledge, London, pp. 40–74.

53. Nicholls, N., Gruza, G.V., Jouzel, J., *et al.* (1996) Observed climate variability and change. Chapter 3 in: *Report of Working Group 1* (eds J.T. Houghton, L.G. Meira-Filho, B.A. Callander, N. Harris, A. Kattenberg and K. Maskell). IPCC Second Assessment Report. Cambridge University Press, Cambridge, pp. 133–192.

54. Bradley, R.S. and Jones, P.D. (1993) Little Ice Age summer temperature variations; their nature and relevance to recent global warming trends. *The Holocene* **3**, 367–376.

55. Bradley, R.S. and Jones, P.D. (1995) Recent developments in studies of climate since AD 1500. In: *Climate Since AD 1500*, 2nd edn (eds R.S. Bradley and P.D. Jones). Routledge, London, pp. 666–679.

56. Jones, P.D. and Bradley, R.S. (1992) Climatic variations over the last 500 years. In: *Climate Since AD 1500* (eds R.S. Bradley and P.D. Jones). Routledge, London, pp. 649–665.

57. De Menocal, P., Ortiz, J., Guilderson, T. and Sarnthein, M. (2000) Coherent high- and low- latitude climate variability during the Holocene warm period. *Science* **288**, 2198–2202.

58. PAGES 2k Consortium (2013) Continental-scale temperature variability during the last two millennia. *Nature Geoscience* **6**, 339–346.

59. Schurer, A.P., Tett, S.F.B. and Hegerle, G.C. (2014) Small influence of solar variability on climate over the past millennium. *Nature Geoscience* **7**, 104–108.

60. D'Arrigo, R., Wilson, R. and Tudhope, A. (2008) The impact of volcanic forcing on tropical temperatures during the past four centuries. *Nature Geoscience* **2**, 51–56.

61. Lavigne, F., Degeal, J.-P., Komorowski, J.-C., Guillet, S., Robert, V., Lahitte, P., *et al.* (2013) Source of the great A.D. 1257 mystery eruption unveiled, Samalas volcano, Rinjani Volcanic Complex, Indonesia. *Proceedings of the National Academy of Sciences* **110** (42), 16 742–16 747.

62. Miller, G.H., Geiersdottir, A., Zhong, Y., Larsen, D.J., Otto-Bliesner, B.L., Holland, M.M., *et al.* (2012) Abrupt onset of the Little Ice Age triggered by volcanism and sustained by sea-ice/ocean feedbacks. *Geophysical Research Letters* **39** (2), doi: 10.1029/2011GL050168.

63. Hu, F.S., Ito, E., Brown, T.A., Curry, B.B. and Engstrom, D.R. (2001) Pronounced climatic variations in Alaska during the last two millennia. Proceedings of the National Academy of Sciences **98** (19), 10 552–10 556.

64. Loso, M.G. (2009) Summer temperatures during the Medieval Warm Period and Little Ice Age inferred from varved proglacial lake sediments in southern Alaska. *Journal of Paleolimnology* **41**, 117–128.

65. Esper, J., Cook, E.R. and Schweingruber, F.H. (2002) Low frequency signals in long tree-ring chronologies for reconstructing past temperature variability. *Science* **295**, 2250–2252.

66. Esper, J., Frank, D.C., Wilson, R.S. and Briffa, K.R. (2005) Effect of scaling and regression on reconstructed temperature amplitude for the past millennium. *Geophysical Research Letters* **32**, L07711.

67. Moberg, A., Sonechkin, D.M., Holmgren, K., Datsenko, N.M. and Karlén, W. (2005) Highly variable Northern Hemisphere temperatures reconstructed from low- and high-resolution proxy data. *Nature* **433**, 613–618.

68. Kirby, J. (2007) Cosmic rays and climate. *Surveys in Geophysics* **28**, 333–375.

69. Ljungqvist, F.C., Krusic, P.J., Brattström, G. and Sundqvist H.S. (2012) Northern Hemisphere temperature patterns in the last 12 centuries. *Climate of the Past* **8**, 227–249.

70. Werner, K., Spielhagen, R.F., Bauch, D., Hass, H.C., Kandiano, E. and Zamelczyk, K. (2011) Atlantic water advection to the eastern Fram Strait – multi-proxy evidence for Holocene variability. *Palaeogeography, Palaeoclimatology, Palaeoecology* **308**, 264–276.

71. Dokken, T., Nisancioglu, K.H., Li, C., Battisti, D.S. and Kissel, C. (2013) Dansgaard-Oeschger cycles: interactions between ocean and sea ice intrinsic to the Nordic Seas. *Paleoceanography* **28** (3), 491–502.

72. Vinther, B.M., Jones, P.D., Briffa, K.R., Clausen, H.B., Andersen, K.K., Dahl-Jensen, D. and Johnsen, S.J. (2010) Climatic signals in multiple highly resolved stable isotope records from Greenland. *Quaternary Science Research* **29**, 522–538.

73. Kobashi, T., Kawamura, K., Severinghaus, J.P., Barnola, J.-M., Nakaegawa, T., Vinther, B.M., *et al.* (2011) High variability of Greenland surface temperature over the past 4000 years estimated from trapped air in an ice core. *Geophysical Research Letters* **38**, L21501.

74. Hendry, E.J., Gagan, M.K., Alibert, C.A., McCulloch, M.T., Lough, J.M. and Isdale, P.J. (2002) Abrupt decrease in tropical Pacific sea surface salinity at end of Little Ice Age. *Science* **295**, 1511–1514.

75. McGann, M. (2008) High resolution foraminiferal, isotopic and trace element records from Holocene estuarine deposits of San Francisco Bay, California. *Journal of Coastal Research* **24** (5), 1092–1109.

76. Zhao, M., Eglinton, G., Read, G.G. and Schimmelmann, A. (2000) An alkenone ($U^{k'}_{37}$) quasi-annual sea surface temperature record (AD1440 to 1940) using varved sediments from the Santa Barbara Basin. *Organic Geochemistry* **31**, 903–917.

77. Mayewski, P.A. and Maasch, K.A. (2006) Recent warming inconsistent with natural association between temperature and atmospheric circulation over the last 2000 years. *Climate of the Past Discussions* **2**, 1–29.

78. Mayewski, P.A., Meredith, M.P., Summerhayes, C.P., Turner, J., Worby, A., Barrett, P.J., *et al.* (2009) State of the Antarctic and Southern Ocean climate system. *Reviews of Geophysics* **47**, RG1003.

79. Turner, J., Bindschadler, R.A., Convey, P., Di Prisco, G., Fahrbach, E., Gutt, J., *et al.* (2009) *Antarctic Climate Change and the Environment*. Scientific Committee on Antarctic Research, Cambridge.

80. Mayewski, P.A., Maasch, K.A., Dixon, D., Sneed, S.B., Oglesby, R., Korotkikh, E., *et al.* (2013) West Antarctica's sensitivity to natural and human forced climate change over the Holocene. *Journal of Quaternary Science* **18** (1), 40–48.

81. Mulvaney, R., Abram, N.J., Hindmarsh, R.C.A., Arrowsmith, C., Fleet, L., Triest, J., *et al.* (2012) Recent Antarctic Peninsula warming relative to Holocene climate and ice-shelf history. *Nature* **489**, 141–145.

82. Etourneau, J., Collins, L.G., Willmott, V., Kim, J.-H., Barbara, L., Leventer, A., *et al.* (2013) Holocene climate variations in the western Antarctic Peninsula: evidence for sea ice extent predominantly controlled by changes in insolation and ENSO variability. *Climate of the Past* **9**, 1431–1446.

83. Pike, J., Swann, G.E.A., Leng, M.J. and Snelling, A.M. (2013) Glacial eischarge along the West Antarctic Peninsula during the Holocene. *Nature Geoscience* **6**, 199–202.

84. Cowie, J. (2013) *Climate Change: Biological and Human Aspects*, 2nd edn. Cambridge University Press, Cambridge.

85. Lund, D.C., Lynch-Stieglitz, J. and Curry, W.B. (2006) Gulf Stream density structure and transport during the past millennium, *Nature* **444**, 601–604.

86. Schlesinger, M.E. and Andronova, N.G. (2004) Has the sun changed climate? Modeling the effect of solar variability on climate. In: *Solar Variability and its Effects on Climate* (eds J.M. Pap and P. Fox). Geophysical Monographs 114. American Geophysical Union, Washington, DC, pp. 261–282.

87. Kobashi, T., Shindell, D.T., Kodera, K., Box, J.E., Nakaegawa, T. and Kawamura, K. (2013) On the origin of multidecadal to centennial Greenland temperature anomalies over the past 800 yr. *Climate of the Past* **9**, 583–596.

88. Lockwood, M. (2010) Solar change and climate: an update in the light of the current exceptional solar minimum. *Proceedings of the Royal Society A* **466**, 303–329.

89. Hansen, J., Sato, M. and Ruedy, R. (2013) Global temperature update through 2012. Available from www.nasa.gov/pdf/719139main_2012_GISTEMP_summary.pdf (last accessed 29 January 2015).

90. Bard, E. and Delaygue, G. (2008) Comment on 'Are there connections between the Earth's magnetic field and climate?' by V. Courtillot, Y. Gallet, J.-L. Le Mouël, F. Fluteau, A. Genevey EPSL 253, 328, 2007. *Earth and Planetary Science Letters* **265**, 302–307.

91. Lockwood, M. and Fröhlich, C. (2007) Recent oppositely directed trends in solar climate forcings and the global mean surface air temperature. *Proceedings of the Royal Society A* **471**(2175), doi:10.1098/rspa.2007.0347.

92. Trenberth, K.E. and Fasullo, J.T. (2013) An apparent hiatus in global warming. *Earth's Future* **1** (1), 19–32.

93. Budyko, M.I. (1977) On present-day climatic changes. *Tellus* **29**, 193–204.

94. Amman, C.M., Joos, F., Schimel, D.S., Otto-Bliesner, B.L. and Tomas, R.A. (2007) Solar influence on climate during the past millennium: results from transient simulations with the NCAR climate system model. *Proceedings of the National Academy of Sciences* **104** (10), 3713–3718.

95. Miller, G.H., Alley, R.B., Brigham-Grette, J., Fitzpatrick, J.J., Polyak, L., Serreze, M.C. and White, J.W.C. (2010) Arctic amplification: can the past constrain the future? *Quaternary Science Reviews* **29** (15–16), 1779–1790.

96. Polyakov, I.V., Bekryaev, R.V., Alekseev, G.V., Bhatt, U.S., Colony, R.L., Johnson, M.A., *et al.* (2003) Variability and trends of air temperature and pressure in the maritime Arctic, 1875–2000. *American Meteorological Society* **16**, 2067–2077.

97. Frolov, I.E., Gudkovich, Z.M., Karklin, V.P., Kovalev, E.G. and Smolyanitsky, V.M. (2009) *Climate Change in Eurasian Arctic Seas: Centennial Ice Cover Observations*. Praxis, Chichester.

98. Wood, K.R. and Overland, J.E. (2010) Early 20th century Arctic warming in retrospect. *International Journal of Climatology* **30**, 1269–1279.

99. Overland, J.E., Wood, K.R. and Wang, M. (2011) Warm Arctic – cold continents: climate impacts of the newly open Arctic Sea. *Polar Research* **30**, 15 787.

100. Polyakov, I.V., Walsh, J.E. and Kwok, R. (2012) Recent changes of Arctic multiyear sea ice coverage and the likely causes. *Bulletin of the American Meteorological Society* **93**, 145–151.

101. Kinnard, C., Zdanowicz, C.M., Koerner, R.M. and Fisher, D.A. (2008) A changing Arctic seasonal ice zone: observations from 1870–2003 and possible oceanographic consequences. *Geophysical Research Letters* **35**, L02507.

102. Kinnard, C., Zdanowicz, C.M., Fisher, D.A., Isaksson, E., de Vernal, A. and Thompson, L.G. (2011) Reconstructed changes in Arctic sea ice over the past 1,450 years. *Nature* **479**, 509–512.

103. Tingley, M.P. and Huybers, P. (2013) Recent temperature extremes at high northern latitudes unprecedented in the past 600 years. *Nature* **496**, 201–205.

104. Kaufman, D., Schneider, D.P., McKay, N.P., Ammann, C.M., Bradley, R.S., Briffa, K.R., *et al.* (2009) Recent warming reverses long-term Arctic cooling. *Science* **325**, 1236–1239.

105. Walsh, J.E. (2013) Melting ice – what is happening to Arctic sea ice, and what does it mean for us? *Oceanography* **26** (2), 171–181.

106. Bjune, A.E., Seppä, H. and Birks, H.J.B. (2009) Quantitative summer-temperature reconstructions for the last 2000 years based on pollen-stratigraphical data from northern Fennoscandia. *Journal of Paleolimnology* **41**, 43–56.

107. Miller, G.H., Lehman, S.J., Refsnider, K.A., Southon, J.R. and Zhong, Y. (2013) Unprecedented recent summer warmth in Arctic Canada. *Geophysical Research Letters* **40**, 5745–5751.

108. Esper, J., Frank, D.C. Timonen, M., Zorita, E., Wilson, R.J.S., Luterbacher, J., *et al.* (2012) Orbital forcing of tree-ring data. *Nature Climate Change* **2**, 862–866.

109. Thompson, L.G., Mosley-Thompson, E., Brecher, H., Davis, M., León, B., Les, D., *et al.* (2006) Abrupt tropical climate change: past and present. *Proceedings of the National Academy of Sciences* **103**, 10 536–10 543

110. Büntgen, U., Tegel, W., Nicolussi, K., McCormick, M., Frank, D., Trouet, V., *et al.* (2011) 2500 years of European climate variability and human susceptibility. *Science* **331**, 578–582.

111. Büntgen, U., Kyncl, T., Ginzler, C., Jacks, D.S., Esper, J., Tegel, W., *et al.* (2013) Filling the eastern European gap in millennium-long temperature reconstructions. *Proceedings of the National Academy of Sciences* **110** (5), 1773–1778.

112. Brohan, P., Kennedy, J.J., Harris, I., Tett, S.F.B. and Jones, P.D. (2006) Uncertainty estimates in regional and global observed temperature changes: a new dataset from 1850. *Journal of Geophysical Research* **111** (D12), doi: 10.1029/2005JD006548.

113. Marcott, S.A., Shakun, J.D., Clark, P.U. and Mix, A.C. (2013) A reconstruction of regional and global temperature for the past 11,300 years. *Science* **339**, 1198–1201.

114. Friedli, H., Lötscher, H., Oeschger, H., Siegenthaler, U. and Stauffer, B. (1986) Ice core record of the $^{13}C/^{12}C$ ratio of atmospheric CO_2 in the past two centuries. *Nature* **324**, 237–241.

115. Wolff, E.W. (2011) Greenhouse gases in the Earth System: a palaeoclimate perspective. *Philosophical Transactions of the Royal Society A* **369**, 2133–2147.

116. Crutzen, P.J. and Stoermer, E.F. (2000) The 'Anthropocene'. *International Geosphere and Biosphere Newsletter* **41**, 17–18.

117. Met Office (2013) *The Recent Pause in Global Warming (2): What are the Potential Causes?* Met Office, Exeter.

118. Kosaka, Y. and Xie, S.-P. (2013) Recent global-warming hiatus tied to equatorial Pacific surface cooling. *Nature* **501**, 403–407.

119. England, M.H., McGregor, S., Spence, P., Meehl, G.A., Timmermann, A., Cai, W., *et al.* (2014) Recent intensification of wind-driven circulation in the Pacific and the ongoing warming hiatus. *Nature Climate Change* **4**(3), 222–227.

120. Schmidt, G.A., Shindell, D.T. and Tsigaridis, K. (2014) Reconciling warming trends. *Nature Geoscience* **7**, 158–160.

121. Santer, B.D., Bonfils, C., Painter, J.F., Zelinka, M.D., Mears, C., Solomon, S., *et al.* (2014) Volcanic contribution to decadal changes in tropospheric temperature. *Nature Geoscience* **7**, 185–189.

122. Solomon, S., Daniel, J.S., Neely, R.R. III,, Vernier, J.-P., Dutton, E.G. and Thomason, L.W. (2011) The persistently variable 'background' stratospheric aerosol layer and global climate change. *Science* **333**, 866–870.

123. Kaufmann, R.K., Kauppi, H., Mann, M.L. and Stock, J.H. (2011) Reconciling anthropogenic climate change with observed temperature 1998-2008. *Proceedings of the National Academy of Sciences* **108** (29), 11 790–11 793.

124. Chen, X. and Tung, K.-K. (2014) Varying planetary heat sink led to global-warming slowdown and acceleration. *Science* **345**, 897–903.

125. Cowtan, K. and Way, R. (2013) Coverage bias in the HadCRUT4 temperature series and its impact on recent temperature trends. *Quarterly Journal of the Royal Meteorological Society B* **140** (683), 1935–1944.

126. Mann, M.E., Bradley, R.S. and Hughes, M.K. (1998) Global-scale temperature patterns and climate forcing over the past six centuries. *Nature* **392**, 779–787.

127. Hughes, M.K. and Diaz, H.F. (1994) Was there a 'Medieval Warm Period', and if so, where and when? *Climate Change* **26**, 109–142.

128. Mann, M.E., Bradley, R.S. and Hughes, M.K. (1999) Northern Hemisphere temperatures during the past millennium: inferences, uncertainties, and limitations. *Geophysical Research Letters* **26** (6), 759–762.

129. Folland, C.K., Karl, T.R., Christy, J.R., Clarke, R.A., Gruza, G.V., Jouzel, J., *et al.* (2001) Observed climate variability and change. Chapter 2 in: *Climate Change 2001, the Scientific Basis* (eds J.T. Houghton, Y. Ding, D.J. Griggs, M. Noguer, P.J. Van der Linden, X. Dai, *et al.*). Contribution of Working Group 1 to the IPCC Third Assessment Report. Cambridge University Press, Cambridge, pp. 101–181.

130. McIntyre, S. and McKitrick, R. (2003) Corrections to the Mann et al. (1998) proxy database and Northern Hemisphere average temperature series. *Energy and Environment* **14** (6), 751–771.

131. McIntyre, S. and McKitrick, R. (2005) Hockey sticks, principal components, and spurious significance. *Geophysical Research Letters* **32** (2), L03710.

132. Montford, A.W. (2010) *The Hockey Stick Illusion: Climategate and the Corruption of Science.* Stacey International, London.

133. Mann, M.E. (2012) *The Hockey Stick and the Climate Wars: Dispatches from the Front Lines.* Columbia University Press, New York.

134. North, G.R., Biondi, F., Bloomfield, P., Christy, J.R., Cuffey, K.M., Dickinson, R.E., *et al.* (2006) *Surface Temperature Reconstructions for the Last 2000 Years.* National Academies Press, Washington, DC.

135. Mann, M.E. and Jones, P.D. (2003) *2,000-Year Hemispheric Multi-proxy Temperature Reconstructions.* IGBP PAGES/World Data Center for Paleoclimatology Data Contribution Series 2003-051. NOAA/NGDC Paleoclimatology Program, Boulder, CO.

136. Moberg, A., Sonechkin, D.M., Holmgren, K., Datsenko, N.M. and Karlén, W. (2005) *2,000-Year Northern Hemisphere Temperature Reconstruction.* IGBP PAGES/World Data Center for Paleoclimatology Data. Contribution Series 2005-019. NOAA/NGDC Paleoclimatology Program, Boulder, CO.

137. Jansen, E., Overpeck, J., Briffa, K.R., Duplessy, J.-C., Joos, F., Masson-Delmotte, V., *et al.* (2007) Palaeoclimate. In: *Climate Change 2007, the Physical Science Base* (eds S. Solomon, D. Qin, M. Manning, M. Marquis, K. Averyt, M.M.B. Tignor, *et al.*). Contribution of Working Group I to the IPCC Fourth Assessment Report. Intergovernmental Panel on Climate Change, Geneva, pp. 433–497.

138. D'Arrigo, R.S., Wilson, R., Liepert, B. and Cherubini, P. (2008) On the 'Divergence Problem' in northern forests: a review of the tree-ring evidence and possible causes. *Global and Planetary Change* **60**, 289–305.

139. Porter, T.J. and Pisaric, M.F.J. (2012) 'Divergence' and implications for reconstructing warm season temperatures from White Spruce tree-rings in Old Crow Plats, Yukon Territory, and neighbouring N.W. North America. Available from http://132.246.11.198/2012-ipy/Abstracts_On_the_Web/pdf/ipy2012arAbstract00330.pdf (last accessed 29 January 2015).

140. Mann, M.E., Zhang, Z., Hughes, M.K., Bradley, R.S., Miller, S.K., Rutherford, S. and Ni, F. (2008) Proxy-based reconstructions of hemispheric and global temperature

variations over the past two millennia. *Proceedings of the National Academy of Sciences* **105** (36), 13 252–13 257.

141. Mann, M.E., Zhang, Z., Rutherford, S., Bradley, R.S., Hughes, M.K., Shindell, D., *et al.* (2009) Global signatures and dynamical origins of the Little Ice Age and Medieval Climate Anomaly. *Science* **326**, 1256–1260.

142. Wanner, H., Mercolli, L., Grosjean, M. and Ritzl, S.P. (2014) Holocene climate variability and change; a data based review. *Journal of the Geological Society of London.* Published online 17 October 2014. Available from: http://jgs.lyellcollection.org/content/early/2014/10/10/jgs2013-101.abstract (last accessed 29 January 2015).

143. Goosse, H., Mann, M.E. and Renssen, H. (2009) Climate of the past millennium: combining proxy data and model simulations. In: *Natural Climate Variability and Global Warming: A Holocene Perspective* (eds R.W. Battarbee and H.A. Binney). Wiley-Blackwell, Chichester and Oxford, pp. 163–188.

144. Jevrejeva, S., Moore, J.C., Grinsted, A. and Woodworth, P.L. (2008) Recent global sea level acceleration started over 200 years ago. *Geophysical Research Letters* **35**, L08715.

145. Church, J.A., Roemmich, D., Domingues, C.M., Willis, J.K., White, N.J., Gilson, J.E., *et al.* (2010) Ocean temperature and salinity contributions to global and regional sea-level change. In: *Understanding Sea-Level Rise and Variability* (eds J.A. Church, P.L. Woodworth, T. Aarup and W.S. Wilson). Wiley-Blackwell, Chichester and Oxford, pp. 143–176.

146. Steffen, K., Thomas, R.H., Rignot, E., Cogley, J.G., Dyurgerov, M.B., Raper, S.C.B., *et al.* (2010) Cryospheric contributions to sea-level rise and variability. In: *Understanding Sea-Level Rise and Variability* (eds J.A. Church, P.L. Woodworth, T. Aarup and W.S. Wilson). Wiley-Blackwell, Chichester and Oxford, pp. 177–225.

147. Church, J.A., Aarup, T., Woodworth, P.L., Wilson, W.S., Nicholls, R.J., Rayner, R., *et al.* (2010) Sea-level rise and variability: synthesis and outlook for the future. In: *Understanding Sea-Level Rise and Variability* (eds J.A. Church, P.L. Woodworth, T. Aarup and W.S. Wilson). Wiley-Blackwell, Chichester and Oxford, pp. 402–419.

148. Church, J.A., White, N.J., Aarup, T., Wilson, W.S., Woodworth, P.L., Domingues, C.M., *et al.* (2008) Understanding global sea levels: past, present and future. *Sustainability Science* **3**, 9–22.

149. Gehrels, R. (2010) Sea-level changes since the Last Glacial Maximum: an appraisal of the IPCC Fourth Assessment Report. *Journal of Quaternary Science* **25** (1), 26–38.

150. Lambeck, K., Woodroffe, C.D., Antonioli, F., Anzidei, M., Gehrels, W.R., Laborel, J. and Wright, A.J. (2010) Paleoenvironmental records, geophysical modeling, and reconstruction of sea-level; trends and variability on centennial and longer timescales. In: *Understanding Sea-Level Rise and Variability* (eds J.A. Church, P.L. Woodworth, T. Aarup and W.S. Wilson). Wiley-Blackwell, Chichester and Oxford, pp. 61–121.

151. Woodworth, P.L., Gehrels, W.R. and Nerem, R.S. (2011) Nineteenth and twentieth century changes in sea level. *Oceanography* **24** (2), 80–93.

152. Jevrejeva, S., Grinsted, A. and Moore, J.C. (2009) Anthropogenic forcing dominates sea level rise since 1850. *Geophysical Research Letters* **36**, 20.

153. Kemp, A.C., Horton, B.J., Donnelly, J.P., Mann, M.E., Vermeer, M. and Rahmstorf, S. (2011) Climate-related sea-level variations over the past two millennia. *Proceedings of the National Academy of Sciences* **108** (27), 11 017–11 022.

154. Rahmstorf, S. (2007) A semi-empirical approach to projecting future sea-level rise. *Science* **315**, 368–370.

155. Vermeer, M. and Rahmstorf, S. (2009) Global sea level linked to global temperature. *Proceedings of the National Academy of Sciences* **106**, 21 527–21 532.

156. Grinsted, A., Moore, J.C. and Jevrejeva, S. (2010) Reconstructing sea level from paleo and projected temperatures 200 to 2100 AD. *Climate Dynamics* **34** (4), 461–472.

157. Jones, P.D. and Mann, M.E. (2004) Climate over past millennia. *Reviews of Geophysics* **42**, RG2002.

158. Gregory, J.M., Lowe, J.A. and Tett, S.F.B. (2006) Simulated global-mean sea level changes over the last half-millennium. *Journal of Climate* **19**, 4576–4591.

159. Fairbanks R.G. (1989) A 17,000-year glacio-eustatic sea level record: influence of glacial melting rates on the Younger Dryas event and deep-ocean circulation. *Nature* **342**, 637–642

160. Rohling, E.J., Grant, K., Hemleben, C., Siddall, M., Hoogakker, B.A.A., Bolshaw, M. and Kucera, M. (2008) High rates of sea-level rise during the last interglacial period. *Nature Geoscience* **1**, 38–42.

161. Siegenthaler, U., Stocker, T.F., Monnin, E., Lüthi, D., Schwander, J., Stauffer, B., *et al.* (2005) Stable carbon cycle-climate relationship during the late Pleistocene. *Science* **310**, 1313–1317.

162. Foster, G.L. and Rohling, E.J. (2013) Relationship between sea level and climate forcing by CO_2 on geological timescales. *Proceedings of the National Academy of Sciences* **110** (4), 1209–1214.

163. Hansen, J.E. and Sato, M. (2012) Paleoclimate implications for human-made climate change. In: *Climate Change: Inferences from Paleoclimate and Regional Aspects* (eds A. Berger, F. Mesinger and D. Šijački). Springer, Dordrecht, pp. 21–48.

164. Pfeffer, W.T., Harper, J.T. and O'Neel, S. (2008) Kinematic constraints on glacier contributions to 21st century sea level rise. *Science* **321**, 1340–1343.

165. Jacobs, S.S., Jenkins, A., Giulivi, C. and Dutrieux, P. (2011) Stronger ocean circulation and increased melting under Pine Island Glacier ice shelf. *Nature Geoscience* **4**, 519–523.

166. Mengel, M. and Levermann, A. (2014) Ice plug prevents irreversible discharge from East Antarctica. *Nature Climate Change* **4**, 451–455.

167. Gardner, A.S., Moholdt, G., Wouters, B., Wolken, G.J., Burgess, D.O., Sharp, M.J., *et al.* (2011) Sharply increased

mass loss from glaciers and ice caps in the Canadian Arctic Archipelago. *Nature* **473**, 357–360.

168. Kopp, R.E., Simons, F.J., Mitrovica, J.X., Maloof, A.C. and Oppenheimer, M. (2009) Probabilistic assessment of sea level during the last interglacial stage. *Nature* **462**, 863–868.

169. Stocker, T., Dahe, Q., Plattner, G.-K., Tignor, M., Allen, S. and Midgley, P. (2010) *IPCC Workshop on Sea Level Rise and Ice Sheet Instabilities.* Report of IPCC Workshop, Kuala Lumpur, Malaysia, 21–24 June 2010. IPCC, WG1 Technical Support Unit, University of Bern, Bern.

170. Church, J.A., Gregory, J.M., White, N.J., Platten, S.M. and Mitrovica, J.X. (2011) Understanding and projecting sea level change. *Oceanography* **24** (2), 130–143.

171. Rignot, E., Velicogna, I., van den Broeke, M.R., Monaghan, A. and Lenaerts, J.T.M. (2011) Acceleration of the contribution of the Greenland and Antarctic ice sheets to sea level rise. *Geophysical Research Letters* **38**, L05503.

172. Sapart, C.J., Monteil, G., Prokopiou, M., van de Wal, R.S.W., Kaplan, J.O., Sperlich, P., *et al.* (2012) Natural and anthropogenic variations in methane sources during the past two millennia. *Nature* **490**, 85–88.

16

Putting It All Together

16.1   A Fast-Evolving Subject

Following the explorers of Earth's climate as they progressively uncovered more and more of its complex workings has been a fascinating journey. Starting in the late 18th century, this field of enquiry expanded slowly at first, and then with gathering pace, as scientists began to find ways of measuring greenhouse gases not only in our present atmosphere, but also in fossil air from ice cores dating back 800 000 years. By various clever means, they also found out how to estimate the likely CO_2 content of past atmospheres for millions of years back beyond the reach of ice cores.

Establishing this atmospheric history and its relation to our climate demanded that we develop new ways of sampling the Earth, including the record concealed in sediments beneath the 72% of the planet covered by ocean floor and in the ice of polar regions. We benefited from the invention of new seabed sampling technologies in the form of cores and drills, of ice drilling technology and of new analytical techniques in the laboratory. Ratios of one chemical element or isotope to another added to the tools we could use as proxies for past climate conditions. Fossil remains were our initial tools, along with climate-sensitive sediments. Our fossil armoury expanded through the discovery of mathematical transfer functions that relate the assemblage of marine plankton to the characteristics of the water mass from which it came. Numerical models proved useful in stimulating our imaginations, in linking widely separated data points and in enabling us to test ideas about past climates. Thanks to advances in dating Earth materials, we can now analyse climate change at ever-smaller intervals of time.

In this unfolding of the flower of climate change, whole new subdivisions of science arose – palaeoclimatology, palaeoceanography and biogeochemistry – along with new journals to house the research papers emerging from the growing communities of scientists in these new fields. As a result, we now know a great deal more than we did even a decade ago about what makes our climate variable, and why.

Not only the scope, but also the rate of discovery continues to increase. Palaeoclimate studies began to rise exponentially in the 1940s. In this book, I refer to no more than 10 research papers or books per decade from 1750 to 1940. The numbers then rose fast: to 383 for the decade 2000–09, and to 291 by mid 2014 for the decade 2010–19, when it was only half over. These, the ones I refer to, are just a subset of the publications on past climates, but the pattern is likely to be representative.

Much of the rise in output reflects the growing application in the Second World War and afterwards of echo-sounding to map the deep-ocean floor, the invention of the piston corer in 1947, the development of palaeomagnetic analyses and magnetic mapping of the deep-ocean floor in the early 1950s, the expansion of interest in the workings of the planet from the International Geophysical Year of 1957–58 onwards, including the opening up of Antarctica to research, the evolution of plate tectonic theory in the early 1960s, the advent of deep-ocean drilling in 1968 and widespread drilling into ice sheets and glaciers from about 1980 on.

As the opportunities to study newly sampled materials grew, so did the numbers of young scientists keen to make a name for themselves by working at this intellectual frontier. Yet more people became involved as the number

Earth's Climate Evolution, First Edition. Colin P. Summerhayes.
© 2015 John Wiley & Sons, Ltd. Published 2015 by John Wiley & Sons, Ltd.

of analytical techniques required to unravel the strands of climate change complexity grew. Prior to 1970, each palaeoclimate paper to which I refer averaged 1.2 authors. In the 1970s, it doubled. By the beginning of the 21st century, it had doubled again. My calculations suggest that there were 30 times as many scientists publishing on past climate change in the early 2000s as in the 1950s. This accelerating rate reflects not only the inherent intellectual fascination of the topic, but also the growing need to understand just how variable Earth's climate is, as background for making predictions about its future behaviour. Research on past climate change now occupies well over a thousand scientists spread across the globe.

The pattern of the science has changed, too. Nowadays, we take a much more integrative, holistic approach to understanding climate change. Humboldt would be pleased. The study of past climates brings together researchers from a wide range of disciplines: astronomers, calculating Earth's orbit and axial tilt; astrophysicists, studying the behaviour of the Sun; palaeobotanists and micropalaeontologists, studying how organisms and their ecology change with climate and time; geochemists, with their battery of elements and isotopes; geophysicists, determining where land and ocean were in relation to one another through time; and oceanographers, meteorologists and plain old geologists, with their intimate knowledge of Earth's surface processes – volcanic activity, weathering, sedimentation and the endless recycling implicit in plate tectonics, mountain building and continental drift. It is not enough to consider just one or two variables in isolation, such as the relation between temperature and CO_2. Many factors can affect this interaction. All have to be considered as we try to work out how Spaceship Earth operates. Otherwise, one day, we may find it uninhabitable.

The scientists whose work I reviewed in this book did not take statements about climate change on trust. They produced, examined and interpreted the evidence from the geological record. Many of their findings are so recent that they have not yet made it into standard undergraduate textbooks or teaching courses. They tend to be available only through scientific journals that are not readily accessible by the general public. Do we know everything we need to know? No. As Wally Broecker reminds us, science is *'a constant struggle to understand more fully and more accurately how the world really works'*[1]. There is always more to do. But we do know much more now than we did in any previous decade about how our climate system works.

16.2   Natural Envelopes of Climate Change

What has our journey uncovered? We now know that because natural gases like CO_2, water vapour, CH_4 and N_2O both absorb and re-emit infrared radiation, the planetary atmosphere containing them benefits from a natural greenhouse effect that keeps the Earth some 32 °C warmer than it would be otherwise. If we add more of any one of these gases to the atmosphere, it will enhance that effect. CO_2 can be added by out-gassing from volcanoes and metamorphism, by the decomposition of organic matter, by the rising to the surface of deep-ocean water rich in CO_2 and by the warming of the ocean – as warm water holds less CO_2 than cold water. It can be extracted from the atmosphere by the chemical weathering of silicate and carbonate minerals in rocks, by photosynthesis, by dissolving in cold surface water, by the sinking of that water into the deep ocean, by the marine biological pump (in which dead and decomposing plankton release into deep water the CO_2 absorbed near the ocean's surface) and by eventual incorporation into carbonate sediment and sedimentary organic matter. The effects of CO_2 in the air are exacerbated by water vapour, which increases through evaporation as temperature increases, providing a positive feedback, and which decreases as temperature decreases. Warmth reduces the proportion of CO_2 the ocean can hold and increases the proportion of water vapour the air can carry. Disconnects between temperature and CO_2 can arise, for example when volcanoes make CO_2 abundant in the absence of mountains that might otherwise be subject to chemical weathering, or when sea ice prevents exchange of CO_2 between the ocean and atmosphere, or where the melting of large volumes of land ice soaks up heat from the atmosphere and keeps the local environment cold regardless of CO_2 in the air.

Since the beginning of the Industrial Revolution, the burning of fossil fuels in many industrial processes, in agriculture and in our means of transportation has produced CO_2 as a byproduct. Some of these processes also produced CH_4 and N_2O. The warming generated by these additions caused more water vapour to be evaporated from the oceans, adding to greenhouse warming via positive feedback. More positive feedback comes from the melting of sea ice in the warming Arctic, which decreases the Earth's albedo. As far as I am aware from my contacts within their community and from my reading of their literature, knowledgeable self-styled 'global warming sceptics' accept these basic facts. Disagreement between them and mainstream climate scientists centres on the extent of the likely warming, not on the fact of warming, although some contrarians still misguidedly dispute the notion that humans have anything to do with global warming.

We also learned that for the past 450 Ma, Earth's climate oscillated within certain boundaries, forming a natural envelope. Within that envelope, Earth experienced alternating warm and cold periods driven by internal or external forces. Ultimately, the output from our star, the Sun, drives our planetary climate. But, that accepted, the variations in our climate have several other sources. The primary modifier is the balance, on a time scale of millions of years, between the supply of CO_2 provided by volcanic activity stimulated by plate tectonic processes and the extraction of CO_2 by weathering and its sequestration in sedimentary carbonates and organic matter. Second, superimposed on that system is another natural envelope of climate change governed by the variations in insolation caused by the subtle but regular changes in the orbit of the Earth around the Sun and the tilt of the Earth's axis, operating on a time scale of tens of thousands of years. Third, superimposed on that system is another natural envelope of climate change, driven by fluctuations in solar energy caused by changes within the Sun. These are also regular, on a centennial to decadal time scale. Fourth, superimposed on those systems are smaller internal oscillations occasioned by internal transfers of energy by wind and ocean currents, causing mostly regional changes at the multiannual to multidecadal scale and ranging from El Niño events to events like the Pacific Decadal Oscillation. The ocean accentuates the effects of these various changes in two ways: it transfers heat slowly around the globe, and, with its vast store of dissolved CO_2, it plays an important feedback role in releasing small amounts of CO_2 as its waters warm and in extracting it from the air as its waters cool.

These natural operations are interrupted from time to time by 'catastrophic' events – mostly short-term major volcanic eruptions like that of Mount Pinatubo in 1991, which cooled the climate for 1–2 years. At a larger scale, over the past 450 Ma, they included rare asteroid impacts and more common massive outpourings of flood basalts over a small number of millions of years, which caused massive extinctions of life at intervals of several tens of millions of years. Catastrophic events have been both irregular in time and unpredictable, unlike the present mass extinction documented by Elizabeth Kolbert in her 2014 book *The Sixth Extinction: An Unnatural History*[2].

Plate tectonics played an important role. The changing positions of the continents through time modulated the record of change within the natural envelope, most notably by changing the positions of the seaways that provided routes for ocean currents and local sources of moisture. Sea level rose and fell as mid-ocean ridges grew and declined at the sites of seafloor spreading. These ridges governed the output of CO_2 through time, while its extraction from the air was controlled by the related formation of mountains and plateaus where plates collided. The rises and falls in atmospheric CO_2 were manifest through ocean chemistry, because a rise in atmospheric CO_2 increases the concentration of CO_2 in the ocean. This makes the oceans more acidic, causes deep-ocean carbonate sediments to dissolve and hence raises the level of the carbonate compensation depth (CCD).

16.3   Evolving Knowledge

Many theories explaining the workings of the climate system through the ages have come and gone. Agassiz's ice sheet theory displaced Lyell's iceberg theory. Yet we now know that armadas of icebergs did play a key role in millennial change, especially in the North Atlantic during the Ice Age.

The savants of the 18th century and the geologists of the early 19th century were struck by evidence suggesting that the Earth's climate had cooled with time. They didn't know what had caused this 'Great Cooling', but for them it characterised what they called Tertiary time and we now call the Cenozoic Era – the past 65 Ma. Charles Lyell thought that the cooling might have come about through the movement of the continents into polar regions, but we now know that was not enough: Antarctica reached the South Pole 90 Ma ago but remained free of an ice sheet for 56 Ma!

Wegener gave us moving continents, but most geologists dragged their feet on his idea until palaeomagnetism came of age in the early 1950s. Mapping continental positions following plate tectonic theory helped to put palaeoclimatology on a firmer footing, but did not entirely help to explain why the Cretaceous was hot and the late Cenozoic was cold. In the late 1800s, Arrhenius offered an elegant explanation for the warming and cooling of the Ice Age, based on Tyndall's discoveries about the absorptive properties of CO_2. Chamberlin added a new dimension to Arrhenius's theory by observing that the amount of CO_2 in the atmosphere would fluctuate depending on the temperature of the sea's surface, since cold water held much more dissolved gas than warm water. Arrhenius had considered only the air. As ever, the science progressed.

Chamberlin extended that analysis to cover much of geological time. We had a theory: adding or subtracting CO_2 could change the climate. But did it do so in fact? In the absence of fossil CO_2, geologists once again dragged their feet.

The necessary tests would have to wait until advances in spectroscopy gave us the complete spectra of the main greenhouse gases, CO_2 and water vapour – something Arrhenius lacked. We also needed to know how both gases were distributed in the air, and how CO_2 was distributed in the ocean and exchanged with the atmosphere. Much of this knowledge had appeared by the 1950s, enabling Plass to resurrect Chamberlin's theory. But in the absence of hard data on past levels of CO_2, geologists stayed in their conservative bunker.

Things began to change at the end of the 1970s. The Soviet geologist Alexander Ronov and climatologist Mikhail Budyko showed that periods of warm climate occurred when volcanic activity was common and CO_2 was likely to have been abundant in the air. At almost the same time, Hans Oeschger's team found CO_2 in bubbles of fossil air trapped in ice cores and Minze Stuiver used carbon isotopes from tree rings to estimate likely past levels of atmospheric CO_2 for the Holocene.

By the late 1980s, we knew that the numbers of pores (stomata) on leaves declined as the percentage of CO_2 increased in the atmosphere. David Beerling, Dana Royer and others confirmed that there is a rather good correlation between CO_2 and temperature over the past 450 Ma. CO_2 was abundant in the warm early Palaeozoic, declined to near present-day levels in the late Carboniferous (when there was a major glaciation), rose again during the Mesozoic, then declined to the levels typical of the Pleistocene, our latest Ice Age. Independent confirmation came from the distribution of carbon isotopes in marine fossils and from analyses of alkenones. Like temperature, CO_2 has fluctuated within a fairly narrow natural envelope.

By the early 1980s, geochemists like Bob Garrels, James Walker and Bob Berner took advantage of the increasing power of computers to model the behaviour of the carbon cycle. Refined over the years, their models confirm the interdependence of CO_2 and temperature back through time.

At about the same time, these various developments led Al Fischer to identify two great tectono-climatic cycles with a periodicity of about 300 Ma, driven by convection in the Earth's mantle, which led to cyclic changes in the abundance of CO_2 in the air between a 'greenhouse state', when CO_2 was abundant, as in the

Cretaceous and early Cenozoic, and an 'icehouse state', when CO_2 was depleted, as in the late Carboniferous and late Cenozoic. The great cooling of the Cenozoic took place when volcanism declined, sea level fell, chemical weathering and mountain uplift increased and CO_2 therefore declined. Fischer's tectono-climatic cycle concept gave the Arrhenius–Chamberlin–Plass model a meaningful geological context and a significant degree of respectability. Other geologists rapidly followed Fischer's lead – among them Thomas Worsley, who observed that the correlation in time between prolonged episodes of mountain building and icehouse climate '*is prima facie evidence for orogenically driven CO_2 drawdown and carbon burial*'[3].

Focusing on the cooling of the Cenozoic Era, Dennis Kent and Giovanni Muttoni devised an elegant explanation based on an enhanced supply of CO_2 as plate tectonics moved India north after the breakup of Gondwana to collide with Asia about 50 Ma ago, followed by a decline in CO_2 as the collision caused mountains to rise and to become chemically weathered.

By 34 Ma ago, CO_2 in the atmosphere had declined to the extent that an ice sheet formed on Antarctica. The growing ice sheet enhanced the Earth's albedo, cooling the climate further. Formation of the ice sheet was encouraged by the increasing thermal isolation of Antarctica, when Australia and Tierra del Fuego moved away to the north. Positive feedback also played a role, in that, as declining CO_2 caused temperatures to fall, the Southern Ocean cooled to the point where it could take up more CO_2 from the air. Cooling also meant that less water vapour evaporated from the ocean, reducing the amount of yet another greenhouse gas. The cooling of the Southern Ocean made the surface waters dense enough to sink, taking newly dissolved CO_2 with them into the ocean depths, drawing down yet more CO_2 in a slow runaway process that led in due course to the latest Ice Age.

One of the most convincing examples of the link between CO_2 and temperature comes from the Palaeocene–Eocene boundary 55 Ma ago, when a major injection of carbon into the air made temperatures rise suddenly by 4–6 °C. The amount of CO_2 emitted was sufficient to acidify the ocean, dissolve deep-ocean carbonates and kill off benthic organisms. The Earth system took 100 000 years to recover. This was a time of massive eruption of flood basalts in the northeast Atlantic volcanic province. A magmatic event on that scale, especially the intrusion of basalt sills into carbon-rich rocks, could have supplied most, if not all, of the CO_2. Earthquakes associated with eruptions could have destabilised the continental slope, releasing CH_4 previously trapped as methane hydrates. Methane is a powerful greenhouse gas in its own right, but is rapidly oxidised to CO_2 in the atmosphere.

Closer to the present, we also find warming associated with elevated levels of CO_2 in the air in the mid Pliocene, when CO_2 rose to about 450 ppm – a little more than today's level. Temperatures rose globally by 2–3 °C, and by up to 18 °C in the Arctic. The Southern Ocean was about 5 °C warmer than it is today, and Antarctica's huge Ross Ice Shelf disappeared. Sea level may have risen by 22 ± 10 m on average. As there was limited land ice in the Northern Hemisphere, much of West Antarctica must have melted away, along with parts of East Antarctica. Quite possibly, this warming arose when warm equatorial waters that had formerly flowed from the Atlantic into the Pacific were diverted north towards Greenland by the gradual closure of the central American seaway between 4.7 and 2.5 Ma ago. A generally warmer ocean absorbed less CO_2. More CO_2 in the air kept the planet warm.

The mid Pliocene warming did not last long. Two possible reasons for this spring to mind. For the Northern Hemisphere, Michael Sarnthein suggested that closing the Panama seaway made sea level rise in the Pacific. That pushed cold North Pacific water into the Arctic, where it helped to thermally isolate Greenland and encourage the development of the Greenland Ice Sheet. At the same time, the northward-moving warm water in the Atlantic made more rain fall in northernmost Europe. Runoff formed a freshwater lid on the Arctic Ocean, encouraging the growth of seasonal sea ice. An increase in ice on Greenland and over the Arctic Ocean increased albedo, cooling the region. Sea ice further prevented the exchange of CO_2 and water vapour with the atmosphere, cooling it more. Meanwhile, in the Southern Hemisphere, New Guinea impinged on the Indonesian islands, restricting the supply of warm water from the Pacific to the Atlantic, which helped to cool the ocean, thus enhancing absorption of CO_2 and further cooling the atmosphere.

These various case studies underline the importance of plate tectonic processes as modifiers of the effects of the primary driver for our climate, the Sun. Such modifications depended to a large extent on changes to the supply and demand for CO_2, and hence water vapour. More CO_2 produces more warming, which produces more water vapour, which produces further warming and so on. The climate and the carbon cycle are intimately linked. As Mike Leeder pointed out, we are dealing with a 'Cybertectonic

Earth', and the '*combination of tectonics and biogeochemistry is the great fulfilment of the Huttonian philosophical scheme*'[4].

Just as changes in the balance of CO_2 affect temperature, so can changes in temperature affect CO_2. Fluctuations in the amount of energy received from the Sun also cause climate change. They are of two kinds: cycles in the Earth's orbit and axis and cycles in actual solar output. Each of these operates within the fairly narrow limits forming the natural envelope of the climate system.

Lyell's friend, the astronomer John Herschel, was the first to suggest that periodic fluctuations in the Earth's orbit, along with the precession of the equinoxes, changed the amount of solar radiation (insolation) received at any one place on the Earth's surface, which might account for periodic changes in the climate. James Croll developed that idea to explain the periodic fluctuations of the Ice Age. Once again, geologists tended to drag their feet, not least because they found it difficult to subdivide Pleistocene geological sequences on land with sufficiently high resolution to test his ideas. Using improved calculations, Milutin Milankovitch cleared up the mystery by showing that the climate of the Northern Hemisphere varied in concert with changes in insolation at 55 and 65° N in summer. If insolation was high enough, summers were sufficiently warm to melt winter snow, preventing its accumulation as ice. André Berger refined Milankovitch's calculations in the 1970s, and by 1976 Hays, Imbrie and Shackleton had discovered that ocean sediments faithfully recorded the orbital beat. Milankovitch's prediction was correct: ice grew when insolation was weak, and melted when insolation was strong.

We live in the Holocene, one of the short warm interludes of a 2.6 Ma-long Ice Age that is predominately cold – some 4–5 °C colder than today's average. Following the Last Glacial Maximum about 20 Ka ago, a rise in insolation warmed the planet, melting the great Northern Hemisphere ice sheets. Their continued melting while insolation was high at the start of the Holocene absorbed solar energy, preventing the Northern Hemisphere from warming to the extent suggested by the amount of solar energy received. As a result, the warming signal did not emerge there until 6000–7000 years ago, when the ice sheets melted away. By then, the insolation was already in decline. So, after a relatively short mid Holocene warm period (a climatic optimum), temperatures fell towards a neoglacial period that culminated in the cold temperatures of the Little Ice Age. Celestial mechanics tells us that low insolation (hence, cool conditions) should continue unabated for at least another 1000 years.

With insolation providing the primary driving force for glacial–interglacial cycles, what was the role of CO_2? Arrhenius, Chamberlin and Plass all thought that the greenhouse gas properties of CO_2 meant that it must have played an important role. But it did not come to the fore as a possible controller of Ice Age climate until Oeschger's team found CO_2 in bubbles of fossil air in the upper parts of ice cores in 1978. By 1987, that conclusion was shared by French scientists analysing fossil air from the Vostok ice core in Antarctica. They thought that CO_2 linked the timing of glacial–interglacial change between hemispheres with opposite patterns of insolation. By 2006, the correlation between CO_2 and temperature found in the Vostok core over the past 400 Ka was extended to a full 800 Ka in the Dome C core. Carbon isotopes enabled estimates of atmospheric CO_2 to be made from marine sediment cores, which showed the same pattern. Palaeoclimate scientists agreed that insolation at high latitudes warmed the ocean, which then emitted CO_2, further enhancing temperature. The warming enhanced evaporation, supplying water vapour, causing further warming and so on. CO_2 must be one of the forcing factors for ice volume. As far as cooling was concerned, Wally Broecker thought that falling insolation cooled the polar regions, increased the strength of winds and so stimulated the biological pump that drew CO_2 into the ocean interior. John Martin took that notion one step further, proposing in 1990 that the stronger winds transported iron-rich dust, which enhanced oceanic productivity, drew CO_2 out of the atmosphere and hence accelerated cooling into glacial periods.

The correlation between CO_2 and temperatures is not one-to-one. While CO_2 and temperature rise more or less together as ice sheets decay into interglacials, temperatures fall more rapidly than atmospheric CO_2 when ice sheets begin to grow. There is an obvious explanation. As insolation decreases and temperatures fall, sea ice grows over the polar ocean, increasing the Earth's albedo and exacerbating cooling. Even though a colder ocean absorbs CO_2 from the atmosphere, the fall in sea level and the growth in sea ice reduce the area of sea available for that task. Meanwhile, on land, terrestrial vegetation declines as snow and ice grow, releasing CO_2 to the air through decomposition. So, while temperature could keep up with declining insolation, CO_2 could not. CO_2 is part of a hysteresis loop in which there is a dynamic lag between input and output. It moves with temperature on warming, but lags temperature on cooling.

As CO_2 emerges from the ocean when it warms, it was difficult to understand why ice cores showed an apparent time lag of about 1000 years between warming and the rise in CO_2. This mystery was solved in 2013, when Frédéric Parrenin used new isotopic techniques to establish that in Antarctica the abundance of CO_2 in the atmosphere increased at the same time as the temperature. The major source for this CO_2 was the Southern Ocean. Northern seas did not warm at the same time, because the energy supplied by rising insolation there was spent mostly on melting the Northern Hemisphere ice sheets, a process that kept the northern regions cold despite the rise in global CO_2 exhaled from the Southern Ocean.

It had long seemed odd that the ice sheets of the Northern and Southern Hemispheres waxed and waned more or less in concert despite opposing patterns of average insolation. However, it is not the average insolation that matters. The intensity of northern summer insolation correlates not only with the length of the summer period in the south, but also with the intensity of spring insolation there. We do not need to call on forcing by Northern Hemisphere insolation to explain the link in climate response between the two hemispheres.

The last glacial period was punctuated by small periodic warmings and coolings of millennial duration that were gentler in the south and larger and more abrupt in the north. The Antarctic warm events preceded the Greenland ones, reversing as the northern events (known as Dansgaard–Oeschger events) began. Eric Wolff argued that terminations of the main glacial periods were like Antarctic warm events in which the following warm event in Greenland failed to develop. This allowed the Antarctic warming to continue, to the point that deglaciation began in the north to match that taking place in the south. The north was preconditioned not to develop a Dansgaard–Oeschger warming event, through a combination of low insolation and maximal land ice volume and extent in the Northern Hemisphere. If Wolff is right, then glacial terminations are periods of runaway southern warming, and the Southern Hemisphere is in the driving seat.

The Northern and Southern Hemispheres are connected through both the atmosphere and the ocean. The rapid circulation of the air means that both polar regions get the same signals of change in CO_2 or CH_4 at the same time. The slow circulation of the ocean connects the poles via the Atlantic Meridional Overturning Circulation, which is the Atlantic branch of the global Thermohaline Conveyor. Rising summer insolation in the north eventually triggers the melting of the Laurentide Ice Sheet, which at its maximum extent becomes lower, due to depression of the Earth's crust, caused by isostatic adjustment, and wider, due to ice flow, making it more susceptible to orbitally induced warming. The Scandinavian Ice Sheet melts faster because it is smaller and thinner. The fast retreat of both ice sheets covers the northern North Atlantic with cold freshwater, on which sea ice forms. That inhibits the restart of the Atlantic Meridional Oceanic Circulation. Temperatures stay low, despite the rising insolation and the rising CO_2 supplied from the south, because most of the available solar energy that is not reflected goes to melt northern land ice. Where sea ice is at its maximum extent, it reaches a line from Brest to Newfoundland, causing winter temperatures to fall as low as those in central Siberia. Increasing insolation and the import of heat from the south eventually melt northern sea ice, allowing the Atlantic Meridional Overturning Circulation to restart. That transports CO_2-rich deep water to the south, where the old CO_2 emerges through the upwelling of Circum-Polar Deep Water, to be exchanged with the atmosphere and further reinforce global warming. Under this scenario, the Atlantic Meridional Overturning Circulation helps to maintain glacial (AMOC-off or reduced) states versus interglacial (AMOC-on) states, but is not in the driving seat for deglaciation. The seesaw between the two poles at the millennial scale formed the Dansgaard–Oeschger events and their Antarctic counterparts during glacial times.

Clearly, then, insolation is the primary driver for glacial–interglacial climate change, while CO_2 plays a critical role as the primary amplifier of its effect, being emitted from the Southern Ocean directly as temperatures increase. CO_2 then increases temperatures beyond what insolation would have achieved, thereby causing evaporation, the supply of water vapour, plus further warming and the supply of yet more CO_2 and H_2O. Methane plays an ancillary role, being given off by wetlands as monsoonal activity grows in interglacial times. CO_2, in this context, is no longer the handmaiden of temperature, following where temperature leads, as was once thought – it is at least an equal partner.

Given that insolation oscillates between well-defined limits, and that the climate system marches to the insolation beat during the Ice Age, temperature, too, oscillates between well-defined limits. So too do the greenhouse gases, like CO_2 and methane. The natural envelope of CO_2 during the Ice Age is 180–280 ppm. The rise of CO_2 to consistent levels of 400 ppm in May 2014 is so far outside this natural envelope that it should give us pause for thought. However, it is still within the natural envelope

of the past 390 Ma (200–1000 ppm). While it has been suggested that the slight rise in CO_2 that began about 8 Ka ago might have been caused by early human land clearances, the jury is still out on that question. In any case, it was not enough of a rise to significantly offset the cooling driven by the decline in insolation that led to the neoglacial conditions of the late Holocene.

Changes in sea level associated with melting ice reinforced warming as sea level rose and cooling as sea level fell. Rising sea level increases the area of ocean that can exchange CO_2 with the atmosphere, while falling sea level decreases it. A wider ocean area provides a larger surface for the evaporation of water vapour.

Volcanic eruptions played a minor role in the Ice Age climate story. They tended to increase as the land rose when ice was removed, and should have decreased as ice was added. Similarly, chemical weathering on land played a minor part in the Ice Age, with its influence being more regional than global. For example, cool periods in western Europe over the Holocene were associated with increased rainfall, which implies more chemical weathering there. While weathering may have played a role in CO_2 drawdown within the oscillations of the Ice Age, its contribution would have been dwarfed by ocean–atmosphere interactions, because of the large size of the ocean CO_2 reservoir.

Superimposed on the multi-thousand-year variations in insolation caused by regular changes in the Earth's orbit and axis are much shorter centennial to millennial variations in solar output. These changes are shorter in duration and smaller than those caused by orbital variations. Orbital change induced 4–5 °C of temperature change between glacial and interglacial times, while changes in solar output induced changes of about 0.5–1.0 °C, mostly at the regional scale.

Like orbital variability, solar variability operates within a narrow natural envelope. A key modifier of the 11-year sunspot cycle is the ~208-year Suess–De Vries cycle, whose variation seems to have driven the development of the grand solar minima of the past 2000 years. It is likely that most of the centennial cool periods of the Holocene reflect weak solar output leading to more rainfall and higher lake levels in western Europe, decreased monsoonal activity in the tropics and increased drift ice in the north Atlantic. Drift ice cycles about 1500 years long probably arose from combinations of different solar cycles, mediated by patterns of ocean circulation.

A solar maximum peaking at about 1100 AD lay behind the development of the Medieval Warm Period (1000–1200 AD), which was followed by the Little Ice

Age. The Little Ice Age was not ubiquitously cold, but experienced warm periods nearly as warm as the Medieval Warm Period in the 1630s, 1730s, 1770s and 1840s, associated with solar maxima. The coldest periods of the Little Ice Age occurred at times of grand solar minima, and it was coldest in winter. Its summers were quite warm.

The effects of solar variability on the North Atlantic are stronger than they are elsewhere, due to interactions between the Atlantic and the ice-covered Arctic. As a result, the signal of the Medieval Warm Period is stronger there than elsewhere. Indeed, it is very difficult to identify either the Medieval Warm Period or the Little Ice Age in and around the Southern Ocean. The oceanographic signals of those events in the north are transmitted south over periods of some hundreds of years through the ocean's subsurface via the Atlantic Meridional Overturning Circulation. That makes it difficult to find ubiquitous and distinct global records of any short-term event with a small solar signal like the Medieval Warm Period, which lasted no more than about 200 years. One Holocene signal that is completely global is the short, sharp cooling at 8.2 Ka ago, which was probably caused by the draining of a vast glacial lake. Its global nature suggests that its effects were transmitted rapidly from the North Atlantic to the Southern Ocean via the air.

Solar variability, like orbital insolation, fluctuates within a natural envelope. Its thermal effects range by small amounts on either side of the global temperatures imposed by the slow variations in orbital insolation. Accepting this premise, and knowing that orbital insolation has been low for the past 2000 years and will remain so for the next 1000, it seems likely that solar output will not push global temperatures above the levels reached over the past 1000 years.

Roughly the same intensity of solar activity characterised the Medieval Warm Period and the mid 20th century. Records of past temperature connecting through to those of today show that, while both global temperature and sunspot activity rose from 1900 to about 1940, the two then diverged. Sunspot activity rose to a maximum in 1960–90, then fell to intermediate levels, while temperatures fell to a low in 1950 and stayed low until about 1965, before rising to the high levels typical of the 2000s. The two signals became disconnected in about 1940. Soviet data suggest that the connection was broken by the rise in aerosols from increasing industrial activity. I take the peak of solar activity between 1960 and 1990 to represent the most recent maximum in the Suess–De Vries solar cycle.

The next minimum in that cycle should occur around 2074; the next maximum is likely due in roughly 2174.

In the 11 years before 2010, CO_2 rose by 20 ppm. The last time it rose that much was over a period of 1000 years during the last deglaciation. It is now rising 100 times as fast as it did then. The dramatic rise in CO_2 from the 1800s onwards made it an increasingly important agent of climate change in the late 20th century, as solar output flattened and legislation cut the output of aerosols that would otherwise have reflected solar radiation back into space. The United Kingdom's Clean Air Act of 1956 was promulgated in response to the London smog of 1952 (more on that later) and was updated in 1968 and 1993. The US Clean Air Act was promulgated in 1970 and amended in 1977 and 1990. Other countries have enacted their own clean air acts, including, for example, New Zealand in 1972. Collectively, by the late 1960s to early 1970s, these acts had significantly cut the global aerosol load that cooled the atmosphere and hid the warming that should have appeared due to rising CO_2. The removal of shading aerosols explains part of the global rise in temperature since about 1970. An increase in aerosols from the growing burning of coal in the industrial revolutions of China and India may help to hide the warming that should have continued during the early part of the present century due to growing outputs of greenhouse gases.

The large additional climate forcing provided by increasing CO_2 and water vapour recently shrank the Baffin Island ice cap that grew 1600 years ago, exposing vegetation that last saw the light of day before the Medieval Warm Period. Other data on plant roots now being uncovered by melting ice in the Canadian Arctic suggest that conditions are now the warmest for the Holocene. Climate signals from across the Arctic confirm that conditions are the warmest they have been for the past 2000 years. The one exception is Greenland, probably because its climate is partly insulated from the warming of the Norwegian-Greenland Sea by the cold East Greenland Current.

Confirmation that the Arctic is changing radically also comes from its sea ice. The duration and magnitude of its decline is unprecedented since records began ~1500 years ago. In contrast, sea ice around Antarctica has remained either fairly stable or very slightly increased in recent decades. The slight increase in sea ice is mostly controlled by the strengthening and southward displacement of the westerly wind belt as a result of global warming. The growth of the ozone hole since the late 1970s has increased the strength of the Southern Ocean westerly winds, exacerbating this effect[5]. In addition, as Jinlun

Zhang of the Polar Science Center of the Applied Physics Laboratory at the University of Washington pointed out in 2007, warming the surface of the Southern Ocean temporarily increases its potential to freeze[6]. This paradox arises because warming the ocean increases its stratification, which prevents warm deep-ocean water from rising to melt sea ice at the surface, helping to increase sea ice production. Stronger winds now push floes together in areas where winds converge, causing sea ice to thicken in the Weddell, Bellingshausen, Amundsen and Ross Seas, thus increasing sea ice volume[7]. The net result is the opposite of what we see in the Arctic. The changes in wind patterns have forced Antarctic sea ice to grow in the Indian Ocean sector and shrink in the Pacific Ocean sector. This regional disparity tells us that we cannot make the simplistic assumption that the very slight overall growth of Antarctic ice tells us that global warming is not working! Like many things in nature, the sea ice story is complicated.

Consistent with the current warming, satellite data show that both Greenland and Antarctica are losing land ice at increasing rates. The loss in Greenland is mostly a result of warmer air temperatures and surface melt. That in Antarctica is mostly a result of the erosion of ice shelves from beneath by warm water welling up on to the continental shelf, especially in the Amundsen Sea.

Similar signs that the warming at the end of the 20th century exceeded that of the previous 2000 years come from studies of tree rings in Europe, from global collections of marine sediment cores, from a variety of other Earth materials collected globally and from the shrinkage of the Quelccaya ice cap in the Peruvian Andes, which is now reversing changes that began some 5000 years ago.

These changes also appear in numerical palaeoclimate models fed with data on insolation, solar output, volcanic activity and greenhouse gases. Their outputs show that the decline in Holocene insolation, plus the cooling effects of occasional large volcanic eruptions like that of Mount Pinatubo in 1991, plus the ups and downs of the solar cycle, should have cooled the climate slightly since 1900. Instead, it has warmed. The only plausible explanation is the addition of greenhouse gases, particularly CO_2 and methane from human activities, plus water vapour evaporated from the warming ocean, plus the effect of decreasing albedo attributable to the loss of Arctic sea ice and the greening of the Arctic. We do not need the models to tell us this. They merely confirm what is obvious from the palaeoclimate and modern data.

16.4   Where is Climate Headed?

My analysis implies that the Little Ice Age has not yet come to an end. It is a manifestation of the Holocene neoglacial, and should extend for at least another 1000 years. What might we expect if the present decline in solar activity leads to another period like the Maunder Minimum, with no sunspots, somewhere around 2074? Based on observations by Moberg, the temperature of the Northern Hemisphere during such a minimum might fall by up to ~0.7 °C at most. Global temperature would fall much less, perhaps by 0.2–0.3 °C. What effect would that have on the forecast global temperature rise? According to the latest calculations by the Intergovernmental Panel on Climate Change (IPCC), the effect of rising CO_2 might be to increase the global temperature in 2070 by between 1 and 3 °C[8]. Subtracting 0.3 °C caused by a possible Maunder Minimum would reduce that rise to between 0.7 and 2.7 °C, values that are still substantial. The actual fall might be somewhat less than 0.3 °C, given that the other grand solar minima of the past 2000 years were not as deep as the Maunder Minimum. Even such a large minimum would not stop global warming. Gerald Meehl of the US National Centre for Atmospheric Research and colleagues reached much the same conclusion in May 2013[9].

Beside the effects of variations in solar output, we would expect the future trajectory of global warming to be modified slightly upward or downward by internal oscillations like that of the El Niño (warm)–La Niña (cool) couplet or the Pacific Decadal Oscillation (positive phase = warm; negative phase = cool), as well as by cooling attributable to the occasional volcano capable of putting large volumes of reflective dust and sulphuric acid into the stratosphere. In addition, we are likely to see some slight cooling from any increase in the burning of coal and the associated output of aerosols resulting from continued industrial expansion and car use in the developing world and the BRIC countries (Brazil, Russia, India and China).

The Pacific Decadal Oscillation will likely change to its warm phase within the next decade. Its current negative phase has distorted the global warming average, making it seem that global warming may be coming to a halt [a]. In reality, warming away from the Pacific has continued to increase, which is why glaciers and ice sheets continue to

---

[a] Note added in press: In fact, it now seems likely that the current slowdown in global warming has more to do with increased sequestration of heat in the deep Atlantic and Southern Oceans. Slowing of that cyclic process will eventually lead to a recovery (acceleration) in surface warming (see footnote b in Chapter 15).

melt, the ocean's heat content has gone on rising, as has sea level, and Arctic sea ice has continued to shrink and thin. **These are the undistorted signals of global warming and should take precedence over just the average global temperature signal in our analysis of how the climate is changing.**

By how much might sea level rise as the ocean warms and land ice melts? The IPCC suggests 25–95 cm by 2100[8]. The geological evidence shows that with the rise in temperature of 2–3 °C above today's levels in recent interglacial periods, sea level was between 4 and 9 m higher than it is today. And when warming persisted for much longer than the length of an interglacial period, sea levels rose by up to ~15 m during the Palaeocene–Eocene Thermal Maximum, and possibly by ~22 m in the mid Pliocene warm period, which saw comparable rises in temperatures. With a rise in CO_2 to levels of 400–450 ppm and a warming of 2 °C, we could see a long-term rise in sea level of >9 m above the present in the 200–500-year timeframe[10]. Sea level rises of the order of 9 m would require the melting of significant volumes of land ice. Already there are signs that both the West Antarctic Ice Sheet in the vicinity of the Amundsen Sea and the East Antarctic Ice Sheet in the Wilkes Basin are at risk of substantial decline[11, 12].

As a side effect, the rise in CO_2 is also gradually acidifying the ocean, which will lead to the $CaCO_3$ content of the surface ocean falling below the level needed to sustain the building of aragonite skeletons by marine plankton. This will affect the Southern Ocean first, to the detriment of organisms like pteropods at the base of the food chain[13]. The deleterious effects of ocean acidification are evident from the Palaeocene–Eocene Thermal Maximum 55 Ma ago, an event from which it took the world 100 000 years or more to recover.

The prospect of rapid climate change is of significant concern to scientists and policy makers alike. In 2013, James C. White of the University of Colorado at Boulder led a team of experts to evaluate the need to understand and monitor abrupt climate change and its impacts, on behalf of the US National Academy of Science[14]. And in 2011, on behalf of the Geological Society of London, I attended two meetings of experts convened by the UK government's chief scientific advisor, Sir John Beddington, to consider the current evidence for and views on potential thresholds or 'tipping points' in the climate system. Both the US and the UK groups agreed with the general consensus in the wider climate science community that there are likely to be tipping points in the climate system beyond which the rate of change may accelerate, but that there is a lot of uncertainty about what they are, when they might occur and whether or not they might be preceded by warning signals.

Is the climate record any guide as to what these tipping points might be? Possibly not, because the world now is not the same as it was when rapid changes happened in the past. For example, the rapid changes typical of Dansgaard–Oeschger events took place during the intermediate stages of glaciations, when there were major ice sheets on North America and Scandinavia. Will further warming switch off the Thermohaline Conveyor, which brings warm air to western Europe via the Gulf Stream and its northern branches? According to Richard Alley[15], most numerical models suggest that this circulation may slow but will not stop. Some models, however, suggest that the melting of the Greenland ice could put a freshwater lid on the northern North Atlantic that would make the conveyor stop. Alley suggests that '*even though a large, rapid, high-impact event seems unlikely based on most of the literature, the nonzero possibility and the potentially large impacts motivate further research*'[15].

One of the worries about continued global warming is its effect on possible supplies of CH_4, which is a more potent greenhouse gas than CO_2. Methane is abundant on the deep-sea floor, where it is trapped in cages of ice known as clathrates (see Chapter 10). While these clathrates are currently stable, their stability depends on the relation between their pressure and their temperature. As the deep ocean warms, the stability of the methane clathrate field may be compromised[16]. Melting clathrates would destabilise the sediments in which they sit, causing submarine slides and slumps that would release large quantities of methane. The volumes of gas trapped in these sediments are enormous, amounting to several trillion tonnes. However, the likelihood of marine clathrates melting over the next couple of hundred years is remote[16].

More likely is the emission of methane from the melting of Arctic permafrost, which is thought to contain about 1400 Gt of carbon[17]. Arctic expeditions on land and at sea find methane bubbling up from melting permafrost[17]. When we examine ice cores, we find that methane values ranged from 350–400 ppb during glacials to 600–800 ppb during interglacials[18]. These limits define methane's natural envelope for the Ice Age. Even though past interglacials were 2–3 °C warmer than the present, their methane values did not rise above 800 ppb. We could infer that a comparable warming in the near future might not be accompanied by significantly increased emissions of methane[18]. Nevertheless, present CH_4 values (~1800 ppb) are already double the highest levels of past interglacials. Continued emissions of CH_4 from our activities, added to a likely increase in natural emissions from melting permafrost, must be of concern. The potential size of this concern is underlined by the fact that about 25% of the land in the Northern Hemisphere is underlain by permafrost of a few meters to more than 1 km in thickness. In some places in Siberia, natural gas is trapped beneath the permafrost and might be released if it melted.

Will the ice sheets on Greenland and West Antarctic melt rapidly? As we saw earlier, parts of the West Antarctic Ice Sheet are now in a state of irreversible decline. There is also a risk that, if sea level rises a few centimetres more, we might lose ice from the Wilkes Basin of East Antarctica. We know that the Ross Ice Shelf melted away during the warm Pliocene. Numerical models suggest that it may also have done so during warm interglacials, along with much of West Antarctica[19]. But the time scale for such melting is likely to be a few hundred years, not decades. Even so, there is no reason to be complacent, since the melting of all Greenland's ice would raise sea level by about 7 m, while the melting of West Antarctica could add a further 5–6 m. We appear to be looking at a maximum rise of 1–2 m by 2100.

Taking into consideration what we know of past climate, my view is that we are not in for rapid change in the next few decades. We face slow, incrementally accelerating change, which to most people will seem imperceptible – at least for a while. If nothing is done, we will be subject to what I call a 'creeping catastrophe'. By that, I mean that the effect by 2100 would be the equivalent of a catastrophe if it had happened all at once. Global warming is not like a tsunami.

Will we get over global warming quickly if we stop emissions now, or soon? No.

There is a popular misconception that our additions of CO_2 to the air fall out within just a few years. In fact, David Archer of the University of Chicago calculates that '*The carbon cycle of the biosphere will take a long time to completely neutralize and sequester anthropogenic CO_2*'[20]. If we stopped putting CO_2 into the atmosphere tomorrow, '*17–33% of the fossil fuel carbon will still reside in the atmosphere 1 kyr from now, decreasing to 10–15% at 10 kyr, and 7% at 100 kyr. The mean lifetime of fossil fuel CO_2 is about 30–35 kyr*'[20]. This is comparable to '*the 10 kyr lifetime of nuclear waste … [which] seems quite relevant to public perception of nuclear energy decisions today. A better approximation of the lifetime of fossil fuel CO_2 for public discussion might be "300 years, plus 25% that lasts forever"*'[20]. Along with Andrey Ganopolski of the Potsdam Institute for Climate Impact Research, Archer took that analysis one step further. Estimating that 25% of present anthropogenic emissions will remain in the atmosphere for thousands of years and that about 7% will remain beyond 100 Ka, they concluded that it may take 500 Ka before the CO_2 falls low enough for the next glacial period to start[21].

16.5 Some Final Remarks

As my colleagues and I pointed out in writing the 2013 addendum to the statement on climate change of the Geological Society of London, '*These various geologically based considerations lend strength to the argument that continued emissions of CO_2 will drive further rises in both temperature and sea level. Given that the Earth system takes a long time to reach equilibrium in the face of change, the present changes are likely to continue long beyond 2100*'[22]. Furthermore, to end with the concluding phrase of the original Geological Society statement, '*In the light of the evidence presented here it is reasonable to conclude that emitting further large amounts of CO_2 into the atmosphere over time is likely to be unwise, uncomfortable though that fact may be*'[22].

We do not regard that phrase as alarmist. It is a simple statement of fact. Nevertheless, there is a significant implication in our continued and indeed expanding use of fossil fuels and in the resulting expansion in emissions of CO_2 and their accumulation in the air and in the ocean. This book is not the place for a comprehensive review of potential impacts, nor of the means of adapting to or mitigating them. Nevertheless, some comments seem in order based on what the past tells us. If we continue to emit CO_2 in growing amounts, the world will warm over the next 50–100 years, even if some of that warming is offset by a decrease in solar output that will, by its nature, be temporary. Even if such a temporary cooling effect should arise, it will not stop CO_2 from further increasing ocean acidification, with deleterious effects on the base of the food chain.

Continued warming will melt more glaciers, ice caps and parts of the big ice sheets, raising sea level further. Low-lying coastal communities will be affected. Many coastal cities may find themselves having to seek advice from the Dutch to stop the sea invading public places. Governments are already considering what coastal engineers must do to protect against ongoing sea level rise in vulnerable areas. They can't all be protected, and some coastal land will be lost. In the developing world, where coastal engineering is unaffordable, migration seems inevitable.

The effects of warming on land areas will be diverse. As we saw earlier, cooling brings rain to middle latitudes. Warming does the opposite. Areas that are now dry may get dryer, while wet areas may get wetter. The reverse tends to be true in monsoonal areas. That geologically based conclusion is echoed by the reports of the IPCC and other organisations studying potential impacts. The drying of already dry areas like Australia, southern Africa, the southwestern United States and the Mediterranean has begun. It will have deleterious effects on water supplies and agriculture, leading to migration on a scale that the military sees as a potential threat[23]. Warming also has an upside. It will push north the potential for wheat growing in Canada and Siberia. More CO_2 in the air will promote faster growth for some plants, but the accompanying warming may prove deleterious to them once specific thresholds are passed. How plants respond depends not just on CO_2, but also on heat and water.

My conclusions are based on the work of a great many palaeoclimatologists from a wide range of countries, as should be evident from the multitude of references to the scientific literature in each chapter. They tell us that the argument by a few scientists that the present warming is a natural response of the climate system to the end of the Little Ice Age is wrong. Among those sceptics of the human influence on global warming is the Australian geologist Dr Robert Carter. Carter argues that we should accept what he calls his 'null hypothesis' of climate change, which is that '*global climate changes are presumed to be natural unless and until specific evidence is forthcoming for human causation*'[24]. We now know, from the palaeotemperature profiles emerging from the outstanding and comprehensive studies of the likes of Vinther, Wanner, Mayewski, Moberg, Esper, Gifford Miller, Marcott and others, that it is highly improbable that the rapid rise in temperature since 1900 to the levels seen during the last decades of the 20th century and those of the beginning of the 21st is the natural result of a supposed 'recovery' to supposed 'normal values' from the cold depths of the Little Ice Age. The rise has been greater and faster than any seen during the 11 700 years of the Holocene, apart from the 'return to normal' after the short-lived 8.2 Ka cold event. What should be considered 'normal' in the present case, as pointed out by Esper and several others, are the underlying cool temperatures of the past 2000 years of the Holocene neoglacial.

Those temperatures are the core that runs through the various data within a range (the 'natural envelope') extending from the peak of medieval warming to the cold of the Maunder Minimum. This core temperature is what I would call the 'norm', and it is a lot colder than what we are experiencing now. We have moved away from that average, and out of the influence of the natural envelope of orbital insolation for the late Holocene, out of the natural envelope of solar output for the past 2000 years and out of the natural envelope of variability in atmospheric CO_2 for the past 800 Ka, into a new domain, driven by our own emissions. Under these circumstances, the correct null hypothesis is that **the present global change should be presumed to be unnatural until specific evidence is forthcoming that this is not the case**.

If I had been writing a book just about the climate of the Holocene, I could have gone into much greater detail about how unstable the hydrological cycle was over the past 11 700 years and how that impacted upon early civilisations. Drought may well have been a contributory factor, for example, in the collapse of the Mayan civilisation in central America. That level of detail was not my aim. It is well covered elsewhere[25]. Here, it is sufficient for me to remind you of Magny's research linking excessive rainfall in western Europe to the lows in solar activity that caused widespread cooling. Long periods of floods or droughts will have been inimical to the development of populations anywhere on the globe. They have happened in the past and will undoubtedly happen again, and they will be just as disruptive or even more so, given that they are likely to affect much larger populations than those of the Maya. Many of those affected will not be city dwellers, but small-scale agriculturalists. There will be little or no opportunity for them to move to greener pastures. As Gwynne Dyer pointed out, conflicts will become more likely[23].

Before the Pleistocene Ice Age, there were larger fluctuations in CO_2 than we have experienced in the past 2.6 Ma. These too were constrained within a natural envelope since the rise of the land plants some 450 Ma ago. During most of that time, levels of CO_2 in the air did not rise much above about 2000 ppm. At the low end, given the continuity of photosynthesising plants, they most likely did not fall below about 180 ppm, the lowest values of the Ice Age. This broad natural envelope reflects the interaction between changing rates of supply of CO_2 – from volcanism and related processes in the Earth's interior – and its rates of extraction by chemical weathering, photosynthesis and the sedimentation of marine carbonates. Lovelock's notion that Gaian processes maintain Earth's temperature at a suitable level (within a natural envelope) may well apply also in the broad sense to the abundances of the atmospheric gases, although he argued that life was providing the main control, whereas my research suggests we may be dealing more with a largely inanimate geochemical thermostat.

The concept of natural envelopes for climate that I promote here fits well with the notions of Hutton and Lyell that Earth cycles fall within certain well-defined limits. But there is also plenty of room for Cuverian catastrophes, notably in the form of the lengthy and massive eruptions of flood basalts in Large Igneous Provinces from time to time, and the occasional asteroid impact. While these pushed Earth's climate out of its natural climate envelope and caused mass extinctions of life, things returned to 'normal' afterwards.

16.6   What Can Be Done?

Will we be able to do anything about the changes we are now making to the climate system? We did successfully take a stand against the emission of ozone-destroying substances, but they were mainly produced by a very few chemical companies in a very few countries, and it turned out after much discussion that the companies had alternatives on the production line. Problem solved. An easy win.

Global warming is not like that. We all contribute to it, by heating or cooling our homes, driving cars, travelling on planes and trains and consuming the products of manufacture and agriculture from a globalised industry. The energy consumed comes not from a few chemical companies in a few countries but from globally distributed gas, oil and coal companies located in many countries and managed either privately or nationally. These activities are supported by a massive infrastructure stretching from coalmines and oil and gas wells through pipelines, tankers and bulk carriers to power stations and petrol station forecourts. This complexity and ubiquity is what makes the solution so much more difficult than closing the ozone hole.

The general tendency of governments faced with environmental problems is to wait until a catastrophe happens, at which point people clamour for change and governments finally act. Take the River Thames, for example. For centuries, the river was used as London's sewer[26]. This became more and more of a problem as the city's population grew, until, in the words of Prime Minister

Benjamin Disraeli (1804–1881), the Thames had become '*A Stygian pool reeking with ineffable and unbearable horror*'[26]. The year 1858 was labelled the 'Great Stink'. Members of Parliament contemplated moving the House of Commons upstream to Hampton Court, and London's law courts prepared to move out to Oxford. Something had to be done. Parliament already had a solution to hand, having set up a Metropolitan Board of Works in 1855 to consider how to clean up the river. Civil engineer Joseph Bazalgette (1819–1891) devised a scheme to take sewage through tunnels to two main outfalls well downstream. His network of 1100 miles of local sewers and 165 miles of main sewers was described by the *Observer* newspaper of the time as '*the most extensive and wonderful work of modern times*'[26]. It solved the problem, and is still in use today. Peter Ackroyd describes the enormous task of disposing of London's human waste as '*the city's secret industry*'[26]. That industry is now at work everywhere in all cities, ensuring public health. We do not think twice about it. Out of sight, out of mind! But it took a virtual catastrophe to start the ball rolling. And no doubt there was a large initial investment.

The story of smog is much the same. The burning of coal in London polluted the city with smoke and was a cause of complaint to Elizabeth I, among others. In 1661, the diarist and writer John Evelyn (1620–1706) wrote a treatise on the problem, entitled *Fumifugium, or The Inconvenience of the Aer and Smoak of London*, lamenting the condition of the city covered by '*a Hellish and dismal Cloud of SEA-COAL*'[26]. Ackroyd reminds us that '*Victorian fog is the world's most famous meteorological phenomenon. It was everywhere, in Gothic drama and in private correspondence, in scientific correspondence and in "Bleak House" (1852–53)*'[26] *... Gas lights were turned on throughout the day in order to afford some interior light ... The street lamps seemed points of flame in the swirling miasma*'[26]. The worst of the fogs were the London 'smogs' of the 1950s, which I remember well. You could hardly see your hand in front of your face at midday on occasion. Thousands died of respiratory problems. Although public disquiet led to the United Kingdom passing a Clean Air Act in 1956, a severe smog in 1962 '*killed sixty people in three days; there was "nil visibility" on the roads, shipping "at a standstill", trains cancelled*'[26]. Something had to be done – and it was, by means of a more extensive Clean Air Act in 1968. Further improvements followed. The costs were accepted. Smog was stopped.

These simple examples from one city in one country represent many similar public health initiatives around the world. They illustrate the truth of the words of Jorgen Randers in his global forecast for the next 40 years to 2052: '*Experience shows that it is hard for democratic, free-market economies to make proactive decisions to increase voluntary investments before they are unavoidable. It is much simpler [to do so] after crisis has struck and there is an externally imposed threat of destroyed infrastructure and livelihoods ... Solutions will come on line much later than optimal – at least in those parts of the world where the majority favors the market. Collective solutions will not be used until it is overabundantly clear that private solutions (based on individual initiatives in an unrestrained market) will not suffice*'[27]. Perceptive words.

We are now living through the global equivalent of London's 'Great Stink' or the vile smogs that preceded the Clean Air Acts of the 1950s and 1960s. The difference is that the scale is now global, not local, and that the CO_2 that we are adding to the atmosphere is colourless and odourless. Not being able to see it or smell it, it's hard to believe it's really there or that it can be in any way a danger to us. CO_2 is an integral component of the natural environment, so can it be classified as a pollutant? Yes, in the same way that many toxic chemicals, like arsenic or mercury, are naturally present in the environment in small amounts but become pollutants when we dump concentrations of them into small areas. In the case of CO_2, that area is the whole atmosphere and the upper layers of the ocean. I use the word 'pollutant' here in the sense of a substance that has undesirable side effects when added to the environment. Note the use of the word 'added'. It is only our *additions* of CO_2, not the natural background of CO_2, that pollute.

While a target of no more than 450 ppm of combined greenhouse gas emissions in the atmosphere has been deemed likely to hold global warming to an average of 2 °C above preindustrial levels, this could mean an 80% cut in CO_2 emissions by 2050. As Randers pointed out[27], given the CO_2 levels we have today (he meant 2012), it seems highly unlikely that we shall meet that target. CO_2 has since gone on climbing, reaching a persistent peak of over 400 ppm in May 2014; 450 ppm is just around the corner. Nevertheless, improvements are being made. Energy efficiency is likely to increase by a further 30% by 2050, but by then there will be 2 billion more people on the planet, wanting to use progressively more energy[27]. Randers forecasts that energy use will grow by 50% in that time period. If there were no change in energy sources,

that would imply an increase in CO_2 emissions by 50%. But Randers expects a levelling off in the use of oil, which should peak by 2025 and then decline, and a similar peak in the use of coal and gas by 2040, followed by a decline.

Energy demand will increasingly be met by renewable energy sources, because they are technically feasible and their costs continue to fall, making them increasingly competitive. As an example of the rapid rate of falling costs, Randers quotes solar panels, which declined from $108/W peak capacity in 1975 to $1.3/W by 2010. In due course, solar panels may give way to solar roof tiles, which could be fitted to every house[28]. If Randers is right, the world's consumption of fossil fuels could be in steep decline by 2052[27]. Others, like Jeremy Leggett, broadly agree[28]. In the short term, Randers sees a major swing to gas, which has already begun in the United States and western Europe. Replacing coal with natural gas in power stations will reduce the amount of CO_2 emitted by two-thirds, and is a big step towards a low-carbon future.

While this is a good thing in the short-term, it postpones the inevitable shift away from burning fossil fuels to eventual use of direct or indirect solar power. The road to moving there directly will remain blocked for some time by a combination of costs, vested interests, the emphasis on short-term profits and the weakness of governments in the face of powerful lobbies[27, 28]. Long-term thinking is rare in our predominantly market-driven modern economies. Long-term action is even rarer. By mid century, however, the problems caused by global warming will loom much larger, alternative (renewable energy) technologies will be much cheaper and carbon capture and storage may have begun, which might lead governments to take more serious action than now seems possible.

As the US economist William Nordhaus, of Yale University, points out, the economic case for action seems beyond dispute[29]. Nordhaus accepts that we do not know everything about global warming with 100% certainty, but good scientists are never 100% sure about any empirical phenomenon. '*The advice of climate science contrarians is to ignore the dangers in the Climate Casino. To heed that advice is a perilous gamble ... Those who burn fossil fuels are enjoying an economic subsidy – in effect, they are grazing on the global commons and not paying for what they eat. Raising the carbon price ... would correct for the implicit subsidy on the use of carbon fuels*'[29]. Companies who think otherwise, he opines, '*are really looking out for their profits and not for the public welfare*'[29]. Much of the confusion about global warming in the public's mind, he reminds us, comes from disinformation supplied by lobbyists keen to 'sow doubt', a tactic well-documented by Naomi Oreskes and Erik Conway[30], and used in the past by those keen to prevent the public from appreciating the strong statistical link between cigarette smoking and cancer.

I hope that *Earth's Climate Evolution* contributes constructively to the discussion of these major issues by bringing to the fore what we have learned from the 200-year-old science of palaeoclimatology, the study of the history of climate change. It has a great deal to tell us about what our future may look like as we continue to spew CO_2 into the atmosphere. One can always argue with particular pieces of evidence and with the outputs of computer models of the climate system, but as my friend Bryan Lovell is fond of pointing out, 'you can't argue with a rock'. As long as he is prepared for ice to be included in the definition of rock, I wholeheartedly agree. Rocks carry an integral climate message from the past. They have a stirring tale to tell that should make us pause in our onward rush. They happen, not by accident, to carry the same message that we hear, through other routes and based on other sciences, from the IPCC. Should we be alarmed? The house is not yet on fire. We do not have to jump out of the window. But it seems to me that we are in rapidly growing need of insurance against a sea of future troubles.

Aside from examining the geological evidence to see if current theories of climate change are making testable predictions, we also have to challenge the global warming sceptics to present coherent and comprehensive hypotheses of their own, the predictions of which can be tested against the observable record. For example, if anyone disagrees with the conclusion that a rise in CO_2 caused the rise in temperature at the Palaeocene–Eocene boundary, or that a fall in CO_2 led to the cooling of the Cenozoic, they must come up with plausible and testable alternatives. As Nate Silver reminds us, we cannot just '*rummage through fact and theory alike for argumentative and ideological convenience*'[31]. Rather, we must '*weigh the strength of the new evidence against the overall strength of the theory*'[31]. Cherry-picking the data to support a particular argument is not on. And belief is out of the question.

To end, I return to the two questions I posed in the Introduction: **Can what we see of climate in the geological record tell us anything about what might happen if we go on emitting more and more carbon dioxide and other greenhouse gases into the atmosphere?** and **What are the chances that our increasing use of fossil fuels will drive Earth's climate out of the icehouse, where it has been stuck for several million years, and back into the greenhouse – the dominant climate mode for much of the past 450 million years?** The evidence I provide from the geological record shows that the answer to the first of these is **Yes** and a logical answer to the second is **High**.

References

1. Broecker, W. (2010) *The Great Ocean Conveyor: Discovering the Trigger for Abrupt Climate Change*. Princeton University Press, Princeton, NJ.
2. Kolbert, E. (2014) *The Sixth Extinction: An Unnatural History*. Bloomsbury, London.
3. Kidder, D.L. and Worsley, T.R. (2012) A human-induced hothouse climate. *GSA Today* **22** (2), 4–11.
4. Leeder, M. (2007) Cybertectonic Earth and Gaia's weak hand: sedimentary geology, sedimentary cycling and the Earth System. *Journal of the Geological Society of London* **164**, 277–296.
5. Maksym, T., Stammerjohn, S.E., Ackley, S. and Massom, R. (2012) Antarctic sea ice – a polar opposite. *Oceanography* **25** (3), 140–151.
6. Zhang, J. (2007) Increasing Antarctic sea ice under warming atmospheric and oceanic conditions. *Journal of Climate* **20**, 2525–2529.
7. Zhang, J. (2014) Modeling the impact of wind intensification on Antarctic sea ice volume. *Journal of Climate* **27**, 202–214.
8. IPCC (2013) Technical summary. In: *Climate Change 2013: The Physical Science Basis*. Contribution of Working Group I to the IPCC 5th Assessment Report. Intergovernmental Panel on Climate Change, Paris.
9. Meehl, G.A., Arblaster, J.M. and March, D.R. (2013) Could a future 'Grand Solar Minimum' like the Maunder Minimum stop global warming? *Geophysical Research Letters* **9**, 1789–1793.
10. Forster, G.L. and Rohling, E.J. (2013) Relationship between sea level and climate forcing by CO_2 on geological timescales. *Proceedings of the National Academies of Science* **110** (4), 1209–1214.
11. Pritchard, H.D., Ligtenberg, S.R.M., Fricker, H.A., Vaughan D.G., Van den Broeke, M. R. and Padman, L. (2012) Antarctic ice-sheet loss driven by basal melting of ice shelves. *Nature* **484**, 502–505.
12. Mengel, M. and Levermann, A. (2014) Ice plug prevents irreversible discharge from East Antarctica. *Nature Climate Change* **4**, 451–455.
13. Orr, J.C., Fabry, V.J., Aumont, O., Bopp, L., Doney, S.C., Feely, R.A., *et al.* (2005) Anthropogenic ocean acidification over the twenty-first century and its impact on calcifying organisms. *Nature* **437**, 681–686.
14. White, J.C., Alley, R.B., Archer, D.E., Barnosky, A.D., Foley, J., Fu, R., *et al.* (2013) *Abrupt Impacts of Climate*

change – Anticipating Surprises. National Academies Press, Washington, DC.

15. Alley, R.B. (2007) Wally was right: predictive ability of the North Atlantic 'Conveyor Belt' hypothesis for abrupt climate change. *Annual Reviews in Earth and Planetary Science* **35**, 241–272.

16. Hay, W.W. (2013) *Experimenting on a Small Planet: A Scholarly Entertainment*. Springer, New York.

17. Shakhova, N., Semiletov, I., Salyuk, A. and Kosmach, D. (2008) Anomalies of methane in the atmosphere over the East Siberian Shelf: is there any sign of methane leakage from shallow shelf hydrates? *Geophysical Research Abstracts* **10**, EGU2008-A-01526.

18. Wolff, E.W. (2011) Greenhouse gases in the Earth System: a palaeoclimate perspective. *Philosophical Transactions of the Royal Society A* **369**, 2133–2147.

19. Pollard, D. and De Conto R.M. (2009) Modelling West Antarctic ice sheet growth and collapse through the past five million years. *Nature* **458**, 329–333.

20. Archer, D. (2005) Fate of fossil fuel CO_2 in geologic time. *Journal of Geophysical Research* **110**, C09S05.

21. Archer, D. and Ganopolski, A. (2005) A movable trigger: fossil fuel CO_2 and the onset of the next glaciation. *Geochemistry, Geophysics, Geosystems* **6** (5), doi:10.1029/2004GC000891.

22. Geological Society of London (2013) Addendum to geological society statement on climate change. Available from http://www.geolsoc.org.uk/climatechange (last accessed 29 January 2015).

23. Dwyer, G. (2008) *Climate Wars: The Fight for Survival as the World Overheats*. Random House, London.

24. Carter, R.M. (2010) *Climate: The Counter Consensus*. Stacey International, London.

25. Diamond, J. (2005) *Collapse: How Societies Choose to Fail or Survive*. Allen Lane, London.

26. Ackroyd, P. (2001) *London: The Biography*. Vintage, London.

27. Randers, J. (2012) *2052: A Report to the Club of Rome Commemorating the 40th Anniversary of 'The Limits to Growth'*. Chelsea Green Publishing, White River Junction, VT.

28. Leggett, J. (2014) *The Energy of Nations: Risk Blindness and the Road to Renaissance*. Earthscan, London.

29. Nordhaus, W. (2013) *The Climate Casino: Risk, Uncertainty, and Economics for a Warming World*. Yale University Press, New Haven, CT.

30. Oreskes, N. and Conway, E.M. (2010) *Merchants of Doubt*. Bloomsbury, London.

31. Silver, N. (2013) *The Signal and the Noise: The Art and Science of Prediction*. Penguin, London.

Appendix A

Further Reading

Alley, R.B. (2000) *The Two-Mile Time Machine: Ice Cores, Abrupt Climate Change, and our Future*. Princeton Universtiy Press, Princeton, NJ.

Alley, R.B. (2011) *Earth: The Operators' Manual*. Norton, New York.

Alverson, K., Bradley, R.S. and Pedersen, T. (2003) *Paleoclimate, Global Change and the Future*. Springer, London.

Archer, D. and Pierrehumbert, R. (2011) *The Warming Papers: The Scientific Foundation for the Climate Change Forecast*. Wiley-Blackwell, Chichester.

Battarbee, R.W. and Binney, H.A. (2008) *Natural Climate Variability and Global Warming: A Holocene Perspective*. Wiley-Blackwell, Chichester.

Broecker, W. (2010) *The Great Ocean Conveyor: Discovering the Trigger for Abrupt Climate Change*. Princeton University Press, Princeton, NJ.

Cowie, J. (2013) *Climate Change: Biological and Human Aspects*, 2nd edn. Cambridge University Press, Cambridge.

Cronin, T.M. (2010) *Paleoclimates*. Columbia University Press, New York.

Hay, W.W. (2013) *Experimenting on a Small Planet: A Scholarly Entertainment*. Springer, London.

Jouzel, J., Lorius, C. and Raynaud, D. (2013) *The White Planet: The Evolution and Future of our Frozen World*. Princeton University Press, Princeton, NJ.

Köppen, W. and Wegener, A. (1924) *Die Klimate der geologischen Vorzeit*. Borntraeger, Berlin. Reprinted in 2015 by the same publishing house and translated into English by Bernard Oelkers (Bremen).

Macdougall, D. (2006) *Frozen Earth: The Once and Future Story of Ice Ages*. California University Press, Berkeley, CA.

Mann, M.E. (2012) *The Hockey Stick and the Climate Wars: Dispatches from the Front Lines*. Columbia University Press, New York.

Masson-Delmotte, V. (2011) *Climat: Le Vrai et le Faux*. Éditions Le Pommier, Paris.

Mayewski, P.A. and White, F. (2002) The Ice Chronicles: The Quest to Understand Global Climate Change. *University Press of New England*, Hanover, NH.

North, G.R., Biondi, F., Bloomfield, P., Christy, J.R., Cuffey, K.M., Dickinson, R.E., *et al.* (2006) *Surface Temperature Reconstructions for the Last 2,000 Years*. National Academies Press, Washington, DC.

Pollack, H. (2009) *A World Without Ice*. Avery-Penguin, New York.

Ruddiman, W.F. (2005) *Plows, Plagues and Petroleum*. Princeton University Press, Princeton, NJ.

Turner, J. and Marshall, G.J. (2011) *Climate Change in the Polar Regions*. Cambridge University Press, Cambridge.

Zalasiewicz, J. and Williams, M. (2012) *The Goldilocks Planet: The Four Billion Year Story of Earth's Climate*. Oxford University Press, Oxford.

Earth's Climate Evolution, First Edition. Colin P. Summerhayes.
© 2015 John Wiley & Sons, Ltd. Published 2015 by John Wiley & Sons, Ltd.

Appendix B

List of Figure Sources and Attributions

Figure 2.1 Extract from a portrait by Sir Henry Raeburn (1756–1823) in the Scottish National Portrait Gallery. Reproduced by permission of the British Geological Survey, © NERC, all rights reserved; CP14/049.

Figure 2.2 From Anon (1833) Portrait of Baron Georges Cuvier. In The Gallery of Portraits: With Memoirs, vol. 2, C. Knight, London.

Figure 2.3 Photograph of a painting by Friedrich Georg Weitsch (1758–1828). From McCrory, D. (2010) Nature's Interpreter – The Life and Times of Alexander von Humboldt. Lutterworth Press, Cambridge. Reproduced with permission of Lutterworth Press.

Figure 2.4 Extract from the Frontispiece to Bonney, T.G. (1895) Charles Lyell and Modern Geology. Macmillan, New York. Published in compliance with the conditions specified by Project Gutenberg; http://www.gutenberg.org/files/34502/34502-h/34502-h.htm.

Figure 2.5 From Figures 14 (A) and 15 (B) in Lyell, C. (1875) Principles of Geology. 12th Ed., v.1, John Murray, London; (also reprinted as Figure 1 in Summerhayes, C.P. (1990) Palaeoclimates. J. Geol. Soc. London 147, 315–320).

Figure 2.6 Published with permission from the Oxford University Museum of Natural History.

Figure 2.7 (a) and (b) Photographed by the author.

Figure 2.8 Published with permission from the Archives of the Museum of Comparative Zoology, Ernst Mayr Library, Harvard University.

Figure 2.9 Reproduced by permission of the British Geological Survey, © NERC, all rights reserved; CP14/049.

Figure 2.10 Reproduced by permission of the British Geological Survey, © NERC, all rights reserved; CP14/049.

Figure 3.1 Extract from a photograph in Irons, J.C. (1896) Autobiographical Sketch of James Croll, with Memoirs of his Life and Work. Edward Stanford, London.

Figure 3.2 From Croll, J. (1875) Climate and Time in their Geological Relations: a Theory of Secular Changes of the Earth's Climate. Appleton and Co., New York; following p. 312.

Figure 4.1 From Figure 6, in Bard, E., (2004) Greenhouse effect and ice ages: historical perspective. Comptes Rendus Geoscience **336**. Reproduced with permission of Elsevier.

Figure 4.2 Licensed under the Creative Commons Attribution-Share Alike 2.0 Generic license.

Figure 4.3 Source unknown.

Figure 4.4 From Figure 2 in Tyndall, J. (1864) The Absorption and Radiation of Heat by Gaseous and Liquid Matter. London, Edinburgh and Dublin Philosophical Magazine and Journal of Science vol. XXVIII, 81–106, (reproduced as Figure 14 in Part V of Tyndall, J. (1872) Contributions to Molecular Physics in the Domain of Radiant Heat – a Series of Memoirs Published in the 'Philosophical Transactions' and 'Philosophical Magazine', with Additions. Longmans, Green and Co., London).

Figure 4.5 Extract from a copy by Tägtström of an original oil painting by R.Berg. Reproduced with permission from the Centre for History of Science of the Royal Swedish Academy of Science.

Figure 4.6 Reproduced with permission from the Archives and Records Management Office, University of Wisconsin - Madison.

Figure 4.7 Reproduced with permission from the Geological Survey of Austria.

Figure 5.1 Reproduced with permission from the Alfred Wegener Institute for Marine and Polar Research.

Figure 5.2 From Figure 23 in Wegener, A. (1920) "Die Entstehung der Kontinente und Ozeane". 2nd ed., Die Wissenschaft, Band 66. Vieweg und Sohn, Braunschweig.

Figure 5.3 From Figure 4 in Köppen, W., and Wegener, A. (1924) Die Klimate der Geologischen Vorzeit. Borntraeger, Berlin. Reproduced with permission from E. Schweizerbart'sche Verlagsbuchhandlung OHG.

Figure 5.4 Reproduced with permission from the School of Geosciences, University of Edinburgh.

Figure 5.5 Reproduced with permission from the Department of Geosciences, Princeton University.

Figure 5.6 Reproduced with permission from the University of Toronto Mississauga archives.

Figure 5.7 Reproduced with permission from the United States Geological Survey.

Figure 5.8 From Maps 31 and 16 in Smith, D.G., Smith, A.G., and Funnell, B.M. (1994) Atlas of Mesozoic and Cenozoic Coastlines. Reproduced with permission from Cambridge University Press.

Figure 5.9 Prepared by Robert A Rohde and made available through Global Warming Art at http://commons.wikimedia.org/wiki/File:Phanerozoic_Sea_Level.png. Licensed under the Creative Commons Attribution-GNU Free documentation License.

Figure 6.1 From Figure 8, in Köppen, W., and Wegener, A. (1924) Die Klimate der Geologischen Vorzeit. Borntraeger, Berlin. Reproduced with permission from E. Schweizerbart'sche Verlagsbuchhandlung OHG.

Figure 6.2 From Figure 4 in Lewis, J.M. (1996) Winds over the world sea: Maury and Köppen. Bulletin of the American Meteorological Society **77** (5),

935–952. Reproduced with permission from the American Meteorological Society.

Figure 6.3 From Figure 7 in Köppen, W. (1931) Grundriss der Klimakunde. Walter de Gruyter and Co., Berlin and Leipzig.

Figure 6.4 From Figure 2 in Summerhayes, C.P. (1990) Palaeoclimates. J Geol. Soc. London, **147**, 315–320, based on Figure 1 in Scotese, C.R., and Summerhayes, C.P. (1986) Computer model of paleoclimate predicts coastal upwelling in the Mesozoic and Cenozoic. Geobyte **1** (3), 28–44 and 94. Reproduced with the permission of the Geological Society of London.

Figure 6.5 From the portrait by Paja Jovanovic, 1943, reproduced with the permission of John Imbrie, Brown University.

Figure 6.6 From Figure 3 in Grubic, A. (2006) The astronomic theory of climatic changes of Milutin Milankovitch. Episodes **29** (3), 197–203.

Figure 6.7 From Figure 1a in Robinson, P.L. (1973) Palaeoclimatology and continental drift. In Implications of Continental Drift to the Earth Sciences, Vol. 1 (eds. D.H. Tarling and S.K. Runcorn). Academic Press, London, pp 451–476. Reproduced with permission from Elsevier.

Figure. 6.8. From Figure 6 in Parrish, J.T., Bradshaw, M.T., Brakel, A.T., et al. (1996) Palaeoclimatology of Australia during the Pangean interval. Palaeoclimates, **1**, 241–281 (Adapted from Parrish, J.T., and Curtis, R.L. (1982) Atmospheric circulation, upwelling and organic-rich rocks in the Mesozoic and Cenozoic. Palaeogeography, Palaeoclimatology and Palaeoecology **40**, 31–66). Published with the permission of Taylor and Francis.

Figure 6.9 From Figure 3, in Parrish, J.T., and Curtis, R.L. (1982) Atmospheric circulation, upwelling and organic-rich rocks in the Mesozoic and Cenozoic. Palaeogeography, Palaeoclimatology and Palaeoecology **40**, 31–66). Published with the permission of Elsevier.

Figure 6.10 Photograph supplied by Jane Francis.

Figure 6.11 Painting by Robert Nicholls. From Figure 3, in Francis, J.E., Ashworth, A., Cantrill, D.J., et al (2008) 100 million years of Antarctic climate evolution: evidence from fossil plants. In: Antarctica: A Keystone in a Changing World (eds. A.K. Cooper, P.J. Barrett, H. Stagg, B. Storey, E. Stump and W. Wise). National Academies

Press, Washington DC, pp 19–27). Reproduced with the permission of Robert Nicholls.

Figure 6.12 Reproduced with the permission of the Integrated Ocean Drilling Program.

Figure 7.1 Photograph supplied by Jim Kennett.

Figure 7.2 Photograph by Robert Ginsburg. Reproduced with permission from the Archives of the Rosenstiel School of Marine and Atmospheric Science, University of Miami.

Figure 7.3 Reproduced with the permission of I.N. McCave, Cambridge University.

Figure 7.4 Reproduced with the permission of Tony Phillips, British Antarctic Survey.

Figure 7.5 Photograph supplied by Peter Barrett.

Figure 7.6 From Figure 8.12 in Francis, J.E., Marenssi, S., Levy, R., et al (2009) From Greenhouse to Icehouse – the Eocene/Oligocene in Antarctica. In: Antarctic Climate Evolution (eds. F Florindo and M. Siegert). Developments in Earth and Environmental Sciences **8**. Elsevier, Amsterdam, pp 309–368. Published with the permission of Elsevier.

Fig. 7.7 From Figure 9, in Cramer, B.S., Miller, K.G., Barrett, P.J., and Wright, J.D. (2011) Late Cretaceous-Neogene trends in deep ocean temperature and continental ice volume: reconciling records of benthic foraminiferal geochemistry (∂^{18}O and Mg/Ca) with sea level history. Journal of Geophysical Research **116**, C12023. Published with the permission of Wiley.

Figure 8.1 Source unknown.

Figure 8.2 Published with the permission of Special Collections and Archives, University of California, San Diego.

Figure 8.3 Published with the permission of Special Collections and Archives, University of California, San Diego.

Figure 8.4 Published with the permission of the University of East Anglia.

Figure 8.5 Photograph provided by Professor Valentin Meleshko.

Figure 8.6 From Figure 5 in Budyko, M.I. (1977) On present-day climatic changes. Tellus **20**, 193–204. Published with the permission of Wiley.

Figure 8.7 Reproduced with the permission of Dr W.S. Broecker.

Figure 8.8 From Figure 1 in Berner, R.A. (1999) A new look at the long-term carbon cycle. GSA Today **11** (9), 1–6. Published with the permission of GSA Today.

Figure 9.1 Reproduced with the permission of the Tyler Prize for Environmental Achievement.

Figure 9.2 Reproduced with the permission of Eric Wolff, Department of Earth Sciences, University of Cambridge.

Figure 9.3 From Figure 3, in Royer, D.L., Berner, R.A., Montañez, I.P., et al (2004) CO_2 as a primary driver of Phanerozoic climate. GSA Today **14** (3), 4–10. Reproduced with the permission of GSA Today.

Figure 9.4 Figure 1 in Beerling, D.J., and Royer, D.L. (2011) Convergent Cenozoic CO_2 history. Nature Geoscience **4**, 418–420. Reproduced with the permission of Nature.

Figure 9.5 Reproduced with the permission of the Huntsman Foundation and R A Berner.

Figure 9.6 Reproduced with the permission of Bill Ruddiman.

Figure 9.7 From Figure 2 in Royer, D.L., Pagani, M., ad Berling, D.J., (2012) Geobiological constraints on Earth system sensitivity to CO_2 during the Cretaceous and Cenozoic. Geobiology **10** (4), 298–310. Published with the permission of Wiley.

Figure 10.1 Photograph supplied by Jim Kennett.

Figure 10.2 From Figure 1 in Vaughan, A.P.M. (2007) Climate and geology – a Phanerozoic perspective. In: Deep-Time Perspectives on Climate change: Marrying the Signal from Computer Models and Biological Proxies (eds. M. Williams, A.M. Haywood, F.J. Gregory, and D.N. Schmidt). Micropalaeontological Society Special Publication. The Geological Society of London, pp. 5–59. Reproduced with the permission of the Geological Society of London.

Figure 10.3 From Figure 1 in Van de Wal, R.S.W., De Boer, B., Lourens, L.J., et al (2011) Reconstruction of a continuous high-resolution CO_2 record over the past 20 million years. Climate of the Past **7**, 1459–1469. Published with permission of Climate of the Past.

Figure 10.4 From Figure 1 in Courtillot, V.E. and Renne, P.R. (2003) On the ages of flood basalt events. C. R. Geoscience 335, 113–140. Published with the permission of Elsevier.

Figure 11.1 From Figure 6 in Barron, E.J., Fawcett, P.J., Peterson, W.H., et al (1995) A 'simulation' of mid-Cretaceous climate. Paleoceanography **10** (2) 953–962. Published with permission of AGU/Wiley.

Figure 11.2 From Figure 6 in Kent, D.V., and Muttoni, G., (2013) Modulation of late Cretaceous and Cenozoic climate by variable drawdown of atmospheric pCO_2 from weathering of basalt provinces on continents drifting through the equatorial humid belt. Climate of the Past **9**, 525–546. Published with the permission of Climate of the Past.

Figure 11.3 From Figure 7 in Kent, D.V., and Muttoni, G., (2013) Modulation of late Cretaceous and Cenozoic climate by variable drawdown of atmospheric pCO_2 from weathering of basalt provinces on continents drifting through the equatorial humid belt. Climate of the Past **9**, 525–546. Published with the permission of Climate of the Past.

Figure 11.4 From Figure 2 in Pollard, D. and DeConto, R.M. (2005) Hysteresis in Cenozoic Antarctic ice-sheet variations. Global and Planetary Change **45**, 9–21. Published with permission of Elsevier.

Figure 11.5 From Figure 6.2 in Jansen, E., Overpeck, J., Briffa, K., et al. (2007) Paleoclimate. Chapter 6 in the 4th Assessment Report of the IPCC. Published with the permission of the IPCC Secretariat.

Figure 11.6 From Figure 14 in Sarnthein, M., Bartoli, G., Prange, M., et al. (2009) Mid-Pliocene shifts in ocean overturning circulation and the onset of Quaternary-style climates. Climate of the Past **5**, 269–283. Published with the permission of Climate of the Past.

Figure 12.1 Reproduced with the permission of André Berger.

Figure 12.2 Published with the permission of John Imbrie.

Figure 12.3 From Figure 4 in Lisiecki, L.E., and Raymo, M.E. (2005) A Pliocene-Pleistocene stack of 57 globally distributed benthic $\partial^{18}O$ records. Paleoceanography **20**, PA1003. Published with the permission of Wiley.

Figure 12.4 From http://www.grida.no/graphicslib/detail/world-ocean-thermohaline-circulation_79a9. Reproduced with permission of Hugo Ahlenius/GRID-Arendal.

Figure 13.1 From Figure 1 in Carlson, W.E., and Winsor, K. (2012) Northern Hemisphere ice-sheet responses to past climate warming. Nature Geoscience **5** (9), 607–613. Reprinted with permission of Nature.

Figure 13.2 Reproduced with the permission of the Tyler Prize for Environmental Achievement.

Figure 13.3 From Figure 1 in Ahn, J., and Brook, E.J. (2007) Atmospheric CO_2 and climate from 65 to 30 ka B.P. Geophysical Research Letters **34** L10703, GL029551. Published with the permission of Wiley.

Figure 13.4 Reproduced with the permission of the Tyler Prize for Environmental Achievement.

Figure 13.5 From Figure 3 in Petit, J.R., Jouzel, J., Raynaud, D., et al (1999) Climate and atmospheric history of the past 420,000 years from the Vostok ice core, Antarctica. Nature **399**, 429–436. Reprinted with the permission of Nature.

Figure 13.6. From Figure 2 in Lüthi, D., Le Floch, M., Bereiter, B., et al (2008) High resolution carbon dioxide concentration record 650,000–800,000 years before present. Nature **453**, 379–382. Reprinted with the permission of Nature.

Figure 13.7 From Figure 3 in Pedro, J.B., Rasmussen, S.O., and Van Ommen, T.D (2012) Tightened constraints on the time-lag between Antarctic temperature and CO_2 during the last deglaciation. Climate of the Past **8**, 1213–1221. Published with permission from Climate of the Past.

Figure 14.1 From Figure 5 in Wanner, H., Solomina, O., Grosjean, M., et al (2011) Structure and origin of Holocene cold events. Quaternary Science Reviews **30**, 3109–3123. Published with the permission of Elsevier.

Figure 14.2 From Figure 2 in Vinther, B.M., Buchardt, S.L., Clausen, H.B., et al (2009) Holocene thinning of the Greenland Ice Sheet. Nature **461**, 385–388. Reprinted with the permission of Nature.

Figure 14.3 From Figure 3 in Mulvaney, R., Abram, N.J., Hindmarsh, R.C.A., et al (2012) Recent Antarctic Peninsula warming relative to Holocene climate and ice-shelf history. Nature **489**, 141–145. Reprinted with the permission of Nature.

Figure 14.4 From Figure 4 in Crosta, X., Debret, M., Denis, D. et al (2007) Holocene long- and short-term climate change off Adélie Land, East Antarctica. Geochemistry, Geophysics, Geosystems **8** (11), doi:10.1029/2007GC001718. Published with permission from Wiley.

Figure 15.1 From Figure 3, in Steinhilber, F., Abreu, J.A., Beer, J., et al (2012) 9,400 years of cosmic radiation and solar activity from ice cores and tree rings. Proceedings of the National Academy of Sciences **109** (16), 5967–5971. Published with the permission of the National Academy of Sciences of the U.S.A.

Figure 15.2 From Figure 6, in Magny, M. (2007) West-Central Europe. In: Lake Level Studies, Encyclopedia of Quaternary Science. Elsevier, Berlin, pp 1389–1399. Published with the permission of Elsevier.

Figure 15.3 From Figure 10, in Lamb, H.H. (1966) Britain's climate in the past. In: The Changing Climate: Selected Papers by H.H. Lamb. Methuen, London, pp. 170–195. Published with the permission of Taylor and Francis.

Figure 15.4 From Figure 7.1(c) in Folland, C.K., Karl, T.R., and Vinnikov, K.Y. (1990) Observed climate variations and change. In: Report of Working Group 1, Climate change (eds. J.T. Houghton, G.J. Jenkins and J.J. Ephraums). The IPCC First Assessment Report. Cambridge University Press, Cambridge, pp 195–238. Published with the permission of the IPCC Secretariat.

Figure 15.5 From Figure 4, in Ljungqvist, F.C., Krusic, P.J., Brattström, G., and Sundqvist, H.S. (2012) Northern Hemisphere temperature patterns in the last 12 centuries. Climate of the Past **8**, 227–249. Published with the permission of Climate of the Past.

Figure 15.6 From Figure 4 in Mayewski, P.A., and Maasch, K.A., (2006) Recent warming inconsistent with natural association between temperature and atmospheric circulation over the last 2000 years. Climate of the Past Discussions **2**, 1–29. Published with permission from Climate of the Past.

Figure 15.7 From Figure 4 in Mulvaney, R., Abram, N.J., Hindmarsh, R.C.A., et al (2012) Recent Antarctic Peninsula warming relative to Holocene climate and ice-shelf history. Nature **489**, 141–145. Reprinted with the permission of Nature.

Figure 15.8 From Figure 3 in Bard, E., and Delague, G. (2008) Comment on 'Are there connections between the Earth's magnetic field and climate?' by V Courtillot, Y. Gallet, J.-L. Le Mouël, F. Fluteau, A. Genevey, EPSL 253, 328, 2007. Earth and Planetary Science Letters **265**, 302–307. Published with permission of Elsevier.

Figure 15.9 From Figure 2 in Esper, J., Frank, D.C., Timonen, M., et al (2012) Orbital forcing of tree-ring data. Nature Climate Change **2**, 862–866. Reprinted by permission of Nature.

Figure 15.10 From Figure 2.20, in Folland, C.K., Karl, T.R., Christy, J.R., et al (2001) Observed climate variability and change. In: Climate Change 2001, the Scientific Basis (eds. J.T. Houghton, Y. Ding, D.J. Griggs et al). Contribution of Working Group 1 to the IPCC Third Assessment Report. Cambridge University Press, Cambridge, pp. 101–181. Reproduced with the permission of the IPCC Secretariat.

Figure 15.11 From Figure 6.10 in Jansen, E., Overpeck, J., Briffa, K.R., et al (2007) Palaeoclimate. In: Climate Change 2007, the Physical Science Base (eds. S. Solomon, D. Qin, M. Manning et al). Contribution of Working Group 1 to the IPCC Fourth Assessment Report. Intergovernmental Panel on Climate Change, Geneva, pp 433–497. Published with the permission of the IPCC Secretariat.

Figure 15.12 From Figure 5 in Wanner, H., Mercolli, L., Grosjean, M., and Ritzl, S.P. (2015) Holocene climate variability and change: a data-based review. Journal of the Geological Society of London **172**, 251–253. Published with the permission of the Geological Society of London.

Fig 15.13 From Figure 1 in Jevrejeva, S., Moore, J.C., Grinsted, A., and Woodworth, P.L. (2008) Recent global sea level acceleration started over 200 years ago. Geophysical Research Letters **35**, L08715. Published with the permission of Wiley.

Index

Printed and bound by CPI Group (UK) Ltd, Croydon, CR0 4YY